BACTERIAL DISEASES OF CROP PLANTS

BACTERIAL DISEASES OF CROP PLANTS

Suresh G. Borkar
Rupert Anand Yumlembam

CRC Press
Taylor & Francis Group
Boca Raton London New York

CRC Press is an imprint of the
Taylor & Francis Group, an **informa** business

CRC Press
Taylor & Francis Group
6000 Broken Sound Parkway NW, Suite 300
Boca Raton, FL 33487-2742

First issued in paperback 2021
First issued in hardback 2019

ISBN 13: 978-1-03-209754-1 (pbk)
ISBN 13: 978-1-4987-5598-6 (hbk)

This book contains information obtained from authentic and highly regarded sources. Reasonable efforts have been made to publish reliable data and information, but the author and publisher cannot assume responsibility for the validity of all materials or the consequences of their use. The authors and publishers have attempted to trace the copyright holders of all material reproduced in this publication and apologize to copyright holders if permission to publish in this form has not been obtained. If any copyright material has not been acknowledged please write and let us know so we may rectify in any future reprint.

Publisher's Note
The publisher has gone to great lengths to ensure the quality of this reprint but points out that some imperfections in the original copies may be apparent.

Library of Congress Cataloging-in-Publication Data

Names: Borkar, S. G. (Suresh Govindrao), 1956- author. | Yumlembam, Rupert
Anand, author.
Title: Bacterial diseases of crop plants / authors: Suresh G. Borkar and
Rupert Anand Yumlembam.
Description: Boca Raton, FL : CRC Press, Taylor & Francis Group, 2017. |
Includes bibliographical references and index.
Identifiers: LCCN 2016008534 | ISBN 9781498755986 (alk. paper)
Subjects: LCSH: Bacterial diseases of plants.
Classification: LCC SB734 .B67 2017 | DDC 632/.32--dc23
LC record available at http://lccn.loc.gov/2016008534

**Visit the Taylor & Francis Web site at
http://www.taylorandfrancis.com**

**and the CRC Press Web site at
http://www.crcpress.com**

Contents

Preface

Food and agriculture are vital components in the development and survival of civilizations. Around half of the world's population and its economy are influenced by agricultural farm production. Plant diseases take as much as a 30% toll of crop harvest if not managed properly and efficiently.

Bacterial diseases of crop plants are significant threats in the plant disease scenario around the world and are observed on all kinds of cultivated and commercial-value plants, including cereals, pulses, oilseeds, fruits, vegetables, cash crops, plantation crops, spices, ornamentals and flowering plants, forage crops, forest trees, and lawn grasses. The impact is evident from the losses these bacterial diseases cause around the world per se. In the case of fire blight in apples and pears since 1995, the Italian government has destroyed 0.5 million pear trees in an attempt to eradicate infection of the bacterium *Erwinia amylovora* on this crop in Italy alone. The USDA, the Animal and Plant Health Inspection Service, and the Division of Plant Industry have formed a joint state/federal eradication campaign to eliminate citrus canker disease. An average of US $50 million per year and 600 personnel have been dedicated to this program in Florida alone. From 1995 through 2005, the compensation paid to commercial growers and homeowners for residential citrus destroyed amounted to US $1 billion. Cassava bacterial blight and banana wilt caused starvation of families involved in the farming of these crops in the African continent. Bacterial blight in pomegranates in recent years caused losses of around INR 23,183 million to pomegranate farmers in the Indian peninsula. There have been several similar cases around the world.

Bacterial diseases of crop plants are often difficult to identify and diagnose, perhaps due to less trained scientific manpower in this branch of plant pathology. Bacterial diseases spread quickly and are difficult to control due to the scarcity of targeted pesticides having label claim (chemical recommended for a particular bacterial disease on the concerned crop—mentioned on the label of the packet/bottle of pesticide) for these. In this scenario, a concise source of information on each bacterial disease of cultivated and commercial-value plants is necessary for all concerned with food and agriculture, including students, farmers and traders, farm extension workers, and the pesticide industry. *Bacterial Diseases of Crop Plants* is a step forward in this direction.

Prof. Suresh G. Borkar

Acknowledgments

The authors acknowledge the scientific contribution of all those whose work and references are quoted in this book. Similarly, some photographs of bacterial diseases were taken from the Internet and the authors gratefully acknowledged the respective contributors, whose details are included in the captions.

The authors thank Kavita Sandip Dhawade for her painstaking effort in the typing and manuscript setting work.

Prof. Suresh G. Borkar

Authors

Prof. Suresh G. Borkar, is university head of department in the Department of Plant Pathology and Agricultural Microbiology since 2005. He graduated from Dr. Punjabrao Deshmukh Krishi Vidyapeeth, Akola, in 1977, and earned his MSc and PhD from IARI, New Delhi, in 1979 and 1983, respectively. He did his postdoctorate from INRA, Angers, France, in 1984 and DSc from Washington International University in 1999. He is a fellow of the Indian Phytopathological Society and the Eurasian Academy of Environmental Sciences.

After returning from France, Prof. Borkar joined his first appointment as assistant professor of plant pathology at Jawaharlal Nehru Krishi Vishwavidyalaya, Jabalpur (Madhya Pradesh) and served the university from January 1985 to December 1989. He was selected as associate professor of plant pathology at Mahatma Phule Krishi Vidyapeeth, Rahuri, and joined the position in December 1989. Later, he was appointed professor of plant pathology in 1994 by the Maharashtra Council of Agriculture Education and Research for Mahatma Phule Krishi Vidyapeeth, Rahuri, and joined the position in May 1994. He has also served as dean, Post Graduate Institute of Mahatma Phule Krishi Vidyapeeth, Rahuri, during 2012–2013.

Prof. Borkar has published more than 88 research papers in around 25 national and 7 foreign journals. He has guided around 25 MSc and PhD students in their research. Most of his students are serving as ICAR scientists and are in agricultural universities. He has received 13 awards from scientific societies and social organizations within the country and abroad. He has developed four wheat varieties, several technologies and recommendations, new strains of beneficial microbes, and published three books. He holds six patents. He has chaired several sessions in ICAR workshops and national seminars. He has visited several universities abroad—France, Greece, the United Kingdom, and Nepal. He is on the selection committee of different agricultural universities in the India. He is a well-known teacher and scientist at the national and international level.

Rupert Anand Yumlembam has completed his MSc from Mahatma Phule Krishi Vidyapeeth, Rahuri, Maharashtra (India), in 2011 and PhD from Bidhan Chandra Krishi Viswavidyalaya (BCKV), Mohanpur, West Bengal (India), in 2015, both in plant pathology. He has worked extensively in the field of plant bacteriology for these degrees. He has published five-research articles in national journals and a book chapter. His area of interest is plant bacteriology identification of new anti-bacterial molecules of plant origins. He has filed one patent on the new bactericidal molecule 4-(4-chlorophenyl) pyridine. He has participated actively in seminars, conferences, and symposiums held by professional societies.

1 Global Overview
Economic Losses due to Bacterial Plant Pathogens and Diseases

1.1 INTRODUCTION

Bacteria appeared on planet Earth long before plants, birds, water- or land-based animals, and human beings, and they have exerted harmful or beneficial effects on these hosts ever since. Bacteria are present everywhere in the environment, ranging from the surface of living beings and dead material to the rhizosphere, and are involved in various kinds of activities.

Bacteria as causative agents of plant diseases were first demonstrated by Dr. T.J. Burrill of Illinois University in 1873, who observed that fire blight in apple and pear in the United States and Europe was caused by bacteria. This bacterium was later identified as *Erwinia amylovora*. Once bacteria were established as causal agents of plant diseases, many more bacterial plant diseases came to light and were studied in detail, including measures to control them.

Among the various genera of bacteria, *Agrobacterium* (causing galls on plants), *Ralstonia* (causing wilts), *Pseudomonas*, *Xanthomonas* or Pantoea (causing leaf spot, blight, blast, cankers, and wilt), *Corynebacterium* or *Curtobacterium* (causing leaf spot, fruit spot, and wilt), and *Erwinia* or *Dickeya* (causing soft rot or wilt) are known to be plant pathogenic bacteria.

The diseases caused by plant bacteria in important crop plants cause varying degrees of economic hardship. These not only cause severe losses in production and economies of cultivators, as in the case of bacterial blight of rice in Asian countries, but also, at times, cause malnutrition, hunger, and despair as reported in the case of banana wilt pathogen and cassava bacterial pathogen in African countries.

The burden of combating these economically significant bacterial diseases falls not only on cultivators but also on governments. The U.S. government spent US$1 billion to eradicate citrus canker disease during 1995–2005, while the Indian government spent INR 10,000 million to combat bacterial blight of pomegranate during 2003–2008, besides losses of INR 23,183 million to the farmers.

This shows the severity and gravity of bacterial plant diseases in crop production systems around the world.

1.2 OVERVIEW OF LOSSES DUE TO BACTERIAL PLANT DISEASES

Table 1.1 lists some of the major bacterial plant diseases that affect various crops in different countries, causing significant losses, and those on which individual governments have spent significant amounts of money to combat.

Bacterial diseases (and the resulting losses) vary from country to country. Per se, bacterial blight of rice occurs mostly in Asian countries and banana wilt and cassava bacterial blight (CBB) in African countries, while bacterial diseases of pome fruits occur in Europe and the United States. The losses due to bacterial diseases in the United States are enumerated in Table 1.2.

The top 10 bacterial plant pathogens that impact world agriculture and farm economics are listed in Table 1.3.

TABLE 1.1

Overview of Losses due to Bacterial Plant Diseases

Sr. No.	Name of Disease	Country	Annual Loss in Production or the Amount Spent to Combat the Disease	Year
1.	Bacterial leaf blight of rice	Japan	22,000–110,000 tons	1954
		Philippines	22%	1973
2.	Brown sheath rot of rice	Indonesia	72%	1997
		Madagascar	100%	1987
3	Gummosis disease of sugarcane	Australia	30%–40% loss in cane tonnage	1893
			9%–17% loss in sugar content	1899
		Mauritius	Heavy losses cv. abandoned	1890
4.	Sugarcane leaf scald	Mauritius	Sugar yield loss 31%	1990
5.	Bacterial canker of hazelnut	Italy	1,000 ha area affected. Loss of US$ 1.5 million	1990
6.	Bacterial blight of grapevine	South Africa	Harvest losses of 70% or more	1940
7.	Banana *Xanthomonas* wilt	Uganda	Loss of US$4 billion; 30%–52% reduction in yield	2001–2006
8.	Moko disease of banana	Guyana	74% reduction in yield	
9.	Citrus canker	United States	US$1 billion for eradication	1995–2005
10.	Bacterial fruit blotch of watermelon	United States	Widespread outbreaks	1994
11.	Bacterial wilt and canker of tomato	North Carolina	70% reduction in yield	
		France	30% reduction in yield	1991
12.	Black rot of cabbage	United States	US$1 million	1976
		Russia	57% loss	1992
		Mozambique	70% loss	2009
		India	50% loss	1991
13.	Bacterial blight of pea	Australia (Victoria)	70% yield loss	1991
14.	Fasciation of chrysanthemum	United States	US$1 million	1934
15.	Bacterial blight of cassava	Zaire	75% tuber loss	1970
		Brazil	50% tuber loss	1974
		Nigeria	75% tuber loss	1973
		Uganda	100% tuber loss	1980
		Colombia	80%–92% tuber loss	1983
16.	Bacterial blight of pomegranate	India	INR 10,000 million spent by the government to combat the disease INR 23,183 million loss to farmers	2003–2008

The important host range of these pathogens, their geographical distribution, and the diseases they cause that have economic implications are summarized in the following sections.

1.3 MAJOR BACTERIAL PLANT PATHOGENS INFLUENCING WORLD AGRICULTURAL CROP PRODUCTION

1.3.1 *Pseudomonas syringae* pathovars

The bacterium *P. syringae* and its pathovars infect a majority of temperate fruit crops like apricot, plum, apple, cherry, peaches, almond, chestnut, hazelnut, raspberry, avocado, kiwi, and oranges.

TABLE 1.2

Losses due to Bacteria in the United States in 1976

Pathogens	Losses (in Million U.S. Dollars)
A. tumefaciens	23
Clavibacter michiganensis subsp. *insidiosus*	17
C. m. subsp. *nebraskensis*	3
C. m. subsp. *sepedonicus*	1.8
E. amylovora	5
E. chrysanthemi	2.3
Erwinia soft rots	14
Xy. fastidiosa (phony peach)	20
Xy. fastidiosa (pierce's disease)	3
P. syringae pv. *glycinea*	64
P. s. pv. *phaseolicola*	2
P. s. pv. *syringae*	18
R. solanacearum	9
Clavibacter xyli (ratoon stunt of sugarcane)	10
Xanthomonas arboricola pv. *juglandis*	2.2
X. a. pv. *pruni*	2
X. axonopodis pv. *malvacearum*	5
X. a. pv. *phaseoli*	5
X. campestris pv. *campestris*	1
Xanthomonas translucens pv. *translucens*	1

Source: Kennedy, B.W. and Alcorn, S.M., *Plant Dis.*, 64, 674, 1980.

TABLE 1.3

Top 10 Bacterial Plant Pathogens as Voted for and Ranked by Plant Bacteriologists Associated with the Journal *Molecular Plant Pathology*

Bacterial Pathogen

1. *P. syringae* pathovars
2. *R. solanacearum*
3. *A. tumefaciens*
4. *X. oryzae* pv. *oryzae*
5. *X. campestris* pathovars
6. *X. axonopodis* pv. *manihotis*
7. *E. amylovora*
8. *Xy. fastidiosa*
9. *Dickeya* (*dadantii* and *solani*)
10. *P. carotovorum* (and *P. atrosepticum*)

Source: Mansfield, J. et al., *Mol. Plant Pathol.*, 13, 614, 2012.

Other host plants include lilac, conifer, wheat, and pea and vegetable crops like tomato, capsicum, beans, celery, leek, parsley, cucumber, sugarbeet, spinach, and coriander.

This bacterium is found in different geographical regions of the United States, Greece, Mexico, France, New Zealand, Australia, Israel, Italy, Europe, Asia, Africa, Oceania, Japan, etc.

The major diseases caused by this bacterium and its pathovar are bacterial cankers, leaf spot, and blight or halo blight depending on the host plant.

P. syringae scores heavily on all counts. The economic impact of *P. syringae* is increasing, with a resurgence of old diseases, including bacterial speck of tomato (pv. tomato; Shenge et al., 2007), and the occurrence of other significant infections worldwide, such as bleeding canker of horse chestnut (pv. *aesculi*; Green et al., 2010). The *European Handbook of Plant Diseases* (Smith et al., 1988) describes 28 pathovars, each attacking a different host species. We can now add pv. *aesculi* to this list. Several pathovars cause long-term problems in trees, often through the production of distortions and cankers (e.g., pathovars *savastanoi* and *morsprunorum*). Infections of annual crops are more sporadic, and outbreaks are often caused by sowing contaminated seed. Many reports highlight the seed-borne nature of *P. syringae*, but it is a remarkably adaptive pathogen, occuring in some apparently bizarre sites, such as snow melt waters (Morris et al., 2007). Once new infections have established, given favorable conditions of rainfall and temperature, disease outbreaks are often devastating, as observed with bean halo blight caused by pv. *phaseolicola* (Murillo et al., 2010).

Bacterial canker of plum caused by *P. syringae* pv. *syringae* (Pss) causes annual tree mortality rates as high as 30% in Germany (Hinrichs-Berger, 2004), and bacterial canker of hazelnut has resulted in the mortality of large numbers of trees in Italy and other European countries (Scortichini, 2002). Yield losses due to lesions on fruit are typically more sporadic and variety dependent, but significant yearly losses have been reported in some instances. In Turkey, bacterial canker affected nearly 80% of apricot trees in the provinces of Erzurum, Erzincan, and Artvin (Kotan and Sahin, 2002) and 20% in Malatya (Donmez et al., 2010). Also, leaf-bud and fruit-bud death and branch diebacks caused by *P. syringae* were observed in about 10% of peach trees grown near Izmir (Ozaktan et al., 2008). In central Italy, the presence of severe twig dieback and death of whole plants induced by Pss was observed in almost 30% of 1-year-old apricot orchards (Scortichini, 2006). In Oregon, United States, sweet cherry infections by Pss were reported (Spotts et al., 2010).

1.3.2   *Ralstonia solanacearum*

The bacterium *R. solanacearum* (syn. *Pseudomonas solanacearum or Burkholderia solanacearum* [Smith]) and its pathotypes infect plants of the Solanaceae family, particularly potato, tomato, tobacco, eggplant, banana, and plantain, making them the major hosts, while peanut, bell pepper, cotton, sweet potato, cassava, castor, ginger, and other solanaceous weeds are also affected.

This bacterium is found in different regions of the Philippines, Hawaii, Queensland, Australia, Central and South America, India, Africa (Angola, Burkina Faso, Burundi, Congo, Ethiopia, Gabon Gambia, Kenya, Madagascar, Malawi, Mauritius, Mozambique, Nigeria, Reunion, Rwanda, Senegal, Seychelles, Sierra Leone, Somalia, South Africa, Swaziland, Tanzania, Tunisia, Uganda, Zaire, Zambia, and Zimbabwe), Western Europe, and Mexico.

The major diseases caused by this pathogen and its pathotypes are wilt on solanaceous crop and moko disease of banana.

R. solanacearum causes a vascular wilt disease and has been ranked as the second most significant bacterial pathogen. It is one of the most destructive pathogens identified to date because it induces rapid and fatal wilting symptoms in host plants. The host range is extensively wide, more than 200 species, and the pathogen is distributed worldwide and induces a destructive economic impact (Kelman, 1998). Direct yield losses by *R. solanacearum* vary widely according to the host, cultivar, climate, soil type, cropping pattern, and strain. For example, yield losses vary from 0% to

91% in tomatoes, 33% to 90% in potatoes, 10% to 30% in tobacco, 80% to 100% in bananas, and up to 20% in groundnuts (Elphinstone, 2005). In potatoes alone, it is responsible for an estimated loss of US$1 billion each year worldwide (Elphinstone, 2005). Difficulties are associated with controlling this pathogen due to its ability to grow endophytically, survive in soil, especially in the deeper layers, and travel along water, and its relationship with weeds (Wang and Lin, 2005).

1.3.3 AGROBACTERIUM TUMEFACIENS

The bacterium *A. tumefaciens* infects almond, apricot, cherry, peach, plum, apple, grape, poplar, raspberry, rose, walnut, willow, *Rubus*, chrysanthemum, citrus, *Ficus*, olive, pear, wild blackberries, marigold, clematis, gypsophila, lilac, macadamia, quince, and wisteria. It occasionally infects aster, birch, blueberry, dahlia, *Gossypium*, hydrangea, impatiens, maple, rhododendron, Sequoia, and spruce.

This bacterium is distributed worldwide but causes serious damages in the United States, Europe, and Australia.

The major diseases caused by this pathogen are crown galls and root galls on its host plants.

Infected plants, raised particularly from grafting, layering, and cutting, and those that have tumors on the main roots and collars are unfit for marketing. The highest losses occur in young plants, that is, those that are still in the nursery. Even though all plant materials with visible crown galls are eliminated in any given nursery, healthy looking but still infected trees from the same nursery can be purchased and planted by growers. This contributes to the spread of crown gall. The disease seldom kills plants, but it can elicit lack of vigor and reduced growth. Diseased crops may be poorly productive and exhibit a marked dwarfing, as observed in almond (Flint, 2002). Sobiczewski et al. (1991) reported that in water-deficient conditions, 1-year-old shoots of crown galled mazzard cherries were 50% shorter and the crown diameter was 25% smaller than in healthy plants. In China, tumorigenic *Agrobacterium* caused considerable economic loss in both the nursery industry and commercial stone fruit orchards. The incidence of disease in cherry (*Prunus avium* L.) and peach (*Prunus persica* L.) orchards reached 20%–90% (Li, 2008; Liu et al., 2009) in major cultivation areas, and major reductions in yield occurred in regions of the North China Plain and areas along the Yangtze River (Ma and Wang, 1995). There is an urgent need for prediction and management of crown gall disease in cherry and peach production.

1.3.4 XANTHOMONAS ORYZAE

The principal host of this bacterium is rice. This bacterium also attack a number of wild or minor cultivated grasses and weeds that belong to the family Poaceae.

The bacterium *X. oryzae* is geographically distributed in Russia, Asia (Bangladesh, Cambodia, China, India, Indonesia, Japan, Korea, Lao, Malaysia, Myanmar, Nepal, Pakistan, the Philippines, Sri Lanka, Taiwan, Thailand, Vietnam), Africa (Burkina Faso, Cameroon, Gabon, Mali, Niger, Senegal, Togo), North America (Mexico), the United States (Louisiana, Texas), and South America (Bolivia, Colombia, Ecuador, Venezuela, Costa Rica, El Salvador, Honduras, Nicaragua, Panama), and Australia.

Bacterial leaf blight (BLB) is the most serious disease of rice in Southeast Asia, particularly due to the widespread cultivation of dwarf, high-yielding cultivars (Ray and Sengupta, 1970; Feakin, 1971). In 1954, in Japan, 90,000–150,000 ha was affected and annual losses were estimated at 22,000–110,000 tons. The disease was first reported in India in 1951, but it was not until 1963 that an epiphytotic occurred. In the Philippines, present losses are of the order of 22.5% in wet and 7.2% in dry seasons in susceptible crops and 9.5% and 1.8%, respectively, in resistant crops (Exconde, 1973). Nitrogen fertilization considerably increases susceptibility. Losses are generally insignificant in the less fertile soils and in summer-grown crops (December–April). Transplanted autumn (May–September) and winter (July–December) crops, however, suffer considerable losses. Diseased crops contain a high proportion of chaffy grains. This global situation has been reviewed by an international workshop held in Manila, the Philippines (Banta, 1989). Yield losses of 10%–50% from BLB have been reported (Ou, 1972). Outbreaks of bacterial leaf blight (BLB) are most common during the monsoon season in Southeast Asia and India (Mew et al., 1993).

1.3.5 XANTHOMONAS CAMPESTRIS PATHOVARS

The major host of *X. campestris* pathovar includes cauliflower, cabbage, broccoli, brussel, kohlrabi, Chinese cabbage, rutabaga, turnip, cole and rape, charlock, mustard, radish, and many other hosts.

This bacterium is geographically distributed worldwide and is present in Africa, Asia, Australia, Oceania, Europe, North America, Central America, West Indies, and South America. It is also present in the Philippines, Norway, Mauritius, Bulgaria, Holland, South Africa, Rhodesia, Queensland, Mozambique, Greece, Israel, Yugoslavia, Hawaii, Thailand, Basilicata, and India.

The major disease caused by this pathogen is blackleg of cruciferous crops and *Xanthomonas* wilt of banana, besides leaf spot or blight diseases on various other crops.

Pathovars of *X. campestris* cause diseases of agronomic importance throughout the world. Among the most notable of these pathogens are *X. campestris* pv. *campestris* (Xcc), the causal agent of black rot of crucifers that affects all cultivated brassicas; *X. campestris* pv. *vesicatoria* (Xcv), now reclassified as *Xanthomonas euvesicatoria*, the causal agent of bacterial spot of pepper and tomato; and *X. campestris* pv. *musacearum*, the causal agent of enset wilt.

These diseases are particularly severe in regions with a warm and humid climate, although black rot has economical implication in temperate regions as well, for example, in Cornwall and other western areas of the United Kingdom.

1.3.5.1 *Xanthomonas campestris* pv. *musacearum*

X. campestris pv. *musacearum* is reported to cause sporadic, but often considerable, losses in the genus *Ensete* in Ethiopia, sometimes causing farmers to abandon severely affected gardens for up to 5 years (Yirgou and Bradbury, 1968). It is likely that the economic and social impacts of *X. campestris* pv. *musacearum* in Uganda will be devastating, given the importance of banana both as a source of food and income. Preliminary evidence suggests that production from affected cultivations can be reduced by more than 90% in less than a year of the first appearance of the disease. Although many stools survive initial infection and may continue to produce suckers, these suckers often wilt before they flower, and the fruit produced is usually affected and inedible. Further spread of the disease in East Africa and beyond would thus be of very considerable significance to both the economy and food security of the region (CABI, 2015). The disease caused US$4 billion losses in Uganda during the period 2001–2006 with 32%–52% decrease in yield, a substantial cause of poverty in marginal and poor farmers.

1.3.5.2 *Xanthomonas campestris* pv. *campestris*

Black rot is considered a major disease of crucifers worldwide. For many years the disease was considered of relatively minor importance to crucifer growers in the major northern production areas of the Unites States and Western Europe. Outbreaks of the disease were sporadic and limited. During the late 1960s and early 1970s and 1990s, the frequency and severity of the disease increased. Approximately 70% of several million transplants from one single seedbed were systemically infected in the United States in 1973 (Williams, 1980). In 1976, losses of US$1 million were estimated (Kennedy and Alcorn, 1980). In Canada, rutabaga (swede) producers lost up to 60% of their crops to black rot during the winter of 1979–1980 (McKeen, 1981).

Nemeth and Laszlo (1983) reported black rot as the cause of considerable damage in cabbage and cauliflower in Hungary. Radunovic and Balaz (2012) reported the presence of black rot in cabbage, kale, broccoli, and collard crops. In some regions of Russia, black rot caused 23%–57% losses in susceptible cabbage cultivars (Ignatov, 1992). Recurrent black rot epidemics have been reported from Italy during 1992–1994 and 1997 (Caponero and Iacobelis, 1994; Scortichini, 1994; Catara et al., 1999). In Israel, black rot causes major economic losses in cabbage, cauliflower, radish, and kohlrabi, especially during the winter season (Kritzman and Ben-Yephet, 1990). In Korea, black rot is considered a major disease in cabbage (Kim, 1986). Surveys conducted in 25 crucifer fields in 8 provinces in Thailand, where black rot is known to cause severe losses, revealed the presence of the black rot organism from plants showing disease symptoms in 21 fields (Schaad and Thaveechai, 1983). Infected seed lots were reported from commercial seed plots in Japan (Shiomi, 1992), and in 1997–1998, black rot infected 50%–90% of plants of susceptible cabbage cultivars grown in three prefectures of Japan (Ignatov et al., 1997). During 1989–1992, *X. campestris* pv. *campestris* caused seed yield reductions in cauliflower in India (Shyam et al., 1994). In Himachal Pradesh (India), curd rot of cauliflower has been a menace to the seed crop and is the cause of huge losses to farmers (Shyam et al., 1994). Black rot appears annually in Manipur (India) near the end of February. Its effects are more severe (up to 50% losses) in susceptible cultivars (Gupta et al., 1991). The widespread occurrence of black rot in Rajasthan, with a high incidence of seed infection, can be the cause of severe losses (Sharma et al., 1992).

In Kenya, black rot is endemic and causes much damage (Onsando, 1988, 1992). The disease is considered of intermediate economic importance in Mozambique (Plumb-Dhindsa and Mondjane, 1984). Black rot is widespread in Zimbabwe, where it is considered the most significant disease in brassicas (Mguni, 1987, 1995). The pathogen was found in crucifer crops from the five agroecological regions of the country. Disease incidence was higher during 1994 (10%–80%) than during 1995 (10%–50%) (Mguni, 1996). In Mozambique, the disease was reported in the southern districts of Boane, Mahotas, and Chokwe. In Boane, the highest incidence of black rot was recorded on the Copenhagen Market variety (70%), followed by Starke (67.9%) and Glory F1 (67.3%).

1.3.5.3   *Xanthomonas campestris* pv. *vesicatoria*

This bacterium is known to cause infection and damage in capsicum and tomato plants; however, a severe epidemic of the bacterial infection in tomato was reported in 1992 in Nashik district of India (Borkar, 1997). Tomato cultivation in Nashik occupies over 14,000 ha of land around the year with a net profit of INR 0.1 million/ha, as 60% of tomatoes are transported to other Indian states and abroad. Infection caused by *X. campestris* pv. *vesicatoria* was consistently appearing in tomato cultivation in that area since 1990, but during kharif 1992, tomato crop succumbed heavily to the disease throughout the cultivated area due to favorable epiphytotic conditions (26°C–30°C temperature, 93%–97% RH (relative humidity) coupled with either rain/downpours and cloudiness), and within 10–15 days, the entire tomato cultivation was devastated, and the farmers had only one picking compared to four to five normal picking. Most of the tomato hybrids under cultivation were susceptible to the disease. Market surveys revealed that there was a 40% decline in tomato supply to the market, and the losses due to the disease caused by this pathogen were estimated at INR 190 million.

1.3.6   *Xanthomonas axonopodis*

The bacterium *X. axonopodis* and its pathovars majorly infect citrus plants, pomegranate, cassava, and beans.

This bacterium is geographically distributed in Asia (India, Indonesia, Japan, Malaysia, the Philippines, Southeast Asia, Taiwan, Thailand), Africa (Benin, Cameroon, Comoros, Congo, Cote d'Ivore, Ghana, Madagascar, Malawi, Mali, Mauritius, Niger, Nigeria, Rwanda, South Africa, Sudan, Tanzania, Togo, Uganda, Zaire), America (Argentina, Barbados, Brazil, Caribbean, Cuba, Dominican Republic, French Guiana, Mexico, Nicaragua, Panama, South America, Trinidad and Tobago, Venezuela), and Oceania (Guam, Micronesia, Palau).

The major diseases caused by this pathogen is citrus canker, oily spot of pomegranate, and bacterial blight of cassava.

The genus *Xanthomonas* currently consists of 20 species, including *X. axonopodis* (Vauterin et al., 2000). Six distinct genomic groups have been defined within *X. axonopodis*, with many pathovars causing economically significant diseases on different host plants of agronomic significance (Rademaker et al., 2005; Young et al., 2008).

The most important pathovars of *X. axonopodis* causing heavy losses are as follows.

1.3.6.1   *Xanthomonas axonopodis* pv. *manihotis*

Cassava (*Manihot esculenta*) is the staple food of nearly 600 million people in the world's tropical regions. *X. axonopodis* pv. *manihotis* (Xam) is the causal agent of Cassava Bacterial Blight (CBB), a major disease, endemic in tropical and subtropical areas. Xam induces a wide combination of symptoms, including angular leaf lesions, blight, wilt, stem exudates, and stem canker. This foliar and vascular disease severely affects cassava production worldwide. Losses between 12% and 100% affect both yield and planting material (Lozano, 1986; Verdier et al., 2004). Over recent years, significant recurrence of the disease has been reported in different regions of Africa and Asia. Host resistance is still the most effective way to control this disease; however, no breeding strategy has been developed.

1.3.6.2   *Xanthomonas axonopodis* pv. *phaseoli*

X. axonopodis pv. *phaseoli* causes the most severe disease under fairly high temperature conditions (25°C–35°C), coupled with high rainfall and humidity. In the EPPO (European and Mediterranean Plant Protection Organization) region, (refer to 15.4.5) it is mainly present in eastern and southern areas, where it has a rather variable impact. As early as 1918, 75% of the fields in New York City (USA) were affected and serious losses occurred. In the following years, losses of 20%–50% were recorded. In 1953, the disease was widespread in western Nebraska (USA) and the loss caused was estimated to be more than US$1 million. In 1976, it was the most economically significant bacterial disease of beans in the United States, causing an estimated loss of US$4 million (Kennedy and Alcorn, 1980). Losses in the field bean crop in Ontario (Canada) varied from a maximum of 1,251,913 kg in 1970 to a minimum of 217,724 kg in 1972. In Romania, between 1962 and 1969, 45% of bacterial diseases of bean were caused by *X. axonopodis* pv. *phaseoli*, while in Hungary in 1974, only 4% were caused by this pathogen.

1.3.6.3   *Xanthomonas axonopodis* pv. *punicae*

Pomegranate is an important crop of the Asian continent, some states of Russia, tropical Africa, Latin America, and the United States, particularly California and Arizona.

In India, particularly in peninsular India comprising states of Maharashtra, Karnataka, and Andhra Pradesh, the crop is grown in 112,000 ha with a production of 792,000 metric tons of pomegranate fruits and a revenue of INR 15,840–63,360 million (US$316–US$1267 million); this was threatened in 2002 by a bacterial disease known as oily spot of pomegranate caused by *X. axonopodis* pv. *punicae* and spread to assumed epidemic form by 2007. The bacterium is so destructive that it devastated the entire orchards of the cultivators within one crop season, causing 80%–100% yield losses, under favorable conditions. The monetary losses per hectare ranges from INR 250,000 to 500,000, and the total losses for pomegranate cultivation in pomegranate growing states were around INR 23,183 million, indicating the seriousness and the devastating ability of this pathogen.

Considering the devastating nature of the disease, the Government of India has provided INR 50,000 per orchard to the farmers to combat the disease situation with the approved total outlay of INR 10,000 million for combating the situation in epidemic/nonepidemic areas in 2 years. However, due to the formation of streptocycline resistance in this disease-causing bacterium, control of the disease in infected fields was nonsatisfactory. At Mahatma Phule Agricultural University, Rahuri, Maharashtra, India, the strategies to combat this pathogen were studied by Professor S.G. Borkar, head, Department of Plant Pathology, with Dr. K.S. Raghuwanshi, associate professor in the department, who searched for resistant cultivars and effective chemical formulations and found 2 out of 29 formulations to be effective.

Based on laboratory results, a protocol known as MPKV protocol was formulated and tested in 10 heavily infected orchard farms of epidemic areas of Nashik in Maharashtra on 25-acre fields. The fields where MPKV protocol was not followed were found heavily infected by the disease with 100% fruit infection, whereas in the fields where MPKV protocol was followed, there was no infection and the orchards were free from this disease. With this protocol, virtually the infected gardens were rejuvenated with no disease symptoms. Therefore, based on the field research trial results, it has been recommended to follow MPKV protocol for the control/eradication of oily spot disease of pomegranate. The farmers who followed this protocol observed no infection of this dreaded bacterial disease in their fields and thus eradicated the disease from their orchards.

1.3.7 *ERWINIA AMYLOVORA*

The bacterium *E. amylovora* infects apple, pear, June berry, quince, medlar, firethorn, mountain ash, white beam, loquat, *Photinia* sp., hawthorn, *Cotoneaster*, and raspberry as its major host plants.

This bacterium is geographically distributed in North America (Hudson Valley [New York], California) Europe (Italy, Ireland), the Middle East, New Zealand, Australia, and Northern Japan.

The major disease caused by this pathogen is fire blight of apple and pear.

The global economical impact of fire blight is difficult to assess as often fire blight incidences, other than major outbreaks, are not reported. When reported, they relate to the loss of crop for a year, while the economic impact may last for up to 7 years (time required for new trees to be in full production). Furthermore, outbreaks are irregular and mostly unpredictable. Few publications assess the cost of fire blight. In their review, Bonn and van der Zwet (2000) give the cost of the reported cases of fire blight, for different parts of the world.

It is generally considered that fire blight—with exceptions for particular years and cultivars—is unlikely to cause severe damage in Northern Europe (the United Kingdom, Sweden, Norway, and Denmark). In contrast, the threat is high for susceptible cultivars (not only pear but also a number of recently released apple varieties and a number of ornamentals) in Southern and Central Europe.

Conversely, fire blight may be considered a disease of usually minor direct influence for a number of apple cultivars (such as Golden Delicious) and ornamentals in most areas. But the risk of unusual climatic conditions conducive to disease activity is inevitable. This is illustrated, for example, by unexpected infections in UK cider apples in 1980 and 1982 (Gwynne, 1984); in pears and apples in Aquitaine, Anjou, and Paris in France in 1978 and 1984, (Lecomte and Paulin, 1989); and in apple and pear flowers in Switzerland in 1995 (Mani et al., 1996).

In Crimea, regional pear varieties are susceptible to *E. amylovora*, which can cause losses of 60%–90% of flowers and buds in some years and reduce yields by a factor of 8–10 (Kalinichenko and Kalinichenko, 1983).

In Egypt, the first fire blight outbreaks since 1962 were recorded in 1982 and were associated with heavy rainfall during bloom. Outbreaks also occurred in 1983 and 1984, which were associated with rainfall combined with windstorms during bloom and 2 days of continuous rainfall during bloom, respectively. The severe occurrence of the disease was expressed mainly as flower blight and caused a loss of 10%–75% flowers per tree (van der Zwet, 1986).

Fire blight is said to be the reason for decreasing pear production in the eastern United States (van der Zwet and Keil, 1979). Similarly it caused, directly or indirectly (following restrictive regulations), the progressive suppression of a number of cultivars in Europe: Laxton's Superb, Beurré Durondeau, Passe Crassane for pears, James Grieves and several cider varieties for apples, and the ornamentals *Cotoneaster salicifolius* and *Pyracantha atalantioides* (Paulin, 1996).

In the state of New York, a potential economic loss of the rootstock phase in fire blight was estimated at $8818/ha based on a 10% tree loss in high-density apple orchards (Momol et al., 1999).

An epidemic in southwest Michigan apple orchards in 2000 was particularly severe and followed unusually warm, humid, wet weather in May. It was estimated that between 350,000 and 450,000 apple trees will be killed and 1500–2300 acres of apple orchards will be lost; the development costs of these orchards were more than $9 million. Apple yields will be reduced by 35% over the region and some growers will experience losses of 100%. Out of the normal 4.5–7 million bushels produced in the region, the expected crop loss was 2.7 million bushels worth an estimated $10 million. It was estimated that the cumulative loss of yield will be $36 million since 5 years will be required for the region to recover. This will bring the total economic loss in the region to ca. $42 million (Longstroth, 2001). The total loss might have surpassed those estimates that were made in 2000, as growers found out that partly infected orchards did not prove economical.

1.3.8 *Xylella fastidiosa*

The bacterium *Xy. fastidiosa* (Xanthomonadales, Xanthomonadaceae) infects grapevine, citrus, almond, elm, oak, oleander, maple, sycamore, coffee, peach, mulberry, plum, pear, pecan, periwinkle, and so on as its host plants.

This bacterium is geographically distributed in Asia (India, Taiwan), North America (Mexico, the United States), Central America (Costa Rica), and South America (Argentina, Brazil, Venezuela).

The major diseases caused by this pathogen are Pierce's disease (PD) of grape, citrus variegated chlorosis (CVC), and almond leaf scorch disease (ALSD).

Xy. fastidiosa is the only species in the genus, but different strains have been well characterized as pathotypes, with cross-infections among different hosts and strains having been reported, but without the development of disease symptoms.

In the United States, within the main areas where *Xy. fastidiosa* occurs naturally (coastal plains of the Gulf of Mexico), *Vitis vinifera* and *Vitis labrusca* cannot be cultivated because they are rapidly infected due to high rates of natural spread. As a consequence, only selections of *V. rotundifolia* (muscadine) and specially bred resistant hybrids can be cultivated. The same situation exists throughout tropical America. In California, however, *Xy. fastidiosa* occurs only in "hot spots." *V. vinifera* has to be cultivated outside these hot spots. There have been considerable losses in the past before this situation was clarified. Control is based essentially on the principle of locating and delimiting hot spots (Goheen and Hopkins, 1988) by trapping insect vectors and testing wild hosts serologically. Vector habitats can be eliminated as a preventative measure, but this is not possible in all situations. Control has been attempted by antibiotic treatment of grapevines against *Xy. fastidiosa* and by insecticide treatment against its vectors, but with only partial success. These methods are little used in practice.

Pierce's disease is thus a major constraint in grapevine production in the United States and tropical America. However, it does not occur in all grapevine-producing areas of the United States and has apparently no tendency to spread to uninfested areas. The distribution of *Xy. fastidiosa* thus appears to be limited by climatic constraints, affecting the bacterium itself or its vectors.

By contrast, in peaches, phony disease does not kill trees or cause dieback, but it does significantly reduce the size and number of fruits. An analysis of its biophysical effects on peach trees was done by Anderson and French (1987). The disease was widespread in the southeastern United States in the 1940s, when 5-year-old orchards were often found to be 50% affected and older orchards entirely so. However, the efficient control methods now available (insecticides, destruction of infected trees, elimination of wild host plants around orchards) allow a high degree of control, except in areas with high incidence.

In citrus, variegated chlorosis has been described by Roistacher as "the world's most destructive disease of sweet orange." It has spread rapidly through large areas of southern Brazil and appears to be unstoppable. In a June survey made in 1990, it was reported that 13 sites out of 920 examined in São Paulo had infected trees. In a follow-up survey in August 1991, 72 out of the 920 sites were infected (a fivefold increase in a period of just 14 months). A total of 1.8 million trees are now infected, and some growers in São Paulo are now planting mangoes instead of citrus (EPPO/OEPP, 1992).

1.3.9 *Dickeya* (*dadantii* and *solani*)

The bacterium *Dickeya* infects potato and hyacinth bulb as its major host plants.

The bacterium *Dickeya solani* is geographically distributed in Europe, Israel, England, Wales, Scotland, Belgium, Finland, France, Poland, the Netherlands, and Spain.

The major disease caused by this pathogen is *D. solani* rot of potato.

Known earlier as *Erwinia chrysanthemi* (Echr), it is a complex of different bacteria now reclassified as species of *Dickeya*. While *Dickeya dadantii* and *Dickeya zeae* (formerly Echr biovar 3 or 8) are pathogens of potato in warmer countries, *Dickeya dianthicola* (formerly Echr biovar 1 and 7) appears to be spreading to potatoes in Europe.

The bacterial pathogen *D. solani* emerged as a major threat to potato (*Solanum tuberosum*) production in Europe in 2004 and has spread to many potato-growing regions via international trade. In December 2013, soft rot symptoms were observed in hyacinth (*Hyacinthus orientalis*) bulbs imported from the Netherlands into China at Ningbo Port.

A survey was carried out in the potato-growing regions of Zimbabwe in April 2009 to assess the prevalence of bacterial soft rot. A total of 125 tubers with soft rot symptoms were collected. The disease caused severe economic losses ranging from 20% to 60% on tubers in the field and in storage. Affected tubers had symptoms that ranged from light vascular discoloration to complete seed piece decay. Infected tuber tissue was often cream colored and soft to touch. In the field, plants showed severe wilting, often accompanied by a slimy brown necrosis of the lower stems (Ngadze et al., 2010).

In Israel, yield reductions of 20%–25% resulting from *Dickeya* infections have been recorded on various potato cultivars, where disease incidence was greater than 15% (Tsror et al., 2009). In Finland, in a 1-year field trial, a comparison was made between direct losses caused by *D. dianthicola* and *D. solani* measured by tuber and stem rot (Laurila et al., 2008). On average, no difference was found in the percentages of tuber decay (5%–6%) between these pathogens, but the percentage of stem rot was much higher for *D. dianthicola* (73% vs. 20%). However, most direct losses to potato production in Europe caused by *Dickeya* occurred as a result of downgrading or rejection of potatoes during seed tuber certification. Since national certification tolerances differ, the economic impact varies from country to country. Strict tolerances in the Netherlands have led to increased direct losses of up to €30M annually (Prins and Breukers, 2008) as a result of downgrading and rejection of seed tuber stocks caused by blackleg. However, it is not possible to differentiate losses caused by *Pectobacterium* and *Dickeya*. The appearance of symptoms during seed tuber certification and growing-crop inspections depends largely on the prevailing climate, especially during the early part of the season. The inoculum level on seed and the cultivar used may also play a role in symptom development, although inoculation of tubers with as little as 40 cells per gram of potato peel was sufficient to cause 30% and 15% diseased plants in field experiments in the Netherlands in 2005 and 2006, respectively (van der Wolf and De Boer, 2007). During official inspections in England and Wales, some potato crops found to be affected by either *D. dianthicola* or *D. solani* showed foliar symptoms ranging from <1% to 30% (Toth et al., 2011).

Crop losses can occur for seed tuber growers, suppliers, and exporters if these tubers are exported to warmer climates and break down as a result of the presence of *Dickeya* spp. With the emergence and spread of *D. solani*, together with the effects of climate change, more frequent losses of this kind can be expected. Data from Israel indicate the scale of potential losses that can be expected when seed tubers latently infected with *D. dianthicola* or *D. solani* are grown in warmer climates (Lumb et al., 1986; Tsror et al., 2006). For example, in the spring of 2005, seed tubers imported to Israel from Europe caused a severe outbreak of disease (subsequently found to be caused by *D. solani*) in more than 200 ha across different locations. Five cultivars were involved, with disease incidence ranging from 5% to 30% (8.2% average). In addition to foliar wilting symptoms, rotted progeny tubers were also observed in the field. In one cultivar, a high wilt incidence (30%) was observed, and when visually healthy progeny tubers were replanted, a further wilt incidence of

10%–15% was observed in the following autumn–winter season. In the spring of 2006, the disease was again observed in more than 260 ha in seven cultivars, with disease incidence ranging from 2% to 30% (10% average) (Tsror et al., 2009). In the spring of 2009, 12 cultivars were involved with a disease incidence of 0.5%–30%. When healthy potato seed tubers were planted in two locations in 2005, following the original diseased crop, no transmission of the disease to the healthy crop was observed and the pathogen was not detected on the progeny tubers. It was thus concluded that the pathogen is not significantly soil borne and its primary means of dispersal is by seed tubers.

1.3.10 *PECTOBACTERIUM CAROTOVORUM* (AND *PECTOBACTERIUM ATROSEPTICUM*)

The bacterium *Pectobacterium carotovora* infects many vegetable crops like tomato, chili, eggplant, cabbage, celery, onion, garlic, leek, sugarbeet, cauliflower, pepper, melon, chicory, cucurbits, carrot, yam, sweet potato, lettuce, cassava, banana, tobacco, orchids, beans, potato, and opium poppy.

This bacterium is geographically distributed worldwide, particularly in Europe, Asia, Africa, North America, South America, and Oceania.

The major disease caused by this pathogen is soft rot in its host plants.

The mean incidence of soft rot ranged from 10% to 100% in 34 potato varieties and selections evaluated for resistance to *Erwinia carotovora* subsp. *carotovora* and *E. carotovora* subsp. *atroseptica* (Reeves et al., 1999). The mean reduction of tuber weight ranged from 0.13% to 16.43%, and a significant correlation coefficient of soft rot incidence and reduction of tuber weight (%) was obtained. In tobacco, yield losses due to barn rots (*E. carotovora* subsp. *carotovora* and *Sclerotinia sclerotiorum*) and hollow stalk (*E. carotovora* subsp. *carotovora*) were 16.53 and 24.7 kg/ha, respectively, recorded in Japan in 1980, which were correlated with meteorological factors (Itagaki et al., 1983).

REFERENCES

Anderson, P.C. and W.J. French. 1987. Biophysical characteristics of peach trees infected with phony peach disease. *Physiol. Mol. Plant Pathol.*, 31, 25–40.

Banta, S.J. 1989. *Bacterial Blight of Rice*. IRRI, Manila, Philippines.

Bonn, W.G. and T. van der Zwet. 2000. Distribution and economic importance of fire blight. In: Vanneste, J.L. (ed.), *Fire Blight, the Disease and Its Causative Agent, Erwinia amylovora*. CABI, Wallingford, U.K., pp. 37–53.

Borkar, S.G. 1997. First epidemic of bacterium *Xanthomonas campestris*. pv. *vesicatoria* on tomato. In: *Proceedings of the International Conference on Integrated Plant Disease Management for Sustainable Agriculture*, November 10–15, 1997. IARI, New Delhi, pp. 310.

CABI. 2015. *Xanthomonas campestris* pv. *musacearum* (banana *Xanthomonas* wilt (BXW)). (n.d.). Retrieved June 30, 2015, from http://www.cabi.org/isc/datasheet/56917.

Caponero, A. and N.S. Iacobellis. 1994. Foci of black rot on cauliflower in Basilicata. *Informatore Agrario.*, 50, 67–68.

Catara, V., Branca, F., and P. Bella. 1999. Outbreak of 'black rot' of Brassicaceae in Sicily. *Informatore Fitopatologico*, 49, 7–10.

Donmez, M.F., Karlidag, H., and A. Esitken. 2010. Identification of resistance to bacterial canker (*Pseudomonas syringae* pv. *syringae*) disease on apricot genotypes grown in Turkey. *Eur. J. Plant Pathol.*, 126, 241–247.

Elphinstone, J.G. 2005. The current bacterial wilt situation: A global overview. In: Allen, C., Prior, P., and A.C. Hayward (eds.), *Bacterial Wilt Disease and the Ralstonia solanacearum Species Complex*. American Phytopathological Society Press, St Paul, MN, pp. 9–28.

EPPO/OEPP. Xylella fastidiosa. EPPO data sheets on quarantine organisms No. 166. EPPO Reporting Service 500/02, 505/13 and 1998/9.

Exconde, O.R. 1973. Yield losses due to bacterial leaf blight of rice. *Philippines Agric.*, 57, 128–140.

Feakin, S.D. 1971. Pest control in rice. *PANS Manual*, 3, 69–74.

Flint, M.L. 2002. *Integrated Pest Management for Almonds*, 2nd ed. State Wide Integrated Pest Management Project, Division of Agriculture and Natural Resources, University of California, Oakland, CA.

Goheen, A. C. and D. L. Hopkins. 1988. Pierce's Disease. Pp. 44–45 In: *Compendium of Grape Diseases*, R.C. Pearson and A.C. Goheen, eds. St. Paul, MN: American Phytopathological Society.

Green, S., Studholme, D.J., Laue, B.J. et al. 2010. Comparative genome analysis provides insights into the evolution and adaptation of *Pseudomonas syringae* pv. *aesculi* on *Aesculus hippocastanum*. *PLoS ONE*, 5, e10224.

Gupta, M., Parmar, Y.S., Bharat, N., Chauhan, A., and A. Vikram. 1991. First report of bacterial leaf spot of coriander caused by *Pseudomonas syringae* pv. *coriandricola* in India. *Plant Dis.*, 97(3), 418.

Gwynne, D.C. 1984. Fire blight in perry pears and cider apples in the South West of England. *Acta Hortic.*, 151, 41–47.

Hinrichs-Berger, J. 2004. Epidemiology of *Pseudomonas syringae* pathovars associated with decline of plum trees in the Southwest of Germany. *J. Phytopathol.*, 152, 153–160.

Ignatov, A., Vicente, J.G., Conway, J., Roberts, S.J., and J.D. Taylor. 1997. Identification of *Xanthomonas campestris* pv. *campestris* races and sources of resistance. In: *ISHS Symposium on Brassicas. 10th Crucifer Genetics Workshop*, Rennes, France, September 23–27, 1997, p. 215.

Ignatov, A.N. 1992. Resistance of head cabbage to black rot. PhD (Candidate of Sciences) thesis, TSKHA, Moscow, Russia, 25pp. (in Russian).

Itagaki, R., Araki, M., and I. Suzuki. 1983. Analysis of tobacco production under the abnormal weather of 1980 in the Northeastern District of Japan. IV. Relationship between weather conditions and losses in yield of tobacco caused by diseases and insect pests. *Bull. Morioka Tobacco Exp. Station*, 17, 53–62.

Kalinichenko, G.V. and R.I. Kalinichenko. 1983. Characteristics of fruiting of pear under bacterial infection in the Crimea. *Sel'skokhozyaistvennaya Biologiya*, 4, 68–71.

Kelman, A. 1998. One hundred and one years of research on bacterial wilt. In: Prior, P.H., Allen, C., and J. Elphinstone (eds.), *Bacterial Wilt Disease: Molecular and Ecological Aspects*. Springer, Heidelberg, Germany, pp. 1–5.

Kennedy, B.W. and S.M. Alcorn. 1980. Estimates of U.S. crop losses to prokaryote plant pathogens. *Plant Dis.*, 64, 674–676.

Kim, B.S. 1986. Testing for detection of *Xanthomonas campestris* pv. *campestris* in crucifer seeds and seed disinfection. *Korean J. Plant Pathol.*, 2, 96–101.

Kotan, R. and F. Sahin. 2002. First record of bacterial canker caused by *Pseudomonas syringae* pv. *syringae*, on apricot trees in Turkey. *Plant Pathol.*, 51, 798.

Kritzman, G. and Y. Ben-Yephet. 1990. Control by metham-sodium of *Xanthomonas campestris* pv. *campestris* and the pathogen's survival in soil. *Phytoparasitica*, 18(3), 217–227.

Laurila, J., Ahola, V., Lehtinen, A. et al. 2008. Characterisation of *Dickeya* strains isolated from potato and river water samples in Finland. *Eur. J. Plant Pathol.*, 122, 213–225.

Lecomte, P. and J.P. Paulin. 1989. Disease control in apple and pear orchards. *Phytoma*, 408, 22–26.

Li, Y.Y. 2008. Research of cherry crown gall disease in region of Dalian. Master dissertation, Liaoning Normal University, Liaoning, China.

Liu, C.H., Jia, K.G., Zhu, L.X., Li, H., Ye, H., and Y.C. Wang. 2009. Resistance of *Prunus persica* to crown gall disease. *J. China Agric. Univ.*, 14, 68–71.

Longstroth M, 2001. The 2000 fire blight epidemic in southwest Michigan apple orchards. The Compact Fruit Tree 34, 16–19. Lopes CA, Stall RE, Bartz JA, 1986.

Lozano, J.C. 1986. Cassava bacterial blight: A manageable disease. *Plant Dis.*, 70, 1089–1093.

Lumb, V.M., Pèrombelon, M.C., and M.D. Zutra. 1986. Studies of a wilt disease of the potato plant in Israel caused by *Erwinia chrysanthemi*. *Plant Pathol.*, 35, 196–202.

Ma, D.Q. and H.M. Wang. 1995. Fruit crown gall disease and biological control. *China Fruits*, 2, 42–44.

Mani, E., Hasler, T., and J.D. CharriFre. 1996. How much do bees contribute to the spread of fire blight? *SchweizerischeBienen-Zeitung*, 119(3), 135–140.

Mansfield, J., Genin, S., Magori, S. et al. 2012. Top 10 plant pathogenic bacteria in molecular plant pathology. *Mol. Plant Pathol.*, 13, 614–629.

McKeen, W.E. 1981. Black rot of rutabaga in Ontario and its control. *Can. J. Plant Pathol.*, 3, 244–246.

Mew, T., Alvarez, A., Leach, J., and J. Swings. 1993. Focus on bacterial blight of rice. *Plant Dis.*, 77, 5–12.

Mguni, C.M. 1987. Diseases of crops in the semiarid areas of Zimbabwe: Can they be controlled economically. In: Cropping *in the Semiarid Areas of Zimbabwe. Workshop Proceedings*, Vol. 2. Agritex, Harare, Zimbabwe, pp. 417–433.

Mguni, C.M. 1995. Cabbage research in Zimbabwe. In: *Brassica Planning Workshop for East and South Africa Region*, May 15–18, 1995, Lilongwe, Malawi. GTZ/IPM Horticulture, Nairobi, Kenya, p. 31.

Mguni, C.M. 1996. Bacterial black rot (*Xanthomonas campestris* pv. *campestris*) of vegetable brassicas in Zimbabwe. PhD thesis, The Royal Veterinary and Agricultural University, Copenhagen, Denmark, 144pp.

Momol, M.T., Norelli, J.L., Aldwinckle, H.S., and D.I. Breth. 1999. Internal movement of *Erwinia amylovora* from infection in the scion and economic loss estimates due to the rootstock phase of fire blight of apple. *Acta Hortic.*, 489, 505–507.

Morris, C.E., Kinkel, L.L., Xiao, K., Prior, P., and D.C. Sands. 2007. Surprising niche for the plant pathogen *Pseudomonas syringae*. *Infect. Genet. Evol.*, 7, 84–92.

Murillo, J., Bardaji, L., and E. Führer. 2010. La grasa de lasjudías, causadapor la bacteria *Pseudomonas syringae* pv. *phaseolicola*. *Phytoma*, 224, 27–32.

Nemeth, J. and E.M. Laszlo. 1983. Bacterial black rot (*Xanthomonas campestris* (Pammel) Dowson) of *Brassica species*. *Novenyvedelem*, 19, 391–397.

Ngadze, E., Coutinho, T.A., and J.E. van der Waals. 2010. First report of soft rot of potatoes caused by *Dickeya dadantii* in Zimbabwe. *Plant Dis.*, 94, 1263. http://dx.doi.org/10.1094/PDIS-05-10-0361.

Onsando, J.M. 1988. Management of black rot of cabbage caused by *Xanthomonas campestris* pv. *campestris* in Kenya. *Acta Hortic.*, 218, 311–314.

Onsando, J.M. 1992. *Black Rot of Crucifers*. Prentice Hall, Englewood Cliffs, NJ, pp. 243–252.

Ou, S.H. 1972. *Rice Diseases*. Commonwealth Mycological Institute, Kew, Surrey, U.K., p. 368.

Ozaktan, H., Akkopru, A., Bozkurt, A., and M. Erdal. 2008. Information on peach bacterial canker in Aegean Region of Turkey. In: *Proceedings of STF Meeting on "Determination of the incidence of the different pathovars of Pseudomonas syringae in stone fruits" COST Action 873 "Bacterial diseases of stone fruits and nuts,"* at 27th-28th March 2008, Skierniewice, Poland, p. 8.

Paulin, J.P. 1996. Control of fire blight in European pome fruits. *Outlook Agric.*, 25(1), 49–55.

Plumb-Dhindsa, P. and A.M. Mondjane. 1984. Index of plant diseases and associated organisms of Mozambique. *Trop. Pest Manage.*, 30, 407–429.

Prins, H. and A. Breukers. 2008. *In de Puree? De Gevolgen van Aantasting door Erwinia voor de Pootaardappelsector in KaartGebracht*. LEI, Den Haag, the Netherlands.

Rademaker, J.L.W., Louws, F.J., Schultz, M.H. et al. 2005. A comprehensive species to strain taxonomic framework for *Xanthomonas*. *Phytopathology*, 95, 1088–1111.

Radunovic, D. and J. Balaz. 2012. Occurrence of *Xanthomonas campestris* pv. *campestris* (Pammel) Dowson 1939, on Brassicas in Montenegro. *Pestic. Phytomed. (Belgrade)*, 27(2), 131–140. Ignatov doi: 10.2298/PIF1202131R.

Ray, P.R. and T.K. Sengupta.1970. A study on the extent of loss of yield in rice due to bacterial blight. *Indian Phytopathol.*, 23, 713–714.

Reeves, A.F., Olanya, O.M., Hunter, J.H., and J.M. Wells. 1999. Evaluation of potato varieties and selections for resistance to bacterial soft rot. *Am. J. Potato Res.*, 76, 183–189.

Schaad, N.W. and N. Thaveechai. 1983. Black rot of crucifers in Thailand. *Plant Dis.*, 67, 1231–1234.

Scortichini, M. 1994. Occurrence of *Pseudomonas syringae* pv. *actinidiae* on kiwifruit in Italy. *Plant Pathol.*, 43, 1035–1038.

Scortichini, M. 2002. Bacterial canker and decline of European hazelnut. *Plant Dis.*, 86, 704–709.

Scortichini, M. 2006. Severe outbreak of *Pseudomonas syringae* pv. *syringae* on new apricot cultivars in Central Italy. *J. Plant Pathol.*, 88, 65–70.

Sharma, J., Agrawal, K., and D. Singh. 1992. Detection of *Xanthomonas campestris* pv. *campestris* (Pammel) Dowson infection in rape and mustard seeds. *Seed Res.*, 20, 128–133.

Shenge, K.C., Mabagala, R.B., Mortensen, C.N., Stephan, D., and K. Wydra. 2007. First report of bacterial speck of tomato caused by *Pseudomonas syringae* pv. *tomato* in Tanzania. *Plant Dis.*, 91, 462.

Shiomi, T. 1992. Black rot of cabbage seeds and its disinfection under a hot-air treatment. *Japan. Agric. Res. Q.*, 26, 13–18.

Shyam, K.R., Gupta, S.K., and R.K. Mandradia. 1994. Prevalence of different types of curd rots and extent of yield loss due to plant mortality in cauliflower seed crop. *Indian J. Mycol. Plant. Pathol.*, 24(3), 172–175.

Smith, I.M., Dunez, J., Lelliott, R.A., Phillips, D.H., and S.A. Archer. 1988. *European Handbook of Plant Diseases*. Blackwell Scientific Publications, Oxford, U.K.

Sobiczewski, P., Karczewski, J., and S. Berczynski. 1991. Biological control of crown gall *Agrobacterium tumefaciens* in Poland. *Fruit Sci. Rep.*, 18, 125–132.

Spotts, R.A., Wallis, K.M., Serdani, M., and A.N. Azarenko. 2010. Bacterial canker of sweet cherry in Oregon—Infection of horticultural and natural wounds, and resistance of cultivar and rootstock combinations. *Plant Dis.*, 94, 345–350.

Toth, I.K., van der Wolf, J.M., and G. Saddler. 2011. *Dickeya* species: An emerging problem for potato production in Europe. *Plant Pathol.*, 60, 385–399.

Tsror, L., Erlich, O., Lebiush, S., Zig, U., and J.J. van de Haar. 2006. Recent outbreak of *Erwinia chrysanthemi* in Israel—Epidemiology and monitoring in seed tubers. In: Elphinstone, J.G., Weller, S., Thwaites, R., Parkinson, N., Stead, D.E., and G. Saddler (eds.), *Proceedings of the 11th International Conference on Plant Pathogenic Bacteria*, July 10–14, 2006, Edinburgh, Scotland, p. 70.

Tsror, L., Erlich, O., Lebiush, S. et al. 2009. Assessment of recent outbreaks of *Dickeya* sp. (syn. *Erwinia chrysanthemi*) slow wilt in potato crops in Israel. *Eur. J. Plant Pathol.*, 123, 311–320.

van der Wolf, J.M. and S.H. De Boer. 2007. Bacterial pathogens of potato. In: Vreugdenhil, D. (ed.), *Potato Biology and Biotechnology: Advances and Perspectives*. Elsevier, Oxford, U.K., pp. 595–617.

Vauterin, L., Rademaker, J., and J. Swings, 2000. Synopsis on the taxonomy of the genus *Xanthomonas*. *Phytopathol.*, 7, 677–682.

Verdier, V., Restrepo, S., Mosquera, G., Jorge, V., and C. Lopez. 2004. Recent progress in the characterization of molecular determinants in the *Xanthomonas axonopodis* pv. *manihotis–cassava* interaction. *Plant Mol. Biol.*, 56, 573–584.

Wang, J.F. and C.H. Lin. 2005. *Integrated Management of Tomato Bacterial Wilt*. AVRDC-The World Vegetable Center, Tainan, Taiwan.

Williams, P.H. 1980. Black rot: A continuing threat to world crucifers. *Plant Dis.*, 64, 736–745.

Yirgou, D. and J.F. Bradbury. 1968. Bacterial wilt of enset (*Ensete ventricosum*) incited by *Xanthomonas musacearum* sp. nov. *Phytopathology*, 58, 111–112.

Young, J.M., Park, D.C., Shearman, H.M., and E. Fargier. 2008. A multilocus sequence analysis of the genus *Xanthomonas. Syst. Appl. Microbiol.*, 5, 366–377.

Zwet T. van der. 1986. Identification, symptomatology, and epidemiology of fire blight on Le Conte pear in the Nile Delta of Egypt. *Plant Dis.*, 70(3), 230–234.

Zwet T. van der and H.L. Keil. 1979. Fire blight, a bacterial disease of rosaceous plants. Agriculture Handbook No. 510. Science and Education Administration of USDA, Beltsville, MD, 200pp.

SUGGESTED READING

Golkhandan, E., Kamaruzaman, S., Sariah, M., Zainal Abidin, M.A., and A. Nasehi. 2013. Characterisation of *Pectobacterium carotovorum* causing soft rot on *Kalanchoe gastonis-bonnierii* in Malaysia. *Archiv. Phytopathol. Plant Protect.*, 46(15), 1809–1815. http://dx.doi.org/10.1080/03235408.2013.778452.

Hayward, A.C. 1983. *Pseudomonas solanacearum*: Bacterial wilt and moko disease. In: Fahy, P.C. and G.J. Persley (eds.), *Plant Bacterial Diseases*. Academic Press, Sydney, New South Wales, Australia, pp. 129–135.

Vidaver, A.K., and Lambrecht, P.A. 2004. Bacteria as plant pathogens. The Plant Health Instructor. On-line DOI:10.1094/PHI-I-2004-0809-01.

Vidhyasekaran, P. 2002. *Bacterial Disease Resistance in Plants. Molecular Biology and Biotechnological Applications*. The Haworth Press, Binghamton, NY, 452pp.

Yabuuchi, E., Kosako, Y., Yano, I., Hotta, H., and Y. Nishiuchi. 1995. Transfer of two *Burkholderia* and an *Alcaligenes* species to *Ralstonia* gen. nov.: Proposal of *Ralstonia pickettii* (Ralston, Palleroni and Doudoroff 1973) comb. nov., *Ralstonia solanacearum* (Smith 1896) comb. nov. and *Ralstonia eutropha* (Davis 1969) comb. nov. *Microbiol. Immunol.*, 39, 897–904.

2 Bacterial Diseases of Cereal Crops

2.1 WHEAT

2.1.1 Black Chaff of Wheat

Pathogen: *Xanthomonas translucens* pv. *translucens* (Jones et al.) Vauterin et al.

Synonyms: *Xanthomonas campestris* pv. *translucens* (Jones et al.) Dye
X. campestris pv. *cerealis* (Hagborg) Dye
X. campestris pv. *hordei* (Hagborg) Dye
X. campestris pv. *secalis* (Reddy et al.) Dye
X. campestris pv. *undulosa* (Smith et al.) Dye

Common names: Black chaff, leaf streak, bacterial stripe, or bacterial blight of cereals and grasses

Symptoms

Infected leaves show narrow, water-soaked streaks (yellowish in barley and triticale), necrotic at the center with a rust-colored margin (in wheat). On young leaves, generally, small, water-soaked leaf spots appear that later develop into longitudinal yellow to brown streaks (Smith, 1917; El Banoby and Rudolph, 1989; Duveiller and Maraite, 1993b). Leaves may wilt. In wet weather, small droplets of yellowish bacterial slime can be observed on the lesions (Jones et al., 1917) (Photo 2.1). Symptoms often develop in the middle of the leaf, where dew remains longer in the morning. Seedlings hardly show any symptoms. Glumes and seeds show "black chaff" symptoms, with purple-black discoloration of the surface. Symptoms take 10–14 days to appear. Streaks are more usual on triticale than on wheat. Culms, leaves, rachis, glumes, and awns may become infected, and symptoms on wheat have been reported to vary with the environment, variety, disease severity, and interaction with fungi (Bamberg, 1936; Boosalis, 1952). On the stem, brown to black stripes may be formed. On the chaff of many hosts, water-soaked spots are formed that later develop into brown to black spots. Ears may show brown discoloration and distortion. Seeds may be black and shriveled. The heavily infected crops show water-soaked to yellow-brown streaks, wilting, and complete necrosis of leaves.

X. translucens pv. *translucens* has ice-nucleating activity (Kim et al., 1987) and may therefore be associated with frost injury (Sands and Fourrest, 1989). Dissemination of the bacterium may in turn be favored by this injury, since symptoms tend to appear after periods of subfreezing temperatures. However, ice nucleation is not a necessary condition for the induction of an epidemic (Duveiller et al., 1991).

Strains of *X. translucens* pv. *translucens* have been found to be host specific, and the original *formae speciales*, later pathovars, were defined in this way: *hordei* (= *translucens*) on barley, *secalis* on rye, and *undulosa* on wheat and triticale. pv. *cerealis* had a relatively wide host range. Duveiller (1989) noted that recent isolates from wheat, rye, and triticale were not host specific. The fact that specific and nonspecific strains can be found tends to support the use of a broad concept of the pathovar (Paul and Smith, 1989).

Host

Primary host: barley (*Hordeum vulgare*), rye (*Secale cereale*), wheat (*Triticum* spp.), and triticale (*Triticum* × *secale*).

Secondary host: *Bromus* spp., *Phalaris* spp., *Elymus repens*, and on other Poaceae by inoculation.

PHOTO 2.1 Symptoms of bacterial leaf streak on wheat. (Courtesy of Mary Burrows, Montana State University, Bozeman, MT, Bugwood.org.)

Geographical Distribution

EPPO region: Belgium (found but not established), Bulgaria (Koleva, 1981; but the bacterium is now declared absent), France (found but not established), Israel, Libya, Morocco, Romania, Russia (southern, Caucasus), Spain (single outbreak; Noval, 1989), Sweden (found but not established), Syria (Mamluk et al., 1990), Tunisia, Turkey (Demir and Ustun, 1992), Ukraine.

Asia: Azerbaijan, China (Henan, Xinjiang), Georgia, India (Delhi), Iran (Alizadeh and Rahimian, 1989), Israel, Japan, Kazakhstan, Malaysia (Sabah), Pakistan, Russia (Siberia), Syria, Turkey, Yemen.

Africa: Ethiopia, Kenya, Libya, Madagascar, Morocco, Senegal (probably a misidentification, since record is on rice), South Africa, Tanzania, Tunisia, Zambia.

North America: Canada (Alberta, Manitoba, New Brunswick, Quebec, Saskatchewan), Mexico, the United States (Arkansas, Colorado, Illinois, Indiana, Iowa, Kansas, Michigan, Minnesota, Mississippi, Missouri, Nebraska, North Carolina, North Dakota, Ohio, Oklahoma, South Dakota, Texas, Utah, Virginia, Washington, Wisconsin).

South America: Argentina, Bolivia, Brazil (Mato Grosso do Sul, Paranà), Paraguay, Peru, Uruguay.
Oceania: Australia (New South Wales).

Disease Cycle

X. translucens pv. *translucens* is a seed-borne pathogen. The transmission rate is very low but causes serious outbreaks in the field under suitable conditions. The pathogen is disseminated by seed on a large scale (Sands and Fourest, 1989). On a local scale, bacteria are transmitted by rain, dew, and contact between plants (Boosalis, 1952). Aphids trapped in sticky exudates may carry the bacterium and transmit it to wheat and barley, thus enabling long-distance dissemination (Boosalis, 1952). Milus and Mirlohi (1995) concluded that other means of overwintering were insignificant by comparison with survival on the seed.

  X. translucens pv. *translucens* is a true parenchymatous pathogen. Intercellular invasion occurs after entry through stomata. The spread from a single plant can affect up to 30 m^2 during a growing season. However, movement in space is usually more limited. Infection cycles can be as short as 10 days (Hall et al., 1981). Bacteria may survive in seeds longer than 63 months (Forster and Schaad, 1990). However, recovery is greatly decreased after some months of storage. Survival in the field does not only depend on the infection of host plants, as epiphytic populations may survive on non-host species as well (Timmer et al., 1987). Moreover, bacteria may overwinter on perennial hosts or

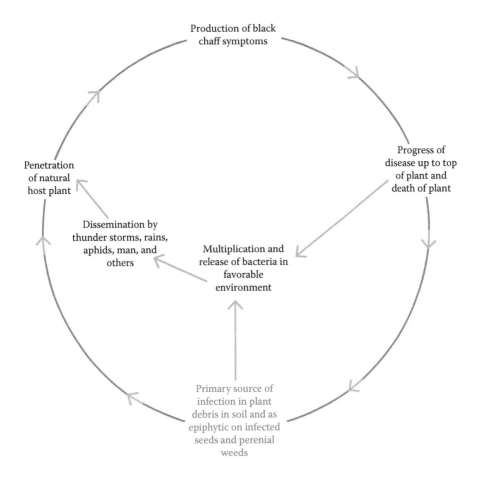

FIGURE 2.1   Disease cycle of black chaff of wheat caused *by X. translucens* pv. *translucens.*

on crop debris in the soil (Boosalis, 1952; Mehta, 1986a). The disease cycle of black chaff of wheat is illustrated in Figure 2.1.

Outbreaks of the disease are sporadic and more frequent on breeder's plots. They are prevalent during wet seasons. Inoculation experiments (Sands et al., 1986) show that plants are most readily infected under wet conditions (rain or sprinkler irrigation). However, the importance of dew, rainfall, or irrigation has not yet been documented, and it is not certain whether free water is needed. The bacterium tolerates a wide range of temperatures (15°C–30°C) (Duveiller et al., 1991), its optimal temperature being around 22°C. The pathogen grows best when relative humidity is high. The disease is favored by warm, moist conditions (26%–30%), especially at heading.

Economic Impact

Little quantitative information is available on losses caused by this disease (Duveiller, 1994a). Direct yield losses have been evaluated from 10% or less up to 40%. Duveiller and Maraite (1993a) have developed a system for forecasting losses from infection levels earlier in the season. Further Duveiller (1994b) has proposed a disease assessment key. The pathogen may also cause sterility of wheat spikes (Forster and Schaad, 1988). Finally, high infection levels may lead to a 10%–30% decrease in kernel weight (Shane et al., 1987). Both durum and bread wheats can be severely affected, while triticale and barley are less frequently affected. There seems to be little information on this disease on rye.

Control Measures
No known control measures exist for the disease in the field (Duveiller, 1994a). Chemical control focuses on seed treatments. Organomercurial seed treatments were used in the past and were mostly considered effective. The recent resurgence of the disease has been linked to the withdrawal of this group of pesticides. Duveiller (1994a), however, queries this and attributes the recent developments to other causes: cultivation of cereals in new areas, favorable conditions for the disease, susceptible cultivars, and so on. Various other treatments are now applied to seed lots to eliminate bacteria, especially cupric acetate (Schaad et al., 1981), formalin (Duveiller, 1989), and guazatine (Mehta, 1986b), but these treatments are phytotoxic. Panoctine 30 was 95% effective, without phytotoxicity. An alternative treatment is dry heat at 72°C for 7 days, as proposed by Fourest et al. (1990), but the effectiveness of this treatment remains to be confirmed.

Resistant cultivars are available for many cereals and the level of resistance depends on the cereal involved (Milus and Mirlohi, 1994). In the absence of any really effective seed treatment, control should center on pathogen-free seed certification (Duveiller, 1994a).

The use of healthy seeds through seed certification programs (and eventually by decontamination with acidic cupric acetate, but this is not 100% effective) and of varieties with low susceptibility or high resistance is the only way to prevent or control diseases caused by *X. translucens*.

2.1.2 BASAL GLUME ROT OF WHEAT

Pathogen: *Pseudomonas syringae* pv. *atrofaciens* (McCulloch) Young, Dye and Wilkie

Synonyms: *Pseudomonas atrofaciens and Phytomonas atrofaciens*
          Bacterium atrofaciens McClloch
          Ph. atrofaciens (McCulloch) Bergey et al.
          P. atrofaciens (McCulloch) Stevens

Common names: Basal glume rot, basal bacteriosis of wheat
Symptoms
Infected leaves have small, dark, water-soaked spots. The spots eventually elongate and become yellow, then necrotic as the tissue dies. Symptoms occur during wet weather, particularly at heading time. The main symptom is a brown, discolored area at the base of the glumes (Photo 2.2) that cover at the kernel. This discoloration is darker on the inside than on the outside of the glume. Usually, only one-third of the glume is discolored but sometimes the entire glume may be affected. Infrequently, the only sign of disease is a dark line at the attachment of the glume to the spike.

Severely diseased spikelets are slightly dwarfed and lighter in color than healthy ones. The base of a diseased kernel has a brown to black discoloration.

Host
Wheat, barley, oats, rye.

Geographical Distribution
Australia, Bulgaria, Canada, former Czechoslovakia, Morocco, New Zealand, Romania, Russia (Siberia), South Africa, Ukraine, the United States, and Zimbabwe (CMI, 1982); reported as a new record on durum wheat in Syria (Mamluk et al., 1990).

Disease Cycle
Basal bacteriosis actively develops in cool and damp years, particularly during cold, damp spring. Its distribution is promoted by low monthly average temperature (15°C–18°C) during the beginning of heading till maturing, with increased air humidity (60%–65% and more) and excessive precipitations just before grain ripening. The optimum temperature for basal bacteriosis development is 23°C–25°C.

P. syringae pv. *atrofaciens* survives epiphytically on seed or in infested soil residue. This bacterium is disseminated with dust particles by wind and becomes entrapped in the water in grooves and small

PHOTO 2.2   Field symptoms of bacterial infection on wheat glumes. (Courtesy of Mary Burrows, Montana State University, Bozeman, MT, Bugwood.org.)

spaces of the spikelets. It is also disseminated by insects and splashing water. It multiplies near glume joints in the presence of moisture but remains dormant when conditions are dry.

Economic Impact

In nature, the pathogen of basal bacteriosis of wheat also attacks rye, barley, and oats. Depending on the zone of wheat cultivation and weather conditions favorable for bacteriosis development, this disease can infect 10%–80% of ears of wheat plants during epiphytotics. In the Central Black Earth Region, basal bacteriosis severity ranges from 1% to 50% (at development 0.3%–25.3%), depending on a variety of spring wheat and conditions of cultivation; some varieties of winter wheat in Voronezh and Lipetsk regions can be infected up to 72%. In Krasnodar Region, about 36% of plants are affected by this disease with a severity of 50%–100%.

Control Measures

- Seed should be thoroughly cleaned and treated with a seed-protectant fungicide.
- Rotate wheat/barley with resistant crops, such as legumes.
- Destroy residue.
- Other control measures include optimal agriculture, maintenance of crop rotation, cultivation of relatively resistant varieties, careful removal of plant residues, separating seeds from shrunken grains, pesticide treatment of seeds before sowing, and treatment of plants by pesticides during vegetation period.

2.1.3   BACTERIAL LEAF BLIGHT OF WHEAT

Pathogen: *Pseudomonas syringae* pv. *syringae* van Hall

Synonyms: *P. syringae*

Common names: Bacterial blotch disease

Symptoms

Bacterial leaf blight develops on the uppermost leaves after plants reach the boot stage. Initially, small, water-soaked lesions appear and may expand and coalesce into irregular streaks or blotches within 2–3 days under cool, wet conditions. Lesions turn from grayish green to tan or white as tissues become necrotic. Ears and glumes may occasionally be infected, resulting in tan to brown necrotic spots with distinct margins.

Host

The pathogen has a very wide host range. Commonly attacked hosts are apple (*Malus* spp.), apricot (*Prunus armeniaca*), kidney bean (*Phaseolus vulgaris*), European bird cherry (*Prunus padus*), hawthorn (*Crataegus* spp.), lilac (*Syringa vulgaris*), orange (*Citrus sinensis*), peach (*Prunus persica*), plum (*Prunus domestica* and Japanese plum), poplar (*Populus* spp.), sweet cherry (*Prunus avium*), sour cherry (*Prunus cerasus*), and wheat (*Triticum aestivum*). Some host specialization has been reported (Little et al., 1998). *P. s.* pv. *syringae* can also cause stem canker and dieback in conifers (*Pinus radiata*; see Dick, 1985).

Control Measures

Hot-water treatment of seed at 53°C for 30 min and seed treatment with phytobactiomycin, quinolate, falisan, and carboxin are reported to be effective (Koleva, 1981).

2.1.4 Yellow Ear Rot of Wheat

Pathogen: *Rathayibacter tritici* (ex Hutchinson) Zgurskaya et al. (1993).

Synonyms: *Clavibacter tritici* (Carlson and Vidaver 1982) Davis et al.
             Corynebacterium tritici

Common names: Yellow ear rot of wheat, bacterial rot of wheat ears, tundu disease

Symptoms

The "tundu" disease is characterized by the twisting of the stem, distortion of the earhead, and rotting of the spikelets with a profuse oozing liquid from the affected tissues. Hence, the name of the disease is yellow ear rot. The ooze contains masses of bacterial cell and the nematode *Anguina tritici*. The nematode alone causes wrinkling, twisting, and various other distortions of the leaves and stems. Infected plants are shorter and thicker than healthy plants. In the distorted earheads, dark galls are found in place of kernels (Singh et al., 1959; Gupta, 1966; Midha, 1969). When the bacterium is associated with the nematode, the disease symptoms are intensified at the flowering stage and the yellow ear rot sets in. Due to the combined action of the nematode and the bacterium, the earhead becomes chaffy and the kernels are replaced by dark nematode galls, which are also contaminated with the bacterium. The disease is severe during the rainy season.

   The infected plants produce more tillers than do the healthy ones. Another interesting feature is the early emergence of ears in nematode-infected plants, which is about 30–40 days earlier than in the healthy ones. Suryanarayana and Mukhopadhyaya (1971) observed that kernels are modified into galls and bear a superficial resemblance to normal kernels. The number of nematode larvae per gall may vary from 80 to 32,000.

Host

Wheat, *Agrostis avenacea*, and *Polypogon monspeliensis*

Geographical Distribution

It was first reported in Punjab in 1917. The disease also occurs in Egypt, China, Australia, Cyprus, and Canada. Similar diseases occur on a number of grasses in many European countries.

Disease Cycle

The disease starts from seeds contaminated with the nematode galls. When such contaminated seeds are sown in the field, they absorb moisture from the soil. The larvae escape from the galls and shortly climb up the young wheat plants. The nematodes have a tendency to seek the tender growing points of the plants where they remain as ectoparasites. After the plants flower, the nematodes enter the floral parts and form galls in the ovaries. While within the galls, sexual differentiation occurs, copulation takes place, and the female lays eggs, up to 2000 in number. The eggs hatch into larvae under favorable conditions and become active parasites. Once the nematode enters the tissue of the ovary, the bacterium becomes active and causes rotting. The yellow ooze coming out of the rotting earhead provides the inoculum for the secondary spread of the disease, which is favored by wind and rain. The nematode probably functions as a vector, transporting the bacterium to the otherwise inaccessible meristematic regions of the host. The nematode secretes some substance in the presence of which the bacterium can cause the disease. Chand (1967) observed that the bacterium *C. tritici* survived in soil debris under laboratory conditions for about 7 months and concluded that the diseased debris probably does not take any part in primary infection because of the limited survival of bacterium in the soil.

Mathur and Ahmad (1964) reported that bacteria remained viable for at least 5 years in the galls of *Anguina tritici*. The nematode galls are reported to remain in the soil for 20 years or more and perhaps the bacterium can survive inside the gall for a fairly long time. Midha et al. (1971) have shown that the galls serve as nutrient sources to the bacterium. Pathak and Swarup (1984) have shown that the bacterium does not survive in the free state in the soil, nor do the nematodes separately carry the bacterium on their bodies. Nematode galls show the presence of bacterium in up to 40%–50% of cases, and it is believed that the main sources of survival of the pathogen are the nematode galls.

Control Measures
The sowing of gall-free seeds in noninfested soil will help to reduce the incidence of disease. The seeds can be freed from galls by being floated in brine, at 160 g of sodium chloride per liter of water. Wheat, barley, oats, or other susceptible crops should not be sown in infested soil. Solar heat during May and June in the northern wheat belt of India can be utilized for destroying galls and nematodes. Infected plants should be carefully taken out and burnt. Seeds for sowing purposes should be taken from disease-free areas only.

2.1.5 BACTERIAL MOSAIC OF WHEAT

Pathogen: *Clavibacter michiganensis* subsp. *tessellarius*

Synonyms: *C. m.* subsp. *michiganensis* and *C. m.* subsp. *nebraskensis*

Symptoms
The disease is characterized by small, yellow lesions with undefined margins. The lesions are more or less distributed uniformly over the whole leaf. Individual lesions may resemble the hypersensitive reaction of rust. Water soaking and bacterial oozing from lesions are not observed.

Geographical Distribution
Illinois (Chang et al., 1990), Alaska, Iowa, the United States, and Canada

Disease Cycle
C. michiganensis subsp. *tessellarius* is seed borne (McBeath and Adelman, 1986; McBeath et al., 1988).

Economic Impact
The economic significance of this disease is unknown (Carlson and Vidaver, 1982a). However, since the disease can destroy flag leaves in severely infected plants, it may cause significant losses (McBeath, 1993).

Control Measures
Control methods include discarding of contaminated seed and using varieties that are resistant to the disease. There seems to be a wide range of host response to the disease among spring wheat genotypes, indicating that genetic improvement to achieve disease resistance trait is possible.

2.1.6   Gumming Disease of Wheat

Pathogen: *Clavibacter iranicus*
The bacterium has only been reported in Iran (Scharif, 1961), where it causes a gumming disease on wheat spikes. Recently, Zgurskaya et al. (1993) suggested renaming this pathogen as *Rathayibacter iranicus*. This bacterium should be considered as being different from *C. tritici*.

Symptoms
Disease symptoms include gumming of wheat spikes.

Economic Impact
Its economic impact has not been estimated yet.

Control Measures
Not worked out.

2.1.7   Bacterial Sheath Rot of Wheat

Pathogen: *Pseudomonas fuscovaginae*

Symptoms
The symptoms include irregular, angular, blackish brown lesions bordered by a purple-black water-soaked area. Infection usually occurs on the adaxial side of the sheath, where moist dews are retained. The most severe infection results in poor spike emergence and sterility (Duveiller et al., 1988; Duveiller and Maraite, 1990).

Disease Cycle
Not much is known about environmental conditions favorable for the bacterial sheath rot, but low temperatures and high humidity at the booting–heading stage are considered favorable.

Control Measures
No control measures are reported, except the use of less susceptible genotypes and discarding infected seed. In Nepal, genotypes Annapurna-1, Annapurna-2, Annapurna-3, and WK685 have shown high levels of infection; however, on RR21, they showed less disease incidence (Anonymous, 1995).

2.1.8   Stem Melanosis of Wheat

Pathogen: *Pseudomonas cichorii* (Swingle, 1925; Stapp, 1928)

Synonyms: *Pseudomonas endiviae* (Kotte, 1930)
Bacterium cichorii (Swingle) (Elliott, 1930)
Bacterium endiviae (Kotte, 1930)
Bacterium formosanum (Okabe, 1935)
Chlorobacter cichorii (Swingle, 1925; Patel and Kulkarni, 1951)
Phytomonas cichorii (Swingle, 1925)
Phytomonas endiviae (Kotte) (Clara, 1934)
Pseudomonas formosanum (Okabe) (Krasil'nikov, 1949)
Pseudomonas papaveris (Lelliott and Wallace, 1955)
P. papaveris (Takimoto) (Okabe and Goto, 1955)

Symptoms
Symptoms are first observed at the milky ripe stage of the plant where small, light-brown lesions develop beneath the lower two nodes. Later these lesions darken and coalesce on the stem, rachis,

and peduncle, with occasional mottling on the glumes. The rachis, upper portion of the peduncle, and portions of the stems immediately below the nodes turn dark. Heads remain bleached with shriveled grain.

Host

Capsicum annuum, Cichorium endivia var. *crispum, Ci. endivia* var. *latifolia, Lactuca sativa*

Geographical Distribution

Europe: Belgium, Bulgaria, France, Germany, Greece, Italy, Macedonia, Portugal, Serbia and Montenegro, Spain, the United Kingdom, Ukraine

Asia: China—Hebei, Neimenggu; India—Delhi; Iran; Japan—Hokkaido, Shikoku; Korea Republic; Taiwan; Turkey

Africa: Burundi, South Africa, Tanzania

North America: Canada—Alberta, Ontario; Mexico; USA—Alabama, California, Florida, Georgia, Illinois, Louisiana, Montana, New York, Washington
Central America and Caribbean: Barbados, Cuba, Puerto Rico

South America: Argentina; Brazil—Minas Gerais, Parana, Rio de Janeiro, Rio Grande do Sul, Sao Paulo; Chile; Colombia

Oceania: Australia—New South Wales, Queensland, New Caledonia; New Zealand

Disease Cycle

The epidemiology of stem melanosis is not worked out, but the combination of high humidity and high temperature conditions (29°C) promotes its spread. Stem melanosis has been found to be associated with copper-deficient soils (Piening et al., 1987, 1989).

Control Measures

The application of Cu chelate at 2–4 kg/ha to copper-deficient soils where the disease is generally observed is recommended. Cu reduces stem melanosis and increases grain yield (Piening et al., 1987, 1989).

2.1.9 PINK SEED OF WHEAT

Pathogen: *Erwinia rhapontici* (Millard) Burkholder

Synonyms: *Erwinia carotovora* pv. *rhapontici* (Millard)

Symptoms

Seeds harboring the pathogen/parasite have a light-pink appearance due to the presence of a diffusible pigment produced by the bacterium in the testa. These symptoms are similar to the coloration of seeds treated with fungicides containing dyes and to seeds from plants infected by *Fusarium* species, the causal agents of scab or head blight. Wheat spikes artificially inoculated with the pathogen develop a maroon color in the spikelet tissue surrounding the inoculation point.

Hosts

Wheat, hyacinth, onion, rhubarb, and pea are also affected. It can be considered as an opportunistic pathogen that invades injured kernels (Wiese, 1987).

Geographical Distribution

The United States (Idaho, North Dakota), Canada, England, France, Russia, Ukraine (Roberts, 1974; Sellwood and Lelliott, 1978; CMI, 1981; McMullen et al., 1984; Forster and Bradbury, 1990; Huang et al., 1990).

Disease Cycle

Seed borne.

Economic Impact
Pink seed causes cosmetic effects and is considered inconsequential (Wiese, 1987).

Control Measures
Currently there are no known control measures.

2.2 RICE

2.2.1 BACTERIAL LEAF BLIGHT OF RICE

Pathogen: *Xanthomonas oryzae* pv. *oryzae* (Ishiyama) Swings et al.

Synonyms: *Pseudomonas oryzae* Ishiyama
        X. campestris pv. *oryzae* (Ishiyama) Dye
        Xanthomonas itoana (Tachinai) Dowson, *Xanthomonas kresek* Schure
        X. oryzae (Ishiyama) Dowson
        X. translucens f. sp. *oryzae* (Ishiyama) Pordesimo (this name has also, incorrectly, been used for pv. *oryzicola*)

Common names: Bacterial leaf blight (BLB), kresek disease, BLB (English), Maladie bactérienne des feuilles du riz (French), Bakterielle Weissfleckenkrankheit, bakterieller Blattbrand (German), Enfermedad bacteriana de las hojas del arroz (Spanish)

Symptoms
Bacterial leaf blight appears on the leaves of young plants as pale-green to gray-green water-soaked streaks near the leaf tip and margins. These lesions coalesce and become yellowish white with wavy edges (Photo 2.3). Eventually, the whole leaf may be affected, becomes whitish or grayish, and then dies. Leaf sheaths and culms of the more susceptible cultivars may be attacked. Systemic infection, known as kresek (Reddy, 1984), results in desiccation of leaves and death, particularly of young transplanted plants. In older plants, the leaves become yellow and then die. In later stages, the disease may be difficult to distinguish from bacterial leaf streak. For more information, see Bradbury (1970a,b), Feakin (1971), and Ou (1972).

Differentiation Test for *Xanthomonas oryzae* pv. *oryzae* Infection
A simple test to differentiate leaf blight consists of immersing the cut end of the basal part of an infected leaf in a dilute solution of basic fuchsine for 1–2 days (Goto, 1965). The area of latent

PHOTO 2.3 Typical symptoms (rapid yellowing, wilting, and dying) of kresek disease in rice. (Courtesy of International Rice Research Institute (IRRI), Los Baños, Philippines, flickr.com.)

bacterial infection beyond the visible lesion remains unstained and appears as green spots with undulate margins clearly separated from the stained healthy part of the leaf. This reaction only occurs with *X. oryzae* pv. *oryzae* and not with other blight organisms. A limitation of this technique is that young flag leaves and short, old leaves do not stain well.

Colonies of *X. oryzae* pv. *oryzae* on incubation at 25°C–28°C in 1% dextrose nutrient agar or Wakimoto agar (Reddy and Ou, 1974) are slow growing, mucoid, and straw colored to yellow in color. The bacteria are Gram-negative rods, capsulated and motile with a polar flagellum. Dimensions are 1.1–2.0 × 0.4–0.6 μm.

Host
The principal host is rice. The bacteria also attack a number of wild or minor cultivated Poaceae (*Leersia* spp., *Leptochloa* spp., *Oryza* spp., *Paspalum scrobiculatum*, *Zizania*, *Zoysia* spp.), including poaceous weeds, which may act as carriers (Li et al., 1985). Many more were found to be susceptible by artificial inoculation (Bradbury, 1970a,b).

Geographical Distribution
EPPO A1 quarantine pest

EPPO region: Russia (Far East, Southern Russia; found but not established); Ukraine, found on rice seeds by Koroleva et al. (1985); but the status of this record is uncertain.

Asia: Bangladesh, Cambodia, China (widespread), India (widespread), Indonesia (widespread), Japan (Honshu, Kyushu), Democratic People's Republic of Korea, Lao, Malaysia (Peninsular, Sabah, Sarawak), Myanmar, Nepal, Pakistan, the Philippines, Sri Lanka, Taiwan, Thailand, Vietnam.

Africa: Burkina Faso, Cameroon, Gabon, Madagascar (probably not, but suspect symptoms seen by Buddenhagen [1985]), Mali, Niger, Senegal, Togo. See John et al. (1984).

North America: Mexico, the United States (Louisiana, Texas).

Central America and Caribbean: Costa Rica, El Salvador, Honduras, Nicaragua (unconfirmed), Panama.

South America: Bolivia, Colombia, Ecuador, Venezuela.

Oceania: Australia (Northern Territory, Queensland).

EU: Absent.

Disease Cycle
The bacterium enters through hydathodes and wounds on the roots or leaves. Penetration may also occur via stomata, with a resultant buildup of bacteria that subsequently exude onto the leaf surface and reenter the plant through the hydathodes. Once inside the vascular system, the bacterium multiplies and moves in both directions. Spreading takes place through wind and rain but primarily through flood and irrigation water (Dath and Devadath, 1983). The spread pattern in a rice field has been analyzed by Nayak and Reddy (1985).

Potential inoculum sources include infected planting material, volunteer rice plants (Durgapal, 1985), infected straw or chaff (Devadath and Dath, 1985), and weed hosts, although the exact role of these sources in nature is poorly understood. Seed transmission is generally thought to occur to a certain extent (Hsieh et al., 1974), but Murty and Devadath (1984) had difficulty in demonstrating this experimentally. Infected seeds did not give rise to infected seedlings but did introduce the bacterium into the soil. Singh et al. (1983), however, observed seed transmission regularly in growth chambers when using heavily infested samples. Singh (1971) found that the bacterium cannot survive in unsterilized soil but can survive for only 15–38 days in field and pond water, but Murty and Devadath (1982) showed that this depends on the soil type. Raj and Pal (1988) failed to observe overwintering in soil or seed and found survival only in infected leaves. Reddy (1972) stated that

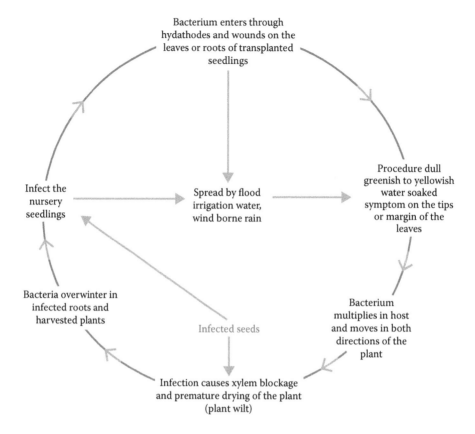

FIGURE 2.2   Disease cycle of bacterial leaf blight of rice caused by *X. oryzae* pv. *oryzae*.

X. oryzae pv. *oryzae* survives for 7–8 months in seed, but for only 3–4 months in straw and stubble; Kauffman and Reddy (1975) reported that although glumes were readily infected, viable bacteria could not be detected on seeds stored for 2 months. It is thought that bacteriophages play a role in reducing bacteria in germinating seeds. In general, it is clear that present results on survival and sources of inoculum are contradictory. The disease cycle of bacterial leaf blight of rice is illustrated in Figure 2.2.

Different races or pathotypes of the bacterium exist and can be distinguished by their behavior on differential cultivars (Mew, 1987). New races appear frequently and the bacterium is highly variable in virulence.

Economic Impact

Bacterial leaf blight is the most serious disease in rice in Southeast Asia, particularly with the widespread cultivation of dwarf, high-yielding cultivars (Ray and Sengupta, 1970; Feakin, 1971). In 1954, in Japan, 90,000–150,000 ha was affected and annual losses were estimated at 22,000–110,000 tons. The disease was first reported in India in 1951, but it was not until 1963 that an epiphytotic occurred. In the Philippines, present losses are of the order of 22.5% in wet to 7.2% in dry seasons in susceptible crops and 9.5% and 1.8%, respectively, in resistant crops (Exconde, 1973). Nitrogen fertilization considerably increases susceptibility. Losses are generally less significant in the less fertile soils and in summer-grown crops (December–April). Transplanted autumn (May–September) and winter (July–December) crops, however, suffer considerable losses. Diseased crops contain a high proportion of chaffy grains. This situation has been reviewed by an international workshop held in Manila, Philippines (Banta, 1989).

Control Measures

Careful crop management (Padmanabhan, 1983), use of resistant cultivars, and seed treatments (Singh and Monga, 1985) help reduce the incidence of the disease. Dipping rice seedlings in an antibiotic before transplanting has been proposed (Buddenhagen, 1983; Durgapal, 1983; Kaul and Sharma, 1987; Mew 1987). Bacterization of seeds with fluorescent pseudomonads has been tried as a biological control method (Anuratha and Gnanamanickam, 1987).

Since the increase in disease severity in the 1970s and 1980s, varietal resistance has become a very important consideration and there is a large volume of literature on breeding and screening for resistance to leaf blight. The existence of different races makes it important to obtain stable (Nayak and Chakrabarti, 1986) or adult-plant resistance (Qi and Mew, 1985). Ajaya, Asha, Biraj, CO-43, Gobind, IR-64, Janaki, PR-4141, Radha, Sona Mahsuri, Sujata, Suraj, Swarna, Udaya, and so on, are some of the resistance varieties available in India.

An ideal agent for chemical control is one that functions at low concentrations by either killing or inhibiting the multiplication of pathogens by blocking their metabolic pathway. BLB has been controlled by chemicals like Bordeaux mixture with or without sugar, copper–soap mixture, and copper–mercury fungicides. Spraying copper oxychloride was recommended to control rice bacterial blight disease (Sulaiman and Ahmed, 1965). Chlorination of field water with stable bleaching powder also effectively reduced disease severity in India (Chand et al., 1979; Sivaswamy and Mahadevan, 1986). Synthetic organic bactericides such as nickel dimethyl dithiocarbamate, dithianon, phenazine, and phenazine N-oxide were also recommended (Fukanaga, 1966; Takahi, 1985).

A foliar spray of cowdung extract (20 g/L) was reported to suppress BLB development in the state of Kerala in southern India (Mary et al., 1986). Also, dithiocarbamate fungicides were reported to inhibit the growth of *Xoo* by arresting fatty acid (Yoneyama et al., 1978) and lipid biosynthesis (Yoneyama and Misato, 1978). A few antibiotics like streptocycline and fungicides, such as zineb and carbendazim, inhibited the pathogen *in vitro* (Hoa et al., 1984; Mahto et al., 1988).

However, an effective and economical chemical control is yet to be developed for BLB disease. This may be because the pathogen population is highly variable in its sensitivity to the chemicals used for disease control. The existence and development of drug-resistant strains also pose serious problems in formulating foolproof control agents.

2.2.2 BACTERIAL LEAF STREAK OF RICE

Pathogen: *Xanthomonas oryzae* pv. *oryzicola* (Fang et al.) Swings et al.

Synonyms: *X. campestris* pv. *oryzicola* (Fang et al.) Dye
  Xanthomonas oryzicola (Fang et al.)
  X. translucens f.sp. *oryzicola* (Fang et al.) Bradbury

Common names: Bacterial leaf streak (BLS, English), Brûlure bactérienne, stries bactériennes, du riz (French), Quemaduras bacterianas, estrías bacterianas, del arroz (Spanish)

Symptoms

Narrow, dark greenish, water-soaked, interveinal streaks of various lengths appear, initially restricted to the leaf blades. The lesions enlarge, turn yellowish orange to brown depending on the cultivar, and eventually coalesce (Photo 2.4). Tiny amber droplets of bacterial exudate are often present on the lesions. In its advanced stages, the disease is difficult to distinguish from that caused by *X. oryzae* pv. *oryzae* but lesion margins remain linear rather than wavy as in the latter case. Direct observation of the bacterium may be necessary for confirmation. Damage is often associated with lepidopterous leaf rollers, leaf folders, and hispa beetles, since bacteria readily enter the damaged tissue resulting from insect infestation.

Colonies of *X. oryzae* pv. *oryzicola* on incubation at 25°C–28°C in 1% dextrose nutrient agar or Wakimoto agar (Reddy and Ou, 1974) are fairly slow growing, usually pale yellow, round, smooth,

PHOTO 2.4   Field symptoms of bacterial leaf streak disease of rice. (Courtesy of International Rice Research Institute (IRRI), Los Baños, Philippines, flickr.com.)

entire, domed, and mucoid. The bacteria are Gram-negative rods, capsulated, and motile with a polar flagellum. Dimensions are 1.0–2.5 × 0.4–0.6 μm.

Host

The principal host of the pathogen is rice. *Oryza sativa* subsp. *japonica* is usually more resistant than subsp. *indica* to pv. *oryzicola*. The bacteria also attack a number of wild or minor cultivated Poaceae (*Leersia* spp., *Leptochloa* spp., *Oryza* spp., *Paspalum scrobiculatum*, *Zizania*, *Zoysia* spp.), including poaceous weeds that may act as carriers (Li et al., 1985). Many more were found to be susceptible by artificial inoculation (Bradbury, 1970a,b).

Geographical Distribution

EPPO region: Absent.

Asia: Confined to tropical areas including Bangladesh, Cambodia, China (Fujian, Guangdong, Hainan), India (Bihar, Karnataka, Maharashtra, Madhya Pradesh, Uttar Pradesh), Indonesia (widespread), Lao, Myanmar, Nepal, Pakistan, the Philippines, Thailand, Malaysia (Peninsular, Sabah, Sarawak), Vietnam.

Africa: Madagascar (Buddenhagen, 1985), Nigeria, Senegal.

Oceania: Australia (Northern Territory).

EU: Absent.

Disease Cycle

The bacterium enters the leaf through stomata or wounds. Spread within a crop occurs by mechanical contact and via rain and irrigation water; under favorable warm, wet conditions, rapid and severe disease development can occur. The role of seed transmission in the perpetuation of the disease is recognized, but the part played by weeds is little understood. Rao (1987) suggested that seed transmission can occur from one summer season to the next, but not if summer seed is sown in the winter season, as the pathogen cannot establish in cool, dry winter weather. The bacterium can persist from season to season on infected leaves and leaf debris but is unable to survive in nonsterile soil (Devadath and Dath, 1970).

The bacteria can only move short distances in infected crops. The only means of long-distance dispersal is in infected rice seeds. The bacteria are usually found in the glumes, but they may also penetrate the endosperm. For *X. oryzae* pv. *oryzicola*, the planting of disease-free seed is considered of utmost importance in control.

Economic Impact

Bacterial leaf streak is only of importance in some areas during very wet seasons and where high rates of nitrogen are used. It does not usually reduce yields if low N rates are applied. Losses of 5%–30% have been reported from India, while in the Philippines losses were not considered significant in either wet or dry seasons (Opina and Exconde, 1971). In general, bacterial leaf streak is a much less important disease than bacterial leaf blight.

Phytosanitary Risk

Both *X. oryzae* pv. *oryzae* and *X. oryzae* pv. *oryzicola* are EPPO A1 quarantine pests (OEPP/EPPO, 1979, 1980). The former is also of quarantine significance for COSAVE, CPPC, IAPSC, NAPPO, IAPSC, has been defined as Comite de Sanidad Vegetal del Cono Sur (COSAVE), The Caribbean Plant Protection Commission (CPPC), The Inter-African Phytosanitary Council (IAPSC), The Inter-African Phytosanitary Council (IAPSC) and The North American Plant Protection Organization (NAPPO) (refer to 15.4), while the latter is listed only by COSAVE and IAPSC. Bacterial leaf blight is an extremely severe disease, causing extensive problems in the Far East, but it is absent from the European rice-growing areas. Its existing distribution suggests that it could survive in Mediterranean countries, and it clearly presents a serious risk for the EPPO region. Bacterial leaf streak is a much less important disease, with a more tropical distribution and correspondingly lower probability of establishment in the EPPO region. Its quarantine status for the EPPO region is arguable, but in any case, the measures taken against *X. oryzae* pv. *oryzae* will also exclude it.

Control Measures

The bacterial leaf streak pathogen hardly requires any particular control measures except the use of healthy seed. Neither treatments nor resistance is mentioned to any significant extent in the literature.

2.2.3   Sheath Brown Rot of Rice

Pathogen: *Pseudomonas fuscovaginae*

Synonyms: *Sarocladium oryzae*

Common names: Sheath brown rot of rice (English), Burundi on rice (Japan), fluorescent bacteria (Brazil)

Symptoms

Sheath brown rot symptoms appearing at seedling and later stages include discoloration and rotting of sheath (Photo 2.5). After transplanting, infected seedlings initially show yellow-brown discolorations on their lower leaf sheaths. Later the discolorations turn gray brown to dark brown. Ultimately, rot results in the death of the infected seedlings. The infected flag leaf sheath becomes water soaked and necrotic. Other sheaths also exhibit lesions. Spikelets of emerging panicles are discolored, are sterile, or may be symptomless except for small brown spots. In acute infections, the sheaths turn

PHOTO 2.5   Field symptoms of bacterial sheath brown rot of rice. (Courtesy of International Rice Research Institute (IRRI), Los Baños, Philippines, flickr.com.)

completely grayish brown or dark brown and panicles shrivel and dry (Tanii et al., 1976; Miyajima, 1983; Zeigler and Alvarez, 1987a,b; Rott et al., 1989; Webster and Gunnell, 1992; Cottyn et al., 1994).

Pathogen

The bacterial colonies of *P. fuscovaginae* are creamy white, circular, raised, and butyrous with entire margins, while some colonies are flat and translucent.

Host

Major hosts: *O. sativa* (rice), *T. aestivum* (wheat)

Minor hosts: *Agrostis* (bent grasses), *Avena sativa* (oats), *Bromus marginatus* (mountain brome [grass]), *H. vulgare* (barley), *Lolium perenne* (perennial ryegrass), *Poa pratensis* (smooth-stalked meadow grass), *S. cereale* (rye), *Sorghum bicolor* (sorghum), triticale, *Zea mays* (maize).

This pathogen was first reported in rice (Tanii et al., 1976), but it has been found on other cereals as well, including wheat (Duveiller et al., 1989; Duveiller and Maraite, 1990).

Geographical Distribution

Europe: Russia, Yugoslavia (Fed. Republic)

Asia: Indonesia, Japan, Hokkaido, Honshu; the Philippines (Taniii et al., 1976; Miyajima, 1983; Cottyn et al., 1996)

Africa: Burundi, Madagascar, Rwanda, Tanzania, Zaire

North America: Mexico (Duveiller et al., 1988; Rott et al., 1989)

Central America and Caribbean: Costa Rica, Cuba, Dominican Republic, El Salvador, Guatemala, Jamaica, Nicaragua, Panama, Trinidad, Tobago

South America: Argentina; Bolivia; Brazil—Sao Paulo; Chile, Colombia, Ecuador, Peru, Suriname, Uruguay (Zeigler and Alvarez, 1987a).

Disease Cycle

The spread of the disease is expected to be very quick due to the use of infected rice seeds for planting in almost all areas. This could simply happen because sheath brown rot disease is confirmed as a seed-borne disease (Duveiller et al., 1988; Xie et al., 2002). Rice seedlings grown in steam-sterilized soil from discolored seeds obtained from tillers infected with *P. fuscovaginae* showed symptoms on sheaths and leaves at 10–20 days after planting (Zeigler and Alvarez, 1987b). Bacterial sheath brown rot symptoms are observed in direct-seeded and transplanted rice raised from naturally and artificially infected seed. Disease symptoms are recorded 30 days after transplanting: the intensity of the disease increased along with the age of plants (Cahyaniati and Mortensen, 1995). Bacteria are detected at a concentration of 4,000–80,000 cells per grain, after storage of rice seeds for about 6 months. The pathogen survived in dry grains until the next autumn at the longest. The disease developed after these infected seeds were sown in sterilized soils (Miyajima, 1983).

Economic Impact

Yield losses as high as 72.2% is reported in Indonesia (Cahyaniati and Mortensen, 1997). Severe losses caused by the disease is also reported (Notteghem, 1998) in rice production areas at Central Africa and Madagascar. Under a very severe infection, total yield loss (almost 100%) in Madagascar was reported by Rott (1987).

Control Measures

Cultural and Sanitary Methods

Eradication and burning of regrowths and plant litter immediately after harvest and off-season cultivation of a crop (such as potatoes or lupins [*Lupinus* sp.]) that is not attacked by the bacterium may be used to control *P. fuscovaginae* (Rott, 1987). Results of studies with irrigated rice demonstrated the need for cultivars with cold tolerance, early maturity, and pest and disease resistance. Sowing time should be adjusted to avoid low temperatures. After this date, the area had phosphate deficiency problems. Farmers need to be encouraged to use seedlings that are 20–30 days old rather than older seedlings (Macapuguay and Mnzaya, 1988). In Madagascar, bacterial sheath brown rot has not yet been found on rainfed rice (Rott, 1987).

Chemical Control

Antibiotics such as streptomycin, alone or in combination with oxytetracycline, can effectively control sheath brown rot if applied at or a few days after panicle emergence (Webster and Gunnell, 1992).

2.3   MAIZE

2.3.1   Goss's Bacterial Wilt and Blight of Maize

Pathogen: *Clavibacter michiganensis* subsp. *nebraskense* (Vidaver and Mandel, 1974; Davis et al., 1984)

Synonyms: *Corynebacterium nebraskense*

Common names: (Nebraska) leaf freckles and wilt of maize, Goss's bacterial wilt, and blight of corn

Symptoms

The first symptoms are water-soaked leaf spots (on seedlings or older plants) that develop into longitudinal streaks parallel to the main veins. Sometimes drops of bacterial slime are present. In later stages of infection, plants wilt, show leaf blight, and may be stunted. Leaf streaks turn gray to light green or yellow to red stripes with wavy or irregular margins, which follow leaf veins (confusion with leaf symptoms of *Erwinia stewartii* and cold damage is possible). Discrete, water-soaked spots (freckles) along the veins are characteristic of the disease (Photo 2.6). These spots are dark green to blackish in appearance and look like freckles (hence the name) when infected leaves turn brown. When streaks coalesce, leaf scorch is seen, which may be confused with symptoms of drought. Systemically infected plants may show yellowish discolored vascular bundles, and a dry or water-soaked to slimy-brown rot of the roots and lower stalk may occur. From these infected stalks and roots, leaves can become reinfected. Plants at any stage of development can become infected, wilt, and die.

PHOTO 2.6 *C. michiganensis* subsp. *nebraskense* infection on maize leaf. (Courtesy of Dean Malvick, University of Minnesota, St. Paul, MN.)

Host
Maize (*Z. mays*), *Saccharum officinarum* (sugarcane), *Sorghum bicolor* (common sorghum), *Sorghum sudanense* (Sudan grass), and *Zea mexicana* (teosinte) have also been reported as natural hosts. *Echinochloa crus-galli* (barn grass), triticale, and *T. aestivum* (wheat) have been reported as secondary hosts.

Geographical Distribution
Canada, the United States.

Disease Cycle
The main source of inoculum appears to be maize crop residues that have been left over in winter. Seed infection (internal and external) and transmission occurs but appears to be minor as compared to residue transmission (Schuster, 1975).

Control Measures
Control measures are crop rotation, destruction of maize debris, and general hygiene, as well as testing of seed and planting of seed lots that tested free from the pathogen. Planting of resistant varieties is sometimes possible. High resistance in dent maize hybrids and inbreeds has been found, but sweet corn varieties appear to be less resistant (Schuster, 1975; Wysong et al., 1981; Smidt and Vidaver, 1986).

2.3.2 STEWART'S DISEASE OF MAIZE

Pathogen: *Pantoea stewartii* subsp. *stewartii* (Smith) Mergaert et al.

Synonyms: *Aplanobacter stewartii* (Smith) McCulloch
  Bacterium stewartii (Smith) Smith
  Erwina stewartii (Smith) Dye
  Pantoea stewartii subsp. *stewartii* (Smith) Mergaert et al.
  Pseudomonas stewartii Smith
  Xanthomonas stewartii (Smith) Dowson

Common names: Stewart's disease, bacterial wilt of maize, bacterial leaf blight of maize

Symptoms
Short to long pale-green to white-yellow streaks on the leaves are formed (Photo 2.7) and sweet corn hybrids may show rapid wilting in an early stage of the disease. Streaks turn brown in later stages

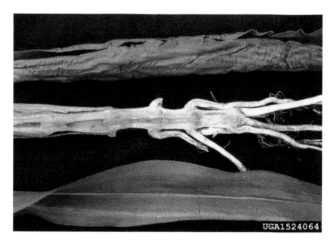

PHOTO 2.7 Typical yellowish leaf streaks of Stewart's disease on maize leaf. (Courtesy of Department of Plant Pathology, North Carolina State University, Raleigh, NC, 2ugwood.org.)

of the disease. The entire vascular tissue may be infected and bacteria can be found in roots, stalks, leaves, tassels, cobs, husks, and kernels.

Host

Primary host: Maize (Z. *mays*), especially sweet corn, *Tripsacum dactyloides* (eastern gamma grass), and Z. *mexicana* (a wild relative of maize)

Secondary host: *Agrostis gigantea* (bench couch grass), *Dactylis glomerata* (orchard grass), *Panicum* spp. (panic grass), *Poa pratensis* (June grass), *Setaria lutescens* (foxtail grass), and *S. sudanense* (Sudan grass).

Geographical Distribution

Europe: Austria, Greece, Poland, Romania, Russia

America: Bolivia, Brazil, Canada, Central America, Costa Rica, Guyana, Mexico, Peru, Puerto Rico, the United States

Disease Cycle

Bacteria may exude as droplets of slime from the husks. Bacteria penetrate the seed, but not the embryo. The disease may be confused with other leaf blights such as those caused by the bacteria *Acidovorax avenae* subsp. *avenae*, *Burkholderia andropogonis*, *C. michiganensis* subsp. *nebraskensis*, or fungi, such as *Setosphaeria turcica* and *Cochliobolus heterostrophus*. The pathogen survives in living host plants and seed and may be also spread by seed. Most important dispersal (in the field), however, is by an efficient insect *vector*, the corn flea beetle *Chaetocnema pulicaria*. *P. s.* pv. *stewartii* overwinters in the alimentary tract of *Ch. pulicaria*, which emerges from hibernation and feeds on young maize plants (Pepper, 1967). Overwintering beetles cause the first infections, but also the second (summer) generation is active in transmission. Other (minor) vectors have been described in North America. The vectors are absent in Europe.

Control Measures

Use of healthy seeds and low susceptibility or resistant varieties and early spraying with insecticides to reduce vector populations are the most important ways to prevent or control disease caused by *P. stewartii* subsp. *stewartii*. Stewart's wilt can be predicted on the basis of the average daily temperature in December, January, and February, which relates to the survival of *Ch. pulicaria*. If the average daily temperature during these months is above freezing, flea beetles survive and Stewart's wilt is likely to be severe on susceptible hybrids. At average daily temperatures of lower than $-3°C$, most flea beetles do not survive and it is unlikely that Stewart's wilt will be severe (Esker and Nutter, 2002). There is sufficient resistance present in maize hybrids (Michener et al., 2003).

2.3.3   BACTERIAL LEAF BLIGHT AND STALK ROT OF MAIZE

Pathogen: *Acidovorax avenae* subsp. *avenae* (Manns) Willems et al.

Synonyms: *Phytomonas setariae* (Okabe) Burkholder
            Pseudomonas alboprecipitans Rosen
            Pseudomonas avenae Manns
            Pseudomonas rubrilineans (Lee et al.) Stapp
            Pseudomonas setariae (Okabe) Savulescu
            Xanthomonas rubrilineans (Lee et al.) Starr et Burkholder

Common names: Bacterial leaf blight of maize, bacterial leaf stripe, stalk rot of maize

Symptoms

Disease symptoms on leaves are severe after heavy rains. Leaves emerging from whorls have soaked, brown lesion that become elliptical and gray or white. Lesions may coalesce during

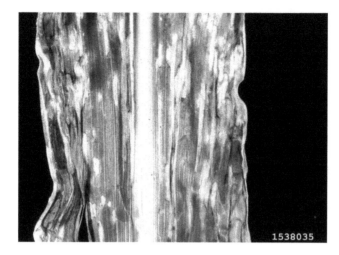

PHOTO 2.8 Field symptoms of bacterial leaf blight and stalk rot of maize. (Courtesy of University of Georgia Plant Pathology, Athens, GA, Bugwood.org.)

wet weather and form large necrotic areas that cause leaves to shred when diseased plants are buffeted by the wind (Photo 2.8).

Plant tops become gray to brown and have shortened internodes. The outside of the stalk may be brown to black and appear water soaked. The inside of the diseased stalk is brown and slimy and has a foul odor that somewhat resembles that of silage. Diseased plants may be stunted.

Host

Euchlaena mexicana, *Sa. officinarum*, *Z. mays*, Poaceae

Geographical Distribution

Asia: Cambodia, China, India, Indonesia, Japan, Malaysia, Nepal, Pakistan, the Philippines, Sri Lanka, Taiwan, Vietnam

Africa: Ethiopia, Kenya, Madagascar, Mauritius, Mozambique, Nigeria, Reunion, South Africa, Tanzania, Uganda, Zaire

America: Argentina, Barbados, Brazil, Colombia, Costa Rica, Cuba, Dominican Republic, El Salvador, French Guiana, Guadeloupe, Guyana, Honduras, Jamaica, Martinique, Mexico, Nicaragua, Puerto Rico, Suriname, Trinidad and Tobago, the United States, Venezuela

Oceania: Australia, Fiji, Guam

Disease Cycle

Ac. avenae subsp. *avenae* does not survive well in infected residue or soil. Vasey grass, *Paspalum urvillei*, is considered a primary means of inoculums in Florida. Contaminated farm equipment is a primary means of dissemination within fields and, likely, between fields.

Infection typically occurs through stomata of leaves that are in whorls and harbor large populations of *Ac. avenae* subsp. *avenae* and other epiphytic bacteria. Moisture in the whorl is considered necessary for infection to occur. Bacteria also are splashed and blown onto the plant and enter the stalk through hail wounds and other injures. Warm, rainy weather favors disease development but leaf blight can occur below 18°C also.

Control Measures

Grow the most resistant hybrid. Full season field maize hybrids are more susceptible than short season hybrids and sweet maize. Bacterial leaf blight and stalk rot is not considered a serious disease.

2.3.4 *Xanthomonas* Leaf Spot of Maize

Pathogen: *Xanthomonas campestris* pv. *holicola*

Synonyms: *X. campestris* pv. *zeae*

Common names: Bacterial leaf spot

Symptoms

Narrow, water-soaked streaks (33 × 20 mm long) occur on diseased leaves. Reddish-brown streaks covered by dried bacterial exudates coalesce and form large, irregular-shaped, necrotic areas that destroy most of the leaves.

Control Measures

No known control measures are available in the literature.

2.3.5 Bacterial Stalk and Top Rot of Maize

Pathogen: *Erwinia carotovora* subsp. *carotovora* and *Erwinia chrysanthemi* pv. *zeae*

Synonyms: *E. carotovora* pv. *zeae* and *Erwinia chrysanthemi*

Common names: Bacterial stalk and top rot

Symptoms

Initially, water-soaked lesions in diseased leaf sheath extend, as streaks, into leaf laminae and cause decay in pith. Diseased nodes are tan to brown, water soaked, and soft and slimy and have a foul odor that resembles spoiled silage. At midseason, plants fall as one to several diseased internodes twist and collapse (Photo 2.9). The fallen plant may remain green for several days because the vascular strains remain intact and so do not decay.

Disease Cycle

E. carotovora subsp. *carotovora* and *E. chrysanthemi* pv. *zeae* live saprophytically on infested residue in soil and also are seed borne. Bacteria disseminated by wind or splashed onto host plants

PHOTO 2.9 Stalk rot and top rot symptoms of maize due to *Erwinia*. (Courtesy of Gregory E. Shaner, Botany & Plant Pathology Department, Purdue University, West Lafayette, IN.)

during wet conditions caused by rainfall or overhead irrigation enter the host plant through hydathodes, stomata, or wounds on leaves and stalks.

Disease development is aided by high temperatures (30°C–35°C) and poor air circulation. The enzyme xylanase is produced by *E. chrysanthemi* pv. *zeae* and has been shown to kill cells and macerate tissue in monocotyledonous plants.

Control Measures

No disease management is practical, but chlorine in irrigation water has been reported to reduce disease incidence and severity.

2.3.6 *Holcus* Spot of Maize

Pathogen: *Pseudomonas syringae* pv. *syringae*

Synonyms: *Pseudomonas holci*

Common names: *Holcus* spot

Symptoms

Lesions are more numerous toward the tips of diseased lower leaves and initially are dark green with water-soaked margins. Later, lesions dry up and become tan to brown, round- to elliptical-shaped (2–10 mm in diameter) areas with reddish to brown margins, depending on the maize hybrid. Larger lesions may be surrounded by a yellowish halo (Photo 2.10).

Holcus spot initially may appear to be a serious disease on young plants, but warm and dry weather conditions arrest bacterial spread on the plant and little damage usually occurs. *Holcus* spot has little or no effect on yields.

Disease Cycle

The bacterium overwinters in infected residue. During temperatures of 25°C–30°C and wet, windy weather early in growing season, bacteria are splashed or blown onto host plant, where they invade the leaf either through stomata or through small injuries.

Control Measures
- Grow hybrids that are more resistant than others.
- Rotate maize with resistant crops.

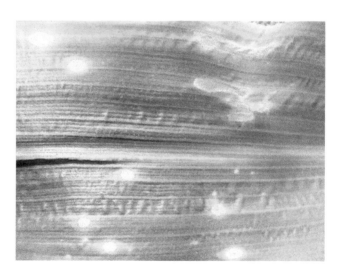

PHOTO 2.10   Symptoms of *Holcus* spot on maize leaf. (Courtesy of Tex Young, www.greatlakeshybrids.com.)

2.3.7   BURKHOLDERIA LEAF SPOT OF MAIZE

Pathogen: *B. andropogonis* (Smith) Gillis et al.

Synonyms: *Phytomonas stizolobii* (wolf) Bergey
Pseudomonas andropogonis (Smith) Stapp
Pseudomonas stizolobii (wolf) Stapp
Pseudomonas woodsii (Smith) Stevens

Common names: Bacterial leaf spot of maize, bacterial stripe of maize, bacterial stripe of sorghum

Symptoms
Lesions occur first on lower leaves and progress up the plant. These are initially water soaked, long, and narrow with parallel sides and olive to amber in color. The lesions may also be pale yellow or greenish white in color. The lesions eventually enlarge and collapse to involve a large area on the infected leaf.

Host
Sorghum vulgare, Z. mays

Geographical Distribution
Europe: Bulgaria, Hungary, Italy, Russia

Asia: China, Iraq, Japan, Pakistan, the Philippines, Southeast Asia, Taiwan, Thailand

Africa: Africa, Ethiopia, Kenya, Mauritania, Nigeria, Rwanda, South Africa, Sudan, Togo, Uganda, Zambia, Zimbabwe

America: Argentina, Brazil, Canada, Colombia, Costa Rica, Cuba, El Salvador, Haiti, Honduras, Mexico, the United States, Uruguay

Oceania: Australia, Micronesia, New Zealand

Control Measures
Not worked out.

2.4   BARLEY

2.4.1   BASAL KERNEL BLIGHT OF BARLEY

Pathogen: *Pseudomonas syringae* pv. *syringae*

Common names: Barley basal kernel blight, bacterial leaf blight

Symptoms
Infected kernels have tan to dark brown spots (approximately 2 mm in diameter) with distinct margins that develop at the embryo end on the posterior surface of the kernel. A lemma spot symptom caused by *P. syringae* pathovars is a well-defined tan to brown discoloration of the kernel lemma. Bacteria are found in intercellular spaces of the aleurone layer, within amyloplast cells and in vascular bundles.

Geographical Distribution
North America.

Disease Cycle
P. syringae pv. *syringae* survives on infested residue and leaf surfaces of nonhost plants and is seed borne. Kernel infection occurs before the milky dough stage. The infection process starts with an epiphytic inoculums buildup on the kernel surface. Bacteria enter kernels primarily through stomata located on the inner epidermis of the barley lemma before the lemma becomes attached to the caryopsis. At the soft drought stage, *P. syringae* might enter kernels through pits or wounds in the external epidermis.

Control Measures
- Destruction of infected residue.
- Treatment of seed with seed-protectant copper fungicide.

2.4.2 BACTERIAL STRIPE BLIGHT OF BARLEY

Pathogen: *Pseudomonas syringae* pv. *striafaciens* Elliott

Synonyms: *P. striafaciens*

Common names: Bacterial stripe blight

Symptoms
Initially, sunken, water-soaked dots appear on dispersed leaves and, if dots are numerous, the leaves may die. The dots enlarge into water-soaked stripes or blotches with narrow yellowish margins that extend the length of the leaf blade. Stripes become a translucent rusty brown as they age. Droplets of bacteria exude from stripes (Photo 2.11) in moist weather and later dry up and form white scales on the surfaces. Bacterial strip occurs infrequently.

Geographical Distribution
Australia, Europe, North America, and South America.

Disease Cycle
Bacterial stripe blight is seed borne and survives in infested crop refuse for at least 2 years. Bacteria are splashed or blown onto leaves during cool, wet weather. Warm, dry weather stops the spread of the bacteria.

Control Measures
- Destroy infected residue.
- Treat seeds with a seed-protectant copper fungicide.
- Grow resistant cultivars.

2.4.3 BASAL GLUME ROT OF BARLEY

Pathogen: *Pseudomonas syringae* pv. *atrofaciens* (McCulloch) Young et al.

PHOTO 2.11 Symptoms of bacterial stripe blight disease on barley leaf. (Courtesy of Clemson University— USDA Cooperative Extension Slide Series, Clemson, SC, Bugwood.org.)

Symptoms

Infections begin as small, dark-green, water-soaked lesions that turn dark brown to blackish in color. On the spikelets, lesions generally start at the base of the glume and may eventually extend over the entire glume. Diseased glumes have a translucent appearance when held toward the light. Dark brown to black discoloration occurs with age. The disease may spread to the rachis, and lesions may also develop on the kernels. Under wet or humid conditions, a whitish gray bacterial ooze may be present. Stem infections result in dark discoloration of the stem. Leaf infections result in small, irregular, water-soaked lesions. Symptoms can be confused with those of other bacterial diseases, genetic melanism, diseases caused by *Septoria nodorum*, and frost damage. The disease causes brown spotting of leaves, stem, glumes, and other aerial parts of the plants and finally damages them.

Host

Wheat, barley, and small grain cereal crops.

Geographical Distribution

The disease occurs in USSR, Europe, and America.

Disease Cycle

P. syringae pv. *atrofaciens* is seed borne and also may persist in soil. The pathogen multiplies near glume joints and lies dormant when moisture is limiting. The pathogens survive on crop debris as well as on various grass hosts. It is disseminated by splashing rain or by insects.

Control Measures

Use of clean seed, crop rotation, avoiding overhead irrigation, and eliminating crop residue.

2.5 OATS

2.5.1 HALO BLIGHT OF OATS

Pathogen: *Pseudomonas coronafaciens* (Elliot, 1920; Schaad and Cunfer, 1979)

Synonyms: *P. syringae* pv. *coronafaciens* (formerly connected to halo blight of oats)
P. syringae pv. *striafaciens* (formerly connected to bacterial stripe blight of oats and barley)

Common names: Halo blight of oats and bacterial stripe blight of oats and barley

Symptoms

1. *Halo blight*: Symptoms are characteristic small, pale green, oval leaf spots with slightly sunken centers. Around the spot, a yellow to red-brown halo usually develops that can be more than 1 cm in diameter or elongate along the entire leaf length. On chaff, small spots with a yellow halo may also occur, and on glumes, yellow translucent spots between the veins can be observed.
2. *Bacterial stripe*: Symptoms are small water-soaked leaf spots, often in rows (stomatal infection). When these spots coalesce, they form narrow water-soaked stripes, sometimes surrounded by a narrow yellow halo. Under humid conditions, bacterial slime can sometimes be observed. Spots and stripes turn red to brown black in later stages (Photo 2.12) (Schaad and Cunfer, 1979).

Host

Primary host: Oats (*Av. sativa*) and barley (*H. vulgare*).

Secondary host: *S. cereale* (rye), *T. aestivum* (wheat), *Avena byzantina* (red oats, Algerian oats), *Agropyron repens* (quack grass), *Bromus* spp. (brome grass), and *Phleum pratense* (timothy).

PHOTO 2.12 Typical field symptoms of halo blight in oats, caused by *Pseudomonas coronafaciens*. (Courtesy of floridadownunder.files.wordpress.com.)

Geographical Distribution
Described for the first time by Elliot (1920) from the United States. Widespread in most countries producing oats and other grains. Although devastating infections were described in the United States in the 1920s, the disease is now sporadic and generally causes no economic losses, except for some sporadic outbreaks. Cold, wet weather conditions or aberrant rain and irrigation conditions favor these outbreaks.

Disease Cycle
P. coronafaciens can be transmitted by seed and can survive in crop debris for several years (Kleinhempel et al., 1989). In the field, bacteria are spread by wind-driven rain. The role of insects and surface water is unknown.

Control Measures
Crop rotation, destruction of crop debris, and the use of certified seed are the only possible and usually also the only necessary means for control. Copper-based pesticides are sometimes used in the United States for preventive spraying or seed treatment.

2.5.2 BACTERIAL STRIPE BLIGHT OF OATS

Pathogen: *Pseudomonas syringae* pv. *striafaciens*

Symptoms
Symptoms appear as small, water-soaked lesions on leaves. These lesions may coalesce into stripes or blotches, which may extend the length of the leaf blade (Photo 2.13). These stripes often have narrow, yellowish margins. As the stripes age, they turn a translucent rusty brown.

Disease Cycle
The bacterium is seed borne and can survive in infected crop stubble for up to 2 years. During periods of cool wet weather, the bacteria are blown or splashed onto leaves.

Economic Impact
Bacterial stripe is widely distributed in oats but rarely causes serious problems.

PHOTO 2.13 Typical symptoms of bacterial stripe blight on oat leaves due to *P. syringae* pv. *striafaciens.* (Courtesy of Department of Agriculture and Food. Copyright Government of Western Australia, agric.wa.gov.au.)

Control Measures
Warm dry weather prevents further spread. If control measures are needed, crop rotation, destruction of crop debris, and using certified seed should lessen disease.

2.5.3 BACTERIAL LEAF STRIPE OF OATS

Pathogen: *Pseudomonas avenae* Manns

Symptoms
Water soaking occurs with advancing of interveinal lesions that vary in length from several to more than 25 cm. Older lesions are usually light brown.

Host
Pearl millet, corn, sorghum, and sugarcane in primary citation. Additional hosts cited are barley, wheat, oats, Italian millet, barnyard millet (*Panicum crus-galli* var. *frumentaceum*), proso millet, foxtail millet, finger millet, rice, rye, and Vasey grass (*Pas. urvillei*).

Control Measures
Crop rotation, destruction of crop debris, and the use of certified seed are the only possible and usually also the only necessary means for control. Copper-based pesticides are sometimes used as preventive spraying and seed treatment.

2.6 SORGHUM

2.6.1 BACTERIAL LEAF STREAK OF SORGHUM

Pathogen: *Xanthomonas vasicola* pv. *holcicola* (Elliot, 1930; Vauterin et al., 1995)

Synonyms: *X. holcicola*
              X. campestris pv. *holcicola*

Common names: Bacterial leaf streak

Symptoms
Small water-soaked leaf spots are the first symptom of the disease. These spots later enlarge and elongate, become reddish brown with a pale brown center, and may cover a large part of the leaf. In a final or severe stage, leaves wither and drop early. Yellowish bacterial slime is exuded from

the spots under humid conditions and dries into a white papery film in dry weather. Symptoms can be confused with those caused by *B. andropogonis* (bacterial leaf stripe). Stripes caused by the latter bacterium, however, are more uniformly reddish brown. Symptoms may also be confused with those caused by *P. syringae* pv. *syringae* (bacterial eye spot or leaf spot). Here, spots remain round to ellipsoid and bacterial slime is not observed.

Host

Sorghum or Indian millet (*Sorghum bicolor = S. vulgare = Andropogon sorghum*) and its varieties *caffrorum* (= *S. caffrorum*), *durra* (= *S. durra*), and *technicus* (= *S. technicum*). *S. almum* (Columbus grass or 5-year sorghum), *S. halepense* (Arabian millet or Aleppo grass), *S. sudanense* (Sudan grass), *Panicum miliaceum* (broomcorn millet), and *Z. mays* (maize) have also been reported to be natural hosts.

Geographical Distribution

First described by Elliot (1930) from the United States.

It is present in Argentina, Australia, Ethiopia, Gambia, India, Israel, Mexico, New Zealand, Niger, the Philippines, South Africa, Thailand, Romania, S. Russia, Ukraine, the United States. Never observed in Europe.

Economic Importance

The disease has only limited importance; it is important only occasionally in springtime under warm weather conditions and becomes less serious during hot and dry summer months, and there is considerable variation in resistance.

Control Measures

Use of healthy seeds and low susceptibility or resistant varieties is the main way to prevent or control diseases caused by *X. vasicola* pv. *holcicola*. Furthermore, rotations with nongrasses or grain crops in a 1:3 scheme, removal of crop residue, controlling weeds, and (in the case of fodder grasses) sowing of mixtures are useful in reducing risk of disease. In some cases, burning of old grass in spring may reduce initial inoculum.

2.6.2 Bacterial Leaf Stripe of Sorghum

Pathogen: *Pseudomonas andropogonis*

Synonyms: *P. syringae* subsp. *andropogonis*

Common names: Bacterial stripe

Symptoms

Lesions occur first on lower leaves and progress up the plant, but leaves above the ears are rarely infected. Lesions (sometimes described as stripes) initially are water soaked, long and narrow with parallel sides, and olive to amber in color. Some investigators have described lesion color as pale yellow or greenish white. Lesions eventually enlarge, or elongate further, and coalesce with other lesions to involve a large area on the infected leaf. As lesions enlarge and coalesce, their color becomes lighter. Severely diseased leaves are easily shred by wind.

Although upper leaves are normally not diseased, they become completely white, a secondary effect of the infection of the lower leaves. In susceptible inbred lines, most leaves below the ear may be killed.

Other foliar symptoms are slightly sunken and circular to ellipsoidal shaped (1–4 mm in diameter) spots are tan to brown with one or more darker brown rings and irregular margins. Some spots are surrounded by chlorotic rings 1 mm wide. Some spots coalesce into elongated blotches (Photo 2.14).

Host

Sorghum, Sudan grass, Broom corn, as well as several other related grasses.

PHOTO 2.14   Natural symptoms of bacterial leaf stripe on sorghum. (Courtesy of Scot Nelson, flickr.com.)

Disease Cycle

P. andropogonis is seed borne and also overwinters in residue. During extended periods of warm and wet weather, bacteria enter the leaves through stomata and possible leaf wounds caused by different means. Disease development is favored by a period of warm and moist weather at a temperature of 22°C–28°C.

Control Measures

Grow resistant hybrids and varieties. Bacterial stripe is not normally economically important but has been severe on a few susceptible inbred lines.

2.7   PEARL MILLET

2.7.1   Bacterial Spot of Pearl millet

Pathogen: *Pseudomonas syringae* van Hall

Symptoms

Bacterial spot is also known as *Holcus* spot of pearl millet. The disease spots are round, oblong, linear, or irregular (Photo 2.15). Water-soaked leaf spots expand to form oval to elongated tan necrotic lesions with a thin dark brown margin (Jensen et al., 1991).

Pathogen and Disease Development

Temperature for growth of the pathogen ranges from 0°C to 35°C, with optimum temperatures between 25°C and 30°C. The pathogen is resistant to freezing in water. The disease development is noticed within these temperature ranges.

Host

Pearl millet, Napier grass, sorghum, Sudan grass, Johnson grass, foxtail, and maize.

Geographical Distribution

Iowa (United States) and New South Wales (Australia).

Transmission

Unknown for pearl millet, but is transmitted in seeds in Napier grass (Richardson, 1979). The pathogen is susceptible to desiccation on grass but resistant on sorghum seed.

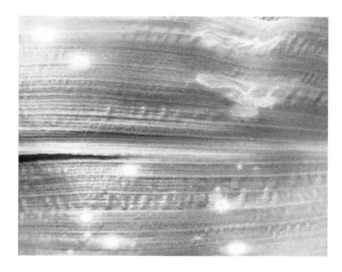

PHOTO 2.15   Leaf symptoms of *P. syringae* on pearl millet. (Courtesy of greatlakeshybrids.com.)

Control Measures
Spray of copper fungicides and antibiotics depending on the severity of the disease.

2.7.2   Bacterial Leaf Streak of Pearl Millet

Pathogen: *Xanthomonas campestris* pv. *pennamericanum* (Pammel) Dowson

Symptoms
Symptoms are not clearly defined in the literature. These are similar to bacterial leaf stripe and streak of sorghum.

Host
Pearl millet and proso millet (*Pa. miliaceum*).

Geographical Distribution
Nigeria, Senegal, and Niger

Control Measures
As in bacterial spot of pearl millet.

2.7.3   Bacterial Leaf Stripe of Pearl Millet

Pathogen: *Pseudomonas avenae* Manns

Symptoms
Water-soaked interveinal lesions on leaves that vary in length from several to more than 25 cm. Older lesions are usually light brown.

Host
Pearl millet, corn, sorghum, and sugarcane are the primary hosts. Additional hosts are barley, wheat, oats, Italian millet, barnyard millet (*Pa. crus-galli* var. *frumentaceum*), proso millet, foxtail millet, finger millet, rice, rye, and Vasey grass (*Pas. urvillei*).

Geographical Distribution
Nigeria.

Transmission
Not known to be transmitted through seed in pearl millet but has been demonstrated to be seed borne in rice and Vasey grass.

Control Measures
As in bacterial spot of pearl millet.

2.7.4 PANTOEA DISEASE OF PEARL MILLET

Pathogen: *Pantoea agglomerans* (Ewing and Fife)

Symptoms
Straw-colored lesions with a chlorotic edge often extending the length of the leaf. Water soaking at leaf tips and margins occur in seedlings.

Host
Pearl millet.

Geographical Distribution
Zimbabwe and possibly India.

Transmission
Not known to be seed transmitted.

Control Measures
As in bacterial spot of pearl millet.

2.8 MINOR MILLET

2.8.1 RYE

2.8.1.1 Bacterial Streak (Black Chaff) of Rye

Pathogen: *Xanthomonas campestris* pv. *translucens*

Common names: Bacterial streak (black chaff) of rye

Symptoms
The symptoms of the disease first appear as small water-soaked leaf spots enlarging to yellowish or brownish blotches or irregular stripes (Photo 2.16). Conspicuous bacterial exudations appear scattered over the lesions.

PHOTO 2.16 Symptoms of bacterial streak on rye leaf. (Courtesy of University of Georgia, College of Agricultural and Environmental Sciences, Athens, GA, plantpath.caes.uga.edu.)

2.8.1.2   Bacterial Halo Blight of Rye

Pathogen: *Pseudomonas syringae* pv. *coronafaciens* (Elliott, 1920; Young et al., 1978)

Common names: Bacterial halo blight of rye

Symptoms
The symptoms of the disease first appear as small water-soaked leaf spots enlarging to yellowish or brownish blotches or irregular stripes (Photo 2.17). Conspicuous bacterial exudations appear scattered over the lesions.

Control Measures
No information is available.

2.8.2   Finger Millet

2.8.2.1   Leaf Spot of Finger Millet

Pathogen: *Xanthomonas eleusineae*

Symptoms
The spots appear on both surfaces of the leaf blade. They are linear and narrow, spreading along the veins and are about 2–5 mm long. In the beginning, the spots are light yellowish brown, but with age become dark brown. In advanced stages, the leaf splits along the streak, shredding the blade. All leaves, including the tender shoots in a plant are affected by the pathogen. Though the bacterium affects mainly the leaves, at times characteristic streaks are also found on the peduncle of the ear-head, and such streaks are narrow and 5–10 mm in length and appear subcuticular.

Disease Cycle
The organism is present chiefly in the vascular bundle and seems to spread in both directions in the bundles.

Control Measures
No information available.

PHOTO 2.17   Symptoms of bacterial halo blight on rye leaves. (Courtesy of T.T. Tsukiboshi, National Agriculture and Food Research Organization (NARO), Ibaraki, Japan, naro.affrc.go.jp.)

2.8.2.2   Bacterial Blight of Finger Millet

Pathogen: *Pseudomonas eleusineae*

Symptoms

The disease affects the older leaves in the initial stage followed by younger leaves. Brown discoloration of the leaf sheath from the base upward is seen. The infected portion of the lamina invariably involves the midrib and appears straw colored. The symptom spreads to about three-fourths of lamina and then abruptly stops, or in some cases reaches up to the leaf tip. The disease proceeds from the margin toward the midrib, as a result of which the midrib and the adjacent area remain healthy while the other portions are infected. The presence of bacterial cells is easily detected in the phloem vessels. Due to the characteristic drooping of the leaves, infected plants can be detected even from a distance.

Infected culms show light brown discoloration along one side. Sometimes the discoloration begins from the base but mostly it begins 2–3 in. above the base and extends up to the leaf sheath. However, there is no apparent reduction in girth or turgidity of the affected culms as compared to healthy ones. Plants less than a month old are usually free from the disease.

Disease Cycle

The bacterium is systemic and soil-borne. No more information is available on life cycle and control measures.

Control Measures

No information available.

REFERENCES

Alizadeh, A. and H. Rahimian. 1989. Bacterial leaf streak of Gramineae in Iran. *OEPP/EPPO Bull.*, 19, 113–117.

Anonymous. 1995. Occurrence of sheath rot, a new disease in wheat. *Lumle Newsl.*, 1(4), 3.

Anuratha, C.S. and S.S. Gnanamanickam. 1987. *Pseudomonas fluorescens* suppresses development of bacterial blight symptoms. *Int. Rice Res. Newsl.*, 12, 1, 17.

Bamberg, R.H. 1936. Black chaff disease of wheat. *J. Agric. Res.*, 52, 397–417.

Banta, S.J. 1989. *Bacterial Blight of Rice*. IRRI, Manila, Philippines.

Boosalis, M.G. 1952. The epidemiology of *Xanthomonas translucens* on cereals and grasses. *Phytopathology*, 42, 387–395.

Bradbury, J.F. 1970a. *Xanthomonas oryzae*. CMI Descriptions of Pathogenic Fungi and Bacteria No. 239. CAB International, Wallingford, U.K.

Bradbury, J.F. 1970b. *Xanthomonas oryzicola*. CMI Descriptions of Pathogenic Fungi and Bacteria No. 240. CAB International, Wallingford, U.K.

Buddenhagen, I.W. 1983. Disease resistance in rice. In: Lamberti, F., Waller, J.M., and N.A. van derGraaff (eds.), *Durable Resistance in Crops*. Plenum, New York, pp. 401–428.

Buddenhagen, I.W. 1985. Rice disease evaluation in Madagascar. *Int. Rice Comm. Newsl.*, 34, 74–78.

Cahyaniati, A. and C.N. Mortensen. 1995. Bacterial sheath brown rot of rice (*Pseudomonas fuscovaginae*) grown in Indonesia. In: *International Seed Testing Association (ISTA) Pre-Congress Seminar on Seed Pathology*, Copenhagen, Denmark, June 6, 1995.

Cahyaniati, A. and C.N. Mortensen. 1997. Bacterial sheath brown rot of rice (*Pseudomonas fuscovaginae*) grown in Indonesia. *Seed Health Testing in the Production of Quality Seed*, 195pp.

Carlson, R.R. and A.K. Vidaver. 1982a. Bacterial mosaic, a new corynebacterial disease of wheat. *Plant Dis.*, 66, 76–79.

Carlson, R.R. and A.K. Vidaver, 1982b. Taxonomy of *Corynebacterium* plant pathogens, including a new pathogen of wheat, based on polyacrylamide gel electrophoresis of cellular proteins. *Int. J. Syst. Bacteriol.*, 32, 315–326.

Chand, J.N. 1967. Longevity of *Corynebacterium tritici* (Hutchinson) Burk, causing "Tundu", disease of wheat in Haryana. *Sci. Cult.*, 33, 539.

Chand, T., Singh, N., Singh, H., and B.S. Thind. 1979. Field efficacy of stable bleaching powder to control bacterial leaf blight. *Int. Rice Res. Newsl.*, 4, 12.

Chang, R.J., Ries, S.M., Ervings, A.D., and C.J. D'Arcy. 1990. Bacterial mosaic of wheat in Illinois. *Plant Dis.*, 74, 1037.

Clara, F.M. 1934. A comparative study of the green-fluorescent bacterial plant pathogens. *Cornell Agr. Exptl. Sta. Mem.* 159: 1–36.

CMI. 1981. *Distribution Maps of Plant Diseases No. 544*, 1st ed. CAB International, Wallingford, U.K.

CMI. 1982. *Distribution Maps of Plant Diseases No. 109*, 5th ed. CAB International, Wallingford, U.K.

Cottyn, B., Cerez, M.T., Van Outryve, M. F., Barroga, J., Swings, J., and T.W. Mew. 1996. Bacterial diseases of rice. I. Pathogenic bacteria associated with sheath rot complex and grain discoloration of rice in the Philippines. *Plant Dis.*, 80, 429–437.

Dath, A.P. and S. Devadath. 1983. Role of inoculum in irrigation water and soil in the incidence of bacterial blight of rice. *Indian Phytopathol.*, 36, 142–144.

Davis, M.J. Jr., Gillaspie, A.G., Vidaver, A.K., and R.W. Harris. 1984. *Clavibacter*: A new genus containing some phytopathogenic coryneform bacteria, including *Clavibacter xyli* subsp. *xyli* sp. *nov.*, subsp. *nov.* and *Clavibacter xyli* subsp. *cynodontis* subsp. *nov.*, pathogens that cause ratoon stunting disease of sugarcane and Bermuda grass stunting disease. *Int. J. Syst. Bacteriol.*, 34(2), 107–117.

Demir, G. and N. Ustun. 1992. Studies on bacterial streak disease (*Xanthomonas campestris* pv. *translucens*) of wheat and other Gramineae. *J. Turkish Phytopathol.*, 21, 33–40.

Devadath, S. and A.P. Dath. 1970. Epidemiology of *Xanthomonas translucens* f.sp. *oryzae*. *Oryza*, 7, 13–16.

Devadath, S. and A.P. Dath. 1985. Infected chaff as a source of inoculum of *Xanthomonas campestris* pv. *oryzae* to the rice crop. *Zeitschrift für Pflanzenkrankheiten und Pflanzenschutz*, 92, 485–488.

Dick, M. 1985. Bacterial stem canker. New Zealand Forest Service, Forest Pathology in New Zealand No. 10.

Durgapal, J.C. 1983. Management of bacterial blight of rice by nursery treatment—Preliminary evaluation. *Indian Phytopathol.*, 36, 146–149.

Durgapal, J.C. 1985. Self-sown plants from bacterial blight-infected rice seeds—A possible source of primary infection in northwest India. *Curr. Sci.*, 54, 1283–1284.

Duveiller, E. 1989. Research on "*Xanthomonas translucens*" at CIMMYT. *OEPP/EPPO Bull.*, 19, 97–103.

Duveiller, E. 1994a. Bacterial leaf streak or black chaff of cereals. *OEPP/EPPO Bull.*, 24, 135–158.

Duveiller, E. 1994b. A pictorial series of disease assessment keys for bacterial leaf streak of cereals. *Plant Dis.*, 78, 137–141.

Duveiller, E. and H. Maraite. 1990. Bacterial sheath rot of wheat caused by *Pseudomonas fuscovaginae* in the highlands of Mexico. *Plant Dis.*, 74, 932–935.

Duveiller, E. and H. Maraite. 1993a. Study on yield loss due to *Xanthomonas campestris* pv. *undulosa* in wheat under high rainfall temperate conditions. *Zeitschrift für Pflanzenkrankheiten und Pflanzenschutz*, 100, 453–459.

Duveiller, E. and H. Maraite. 1993b. *Xanthomonas campestris* pathovars on cereals: Leaf streak or black chaff diseases. In: Swings, J. (ed.), *Xanthomonas*. Chapman & Hall, London, U.K., Chapter 1, pp. 76–79.

Duveiller, E., Miyajima, K., Snacken, F., Autrique, A., and H. Maraite. 1988. Characterization of *Pseudomonas fuscovaginae* and differentiation from other fluorescent pseudomonads occurring on rice in Burundi. *J. Phytopathol.*, 122, 97–107.

Duveiller, E., Bragard, C., and H. Maraite. 1991. Bacterial diseases of wheat in the warmer areas—Reality or myth? In: Saunders, D. (ed.), *Wheat for the Non-traditional Warm Areas. Proceedings of the International Conference*, Iguazu Falls, Brazil. CIMMYT, Mexico, DF, pp. 189–202.

El Banoby, F.E. and K. Rudolph. 1989. Multiplication of *Xanthomonas campestris* pvs. *secalis* and *translucens* in host and nonhost plants (rye and barley) and development of water-soaking. *OEPP/EPPO Bull.*, 19, 104–111.

Elliot, C. 1920. Halo blight of oats. *J. Agric. Res.*, 19, 139–172.

Elliot, C. 1930. Bacterial streak disease of sorghum. *J. Agric. Res.*, 40, 963–976.

Esker, P.D. and F.W. Nutter Jr. 2002. Assessing the risk of Stewart's disease of corn through improved knowledge of the role of the corn flea beetle vector. *Phytopathology*, 92, 668–670.

Exconde, O.R. 1973. Yield losses due to bacterial leaf blight of rice. *Philippines Agric.*, 57, 128–140.

Feakin, S.D. 1971. Pest control in rice. *PANS Manual*, 3, 69–74.

Forster, R.L. and J.F. Bradbury. 1990. Pink seed of wheat caused by *Erwinia rhapontici* in Idaho. *Plant Dis.*, 74, 81.

Forster, R.L. and N.W. Schaad. 1988. Control of black chaff of wheat with seed treatment and a foundation seed health program. *Plant Dis.*, 72, 935–938.

Forster, R.L. and N.W. Schaad. 1990. Longevity of *Xanthomonas campestris* pv. *translucens* in wheat seeds under storage conditions. In: *Proceedings of the Seventh International Conference on Plant Pathogenic Bacteria*, Vol. 1. Akademiai Kiado, Budapest, Hungary, pp. 329–331.

Fourest, E., Rehm, L.D., Sands, D.C., Bjarko, M., and R.E. Lund. 1990. Eradication of *Xanthomonas translucens* from barley seed with dry heat treatments. *Plant Dis.*, 74, 816–818.

Fukanaga, K. 1966. Antibiotics and new fungicides for control of rice diseases. In: *Symposium on Plant Diseases in the Pacific. 11th Pacific Science Congress*, Tokyo, Japan, pp. 170–180.

Goto, M. 1965. A technique for detecting the infected area of bacterial leaf blight of rice caused by *X. oryzae* before symptom appearance. *Ann. Phytopathol. Soc. Jpn.*, 30, 37–41.

Gupta, P. 1966. Studies on ear-cockle and 'tundu' disease of wheat. PhD thesis, Indian Agricultural Research Institute, New Delhi, India.

Hall, V.N., Kim, H.K., and D.C. Sands. 1981. Transmission and epidemiology of *Xanthomonas translucens*. *Phytopathology*, 71, 878.

Hoa, T.T.C., Binh, T.C., Kandaswamy, T.K., and N. Van Luat. 1984. Prophylactic chemical treatments for control of bacterial blight (BB). *Int. Rice Res. Newsl.*, 9(3), 14.

Hsieh, S.P.Y., Buddenhagen, I.W., and H.E. Kauffman. 1974. An improved method for detecting the presence of *Xanthomonas oryzae* in rice seed. *Phytopathology*, 64, 273–274.

Huang, H.C., Phillippe, R.C., and L.M. Phillippe. 1990. Pink seed of pea: A new disease caused by *Erwinia rhapontici*. *Can. J. Plant Pathol.*, 12, 445–448.

Jensen, S.G., P. Lambrecht, G.N. Odovody, and A.K. Vidaver. 1991. A leaf spot of pearl millet caused by *Pseudomonas syringae*. *Phytopathology*, 81, 1193 (abstract).

John, V.T., Dobson, R., Alluri, K., Zan, K., Efron, Y., Wasano, K., Thottapilly, G., Gibbons, J.W., Rossel, H.W. (1984) Rice: pathology, virology. In: *Annual Report, International Institute of Tropical Agriculture 1983, 1984*, pp. 19–22. IITA, Ibadan, Nigeria.

Jones, L.R., Johnson, A.G., and C.S. Reddy. 1917. Bacterial-blight of barley. *J. Agric. Res.*, 11, 625–643.

Kauffman, H.E. and A.P.K. Reddy. 1975. Seed transmission studies of *Xanthomonas oryzae* in rice. *Phytopathology*, 65, 663–666.

Kaul, M.L.H. and K.K. Sharma. 1987. Bacterial blight of rice—A review. *Biologisches Zentralblatt*, 106, 141–167.

Kim, H.K., Orser, C., Lindow, S.E., and D.C. Sands. 1987. *Xanthomonas campestris* pv. *translucens* strains active in ice nucleation. *Plant Dis.*, 71, 994–997.

Kleinhempel, H., Naumann, K., and D. Spaar. 1989. *Bakterielle Erkrankungen der Kulturpflanzen*. Gustav Fischer Verlag, Jena, Germany.

Koleva, N. 1981. Bacteriosis of winter wheat. *Rastitelna Zashchita*, 29, 15–17.

Koroleva, I.B., Gvozdyak, R.I., and L.A. Pasichnik. 1985. *Xanthomonas campestris* pv. *oryzae*, causal agent of a bacterial disease of rice seed in Ukraine. *Mikrobiologicheskii Zhurnal*, 47, 93–95.

Kotte, W. 1930. Eine bakterielle Blattfaule der Winter-Endivie (*Cichorium endiviae* L.). *Phytopath. Zeitschr.*, 1, 605–613.

Li, Z.Z., Zhao, H., and X.D. Ying. 1985. The weed carriers of bacterial leaf blight of rice. *Acta Phytopathol. Sin.*, 15, 246–248.

Little, E.L., Bostock, R.M., and B.C. Kirkpatrick. 1998. Genetic characterization of *Pseudomonas syringae* pv. *syringae* strains from stone fruits in California. *Appl. Environ. Microbiol.*, 10, 3818–3823.

Macapuguay, F. and M. Mnzaya. 1988. Progress on irrigated rice agronomy at Usangu village—M'beya zone—Tanzania. *Int. Rice Comm. Newsl.*, 37, 40–44.

Mahto, B.N., Singh, R.N., and G.P. Singh. 1988. Response of rice bacterial blight pathogens *in vitro* to antibiotics and fungi toxicants. *Int. Rice Res. Newsl.*, 13, 23.

Mamluk, O.F., Al Ahmed, M., and M.A. Makki. 1990. Current status of wheat diseases in Syria. *Phytopathologia Mediterranea*, 29, 143–150.

Mary, C.A., Dev, V.P.S., Karunakaran, S., and N.R. Nair. 1986. Cowdung extract for controlling bacterial blight. *Int. Rice Res. Newsl.*, 11, 19.

Mathur, R.S. and Z.U. Ahmad. 1964. Investigations on the nature of association between *Anguina tritici* and *Corynebacterium tritici*. *Proc. Natl. Acad. Sci., India*, 34(B), 335–336.

McBeath, J.H. 1993. Other bacterial diseases. In: Mathur, S.B. and B.M. Cunfer (eds.), *Seed-Borne Diseases and Seed Health Testing of Wheat*. Jordbrugsforlaget, Frederiksberg, Denmark, pp. 137–146.

McBeath, J.H. and M. Adelman. 1986. Detection of *Corynebacterium michiganense* subsp. *tessellarius* in seeds and wheat plants. *Phytopathology*, 76, 1099 (abstract).

McBeath, J.H., Adelman, M., and L. Jackson. 1988. Screening wheat germplasm for *Corynebacterium michiganense* subsp. *tessellarius*. *Phytopathology*, 78, 1566 (abstract).

McMullen, M.P., Stack, R.W., Miller, J.D., Bromel, M.C., and V.L. Youngs. 1984. *Erwinia rhapontici*, a bacterium causing pink wheat kernels. *Proc. North Dakota Acad. Sci.*, 38, 78.

Mehta, Y.R. 1986a. Survival of *Xanthomonas campestris* pv. *undulosa* in field conditions. In: *Reunião Nacional de Pesquisa em Trigo*, Londrina, Brazil, p. 55.

Mehta, Y.R. 1986b. Effects of guazatin plus to control *Xanthomonas campestris* pv. *undulosa* in wheat. In: *Reunião Nacional de Pesquisa em Trigo*, Londrina, Brazil, p. 56.

Mew, T.W. 1987. Current status and future prospects of research on bacterial blight of rice. *Annu. Rev. Phytopathol.*, 25, 359–382.

Michener, P.M., Freeman, N.D., and J.K. Pataky. 2003. Relationships between reactions of sweet corn hybrids to Stewart's wilt and incidence of systemic infection by *Erwinia stewartii*. *Plant Dis.*, 87, 223–228.

Midha, S.K. 1969. Studies on the ear-cockle nematode and the development of ear cockle and 'tundu' disease of wheat. PhD thesis, Indian Agricultural Research Institute, New Delhi, India.

Midha, S.K., Chatrath, M.S., and G. Swarup. 1971. On the feeding of *Anguina tritici* on growing point of wheat seedlings. *Indian J. Nematol.*, 1, 93–94, 120.

Milus, E.A. and A.F. Mirlohi. 1994. Use of disease reactions to identify resistance in wheat to bacterial streak. *Plant Dis.*, 78, 157–161.

Milus, E.A. and A.F. Mirlohi. 1995. Survival of *Xanthomonas campestris* pv. *translucens* between successive wheat crops in Arkansas. *Plant Dis.*, 79, 263–265.

Miyajima, K. 1983. Studies on bacterial sheath brown rot of rice plant caused by *Pseudomonas fuscovaginae*, Tanii, Miyajima and Akita [in Japanese, English abstract]. *Rep. Hokkaido Pref. Exp. Stn.*, 43, 1–74.

Murty, V.S.T. and S. Devadath. 1982. Survival of *Xanthomonas campestris* pv. *oryzae* in different soils. *Indian Phytopathol.*, 35, 32–38.

Murty, V.S.T. and S. Devadath. 1984. Role of seed in survival and transmission of *Xanthomonas campestris* pv. *oryzae* causing bacterial blight of rice. *Phytopathologische Zeitschrift*, 110, 15–19.

Nayak, P. and N.K. Chakrabarti. 1986. Stable resistance to bacterial blight disease in rice. *Ann. Appl. Biol.*, 109, 179–186.

Nayak, P. and P.R. Reddy. 1985. Spread pattern of bacterial blight disease in rice crop. *Indian Phytopathol.*, 38, 39–44.

Notteghem, J.L. 1998. Emerging disease problems in intensive rice systems in Africa. In: *The Seventh International Congress of Plant Pathology*, Edinburgh, Scotland, August 9–16, 1998. British Society for Plant Pathology, London, U.K.

Noval, C. 1989. Maladies bactériennes des graminées en Espagne. *OEPP/ EPPO Bull.*, 19, 131–135.

OEPP/EPPO. 1979. Data sheets on quarantine organisms No. 3, *Xanthomonas campestris* pv. *oryzicola*. *OEPP/EPPO Bull.*, 9(2).

OEPP/EPPO. 1980. Data sheets on quarantine organisms No. 2, *Xanthomonas campestris* pv. *oryzae*. *OEPP/ EPPO Bull.*, 10(1).

Okabe, N. 1935. Bacterial diseases of plants occurring in Taiwan (Formosa). V. *Journal of the Society of Tropical Agriculture* 7: 57–66 pp.

Opina, O.S. and O.R. Exconde. 1971. Assessment of yield loss due to bacterial leaf streak of rice. *Philippine Phytopathol.* 7, 35–39.

Ou, S.H. 1972. *Rice Diseases*. CAB International, Wallingford, U.K.

Padmanabhan, S.Y. 1983. Integrated control of bacterial blight of rice. *Oryza*, 20, 188–194.

Pathak, K.N. and G. Swarup. 1984. Incidence of *Corynebacterium michiganense* pv. *tritici* in the ear-cockle nematode (*Anguina tritici*) galls and pathogenicity. *Indian Phytopathol.*, 37, 267–270.

Paul, V.H. and I.M. Smith. 1989. Bacterial pathogens of Gramineae: Systematic review and assessment of quarantine status for the EPPO region. *OEPP/EPPO Bull.*, 19, 33–42.

Pepper, E.H. 1967. *Stewart's Bacterial Wilt of Corn*. Monograph 4. American Phytopathological Society, St. Paul, MN.

Piening, L.J., MacPherson, D.J., and S.S. Mahli. 1987. The effect of copper in reducing stem melanosis of Park wheat. *Can. J. Plant Sci.*, 67, 1089–1091.

Piening, L.J., MacPherson, D.J., and S.S. Mahli. 1989. Stem melanosis of some wheat, barley, and oat cultivars on a copper deficient soil. *Can. J. Plant Sci.*, 11, 65–67.

Qi, Z. and T.W. Mew. 1985. Adult-plant resistance of rice cultivars to bacterial blight. *Plant Dis.*, 69, 896–898.

Raj, K. and V. Pal. 1988. Overwintering of *Xanthomonas campestris* pv. *oryzae*. *Int. Rice Res. Newsl.*, 13, 22–23.

Rao, P.S. 1987. Across-season survival of *Xanthomonas campestris* pv. *oryzicola*, causal agent of bacterial leaf streak. *Int. Rice Res. Newsl.*, 12, 2, 27.

Ray, P.R. and T.K. Sengupta. 1970. A study on the extent of loss in yield in rice due to bacterial blight. *Indian Phytopathol.*, 23, 713–714.

Reddy, P.R. 1972. Studies on bacteriophages of *Xanthomonas oryzae* and *Xanthomonas translucens* f.sp. *oryzicola*, the incitants of blight and streak diseases of rice. PhD thesis, Banaras Hindu University, Varanasi, India.

Reddy, P.R. 1984. Kresek phase of bacterial blight of rice. *Oryza*, 21, 179–187.

Reddy, P.R. and S.H. Ou. 1974. Differentiation of *Xanthomonas translucens* f.sp. *oryzicola* (Fang et al.) Bradbury, the leaf-streak pathogen, from *Xanthomonas oryzae* (Uyeda and Ishiyama) Dowson, the blight pathogen of rice, by enzymatic tests. *Int. J. Syst. Bacteriol.*, 24, 450–452.

Richardson, M.J. 1979. *An Annotated List of Seed Borne Disease*, 3rd ed. International Seed Testing Association, Zurich, Switzerland.

Roberts, P. 1974. *Erwinia* (Millard) Burkholder associated with pink grain of wheat. *J. Appl. Bacteriol.*, 37, 353–358.

Rott, P. 1987. Brown rot (*Pseudomonas fuscovaginae*) of the leaf sheath of rice in Madagascar. Institute de Recherches Agronomiques Tropicales et des Cultures Vivrieres, Montpellier, France, 22pp.

Rott, P., Notteghem, J.L., and P. Frossard. 1989. Identification and characterization of *Pseudomonas fuscovaginae*, the causal agent of bacterial sheath brown rot of rice, from Madagascar and other countries. *Plant Dis.*, 73(2), 133–137.

Sands, D.S. and E. Fourest. 1989. *Xanthomonas campestris* pv. *translucens* in North and South America and in the Middle East. *OEPP/EPPO Bull.*, 19, 127–130.

Sands, D.S., Mizrak, G., Hall, V.N., Kim, H.K., Bockelman, H.E., and M.J. Golden. 1986. Seed-transmitted bacterial diseases of cereals: Epidemiology and control. *Arab. J. Plant Protect.*, 4, 127–125.

Schaad, N.W. and B.M. Cunfer. 1979. Synonymy of *Pseudomonas coronofaciens*, *Pseudomonas coronofaciens* pathovar *zeae*, *Pseudomonas coronofaciens* subsp. *atropurpurea* and *Pseudomonas striafaciens*. *Int. J. Syst. Bacteriol.*, 29, 213–221.

Schaad, N.W., Gabrielson, R.L., and M.W. Mulanase. 1981. Hot acidified cupric acetate soaks for eradication of *Xanthomonas campestris* from crucifer seeds. *Appl. Environ. Microbiol.*, 39, 803–807.

Scharif, G. 1961. *Corynebacterium iranicum* sp. *nov.* on wheat (*Triticum vulgare* L.) in Iran and a comparative study of it with *C. tritici* and *C. rathayi*. *Entomologie et phytopathologie appliquées*, 19, 1–24.

Schuster, M.L. 1975. Leaf freckles and wilt of corn incited by *Corynebacterium nebraskense* Schuster, Hoff, Mandel, Lazar, 1972. *Neb. Agric. Exp. Stn. Res. Bull.*, 270, 40pp.

Sellwood, J.E. and R.A. Lelliot. 1978. Internal browning of hyacinth caused by *Erwinia rhapontici*. *Plant Pathol.*, 27, 120–124.

Shane, W.W., Baumer, J.S., and P.S. Teng. 1987. Crop losses caused by *Xanthomonas* streak on spring wheat and barley. *Plant Dis.*, 71, 927–930.

Singh, D., Vinther, F., and S.B. Mathur. 1983. Seed transmission of bacterial leaf blight in rice. *Seed Pathol. News*, 15, 11.

Singh, R., Singh, J., and S.C. Mathur. 1959. Ear-cokle or "Sehun" disease of wheat. *J. Agric. Animal Husb.*, 3, 7–9.

Singh, R.A. and D. Monga.1985. New methods of seed treatment for eliminating *Xanthomonas campestris* pv. *oryzae* from infected rice seed. *Indian Phytopathol.*, 38, 629–631.

Singh, R.N. 1971. Perpetuation of bacterial blight disease of paddy and preservation of its incitant. I. Survival of *Xanthomonas oryzae* in water. II. Survival of *Xanthomonas oryzae* in soil. *Indian Phytopathol.*, 24, 140–144, 153–154.

Sivaswamy, N.S. and A. Mahadevan. 1986. Effect of stable bleaching powder on growth of *Xanthomonas campestris* pv. *oryzae*. *Indian Phytopathol.*, 39, 32–36.

Smidt, M. and A.K. Vidaver. 1986. Population dynamics of *Clavibacter michiganense* subsp. *nebraskense* infield-grown dent corn and popcorn. *Plant Dis.*, 70, 1031–1036.

Smith, E.F. 1917. A new disease of wheat. *J. Agric. Res.*, 10, 51–54.

Stapp, C. 1928. Schizomycetes (Spaltpilze oder Bakterien). In: Sorauer (ed.), *Handbuch der Pflanzenkrankheiten*, 5th edn., Vol. 2. Paul Parey, Berlin, Germany, pp. 1–295.

Sulaiman, M. and L. Ahmed. 1965. Controlling bacterial blight in paddy in Maharashtra. *Indian Farm.*, 15, 27–29.

Suryanarayana, D. and M.C. Mukhopadhaya. 1971. Ear-cockle and 'tundu' disease of wheat. *Indian J. Agric. Sci.*, 41, 407–413.

Swingle, D.B. 1925. Center rot of French endive or wilt of Chicory (*Cichorium intybus* L.). *Phytopathology*, 15, 730.

Takahi, Y. (1985) Shirahagen R-S (tecloftalam). *Japan Pesticide Information* No. 46, pp. 25–30.

Tanii, A., Miyajima, K., and T. Akita. 1976. The sheath brown rot disease of rice plant and its causal bacterium *Pseudomonas fuscovaginae* sp. *nov. Ann. Phytopathol. Soc. Jpn.*, 42, 540–548.

Timmer, L.W., Marois, J.J., and D. Achor. 1987. Growth and survival of Xanthomonads under conditions non-conducive to disease development. *Phytopathology*, 77, 1341–1345.

Vauterin, L., Hoste, B., Kersters, K., and J. Swings. 1995. Reclassification of *Xanthomonas. Int. J. Syst. Bacteriol.*, 45, 472–489.

Vidaver, A.K. and M. Mandel. 1974. *Corynebacterium nebraskense*, a new orange pigmented phytopathogenic species. *Int. J. Syst. Bacteriol.*, 24, 482–485.

Webster, R.K. and P.S. Gunnell (eds.). 1992. *Compendium of Rice Diseases.* APS Press, St. Paul, MN.

Wiese, M.V. 1987. *Compendium of Wheat Diseases.* 2nd Edition. American Phytopathological Society, St. Paul, MN. pp. 5–10.

Wysong, D.S., Doupnik, B., and L. Lane. 1981. Goss's wilt and corn lethal necrosis—Can they become a major problem? In: *Proceedings of the Annual Corn Sorghum Research Conference 36.* American Seed Trade Association, Washington, DC, pp. 104–130.

Xie, G.L., Zhu, G.N., and X.P. Ren. 2002. Diversity of pathogenic bacteria from rice seeds. *Acta Phytopathol. Sinica*, 32, 114–120 (in Chinese with English abstract).

Yoneyama, K., Sekido, S., and T. Misato. 1978. Studies on the fungicidal action of dithiocarbamates and effect of sodium dimethyldithiocarbamate on fatty acid synthesis of *Xanthomonas oryzae. Ann. Phytopathol. Soc. Jpn.*, 44, 313–320.

Zeigler, R.S. and E. Alvarez. 1987b. Bacterial sheath rot of rice caused by a fluorescent *Pseudomonas* in Latin America. *Fitopatol. Brasileira*, 12, 193–198.

Zeigler, R.S. and E. Alvarez. 1987a. Bacterial sheath brown rot of rice caused by *Pseudomonas fuscovaginae* in Latin America. *Plant Dis.*, 71, 592–597.

Zgurskaya, H.I., Evtushenko, L.I., Akimov, V.N., and L.V. Kalakovtskii. 1993. *Rathayibacter* gen. *nov.*, including the species *Rathayibacter rathayi* comb. *nov.*, *Rathayibacter tritici* comb. *nov.*, *Rathayibacter iranicus* comb. *nov.*, and six strains from annual grasses. *Int. J. Syst. Bacteriol.*, 43, 143–149.

SUGGESTED READING

Agrios, G.N. 2005. *Plant Pathology*, 5th edn. Elsevier Academic Press, Burlington, MA, 922pp.

CABI; EPPO. 1997. *Pseudomonas fuscovaginae.* [Distribution map]. Distribution Maps of Plant Diseases 1997 No. December (1st ed.), Map 742.

Carlson, R.R. and A.K. Vidaver. 1982b. Taxonomy of Corynebacterium plant pathogens, including a new pathogen of wheat, based on polyacrylamide gel electrophoresis of cellular proteins. *Int. J. Syst. Bacteriol.*, 32, 315–326.

Duveiller, E., Fucikovsky, L., and K. Rudolph. 1997. *The Bacterial Diseases of Wheat: Concepts and Methods of Disease Management.* CIMMYT, Mexico, DF (Mexico), 84pp.

Dye, D.W., and R.A. Lelliott. 1974. Genus II. Xanthomonas Dowson 1939, 187, p. 243–249. In R. E. Buchanan and N. E. Gibbons (ed.), Bergey's manual of determinative bacteriology, 8th ed. The Williams & Wilkins Co., Baltimore.

Gnanamanickam, S.S. 2009. Biological control of rice diseases. *Progr. Biol. Control*, 8, 67–78.

Janse, J.D. 2005. *Phytobacteriology: Principles and Practice.* Oxford Press, Wallingford, U.K., 360pp.

Peters, R.A., Timian, R.G., and D. Wesenberg. 1983. A bacterial kernel spot of barley caused by Pseudomonas syringae pv. syringae. *Plant Dis.*, 67, 435–438.

Swings, J., Van den Mooter, M., Vauterin, L. et al. 1990. Reclassification of the causal agents of bacterial blight (Xanthomonas campestris pv. oryzae) and bacterial leaf streak (Xanthomonas campestris pv. oryzicola) of rice as pathovars of Xanthomonas orzae (ex Ishiyama 1922) sp. nov., nom. rev. *Int. J. Syst. Bacteriol.*, 40, 309–311.

3 Bacterial Diseases of Pulse Crops

3.1 GREEN GRAM

3.1.1 BACTERIAL BLIGHT OF GREEN GRAM

Pathogen: *Xanthomonas phaseoli*

Symptoms

Leaf spots first appear as small, water-soaked, or light-green areas on leaflets. They later become dry and brown. This disease is characterized by many brown, dry, and raised spots on the leaf surface (Photo 3.1). The spots may join to affect much of leaf surface, eventually killing the leaflet. When the disease is severe, several such spots coalesce and the leaves become yellow and fall off prematurely. The lower surface of the leaf appears red in color due to the formation of raised spots. Similar water-soaked spots develop on pods where the spot margin is with a shade of red. Severely diseased pods shrivel. In humid weather, a yellowish crust of the blight bacteria covers the spot surface.

Disease Cycle

The bacteria are readily spread by water splashes, walking, or working in the field while plants are wet. The bacteria enter the host through the wounds sustained during field operations. Therefore, avoid field operations when the foliage is wet.

Transmission

The bacterium is also seed borne.

Control Measures

Cultural practices are important in controlling blights. Eliminate weeds and volunteer beans and other potential hosts of bean blight, as this will reduce disease incidence. Good weed control will also improve aeration around the crop so that the plants dry faster; this will reduce the chances for bacterial spread and infection.

A rotation of at least 2 years between bean crops will give time for the bacterial population to decline in the debris. Deep plowing will also encourage the breakdown of infected plant debris and distract the debris and stubbles.

The incidence of bean blight can also be reduced if beans are grown with maize rather than in a monoculture. Use disease-free seeds. Soak the seeds in 500 ppm streptocycline solution for 30 min before sowing followed by two sprays of streptocycline combined with 3 g of copper oxychloride per liter at an interval of 12 days.

3.1.2 BACTERIAL BROWN SPOT OF GREEN GRAM

Pathogen: *Pseudomonas syringae* pv. *syringae*

Bacterial brown spot is the most economically significant disease of processing beans in the north central region of the United States. It occurs in other bean-growing areas in the United States and the world.

Symptoms

Small (3–9 mm in diameter), oval necrotic lesions are apparent on leaves. A narrow yellow-green zone of tissue may be seen surrounding the lesions. Bacterial exudates (ooze) and water soaking

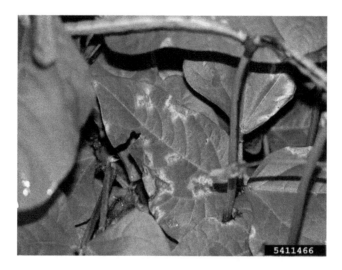

PHOTO 3.1 Leaf infection of bacterial blight pathogen on green gram; angular necrotic lesions of common blight with characteristic yellow border. (Courtesy of Don Ferrin, Louisiana State University Agricultural Center, Baton Rouge, LA, Bugwood.org.)

are rarely observed prior to necrosis. The leaf tissue around the lesion may be puckered. Lesions coalesce and their centers fall out, giving leaves a tattered appearance. On pods, dark-brown lesions are generally small, 1–3 mm in diameter, and necrotic and cause pod malformation by inciting cessation of growth of nearby tissue.

Disease Cycle
Seed transmission is very low and is rarely of significance. Sources of primary infection are usually weed hosts on which the bacterium survives as an epiphyte on leaf surfaces. It can survive in plant debris for 1 year. Spread of the pathogen is by windblown rain or overhead sprinkler irrigation.

Overcast, cloudy, humid weather favors the disease, especially if such conditions immediately follow rain or irrigation. Moderate and warm temperatures are conducive to disease development. Such conditions favor multiplication of the bacterium on leaf surfaces. Plants injured by high winds, hail, or blowing sand are very susceptible to infection.

Control Measures
Crop rotation should be practiced. Spray with copper-containing chemicals weekly after symptoms have been observed, particularly when weather conditions favor disease development. Use resistant cultivars when available.

3.1.3 Bacterial Wilt of Green Gram

Pathogen: *Corynebacterium flaccumfaciens*

Bacterial wilt is of modest importance and uncommon occurrence. However, it has been repeatedly observed in the central United States and has been reported in 14 states and several countries.

Symptoms
Infected plants at first wilt temporarily during the warmest part of the day but regain their normal appearance during cool periods. Eventually a gradual systemic wilting of the plant persists, and the plant dies after turning straw color. A systemic dark-brown to black discoloration is apparent in the vascular tissue inside the root and lower hypocotyls. Stem cankers and water-soaked pods also occur.

Disease Cycle

The bacterium *C. flaccumfaciens* can be borne on seeds, where it can live for many years. It can overwinter in plant debris or on weeds. It has been reported to be spread by surface irrigation water and by hailstorms. The disease is favored by warm temperature of 32°C and dry weather. Since wounds provide entry into the plants, blowing sand or hailstorms may likely result in more disease.

Control Measures

Use disease-free seed, practice crop rotation, and use resistant cultivars.

3.2 BLACK GRAM

3.2.1 BACTERIAL LEAF BLIGHT OF BLACK GRAM

Pathogen: *Xanthomonas campestris* pv. *cassiae*

Symptoms

Small water-soaked lesions develop on leaves with chlorotic haloes that later turn to dark-brown spots. Postemergence seedling rot is also common.

Control Measures

Use disease-free seed.

Destruct debris and stubbles.

Soak the seeds in 500 ppm streptocycline solution for 30 min before sowing followed by two sprays of streptocycline combined with 3 g of copper oxychloride per liter at an interval of 12 days.

3.3 PIGEON PEA

3.3.1 BACTERIAL LEAF SPOT AND STEM CANKER OF PIGEON PEA

Pathogen: *Xanthomonas cajani* sp.

A new bacterial disease of pigeon pea was first observed at Jalgaon and Anand (Bombay provience) in September 1949.

Symptoms

Circular, water-soaked spots (0.5 mm in diameter) appears on the leaves of 1-month-old plants. As they expand, the lesions become quadrilateral (1 mm) and surround by a halo on the upper leaf surfaces. The initially light brown spots later turn dark brown and raised on the upper surface in consequence of the desiccation of the bacterial exudate. Lesions 2 mm in diameter may be formed by coalescence of the spots, which in severe cases occur all over the leaf and petiole and induce general yellowing and eventual shedding. The dark-brown cankers produced on the main stem and lateral branches cause decortication if present in sufficient numbers. Later, rough, raised cankerous lesions appear on stems. Cankers can cause stems to break, but the broken part usually remains attached to the plant. Affected branches dry when infection is severe.

The causal organism *X. cajani* requires the optimum temperature of 30°C for its development and its thermal death point is at 51°C.

Disease Development
- The disease appears during rainy season in India. The bacterium is basically seed-borne pathogen.
- Warm (25°C–30°C) and humid weather favors disease development.
- Incidence is higher in low-lying areas.
- Leaf infection can occur at all stages, while stem infection occurs in younger plants.

Control Measures
- Select well-drained fields.
- Select seeds from healthy plants.
- Spray antibiotics like 100 ppm of streptocycline at 10-day intervals.

3.4 SOYBEAN

3.4.1 BACTERIAL BLIGHT OF SOYBEAN

Pathogen: *Pseudomonas savastanoi*

Symptoms

Brown spots on the margins of the cotyledons characterize plants infected early in the growing season. Young plants may be stunted and if the infection reaches the growing point, they may die. Symptoms in later growth stages include angular lesions, which begin as small yellow-to-brown spots on the leaves (Photo 3.2). The centers of the spots turn dark reddish brown to black and dry out. A yellowish-green "halo" appears around the edge of water-soaked tissue that surrounds the lesions. Eventually the lesions will fall out of the leaf and the foliage will appear ragged (Photo 3.3). Generally young leaves are most susceptible to blight infection. Lesions can also occur on the pods causing the seeds to become shriveled and discolored. However, seeds usually do not show symptoms.

Geographical Distribution

The disease is geographically distributed in the United States.

Disease Cycle

Bacterial blight (*P. savastanoi*) of soybeans is typically an early season disease, which overwinters in the field on plant residue. Initial infection of soybeans occurs when wind or splashing water

PHOTO 3.2 Bacterial blight of soybean. (Courtesy of Clemson University—USDA Cooperative Extension Slide Series, Clemson, SC, Bugwood.org.)

PHOTO 3.3   Trifoliate with ragged appearance due to bacterial blight. (Courtesy of Loren Giesler, University of Nebraska-Lincoln, Lincoln, NE.)

droplets from plant residue on the soil surface carry bacterial cells to the leaves. The bacteria enter the plants through stomata and wounds on leaves. In order for infection to occur, the leaf surface must be wet. Seedlings may be infected through infected seed.

Bacteria that enter the host produce a toxin, which prevents chlorophyll production. Bacteria can also be spread to uninfected leaves when they rub against infected ones during cultivation, rain, and wind.

Favorable Environmental Conditions
Development of bacterial blight is promoted by cool, wet weather (22°C–27°C). Infection can occur early but is most common at mid-season and continues until hot and dry weather limits development. Disease outbreaks often follow windy, rainstorms.

Disease Management
Genetic Resistance
While this disease is not typically yield limiting in Nebraska, producers should consider the following options in problematic fields. The best management tool is to prevent disease establishment. Cultivars that are not highly susceptible to the disease should be considered for planting.

Cultural Practices
Crop rotation can be an effective method to avoid inoculum from a previously infected crop. Incorporating crop residue by tillage will reduce the amount of inoculum available in the spring to infect plants, but there are moisture and erosion issues to be considered. To prevent the spread of disease, limit cultivation to times when the foliage is dry.

Chemical Control
Copper fungicides are labeled for control of bacterial blight on soybeans, but to be effective, they need to be applied early in the disease cycle.

3.4.2   BACTERIAL PUSTULE OF SOYBEAN

Pathogen: *Xanthomonas axonopodis* pv. *glycines*

Symptoms
Symptoms are generally confined to leaves. Firstly, small yellowish-green areas with reddish-brown and elevated centers appear on one or both leaf surfaces. Spots are more conspicuous on the upper leaf surface. Small, raised pustules develop in the center of the lesions, especially on the lower leaf

PHOTO 3.4 Bacterial leaf pustule symptoms on soybean leaf. (Courtesy of Daren Mueller, Iowa State University, Ames, IA, Bugwood.org.)

surface (Photo 3.4). Sometimes similar pustules also develop on pods. Spots may merge and result in large irregular dead areas that sometimes fall off, giving the leaf a ragged appearance. Heavily infected leaves turn yellow and fall off. Heavy incidence on a susceptible plant may cause complete defoliation (Dunleavy, 1966).

In Nebraska, this disease is typically present on the upper leaves. Initial infection results in the development of tiny pale green spots on the new leaves. These spots have raised centers that may develop on either surface of the leaf but are more common on the lower leaf surface. As the disease progresses, small light-colored pustules will form in the center of the spots. These spots may merge together to form irregular areas that appear as lesions. As spots mature, the lesion and pustule turn brown.

This disease can easily be confused with soybean rust. Mature soybean rust pustules have a small opening at the top for spore release. Bacterial pustule lesions lack the opening on top and spores. If an opening is present, it is typically a linear crack across the surface of the pustule. These features can only be seen under magnification (20× or higher recommended). Bacterial pustule symptoms are also similar to those of bacterial blight, but lesions do not appear water soaked and will have raised centers.

Bacterial pustule, which is caused by *X. axonopodis* (syn. *campestris*) pv. *glycines*, can cause premature defoliation and reduced seed size and quantity. This disease has been reported in most parts of the world where soybeans are grown and the climate is warm with frequent rain. It is most prevalent later in the growing season.

Host Plants
Brunnichia cirrhosa (redvine), *Glycine max* (soybean), *Macrotyloma uniflorum* (horse gram) *Phaseolus* (beans), *Phaseolus lunatus* (lima bean), *Phaseolus vulgaris* (common bean), *Vigna unguiculata* (cowpea).

Geographical Distribution
China, USSR, India, the United States, South America, Africa.

Disease Cycle
Like bacterial blight, bacterial pustule overwinters in crop residue and is carried by wind-driven rain or water droplets splashing from the ground to the plant. In addition, the disease can be spread during cultivation while the foliage is wet. The bacterium enters the plant through natural openings and wounds. Warm weather with frequent showers promotes the development of this disease. Unlike bacterial blight, warm temperatures do not limit development of bacterial pustule. Its optimal temperature range for development is 30°C–33°C.

Forecasting and Modeling Systems
Computer-based models have been developed for soybean disease management (Michalski et al., 1983).

Impact

Premature defoliation occurs due to heavy infections, which in turn produces a reduction in seed size and number. Soybean yield losses due to this disease have been reported to be up to 11% (Hartwig and Johnson, 1953). Preflowering appearance of disease causes economic losses in yield (Saxena, 1977). It can be estimated that soybean yield losses of 15%, 21%, 38%, and 53% are encountered at the 10.1%–25%, 25.1%–50%, 50.1%–75%, and >75% infection rates, respectively (Shukla, 1994).

Disease Management

Cultural Control

Pathogen-free healthy seeds should be used for cultivation.

Proper Sanitation

Diseased crop residue/debris should be destroyed by deep plowing and occasionally by roguing. Some weed management practices may also be employed.

Suitable Crop Culture

Cultivation/intercultural practices could be avoided during wet weather conditions (Nyvall, 1979; Pedigo et al., 1981). Potash and phosphorous play an important role in the management of this pathogen (Sinclair and Backman, 1989).

Soybeans could be rotated with nonsusceptible (preferably nonlegume) crops (Welch, 1985; Kennedy and Sinclair, 1989; Von Qualen et al., 1989).

Highly susceptible varieties should not be planted in areas where the disease is a serious problem and disease-resistant/disease-tolerant plants should be used in problematic regions.

Resistance

Sources for resistance to bacterial pustule disease have been identified in a number of studies (Lehman and Poole, 1929; Feaster, 1951; Hartwig and Lehman, 1951; Khare et al., 1976; Tisselli et al., 1980; AVRDC, 1987; Singh and Jain, 1988; Joshi, 1989; Shukla and Prabhakar, 1990). Borkar and Sharma (1990) identified the resistance against the pathogen in soybean cultivars JS-81-1625, JS-81-1668, JS-82-989, JS-81-303, JS-81-283, DS-10, JS-81-607, and JS-81-1515.

Type of Resistance

Monogenic recessiveness was observed in bacterial pustule (Feaster, 1951; Hartwig and Lehman, 1951; Rao and Patel, 1973) caused by the presence of a gene identified as *Rxp* (Kennedy and Sinclair, 1989).

Chemical Control

Due to the variable regulations around (de)registration of pesticides, for the moment, no specific chemical control recommendations are followed. For further information, refer the following resources:

- EU pesticides database (www.ec.europa.eu/sanco_pesticides/public/index.cfm)
- PAN pesticide database (www.pesticideinfo.org)
- Your national pesticide guide

3.4.3 BACTERIAL TAN SPOT OF SOYBEAN

Pathogen: *Curtobacterium flaccumfaciens* pv. *flaccumfaciens* (Hedges) Collins and Jones

Synonyms: *C. flaccumfaciens* (Hedges) Dowson

Common names: Bacterial wilt (*Phaseolus* beans), Bacterial tan spot (soybeans) (English), Flétrissementbactérien du haricot (French), BakterielleWelke (German) and Architezbacteriana de la soja (Spanish)

Symptoms

C. flaccumfaciens pv. *flaccumfaciens* is primarily a vascular invader that causes vascular brown-ing. In beans (*Phaseolus*) and *Zornia* spp., plant parts (or the entire plant) wilt and seedling death is common. In soybeans, cowpeas, and mung beans, leaf chlorosis with spotting and sometimes flower blighting occurs, but there is no wilting. Necrotic spots may appear on the early shield leaves or, more commonly, on the trifoliate leaves. Spots progressively dry out, become bleached to a tan color, and tear in winds, thus giving the leaf a ragged appearance. Plants infected when young remain chlorotic and stunted. Symptoms are accentuated by moisture stress (Conde and Diatloff, 1991).

Pods show brown spotting, with yellow bacterial slime occasionally visible; seeds may be discolored yellow or brown.

Host

G. max (soybean), *Ipomoea* (morning glory), *Lablab purpureus* (hyacinth bean), *Phaseolus coccin-eus* (runner bean), *Ph. lunatus* (lima bean), *Ph. vulgaris* (common bean), *Vigna angularis* (adzuki bean), *Vigna radiata* (mung bean), *V. unguiculata* (cowpea), *Zornia*.

Geographical Distribution

EPPO region: Recorded in Albania, Ukraine. Found but not established in Greece and Hungary; locally established in Bulgaria (unconfirmed), Romania, Tunisia, Turkey (unconfirmed), Russia (Far East, Southern Russia; only on soybean), and Yugoslavia. Reports from Belgium (OEPP/EPPO, 1982), France, Germany, and Switzerland have not been substantiated.

Africa: Mauritius, Tunisia.

North America: Canada (Ontario), Mexico (unconfirmed), the United States (first reported in 1920, especially in irrigated High Plains and Midwest, but not reported since the early 1970s except in Iowa on soybeans (Hall, 1991). Specific records from Colorado, Connecticut, Iowa, Idaho, Michigan, Montana, Nebraska, Ohio, Oregon, Virginia, Wisconsin).

South America: Colombia, Venezuela.

Oceania: Australia (New South Wales, Queensland, South Australia, Victoria).

Disease Cycle

The bacterium can be transmitted both within and on the seed; it is very resistant to drying and can remain viable for up to 24 years in seed stored in the laboratory. In the field, it has been known to survive in soil for at least two winters between bean crops rotated with wheat. There are no reports of vectors, but the nematode *Meloidogyne incognita* may assist entry by providing wounds.

C. flaccumfaciens pv. *flaccumfaciens* can infect in the absence of rain; it has not been observed to enter via stomata. Once within the plant, the bacterium colonizes the vascular tissue.

There is no information on race variation.

For additional information, see also Hedges (1926), Zaumeyer (1932), and Zaumeyer and Thomas (1957).

Impact

C. flaccumfaciens pv. *flaccumfaciens* is considered a serious pathogen of beans (*Phaseolus*) in parts of Europe and North and South America, where it causes death of seedlings and yield loss in surviv-ing plants. It is becoming a serious, but sporadic, pathogen of soybeans, cowpeas, and mung beans in the United States, Russia, and Australia. Sporadic yield losses of up to 19% in soybeans have been recorded in the United States (Dunleavy, 1978, 1983, 1984). Pasture establishment of the legume *Zornia* spp. is seriously affected in Colombia.

Detection and Inspection Methods

Bacteria may be detected beneath the seed coat by means of a combined cultural and slide agglu-tination test. Bean seed from countries where the disease is known to occur should be inspected for discoloration of the seed coat. Immunofluorescence staining can also be used to detect the bacterium in contaminated seed lots (Calzolari et al., 1987). An EPPO quarantine procedure is in preparation.

Means of Movement and Dispersal
In international trade, the disease is liable to be carried on infected soybean seeds.

Economic Impact
Following the first report of its occurrence in 1920, *C. flaccumfaciens* pv. *flaccumfaciens* became one of the most important bacterial diseases of beans in the United States, causing up to almost total losses in some years. More recently, however, it has become very much less important and has indeed not been reported on beans since the early 1970s (Hall, 1991). In soybeans, the disease was not reported in the United States until 1975 and is of rather minor importance (Sinclair and Backman, 1989). In the EPPO region, it is important on beans in Turkey but causes only minor losses in other countries.

Prevention and Control
No effective chemical controls are available against *C. flaccumfaciens* pv. *flaccumfaciens*, although secondary spread can be retarded by bactericidal sprays of copper fungicide. Heat treatment does not eliminate the pathogen from seeds; there are no established seed treatments.

Weed and volunteer plants and plant debris that act as reservoirs of infection should be eliminated. Pathogen-free seeds and resistant cultivars should be planted (resistance is available for beans [*Phaseolus*] and soybeans but not for cowpeas, mung beans, or *Zornia* spp.).

Phytosanitary Risk
EPPO has listed *C. flaccumfaciens* pv. *flaccumfaciens* as an A2 quarantine pest (OEPP/EPPO, 1982), and CPPC and IAPSC also consider it of quarantine significance. Because of its very low current importance in its area of origin, the quarantine status of the pathogen will be reviewed within EPPO. From its existing distribution, the disease seems most likely to be important in the southern part of the EPPO region, where *Phaseolus* spp. are widely grown. It is not present in the western Mediterranean countries and not established in most eastern Mediterranean countries. The disease does not seem important enough on soybeans to merit any special attention on this crop.

Phytosanitary Measures
EPPO recommends that consignments of seeds of *Ph. vulgaris* imported from infested countries should come from an area where the disease does not occur or from a crop that was found free from the disease during the growing season (OEPP/EPPO, 1990). In future, seed-testing methods will almost certainly provide equivalent protection.

3.4.4 Soybean Wilt

Pathogen: *Curtobacterium flaccumfaciens* pv. *flaccumfaciens*
        Ralstonia solanacearum

Three separate bacterial wilt diseases of soybeans have been reported from several locations in the world (Sinclair and Backman, 1989), but they are extremely uncommon in soybean production. These include two wilt-type diseases caused by the seed-borne pathogen *C. flaccumfaciens* pv. *flaccumfaciens* (*Cff*), or *R. solanacearum*, and the third one referred to as bacterial tan spot, also caused by *C. flaccumfaciens* pv. *flaccumfaciens* (Sinclair and Backman, 1989).

During the 2005–2006 growing seasons, soybean plants exhibiting yellowing and wilting and leaf necrosis ("firing") with yellow borders were observed in irrigated western Nebraska production fields. Isolations on nutrient-broth yeast extract medium from symptomatic plants yielded Gram-positive bacteria closely resembling (morphologically, culturally, and biochemically) pathogenic isolates associated with bacterial wilt of dry beans (Harveson et al., 2006).

Four soybean isolates collected from separate western Nebraska soybean fields and counties (Box Butte, Keith, Perkins, and Scotts Bluff) were tested for pathogenicity on soybeans and dry beans.

Symptoms

Symptoms on dry beans inoculated with soybean isolates first appeared within 7 days after inoculation followed by wilting and mortality within 2 weeks. Symptoms on soybeans were slower to develop (3–4 weeks), and isolates from both soybeans and dry beans caused wilting and firing but no plant death. In comparison, a highly virulent Great Northern isolate (positive control) caused wilting and death of dry bean within 10–14 days and wilting of soybean within 18–20 days.

None of the soybean wilt pathogens are commonly found in Midwestern soybean crops. This is the first report of naturally infected U.S. soybean fields in at least 25 years and the first ever for Nebraska (Harveson et al., 2006). This is also the first report of naturally occurring soybean wilt isolates being pathogenic to dry beans. The severity of the damage on dry beans has been well documented (Harveson et al., 2005). However, the potential for future problems in soybean production in Nebraska and elsewhere is unknown. Crop rotation may affect disease severity and pathogen survival in these two crops, since there are several regions in Nebraska where the two crops may overlap. Currently, the source of inoculum is not established, but the unique presence of both dry bean and soybean pathogens in production fields warrants investigation of the relationships among isolates.

Control Measures

Not worked out.

3.4.5 WILDFIRE OF SOYBEAN

Pathogen: *Pseudomonas syringae* pv. *tabaci*

In 2006 and 2007, a new bacterial disease was observed in field-cultivated soybeans in Boeun District and Mungyeong City of Korea.

Symptoms

The disease caused severe blighting of soybean leaves. Soybean leaves in the fields showed yellowish spots with brown centers. Brown and dead areas of variable size and shape were surrounded by wide, yellow haloes with distinct margins. Spots coalesce and affected leaves fell readily.

Seven bacterial strains were isolated from chlorotic areas of soybean leaves and all produced white colonies in Trypticase soy agar. With the Biolog Microbial Identification System, version 4.2, all strains were identified as *P. syringae* pv. *tabaci* with a Biolog similarity index of 0.28–0.52 and 0.48 with *P. syringae* pv. *tabaci* CFB2106 after 24 h. Upon conducting pathogenicity tests, bacterial strains similar to those found in the field were reisolated from inoculated plants.

This is thought to be the first report of *P. syringae* pv. *tabaci* causing wildfire on soybean in Korea.

Control Measures

Not worked out.

3.4.6 BACTERIAL BLIGHT OF SOYBEAN

Pathogen: *Pseudomonas syringae* pv. *glycinea*

It was first reported in Nebraska in 1906 and is now the most common soybean bacterial disease in the state. Although yield loss from bacterial blight is seldom observed, economic losses can occur when soybeans are grown for seed. The disease is most prevalent early in the season.

PHOTO 3.5 Symptoms of *P. syringae* on soybean leaves. (Courtesy of Daren Mueller, Iowa State University, Ames, IA, Bugwood.org.)

Symptoms

Plants infected early in the growing season are characterized by brown spots on the margins of the cotyledons. Young plants may be stunted, and if the infection reaches the growing point, they may die. Symptoms in later growth stages include angular lesions, which begin as small yellow-to-brown spots on the leaves (Photo 3.5). The centers of the spots turn dark reddish brown to black and dry out. A yellowish-green "halo" appears around the edge of the water-soaked tissue that surrounds the lesions. Eventually, the lesions fall out of the leaf and the foliage appears ragged. Generally, young leaves are most susceptible to blight infection. The disease rarely affects seeds, but when lesions do appear on pods, developing seeds may become shriveled and discolored.

Both Septoria leaf spot and soybean rust start in the lower canopy. However, bacterial blight will be in the mid-to-upper canopy and have green leaves below the affected area as the crop goes into the flowering stage of development.

Disease Cycle

Bacterial blight of soybeans is typically an early season disease. It overwinters in the field in plant residue. Initial infection of soybeans occurs when bacterial cells are carried by splashing or wind-driven water droplets from plant residue on the soil surface to the leaves. Disease outbreaks usually follow rainstorms with high winds. The bacteria enter the plants through stomata (natural openings) and wounds on leaves. In order for infection to occur, the leaf surface must be wet. Seedlings may be infected through infested seeds. Infecting bacteria produce a toxin, which prevents chlorophyll production and gives a yellowish halo to the lesion. Bacteria also can be directly spread from infected leaves to uninfected leaves when they rub against one another during cultivation (especially when there is dew), rain, or wind.

Development of bacterial blight is promoted by cool, wet weather (21°C–26°C). Infection may occur at any time during the growing season but is most common early in the growing season and will continue until hot, dry weather limits disease development.

Control Measures

Crop rotation and spray of copper-based fungicides are recommended.

3.5 RAJMA BEAN/KIDNEY BEAN/DRY BEAN

3.5.1 BACTERIAL WILT OF DRY BEANS

Pathogen: *Curtobacterium flaccumfaciens* pv. *flaccumfaciens*

Bacterial wilt of dry beans, caused by *C. flaccumfaciens* pv. *flaccumfaciens* (*Cff*), has been a sporadic but often serious production problem in dry beans throughout the irrigated High Plains since first being reported in South Dakota in 1922. It was first observed in western Nebraska dry bean

PHOTO 3.6   Initial leaf scorching and at later stages complete wilting of dry bean plant due to bacterial wilt pathogen. (Courtesy of Howard F. Schwartz, Colorado State University, Fort Collins, CO, Bugwood.org.)

production fields in the early to mid-1950s and continued to be an endemic, economically important problem throughout the 1960s and early 1970s. The disease then only periodically appeared in seed, but had little detectable effect on yields after the implementation of crop rotation and seed sanitation practices.

The pathogen was again identified in 2003 for the first time in this area in almost 25 years. Over the past 7–8 years, it has fully re-emerged in the Central High Plains (Nebraska, Colorado, and Wyoming) and has now been identified from more than 400 fields. Affected fields were planted with dry beans from multiple market classes and seed sources, including yellows, great northern, pintos, kidneys, cranberries, blacks, navies, pinks, and small reds. Disease incidence in these fields has varied from trace levels to >90% (Harveson, 2011).

Symptoms

Disease symptoms can be observed on disease foliage and bean pods (Photo 3.6). Field symptoms consist of leaf wilting during periods of warm, dry weather or periods of moisture stress. This occurs because of the pathogen's presence within the vascular system, which blocks normal water movement from roots into the foliage. Plants often recover during evening hours when temperatures are lower but wilt again during the heat of the day. Disease severity and plant mortality are often higher on young plants or those growing from infected seed. Seedlings are particularly susceptible, and if attacked when 2–3 in. tall, they usually die. Symptoms on adult plants are less pronounced as the disease generally develops and progresses more slowly.

Infected plants in Nebraska have additionally exhibited symptoms consisting of interveinal, necrotic lesions surrounded by bright yellow borders. These symptoms may be confused with those caused by common bacterial blight, *X. axonopodis* pv. *phaseoli* (synonym *X. campestris* pv. *phaseoli*), but bacterial wilt lesions tend to be more wavy or irregular. Additionally, water soaking of leaves is not usually observed with wilt, as it is with common bacterial blight and halo blight (*P. syringae* pv. *phaseoli*) infections. However, water soaking has commonly been observed on bacterial wilt–infected yellow bean leaves in Colorado.

If plants survive to produce mature seeds, the pathogen will often color or stain seeds, due to the systemic nature of this disease. Seeds may become infected even while pods appear to remain healthy. White-seeded cultivars are particularly prone to quality reductions due to the conspicuously colored seed coats from systemic infections.

Pathogen Color Variants

Colony growth and staining of seeds reported for original isolates of *Cff* were always yellow until orange and purple colored variants were found in western Nebraska. The purple variant maintains a yellow-colored colony in culture but produces an extracellular, water-soluble bright purple pigment that diffuses into growth medium within 2–3 days and also discolors seed. The purple variant, which is very rare, has only been reported once outside of the western Nebraska Panhandle from cull bean seeds in Alberta, Canada, in 2006. The pigment produced by the purple variant is often unstable and inconsistently expressed, which may explain this variant's lower reported incidence in nature.

Since 2005, all three pathogen color variants have been isolated from infected dry bean seeds and plants in western Nebraska fields during the season, with more than 90% of collected isolates during this time consisting of the yellow and orange variants. Following the 2007 growing season, a pink bacterial isolate closely resembling the wilt pathogen was recovered on isolation media from orange-stained seeds (market class great northern) that originated from research plots affiliated with the University of Nebraska–Lincoln Panhandle Research and Extension Center (Scottsbluff Ag Lab) near Mitchell, Nebraska.

Infection and Survival

The pathogen is seed borne, and infected seeds represent the major source of inoculums and means for dispersal, both long and short distances. The pathogen can be transmitted both within and on the outside of seeds. It can overwinter on infected residue or weeds, but survival in soil by itself is poor. Due to a strong resistance to drying, the pathogen can remain viable up to 24 years in seed stored under optimum conditions in the laboratory.

Initial infection occurs when the pathogen enters the vascular system through either infected seeds or wounds on leaves or stems. Disease is not thought to develop via entry into numerous natural pores in leaves (stomata), unlike common bacterial and halo blights. However, the disease develops and spreads more rapidly and becomes more severe following hailstorms or when temperatures exceed 32°C. Wilting of plants is more pronounced during periods of moisture stress, and secondary spread occurs in a similar manner to that of common bacterial and halo blights.

The most important hosts for bacterial wilt are *Phaseolus* spp., especially *Ph. vulgaris* L. (common bean), but wilt isolates can survive and remain pathogenic in soil for at least 2 years. However, the primary mechanism for survival is in crop residues. A comprehensive survey for 4 years has further revealed the presence of bacterial wilt isolates occurring with other crops grown in rotation with dry beans, including soybeans, corn, wheat, sunflower, and alfalfa. These isolates were found in association with other bacterial diseases, suggesting survival in those crops' residues.

Disease Management

Bacterial diseases of dry beans are very difficult to manage due to the lack of information on the effectiveness of chemical management options. Therefore, genetic resistance is generally considered to be the most effective means of disease management.

The resistant cultivar "Emerson" was developed in the early 1970s by the University of Nebraska specifically for controlling bacterial wilt, which demonstrates the importance that this disease once held. "Emerson" is still available today, but it is grown on a limited basis as a specialized variety for targeted markets in Europe. It cannot be produced on all fields where the disease has recently been identified. Thus, an emerging problem that needs addressing is utilizing newly developed resistant cultivars. New resistance studies have begun in Nebraska in which *Phaseolus* germplasm collections are being evaluated for resistance to *Cff* in the ongoing effort to produce new wilt-resistant cultivars adapted for dry bean production in this region.

Seeds from previously blighted fields should not be reused. Certified seeds of disease-resistant cultivars should be planted where possible. Seed treatments of plants with streptomycin help to reduce contamination of the seed coat and establish a vigorous, early season stand. Rotation of beans with other crops for 2–3 years and elimination of bean volunteers during growing season can also be done. Old bean straw from infested crops should not be spread on new fields to be used for bean production. Plantation of beans close to recently blighted fields should be avoided. Stay out of bean fields when wet. Moving through blighted fields when foliage is wet can spread pathogen to other plants. Avoid reusing irrigation water.

Consider preventative sprays of a copper-based bactericide during mid-vegetative or early flowering periods, depending upon weather and potential pathogen involved.

3.5.2 BACTERIAL BROWN SPOT OF DRY BEANS

Pathogen: *Pseudomonas syringae* pv. *syringae*

Bacterial brown spot is a more recently discovered disease of dry beans in Nebraska than either halo blight or common bacterial blight. It was first seen in western Nebraska dry bean fields on a limited basis throughout the late 1960s. Epidemics still occur sporadically, but the pathogen's presence has increased in incidence and damage along with halo blight during the past 20 years in the Central High Plains.

Brown spot–tolerant bean cultivars were first identified and reported by Schuster and Coyne in 1969 from field and greenhouse tests in Nebraska using the dry bean lines US 1141 and GN Nebraska #1 selection 27, and the snap bean cultivar Tempo. When the disease occurs, it can be very damaging due to the lack of resistance in modern commercial cultivars.

Symptoms

Disease symptoms can be observed on disease foliage and on infected beans pods (Photo 3.7). Lesion size can vary, but generally lesions are small, circular, and brown, often surrounded by a yellow zone. As the disease progresses, lesions begin combining to form linear, necrotic streaks bound by leaf veins. If water soaking occurs, it appears as small circular spots on the underside of leaves. Old lesion centers fall out, leaving tattered strips or "shot holes" on affected leaves and evidence of water soaking may be visible in the edge of tissue next to the shot holes. Stem and petiole lesions are occasionally found when the pathogen becomes systemic.

Pod lesions are circular and water soaked initially but later turn brown and become necrotic. If young pods or those in the flat stage become infected, they may be bent or twisted with ring-spots or water-soaked brown lesions.

Like the halo blight pathogen, a cream- to silver-colored bacterial ooze may emerge from stems, pods, or leaves after infection under humid climatic conditions. Seeds may also shrivel or discolor or become infected if lesions penetrate into pod walls.

Pathogen

P. syringae pv. *syringae* (*Pss*) is a strictly aerobic or oxygen-requiring bacterium capable of movement. Like *Psp* (the halo blight pathogen), it produces diffusible fluorescent pigments on iron-deficient media and cream-to-white colonies on standard media. Pathogenic isolates may produce a bacteriocin (compound toxic to other related bacteria) in the host plant.

The pathogen *P. syringae* causes disease on many different plants, but only certain forms of the pathogen cause brown spot. *Pss* has been demonstrated to be extremely variable among

PHOTO 3.7 Early shot-hole lesions on leaves of dry beans due to bacterial brown spot (*P. syringae* pv. *syringae*). (Courtesy of Howard F. Schwartz, Colorado State University, Fort Collins, CO, Bugwood.org.)

leguminous hosts. For example, isolates from cowpeas (black-eyed peas) were shown to be weaker parasites than those from the original New Jersey bean specimens. The isolates originally found in Nebraska dry bean production areas were shown to be less infectious than snap bean isolates from Wisconsin. Isolates from canning peas in Wisconsin were able to produce brown spot symptoms on beans and vice versa. Furthermore, its pathogenic host range is very large, including *Ph. vulgaris* (common bean), *Ph. lunatus* (lima bean), *Pueraria lobata* (kudzu), *G. max* (soybean), *Pisum sativum* (pea), *Vicia faba* (fava bean), *Vigna sesquipedalis* (yard-long bean), *V. unguiculata* (cowpea), and *Vigna sinensis* (mung bean).

Bacterial brown spot, like common bacterial blight, is a warm weather–oriented disease. The pathogen potentially causes the most damage when temperatures range from 27°C to 30°C, as opposed to the cool weather favoring disease development of halo blight from 17°C to 20°C.

Disease Cycle

Conditions for brown spot development in Nebraska often occur during mid-vegetative to early flowering periods of plant growth. The pathogen may only need cool nights to thrive. Like other dry bean bacterial pathogens, the brown spot pathogen survives in bean residue or seeds from the previous year, although seed infections are not thought to be important in the epidemiology of the disease except to introduce the pathogen into new areas. The pathogen has additionally been shown to have a resident phase on both host and nonhost plants. Thus, the pathogen is also capable of growing and surviving as an epiphyte on the surface of both weed and legume crop hosts. Wet weather, hail, violent rain, and windstorms help the pathogen to spread among and between fields.

Disease Management

Chemical Methods

Chemical control results vary depending on pathogen, weather, and disease pressure. Increased yields have been realized for halo blight and brown spot infections through use of copper-based applications 40 days after emergence (mid-vegetative or early flowering periods) and then repeated every 7–10 days for a total of three applications.

Cultural Methods

Seeds from previously blighted fields should not be reused. Certified seeds of disease-resistant cultivars should be planted where possible. Seed treatments of plants with streptomycin help to reduce contamination of the seed coat and establish a vigorous, early season stand. Rotation of beans with other crops for 2–3 years and elimination of bean volunteers during growing season can also be done. Old bean straw from infested crops should not be spread on new fields to be used for bean production. Plantation of beans close to recently blighted fields should be avoided. Stay out of bean fields when wet. Moving through blighted fields when foliage is wet can spread pathogen to other plants. Avoid reusing irrigation water.

3.5.3 Halo Blight of Dry Beans

Pathogen: *Pseudomonas syringae* pv. *phaseolicola*

Halo blight, like common bacterial blight, has been found in Nebraska for more than 70 years. It is considered to be a major problem wherever moderate temperatures occur during bean production. Central High Plains losses have been reduced by planting cultivars with genetic resistance to the pathogen. While several popular cultivars have good levels of resistance to common and halo blights, they are more prone to infection by rust or white mold.

Symptoms

Halo blight symptoms can be observed on infected leaves and bean pods (Photos 3.8 and 3.9). The first symptoms of infection are small water-soaked spots on leaflets. Under arid conditions, the infected tissue dies and turns tan-colored and necrotic. A broad yellow-green halo then develops around

PHOTO 3.8 Halo blight symptoms on leaves of dry bean. (Courtesy of Howard F. Schwartz, Colorado State University, Fort Collins, CO, Bugwood.org.)

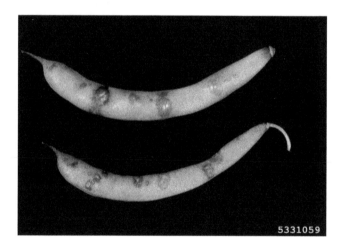

PHOTO 3.9 Halo blight symptoms on pod of dry bean. (Courtesy of Mary Ann Hansen, Virginia Polytechnic Institute and State University, Blacksburg, VA, Bugwood.org.)

necrotic spots, in contrast to the narrow, bright yellow border around lesions characteristic with common blight. The necrotic spots additionally remain very small, unlike that of common blight.

Haloes are not evident in hot weather. If infected systemically, young leaflets become curved and chlorotic and do not show necrotic spots or broad yellow halos. Yellow-green chlorosis becomes more pronounced at temperatures of 17°C–20°C due to the pathogen's production of a toxin.

When the temperature is above 24°C, toxin production often decreases and chlorotic symptoms become less noticeable. Then the symptoms appear similar to brown spot. In severe cases, a general systemic chlorosis may develop. Systemic infections are not commonly encountered but occur more readily in some dry bean market classes such as light red kidneys.

Pod symptoms begin as water-soaked circular spots or streaks along pod sutures. Bacterial ooze may emerge from stems, pods, or leaves 7–10 days after infection, giving lesions a greasy, water-soaked appearance.

Seeds may shrivel or discolor if lesions expand into pod sutures or penetrate young pod walls.

Pathogen

On common isolation media, bacterial colonies appear cream to silvery in color, in contrast to the yellow colonies found with the common blight bacterial pathogen.

P. syringae pv. *phaseolicola* (*Psp*) is an aerobic or oxygen-requiring bacterium. *Psp* can usually be identified as cream-to-white colored colonies on standard media. However, it also produces diffusible, fluorescent pigments in iron-deficient culture media and thus is also classified as a fluorescent pseudomonad, as is the brown spot pathogen.

Maximum growth and production of the bacterial-induced toxin occur at 17°C–20°C, which is responsible for the chlorosis symptoms described earlier.

Halo blight is a low-temperature disease and most destructive in areas where temperatures are moderate. Yield loss potential is greatest at temperatures that enhance production of the toxin (17°C–20°C). It is frequently detected early in the season, shortly after its emergence on seedlings or during mid-vegetative stages of plant growth when conditions are favorable.

Disease Cycle

Like common blight, the halo blight pathogen survives in residue or seeds from the previous year. However, seed transmission these days is very low in the United States, due to the use of western-grown certified seeds grown in arid areas.

However, the pathogen is commonly found on leaves as an epiphyte on legume hosts. Several pathogenic races have been identified, distinguished by host specificity, and substantial variation in virulence has been observed in natural populations. The disease cycle of halo blight of dry beans is illustrated in Figure 3.1.

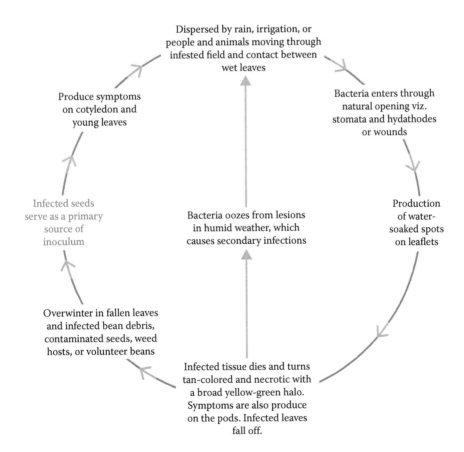

Dispersed by rain, irrigation, or people and animals moving through infested field and contact between wet leaves

Produce symptoms on cotyledon and young leaves

Bacteria enters through natural opening viz. stomata and hydathodes or wounds

Infected seeds serve as a primary source of inoculum

Bacteria oozes from lesions in humid weather, which causes secondary infections

Production of water-soaked spots on leaflets

Overwinter in fallen leaves and infected bean debris, contaminated seeds, weed hosts, or volunteer beans

Infected tissue dies and turns tan-colored and necrotic with a broad yellow-green halo. Symptoms are also produce on the pods. Infected leaves fall off.

FIGURE 3.1 Disease cycle of halo blight of dry beans caused by *P. syringe* pv. *phaseolicola*.

Disease Management

Chemical Methods

Chemical control results vary depending on pathogen, weather, and disease pressure. Increased yields have resulted from treating halo blight and brown spot infections with copper-based applications 40 days after emergence (mid-vegetative or early flowering periods) and repeating application every 7–10 days for a total of three applications.

Cultural Methods

Seeds from previously blighted fields should not be reused. Certified seeds of disease-resistant cultivars should be planted where possible. Seed treatments of plants with streptomycin help to reduce contamination of the seed coat and establish a vigorous, early season stand. Rotation of beans with other crops for 2–3 years and elimination of bean volunteers during growing season can also be done. Old bean straw from infested crops should not be spread on new fields to be used for bean production. Plantation of beans close to recently blighted fields should be avoided. Stay out of bean fields when wet. Moving through blighted fields when foliage is wet can spread pathogen to other plants. Avoid reusing irrigation water.

REFERENCES

AVRDC. 1987. *Soybean Varietal Improvement.* Asian Vegetable Research and Development Center, Shanhua, Taiwan.

Borkar, S.G. and J.K. Sharma. 1990. Source of resistance to bacterial leaf spot pathogen of soyabean. *Oilcrops Newsl.*, (7), 32.

Calzolari, A., Tomesani, M., and U. Mazzuchi. 1987. Comparison of immunofluorescence staining and indirect isolation for the detection of *Corynebacterium flaccumfaciens* in bean seeds. *Bull. OEPP/EPPO Bull.*, 17, 157–163.

Conde, B.D. and A. Diatloff. 1991. Diseases of mungbeans. In: Imrie, B.C. and R.J. Lawn (eds.), *Mungbean— The Australian Experience: Proceedings of the First Australian Mungbean Workshop.* CSIRO Division of Tropical Crops and Pastures, Brisbane, Queensland, Australia, pp. 73–77.

Coyne, D.P. and M.L. Schuster. 1969. Moderate varietal tolerance of beans to brown spot (*Pseudomonas syringae*). *Annu. Rep. Bean Improv. Coop.*, 12, 13.

Dunleavy, J.M. 1966. Factors influencing the spread of brown stem rot in soyabeans. *Phytopathology*, 56, 290–300.

Dunleavy, J.M. 1978. Bacterial tan spot, a new disease of soybeans. *Am. Phytopathol. Soc.*, 70th Annual Meeting (abstract). October 28-November 2, Tucson, AZ.

Dunleavy, J.M. 1983. Bacterial tan spot, a new foliar disease of soybeans. *Crop Sci.*, 23, 473–476.

Dunleavy, J.M. 1984. Yield losses in soybeans caused by bacterial tan spot. *Plant Dis.*, 6, 774–776.

Feaster, C.V. 1951. Bacterial pustule disease on soybean: Artificial inoculation, varietal response, and inheritance of resistance. *Missouri Agric. Exp. Stn Bull.*, 487.

Hall, R.J.B. (ed.). 1991. *A Compendium of Bean Diseases.* American Phytopathological Society, St. Paul, MN.

Hartwig, E.E. and H.W. Johnson. 1953. Effect of the bacterial pustule disease on yield and chemical composition of soybeans. *Agron. J.*, 45, 22–23.

Hartwig, E.E. and S.G. Lehman. 1951. Inheritance of resistance to bacterial pustule disease in soybeans. *Agron. J.*, 43, 226–229.

Harveson, R. M. 2011. Reflections on Bacterial Wilt and a Summary of Studies Conducted in Nebraska. University of Nebraska–Lincoln.

Harveson, R.M., Schwartz, H.F., Vidaver, A.K., Lambrecht, P.A., and K. Otto. 2006. New outbreaks of bacterial wilt of dry beans in Nebraska observed from field infections. *Plant Dis.*, 90, 681.

Harveson, R.M., Vidaver, A.K., and H.F. Schwartz. 2005. Bacterial wilt of dry beans in western Nebraska. NebGuide Series No. G05-1562-A. Cooperative Extension Service, University of Nebraska, Lincoln, NE.

Hedges, F. 1926. Bacterial wilt of beans (*Bacterium flaccumfaciens* Hedges), including comparisons with *Bacterium phaseoli. Phytopathology*, 16, 1–22.

Joshi, S. 1989. Screening AVRDC soybean breeding lines and accessions for soybean rust, bacterial pustule and frog-eye leaf spot. *Soybean Rust Newsl.*, 9, 1–2 (abstract).

Khare, M.N., Agarwal, S.C., and A.C. Jain. 1976. Studies on disease of lentil. In: *Proceedings of the XVI Annual Workshop on Rabi Pulses*, September 27–30, 1976. BHU, Varanasi, pp. 1–4.

Lehman, S.C. and R.F. Poole. 1929. Research in Botany. North Carolina Agricultural Experiment Station Annual Report, Vol. 51, pp. 59–67.

Michalski, R.S., Davis, J.H., Bisht, V.S., and J.B. Sinclair. 1983. A computer-based advisory system for diagnosing soybean diseases in Illinois. *Plant Dis.*, 67, 459–463.

Nyvall, R.F. 1979. Diseases of rye (*Secale cereale* L.). In: *Field Crop Diseases Handbook*. AVI Publishing Company Inc., Westport, CT, pp. 225–240.

OEPP/EPPO. 1982. Data sheets on quarantine organisms No. 48, *Corynebacterium flaccumfaciens. Bull. OEPP/EPPO Bull.*, 12(1): 1–4.

OEPP/EPPO. 1990. Specific quarantine requirements. EPPO Technical Documents No. 1008.

Panhandle Rec, Scottsbluff. http://plantpathology.unl.edu/reflections-bacterial-wilt-and-summary-studies-conducted-nebraska.

Pedigo, L.P., Higgins, R.A., Hammond, R.B., and E.J. Bechinski. 1981. Soybean pest management. In: Pimenter, D. (ed.), *CRC Handbook of Pest Management*, Vol. III. CRC Press, Boca Raton, FL, pp. 417–537.

Rao, M.V.B. and P.N. Patel. 1973. Evaluation of chemical in vitro and in vivo against the pustule pathogen, *Xanthomonas phaseolivar*, sojensis. *Indian Phytopathol.*, 26, 598–599.

Saxena, M.C. 1977. Soybean in India. Country report of All India Coordinator Research Project in Soybean. G.B. Pant University of Agriculture and Technology, Pantnagar, India. Quoted in CAB International. (2000). *The Crop Protection Compendium* [CD-ROM]. CAB International, Wallingford, U.K.

Shukla, A.K. 1994. Pilot estimation studies of soybean yield losses by various levels of bacterial pustule infection. *Int. J. Pest Manag.*, 40, 249–251.

Shukla, A.K. and Prabhakar. 1990. Sources of resistance against bacterial pustule. *Agric. Biol. Res.*, 7, 145–146.

Sinclair, J.B. and P.A. Backman (eds.). 1989. *A Compendium of Soybean Diseases*, 3rd ed. American Phytopathological Society, St. Paul, MN.

Singh, R.B. and J.P. Jain. 1988. Chemical control of bacterial pustule of soybean. *J. Turkish Phytopathol.*, 17, 31–36.

Tisselli, O., Sinclair, J.B., and T. Hymowitz. 1980. Sources of resistance to selected fungal, bacterial, viral, and nematode diseases of soybeans. International Soybean Program INTSOY Series 18. University of Illinois, Urbana, Champaign, IL.

Von Qualen, R.H., Abney, T.S., Huber, D.M., and M.M. Schreiber. 1989. Effect of rotation, tillage, and fumigation on premature dying of soybeans. *Plant Dis.*, 73, 740–744.

Welch, L.F. 1985. Rotational benefits to soybeans and following crops. In: Shibles, R.M. (ed.), *Proceedings of the World Soybean Research Conference III*. Westview Press, Boulder, CO, pp. 1054–1060.

Zaumeyer, W.J. 1932. Comparative pathological history of three bacterial diseases of bean. *J. Agric. Res.*, 44, 605–632.

Zaumeyer, W.J. and H.R. Thomas. 1957. A monographic study of bean diseases and methods for their control. *USDA Tech. Bull.*, (865), 255pp.

SUGGESTED READING

Agrios, G.N. 2005. *Plant Pathology*, 5th edn. Elsevier Academic Press, Burlington, MA, 922pp.

Anonymous. 2010. Common diseases of soybean in the Mid-Atlantic region. Virginia Cooperative Extension—Plant Diseases, Virginia Polytechnic Institute and State University, Blacksburg, VA.

CMI. 1987. *Distribution Maps of Plant Diseases No. 370*, 4th edn. CAB International, Wallingford, U.K.

Dunleavy, J.M. 1963. A vascular disease of soybeans caused by *Corynebacterium* sp. *Plant Dis. Rep.*, 47, 612–613.

EPPO/CABI. 1997. *Quarantine Pests for Europe*. 2nd edition. Edited by Smith IM, McNamara DG, Scott PR, Holderness M. CABI, Wallingford, UK, 1425 pp.

Garrett, K.A. and H.F. Schwartz. 1998. Epiphytic *Pseudomonas syringae* on dry beans treated with copper-based bactericides. *Plant Dis.*, 82, 30–35.

Giesler, L.J. 2013. G2058. Bacterial Diseases of Soybean - University of Nebraska–Lincoln. UNL Extension NebGuide. extensionpublications.unl.edu/assets/pdf/g2058.pdf. https://pdc.unl.edu/agriculturecrops/soybeans/bacterialblight.

Hagedorn, D.J. and D.A. Inglis. 1986. *Handbook of Bean Diseases*. A3374. http://learningstore.uwex.edu/assets/pdfs/A3374.PDF.

Hartman, G.L., Sinclair, J.B., and J.C. Rupe. 1999. *Compendium of Soybean Diseases*, 4th ed. American Phytopathological Society Press, St. Paul, MN.

Hayward, A.C. and J.M. Waterston. 1965. *Corynebacterium flaccumfaciens. CMI Descriptions of Pathogenic Fungi and Bacteria No. 43.* CAB International, Wallingford, U.K.

Hsieh, T.F., Huang, H.C., Erickson, R.S., Yanke, L.J., and H.H. Mundel. 2002. First report of bacterial wilt of common bean caused by *Curtobacterium flaccumfaciens* in western Canada. *Plant Dis.*, 86, 1275.

Kulkarni, Y.S., Patel, M.K., and S.G. Abhankar. 1950. A new bacterial leaf spot and stem canker of Pigeon Pea. *Curr. Sci.*, 19(12), 384.

Miller, P.R. and H.L. Pollard. 1976. *Multilingual Compendium of Plant Diseases.* American Phytopathological Society, St. Paul, MN.

Pacumbaba, R.P. 1987. Outbreak of bacterial blight of soybean at Alabama A&M soybean research field plots and vicinity in 1986: In disease note. *Plant Dis.*, 71, 557.

Schwartz, H.F. 1980. Miscellaneous bacterial diseases. In: Schwartz, H.F. and G.E. Galvez (eds.), *Bean Production Problems.* Centro Internacional de Agricultura Tropical (CIAT), ApartadoAereo 6713, Cali, Columbia, pp. 173–194.

Schwartz, H.F., Franc, G.D., Hanson, L.E., and R.M. Harveson. 2005. Disease management. In: Schwartz, H.F., Brick, M.A., Harveson, R.M., and G.D. Franc (eds.), *Dry Bean Production and Pest Management*, Bull. No. 562A. Colorado State University, Fort Collins, CO, pp. 109–143.

Shukla, A.K. and A.N. Sharma. 1996. Soybean ki kisme and fasal weinurvarak prabandh. Krishi Jagat, Kharif issue (in Hindi).

Shukla, A.K. 1990. Major diseases of soybean. In: Bhatnagar, P.S. and S.P. Tiwari (eds.), *Technology for Increasing Soybean Production in India.* Technical Bulletin No. 1. National Research Center for Soybean, Indore, India, pp. 22–28. Quoted in CAB International. (2000). *The Crop Protection Compendium* [CD-ROM]. CAB International, Wallingford, U.K.

Weller, D.M. and A.W. Saettler. 1976. Chemical control of common and fuscous bacterial blights in Michigan Navy (pea) beans. *Plant Dis. Rep.*, 60(9), 793–797.

Yoshii, K. 1980. Common and fuscous blights. In: Schwartz, H.F. and G.E. Galvez (eds.), *Bean Production Problems: Disease, Insect, Soil and Climatic Constraints of Phaseolus vulgaris.* CIAT, Cali, Colombia, pp. 157–172.

Yoshii, K., Galvez, G.E., and G. Alvarez-Ayala. 1976. Estimation of yield losses in beans caused by common blight. *Proc. Am. Phytopathol. Soc.*, 3, 298–299.

Yu, Z.H., Stall, R.E., and C.E. Vallejos. 1998. Detection of genes for resistance to common bacterial blight of beans. *Crop Sci.*, 38(5), 1290–1296, 36.

4 Bacterial Diseases of Oilseed Crops

4.1 GROUNDNUT

4.1.1 BACTERIAL WILT OF GROUNDNUT

Pathogen: *Ralstonia solanacearum* (Smith) Yabuuchi et al.

Synonyms: *Bacterium solanacearum*
Burkholderia solanacearum
Pseudomonas solanacearum

Bacterial wilt is a major constraint to groundnut production in China, Indonesia, and Vietnam. Yield losses of 10%–30% commonly occur and can reach over 60% in heavily infested fields (Mehan et al., 1994). In China, 0.2 million ha under groundnut have been reported to be infested with bacterial wilt with yield losses of 10%–30% (Liao et al., 1990).

Symptoms
Wilt symptoms can be seen 2–3 weeks after planting. The first sign of disease is a slight drooping or curling of one or more leaves. In more advanced stages, the plants may bend over at the tip, appear dry, and eventually turn brown, wither, and die. Infected plants have discolored and rotten roots and pods.

The diagnostic characteristics of this disease are the dark brown discoloration in the xylem and pith, and the streaming of "bacterial ooze" (Mehan et al., 1994).

Hosts
Arachis hypogaea (groundnut), *Capsicum* spp. (chilies), *Gossypium hirsutum* (Cotton), *Ipomoea batatas* (sweet potato), *Lycopersicon esculentum* (tomato), *Manihot esculenta* (cassava), *Musa* spp. (banana), *Nicotiana* spp. (tobacco), *Solanum melongena* (brinjal), *Solanum tuberosum* (potato), and *Zingiber officinale* (ginger).

Geographical Distribution
Bacterial wilt is global in distribution and is mostly prevalent in several countries of Asia, Africa, and North America.

Pathogen
Ralstonia solanacearum is an aerobic, non-spore-forming, rod shaped, gram-negative bacterium. The bacterial cells measure approximately 0.5×1.5 m. Virulent isolates are mainly nonflagellate and nonmotile. Avirulent isolates usually bear one to four polar flagella and are highly motile. Common fimbriae are often present in both virulent and avirulent isolates. Although it does not produce fluorescent pigments, it produces a brown diffusible pigment on a variety of agar media containing tyrosine.

Disease Cycle
The bacterium grows at a wide range of temperatures from 25°C to 35°C. The bacterium is mainly disseminated through infested soil, water, and infected seed (Anitha et al., 2003).

Detection/Indexing Methods at ICRISAT
Direct plating of 4-week-old leaf-twigs, leaf-bits, and seed on Tetrazolium chloride agar (TZCA) medium is used to detect the wilt pathogen in groundnut (Prasada Rao et al., 2000; Anitha et al., 2004).

Control Measures
Not available.

4.2 SESAMUM

4.2.1 BACTERIAL BLIGHT OF SESAMUM

Pathogen: *Xanthomonas campestris* pv. *sesami*

Symptoms
Plants at all growth stages are affected. Water-soaked, small, and irregular spots form on the leaves, which later increase and turn brown, under favorable disease conditions. Leaves become dry and brittle; severely infected leaves defoliate. Initially water-soaked spots appear on the undersurface of the leaf and then on the upper surface. They increase in size, become angular and restricted by veins, and turn dark brown. Several spots coalesce together, forming irregular brown patches, and cause drying of leaves. The reddish brown lesions may also occur on petioles and stems. In advanced stages of the disease development, the spots form on the twigs, which bear poor capsules. The spots also appear on capsules. The seeds in such capsules do not fill and mature properly.

Disease Cycle
The bacterium survives in the infected plant debris and in seeds. The secondary spread is by rain water. The disease cycle of bacterial blight of sesamum is illustrated in Figure 4.1.

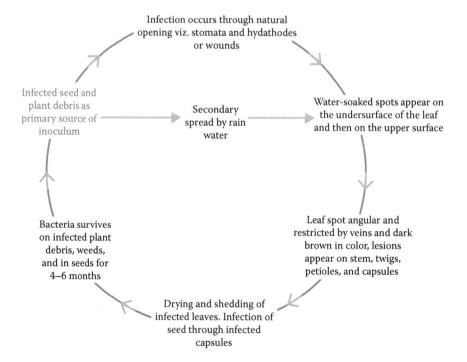

FIGURE 4.1   Disease cycle of bacterial blight of sesamum incited by *X. campestris* pv. *sesame*.

Impact and Losses
In severely affected field, up to 60% losses can occur.

Disease Management
Infected plant debris and crop residues should be removed and burned. Spray streptocycline sulfate or oxytetracycline hydrochloride or streptocycline (500 pm) at 100 g/ha as soon as symptoms are noticed. Continue two more sprays at 15 days interval if necessary. Crop rotation should be done.

Use resistant variety like T-58 and disease-free certified seeds. Deep the seed in Agrimycin-100 (250 ppm) or streptocycline suspension (0.05%) for 30 min. Early planting, that is, immediately after onset of monsoon.

4.2.2 BACTERIAL LEAF SPOT OF SESAMUM

Pathogen: *Pseudomonas syringae* pv. *sesame*

Symptoms
Light-brown angular spots with dark purple margin appear in the leaf veins. Defoliation and death of plant may occur in severe leaf and stem infection.

Sunken and shiny spots appear on the capsules. Early capsule infection render them black and seedless.

Disease Cycle
The bacterium survives in the infected plant debris and in seeds. The secondary spread is by rain water.

Disease Management
Remove and burn infected plant debris and crop residues. Spray streptomycin sulfate or oxytetracycline hydrochloride or streptocyclin (500 ppm) at 100 g/ha as soon as symptoms are noticed. Continue two more sprays at 15 days interval if necessary. Crop rotation should be done.

Use resistant variety like T-58 and disease free certified seeds. Deep the seed in Agrimycin-100 (250 ppm) or streptocycline suspension (0.05%) for 30 min. Early planting, that is, immediately after onset of monsoon.

4.3 SUNFLOWER

4.3.1 APICAL CHLOROSIS OF SUNFLOWER

Pathogen: *Pseudomonas syringae* pv. *tagetis*

Symptoms
Symptoms consisting solely of leaf chlorosis without discernible lesions are observed on sunflower in all vegetative growth stages, although they are more frequent and severe on seedlings. Often only a portion of the initially affected leaf is chlorotic, but subsequently formed leaves are uniformly chlorotic, including the veins. Infected seedlings are frequently stunted and occasionally died; however, infection of older plants rarely result in any stunting. With seedling infection systemic chlorosis lasts up to 8 weeks and spans 8–10 leaves, but chlorosis on older (prebloom) plants is frequently limited to a few leaves. Symptoms are never observed on subapical, fully expanded leaves of plants past the bud stage, nor are chlorotic leaves ever observed to recover. Chlorotic plants are usually scattered throughout a field, occurring singly or in small groups within a row or occasionally spanning two to three rows. The distribution of diseased plants did not appear to be associated with any variation in topography.

Apical chlorosis symptoms first appear as dramatic bleaching or chlorosis of the upper leaves of seedlings. The disease is often confused with iron chlorosis or nitrogen deficiency, but apical chlorosis, unlike mineral deficiencies, is more uniform and the affected plants never recover. Mineral deficiencies are usually accompanied by greening of the leaf veins. Yield is seldom, if ever, affected by the disease in the High Plains. Recovery of individual chlorotic leaves is never observed; a leaf, regardless of the degree of chlorosis, remained chlorotic until senescence. Plants of the more susceptible lines are visibly stunted and systemic chlorosis persisted until flowering on several plants grown to maturity. Chlorosis is never observed on bracts or ray flowers.

Pathogen
Gram-negative, oxidase-negative, fluorescent bacteria can be consistently isolated from leaves and petioles of chlorotic sunflowers and all such strains produced apical chlorosis 4–5 days after inoculation of healthy sunflowers. All pathogenic sunflower strains and the marigold strain of *P.s.* pv. *tagetis* behave identically in biochemical tests and confirmed to previous descriptions (Helimer, 1955; Trimboli et al., 1978), with the exception of an inability to utilize ethanol as a carbon source. Strains from either host produce apical chlorosis in sunflower when inoculated by injection or root immersion. Injection into the growing point or cotyledons produce the most pronounced symptoms and is judged as the most reliable technique. Neither apical chlorosis nor any leaf lesions are produced when bacterial suspensions were sprayed onto sunflower, with or without subsequent incubation in a mist chamber. In contrast, *P.s.* pv. *helianthi* produce angular lesions on sunflower leaves after spray inoculation but no apical chlorosis. With any inoculation method *P.s.* pv. *syringae* is not pathogenic on sunflower.

Host
Sunflower, zinnia, and marigold exhibit apical chlorosis when inoculated with either the sunflower or marigold strains of *P.s.* pv. *tagetis*, as did other species within the Compositae, including chicory, cocklebur, endive, and salsify. Limestone lettuce and Sidwill safflower, also in the Compositae, exhibit only slight chlorosis. None of the other species exhibit chlorosis with either strain, with the exception of tomato, cucumber, and some Brassica species (mustard, rape, and broccoli). Chlorosis on Brassica species, however, was generally limited to one or two leaves and are very mild and ephemeral on tomato and cucumber.

Disease Cycle
Apical chlorosis is caused by the bacterium *P.s.* pv. *tagetis*. Although infected plants are conspicuous in the field, the disease is of little economic importance. The disease is most severe on seedlings during cold, wet weather. The pathogen can be disseminated among plants by splashing rain and irrigation water, by wind as aerosols, and by the movement of contaminated soil. The apical chlorosis pathogen is a very common resident on the leaves of many weeds and other crop plants, but also survives in the soil.

Seed Transmission
Seed transmission is evident as plants from six of seven sunflower seed lots developed symptoms of apical chlorosis. Of 140 plants from the bulk seed sample, 15% were chlorotic 2 weeks after planting. From 0% to 63% of the seedlings derived from single heads developed apical chlorosis; overall, 21% of the 330 plants were systemically chlorotic 4 weeks after planting.

Disease Management
- Practice a 3-year or longer rotation to avoid building up large soil populations of the apical chlorosis bacterium.
- Avoid excessive irrigation and water-logged soils.

- Some hybrids are less susceptible to apical chlorosis than others, but none are completely immune. None of the inbred lines tested exhibit immunity to *P.s.* pv. *tagetis*.
- Seed treatment with antibiotic streptocycline (500 ppm) may reduce the incidence of the disease.

4.3.2 BACTERIAL STALK ROT AND HEAD ROT OF SUNFLOWER

Pathogen: *Erwinia carotovora*

Synonyms: *Pectobacterium carotovorum*, subsp. *carotovorum*
Pectobacterium atrosepticum

The diseases is more common in the U.S. Southern great plain states.

Symptoms

Stalk Rot
Initial symptoms are similar to several fungal stalk rots including water-soaked lesions and streaks that turn dark brown to black. Disease progress causes tissues to soften and emit a characteristic odor of rotten potatoes. Infections begin in wounds as a result of mechanical damage allowing entry of the pathogen. Necrosis and rot of tissues can also move up to stalks to affect developing heads. Infected stalks soften and dry up, becoming dark brown to black and may split open. Plants often lodge under the weight of maturing heads. A foam may appear on infected tissues as a result of bacteria causing fermentation of sugars in plant.

Head Rot
- *Co*alescing lesions develop watery, soft-rot symptoms that become dark brown as disease progresses.
- Heads give off an odor of rotting potatoes, and slimy masses of bacterial growth are present within infected tissues.

Bacterial stalk rot symptoms first appear as a blackening of stems, often centered around a leaf petiole axil. The pith of infected stems is black and watery, and the rot eventually hollows stems. Plants often lodge as stems rot. Heads can be infected occasionally, causing a soft, watery rot that discolors seeds and flowers. The disease can reduce yield and seed quality, but is of little economic concern in the High Plains.

Disease Cycle
Bacterial stalk rot is caused by the bacteria *E. carotovora* subsp. *carotovora* and *E. carotovora* subsp. *atroseptica*. Bacterial stalk rot infection generally occurs late in the season after extended wet periods. Infections occur in areas where water collects, such as leaf petiole axils, and is aided by wounds created by insect feeding (i.e., sunflower budworm), other pathogens or mechanical injury. The bacterial stalk rot pathogens are readily disseminated in and among sunflower fields by splashing rain and irrigation water, as aerosols, and surface irrigation waters. The pathogens survive between sunflower crops in infected crop debris and as pathogens on many other plants, including potato, onion, and carrot.

Factors Favoring Disease Development
Thunderstorms with hail, insect or bird damage to head, warm temperatures with high humidity levels favor the disease development. Mechanical injury (from insects, birds, or hail) is required for infection. Pathogen is found ubiquitously in soil and is spread by rain splashing and driving winds.

Disease Management
Avoid rotations with susceptible crops such as potato and other vegetables. Do not reuse irrigation tail water from vegetable crops if sunflowers are to be irrigated. Practice strict sanitation.

4.3.3 Bacterial Leaf Spot of Sunflower

Pathogen: *Pseudomonas syringae* pv. *helianthi*

Symptoms

Infections begin as small angular to circular necrotic lesions that often are surrounded by a yellow halo. In severe cases the lesions coalesce and may cause death of large areas of the leaf resulting in defoliation.

Control Measures

Infected plant debris and crop residues should be removed and burned. Crop rotation should be followed. Spray copper-containing fungicide (0.2%) as per label or antibiotic streptocycline 500 ppm at 10 days interval up to three sprays.

4.4 SAFFLOWER

4.4.1 Bacterial Leaf Spot and Stem Blight of Safflower

Pathogen: *Pseudomonas syringae*

Symptoms

Disease initially appears as dark, water-soaked lesions on leaves, stems, and petioles. Lesions become red–brown in color and turn necrotic with age, and often develop a pale-yellow margin. Some lesions may have a dark brown to black margin. The terminal bud is often killed, and the interior of petioles may develop a rot that extends below the soil line into roots.

Disease Cycle

Bacterial leaf spot and stem blight of safflower is caused by the bacterium *P. syringae*. Infection occurs when bacterial cells are deposited onto leaves by splashing water, aerosol movement, or from contaminated seed and multiply to form large populations. The bacteria gain entry into plants through natural openings and wounds. Infection occurs most readily during cool (<30°C), wet weather, especially hard wind-driven rains. Bacteria are disseminated within and among fields by splashing water, aerosols, and on contaminated equipment and workers. The pathogen survives between susceptible hosts in and on weeds such as hairy vetch, infested crop debris, and contaminated seed.

Losses

Severe infections can kill plants, but infected plants may recover when hot, dry conditions return. Yield losses can be significant.

Control Measures

Infected plant debris and crop residues should be removed and burned. Seed treatment with streptocycline (0.05%) and spray of copper-containing fungicide may reduce the disease.

4.5 MUSTARD

4.5.1 Bacterial Leaf Spot of Mustard

Pathogen: *Xanthomonas campestris* pv. *armoraceae*

Bacterial leaf spot (*P.s.* pv. *maculicola*; *Psm*) and *Xanthomonas* leaf spot (*X. campestris* pv. *armoraceae*; *Xca*) are important bacterial leaf spots on leaves of Brassica (turnip, mustard, collards, kale, etc.) in Oklahoma. Fungicides that are effective for the control of Cercospora leaf spot, the major fungal leaf spot disease in Oklahoma, are not effective on these bacterial leaf spot. Copper hydroxide (Kocide), copper sulfate (Cuprofix), and the plant defense activator acibenzolar-s-methyl

(Actigard) have been previously evaluated for control of bacterial leaf disease on turnip greens, but have not provided adequate control.

Symptoms
Xanthomonas leaf spot causes light-brown spots on foliage and apparently reduces the photosynthetic area of infected plants. The disease appears to be of little concern.

Disease Cycle
Little is known about the disease cycle, but other plant pathogenic *xanthomonads* are known to be disseminated by splashing rain and irrigation water, workers, equipment, and aerosols. *Xanthomonads* tend to survive poorly in soil, and are usually found in association with infested crop debris, contaminated seed, and weed hosts.

Disease Management
No cultural controls have been reported for *Xanthomonas* leaf spot. However, management strategies common for other diseases caused by *Xanthomonads* include crop rotation to nonhosts, thorough incorporation of plant debris, weed control in and around fields, and furrow irrigation from pathogen-free sources of water such as wells. Copper bactericides may provide some *Xanthomonas* leaf spot suppression if applied preventatively and regularly. John Damicone and Tyler Pierson (2009) reported Kocide and/or actigard to be effective to reduce the incidence of the disease.

4.5.2 BLACK ROT OF MUSTARD

Pathogen: *Xanthomonas campestris* pv. *campestris*
Black rot caused by the bacterium *X. campestris* pv. *campestris* affects cabbage and related crops like brassicas, mustard, and radish worldwide.

Symptoms
On young seedlings a yellowing appears along the margin of the cotyledons, which later shrivels and drops off. On the margins of mature leaves, similar yellowing appears. Initially, a small V-shaped area develops on leaves, but as the diseased area enlarges, the veins become distinctly black (Photo 4.1). In contrast to *Fusarium* yellows the veins are brownish in color. The internal

Typical V shaped lesion

PHOTO 4.1 Bacterial leaf spot symptoms on mustard. (Courtesy of Ashish Kumar, IARI Regional Station, Pusa, India, www.krishisewa.com.)

discoloration of stem occurs. The affected stem, when cut crosswise, shows a characteristic black ring. The bacterium can persist in plant residue for 1–2 years or as long as the plant debris remains intact. In case of severe infection internal discoloration of vegetative organs (black in color) and dry rot occurs. Severely infected plants produce seeds with discoloration and lesions on the seed surface. Plant death can occur in heavily affected crop.

Host
Black rot pathogen affect most of cultivated cruciferous plants and weeds. Cauliflower and cabbage are the most readily affected hosts in the crucifers, although kale is almost equally susceptible. Broccoli and Brussels sprouts have intermediate resistance and radish is quite resistant, but not to all strains. Kohlrabi, Chinese cabbage, rutabaga, turnip, collard, rape, jointed charlock (*Raphanus raphanistrum*), and mustard are also susceptible hosts.

Disease Cycle
The bacteria survive in infected seed, in debris from diseased plants left in the field and in infested soil. Seed-borne bacteria can be disseminated long distances. Many cruciferous weeds can harbor the black-rot bacteria. In a new field, black rot is usually introduced via infected seed or diseased transplants.

Further spread is facilitated by water-splash, running water, and handling infected plants. The bacteria enter the plant mainly through water pores at the edges of leaves. They can also enter through the root system and wounds made by chewing insects. They then move through the water vessels to the stem and the head. Black rot is favored by warm (26°C–30°C) wet conditions.

Disease Management
- Use certified disease-free seed. Hot water seed treatment at 52°C is recommended.
- Mulching of the field crop, where practicable, is highly recommended.
- Avoid wet, poorly drained soils and overhead irrigation.
- Field operations during wet weather should be discouraged.
- Keep the field free of weeds, particularly of the crucifer family. When possible, remove, burn, or plough down all crop debris immediately after harvest to reduce the amount of bacteria in the soil.
- A crop rotation based on at least a 2-year break in cruciferous crops is advocated.
- Use of resistant/tolerant varieties, where commercially available, provides the most effective control of the disease.

4.6 CASTOR

4.6.1 Bacterial Leaf Spot of Castor

Pathogen: *Xanthomonas campestris* pv. *ricinicola*
The disease is prevalent in all the castor-growing areas in India.

Symptoms
The pathogen attacks cotyledons, leaves and veins and produces few to numerous small, round, water-soaked spots which later become angular and dark brown to jet black in color. The spots are generally aggregated toward the tip. At a later stage the spots become irregular in shape particularly when they coalesce and areas around such spots turn pale-brown and brittle.

Bacterial ooze is observed on both the sides of the leaf which is in the form of small shining beads or fine scales.

Pathogen
The optimum temperature for the growth of the bacterium and its thermal death point are 31°C and 51°C, respectively.

Control Measures

- Field sanitation helps in minimizing the yield loss as pathogen survives on seed and plant debris.
- Hot water treatment of seed at 58°C–60°C for 10 min.
- Grow tolerant varieties.

Chemical
Spray copper oxychloride (0.3%) or streptocycline 1 g in 10 L of water or paushamycin (0.025%).

4.7 OIL PALM

4.7.1 BACTERIAL BUD ROT OF OIL PALM

Pathogen: *Erwinia* spp.

Symptoms
Parts of spear leaf petiole or rachi turn brown; discoloration may be associated with a wet rot; spear leaf may be wilted and/or chlorotic. Leaves may be collapsing and hanging from the crown; infection of the bud results in buds becoming rotten and putrid leading to death of the palm.

Geographical Distribution
Disease occurs in oil palm in Colombia, Costa Rica, Democratic Republic of Congo, Ecuador, Nicaragua, Nigeria, Panama, and Southeast Asia.

Disease Management
- Plant Oil palm varieties with resistance to the bacteria.
- Rotting tissue on spear leaves should be removed to prevent bacteria spreading to buds. Palm buds can be protected using copper-based fungicides.

4.8 OLIVE

4.8.1 OLIVE KNOT

Pathogen: *Pseudomonas savastanoi* pv. *savastanoi*
Olive knot disease on olive, *Olea europaea*, is distributed in olive-growing regions worldwide, and reports of the disease in California date back to the late nineteenth century. The disease is caused by *P. savastanoi* pv. *savastanoi* (*Psv*), a bacterial pathogen that is spread short distances during the winter and spring rain events associated with California's Mediterranean climate. Consequently, in California, disease severity is greatest in the northern part of the state where heightened rainfall promotes disease development. Olive knot disease has become more common and serious during the past decade, in part due to increased plantings of the "Manzanillo" olive, a highly susceptible cultivar, and the introduction of super high-density olive plantings for oil production, a system where mechanized cultural practices can promote disease development.

Symptoms
The characteristic symptom of infection is the development of galls, or "knots," at infection sites. Galls are most commonly formed at leaf nodes (sites of bud development), due to infection of leaf scars by the bacterium; however, they also can be formed at other points of pathogen entry, such as pruning wounds or wounds caused by frost damage or hail. Excessive freeze damage increases disease severity even in regions characterized by lower rainfall. The olive knot bacterium produces plant growth regulators at infection sites resulting in plant tissue proliferation and gall development.

Though galls typically form on stems and twigs, galls also have been observed on leaves and fruit. During rain events, bacterial ooze might form on the surface of galls; this ooze is infective and can induce disease when transmitted to uninfected plant parts.

Damage

Galls produced as a result of infection by *Psv* can girdle and kill affected twigs. The death of infected shoots directly reduces yield; however, the disease also affects fruit size and quality. Flavor sensory tests have demonstrated that even trees with few galls can yield off-flavor fruit. The impact of the disease on yield, fruit size, and quality render olive knot of economic importance to both commercial table olive growers and growers of olives for oil. Additionally, severe symptoms of olive knot detract from the aesthetics of olive trees used in commercial and private landscapes.

Disease Cycle

Pseudomonas savastanoi pv. *savastanoi* survives both in gall tissues and as an epiphyte on twigs, leaves, and fruit. Because the pathogen survives better on rough bark surfaces than on foliage, pathogen populations are higher on twigs than on leaves. *Pseudomonas savastanoi* pv. *savastanoi* populations on plant surfaces vary throughout the year, with populations increasing during the rainy season. Disease severity (i.e., the number of galls per tree) is directly related to the magnitude of the epiphytic pathogen population.

Both the epiphytic pathogen population and bacterial ooze emitted from galls can serve as primary inoculum (infectious propagules) for the development of new infections. The pathogen can be transmitted both within a plant and to neighboring plants in wind-blown rain or over larger distances on contaminated pruning tools or infected nursery stock. The pathogen can infect the plant through natural openings that occur when the tree drops its leaves, flowers, or fruit, through wounds resulting from natural events such as frost injury or hail damage or by wounds caused by cultural practices such as pruning and harvesting. Pruning wounds can remain susceptible to infection for at least 14 days. Leaf scars, however, are the most common points of pathogen entry and can remain susceptible to infection for up to 7 days after leaf drop (abscission).

Although olives are evergreen and leaves drop throughout the year, the abscission rate in California is highest during the late spring. Consequently, the tree can be more susceptible to infection during spring rain events as a result of the heightened availability of fresh leaf scars serving as infection courts, or points of pathogen entry. Other factors enhancing leaf drop, such as frost damage or olive leaf spot disease, can increase vulnerability to infection by *Psv*. Generally infections by *Psv* on olive remain localized, resulting in gall formation at the infection site. Secondary galls, although rare, can be initiated by bacterial movement within the xylem vessels of the olive. These secondary galls typically form in close proximity to the primary gall, and the potential for a plant to support secondary gall development can vary by cultivar. Although all olive cultivars are susceptible to the pathogen, disease severity can vary by cultivar, and 1-year-old plants are more susceptible to infection than 3-year-old plants.

Because galls form only when the tree is growing, infections initiated during the winter do not become symptomatic until spring. This latent period between infection and symptom development offers yet another avenue of pathogen transmission, as sale of asymptomatic nursery stock can result in long-distance pathogen spread and introduction of the disease to uninfected landscapes or orchards. Plants containing asymptomatic infections might evade plant health inspectors and allow for international movement of the pathogen.

Disease Management

Although olive knot disease generally is caused by *Psv*, it is important to note that olives also can become infected by a related bacterium, *P. savastanoi* pv. *nerii* (*Psvn*). Both pathogens affect plants in the family Oleaceae; however, *Psvn* is more commonly associated with knot formation on *Nerium oleander*, a disease referred to as oleander knot. While *Psvn* infects both oleander and olive, *Psv* infects only olive. The relative frequency of olive infections caused by *Psvn* in California is unknown. Effective management of both oleander knot and olive knot disease relies on reducing pathogen populations on the plant surface. Though commercial olive growers might utilize copper-based bactericides as a component of an integrated pest management program, products known to be effective aren't registered for use on backyard trees. Bactericidal compounds available to home-owners haven't been evaluated for efficacy in management of *Psv*. For the homeowner, a combination of cultural practices and sanitation is the most appropriate method of disease management.

Exclusion

For landscape plantings of olive, the primary defense against disease is pathogen exclusion. By planting disease-free nursery stock, a homeowner or commercial landscaper might avoid pathogen introduction into a landscape. When purchasing an olive tree, consider the potential for the plant to harbor latent infections. For example, purchasing a plant in winter heightens the potential for asymptomatic infections to evade observation. Gall formation can be complete by late spring, allowing for visual selection of uninfected plants.

Sanitation and Cultural Practices

Galls exude bacterial ooze during rain events; therefore, removing them from infected trees reduces the potential for disease spread. Because galls might form on small branches and twigs as well as large structural branches (scaffolds), tools ranging in size from small pruning shears to pruning saws might be needed to remove affected tissues from the tree. All tools should be routinely steril-ized with a 10% bleach solution to prevent disease transmission both within and between trees. Galls shouldn't be removed during the winter and spring rainy season, because the resulting wounds can serve as new infection courts. Pruning wounds made in the dry summer months aren't suscep-tible to infection, thereby reducing the need to sanitize tools during summer pruning.

REFERENCES

Anitha, K., Chakrabarty, S.K., Girish, A.G. et al. 2004. Detection of bacterial wilt infection in imported groundnut germplasm. *Indian J. Plant Protect.*, 32, 147–148.

Anitha, K., Gunjotikar, G.A., Chakrabarty, S.K. et al. 2003. Interception of bacterial wilt, *Burkholderia solanacearum* in groundnut germplasm imported from Australia. *J. Oilseeds Res.*, 20, 101–104.

Damicone, J. and T. Pierson. 2009. Control of bacterial leaf spot on mustard greens—Spring trial stillwater, OSU Entomology and Plant Pathology. www.ir4.rutgers.edu/FoodUse/PerfData/2750.pdf.

Helimer, E. 1955. Bacterial leaf spot of African marigold (*Tagetis erecta*) caused by *Pseudomonas tagetis* sp. n. *Acta Agric. Scand.*, 5, 185–200.

Liao, B.S., Wang, Y.Y., Xia, X.M., Tang, G.Y., Tan, Y.J., and D.R. Sun. 1990. Genetic and breeding aspects of resistance to bacterial wilt in groundnut. In: Bacterial wilt of groundnut: *Proceedings of an ACIAR/ICRISAT Collaborative Research Planning Meeting*, 18–19 March, 1990, Genting Highlands, Malaysia (Middleton K.J., and Hayward, A.C. (Eds.), pp. 39–43. ACIAR proceedings No. 31. Canberra, Australia. Australian Centre for International Agricultural Research.

Mehan, V.K., Liao, B.S., Tan, Y.J. et al. 1994. Bacterial wilt of groundnut. Information Bulletin No. 35. International Crops Research Institute for the Semi-Arid Tropics, Patancheru, India, 28pp.

Prasada Rao, R.D.V.J., Gunjotikar, G.A., Chakrabarty, S.K. et al. 2000. Detection of *Ralstonia solanacearum* in seeds of wild *Arachis* spp. imported from Brazil. *Indian J. Plant Protect.*, 28, 51–56.

Trimboli, D., Fahy, P.C., and K.F. Baker. 1978. Apical chlorosis and leaf spot of *Tagetes* spp. caused by *Pseudomonas tagetis* Helimers. *Aust. J. Agric. Res.*, 29, 831–839.

SUGGESTED READING

Elliott, M.L., Broschat, T.K., Uchida, J.Y. et al. 2004. *Compendium of Ornamental Palm Disease and Disorders*. American Phytopathological Society Press, St. Paul, MN.

Hewitt, W.B. 1938. Leaf-scar infection in relation to the olive-knot disease. *Hilgardia*, 12, 41–65.

Janse, J.D. 2005. *Phytobacteriology: Principles and Practice*. CABI Publishers, Cambridge, MA, 360pp.

Nega, E., Ulrich, R., Werner, S., and M. Jahn. 2003. Hot water treatment of vegetable seed—An alternative seed treatment method to control seed borne pathogens in organic farming. *J. Plant Dis. Protect.*, 110(3), 220–234.

Onsando, J.M. 1992. Black rot of crucifers. In: Chaube, H.S., Kumar, J., Mukhopadhyay, A.N., and Singh, U.S. (eds.), *Plant Diseases of International Importance*, Vol II: Diseases of Vegetables and Oil Seed Crops. Prentice Hall, Englewood Cliffs, NJ, pp. 243–252.

Penyalver, R., García, A., Ferrer, A. et al. 2006. Factors affecting *Pseudomonas savastanoi* pv. *savastanoi* plant inoculations and their use for evaluation of olive cultivar susceptibility. *Phytopathology*, 96, 313–319.

Quesada, J.M., Penyalver, R., Panadés, J. et al. 2010. Dissemination of *Pseudomonas savastanoi* pv. *savastanoi* populations and subsequent appearance of olive knot disease. *Plant Pathol.*, 59, 262–269.

Schroth, M.N., Osgood, J.W., and T.D. Miller. 1973. Quantitative assessment of the effect of the olive knot disease on olive yield and quality. *Phytopathology*, 63, 1064–1065.

Teviotdale, B.L. and W.H. Krueger. 2004. Effects of timing of copper sprays, defoliation, rainfall, and inoculum concentration on incidence of olive knot disease. *Plant Dis.*, 88, 131–135.

Wilson, E.E. 1935. The olive knot disease: Its inception, development, and control. *Hilgardia*, 9, 233–264.

5 Bacterial Diseases of Cash Crops

5.1 SUGARCANE

5.1.1 Red Stripe and Top Rot of Sugarcane

Pathogen: *Acidovorax avenae* subsp. *avenae* (Manna) Willems et al. (1992)

Synonyms: *A. avenae* (Manns) Willems et al.
A. avenae subsp. *avenae* (Manns) Willems et al.
Phytomonas setariae (Okabe) Burkholder
Pseudomonas alboprecipitans Rosen
Pseudomonas avenae Manns
Pseudomonas rubrilineans (Lee et al.) Stapp
Pseudomonas setariae (Okabe) Savulescu
Xanthomonas rubrilinea (Lee et al.) Starr & Burkhokler

Common Name: Bacterial leaf stripe

Host
Euchlaena mexicana, *Saccharum officinarum*, *Zea mays*, Poaceae.

Symptoms
The disease first makes its appearance on the basal part of young leaves. The stripes appear as water-soaked, long, narrow chlorotic streaks and turn reddish brown within a few days. These stripes are 0.5–1 mm in width and 5–100 mm in length and run parallel to the midrib. The stripes remain confined to lower half of the leaf lamina, and whitish flakes spread to growing points of the shoot and yellowish stripes develop, which later turn reddish brown. Rotting may commence from the tip of the shoot and spreads downward. The core is discolored to reddish brown and shriveled, and cavity forms in the center. Badly affected fields emit a foul and nauseating smell.

Favorable Conditions
Continuous ratooning and prolonged rainy weather with low temperature (25°C) are favorable conditions.

Disease Cycle
The pathogen remains viable in the soil and infected plant residues. The bacterium also survives on sorghum, pearl millet, maize, finger millet, and other species of *Saccharum*. The bacterium primarily spreads through infected canes. The secondary spread is mainly through rain splash, irrigation water, and insects. Infected parenchymatous cells may collapse, and normal functioning of the plant parts may fail. Several grasses, including ragi and bajra, have been reported to be infected by bacteria, and these hosts may also play a role in the perpetuation and spread of the pathogen. The disease cycle of red stripe and top rot of sugarcane is illustrated in Figure 5.1.

Disease Management
- Whenever the disease is noticed, the affected plants should be removed and burned.
- Select sets from the healthy fields.
- Avoid growing collateral hosts near sugarcane fields.

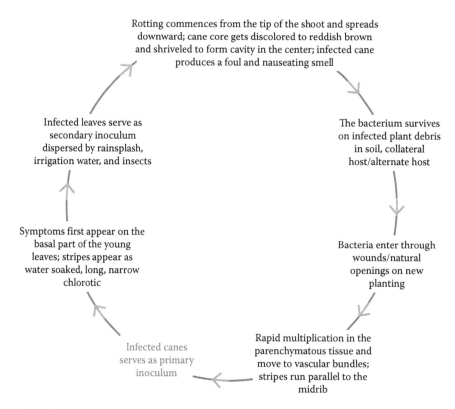

FIGURE 5.1   Disease cycle of red stripe and top rot of sugarcane caused by *Acidovorax avenae* subsp. *avenae*.

- Grow resistant varieties.
- Use of resistant cultivars is the best method to avoid the disease in areas where its occurrence is common. CO cultivars, 6805, 7202, 7321, 7537, 7642, and 8005 and variety Ponda are reported to be resistant.

5.1.2   GUMMOSIS OR COBB'S DISEASE OF SUGARCANE

Pathogen: *Xanthomonas axonopodis* pv. *vasculorum* (ex Cobb, 1893) Vauterin, Hoste, Kersters and Swings (1995)

Synonyms: *Xanthomonas campestris* pv. *vasculorum* (Cobb, 1894; Dye, 1978)
        X. vasculorum (Cobb) Dowson (1939)
        Bacillus vasculorum Cobb (1893)
        Bacterium vasculorum (Cobb) Migula (1900)
        Phytomonas vasculorum (Cobb) Bergey et al. (1923)
        Pseudomonas vasculorum (Cobb) Smith (1901)

Common Names: Sugarcane gumming disease, gumming disease of sugarcane, gummosis, or Cobb's disease of sugarcane

Symptoms

Abnormal colors or necrotic areas develop on leaves, while internal discoloration, internal red necrosis, and ooze are common on stems. There are two distinct phases in the symptom development: the foliar stage and the systemic stage.

On leaves, streaks 3–6 mm wide arise from the leaf margin and develop along the vascular bundles toward the base. Streaks are yellow to orange with red flecks, becoming necrotic and gray with age. The necrosis may involve large areas of leaf. A few short streaks are observed in resistant cultivars or when environmental conditions are not favorable for the disease.

On highly susceptible cultivars and when conditions are favorable, the bacterium progresses down the lamina and the sheath and infects the stalk. The systemic stage of the disease is characterized by a reddish discoloration of the vascular bundles at the nodes and by a bacterial slime. Gum pockets are formed and the slime exudes from the cut surface of the stalks. Stalk deformation and knife-cut lesions due to transverse splits in young elongating tissue can also be observed. When conditions are not favorable for cane growth, the formation of the gum pockets near the apex and in the leaf spindle results in the death of the growing point. Another characteristic of the systemic stage of the disease is the partial or total chlorosis of new leaves in mature canes. Chlorosis can also occur in young ratoons as a result of transmission by contaminated leaves.

Main Host
Areca catechu (betel nut palm), *Bambusa vulgaris* (common bamboo), *Cocos nucifera* (coconut), *Coix lacryma-jobi* (Job's-tears), *Dictyosperma album*, *Panicum maximum* (Guinea grass), *Panicum miliaceum* (millet), *Pennisetum purpureum* (elephant grass), *Roystonea regia* (Cuban royal palm), *S. officinarum* (sugarcane), *Sorghum bicolor* (sorghum), *Sorghum halepense* (Johnson grass), *Sorghum sudanense* (Sudan grass), *Thysanolaena latifolia* (Asian broom grass), *Z. mays* (maize).

Geographical Distribution
Africa: Ghana, Madagascar, Madeira, Malawi, Mauritius, Mozambique, Reunion, South Africa (Natal), Swaziland, Zimbabwe.

Australasia and Oceania: New Guinea.

North America: Mexico.

Central America and West Indies: Belize, Dominica, Dominican Republic, Panama, Puerto Rico, St. Kitts and Nevis.

South America: Argentina (Tucuman), Brazil, Colombia, French Guiana.

Disease Cycle
Plants become infected through wounded roots, wind-driven rain, and insects (flies) as well as infected cuttings play a role in transmission.

Economic Importance
In the past, gumming was a major disease of sugarcane in Australia, Mauritius, and Reunion. Severe epidemics occurred in Australia where reductions of 30%–40% in cane tonnage and 9%–17% in sugar content were reported between 1893 and 1899 (North, 1935). In the 1890s, heavy losses were also observed in Mauritius and susceptible cultivars were abandoned. Besides the losses in the field, gumming causes problems in the milling process, resulting in a lower sugar recovery.

The economic difficulties causes by the disease decreased when the susceptible noble canes were replaced by interspecific hybrids in the 1930s (Ricaud and Autrey, 1989). This led to the eradication of the pathogen from Australia, Fiji, Brazil, and the West Indies. In the Mascareigne islands, there has been evidence of some losses in the hybrids, especially when the cultivars are susceptible to the systemic infection (Ricaud, 1969). During the 1980–1981 epidemic, sugar yields from systemically infected stalks of cv. M377/56 were found to be 19.5% lower than that from stalks with foliar streaks only (Ricaud and Autrey, 1989). A decrease in sugar yield of up to 45% was reported in cv. M377/56 after inoculation in field trials (Autrey et al., 1986).

In the Mascareigne islands, the need to include resistance to gumming in breeding schemes and the replacement of cultivars that show susceptibility increase the economic impact of this pathogen. For instance, in Reunion, the popular cultivar R397 represented 60% of the sugarcane production in 1958 but was then abandoned following a severe epidemic of *X. axonopodis* pv. *vasculorum*.

Control Measures

Use of healthy planting material (cuttings), hygiene and disinfection of cutting knifes, use of low susceptibility or resistant varieties (Ricaud and Autrey, 1989), and early spraying with insecticides to reduce vector populations are the only ways to prevent or control disease caused by *Xanthomonas* pv. *vasculorum*.

Resistance

The degree of cultivar susceptibility is often revealed during an epidemic. Then, the planting of highly susceptible cultivars is prohibited and their replacement is undertaken. The speed of the cultivar replacement depends on the availability of resistant cultivars with an acceptable yield. The breeding schemes in Mauritius and Reunion include the resistance to gumming as an important objective.

Resistance to foliar infection and tolerance to systemic infection appear independently and controlled genetically (North, 1935; Ricaud, 1969). In areas where there is a risk of epidemics, the planting of cultivars with foliar susceptibility should be avoided. When environmental conditions are conducive to epidemics, such cultivars facilitate the increase in inoculum pressure and the selection of strains with higher pathogenicity. The losses encountered during an epidemic are directly related to the tolerance to systemic infection.

Interspecific hybridization has been very successful in the control of many sugarcane diseases, and gumming has been eradicated from areas where resistant cultivars have been grown. Resistance to the disease appears governed by a few major genes rather than a polygenic system (Stevenson, 1965). This could explain the resistance failure of some cultivars after the appearance of new pathogenic strains (Antoine and Pérombelon, 1965; Ricaud and Sullivan, 1974). Evidence of variation in *X. axonopodis* pv. *vasculorum* has accumulated over the years from different parts of the world (Ricaud and Autrey, 1989). Phenotypic differences between the eastern and southern African strains were observed by Hayward (1962) and confirmed at the DNA and membrane protein levels by Qhobela and Claflin (1992). Phenotypic and pathogenic differences were also recorded in Mauritius and in Reunion. It has been suggested that three races of the pathogen have been involved in the epidemics observed in Mauritius. However, the inoculation of the putative differential cultivars in controlled conditions revealed that the different morphological types of strains varied in aggressiveness, not in virulence (Péros, 1988). The variability of the pathogen requires the evaluation of resistance of new genotypes of sugarcane against all prevailing strains.

The screening method developed by North (1935) in Australia is still used in Mauritius. Test varieties are planted between rows of a susceptible cultivar, which is inoculated. The disease spreads from the inoculated plants, and the natural infection of the test varieties is rated during peak infection in the maturing season. Replication of the resistance tests permits an accurate assessment of the behavior of the promising varieties. However, the field trials are space and time consuming and need environmental conditions favorable for the disease to spread. Other methods have been investigated to complement the field trials and for a more rapid screening. The inoculation of detached leaves was found reliable but laborious. The inoculation of potted plants in the greenhouse was performed and produced results in accordance with field behavior (Girard and Péros, 1987). Péros and Lombard (1992) induced the development of the disease in plantlets inoculated *in vitro*. However, the weak expression of the foliar symptoms appeared to limit the use of the *in vitro* test.

Sanitation

The rapid spread of *X. axonopodis* pv. *vasculorum* from infection foci limits the use of control methods based on sanitation (Ricaud and Autrey, 1989). Sanitation control includes the use of seed cane from fields without systemic infection, the eradication of volunteer stools of susceptible cultivars and alternative hosts when replanting, and disinfection of knives and harvesters to avoid mechanical spread. Thermotherapy, which is used to control other pathogens of sugarcane, can also be used. However, the treatment (50°C for 30 min, and 2–3 h the next day at the same temperature) will not completely cure the planting material. Disease-free plantlets can be produced by tissue culture techniques.

5.1.3 SUGARCANE LEAF SCALD

Pathogen: *Xanthomonas albilineans*

Synonyms: *Agrobacterium albilineans*
Bacterium albilineans
Phytomonas albilineans
Pseudomonas albilineans

Symptoms

On leaves, abnormal colors, abnormal forms, necrotic areas with wilting of leaves, and yellowing or death are observed, while on stems, discoloration of bark, internal red necrosis, stunting or resetting, and stems wilt are observed. Whole plant dwarfing and plant death or dieback are common.

Foliar infection induced by the airborne inoculum of *X. albilineans* is characterized by cream-to-yellow stripes starting at the tip or occasionally at the margin of the leaf (Photo 5.1).

Three phases are associated with the symptomatology of the disease: latent, chronic, and acute phases.

5.1.3.1 Latent Phase

Latent infection or the absence of symptoms is a characteristic feature of the disease that occurs in tolerant varieties and under favorable conditions of plant growth.

Chronic Phase

In the chronic phase, a typical white pencil-line stripe (1–2 mm wide) runs from several centimeters to almost the entire length of the leaf. At a later stage, the sharp margins of the stripe may become diffuse, and a red pencil line may be formed in the middle of the stripe. Partial or total chlorosis of leaves occurs and is accompanied by an inward curling of the leaves (scalding).

Acute Phase

Affected stalks may be stunted with the development of axillary buds (side shoots) bearing symptoms of the chronic phase. A longitudinal section of the stalk shows a reddish discoloration of the vascular

PHOTO 5.1   Sugarcane: leaf scald (Pathogen: *Xanthomonas albilineans*). (Courtesy of Scot Nelson, flickr.com.)

bundles, particularly at the nodes. As the disease progresses, lysigenous cavities may be formed and the stalk dies. In the acute phase of the disease, sudden plant death occurs with few or no symptoms.

Host
B. vulgaris (common bamboo), *C. lacryma-jobi* (Job's-tears), *Cymbopogon citratus* (citronella grass), *Imperata cylindrica* (satintail), *P. maximum* (Guinea grass), *Paspalum conjugatum* (sour paspalum), *Paspalum dilatatum* (Dallis grass), *Pe. purpureum* (elephant grass), *Rottboellia cochinchinensis* (itch grass), *S. officinarum* (sugarcane), *Saccharum spontaneum* (wild sugarcane), *So. halepense* (Johnson grass), *Z. mays* (maize).

Geographical Distribution
Africa: Benin, Burkina Faso, Cameroon, Chad, Congo Democratic Republic, Congo, Ghana, Ivory Coast, Kenya, Madagascar, Malawi, Mauritius, Morocco, Mozambique, Nigeria, Réunion, South Africa, Swaziland, Tanzania, Zimbabwe.

North America: the United States (Florida, Hawaii, Louisiana, Texas).

Central America: Barbados, Belize, Cuba, Dominica, Dominican Republic, Grenada, Guadeloupe, Guatemala, Jamaica, Martinique, Mexico (Vera Cruz), Panama, Puerto Rico, St Kitts & Nevis, St Lucia, St Vincent & Grenadines, Trinidad & Tobago.

South America: Argentina, Brazil (São Paulo, Sergipe), Colombia, Ecuador, French Guiana, Guyana, Suriname, Uruguay, Venezuela.

Asia: Cambodia, China, Taiwan, India (Andhra Pradesh, Uttar Pradesh), Indonesia (Java, Sulawesi), Japan, Malaysia, Myanmar, Pakistan, the Philippines, Sri Lanka, Thailand, Vietnam.

Australasia: Australia (New South Wales, Queensland), Fiji, French Polynesia, Papua New Guinea.

Impact
Leaf scald was responsible for heavy losses in sugarcane at the beginning of the century when noble canes, *S. officinarum*, were cultivated (Ricaud and Ryan, 1989). The impact of the disease is reduced by the cultivation of interspecific hybrids but susceptible varieties may be rapidly destroyed. Yield losses of 15%–20% were recorded in sugarcane variety B 69379 in Guadeloupe (Rott et al., 1995). Cochereau and Jean-Bart (1989) estimated a loss of 13 tons of cane per hectare of infected plants compared with healthy cane. In Mexico, leaf scald destroyed 800 ha under variety Mex 64-1487 (Rott et al., 1995), while in Mauritius, a sugar yield-depressing effect of up to 31% was observed in variety M 695/69.

Transmission
These pathogens are primarily transmitted due to infected cuttings and fluffs (mature flowers). The use of contaminated tools and harvesters aids local spread. Airborne infection has occurred and wet and stormy conditions are highly conducive to disease spread. Root-to-root infections have also been recorded (Klett and Rott, 1994).

Since leaf scald is a systemic disease that may be inconspicuous (latent) for lengthy periods of time, infected seed cane is a major cause of disease spread. Cutting knives, including those on machinery, are an important source of infection.

The pathogen can also survive in stubble. The organism does not appear to survive for long periods of time in soil or undecomposed cane trash.

Alternative hosts may offer another means of pathogen survival. *X. albilineans* naturally infects several wild grass weeds, such as elephant grass (*Pe. purpureum*).

Besides transmission by cutting knives, evidence is accumulating to suggest aerial transmission. This may explain, in part, the recent spread of leaf scald.

The amount of damage caused by leaf scald appears to be influenced by environmental conditions. Periods of stress such as drought, waterlogging, and low temperature are reported to increase the severity of the disease. The yield of stalks that are dead or have necrotic tops and leaves with numerous side shoots is decreased to 20%–30% of that of symptomless stalks. At present, there are very few plants showing these symptoms at harvest and leaf scald is not an economic problem at this time in Florida because of resistant varieties.

Prevention and Control
Leaf scald may be controlled effectively by the simultaneous use of several methods. The cultivation of resistant varieties should be supplemented with the use of disease-free cuttings from nurseries. These cuttings should be established using setts that have undergone a cold soak/long hot-water treatment in which the cuttings are soaked in cold running water for 48 h followed by hot-water treatment at 50°C for 3 h.

It is necessary to disinfect cutting instruments regularly when cuttings are prepared and during harvest because *X. albilineans* is transmitted by knives and harvesters. Legislation and quarantine are important to regulate the movement of sugarcane plant material (Frison and Putte, 1993).

5.1.4 Ratoon Stunting of Sugarcane

Pathogen: *Leifsonia xyli* subsp. *xyli* (Davis et al., 1984; Evtushenko et al., 2000)

Synonyms: *Clavibacter xyli* subsp. *xyli* (Davis et al., 1984)
Cl. xyli (Davis et al., 1984)

Common Names: Sugarcane ratoon stunting disease

Symptoms
Stunting is the only overt symptom and can also be caused by a number of other maladies. The degree of stunting resulting from the disease may vary considerably. Yields can be adversely affected even when stunting is not obvious. Disease expression can be enhanced by stress, especially moisture stress. Yield reduction is caused by slower growth of diseased crops with the accompanying production of thinner and shorter stalks and sometimes a reduction in the number of stalks when the disease is severe. In stubble or ratoon crops, diseased plants are slower to initiate growth, and death of individual plants of extremely susceptible cultivars may occur. Some highly susceptible cultivars may show wilting under moisture stress and even necrosis of leaves at the tips and margins.

Diseased stalks of some cultivars may exhibit internal discoloration, but these symptoms are often ephemeral. In mature stalks, vascular bundles at the nodes may become discolored. The intensity of discoloration may vary among cultivars from one time to another, ranging from yellow, orange, pink, salmon, and red to reddish brown. This may be viewed by slicing longitudinally through a node with a knife and examining the fresh cut for discolored vascular bundles in the shape of dots, commas, or short lines. Discoloration does not extend into the internodes. Similar nodal symptoms may be caused by insects or other pathogens but usually extend throughout nodes, instead of being confined to the lower portion of nodes, or extend into the internodes. Juvenile stalk symptoms may be observed in some cultivars by longitudinally slicing through 1- to 2-month-old stalks. The symptoms appear as pinkish discoloration just below the apical meristematic area and may extend downward as much as a centimeter.

Host
S. officinarum (sugarcane)

Geographical Distribution
Europe: Spain, Mainland Spain.

Asia: Bangladesh, China—Guangdong; India—Karnataka, Madhya Pradesh, Punjab, Uttar Pradesh; Indonesia—Java; Japan—Kyushu, Ryukyu Archipelago; Malaysia—Peninsular Malaysia; Myanmar; Pakistan; the Philippines; Sri Lanka; Taiwan; Thailand.

Africa: Burkina Faso, Cameroon, Comoros, Congo, Congo Democratic Republic, Cote d'Ivoire, Djibouti, Egypt, Ethiopia, Kenya, Madagascar, Malawi, Mali, Mauritius, Mozambique, Nigeria, Reunion, Seychelles, Somalia, South Africa, Sudan, Swaziland, Tanzania, Uganda, Zimbabwe.

North America: Mexico; the United States—Florida, Hawaii, Louisiana.

Central America and Caribbean: Antigua and Barbuda, Barbados, Belize, Cuba, Dominican Republic, El Salvador, Guadeloupe, Jamaica, Nicaragua, Panama, Puerto Rico, St Kitts-Nevis, Trinidad and Tobago.

South America: Argentina; Bolivia; Brazil—Rio de Janeiro; Colombia; Guyana; Peru; Uruguay; Venezuela.

Oceania: Australia—New South Wales, Queensland; Fiji.

Impact

Ratoon stunting disease is widely regarded as causing greater economic loss to the cane sugar industry throughout the world than any other disease (Hughes, 1974); yet paradoxically, few other diseases of sugarcane are less conspicuous. Yield losses have frequently been estimated at 5%–10% overall. Losses may be negligible in some years, but in other years, they may be 30% or greater (Steib and Chilton, 1967; Early, 1973; Koike, 1974; Singh, 1974; Béchet, 1976; Liu et al., 1979). The importance of ratoon stunting disease is largely dependent on two factors: varietal susceptibility and disease incidence. In Florida, for example, yield loss from ratoon stunting disease was estimated at $36.8 million for the 1988–1989 crop (Dean and Davis, 1990). This estimate was based on a relatively small average yield reduction of 5% applied to essentially the entire crop in Florida. In South Africa, losses in a normal season have been estimated at 3% of the annual crop (Bailey and Fox, 1984).

Assuming a high disease incidence and a susceptible sugarcane clone, several other factors are important in the severity of yield losses. Moisture stress can increase yield reduction from the disease in susceptible cultivars (Rossler, 1974); heavy losses can occur when sugarcane is grown under dry land conditions and growth is checked by prolonged dry weather (Egan, 1970). Yield reductions are sometimes greater in successive ratoon crops, possibly because of increased disease incidence (Steindl, 1961; Bailey and Béchet, 1986). The presence of other diseases, such as sugarcane mosaic potyvirus (Steib and Chilton, 1967; Koike, 1974), may increase reductions in germination and yield caused by ratoon stunting disease. Weak, diseased crops may invite weed infestation and, thus, be subject to further yield reduction as a result of competition with weeds.

Prevention and Control

Cultural Control and Sanitary Methods

Planting healthy cane can be used to control ratoon stunting disease. Sanitation is important in keeping healthy cane from becoming infected because *L. xyli* subsp. *xyli* is easily transmitted mechanically.

Seed cane can be monitored for freedom from the disease using appropriate diagnostic techniques. In Australia and South Africa, phase-contrast microscopy has been used to screen thousands of samples annually for *L. xyli* subsp. *xyli* (Bailey and Fox, 1984). Continued vigilance in the selection of seed cane over several years has resulted in a reduction in the incidence of ratoon stunting disease in both plantings for seed cane production and commercial crops.

Seed cane can be heat treated to help prevent the spread of *L. xyli* subsp. *xyli* from one geographic area to another and to control ratoon stunting disease within areas where it occurs (Steindl, 1961). Hot-water, hot-air, moist-air, and aerated steam treatments have been used (Benda and Ricaud, 1978). Hot-water treatment at 50°C for 2–3 h has been the most commonly used method (Steindl, 1961; Gul and Hassan, 1995). The application of streptomycin and hot-water treatment at 52°C for 30 min suppressed disease by 22.6% and increased yields of cane and white sugar by 35.54 and 4.55 tons/ha, respectively (Gul and Hassan, 1995). Different cultivars vary in their tolerance to injury by heat. Seed cane from mature plants is usually less affected and generally germinates better after hot-water treatment than that from immature or over-mature plants. Pretreatment of young, heat-sensitive seed cane to increase germination has been reported to reduce the deleterious effects of hot-water treatment (Benda, 1972, 1978). Seed cane was cut 1–5 days before treatment, pretreated at 50°C for 10 min in hot water, and then treated the following day at 50°C for 2–3 h. Hot-air treatment, with inlet air at 58°C for 8 h or at 50°C for 24 h, may be less harmful to immature seed cane than hot-water treatment and is used in some areas (Steindl, 1961). Dipping seed cane in water at ambient temperatures following hot-air treatment may protect germination. Moist-air treatment at 54°C for 7 h in a sealed chamber was developed to overcome the moisture loss associated with hot-air treatment (Shukla et al., 1974). Aerated-steam treatment at 53°C for 4 h uses steam-moistened hot air to reach the necessary internal stalk temperatures more quickly, approaching the treatment time used with hot water (Mayeux et al., 1979).

The two major problems limiting the effectiveness of heat treatment, other than expense, are reductions in germination and lack of complete control. Procedures that have been developed to protect germination in addition to those already discussed include treating seed cane with fungicides or other chemicals during and after heat treatment, careful selection of seed cane for treatment, and leaving leaf sheaths over buds during treatment. Proper functioning of the heating unit and temperature control system and proper loading of the heat chamber can favor both germination and control of ratoon stunting disease.

Under practical conditions, however, heat treatment is often not completely curative for ratoon stunting disease (Damann and Benda, 1983), and germination rates of treated seed cane may be reduced. Moreover, the expense involved in heat treatment is prohibitive. Consequently, heat treatment is often used to establish pathogen-free nurseries that are then used to supply planting material for commercial fields. Even so, the quantity of seed cane produced in such nurseries is usually inadequate, and additional sources of relatively pathogen-free seed cane are required. Seed cane from sources with a recent history of heat treatment can be used to fill the void, especially when adequate sanitary precautions have been taken to prevent spread of the pathogen and ensure that the level of the disease is minimal.

Because *L. xyli* subsp. *xyli* is readily transmitted by mechanical means, sanitation is important in preventing its spread to healthy plants (Comstock et al., 1996). Precautions can be taken to avoid transmission of the bacterium from one field to another on contaminated agricultural equipment. Sugarcane fields that are believed to be free from, or with a lower incidence of, ratoon stunting disease can be harvested first each day. Implements that have been used in diseased sugarcane fields can be cleansed and decontaminated before entering another field. Hot water, steam, flaming, or chemicals can be used to disinfect implements (Gillaspie and Davis, 1992).

Chemical Control
The application of ammonium sulfate to sugarcane crops resulted in 22.89% reduction in disease caused by *L. xyli* subsp. *xyli* and yield increases of 29.09 and 2.91 tons/ha, in cane and white sugar, respectively (Gul and Hassan, 1995).

5.1.5 BACTERIAL MOTTLE OF SUGARCANE

Pathogen: *Pectobacterium chrysanthemi* (Burkholder et al., 1953). Brenner et al. (1973) emend. Hauben et al. (1998).

Symptoms
Bacterial mottle initially produces creamy-white regular stripes, 1–2 mm wide, from the base of leaves toward the leaf tip running parallel to the vascular bundles. The streaks often cease at irregular intervals along the leaf but can extend along the full length of the leaf. As the leaf ages, the streaks develop rusty-red areas. The disease progresses from this initial infection into a systemic stage with varying degrees of chlorotic mottling or complete chlorosis. Chlorotic leaves develop rusty-red flecks and streaks as they age. During warm, humid weather, white globules of bacterial

exudate form on the underside of the leaf and water that accumulates in the spindles and leaves turns milky with bacterial exudate (Steindl, 1964).

Infected shoots are severely stunted and the leaves wither and die prematurely. Diseased plants have excessive tillering and the tillers show the characteristic mottling. Infected stalks often produce stem galls and multiple buds. Internally diseased stalks show brown necrotic areas around the nodes and growing points. These necrotic areas can occasionally form small cavities filled with gummy substance.

Geographical Distribution
Australia.

Epidemiology
Flooding is the main factor associated with bacterial mottle. The disease is more prevalent in years of heavy flooding. Usually the disease only affects a small number of plants in a field and does not spread rapidly.

Economic Importance
Bacterial mottle does not cause significant yield losses because of the restricted spread of the disease.

Control Measures
No active control measures are practiced for bacterial mottle. Some varieties are resistant to the disease. Destruction of infected plants and grasses around sugarcane fields could limit spread, but these practices have not been justified.

5.1.6 MOTTLED STRIPE OF SUGARCANE

Pathogen: *Herbaspirillum rubrisubalbicans* (Christopher and Edgerton, 1930; Baldani et al., 1996)

Synonyms: *Bacterium rubrisubalbicans* (Christopher and Edgerton; Burgvits, 1935)
Phytomonas rubrisubalbicans (Christopher and Edgerton, 1930)
Xanthomonas rubrisubalbicans (Christopher and Edgerton; Savulescu, 1947)
Pseudomonas rubrisubalbicans (Christopher and Edgerton, 1930; Krasil'nikov, 1949)

Common Names: Mottled stripe of sugarcane

Symptoms
Stripe disease primarily affects the leaf blade. Stripes are predominantly red in color, though frequently white areas or white margins occur. This difference in color gives the appearance of red on a white background. Stripes run parallel to the leaf veins and range in length from very short to a meter or more. The width of the stripe is usually 1–4 mm (Christopher and Edgerton, 1930). One to many stripes may occur on the same leaf. When many stripes occur, they sometimes coalesce, forming mottled red and white bands (Christopher and Edgerton, 1930).

Host
S. officinarum (sugarcane), *So. bicolor* (sorghum), *So. halepense* (Johnson grass), *Z. mays* (maize).

Geographical Distribution
Asia: China—Guangdong, Hainan; Indonesia, Japan, Kyushu, Ryukyu Archipelago, Sri Lanka, Thailand.

Africa: Angola, Benin, Burundi, Central African Republic, Cote d'Ivoire, Madagascar, Malawi, Mauritius, Nigeria, Reunion, Tanzania, Togo.

North America: the United States—Florida, Louisiana, Texas.

Central America and Caribbean: Barbados, Cuba, Guadeloupe, Jamaica, Martinique, Nicaragua, Panama, Puerto Rico.

South America: Brazil—Bahia, Mato Grosso, Minas Gerais, Rio de Janeiro; Colombia, Peru, Venezuela.

Oceania: Australia—New South Wales, Queensland; Fiji, New Zealand.

Economic Impact

Mottled leaf stripe is commonly found on sugarcane during summer months but seldom causes sufficient damage to leaf areas that causes any crop loss (Steindl and Edgerton, 1964). There is no evidence that this is a significant disease of sugarcane.

Control Measures

Destruction of infected plants and grasses around sugarcane field could limit spread, but no control measures are practical for mottle stripe.

5.2   TOBACCO

5.2.1   WILDFIRE OF TOBACCO

Pathogen: *Pseudomonas syringae* pv. *tabaci* (Wolf & Foster) Young et al.

Synonyms: *Bacterium tabacum* Wolf & Foster
　　　　　　Phytomonas angulata (Fromme & Murray) Stapp
　　　　　　Pseudomonas angulata (Fromme & Murray) Holland
　　　　　　Pseudomonas tabaci (Wolf & Foster) Stevens
　　　　　　Xanthomonas tabaci

Common names: Angular leaf spot of tobacco, black fire of tobacco, wildfire of tobacco

Wildfire and angular leaf spot can affect tobacco in both the seedbeds/float trays and the field, although wildfire tends to be more of a problem in the seedbed and angular leaf spot in the field. Wildfire and angular leaf spot are not major problems in many tobacco-producing areas, such as the United States, Brazil, and Europe. In Africa, they are diseases of major importance that can cause devastating losses, especially in wet seasons. The bacteria that cause wildfire and angular leaf spot are identical in all respects except that the wildfire bacteria produce a toxin and the angular bacteria do not. Wildfire is therefore caused by the "tox+" strain and angular leaf spot by the "tox–" strain.

Symptoms

The symptoms of the tox+ (toxin-producing) and tox– (non-toxin-producing) forms of this disease are quite different.

Wildfire

(tox+) is characterized by a small brown or black water-soaked lesion, surrounded by a broad chlorotic halo (Photo 5.2). The lesions increase in diameter and may coalesce until the diseased tissue eventually falls out leaving ragged holes. Wildfire can be systemic in seedlings, causing distortion of the apical bud and leaves.

The Angular

(tox–) lesion is brown, dark brown, or black, much larger than the wildfire lesion, has little or no chlorotic halo, and has angular margins because the lesion is confined by the lateral veins (Photo 5.3). In Africa, both diseases tend to be more severe at the top of the plant.

Host

Tobacco, soybean.

Geographical Distribution

Europe: Armenia, Austria, Azerbaijan, Belgium, Bulgaria, France, Georgia, Germany, Greece, Hungary, Italy, Poland, Romania, Russia, Serbia, Spain, Switzerland, Turkey, Ukraine.

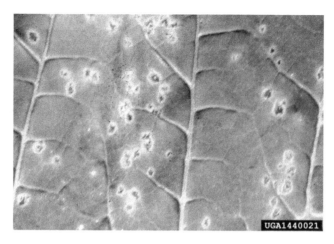

PHOTO 5.2 Symptoms of wildfire on tobacco leaf. (Courtesy of R.J. Reynolds Tobacco Company Slide Set, R.J. Reynolds Tobacco Company, Winston-Salem, NC, Bugwood.org.)

PHOTO 5.3 Angular leaf spot symptoms of wildfire disease on tobacco leaf. (Courtesy of R.J. Reynolds Tobacco Company Slide Set, R.J. Reynolds Tobacco Company, Winston-Salem, NC, Bugwood.org.)

Asia: China, India, Indonesia, Iran, Japan, Kazakhstan, Korea Republic, Lebanon, the Philippines, Taiwan, Vietnam.

Africa: Congo, Kenya, Malawi, Mauritius, Morocco, Mozambique, South Africa, Tanzania, Uganda, Zaire, Zambia, Zimbabwe.

America: Argentina, Brazil, Canada, Colombia, Dominican Republic, Mexico, Puerto Rico, the United States, Venezuela.

Oceania: Australia, New Zealand.

Disease Cycle

The bacterium overwinters in plant debris in the soil, in dried diseased leaves, on seed from infected seed capsules, and on contaminated seedbed covers. From these, the bacteria are carried to the leaves by rain splashes or by wind. They may also be spread during handling of the plants.

The bacteria are spread in wind-driven water droplets, from leaf to leaf and plant to plant within the field, from field to field, and from infected weed hosts or tobacco regrowth. Driving rains and sandblasting winds exacerbate the problem considerably. These diseases can also be seed transmitted. Tobacco regrowth and debris from infected plants should always be destroyed at the end of the season, as they are sources of inoculum to infect overwintering weed hosts. In the semi-tropical areas where these diseases are a problem, winters are seldom cold enough to kill overwintering weeds and tobacco regrowth. Wildfire and angular leaf spots are favored by cloudy wet weather.

High humidity or a film of moisture must be present for infections to occur and for the development of epidemics. Water-soaked areas forming in the leaves during long rainy period or from wind-blown rain are excellent infection courts for the bacterium and result in extensive lesions within 2–3 days. Bacteria enter the leaf through stomata, hydathodes, and wounds. Certain insects such as flea beetles, aphids, and white flies also act as vectors of this pathogen.

Once inside the leaf, the bacteria multiply intercellularly at a rapid rate and secrete the wildfire toxin. The toxin spreads radially from the point of infection and results in the formation of the chlorotic halo, which consists of a zone of cells free of bacteria surrounding the bacteria-containing spot. Variants of the bacterium that do not produce tabtoxin produce a similar disease without halos, known as angular leaf spot or wildfire.

In favorable weather, bacteria continue to spread intercellularly and, through the toxin and enzymes they secrete, cause the breakdown, collapse, and death of the parenchyma cells in which the tissues disintegrate further. Bacteria in the disintegrated areas of the leaf fall to the ground or are carried by air currents and splashing rain to other plants.

Economic Impact

Wildfire of tobacco occurs worldwide. In some regions, it occurs year after year and is very destructive, whereas in other regions, it appears sporadically. In addition to tobacco, the pathogen *P. syringae* pv. *tabaci* also affects soybean.

Wildfire causes losses in both the seedbed and field. Affected seedlings may be killed. In tobacco plants already in the field, wildfire causes large, irregular, dead areas on the leaves, which wither and fall off, making the leaves commercially worthless.

Disease Management

Rotation

Disease spread is reduced by planting earlier fields downwind of later planted fields; the earlier planted fields often serve as an inoculum source. These diseases are generally worse in intensively used fields and can be minimized by suitable rotations.

Removal of Alternate Hosts

Many *solanaceous* weeds are hosts of this pathogen. Examples are Apple of Peru (*Nicandra physalodes*) and Jimson weed/stinkblaar (*Datura stramonium*). Such weeds should be removed from the proximity of the fields and especially seedbeds/greenhouses. This is particularly important in areas that do not have killing winter frosts, where weeds overwinter.

Resistant Varieties

Resistance to wildfire (race 0), derived from *Nicotiana longiflora* via Burley 21, is monogenic, complete, and fully dominant. This resistance has proved very durable in most parts of the world, but in Africa, it broke down in the 1970s, with the emergence of race 1. *Nicotiana rustica*–derived resistance to races 0 and 1 of wildfire and angular leaf spot is also monogenic, complete, and fully dominant. Race 1–resistant varieties were released in Zimbabwe in the early 1990s, but this resistance broke down in a relatively short time, with the emergence of race 2 of both wildfire and angular leaf spot in the late 1990s. No resistance to race 2 has been identified. Polygenic resistance is generally low, but some of the newer multi-resistant Zimbabwean varieties have some polygenic resistance to race 2.

Sanitation

All seedbed tools, particularly those used for clipping/mowing, should be regularly sterilized with bleach or a copper-based compound. Disease spread is minimized if clean fields are reaped before infected ones. Plant debris should be plowed under at the end of the season, and all regrowth destroyed. Any exposed infected plant material may serve as a source of inoculum for next season. To be effective, this should be done by all growers in an area.

Maintenance of Fertility and pH

These diseases are favored by excessive fertility, particularly with high N and low K fertilization. The use of excessive amounts of lime, which interferes with K uptake, can also increase disease severity.

Chemical Control

Streptomycin (e.g., Agri-Mycin 17, Ag streptomycin) is registered for use on tobacco in some countries. However, because of the potential for developing antibiotic resistance and the human/animal health issues, it should be used with extreme caution. Streptomycin is not registered in Africa.

Seedbeds/Float Beds

Fumigation of seedbeds will usually eliminate initial inoculum in the seedbeds. Seedlings should be preventatively sprayed with a combination of a copper-based compound (e.g., Kocide 101, copper oxychloride) and the systemic acquired resistance compound, acibenzolar-S-methyl (A-S-M; e.g., Actigard, Bion), if locally registered. Note that in some countries, A-S-M is not registered for seedlings. As always, use the registered rate, but with A-S-M, it is particularly important not to exceed the recommended rate because this product can be phytotoxic. Young plants are most vulnerable, and burley is more sensitive than other tobacco types. A-S-M should be used on burley seedlings with caution and only when necessary, as it can cause stunting.

Field

Copper sprays are not registered for field use because of residues; copper sprays applied in the field will result in unacceptable residues. The only chemical that can be used to control these diseases in the field is A-S-M (if locally registered). Field sprays are more effective when used in combination with seedbed sprays. A-S-M is most effective when used as a preventative spray, but because of the cost, many growers only use it when symptoms appear.

Scouting

Both the seedbeds/float beds and the field should be scouted regularly. Particular attention should be paid to low-lying areas.

> *Seedbeds/float beds:* Remove and destroy any infected seedlings and all others within 1 m of an infected plant. Drench the surrounding area with disinfectant (e.g., bleach).
> *Field:* A-S-M applications should commence as soon as symptoms are observed if it is not already being used as a preventative measure.

Others

Wildfire and angular leaf spot can be transmitted by seed, so seed should not be collected from infected plants. Seed treatment with silver nitrate should be a routine precaution and is required in many countries where these diseases are a problem. Use of certified seed will minimize the chances of starting a crop with infected seed. Disease severity is increased by any practice resulting in thick, heavy leaves, such as low topping and excessive N, and by any practice that can injure the leaves, such as mowing the grass surrounding tobacco fields.

Integrated Disease Management

An integrated approach to the management and control of wildfire and angular leaf spot includes the following:

- Seed treatment with silver nitrate, ideally as part of the seed certification requirement.
- Use of certified tobacco seed.
- Rotation of seedbed sites.
- Proper fumigation of the seedbed areas.

- Preventative seedbed/float bed sprays, correctly applied, of a combination of a copper-based compound and A-S-M (if locally registered). *Note:* A-S-M should be used on burley seedlings with caution and only when necessary.
- Scouting seedbeds and removing all seedlings within 1 m of an infected plant.
- Eradication of alternate host weeds, particularly those near the seedbeds/float beds.
- Sterilization of seedbed tools.
- Site selection to avoid later planted crops downwind of earlier planted ones.
- Correct fertilization and pH; avoiding excessive N, low K, and high pH.
- Minimizing of leaf injury; avoiding leaf breakage and not mowing too close to the field.
- Correct topping, particularly avoiding low topping.
- Regular scouting, particularly under conducive conditions (wet, cloudy, after rain).
- Field sprays of A-S-M, preventatively or when indicated by scouting.
- Minimizing spread by reaping clean fields before infected ones.
- Destruction of all plant residue and regrowth at the end of the season.

5.2.2 Black Rust of Tobacco

Pathogen: *Pseudomonas pseudozoogloeae*

Synonyms: *P. syringae* pv. *tabaci*
         Bacterium pseudozoogloae

Common names: Black Rust

Symptoms
Symptoms occur on diseased lower leaves as plants approach maturity. The dark green spots that appear initially become necrotic and dark brown in the center and have zonate concentric rings. A dark green margin that surrounds each spot eventually reaches 1–2 cm in diameter. The margins remain dark green after diseased leaves have dried.

Host
Tobacco.

Geographical Distribution
The pathogen is geographically distributed in Indonesia and possibly Japan and Italy.

Disease Cycle
Bacterium pseudozoogloae survives within infested residue and is disseminated by splashing rain.

Control Measures
No management other than to remove and burn the lower diseased leaves.

5.2.3 Blackleg and Hallow Stalk of Tobacco

Pathogen: *Erwinia carotovora* subsp. *carotovora* (Jones) Bergey et al.

Synonyms: *Erwinia carotovora* subsp. *atroseptica* (van Hall) Dye

Symptoms
Blackleg occurs in irregular circular areas approximately 1 m in diameter within plant beds. Rotted petioles and stems become black and eventually spilt open or rot off (Photo 5.4).

Bacterial hollow stalk is first noticed when leaves begin to drop or turn yellow and hang down the stalk. A watery soft rot develops at the base of the affected leaves and often extends up the midrib. The disease may begin at any point of injury, but it often starts at injuries caused by topping or suckering. The pith in the center of the stalk is destroyed, leaving a hollow shell. As the

PHOTO 5.4   Tobacco plant with typical blackleg and hallow stalk of tobacco symptoms. (Courtesy of R.J. Reynolds Tobacco Company Slide Set, R.J. Reynolds Tobacco Company, Winston-Salem, NC, Bugwood.org.)

disease moves down the stalk, the leaves droop. Later they drop off, leaving a bare stalk. Hollow stalk usually appears after topping or suckering and may continue to develop after the tobacco is hung in the barn, causing barn rot. Related bacteria may also cause a soft rot of lower leaves during rainy weather.

Hollow stalk mainly occurs as a slimy, stinking rot of the stalk pith and leaf midribs. This disease is very sporadic in occurrence from season to season and farm to farm. It can cause extensive losses during wet seasons, both in the field and during harvesting and housing. In wet weather, this bacterium usually also becomes established as a saprophyte on tobacco plants and some systemic development occurs in rapidly growing tobacco that is under excessive nitrogen fertilization.

Disease Cycle

Cool and heavy day soils with a neutral pH favor survival of causal bacteria. *E. carotovora* subsp. *carotovora* is disseminated to fields on infected transplants and by any means that moves soil. Bacteria may also be spread within a planting bed by maggot flies of *Hylemya* spp. Host plants become infected by leaves touching infested soil. Bacteria then spread through the petioles and into the stems. Blackleg is most severe during damp, cloudy weather over a wide range of temperatures.

Control Measures

The pathogen requires a wet wound to become established. Consequently, where possible, avoiding wounding of the foliage and stems in wet seasons becomes important. Wounds that were acceptable in a dry season can lead to serious losses in wet seasons. Since wounding is required in topping operations, it should be done in a way that the wound site dries and heals quickly. Consider the following points:

- Taking steps to reduce the amount of wounding that occurs during topping is more important in wet than dry seasons.
- Topping at the recommended stages of growth, rather than waiting until the plants are fully flowering, reduces wounding.
- Plants with fully expanded flower heads require special attention because breaking out such tops usually leaves a wound that will hold water. Tops of plants in full flower should be removed with a knife, with the cut made on an angle sloping toward the sunny side of the plant to aid drying. Knives used for this should be disinfested often by dipping them in either 10% bleach or 70% alcohol.

- Carefully remove any suckers that might be present at topping. Those left on the plant and killed by sucker control chemicals become ideal sites for hollow stalk to begin in wet seasons. When this happens, the rot often takes on a "candy-striping effect," as the bacteria run down the stalk from the rotting sucker and pool in the leaf/stem intersections.
- Topping and suckering during damp or cloudy weather increase hollow stalk activity and therefore should be avoided. Workers rubbing soil on their hands to remove excess gum increases the spread of the hollow stalk pathogen.
- If plants with hollow stalk are encountered during topping, they should not be touched by those doing the topping to reduce the chances of spread.
- Fields under excessive nitrogen fertilization are especially prone to hollow stalk, because the wounds do not heal, so extra care is needed in handling such sites.
- Excessive rates of sucker control chemicals can damage leaves and greatly increase hollow stalk. This is often evident near the ends of rows where ground speeds are lower and higher rates are applied.
- Do not house plants with hollow stalk. Yes, you will need to train the tobacco cutters to leave them standing and provide some type of incentive for them to do this, because the standard incentive to them supports cutting the plant.
- Harvesting immature tobacco can lead to increased bacterial soft rot in the curing barns. Be especially careful not to get these in the upper rails of tobacco where the slimy rotting leaves can drip and fall into lower rails of tobacco.

5.2.4   GRANVILLE WILT OF TOBACCO

Pathogen: *Ralstonia solanacearum*

Synonyms: *Burkholderia solanacearum, Pseudomonas solanacearum*

Symptoms
The first symptom is a wilting on one side of the plant. As the disease progresses, the entire plant wilts and death generally follows. When death does not occur, plants are usually stunted and leaves may be twisted and otherwise distorted (Photo 5.5). The stalk usually turns black, especially at the ground level. At this stage, Granville wilt may be easily confused with other diseases such as black shank. Dark streaks can be seen extending up the plant just beneath the outer bark. Infection may not be noticed immediately because wilting symptoms may not appear until the plant undergoes moisture stress. It is not unusual to observe symptoms several weeks after initial infection.

Geographical Distribution
Bacterial wilt occurs mainly in tropical, semitropical, and warm temperate zones. It is particularly severe in many Asian countries (China, Japan, Malaysia, Pakistan, Thailand, Vietnam), America (the United States, Brazil), and Africa (including South Africa). *R. solanacearum* is quite rare in Europe. It has been reported in Hungary, Italy, and Yugoslavia. It has not been reported currently on tobacco in France. In the countries where *R. solanacearum* is endemic, it is a feared disease that can cause serious damage on tobacco and other crops.

Disease Cycle
The bacterium is spread by anything that moves infested soil or water. Major means of spread include water, infected transplants, and farm vehicles moving from field to field. The motile bacteria gain entry into the plant through natural openings or wounds. Since roots often "wound themselves" as they grow, or are wounded during transplanting, the Granville wilt bacteria have no difficulty in gaining entry into the plant. More extensive root wounding caused by nematodes or root pruning

PHOTO 5.5 Tobacco plant with typical Granville wilt symptoms. (Courtesy of R.J. Reynolds Tobacco Company Slide Set, R.J. Reynolds Tobacco Company, Winston-Salem, NC, Bugwood.org.)

during cultivation provide more paths of entry for the bacteria. High populations (greater than 250,000 bacteria per gram of soil) are usually necessary for infection to occur. The bacteria may also be spread during mechanical topping and harvesting. These bacteria are favored by relatively high soil temperatures and adequate to high moisture levels in the soil. Poor soil drainage and wet, warm growing seasons favor Granville wilt.

Economic Impact

Granville wilt continues to be one of the most destructive diseases of tobacco in North Carolina. It causes losses from over 1% to 2% of the entire tobacco crop, costing growers from $10 to $15 million annually. Granville wilt was first observed in 1880 on a few farms in the northern areas of North Carolina's Middle Belt. During the next 30 years, losses from this disease increased on farms in Granville, Vance, Wake, and Durham counties to the point where it was causing 25%–100% field losses. During that period this disease caused banks to close, farms to be sold, and towns to decline. Today it not only plagues the Middle Belt but is the most chronic disease problem in eastern North Carolina and in the Border Belt.

Pathogen

R. solanacearum is a highly polyphagous bacterium. It includes several strains that differ namely in their ability to metabolize various sugars and denitrify nitrates, as well as in their distinct host range. Based on the latter, it seems that *R. solanacearum* can be separated into five physiological races:

Race 1, with many strains isolated from different regions and hosts
Race 2, adapted to the triploid banana trees in the tropics
Race 3, preferentially attacking the potato, tomato, and eggplant in temperate climates
Race 4, associated with ginger
Race 5, affecting the mulberry tree

There is little information about the "tobacco" strains belonging to these physiological races. It has been reported that some "tobacco" strains do not seem to be able to attack tomato and eggplant, that "potato" strains have little affect on tobacco. "Banana" strains seem unable to infect tobacco.

Diagnostic Method in the Field

A simple diagnostic test for Granville wilt can be done on-farm. When an infected stem segment is suspended in a glass of clear water for a few minutes, bacterial streaming occurs. The bacterial streaming appears as white ooze or a smokey stream, which originates from the cut end of the stem, where the dark streaks are observed under the bark, and slowly moves out into the water.

Disease Management

Crop Rotation

Crop rotation must be the basis on which Granville wilt management programs are established. This practice is perhaps the most essential thing that growers can do to minimize losses due to Granville wilt. In fact, without appropriate crop rotation, it is not possible to manage this disease successfully where the infestation level ranges from moderate to high. Therefore, crop rotation must be the basis on which Granville wilt management programs are established. Crop rotation is effective because the Granville wilt bacteria live in the soil and are not well adapted to survival in the absence of susceptible plant tissue. Thus, their populations decline if a suitable plant such as tobacco is absent for even 1 year. As is true with any other soilborne pathogen, the longer the rotation, the more efficient the control. However, planting a nonhost crop (soybeans, fescue, corn, cotton, milo) just for 1 year will usually significantly reduce the disease loss in the following tobacco crop. Integrating other management practices, such as improved drainage, avoiding late or deep cultivations, stalk and root destruction, and the use of multipurpose fumigants where disease occurred in past years, is better than relying on any one or two practices.

Granville wilt is most damaging in fields where tobacco was grown the previous year, in wet areas in a field, and in years where soil temperatures are normal to above normal. Other plants the bacteria can infect include tomatoes, Irish potatoes, pepper, eggplant, peanuts, and weeds.

Debris Destruction

Roots and stalks from the previous crop should be destroyed as soon as possible after harvest. The decay of old plant residue through stalk and root destruction soon after harvest decreases the number of bacteria present in the soil.

Resistant Varieties

Varieties are available that carry varying levels of resistance to Granville wilt. None of these varieties is immune to this disease and some losses might be expected in severely infested areas with the use of any variety. Nevertheless, growers have an opportunity to select those varieties that will afford them, in most cases, good protection when used in combination with other disease control practices such as stalk and root destruction and crop rotation.

Chemical Control

The fumigants Chlor-O-Pic 100, Telone C-17, Telone C-35, and Pic+ may help control Granville wilt if used in combination with other cultural control practices. All require a 3-week waiting period between time of application and transplanting. Always follow label instructions.

REFERENCES

Antoine, R. and M. Pérombelon. 1965. Cane diseases: 2. Gumming disease. Annual Report 1964, pp. 51–56. Mauritius Sugar Industry Research Institute, Réduit, Mauritius.

Autrey, L.J.C., Dhayan, S., and S. Sullivan. 1986. Effect of race 3 of gumming disease pathogen on growth and yield in two sugarcane varieties. *Proc. Int. Soc. Sugar Cane Technol.*, 19, 420–428.

Bailey, R.A. and G.R. Bechet. 1986. Effect of ratoon stunting disease on the yield and components of yield of sugarcane under rainfed conditions. *Proc. South Afr. Sugar Technol. Assoc.*, 60, 143–147.

Bailey, R.A. and P.H. Fox. 1984. A large-scale diagnostic service for ratoon stunting disease of sugarcane. *Proc. Annu. Congr. South Afr. Sugar Technol. Assoc.*, 58, 204–210.

Baldani, J.I., Pot, B., Kirchhof, G. et al. 1996. Emended description of *Herbaspirillum*: Inclusion of "*Pseudomonas*" *rubrisubalbicans*, a mild plant pathogen, as *Herbaspirillum rubrisubalbicans* comb. nov.; and classification of a group of clinical isolates (EF group 1) as *Herbaspirillum* species 3. *Int. J. Syst. Bacteriol.*, 46, 802–810.

Béchet, G.R. 1976. Ratoon stunting disease and rapid diagnostic techniques. In: *Proceedings of the 50th Annual Congress*, South African Sugar Technologists Association, Mount Edgecombe, South Africa, pp. 65–68.

Benda, G.T.A. 1972. Hot-water treatment for mosaic and ratoon stunting disease control. *Sugar J.*, 34, 32–39.

Benda, G.T.A. 1978. Increased survival of young seedcane after hot-water treatment for RSD control. *Sugar Bull.*, 56(19), 7–8, 13–14.

Benda, G.T.A. and C. Ricaud. 1978. The use of hot-water treatment for sugarcane disease control. *Proc. Int. Soc. Sugar Cane Technol.*, 16, 483–496.

Bergey, D.H., Harrison, F.C., Breed, R.S. et al. 1923. *Bergey's Manual of Determinative Bacteriology*, 1st ed. The Williams and Wilkins Co., Baltimore, MD. pp. 1–442.

Brenner, D.J., Steigerwalt, A.G., Miklos, G.V. et al. 1973. Deoxyribonucleic acid relatedness among *Erwinia* and other Enterobacteriaceae: The soft-rot organisms (Genus *Pectobacterium* Waldee). *Int. J. Syst. Evol. Microbiol.*, 23(3), 205–216. doi:10.1099/00207713-23-3-205.

Christopher, W.N. and C.W. Edgerton. 1930. Bacterial stripe diseases of sugar-cane in Louisiana. *J. Agric. Res.*, 41, 259–267.

Cobb, N.A. 1893. Plant diseases and their remedies. Diseases of the sugarcane. *Agric. Gazette N.S.W*, 4, 1.

Cobb, N.A. 1894. Plant diseases and their remedies. *Agric. Gazette N.S.W. for 1893*, 4, 777–798.

Cochereau, P. and A. Jean Bart. 1989. Etatphytosanitaire des cultures de canne à sucre à Marie-Galante (Guadeloupe). *Bull. Agronomique des Antilles et de la Guyane*, 9, 13–19.

Comstock, J.C., Shine, J.M.J., Davis, M.J. et al. 1996. Relationship between resistance to *Clavibacter xyli* subsp. *xyli* colonization in sugarcane and spread of ratoon stunting disease in the field. *Plant Dis.*, 80(6), 704–708, 23.

Damann, K.E. Jr. and G.T.A. Benda. 1983. Evaluation of commercial heat-treatment methods for control of ratoon stunting disease of sugarcane. *Plant Dis.*, 67(9), 966–967.

Davis, M.J., Gillaspie, A.G. Jr., Vidaver, A.K. et al. 1984. *Clavibacter*: A new genus containing some phytopathogenic coryneform bacteria, Including *Clavibacter xyli* subsp. *xyli* sp. nov., subsp. nov. and *Clavibacter xyli* subsp. *cynodontis* subsp. nov. pathogens that cause ratoon stunting disease of sugarcane and bermudagrass stunting disease? *Int. J. Syst. Bacteriol.*, 34(2), 107–117.

Dean, J.L. and M.J. Davis. 1990. Losses caused by ratoon stunting disease of sugarcane in Florida. *J. Am. Soc. Sugar Cane Technol.*, 10, 66–72.

Dowson, W.J. 1939. On the systematic position and generic names of the Gram negative bacterial plant pathogens. *Zentralblatt fur Bakteriologie. Parasitenkunde und Infektionskrankheiten*, 2(100), 177–193.

Dye, D.W. 1978. Genus IX *Xanthomonas* Dowson 1939. In: Young, J.M., Dye, D.W., Bradbury, J.F., Panagopoulos, C.G., and C.F. Robbs (eds.), A proposed nomenclature and classification for plant pathogenic bacteria. *N.Z. J. Agric. Res.*, 21, 153–177.

Early, M.P. 1973. Ratoon stunting disease of sugar-cane in Kenya. *East Afr. Agric. For. J.*, 39(1), 57–60.

Egan, B.T. 1970. RSD in North Queensland. *Proc. Queensland Soc. Sugar Cane Technol.*, 37, 221–224.

Evtushenko, L.I., Dorofeeva, L.V., Subbotin, S.A. et al. 2000. *Leifsonia poae* gen. nov., sp. nov., isolated from nematode galls on Poa annua, and reclassification of 'Corynebacterium aquaticum' Leifson 1962 as *Leifsonia aquatica* (ex Leifson 1962) gen. nov., nom. rev., comb. nov. and *Clavibacter xyli* Davis *et al.* 1984 with two subspecies as *Leifsonia xyli* (Davis *et al.* 1984) gen. nov., comb. nov. *Int. J. Syst. Evol. Microbiol.*, 50(1), 371–380; 47.

Frison, E.A. and C.A.J. Putte. 1993. Technical guidelines for the safe movement of sugarcane germplasm. FAO/IBPGR, Rome, Italy.

Gillaspie, A.G. Jr. and M.J. Davis. 1992. Ratoon stunting of sugarcane. In: Mukhopadhyay, A.N., Kamar, J., Chaube, H.S., and U.S. Singh (eds.), *Plant Diseases of International Importance. Diseases of Sugar, Forest, and Plantation Crops*, Vol. 4. Prentice Hall, Upper Saddle River, NJ, pp. 41–61.

Girard, J.C. and J.P. Péros. 1987. Uneméthoded'évaluation en serre de la résistance de la canne à sucre à la gommose. *L'Agron. Trop.*, 42, 126–130.

Gul, F. and S. Hassan. 1995. Further studies on the control of ratoon stunting disease of sugarcane in the N.W.F.P. *Pak. J. Phytopathol.*, 7(2), 163–165; 8.

Hauben, L., Moore, E.R.B., Vauterin, L. et al. 1998. Phylogenetic position of phytopathogens within the Enterobacteriaceae. *Syst. Appl. Microbiol.*, 21, 384–397.

Hayward, A.C. 1962. Studies on bacterial pathogens of sugar cane. *Mauritius Sugar Industry Research Institute Occasional* Paper No. 13: 13–27.

Hughes, C.G. 1974. The economic importance of ratoon stunting disease. In: Dick, J. and D.J. Collingwood (eds.), *Proceedings of the XV Congress of the International Society of Sugar Cane Technologists*, Durban, South Africa, June 12–29, 1974. ISSCT, Durban, South Africa, Vol. 1, pp. 213–217.

Klett, P. and P. Rott. 1994. Inoculum sources for the spread of leaf scald disease of sugarcane caused by *Xanthomonas albilineans* in Guadeloupe. *J. Phytopathol.*, 142, 283–291.

Koike, H. 1974. Interaction between diseases on sugarcane: Sugarcane mosaic and ratoon stunting disease. In: *Proceedings XV Congress, International Society of Sugar Cane Technologists*, Durban, South Africa, pp. 258–265.

Liu, L.J., Ramirez Oliveras, G., Serapion, J.L., and C.L. Gonzalez Molina. 1979. Further developments in the study of the ratoon stunting disease of sugarcane in Puerto Rico. *J. Agric. Univ. Puerto Rico*, 63(2), 146–151.

Mayeux, M.M., Cochran B.J., and J.R. Steib. 1979. An aerated steam system for controlling ratoon stunting disease. *Trans. Am. Soc. Agric. Eng.*, 22(3), 653–656.

Migula, W. 1900. *System der Bakterien*, Vol. 2. Gustav Fischer, Jena, Germany.

North, D S (1935). The gumming disease of the sugarcane, its dissemination and control. *Agric. Rep. (Technol.)*, 10., CSR Ltd., Sydney, 149 pp.

Péros, J.P. 1988. Variability in colony type and pathogenicity of the causal agent of sugarcane gumming *Xanthomonas campestris* pv. *vasculorum* (Cobb) Dye. *Z. Pflanzenkr. Pflanzensch.*, 95, 591–598.

Peros, J.P. and H. Lombard. 1992. In vitro evaluation of sugarcane resistance to gumming disease and of *Xanthomonas campestris* pv. *vasculorum* aggressiveness. *Plant Cell Tissue Organ Cult.*, 29, 145–151.

Qhobela, M. and L.E. Claflin. 1992. Eastern and southern African strains of *Xanthomonas campestris* pv. *vasculorum* are distinguishable by restriction fragment length polymorphism of DNA and polyacrylamide gel electrophoresis of membrane proteins. *Plant Pathol.*, 41(2), 113–121.

Ricaud, C. 1969. Investigation on the systemic infection of gumming disease. *Proc. Int. Soc. Sugar Cane Technol.*, 13, 1159–1169.

Ricaud, C. and L.J.C. Autrey. 1989. Gumming disease. In: Ricaud, C., Egan, B.T., Gillaspie, A.G. Jr., and C.G. Hughes (eds.), *Diseases of Sugarcane—Major Diseases*. Elsevier, Amsterdam, the Netherlands, pp. 21–38.

Ricaud, C. and C.C. Ryan. 1989. Leaf scald. In: Ricaud, C., Egan, B.T., Gillaspie, A.G., and C.G. Hughes (eds.), *Diseases of Sugarcane—Major Diseases*. Elsevier, Amsterdam, the Netherlands, pp. 39–58.

Ricaud, C. and S. Sullivan. 1974. Further evidence of population shift in the gumming disease pathogen in Mauritius. *Proc. Int. Soc. Sugar Cane Technol.*, 15, 204–209.

Rossler, L.A. 1974. The effects of ratoon stunting disease on three sugarcane varieties under different irrigation regimes. In: Dick, J. and D.J. Collingwood (eds.), *Proceedings of the XV Congress of the International Society of Sugar Cane Technologists*, Durban, South Africa, June 12–29, 1974, Vol. 1, pp. 250–257.

Rott, P., Soupa, D., Brunet, Y., Feldmann, P., and P. Letourmy. 1995. Leaf scald (*Xanthomonas albilineans*) incidence and its effect on yield in seven sugarcane cultivars in Guadeloupe. *Plant Pathol.*, 44(6), 1075–1084.

Savulescu, T. 1947. Contribution à la classification des bactériacéesphytopathogènes. *Analele Academiei Romane Series III*, 22, 135–160.

Shukla, U.S., Ram, R.S., and R.C. Tripathi. 1974. Effects of moist hot air treatments in controlling GSD and RSD. Annual report of the Indian Institute for Sugarcane Research. Indian Institute for Sugarcane Research, Lucknow, India, pp. 72–74.

Singh, G.R. 1974. Studies on yield of cane and juice quality due to ratoon stunting disease of sugarcane in India. *Indian Sugar*, 24(7), 623–624, 627–629.

Smith, E.F. 1901. The cultural characters of *Pseudomonas hyacinth, Ps. campestris, Ps. phaseoli* and *Ps. stewarti*—Four one-flagellate yellow bacteria parasitic on plants. *U.S.D.A Div. Veg. Phys. Pathol. Bull.*, 28, 1–153.

Steib, R.J. and S.J.P. Chilton. 1967. Interrelationship studies of mosaic and ratoon stunting disease in sugarcane in Louisiana. *Proc. Int. Soc. Sugar Cane Technol.*, 12, 1061–1071.

Steindl, D.R.L. 1961. Ratoon stunting disease. In: Martin, J.P., Abbott, E.V., and C.G. Hughes (eds.), *Sugarcane Diseases of the World*, Vol. I. Elsevier, Amsterdam, the Netherlands, pp. 433–459.

Steindl, D.R.L. 1964. Dwarf. In: Hughes, C.G., Abbott, E.V., and C.A. Wismer (eds.), *Sugar-Cane Diseases of the World*, Vol. 2. Elsevier Publishing Company, Amsterdam, the Netherlands, pp. 158–163.

Steindl, D.R.L. and C.W. Edgerton. 1964. Mottled stripe. In: Hughes, C.G., Abbott, E.V., and C.A. Wismer (eds.), *Sugar-Cane Diseases of the World*, Vol. 2. Elsevier Publishing Company, Amsterdam, the Netherlands, pp. 12–16.

Stevenson, G.C. 1965. *Genetics of Breeding of Sugar Cane*. Longmans, Green & Co. Ltd., London, U.K., 284pp.

Vauterin, L., Hoste, B., Kersters, K., and J. Swings. 1995. Reclassification of Xanthomonas. *Int. J. Syst. Bacteriol.*, 45, 472–489.

Willems, A., Goor, M., Thielemans, S. et al. 1992. Transfer of several phytopathogenic *Pseudomonas* species to *Acidovorax* as *Acidovorax avenae* subsp. *avenae* subsp. nov. comb. nov., *Acidovorax avenae* subsp. *citrulli*, *Acidovorax avenae* subsp. *cattleyae*, and *Acidovorax konjaci*. *Int. J. Syst. Bacteriol.*, 42, 107–119.

SUGGESTED READING

Anne, J. 2013. Ch.15. Wildfire, angular leaf spot. Bacterial Diseases. In: Integrated Disease Management. IPM Taskforce. University of Kentucky, Lexington, KY. www.uky.edu/Ag/Agronomy/IPM_Taskforce/completedDocs/5b-15-AngularWildfire.pdf.

Bailey, R.A. 2004. Diseases. In: *Sugarcane*, 2nd Edition. James, G. (ed). Blackwell Science, Oxford. pp. 338–346.

Birch, R.G. 2001. *Xanthomonas albilineans* and the antipathogenesis approach to disease control. *Mol. Plant Pathol.*, 2, 1–11.

Croft, R., Magarey, R., and P. Whittle. 2000. Disease management. In: Hogarth, M. and P. Allsopp (eds.), *Manual of Cane Growing*. Bureau of Sugar Experimental Stations, Indooroopilly, Brisbane, Australia, pp. 263–289.

Fors, A.L. 1978. Smut in Belize. *Sugarcane Pathol. Newsl.*, 21, 3.

Hoy, J.W. and M.P. Grisham. 1994. Sugarcane leaf scald distribution, symptomatology, and effect on yield in Louisiana. *Plant Dis.*, 78, 1083–1087.

Martin, J.P. and C.A. Wismer. 1961. Red stripe. In: *Sugarcane disease of the world* Vol. I, J.P., Abbott, E.V., and C.G. Hughes (Eds.), pp. 109–126. Amsterdam, The Netherlands, Elsevier Publishing Company.

Martin, J.P. and C.A. Wismer. 1989. Red stripe. In: Ricaud, C., Egan, B.T., Gillaspie, A.G. Jr., and C.G. Hughes (eds.), *Diseases of Sugarcane—Major Diseases*, revised by C.C. Ryan. Elsevier Science Publishers B.V., Amsterdam, the Netherlands, pp. 81–95.

Mila, M. 2011. Bacterial wilt of tobacco: Plant disease fact sheets. Retrieved June 19, 2015, from http://www.cals.ncsu.edu/plantpath/extension/clinic/fact_sheets/index.php?do=disease&id=13.

6 Bacterial Diseases of Fiber Crops

6.1 COTTON

6.1.1 Bacterial Blight of Cotton

Pathogen: *Xanthomonas axonopodis* pv. *malvacearum* (Smith, 1901; Vauterin et al., 1995)

Synonyms: *Xanthomonas malvacearum*

Common name: Angular leaf spot
Tachesangulaires du cotonnier
"Eckige Blattflecken"—Krankheit der Baumwolle

Symptoms
The bacterium is capable of infecting all the portions of the plant that are above the ground. There are four major types of symptoms: (1) seedling infection, (2) angular leaf spot of cotton, (3) blackarm of cotton, and (4) boll rot of cotton.

Seedling Infection
The earliest symptom of the disease is noticed on the cotyledons of the germinating seed. Water-soaked circular lesions appear first on the lower surface of the cotyledons. The infection soon spreads and forms angular patches. The diseased tissues turn brown, distort the shape of the cotyledons, and finally cause drying and withering of the seedling. The whole seedling may collapse and wilt in severe infections.

Leaf Spots
On leaves, spots are usually angular (Photo 6.1). These first appear on the under surface and then on the upper surface. The disease advances along the main veinlets and veins to cause finger-like lesions on the leaf blade, and then the disease is identified as "vein blight" (Photo 6.2). The leaves affected by such vein blights become crinkled and twist inward and dry. The infection may also spread through the leaf blade to the petiole, causing severe defoliation. On the surface of the affected tissues, there may be a profuse exudation of the bacterial slime that may dry and form a thin film.

Blackarm
The pathogen causes dark-brown black lesions on the stem and fruiting branches. The lesions are linear and sunken from the surface, after covering the whole circumference of the stem or branch. Under favorable climatic conditions, most of the branches and the main stem of the plant are affected by the disease. As the disease advances, the branches dry and soon the plant withers.

Boll Lesions
The bacterium affects the boll or fruit also. The first symptoms appear as water-soaked lesions on the surface. Such lesions turn dark brown and finally black. They are invariably sunken on the surface (Photo 6.3). When young balls are attacked with several such lesions, they may drop off from the plant. In severe cases, the bacterial infection spreads inside the boll or the fruit and causes severe reduction in size, quantity, and quality of the lint formed inside. As the disease advances, the seeds invaded by the bacterium suffer a reduction in their size and loss of viability.

The most prominent water-soaking symptoms of the disease were due to the production of extra-cellular polysaccharide produced by the bacterium (Borkar and Verma, 1989b). For induction of

PHOTO 6.1   Angular leaf spot symptoms on cotton leaf. (Courtesy of Jason Brock, University of Georgia, Athens, GA, Bugwood.org.)

PHOTO 6.2   Vein blight symptoms on cotton leaf. (Courtesy of Borkar and Bhosale, PPAM, Mahatma Phule Krishi Vidyapeeth, Rahuri, India.)

water-soaking reaction in cotton *cvs*, 3 mg Extracellular polymeric substances (EPS)/infiltration/spot was found necessary, and concentration below this does not induce any reaction (Borkar, 1982). The process of pathogenesis is completed within 6–8 h of bacterial infection (Borkar and Verma, 1989c); however, the water-soaking symptoms appear after 3 days of bacterial infection.

The susceptibility/resistance of the cotton cultivars is attributed to the quantum of some of the biochemicals present in the plant. The most susceptible cvs acala-44 contained 29% more protein than the most resistant cvs 101-102B, although free amino acid contents of both the cvs were same. During *Xanthomonas axonopodis* pv. *malvacearum* (*Xam*) infection, the protein content increased while free amino acid content decreased in cotton cvs (Borkar and Verma, 1988b). Similarly, the most resistant

PHOTO 6.3   Bacterial infection of cotton on bracts and bolls. (Courtesy of Tom Allen, Mississippi State University, Oktibbeha, MS, mississippi-crops.com.)

cvs contained more tannin than the susceptible and moderately resistance *cvs* (Borkar, 1989). Resistant cotton *cvs* and its intercellular fluid contained a higher amount of sugar compared to susceptible *cvs*. The inoculation of bacterial pathogen in the intercellular fluid of resistant and susceptible cotton *cvs* increased the total sugar content only in the intercellular fluid of susceptible *cvs*, thereby indicating the production of exopolysaccharide, a water-soaking inducing factor (Borkar, 1989a).

Total and dihydroxyphenol contents in the leaves of susceptible *cvs* and its intercellular fluid were low compared to resistant *cvs* and its intercellular fluid, and during Xam–cotton interaction, these increased in resistant *cvs* and its intercellular fluid and decreased in the susceptible *cvs* and its intercellular fluid. A decrease in sugar and an increase in phenol level in resistant *cvs* during Xam–cotton interaction were found responsible for the inhibition of bacterial growth and its multiplication (Borkar and Verma, 1989).

Different types of incompatible reactions induced during Xam–cotton interaction were related to the intensity of dopachrome (browning reaction inducing product of phenol–phenolase interaction) formed. The intensity of dopachrome formation during phenol–phenolase interaction was dependent on the nature of phenolic substrate (specific/nonspecific), position of hydroxyl group in phenol structure, and Cu atoms affecting reagent involved during interaction (Borkar and Verma, 1991). Leaves of most resistant *cvs* contain more phenol (69% more) than the most susceptible *cvs*. Diphenol oxidase (DPO) and its activity were more in the leaves of resistant *cvs* than in the leaves of susceptible *cvs*. DPO of cotton *cvs* was more active (producing more amount of dopachrome) with pyrogallol than that of the susceptible *cvs*, and the formation of dopachrome was dependent on the consumption of molecular oxygen. The presence of higher phenol content and increased DPO activity in the intercellular fluid of resistant *cvs* produced more quickly (than in susceptible cvs) the dopachrome responsible for tissue necrosis/hypersensitive reaction, whereas in the leaves of susceptible *cvs*, low phenol content and slower activity of DPO produced brown pigment or dopachrome more slowly and allowed the Xam to multiply rapidly to produce susceptible water-soaking reaction (Borkar and Verma, 1991b).

Pathogen

The bacterial blight pathogen *Xanthomonas malvacearum* exhibits variability due to its ability to attack one or more resistance genes available in the cotton plant, and thus several races of this pathogen are available in different geographical locations. In India, at least 32 races of the bacterium are present. The molecular studies of the bacterium indicated that the DNA content of the bacterium was dependent on the race number (virulence) of Xcm strains and ranged from 9.62×10^{-7} µg to $11–99 \times 10^{-7}$ µg DNA/cell. Quantitatively, the amount of plasmid DNA was more than the chromosomal DNA in all the races of Xam. The degree of virulence of Xam races was related with the amount of chromosomal DNA but not with plasmid DNA (Borkar and Verma, 1988a). Streptomycin-resistant mutant of Xam synthesized higher amount of DNA than the wild-type streptomycin-susceptible Xam in nature.

Host

Gossypium hirsutum (upland cotton) and *Gossypium barbadense* (sea island or pima cotton); *Gossypium herbaceum*, *Gossypium populifolium* (wild cotton), *Hibiscus vitifolius* (weed in Indian cotton fields), and *Hibiscus rosa-sinensis* (Chase, 1986); and outside the family of Malvaceae *Lochnera pusilla* (weed in Indian cotton fields), *Ceiba pentandra* (kapok tree), and *Jatropha curcas* (purging nut) have been reported as natural hosts.

Geographical Distribution and Importance

The pathogen is widespread in all cotton-growing regions of the world. In Europe, they are reported from Bulgaria, Greece, Italy, Moldavia, Poland, Romania, Russia, Spain, Ukraine, and former Yugoslavia. The disease was a major problem, with high crop losses, after introduction into Africa in the first and the United States and Australia in the second half of the past century. After the development of resistant varieties, the disease has become a more minor problem.

Disease Cycle

The bacterium survives on infected, dried plant debris in soil for several years. The bacterium is also seed borne and remains in the form of slimy mass on the fuzz of seed coat. The bacterium also attacks other hosts like *Thumbergia thespesioides*, *Eriodendron anfructuosum*, and *J. curcas*. The primary infection starts mainly from the seed-borne bacterium. The secondary spread of the bacteria may be through wind, windblown rain splash, irrigation water, insects, and other implements. The disease cycle of bacterial blight of cotton is illustrated in Figure 6.1.

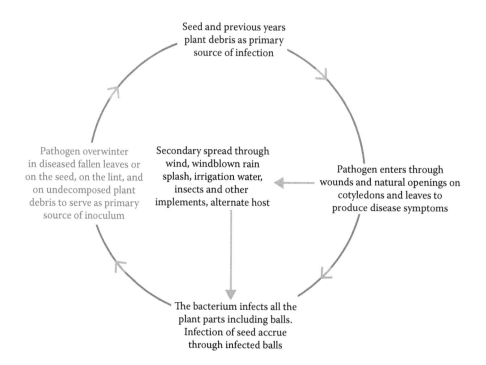

FIGURE 6.1 Disease cycle of bacterial blight of cotton caused by *X. axonopodis* pv. *malvacearum*.

Epidemiology

Optimum soil temperature of 28°C, high atmospheric temperature of 30°C–40°C, relative humidity of 85%, early sowing, delayed thinning, poor tillage, late irrigation, and potassium deficiency in soil are favorable for disease development and spread. Rain followed by bright sunshine during the months of October and November are highly favorable. Wind-driven rain, hail, and sandblasting increase disease severity.

Transmission

Bacteria overwinter on seed and plant debris. Penetration is through stomata or wounds. Bacteria are spread by infected seed and in the field from crop residues and infected plants by water, and wet conditions are necessary for disease development. Wind-driven rain, hail, and sandblasting significantly increase disease severity.

The spotted bollworms (*Earias* spp.) were found to transmit the bacterial blight pathogen. Caterpillars were found to acquire the pathogen by feeding on infected leaves and the bacterial pathogens were detected in the excreta of the caterpillar where it remained infectious for up to 6 days. This excreta help to disseminate the pathogen (Borkar et al., 1980).

Red cotton bugs also help in the spread of the bacterium when the bacterial exudates are available on the infected portions under high humid conditions. Injury caused by the feeding/movement of the bugs on the leaves predisposed them to infection of Xam (Verma et al., 1981).

Losses

Among the plant diseases caused by bacteria, blackarm of cotton is a severe one. It was first described in 1891 by Atkinson from the United States. The disease has been reported from South America, Egypt, Sudan and other African countries, USSR, China, Australia, Sri Lanka, Pakistan, and India. Cotton yield losses of 10% due to this disease have been reported.

Disease Management

Delint the cotton seeds with concentrated sulfuric acid at 100 mL/kg of seed. Treat the delinted seeds with carboxin or oxycarboxin at 2 g/kg or soak the seeds in 1000 ppm streptomycin sulfate overnight.

Remove and destroy the infected plant debris. Rogue out the volunteer cotton plants and weed hosts.

Follow crop rotation with nonhost crops.

Follow early thinning and early earthing up with potash.

Grow resistant varieties like Sujatha, 1412, and CRH 71.

Spray with 100 g of streptomycin sulfate + tetracycline mixture along with copper oxychloride at 1.25 kg/ha.

Variation in *X. malvacearum* plays an important role in the management of this disease. The most virulent race-32 possessed the virulence factor for at least five major genes, that is, B_1, B_2, B_{IN}, B_N, and B_4. The gene combination of $B_1 + B_2$ or $B_2 + B_4$ was immune in certain *Hirsutum cvs* emphasizing the role played by minor genes or the background of the *cvs* and the fact that major genes must be combined with minor genes to obtain immune lines to multiple races of *X. malvacearum* (Verma et al., 1980).

Mixed races of *Xanthomonas* at an infection site produce variable reaction. Even most virulent race when associated with less virulent race does not show translucent water soaking (Verma and Borkar, 1984). This creates a problem to determine the resistance of the cotton *cvs* against bacterial blight pathogen when more than one race is present or inoculated for resistance screening. Therefore, one race-horizontal model (Borkar, 1990) is helpful for screening the varieties for horizontal resistance of cotton lines to bacterial blight pathogen. Use of variable races of the pathogens during screening process and confusing grading obtained, due to mixture of races of Xam, during their interaction on host *cvs* during screening process is avoided by using this model.

Different races behave differently to chemicals used for disease control (Verma et al., 1980) and therefore the presence of races in the geographical location and their susceptibility to the chemical is an important factor in the management of the disease.

6.2 JUTE

6.2.1 BACTERIAL LEAF BLIGHT OF JUTE

Pathogen: *Xanthomonas campestris* pv. *capsularii*
The disease is reported from India.

Symptoms
Small angular brown spots appear on jute leaves and some of the spots are surrounded by yellow halo. In the later stage, the brown spots coalesce leading to the development of larger blighted area on the leaf lamina. The incidence of the disease varies from about 2% to 5%.

Host
Jute.

Geographical Distribution
India.

Disease Cycle
This disease spreads through rain flashes in the field.

Control Measures
Not yet worked out.

REFERENCES

Borkar, S.G. 1982. Compatible/incompatible *Xcm*-cotton interaction. PhD thesis submitted to P.G. School, I.A.R.I., New Delhi, India, p. 171.

Borkar, S.G. 1989. Tannin content of cotton CVS in relation to bacterial blight resistance and its dynamics during different reaction induced by *Xcm*-cotton interaction. *Indian J. Plant Pathol.*, 7(2), 142–144.

Borkar, S.G. 1990. One race-horizontal resistance model. *J. Cotton Res. Dev.*, 4(2), 220–227.

Borkar, S.G. and J.P. Verma. 1988a. DNA content of *Xanthomonas campestris* pv. *malvacearum* and its relation to virulence and drug resistance. *Indian J. Plant Pathol.*, 6(1), 52–55.

Borkar, S.G. and J.P. Verma. 1988b. Quantitative changes in protein and free amino acid during interaction between cotton and *Xanthomonas campestris* pv. *malvacearum*. *Indian J. Plant Pathol.*, 6(1), 46–48.

Borkar, S.G. and J.P. Verma. 1989a. Dynamics of sugars and phenols in the intercellular fluid of susceptible/resistant CVS during infection with *Xanthomonas campestris* pv. *malvacearum*. *Indian J. Mycol. Plant Pathol.*, 19(2), 184–191.

Borkar, S.G. and J.P. Verma. 1989b. Exopolysaccharide, a water soaking inducing factor produced by bacterial blight pathogen *Xanthomonas campestris* pv. *malvacearum*. *Coton et. Fibre Tropicales*, 44(2), 149–153.

Borkar, S.G. and J.P. Verma. 1989c. Chemical inhibition of compatible/incompatible reaction in cotton. *Indian Phytopathol.*, 42(2), 236–240.

Borkar, S.G. and J.P. Verma. 1991a. A basis of different types of incompatible reactions induced during Xcm-cotton interaction. *Indian J. Plant Pathol.*, 9(182), 33–36.

Borkar, S.G. and J.P. Verma. 1991b. Dynamics of phenol and diphenoloxidase contents of cotton CVS during hypersensitive and susceptible reaction induced by *Xanthomonas campestris* pv. *malvacearum*. *Indian Phytopathol.*, 44(3), 281–290.

Borkar, S.G., Verma, J.P., and R.P. Singh. 1980. Transmission of *Xanthomonas malvacearum*, the incitant of bacterial blight of cotton through spotted bollworm. *Indian J. Entomol.*, 42(3), 390–397.

Chase, A.R. 1986. Comparisons of three bacterial leaf spots of *Hibiscus rosa-sinensis*. *Plant Dis.*, 70, 334–336.

Smith, E.F. 1901. The cultural characters of *Pseudomonas hyacinthi*, *Ps. campestris*, *Ps. phaseoli* and *Ps. stewarti*—Four one-flagellate yellow bacteria parasitic on plants. *U.S.D.A. Div. Veg. Phys. Pathol. Bull.*, 28, 151–153.

Vauterin, L., Hoste, B., Kersters, K. et al. 1995. Reclassification of *Xanthomonas. Int. J. Syst. Bacteriol.*, 45, 472–489.

Verma, J.P. and S.G. Borkar. 1984. Reaction of mixed races of *Xanthomonas campestris* pv. *malvacearum. Curr. Sci.*, 53(17), 930–931.

Verma, J.P., Singh, R.P., Borkar, S.G. et al. 1980. Management of bacterial blight of cotton with particular reference to variation in pathogen *X. malvacearum. Ann. Agric. Res.*, 1(1), 98–107.

Verma, J.P., Borkar, S.G., and R.P. Singh. 1981. Transmission of *Xanthomonas campestris* pv. *malvacearum* through red cotton bug. *Ann. Agric. Res.*, 2(1–2), 57–63.

SUGGESTED READING

Janse, J.D. 2005. *Phytobacteriology: Principles and Practice*. CABI Publishers, Cambridge, MA, 360pp.

7 Bacterial Diseases of Fruit Crops

7.1 APPLE AND PEAR

7.1.1 FIRE BLIGHT OF APPLE AND PEAR

Pathogen: *Erwinia amylovora*

Fire blight, also written fireblight, is a contagious disease affecting apples, pears, and some other members of the family Rosaceae. Fire blight is a destructive bacterial disease of apples and pears that kills blossoms, shoots, limbs, and sometimes the entire trees. It poses a serious concern to producers of apples and pears. Under optimal conditions, it can destroy an entire orchard in a single growing season.

Pears are the most susceptible, but apples, loquat, crabapples, quinces, hawthorn, *Cotoneaster*, *Pyracantha*, raspberry, and some other rosaceous plants are also equally vulnerable to fire blight. The disease is believed to be indigenous to North America, from where it spread to rest of the world.

Pathogen

E. amylovora is a native pathogen of wild, rosaceous hosts in eastern North America. These hosts include hawthorn, serviceberry, and mountain ash. Early European settlers introduced apple and pear to North America. The first report of fire blight as a disease of apple and pear occurred in 1780, in the Hudson Valley of New York. In California, the disease was first reported in 1887.

Early nineteenth- and twentieth-century horticultural texts and bulletins recognized fire blight as a serious disease of pear, provided descriptions of symptoms, and outlined pruning practices for control. Nonetheless, in the eastern United States, fire blight proved to be destructively epidemic on pear, limiting the cultivation of this host. Even today, the threat of fire blight restricts commercial production of pear to semi-arid, desert areas west of the Rocky Mountains.

E. amylovora has the distinction of being the first bacterium shown to be a pathogen of plants. Koch's postulates for *E. amylovora* were fulfilled by J.C. Arthur in 1885, but the genesis of the concept that bacteria can be plant pathogens required the contributions of many scientists (notably T.J. Burrill) and growers over a period extending from 1846 to 1901. *E. amylovora* is also one of the first plant pathogens to be associated with an insect vector. In the late 1890s, M.B. Waite linked blossom infection to the movement of the pathogen from flower to flower by pollinating insects.

The disease was introduced to Europe in the 1950s. It has since spread to most countries in Western Europe. Introductions of infested plant material served to establish *E. amylovora* in Europe, the Middle East, and New Zealand. In 1995, fire blight was first observed in the Po River Valley of northern Italy, which is the largest pear production area in the world. Since 1995, the Italian government has destroyed 500,000 pear trees in an attempt to eradicate *E. amylovora*.

Fire blight was first recorded in Ireland in 1986, but to date, the disease has not become established there. In Ireland, *Cotoneaster* appears to be the most susceptible host, especially the larger leaf varieties like *Cotoneaster bullatus*, *Cotoneaster cornubia*, and *Cotoneaster lacteus*.

Fire blight is not believed to be present in Australia, though it might possibly exist there. It has been a major reason for a long-standing embargo on the importation of New Zealand apples to Australia. Japan was likewise believed to be without the disease, but it was discovered in pears grown in northern Japan. Japanese authorities are, however, still denying its existence and the Japanese scientist who discovered it is believed to have committed suicide after his name was leaked to affected farmers.

Pears appear to be more susceptible than apples to fire blight pathogen. This is because pears tend to have more flowers per spur than apples, and these flowers tend to remain open and susceptible for a longer period than those on apple. For example, individual apple flowers stay open for about 80 degree days (DD) above 4°C (44 DD above 4°C), while pear flowers stay open for an average of 120 DD above 4°C (67 DD above 4°C). Because of this longer flower life, nearly 90% of the total flower buds are open at full bloom on pears compared with only 65%–70% of those on apples.

Fire blight is a destructive bacterial disease of apples and pears that kills blossoms, shoots, limbs, and, sometimes, entire trees. The disease is generally common throughout the mid-Atlantic region, although outbreaks are typically very erratic, causing severe losses in some orchards in some years and little or no significant damage in others. This erratic occurrence is attributed to differences in the availability of overwintering inoculum, the specific requirements governing infection, variations in specific local weather conditions, and the stage of development of the cultivars available. The destructive potential and sporadic nature of fire blight, along with the fact that epidemics often develop in several different phases, make this disease difficult and expensive to control. Of the apple varieties planted in the mid-Atlantic region, those that are most susceptible include York, Rome, Jonathan, Jonagold, Idared, Tydeman's Red, Gala, Fuji, Braeburn, Lodi, and Liberty. Stayman and Golden Delicious cultivars are moderately resistant and all strains of Delicious are highly resistant to fire blight, except when tissues are damaged by frost, hail, or high winds.

Host

Amelanchier (June berry), *Malus* (apple), *Chaenomeles* (Quince) *Mespilus* (Medlar), *Pyracantha* (firethorn)
Cotoneaster Crataegus (hawthorn/whitethorn), *Pyrus* (pear)
Cydonia (quince), *Sorbus* (mountain ash, whitebeam)
Eriobotrya (loquat), *Photinia davidiana* (formerly *Stranvaesia davidiana*)

Symptoms

Fire blight is a systemic disease. The term "fire blight" describes the appearance of the disease, which can make affected areas appear blackened, shrunken, and cracked, as though scorched by fire.

Primary infections are established in open blossoms and tender new shoots and leaves in the spring when blossoms are open.

Symptoms of fire blight can be observed on all above-ground tissues, including blossoms, fruits, shoots, branches and limbs, and in the rootstock near the graft union on the lower trunk. Generally, symptoms of fire blight are easy to recognize and distinguishable from other diseases.

On Blossom Clusters

Blossom symptoms are first observed 1–2 weeks after petal fall. The floral receptacle, ovary, and peduncles become water-soaked and dull, grayish green in appearance. Later these tissues shrivel and turn brown to black (Photo 7.1). Similar symptoms often develop in the base of the blossom cluster and young fruitlets as the infection spreads internally. During periods of high humidity, small droplets of bacterial ooze form on water-soaked and discolored tissues. Ooze droplets are initially creamy white, becoming amber tinted as they age.

On blossoms, *E. amylovora* acts as an excellent colonizer of the surfaces of stigmas and, to a lesser extent, the surface of the nectary. This reproduction on blossom surfaces is called epiphytic growth and occurs without the bacterium causing disease. Epiphytic growth of *E. amylovora* on stigmas and movement of the pathogen from blossom to blossom by pollinating insects are two important processes that regulate the incidence of blossom infection. *E. amylovora* also can survive on other healthy plant surfaces, such as leaves and branches, for limited periods (weeks), but colony establishment and epiphytic growth on these surfaces do not occur. Cells of *E. amylovora* excrete large amounts of an extracellular polysaccharide (a major component of bacterial ooze), which creates a

PHOTO 7.1 *E. amylovora*—fire blight on *Malus*. (Courtesy of Charlie, flickr.com.)

matrix that protects the pathogen on plant surfaces. In propagation nurseries, cells of *E. amylovora* surviving on woody surfaces can initiate disease when scions and rootstocks are wounded during grafting. *E. amylovora* can also reside as an endophyte within apparently healthy plant tissue, such as branches, limbs, and budwood. Migration of the pathogen through xylem is one mechanism by which blossom infections of pear/apple can lead to rootstock infections near the graft union.

On Young Shoot
Shoot symptoms are similar to those in blossoms but develop more rapidly. Tips of shoots may wilt rapidly to form a "shepherd's crook." Leaves on diseased shoots often show blackening along the midrib and veins before becoming fully necrotic (Photo 7.2). Numerous diseased shoots give a tree a burnt, blighted appearance, hence the disease name.

Advanced Foliar Symptoms
Infections initiated in blossoms and shoots can continue to expand both up and down larger branches and limbs. Bark on younger branches becomes darkened and water-soaked (Photo 7.3). At advanced stages, cracks will develop in the bark, and the surface will be sunken slightly. Amber-colored

PHOTO 7.2 *E. amylovora*—fire blight on Pyrus calleryana on young shoot, Bradford pear. (Courtesy of Charlie, flickr.com.)

PHOTO 7.3 Advanced foliar symptoms of fire blight on *Pyrus calleryana*, Bradford pear. (Courtesy of Charlie, flickr.com.)

bacterial ooze mixed with plant sap may be present on bark. Wood under the bark will show streaked discolorations.

On Pear and Apple Fruits

Indeterminate, water-soaked lesions form on fruit surface and later turn brown to black. Droplets of bacterial ooze may form on lesions (Photo 7.4), usually in association with lenticels. Severely diseased fruits blacken completely and shrivel.

PHOTO 7.4 Fire blight of apple caused by *E. amylovora*. (Courtesy of Scoth, N.)

Apple Rootstocks

Rootstock infections usually develop near the graft union as a result of internal movement of the pathogen through the tree or from infections through watersprouts or burr knots. The bark of infected rootstocks may show water-soaking, a purplish-to-black discoloration, cracking, and signs of bacterial ooze. Red-brown to black streaking may be apparent in wood just under the bark. Symptoms of rootstock blight can be confused with Phytophthora collar rot. Malling 26 and 9 rootstocks are highly susceptible to fire blight.

Pathogen Biology

E. amylovora is a member of the family Enterobacteriacae. Cells of *E. amylovora* are Gram-negative, rod-shaped, and measure 0.5–1.0 × 3.0 mm, and are flagellated on all sides (peritrichous). Physiologically, *E. amylovora* is classified as a facultative anaerobe. It grows on most standard microbiological media and on several differential media. Optimum temperature for growth is 27°C, with cell division occurring at temperatures ranging from 5°C to 31°C. Identification of *E. amylovora* isolates is based on biochemical and serological tests, inoculation of immature pear fruits and apple seedlings, and DNA hybridization assays.

Disease Cycle

Honeybees and other insects, birds, rain, and wind can transmit the bacterium to susceptible tissue. Injured tissue is also highly susceptible to infection, including punctures and tears caused by plant-sucking or biting insects. Hailstorms aggravate the infection of an entire orchard in a few minutes, and growers do not wait until symptoms appear, normally beginning control measures within a few hours.

Once deposited, the bacterium enters the plant through open stomata and causes blackened, necrotic lesions, which may also produce a viscous exudate. This bacteria-laden exudate can be distributed to other parts of the same plant or to susceptible areas of different plants by rain, birds, or insects, causing secondary infections. The disease spreads most quickly during hot and wet weather and is dormant in the winter when temperatures drop. Infected plant tissue contains viable bacteria, however, and will resume production of exudate upon the return of warm weather in the following spring. This exudate is then the source for new rounds of primary infections. The disease cycle of fire blight of apple and pear is illustrated in Figure 7.1.

The pathogen spreads through the tree from the point of infection via the plant's vascular system, eventually reaching the roots and/or graft junction of the plant. Once the plant's roots are affected, the death of the plant often results. Overpruning and overfertilization (especially with nitrogen) can lead to watersprout and other midsummer growth that leave the tree more susceptible.

Overwintering

E. amylovora overwinters in a small percentage of the annual cankers that were formed on diseased branches in the previous season. These overwintering sites are called "holdover cankers." As temperatures warm in spring, the pathogen becomes active in the margins of holdover cankers. Free bacterial cells are released onto the bark surface, sometimes as visible ooze. Insects attracted to the ooze (e.g., flies) or rain disseminates the bacteria from the canker to blossoms.

Floral Epiphytic Phase

Stigmas, which are borne on the ends of the style, are the principal site of epiphytic colonization and growth by *E. amylovora*. During the floral epiphytic phase, the ultimate population size that the pathogen attains is influenced by temperature, which regulates the generation time of the pathogen, and by the number of blossoms in which the pathogen becomes established, which is facilitated by pollinating insects, honey bees in particular. Under ideal conditions, stigmas of each flower can support 10^6 cells of the pathogen.

Primary Infection in Flowers

Blossom blight is initiated when cells of *E. amylovora* are washed externally from the stigma to the hypanthium (floral cup). On the hypanthium, *E. amylovora* gains entry into the plant through

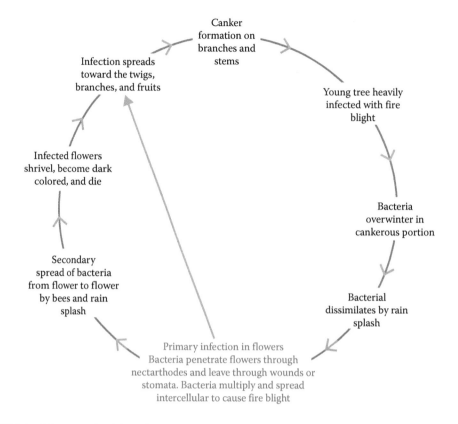

FIGURE 7.1   Disease cycle of fire blight of pear and apple caused by *E. amylovora*.

secretory cells (nectarthodes) located on the surface. In pear, the importance of blossom blight is expanded further by the tendency of this species to produce nuisance, secondary or "rattail" blossoms during late spring and early summer, long after the period of primary bloom.

Secondary Phases

Secondary phases include shoot, fruit, and rootstock blight. These phases are usually initiated by inoculum produced on diseased tissues as a result of blossom infection. Wounds are generally required by *E. amylovora* to initiate shoot and fruit blight. Insects such as plant bugs and psylla create wounds on succulent shoots during feeding. Strong winds, rain, and hail can create numerous, large wounds in host tissues. Infection events induced by severe weather are sometimes called "trauma blight." Rootstock blight of apple can result from shoot blight on watersprouts or from internal translocation of *E. amylovora* from infections higher on the tree.

Canker Expansion

Both primary and secondary infections can expand throughout the summer, with the ultimate severity of an infection being dependent on the host species, cultivar, environment, and age and nutritional status of the host tissues. Young, vigorous tissues and trees are more susceptible to fire blight than older, slower growing tissues or trees. Similarly, trees that have received an excess of nitrogen fertilizer, and therefore are growing rapidly, are more susceptible than trees growing under a balanced nutrient regime. Rates of canker expansion also can be enhanced by a high water status in a tree caused by excessive or frequent irrigation or poorly drained soils. Canker expansion slows in late summer as temperatures cool and growth rates of trees and shoots decline.

Epidemiological Models

Blossom blight is sporadic from season to season owing to the requirement for warm temperatures to drive the development of large epiphytic populations. Several epidemiological models (e.g., Cougarblight and Maryblyt) predict the likelihood of blossom blight epidemics based on observed climatic conditions. The models work by identifying the periods conducive for epiphytic growth of *E. amylovora* on blossoms before infection occurs, and thus are used widely to aid decisions on the need for and timing of chemical applications. Blossom blight risk models accumulate degree units above a threshold temperature of 15.5°C or 18°C. Data on rain or blossom wetness during periods of warm weather are also used in the models to indicate more precisely the timing and likelihood of blossom infection. Other temperature-based models predict the time of symptom expression after an infection event (i.e., the length of the incubation period) based on heat unit sums. These models are used to time orchard inspections and/or pruning activities.

How Does the Disease Spread?

Fireblight can be spread in the following ways:

- *Movement* of infected host plants, which may or may not display symptoms, can transmit fireblight over long distances.
- *Rain splash*: Bacteria emerge from infected material in the form of sticky ooze and droplets that harden on drying but which readily dissolve in water and can be spread by rain splash from an infected host to adjacent host material.
- *Wind*: Fine strands of ooze may be extruded especially from young tissue and these become brittle on drying and are dispersed by wind.
- *Insects* can spread the pathogen from overwintering cankers to early blossoms and between blossoms.

Disease Management

Effective management of fire blight is multifaceted and largely preventative. The grower must utilize a combination of sanitation, cultural practices, and sprays of chemical or biological agents to keep the disease in check.

Action Following Discovery of Fire Blight

All infected plants must be destroyed under the supervision of the Department of Agriculture and Food. It may also be necessary to destroy other adjacent host plants in order to control spread of the disease. The source of the infected material is then traced so that other infected plants may also be destroyed. The plant passport system facilitates the tracing of fireblight host plants traded within the European Union.

Monitoring

Concentrate monitoring in orchard blocks where the disease occurred during the previous season. Observe blighted limbs and shoots for removal during normal pruning operation. There may be a need to remove whole trees on some occasions.

Orchards grown on susceptible rootstocks (M.26, M.9, Mark), where fire blight occurred the previous year in trees showing poor foliage color or dieback, should be examined for rootstock cankers and, if found, removed from the orchard immediately and destroyed. A very important aspect of fire blight management involves monitoring the weather for the specific conditions that govern the buildup of inoculum in the orchard, the blossom infection process, and the appearance of symptoms. A weather station that records the daily minimum and maximum temperatures and rainfall amounts is needed. When 50% of the buds show green tissue, begin keeping a daily record of the cumulative DD greater than 12°C. This information can be used to signal when symptoms are likely to appear in the orchard for blossom blight (103 DD greater than 12°C [57 DD greater than 12.7°C] after infection), canker blight (about 300 DD greater than 12°C [167 DD greater than 12.7°C] after green tip), and early shoot blight (about 103 DD greater than 12°C [57 DD greater than 12.7°C] after blossom blight or canker blight symptoms appear).

Continue to monitor and record the daily minimum and maximum temperatures and rainfall amounts and continue to accumulate DD greater than 12°C (12.7°C). At the full pink stage (i.e., first flower open in the orchard), a record should also be kept of the cumulative degree hours (DH) greater than 18.3°C. Once a total of 200 or more DH greater than 18.3°C (111 DH greater than 18.3°C) has accumulated after the start of bloom, any wetting event caused by rain or heavy dew that wets the foliage is likely to trigger a blossom infection event if the average daily temperature is 15.6°C or more.

This information can be used to schedule streptomycin sprays, which are most effective if applied on the day before or on the day of an infection event. Such sprays protect all flowers open at the time of treatment. However, because other flower buds may open after treatment, reassess the need for additional sprays at 4-day intervals during bloom. Continue to monitor for strikes and remove all blighted limbs.

Monitor the orchard to locate blighted limbs for removal. For the greatest effect on the current season's damage severity, infected limbs should be removed as soon as early symptoms are detected and before extensive necrosis develops. Where the number and distribution of strikes is too great for removal within a few days, it may be best to leave most strikes and cut out only those that threaten the main stem. On young trees, and those on dwarfing rootstocks, early strikes in the tops of the trees often provide inoculum for later infections of shoots and sprouts on lower limbs near the trunk, which may result in tree loss. Give these early strikes a high priority for removal.

Look for symptoms of early tree decline or early fall color in orchards planted on highly susceptible rootstocks (M.26, M.9, Mark) where the disease developed. These symptoms may appear either on one side or throughout individual trees. Examine the rootstock area of these trees just below the graft union for evidence of cankering or bacterial ooze. Remove any tree showing these symptoms during this period.

Elimination of Overwintering Inoculums

Vigilant sanitation through the removal of expanding and overwintering cankers is essential for the control of fire blight in susceptible cultivars. Removal of overwintering ("holdover") cankers is accomplished by inspecting and pruning trees during the winter.

Removing Sources of Infection

Dormant pruning to remove overwintering infections helps reduce inoculum for the next season. Make cuts about 4 in. below any signs of dead bark. Remove pruned material from the orchard. Beginning about 1 week after petal fall, monitor the orchard to locate blighted limbs for removal. For the greatest effect on the current season's damage severity, infected limbs should be removed as soon as early symptoms are detected and before extensive necrosis develops. Where the number and distribution of strikes is too great for removal within a few days, it may be best to leave most strikes and cut out only those that threaten the main stem. On young trees, and those on dwarfing rootstocks, early strikes in the tops of the trees often provide inoculum for later infections of shoots and sprouts on lower limbs near the trunk, which may result in tree loss. Give these early strikes in the tops of trees a high priority for removal. Do not combine the practices of fire blight removal with pruning and training of young, high-density trees.

Midseason Suppression of Established Infections

In summer, established infections are controlled principally by pruning. Effective control through pruning requires that cuts are made 20–25 cm (8–10 in.) below the visible end of the expanding canker and that between cuts the pruning tools are disinfested with a bleach or alcohol solution to prevent cut-to-cut transmission. Repeated trips through an orchard are necessary, as some of the infections are invariably missed and others become visible at later times. Prunings harboring the pathogen are usually destroyed by burning.

In severely affected orchards, cultural practices that slow the growth rate of the tree will also slow the rate of canker development. They include withholding irrigation water, nitrogen fertilizer, and cultivation. Similarly, practices that reduce tree wounding and bacterial movement can reduce secondary infection. They include controlling insects such as plant bugs and psylla, limiting the use of limb spreaders in young orchards, and avoiding the use of overhead sprinklers. Chemicals such as streptomycin or copper can suppress trauma blight if applied immediately after a hailstorm.

Prevention of Blossom Blight

Prevention of blossom infection is important in fire blight management because infections initiated in flowers are destructive and because the pathogen cells originating from blossom infections provide much of the inoculum for secondary phases of the disease, including the infection of shoots, fruits, and rootstocks. Management actions to suppress blossom blight target the floral epiphytic phase. Sprays of antibiotics, streptomycin, or oxytetracycline have effectively suppressed blossom infection in commercial orchards.

Treatment

Sprays of the antibiotics streptomycin or terramycin can prevent new infections. The use of such sprays has led to streptomycin-resistant bacteria in some areas, such as California and Washington. Certain biological controls consisting of beneficial bacteria or yeast can also prevent fire blight from infecting new trees. The only effective treatment for plants already infected is to prune off the affected branches and remove them from the area. Plants or trees should be inspected routinely for the appearance of new infections. The rest of the plant can be saved if the blighted wood is removed before the infection spreads to the roots.

Chemical and Biological Control

Copper compounds also are effective but not used widely because copper can be phytotoxic to the skin of young fruits. *E. amylovora* has become resistant to streptomycin in some production areas, limiting the effectiveness of this chemical. Nonpathogenic, bacterial epiphytes sprayed onto blossoms can preemptively suppress fire blight by colonizing the niche (stigmatic surface) used by *E. amylovora* to increase its epiphytic population size. The bacterium *Pseudomonas fluorescens* strain A506 is registered and sold commercially for this purpose (BlightBan A506).

A copper spray applied at the 1/4 in. green tip stage may reduce the amount of inoculum on the outer surfaces of infected trees. At bloom, antibiotic sprays are highly effective against the blossom blight phase of the disease. These sprays are critical because effective early-season disease control often prevents the disease from becoming established in an orchard. Predictive models, particularly Maryblyt, help to identify potential infection periods and improve the timing of antibiotic treatments, as well as avoid unnecessary treatments. Strains of the pathogen that are resistant to streptomycin are present in some orchards in the eastern United States and are widespread in most apple and pear regions of the western United States.

Fire blight bacteria in Utah County have been tested and are documented to be resistant to streptomycin. Use of streptomycin, to control fire blight, is not recommended in Utah County. Instead, use of compounds containing oxytetracycline is recommended.

Dormant Spray

In orchards with a history of severe fire blight, it is advisable to carefully prune out overwintering cankers and spray with a copper-plus-oil mixture at the delayed dormant stage (silver tip to green tip). Copper compounds can be phytotoxic if they are sprayed much past the bud burst stage (1/2 in. green). This spray is thought to reduce the levels of inoculum in the orchard, but they may also be effective in reducing insect vectors.

Bordeaux mixture (see the following formulation) plus 1 gal of 60–70 s spray oil.

Preparation of Bordeaux mixture: Dissolve 8 lb of crystalline copper sulfate in 100 gal of water in the spray tank. After the copper sulfate is dissolved, add 8 lb of hydrated spray lime (350 mesh), either mixed in water or as powder, to the tank. Constant agitation is needed to thoroughly mix the contents of the tank. Finally, add 1 gal of spray oil. Copper hydroxide plus oil. Copper oxychloride sulfate + basic copper sulfate.

Blossom Sprays

Streptomycin and oxytetracycline, and fixed copper sprays, have proven very effective in reducing fire blight provided they are properly applied at correct times. They are preventive sprays only and must be repeated every 4–5 days as long as new flowers are opening. A biological control compound (a bioantagonistic bacterial competitor) has recently been licensed for the control of fire blight as well (Bloomtime FD [*Pantoea agglomerans*]).

Agricultural antibiotics available for the treatment of fire blight in pears and apples in Utah include Agri-Mycin 17 (streptomycin sulfate [not recommended for use in Utah County]) and/or Mycoshield (oxytetracycline calcium complex).

Fixed Coppers

Copper oxychloride sulfate (C-O-C-S WDG) and Kocide branded products (20/20, 101, 2000, and 3000); read labels as some are phytotoxic and may not be labeled for pear, only for apple.

Timing bactericide applications is critical. A delay of even several hours can reduce the level of control. It is not necessary to spray until mean daily temperature during bloom (average of maximum and minimum temperatures from midnight to midnight) first exceeds 16°C in the spring. Use forecasting models such as Maryblyt or Cougarblight to determine when to spray. Sprays should be repeated at 4- to 5-day intervals throughout the bloom period as long as temperatures are above the mean threshold. Applications are most important on young pears and susceptible varieties of apples.

Fixed coppers may have phytotoxic effects, causing russeting in fruits. Do not use fixed coppers on d'Anjou pears.

Biological control agents, although not widely used, have provided partial control of blossom infections. More effective biological agents are required if their use is to become widespread.

Insect Control

The role of insects in the transmission of fire blight bacteria is well known. It is likely that insects that cause wounds (leafhoppers, plant bugs, pear psylla) can create places for bacteria to enter the tree, and some summer infections (shoot blight) are probably facilitated by insects. Where fire blight is a problem, controlling these insects at levels below their economic injury threshold is advised.

Cultural Practices

Use management systems that promote early cessation of tree growth without adversely affecting tree vigor. Excessive vigor is an important component of orchard risk for fire blight. When tree growth continues past midsummer, it is likely that late-season infections will increase. Orchards should be established on well-drained soils, avoiding low, frost-prone or potentially water-logged areas, and nitrogen fertilizer should be applied based on the analyses of foliage N levels.

Any practice that promotes excessive succulent growth should be avoided. Trees should be fertilized to promote good health, but overfertilization with nitrogen or applications late in the season often cause excessive new growth that is susceptible to fire blight infection. Remove blighted blossoms and twigs as soon as they are evident. The infected blossoms and twigs should be pruned a minimum of 8–12 in. below the obvious infection. Avoid heavy pruning in the early summer because it stimulates succulent growth, which is very susceptible to blight.

During the dormant season, remove any cankers or blighted tissue. Also, remove any suckers growing up from the roots or on the trunk. It is advisable to avoid using pruners on small twigs; use hands to break out the blighted twigs. Pruners should be soaked in a 10% solution (one part bleach and nine parts water) of household bleach, or good surface disinfectant, between cuts when pruning out active fire blight in the summer. Bleach solution is effective but is corrosive to pruning tools. Rinse, dry, and oil the tools several times during the pruning. Pruners need not be sterilized between cuts when pruning is done during the dormant season.

Hosts of fire blight such as *Pyracantha*, hawthorn, *Cotoneaster*, and crabapple growing near the orchard should be eradicated. This will reduce fire blight inoculum in the orchard. Avoid excessive irrigation to reduce humidity in the orchard. If using sprinkler irrigation, do not allow water to wet the foliage since it will act in the same manner as rain in spreading bacteria. No cultivars are completely resistant to fire blight, but some are less susceptible than others. If fire blight is a common problem in your area, plant less susceptible varieties.

Resistant Cultivars

When establishing new orchards, consider susceptibilities of the scion and rootstock to fire blight; although none are immune, there is considerable variation among apple cultivars (and pear cultivars) in susceptibility to fire blight. Some cultivar/rootstock combinations are so susceptible to fire blight that investments in these are extremely high risk. In the eastern United States, Gala on M.26 is a good example. Long-range plans for establishing new orchards with fire blight–susceptible cultivars should include contingency plans for controlling the disease without streptomycin.

Fire blight–resistant cultivars of apple include Red Delicious, Liberty, Prima, Pricilla, Redfree, Spartan, and Sir Prize. There are many moderately resistant cultivars including Honeycrisp, Duchess, Empire, Golden Delicious, Granny Smith, Jonagold, McIntosh, Mutsu, and Winesap. This is not an exhaustive list as there are many cultivars of apples.

Some less-susceptible cultivars of pear include Harrow Delight, Harvest Queen, Kieffer, Moonglow, Seckel, LaConte, and Magness.

Of the pear varieties most commonly grown in the mid-Atlantic region, Bartlett, Bosc, D'Anjou, and Clapp's Favorite are the most susceptible, while Magness, Moonglow, Maxine, and Seckel are highly resistant. All varieties of Asian pears, except Seuri, Shinko, and Singo are moderately to highly susceptible to fire blight.

7.1.2 BLOSSOM BLAST AND DIEBACK OF PEAR

Pathogen: *Pseudomonas syringae*

Blossom blast of pear is caused by *P. syringae* van Hall, an epiphytic phytopathogenic bacterium (Barker and Grove, 1914; Crosse, 1959). Under favorable climatic conditions, the disease has caused serious but sporadic yield reductions in pear-growing regions throughout the world (McKeen, 1955; Dye, 1956; Panagopoulos and Crosse, 1964; Cancino et al., 1974; Sands and Kollas, 1974; Mansvelt and Hattingh, 1986). The most severe and frequent expression of disease symptoms usually follows cool, wet weather (Sands and Kollas, 1974; Waissbluth and Latorre, 1978; Mansvelt and Hattingh, 1987), which favors increases in populations of *P. syringae* (Gross et al., 1983). *P. syringae* is one of the most common ice nucleation–active (INA) bacteria (Maki et al., 1974) and has been shown to increase frost injury to sensitive plants (Klement, 1974; Arny et al., 1976; Weaver, 1978; Anderson et al., 1982; Klement et al., 1984). High populations of *P. syringae* increase the probability of ice nucleation activity (Lindow et al., 1982; Gross et al., 1984) during pear bloom. Although frost damage is often associated with outbreaks of pear blossom blast, the disease can appear without a frost during bloom periods that are cool and wet (Sands and Kollas, 1974; Waissbluth and Latorre, 1978). Pear blossoms at the prebloom through postbloom developmental stages are the most sensitive to frost injury (Proebsting and Mills, 1978; Whitesides and Spotts, 1991). There have been reports of differences in cultivar susceptibility to blossom blast (Panagopoulos and Crosse, 1964; Cancino et al., 1974; Waissbluth and Latorre, 1978), but the susceptibility of many cultivars has not been studied.

Symptoms

Infection may cause blossom blast, leaf spots, dieback of twigs (Photo 7.5) and spurs, dormant-bud death, and bark cankers.

At first, bark cankers appear as light brown, irregular patches on limbs. Later, outer bark and some underlying tissues may wholly or partly slough away. The blossom blast may closely

PHOTO 7.5   Symptoms of blossom blast and dieback on pear. (Courtesy of Part of OSU Extension Plant Pathology Slide Set.)

resemble that of the fire blight, but it is different in that blast seldom extends more than 1–2 in. into the spur and never involves a bacterial exudate. The blossom blast may also resemble the infection of blossom by *Botrytis* on young trees.

Rain and low temperatures, especially frost-inducing temperatures during bloom, increase the incidence of blossom infection. Two common genetic traits increase the bacteria's ability to cause disease. Most produce a powerful plant toxin, syringomycin, which destroys plant tissues as bacteria multiply in a wound. Bacteria also produce a protein that acts as an ice nucleus, increasing frost wounds that bacteria easily colonize and expand.

The varieties show different susceptibility to blossom blight. The Old Home and Asian pear cultivars shows severity on woods whereas Packham's Triumph, Bartlett, Eldorado, Anjou, and Bosc cultivars shows severe blossom blight.

Less severe blossom blight is found on varieties such as Comice, Forelle, red Anjou, and red Bartlett cultivars. Apple is not as susceptible and generally does not need control.

Reports from Pacific Northwest nurseries (Canfield, 1986) help illustrate how variable the severity and host range of *P. syringae* can be. One nursery reported in the spring of 1982 that 30% of 90,000 linden trees showed severe symptoms. Another nursery reported that its linden trees were also severely diseased. However, a third nursery reported that the disease was very serious on Japanese lilac and two cultivars of red maple, Red Sunset and October Glory, but they observed no problem on linden trees. A fourth nursery reported the disease on laburnum trees (golden chain tree) and Bradford pear, with 90% of the latter being killed. At a fifth nursery, damage was light to moderate but mostly light and primarily on the leaves of Thundercloud and Newport plums and on Norway, red, and silver maples.

There is no agreement about the severity of diseases caused by *P. syringae*. Most researchers consider *P. syringae* a weak pathogen, an opportunist that capitalizes on a host weakened by some predisposing condition. A number of factors reportedly make plants more susceptible to infection, the foremost being freeze damage. Freezing wounds the plant, allowing the bacterium to get into and destroy plant cells. Numerous workers have reported that symptom development in the field was related to cold temperatures. Ironically, many strains of *P. syringae* catalyze ice crystal formation on and in the plant tissues (Lindow, 1983). These generally are referred to as INA bacteria. Their presence on the plant serves to raise the freezing temperature above at which sensitive plant tissues would normally freeze. Most frost-sensitive plants have no significant mechanism of frost tolerance and must be protected from ice formation to avoid frost injury.

Ice nucleation activity of *P. syringae* is conferred by a single gene that encodes an outer membrane protein. Individual ice-nucleation proteins do not serve as ice nuclei, but they form large, homogeneous aggregates that collectively orient water molecules into a configuration mimicking the crystalline structure of ice, thereby catalyzing ice formation. Oriented water molecules freeze at temperatures slightly below zero ($-2°C$ to $-10°C$) instead of supercooling.

Predisposing Factors for the Disease

Wounding
Wounding of any kind seems to play a major role in initiating disease development. Wounds may be mechanical or environmental such as frost injury. Wounds have been shown to predispose trees to blossom blight and bacterial canker. Pruning wounds not only allow the bacterium to enter but also aid infection by fungi such as *Cytospora* and *Nectria*.

Plant Dormancy
Dormancy may also predispose susceptible trees to damage from *P. syringae*. Dormant peach trees were reportedly more susceptible to the disease than active ones.

Soil Factors
Factors such as soil pH and mineral nutrition may also predispose trees to *P. syringae* infection.

Dual Infections
Disease severity is greater when the plant is attacked by more than one pathogen. Diagnosis of *P. syringae* symptoms on maple in Oregon nurseries has been complicated by the presence of Verticillium wilt. In maple, symptoms of each disease can be similar, and dual infections have been observed. The relative contribution of each pathogen to the total disease impact upon the tree is unknown.

Sources and Survival of *Pseudomonas syringae*
There are several potential sources of *P. syringae*; however, the relative contribution of each source to disease development remains unknown.

Buds
Buds are considered a major overwintering site of *P. syringae*. This bacterium has been detected inside apparently healthy apple and pear buds during both growing and dormant seasons.

Cankers
Cankers from a previous year's infection have long been thought to be the primary source of inoculum.

Systemic Invasion
P. syringae has been isolated from interior tissues of fruiting cherry trees; most of these trees typically had three to six cankers, but two showed no visible symptoms. Bacteria were detected as far as 20 ft from any obviously diseased tissue. The highest bacterial counts were from the trunk, roots, and lower scaffold limbs. However, of nearly 10,000 bacteria examined, less than 10% were identified as *P. syringae*. In South Africa, bacteria introduced into leaves and leaf petioles during the growing season invaded leaves and shoot of plum and cherry trees and caused disease (Hattingh et al., 1989). In some cases, disease symptoms do not appear until the following spring. Interestingly, *P. syringae* strains isolated from the interior tissues of pears in Oregon were not pathogenic.

Latent Infections

Establishment of *P. syringae* inside symptomless tissues during the summer could represent a very important source of primary inoculum. Systemic invasion of tissues by *P. syringae*, therefore, assumes considerable ecological significance and obviously poses a significant challenge to the idea of controlling the disease by applying protective bactericide sprays to the tree surface.

Epiphytes

P. syringae also exists on the surfaces of many plants and, therefore, is in a position to cause infection, should the right environmental conditions develop. Monitoring surface populations associated with nursery trees in Oregon showed that the population increased rapidly and peaked during the first 2–3 weeks after bud break. The population declined during the summer, increased a small amount in October, and often was undetectable during December through February.

An antibiotic-resistant strain of *P. syringae* could be recovered from symptomless maple trees for up to 10 months after application showing that a particular strain of *P. syringae* can survive successfully over the summer and winter. Epiphytic *P. syringae* may be important for survival and spread during the growing season but they may not be important for survival and overwintering of primary inoculum.

Weeds and Grasses

Weeds also can be hosts for *P. syringae*. Workers in Michigan, Oregon, Poland, and South Africa have reported that *P. syringae* was isolated during the growing season from weeds or grasses.

Soil

It is generally accepted that *P. syringae* survives poorly in soil, but neither the soil phase nor the potential survival of *P. syringae* on roots (the rhizosphere) has been studied in any depth.

Spread of *Pseudomonas syringae*

P. syringae can be moved from place to place by wind, rain, insects, infested budwood, and transportation of infested nursery stock. Mechanical equipment and pruning tools also may be a frequently overlooked means of dispersal or of generating aerosols containing the bacteria. Harvesting alfalfa fields greatly increased the number of INA bacteria captured on Petri plates of agar in nearby citrus groves. *P. syringae* has also been recovered from the air above and next to bean fields.

Control Measures

A variety of methods have been tested for control of *P. syringae* in commercial plantings. They include cultural management, host resistance, biological control with microbial antagonists, and chemical control. Efforts have been targeted primarily either to control disease or to reduce the risk of frost damage from INA *P. syringae*. However, the results of these control efforts have not always been successful.

Cultural Management

Pruning

Pruning in fall and early winter also predisposed the trees to more severe damage from *P. syringae* infections and short-life syndrome (Chandler and Daniell, 1976). Cankers from *P. syringae* infections were longer when twigs were pruned from trees in December than when pruned in January or February. In those cases where trees may be threatened by *Cytospora* sp. and *P. syringae*, pruning in early spring when trees are more resistant to *Cytospora* may be of benefit.

Cauterization

Burning *P. syringae* cankers on limbs of stone fruit trees in New Zealand with a handheld propane burner cauterized the tissues and limited the canker's further spread, so the branch or trunk did not become girdled and killed (Hawkins, 1976). Living tissue surrounding the burned canker area callused quickly, forming an effective barrier against reinfection, and within 2 years, the treated limb showed little evidence of infection. Cauterization was fast and easy to use in the orchard, and a single treatment controlled most cankers. The method was tested successfully on apricots, sweet cherries, and peaches. This method is part of a total control strategy that also includes a spray program of autumn and winter Bordeaux sprays followed by spring sprays of streptomycin.

Biological Control with Bacterial Antagonists
Biological control has been directed almost entirely at frost control using bacterial antagonists to prevent buildup of INA populations of *P. syringae* (Lindow, 1983). Many of these experiments have been successful, but in field tests on apple and pear orchards in Washington, frost control was not achieved by applying either antagonists or chemical bactericides. There may be an intrinsic INA molecule within plant tissue not related to the INA bacteria.

Chemical Controls
Fixed copper compounds (such as Bordeaux and copper hydroxide), streptomycin (an antibiotic), and coordination productions (such as Bravo CM) are registered and have been used to control *P. syringae* with various degrees of success. Adding spreader stickers to these bactericides has gotten longer lasting control under the cool, wet conditions of the Pacific Northwest. Bacterial resistance to these bactericides, which can limit their usefulness, has been documented. Many nurseries alternate sprays of copper and streptomycin or combine them to reduce buildup of resistant strains and avoid copper phytotoxicity in the spring.

A combination of copper–streptomycin spray schedule also was used to control pear blossom blast in California.

Some growers in Oregon have reported poor control of *P. syringae* following either copper or streptomycin sprays. Because fixed copper sprays gave sporadic disease control of bacterial canker in sweet cherry, it is speculated that the poor control was due in large part to systemic infections. Other failures of control may be due to copper-resistant strains, poor chemical coverage, or inadequate timing as well as to reinfection from outside sources.

7.1.3 BLISTER SPOT OF APPLE

Pathogen: *Pseudomonas syringae* pv. *papulans*
Blister spot is a disease of apple fruit caused by the bacterium *P. syringae* pv. *papulans* (Rose) Dhanvantari. Although over 20 cultivars of apple have been reported as susceptible to the bacterium, the disease is usually of economic impact only on Mutsu in New York State. If uncontrolled, the disease generally affects 5%–60% of fruits in an orchard. The bacterium does not cause extensive decay of the fruit but makes it unsuitable for fresh market use.

Symptoms
Blister spot is first observed as small raised green blisters, which develop on fruit from early to mid July. These blisters are associated with stomata and continue to expand during the growing season. Near harvest, they range in diameter from 1 to 5 mm and are purplish-black in color (Photo 7.6). A few to hundreds of lesions may develop per fruit. The decay will rarely extend more than 1–2 mm into the fruit. The bacterium also infects leaves, causing midvein necrosis of leaves.

Host
Malus domestica (apple), *Malus sylvestris* (crab-apple tree), and *Pyrus communis* (European pear).

Pathogen and Disease Cycle
P. syringae pv. *papulans* is very widespread in Mutsu orchards. The bacterium survives the winter in dormant buds. Up to 40% of the dormant buds in an orchard may harbor the pathogen. The bacterium may also overwinter in infected fruit on the orchard floor. Once tissues begin growing in the spring, the pathogen can be detected in high numbers on Mutsu leaves, blossoms, and then fruits, even though no disease may be apparent. High populations of the bacterium have also been detected on Golden Delicious leaves, and infection occurs on Golden Delicious and Jonagold. These infections,

PHOTO 7.6   Symptoms include a midvein necrosis of leaves and blister spot on fruits. (Courtesy of Van der Zwet, T., USDA, Washington, DC.)

however, are much smaller in size and fewer in number than infections on Mutsu and do not cause a serious economic problem. The pathogen also survives on the foliage of some orchard weeds, including *Taraxacum officinale* Weber (dandelion), *Agropyron repens* L. (quackgrass), *Euphorbia esula* L. (leafy spurge), *Trifolium* sp. (clover), and *Malva neglecta* L. (common mallow).

Because of the ubiquitous nature of the pathogen, it can be assumed that it is present at high populations in Mutsu orchards throughout the growing season. The bacterium can be spread to susceptible fruits by insects and rain. The disease cycle of blister spot of apple is illustrated in Figure 7.2.

Fruit Susceptibility
Inoculation studies conducted to determine when during the growing season fruits are susceptible to *P. syringae* pv. *papulans* indicated that the susceptibility period begins from 2 to 21½ weeks after

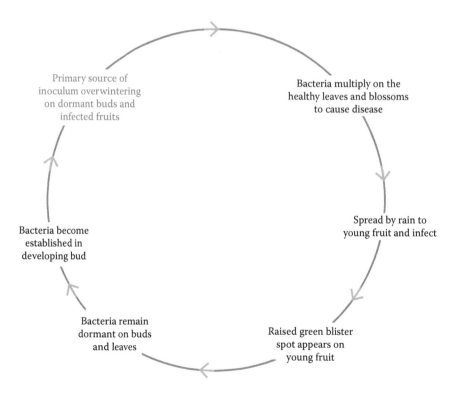

FIGURE 7.2   Disease cycle of blister spot of apple caused by *P. syringae* pv. *papulans.*

petal fall and continues for 4–5 weeks or until stomata on fruit have developed into lenticels. Even during seasons of delayed bloom, this relationship has held. In some years, environmental conditions after petal fall may favor an increased or decreased rate of fruit development, and thus may affect the susceptibility period.

Susceptibility of Varieties

The varieties Cortland, Delicious, Golden Delicious, Idared, McIntosh, and Rome Beauty all developed blister spot lesions when they were artificially inoculated with the pathogen in the orchard. With the exception of Golden Delicious, natural infections have not been observed on these varieties in recent years in New York State. These results do imply, however, that if pathogen populations become high, blister spot might develop on several apple varieties, especially if they are planted in close proximity to Mutsu.

Impact

The economic damage of *P. syringae* pv. *papulans* is not of yield but of cosmetic value. The blisters formed on the fruit seriously reduce the value of apples for the fresh market (Bonn and Fisher, 1989).

The disease has become a serious problem on Mutsu apples in Ontario, Canada, and Michigan and New York, the United States (Dhanvantari, 1977), but on other cultivars the organism is generally considered to be of minor importance.

Control Measures

Blister spot can be controlled using properly timed sprays of streptomycin. Field experiments have shown that it is essential to apply the first of these sprays prior to the onset of increased fruit susceptibility, that is, 2 weeks after petal fall. If the first spray is applied after this critical period, the percent disease control will drop significantly. In orchard field tests, control of blister spot was achieved with a mixture of 0.25 lb (113 g) streptomycin (17% WP) plus 1 pint (0.473 L) of glyodin per 100 gal (379 L) of water applied at a dilute rate of 300 gal of spray per acre. Dilute spray is recommended to ensure thorough coverage of all tissue on which the pathogen may multiply. If growers wish to use a concentrate spray, no less than 100 gal of water should be used. A total of three sprays at weekly intervals is recommended for maximum control (Burr, 1982).

7.1.4   CROWN GALL OF APPLE AND PEAR

Pathogen: *Agrobacterium tumefaciens*

Crown gall is caused by the bacterium *A. tumefaciens*. This bacterium has the widest host range of any plant pathogen. It is capable of causing tumors, or galls, on virtually all plant species, except the monocots (grasses). A similar bacterium, *Agrobacterium rubi*, causes galls on the canes of brambles. The disease is particularly destructive on brambles (raspberries and blackberries) and grapes. It can also cause severe problems on apple, pear, blueberry, all stone fruits, and on ornamentals.

The bacteria induce galls or tumors on the roots, crowns, trunks, and canes of infected plants. These galls interfere with water and nutrient flow in the plants. Seriously infected plants may become weakened, stunted, and unproductive.

A. tumefaciens was originally named *Bacterium tumefaciens* by Cavara in 1897. In 1907, the name was changed to *Phytomonas tumefaciens* (Smith and Townsend) Bergey et al. by E. F. Smith, which was subsequently changed to *A. tumefaciens*, the taxon name of current use. This organism was first isolated from grapevines in 1897 by Fridiano Cavara at the Laboratorio di Botanica del Recherci Instituto Forestale di Vallom Drosa in Naples, Italy. In 1907, in the United States, E. F. Smith and C.O. Townsend isolated the bacterium from chrysanthemum galls.

Symptoms

The disease first appears as small overgrowths or galls on the roots, crown, trunk, or canes. Galls usually develop on the crown or trunk of the plant near the soil line or underground on the roots. Although they can occur, aerial galls are not common on fruit trees.

PHOTO 7.7   Crown galls on apple branch. (Courtesy of Luepschen, N.S., Bugwood.org.)

In early stages of development, the galls appear as tumor-like swellings that are more or less spherical, white or flesh colored, rough, spongy (soft), and wart-like. They usually form in late spring or early summer and can be formed each season. As galls age, they become dark brown to black, hard, rough, and woody. Some disintegrate with time and others may remain for the life of the plant (Photo 7.7).

The tops of infected plants may appear normal. If infection is severe, plants may be stunted, produce dry, poorly developed fruit, or show various deficiency symptoms due to impaired uptake and transport of nutrients and water.

Predisposing Factors

Wounds that commonly serve as infection sites are those made during pruning, machinery operations, freezing injury, growth cracks, soil insects, and any other factor that causes injury to plant tissues. Bacteria are abundant in the outer portions of primary galls, which are often sloughed off into the soil. In addition to primary galls, secondary galls may also form around other wounds and on other portions of the plant in the absence of the bacterium. The bacteria overwinter inside the plant (systemically) in galls, or in the soil. When they come in contact with wounded tissue of a susceptible host, they enter the plant and induce gall formation, thus completing the disease cycle. The bacteria are most commonly introduced into a planting site on or in planting material.

Symptoms become evident 2–4 weeks after infection if temperatures are at or above 20°C, usually coinciding with warmer soil temperatures in May or June. Initially, the galls look like callus outgrowths but then increase rapidly in size and number. Symptom development slows greatly below 15°C and stops. Infection is inhibited above 34°C–35°C. Latent infections are symptomless and usually occur when soils are cool. Gall symptoms typically develop at the infected wound the following season; on rare occasions, galls don't appear until the third growing season.

Some problems can look like crown gall but are not pathogenic. Aerial burrknot on apple tree trunks and branches is a cushion-like assemblage of adventitious roots; it is thought to be caused by genetic reasons rather than due to an infectious agent.

Small galls require careful diagnosis because they may be confused with excessive wound callus or with galls induced by nematodes, fungi, or insects. Thus, pathogenic strains of *Agrobacterium* must be isolated to confirm a crown gall diagnosis.

Nonpathogenic *Agrobacterium* cells are often prevalent in these same tissues and can reach high populations. This makes diagnosis difficult, especially in galls on apple, blueberry, and grapevines where nonpathogens constitute over 99% of the *Agrobacterium* population. DNA probe technology is crucial for the recognition of pathogenic isolates among the general *Agrobacterium* population.

Host

Types of plants affected include shrubs, ornamental flowering trees, fruit trees (temperate to tropical) tubers, grapevines, tobacco, annuals, and members of the coniferae. Table 7.1 gives examples of host plants infected by *Agrobacterium* species and the relative frequency of infection. The frequency of crown gall on plants at any given location varies because the host range of *Agrobacterium* is strain-dependent. Some strains infect a wide range of hosts, others a narrow range. Anderson and Moore (1979) and Bradbury (1986) have listed 391 susceptible plant genera, many of which have many susceptible species. However, crown gall disease has been observed on only a few of these species in their natural habitat. Even the root systems of nonhost plants such as weeds, grasses, and cereals can harbor the pathogen and serve as a reservoir of inoculum.

Pathogen

A. tumefaciens is a rhizoplane bacterium and Gram-negative, strictly aerobic, with bacilliform rods measuring 1×3 μm, whose nutritional requirements are nonfastidious. The rods bear flagella that are arranged subpolarly around the cylindrical circumference of the cell, referred as circumthecal flagellation. When *A. tumefaciens* cells perceive plant phenolic compounds, the virulence genes that are located in the resident Ti (tumor-inducing) plasmid are expressed, resulting in the formation of a long flexuous filament called the T pilus. The activation of VirA also shuts off the motility of the circumthecal flagella, presumably when *A. tumefaciens* cells attach to plant cells. Attachment to the plant cells is a prerequisite for initiating the transfer of the T-DNA into the plant cell. Both the circumthecal flagella and the T pilus play an essential role in virulence, presumably by bringing the bacterial cell to its target followed by attachment to the plant host, respectively.

Based on some distinct phenotypic differences, *A. tumefaciens* isolates were originally classified into three biotypes or biovars (biotypes I, II, and III; biovars 1, 2, and 3). Biotype I or biovar 1 strains produce 3-ketosugars and usually have wide host ranges. Biotype II or biovar 2 strains mainly are classified as the hairy root-forming organism *Agrobacterium rhizogenes*. Biotype III or biovar 3 isolates are mainly confined to grapevines, prefer L-tartaric acid over glucose, and produce polygalacturonase. Because grapevine isolates formed a distinct group verified by DNA homology studies and were frequently limited in host range to grapevines, biovar 3 strains have been reclassified into one species, that is, *Agrobacterium vitis*. *A. rubi* strains infect canes of the genus *Rubus*, representing blackberry and raspberry.

TABLE 7.1
Frequency of *Agrobacterium* Infection in Certain Host Plants

High	Low	Occasional/Rare
Almond	*Chrysanthemum* spp.	Aster
Apricot	*Citrus* sp.	Birch (*Betula*)
Cherry	*Clematis* spp.	Blueberry (*Vaccinium*)
Peach	*Ficus* sp.	Cactus
Plum (*Prunus*)	*Gypsophila* sp.	*Dahlia* spp.
Apple (*Malus*)	Lilac (*Syringae*)	Gossypium
Euonymus spp.	*Macadamia*	Hydrangea
Grape (*Vitis*)	Marigold (*Tagetes*)	Impatiens
Poplar (*Populus*)	Olive (*Olea*)	Incense-cedar (*Calocedrus*)
Raspberry (*Rubus*)	Pear (*Pyrus*)	Maple (*Acer*)
Rose (*Rosa*)	Quince (*Chaenomeles*)	*Rhododendron* spp.
Rubus spp.	*Ribes* spp.	Sequoia
Walnut (*Juglans*)	Wild blackberry (*Rubus*)	Spruce (*Picea*)
Willow (*Salix*)	Wisteria	

Pathogenesis

Role of Ti Plasmid and Virulence Genes

Experimental inoculation of an assay host plant such as Jimson weed (*Datura stramonium*) results in tumor formation within 2 weeks. Virulence and the host range of *A. tumefaciens* are conferred by a large extrachromosomal DNA element designated as the Ti (tumor inducing) plasmid that resides in all virulent strains of this pathogen. The Ti plasmid is approximately 200 kilobases in length and comprises a covalently closed, double-stranded DNA circular molecule.

Upon recognition of plant signals in the form of dimethoxyl phenolic compounds released from plant wounds, virulence gene, *vir*A, encodes histidine kinase (VirA) that phosphorylates the response regulator (VirG), which in turn transcriptionally activates the remaining *vir* genes of the *vir* regulon. The products of the *vir* genes carry out the processing of the T-DNA and functions for T-DNA transfer to the plant cell. Also, perception of the plant phenolic compounds by VirA switches off the cell motility of *A. tumefaciens*. In addition, certain chromosomal genes (e.g., *chv*, *att* genes) of *A. tumefaciens* are also involved in virulence. Most of them have a role in the attachment of *A. tumefaciens* to plant cells in promoting signal molecule recognition and in regulating certain *vir* and T-DNA genes.

T-DNA transfer from *A. tumefaciens* to plants is aided by a long flexuous appendage known as the T pilus. The T-DNA is delivered as a single-stranded DNA molecule coupled to a VirD2 protein into the plant cell. The T-DNA is integrated into the plant chromosome as evidenced by *in situ* hybridization using the T-DNA as the probe. The T-DNA genes are expressed by plant transcriptional machinery.

The products of the T-DNA genes catalyze the formation of auxin, cytokinin, and opines. Profuse production of auxin and cytokinin in the transformed cells results in abnormal cell division, cell enlargement, and uncontrolled growth of the infected plant tissues. Opines are utilized as specific nitrogen and carbon sources by *A. tumefaciens*. The opines produced by the transformed plant cells provide unique nutritional sources for the pathogen, and certain opines promote transfer of the Ti plasmid between *Agrobacterium* strains. Certain opines such as octopine and agrocinopine serve as inducers of the Ti plasmid conjugative transfer system.

Disease Cycle and Epidemiology

A. tumefaciens naturally resides on the rhizoplane of woody and herbaceous weeds. Its presence in soils originates from galls that were broken or sloughed off from infected plants during cultivation practices or disseminated as infected plant material. Irrigation aids in further dissemination of the *A. tumefaciens* bacterial cells. *A. tumefaciens* is also spread by infected and infested planting stocks originating as nursery stock from uncertified sources. Secondary spread then occurs through pruning and cultivation equipment, particularly when galls are removed manually with the same cutting tools used in pruning. Tilling equipment can be contaminated by cutting through galls at or near the base of trunks of infected trees. Roguing (removal) of infected trees and replanting in the same spot where the infected tree had grown is a poor practice because sloughed off galls serve as sources of abundant populations of *A. tumefaciens*.

Although a single cell can be transformed and grow into a gall, the size of the initial tumor depends on the amount of inoculum. Thus, if a large number of cells of *A. tumefaciens* enter a plant wound, disease will be more severe and a large gall will be seen rapidly growing at the infection site. Plants that are systemically infected will harbor *A. tumefaciens* for extended periods of time in the absence of overt symptoms. Tissue injuries induced by accidental wounds made during cultivation practices or by frost will elicit new infections with the appearance of numerous galls along the vascular system of the host.

Disease Management

Pathogen-free plants grown in uninfested soil will not develop crown gall. This emphasizes the importance of planting clean propagating material in clean soil. Good sanitation (Aubert et al., 1982) and cultural practices are important deterrents to crown gall. At harvest, discard all nursery stock showing symptoms to avoid contaminating healthy plants and storage facilities. At harvest, leave noticeably galled plants in the field for later pickup and destruction. If possible, choose a rootstock that is less susceptible (Durgapal, 1977; Zurowski et al., 1985; Jaburek and Holub, 1987; Nesme et al., 1990; Ishizawa et al., 1992; Goodman et al., 1993; Lemoine and Michelesi, 1993; Pierronnet and Eyquard, 1993), avoid planting sites heavily infestated by root-attacking insects and nematodes (Dhanvantari et al., 1975; Tawfik et al., 1986; Vrain and Copeman, 1987), disinfect pruning equipment between trees, and adopt management practices that minimize wounding (Cazelles et al., 1991). Avoid planting into heavy, wet soil. Do not plant trees deeper than they grew in the nursery (Gloyer, 1934). If possible, incubate dormant seedling roots at 22°C–24°C for 10–14 days to heal wounds and reduce susceptibility to *A. tumefaciens* before planting them in wet soil (Moore and Allen, 1986). Use irrigation water from wells, if possible. Avoid planting where galled plants grew in the last 4–5 years (Tawfik et al., 1986); choose fields that were planted recently to vegetables or grain. In summary, avoid exposing plants to pathogenic *Agrobacterium* at any stage of plant production.

Preplanting Management Options

Planting Stocks
Visual examination for crown gall tumors has been the conventional primary screen for diseased material. The method is limited for complete disease control because *A. tumefaciens* can reside on the rhizoplane and systemically in certain host plants such as grapevines, chrysanthemums, and marguerite daisy.

1. Obtain clean (disease-free) nursery stock from a reputable nursery and inspect the roots and crowns yourself to make sure they are free from galls. Avoid planting clean material in sites previously infested with the bacteria.
2. Avoid all unnecessary root, crown, and trunk wounding during cultivation and other machinery operation, and control soil insects. Any practice that reduces wounding is highly beneficial. Preventing winter injury (especially on grapes) is also beneficial.

Planting Site
Site conditions play an important part in the incidence of crown gall. Crown gall is generally much more prevalent in heavy soils or in soil where water stands for a day or so. In New York, crown gall incidence was highest on a heavy clay knoll (15 ft elevation) from which water drained toward flat, loamy portions of the field (Gloyer, 1934). In Oregon, gall incidence on an Old Home × Farmingdale pear rootstock selection was severe (495 of 500 trees infected) in a heavy, wet soil, but in the same field, only 1 of 500 trees was galled outside the wet area.

Cropping history can influence crown gall incidence. Budded apple trees were found to be badly galled in fields where previous nursery crops such as grape, peach, raspberry, and rose had been heavily infected (Gloyer, 1934; Moore and Allen, 1986). This situation is not repeated at every site, but we still recommend avoiding fields with a recent history of crown gall.

Vectors
Nematodes, grubs, and other chewing insects have been implicated as passive carriers of *A. tumefaciens* that subsequently infect the feeding wounds.

Crop Rotation
A crop rotation program employing cereal crops followed by green manuring helps reduce the population size of *A. tumefaciens*.

Natural Resistance
Reports of plant resistance to crown gall are limited and findings are varied. In Britain, apple rootstock Malling Jewel was considerably more resistant than Malling Delight. McIntosh apple trees in Oklahoma, but not in Oregon, were very susceptible to *A. tumefaciens*. In Italy

and the Pacific Northwest, Malling 7 is considered the most susceptible apple rootstock to *A. tumefaciens* (Bazzi, 1983) followed by Malling 9 and 26. M9, however, is reportedly the most susceptible rootstock in Switzerland. Interestingly, the new root-stock Mark, a selection from Malling 9, was much more susceptible to *A. tumefaciens* than Malling 7 (Moore and Canfield, unpublished). This variability is likely because strains of the pathogen are better adapted to one nursery site than to another.

Biological Control

Using *Agrobacterium radiobacter* K84, a biological control agent, has been very effective against crown gall on a number of hosts, but exceptions exist. This biological control is *solely preventative, not curative*; hence, application timing is important to properly protect plant wounds caused at harvest or by pruning. Htay and Kerr (1974) recommend seed and root treatment with K84 for best results. Not all pathogenic strains of *Agrobacterium* are sensitive to K84. For example, most pathogenic agrobacteria isolated from grape tumors are *A. vitis* that are insensitive to K84. Even so, K84 has been effective against several insensitive *Agrobacterium* pathogens in field tests (Vicedo et al., 1993). K84 is compatible with etridiazole alone (Truban), etridiazole plus thiophanate-methyl (Banrot or Zyban), benomyl (Benlate), and Captan fungicides that have been used to protect dormant trees in cold storage (Moore, unpublished).

A new, genetically engineered strain of K84 called K1026 has been patented in the United States. K1026 was developed with a possible view that K84's effectiveness as a biological control of crown gall might break down. K1026 gives the same biological control as K84 and is being used commercially in Australia and Spain (Vicedo et al., 1993). BioCare, Ltd. (Australia) has submitted a registration packet to the U.S. Environmental Protection Agency for approval to market K1026 in this country.

A relatively new biological control agent for crown gall is available for apple, pear, stone fruit, blueberry, brambles, and many ornamentals. It is not effective on grape.

The agent is a nonpathogenic strain of bacterium (*A. radiobacter* strain 84) that protects the plants against infection by the naturally occurring strains of pathogenic bacteria in the soil. Nursery stock is dipped in a suspension of commercially prepared *A. radiobacter* strain 84 at planting time. The antagonistic bacteria act only to protect disease-free plants from future infection by the crown gall bacterium; they cannot cure infected plants.

For the control of crown gall on nursery stock, including apricots, cherries, nectarines, peaches, plums, prunes, and Asian and European pear, Dygall has been temporarily registered as a preventative inoculant. (Note: Dygall has not proven effective on apples and is not registered for use on apples.) Dygall contains the biological control agent *A. radiobacter* strain 84, a bacterium that inhibits the development of crown gall. Dygall must be applied to susceptible plants or cuttings before planting in crown gall–infested soil. Immerse roots or cuttings in the Dygall solution just prior to planting. Keep the treated planting stock cool, prevent any exposure to sunlight, and plant as soon as possible. Follow precautions and directions on the product label.

Chemical Treatments

No registered chemicals that effectively control crown gall are currently available in the markets. Field tests in Oregon and Washington with the antibiotic terramycin and a discontinued copper product, Copac E (Garrett, 1987), consistently gave best results in reducing tumor incidence on apple rootstocks (Canfield and Moore, 1992), but they are not at present registered for public or commercial use. In general, chemical preplant dips or soil drenches have been ineffective (Mirow, 1985).

Eradication of crown gall using creosote-based compounds, copper-based solutions, and strong oxidants such as sodium hypochlorite is transiently effective. The chemical eradicant application procedure is labor-intensive and therefore expensive both monetarily and environmentally. The superficial treatments are ineffective against systemically infected plants. Generally, chemicals are rarely used for control of crown gall.

Genetic Engineering

Transgenic crop plants harboring one or more unique genes tailored to protect the plant from crown gall have been developed. Genes encoding products that degrade or inactivate the T-DNA strand complex when it enters the host cell, which prevents the expression of T-DNA genes encoding indole-acetic acid and cytokinin biosynthesis, and those that prevent the attachment of *A. tumefaciens* to its target are some examples currently being tested. Biotechnology companies, such as DNA Plant Technology (Oakland, CA), are applying sense strand messenger RNA or small-interfering RNAs to develop crown gall–resistant fruit and nut crops.

7.2   APRICOT

7.2.1   Bacterial Canker of Apricot

Pathogen: *Pseudomonas syringae*

Host: Apricots and plums, but all stone fruits are susceptible.

Symptoms

Symptoms are most obvious in spring and include limb dieback with rough cankers and amber-colored gum. There may also be leaf spot and blast of young flowers and shoots. The sour sap phase of bacterial canker may not show gum and cankers, but the inner bark is brown, fermented, and sour smelling. Orange or red flecks and pockets of bacterial invasion under the bark occur outside canker margins. Frequently, trees sucker from near ground level are affected by the bacterial pathogen; but the cankers do not extend below ground.

The bacteria can invade all the above-ground parts of a plant. Symptoms vary widely between hosts and different climatic conditions but are more commonly initiated on green foliage.

Although bacterial canker may affect trees of any age, it is more often associated with death of trees in the first or second year after planting. The tree flowers and begins its growth normally but limbs or whole trees collapse and die within a matter of weeks in spring. Leaves often remain attached to the tree. Death usually occurs in the same season and young trees seldom, if ever, struggle on into the next season. Once the top dies, many shoots arise from just below-ground level because the disease does not invade the roots.

In mature trees, under the right climatic conditions, infection can spread quickly, killing large branches in a matter of weeks.

Areas of infection, particularly under bark of mature branches, are usually sunken. These cankers often exude large amounts of gum and often give off a characteristic vinegar (sour) smell. Slicing bark off the affected area reveals a brown or tan discoloration in the bark (orange in the case of apricots). The outer layers of wood are often stained a brown color in contrast to the normal white, but the inner wood (xylem) is not discolored except in the more severe cases.

Symptoms on leaves vary greatly, but generally first appear as water-soaked spots. Leaf spots are dark brown, circular to angular, and sometimes surrounded with yellow halos. The spots may coalesce to form large patches of dead tissue, especially at margins of leaves, or the centers of the necrotic spots may drop out, resulting in tattered leaves giving a shot hole appearance (Photo 7.8). Infected leaves may abscise during midseason. Infected leaf and flower buds may fail to open in spring, resulting in a condition referred to as dead bud. Small cankers often develop at the bases of these buds. Other infected buds open in spring but collapse in early summer, leaving wilted leaves and dried-up fruit. If blossom infection occurs, whole blossom clusters collapse as infection spreads into the fruit-bearing spurs. Blossom blight and spur blast mostly give it a shot-hole appearance. Infected fruits often develop depressed spots with dark centers and sometimes have underlying gum pockets.

Cankers on trunks, limbs, and branches exude gum during late spring and summer. Leaves on the terminal portions of cankered limbs and branches may wilt and die in summer or early autumn if girdled by a canker (Photo 7.9). Occasionally, large scaffold limbs are killed. Leaf and fruit

PHOTO 7.8   Bacterial canker–infected leaves of apricot. (Courtesy of Keil, H.L., Bugwood.org.)

PHOTO 7.9   Bacterial canker–infected stem of apricot. (Courtesy of Young, J.M., Landcare Research, Auckland, New Zealand.)

infections (Photo 7.10) occur sporadically, but they can be of economic significance in years with prolonged wet, cold weather during or shortly after bloom.

Predisposing Conditions

P. syringae survives on plant surfaces, is spread by splashing rain, and is favored by high moisture and low temperatures in spring. The bacterium is commonly found on healthy as well as diseased plants and becomes pathogenic only on susceptible or predisposed trees.

The disease is found almost exclusively in replanted orchards where ring nematodes flourish or in locations where spring frost is a problem. The disease is worse in low, gravelly, or sandy spots in the orchard. Vigorous trees are less susceptible to bacterial canker, while young trees, 2–8 years

PHOTO 7.10 Bacterial canker–infected fruits of apricot. (Courtesy of Young, J.M., Landcare Research, Auckland, New Zealand.)

old, are most affected. The disease rarely occurs in the first year of planting unless the ground is not fumigated before planting and is uncommon in nurseries.

Bacterial canker pathogen enters the tree through cracks, wounds, leaf scars, and bud scales during cool, wet weather in autumn and winter, though the symptoms will not be obvious until the next spring.

Disease Cycle

The bacteria can survive from one season to the next in bark tissue at canker margins, in apparently healthy buds and systemically in the vascular system. Bacteria multiply within these overwintering sites in the spring and are disseminated by rain to blossoms and to young leaves. Bacteria can live in an epiphytic phase on the surface of symptomless blossoms and leaves from bloom through leaf fall in autumn. After leaves abscise in autumn, the bacteria may enter the tree through fresh leaf scars. Outbreaks of bacterial canker are often associated with prolonged periods of cold, frosty, wet weather late in the spring or with severe storms that injure the emerging blossom and leaf tissues. Freezing can predispose the tissue to infection, but infection depends on the presence of wet weather during the thawing process. Free water on leaf surfaces and high relative humidity are required for at least 24 h before significant leaf infection can occur following violent storms. Symptoms appear about 5 days later at temperatures between 21°C and 27°C.

Control Measures

Firstly, when purchasing trees from nurseries, ensure they are free from bacterial canker. Badly infected trees are a source of infection for adjacent trees and are generally stunted and often die. Remove all diseased trees in young (less than 4 years old) plantings and replace with healthy trees. The bacteria will not infect a tree via its root system, so you can replant in the same spot.

In older plantings, careful pruning is an alternative. Limbs with developing cankers can be cut off and the wounds sealed to prevent further damage; but make sure the limb is cut back to clean wood. Do this only after leaf growth has started and while the tree is actively growing. Dip all pruning tools in an effective disinfectant (e.g., methylated spirits) after each cut and paint all wounds with a copper-based paint or Bordeaux paste.

In difficult situations where the disease is hard to control and particularly in young trees, a cover spray program according to which copper is sprayed during two springs, two autumns, and one winter is suggested. They are applied during early budswell, 7–10 days later, after 25%–50% and 90%–100% of leaf fall, and midwinter. Copper oxychloride, plus oil to act as a sticker, is an acceptable spray to use. Bordeaux mixture is an alternative.

Apricots beyond the pink bud stage are particularly susceptible to copper sprays. Applications of copper-based fungicides cause shot holing of the leaves and can damage fruit. For this reason, copper sprays should not be applied between pink bud and the commencement of leaf fall. Where the severity of disease is such that a copper spray at 25%–50% leaf fall is warranted, a half-strength spray should be used at this timing. Under normal circumstances, a minimum program of at least one early budswell and one autumn leaf fall spray should be sufficient.

The key to bacterial canker management is control of ring nematodes and maintenance of healthy, vigorous trees. Any management practice that improves tree vigor (e.g., lighter, more frequent irrigation with drip or microsprinklers, improved tree nutrition [especially nitrogen]) will help reduce the incidence of this disease.

It is very important to fumigate sandy soils when apricot trees are to be planted following an old apricot, peach, almond, or other *Prunus* spp. orchard. Rootstocks of plum parentage (e.g., Myrobalan, Marianna 2624) are highly susceptible to bacterial canker. Lovell peach and Viking rootstocks are more tolerant than Nemaguard or apricot rootstocks. In soils with high levels of ring nematodes, annual fall treatments with a nematicide are beneficial. There is evidence that pruning during the dormant period may make trees more susceptible than pruning after trees become active in spring or pruning in summer. Copper sprays applied at the beginning and end of leaf fall have been tried with highly variable results.

In light, sandy soils and in some heavy soils, control has been achieved with preplant fumigation for ring nematodes. Nematodes stress trees, which predisposes them to bacterial canker. The benefits of preplant soil fumigation for control of bacterial canker usually last only a few years.

Cultural

Use F 12-1 Mazzard as a rootstock. Use scions or buds from virus-free, canker-free trees. Remove trees with girdled trunks. Prune out branches with cankers. Cankers can be cleaned up by cutting away bark from above and around the edges of the infected area. Cover the wounds with dressing. Sterilize tools between cuts with 10% bleach solution or 70% ethyl alcohol.

Resistance

Resistant cultivars: None.
Intermediate: Corum, Sam, Sue.
Susceptible: Bing, Hardy Giant, Lambert, Royal Anne, Schmidt, Van, Windsor.

Limitations

1. There is no satisfactory control for this disease. It thrives in areas with moist, warm winters such as those found in the coastal region of British Columbia.
2. Locate orchard in an area less likely to be affected by frost.
3. Copper-containing compounds are of limited value because strains of *P. syringae* will develop resistance to them.
4. If copper sprays appear ineffective, check for *Cryptosporiopsis* infection.

7.2.2 Crown Gall of Apricot

Pathogen: *Agrobacterium tumefaciens*

In the winter of 2002, nearly 30% of 2-year-old apricot trees (*Prunus armeniaca* cv. Ninfa) in two commercial orchards in Adana and Mersin, in the eastern Mediterranean region of Turkey, were observed with grown gall symptoms. Tumors and galls were often found at or just below the soil surface on the roots or crown region of the apricot plants as described by Ogawa et al. (1995).

Symptoms

Crown gall disease results in rough, abnormal galls on roots or trunk. Galls are soft and spongy, not hard. The centers of older galls decay. Young trees become stunted and older trees often develop secondary wood rots.

Pathogen

Crown gall bacteria survive in gall tissue and in the soil. They enter the tree only through wounds. Crown gall is most damaging to young trees, either in the nursery or in new orchard plantings.

Disease Management

The incidence of crown gall can be reduced by planting noninfected, "clean" trees. It is also important to carefully handle trees to avoid injury as much as possible, both at planting and during the life of the tree in the orchard. Preplant, preventive dips or sprays with a biological control agent are available and may be helpful in some orchards. Generally, by the time crown gall is evident in an apricot orchard, it is usually best to tolerate the problem for the few remaining years of orchard life, which is about 12–15 years, or just remove the orchard and start anew. When replanting at previously affected site, remove as many of the old tree roots as possible, grow a grass rotation crop to help degrade leftover host material and reduce pathogen levels, and offset the new trees from the previous tree spacing to minimize contact of healthy new roots with any infested roots that may remain.

7.2.3 BACTERIAL LEAF SPOT OF APRICOT

Pathogen: *Xanthomonas arboricola* pv. *pruni* (Smith) Vauterin et al.

Synonyms: *Xanthomonas campestris* pv. *pruni* (Smith) Dye
Xanthomonas pruni (Smith) Dowson

Common names: Bacterial leaf spot, shot hole, black spot (English)
Tache bactérienne, bactériose (French)
Fleckenbakteriose (German)

Bacterial spot caused by *X. arboricola* pv. *pruni* was described for the first time in the United States (Michigan) in 1903 on Japanese plum. The disease is now reported from almost all continents where stone fruits are grown. *X. arboricola* pv. *pruni* attacks only *Prunus* species and particularly the fruit crops. It is best known as a pathogen of plum, nectarine, and peach (Stefani et al., 1989), but it has been reported on apricot (Scortichini and Simeone, 1997), almond (Young, 1977), and cherry as well.

Host

X. arboricola pv. *pruni* attacks only *Prunus* spp., and particularly the fruit crops such as almonds, peaches, cherries, plums, apricots, and *Prunus salicina*. Other exotic or ornamental species of *Prunus* attacked include *Prunus davidiana* and *Prunus laurocerasus*. Cultivars of the Sino-Japanese group (*Prunus japonica* and *Pr. salicina*) are generally more susceptible than European plums (Bazzi and Mazzucchi, 1984; Topp et al., 1989).

Geographical Distribution

X. arboricola pv. *pruni* was first described in North America. It is not clear from the literature whether it has spread from there or naturally has a wider range.

EPPO region: Found in Austria (unconfirmed), Cyprus (unconfirmed), Lebanon, Moldova, the Netherlands (unconfirmed), Switzerland (unconfirmed), Ukraine. Locally established in Bulgaria, Italy, Romania, Russia (European, Far East), Slovakia (unconfirmed), and Slovenia.

Asia: China (widespread), Cyprus (old unconfirmed record, now absent), Hong Kong, India (Himachal Pradesh), Japan, Korea Democratic People's Republic, Korea Republic, Lebanon, Pakistan, Russia (Far East), Saudi Arabia, Taiwan, Tajikistan.

Africa: South Africa, Zimbabwe.

North America: Bermuda, Canada (Manitoba, Nova Scotia, Ontario, Quebec), Mexico, the United States (Alabama, Arkansas, Connecticut, Florida, Georgia, Kentucky, Louisiana, Maryland, Michigan, Missouri, Mississippi, New Jersey, North Carolina, South Carolina, Texas).

South America: Argentina, Brazil (Santa Catarina, São Paulo), Uruguay.

Oceania: Australia (New South Wales, Queensland, Victoria, Western Australia), New Zealand.

EU: Present in European union region.

Disease Symptoms
Symptoms of bacterial spot can be observed on leaves, fruits, twigs, and branches (EPPO, 1997).

Symptoms on Leaves
Infection is first apparent on the lower surface as small, pale-green to yellow, circular, or irregular areas with a light-tan center. These spots soon become evident on the upper surface as they enlarge, becoming angular and darkening to deep-purple, brown, or black. The immediately surrounding tissue may become yellow. The diseased areas drop out, usually after darkening in color, but they may drop out prior to the color change, giving a shot-hole appearance to the leaf. Often, a dark ring of diseased tissue is left with the formation of the shot hole. Spots are usually concentrated toward the leaf tip because the bacteria accumulate in this region in droplets of rain or dew. Bacterial ooze may be associated with the spots. Severely infected leaves turn yellow and drop off. Atypical symptoms reported include a gray leaf spot on the upper surface, and a case in which bacteria infiltrated a large area, giving the leaf a greenish-yellow, translucent appearance. A severe defoliation can occur, leaving a carpet of yellow chlorotic leaves under the trees of susceptible cultivars.

Symptoms on Fruits
On fruits, small circular brown spots appear on the surface. They become sunken, the margins are frequently water-soaked, and there are often light-green haloes that impart a mottled appearance to the fruit. As a result of natural enlargement of the fruit, pitting and cracking occur in the vicinity of the spots. These cracks are often very small and difficult to see, but where heavy infection has occurred on young fruits they can be extensive, severely damaging the fruit surface. Gum flow, particularly after rain, may occur from bacterial wounds; this may easily be confused with damage caused by insects. As a general rule, symptoms on fruits appear 3–5 weeks after petal fall and develop until the skin color changes, when ripening process begins and some physiochemical parameters change. Symptoms often occur after hail damage.

Symptoms on Twigs
On twigs, spring cankers occur on the top portion of overwintering twigs and on watersprouts before green shoots are produced; initially they appear as small, water-soaked, slightly darkened, superficial blisters and extend 1–10 cm parallel to the long axis of the twig and may even girdle it. In this case, the tip of the twig may die, while the tissue immediately below the dead area, in which the bacteria are present, becomes characteristically dark; this is the so-called black tip injury. Twig infections later in the season result in summer cankers, which appear as water-soaked, dark-purplish spots surrounding lenticels. These later dry out and become limited, dark, sunken, circular-to-elliptical lesions with a water-soaked margin. On apricot twigs and branches, cankers are perennial, in contrast to peach, and continue developing in twigs 2 and 3 years old. The inner bark is penetrated, resulting in deep-seated cankers that deform and kill twigs.

Disease Cycle
X. arboricola pv. *pruni* overwinters primarily in the intercellular spaces of the cortex, phloem, and xylem parenchyma toward the tips of twigs produced during the preceding season. On apricot, summer cankers formed in one season continue developing the following spring, so

providing a source of inoculum at this time. Buds and fallen leaves have also been reported as overwintering sites.

In the spring, before host division starts, the bacteria in the intercellular spaces multiply and cause the epidermis to rupture, initiating a visible lesion referred to as a spring canker. Inoculum from these cankers is disseminated in rain and wind and infects new leaf growth via stomata. Lesions developing on the leaf exude bacteria that bring about secondary infections. Du Plessis (1983, 1987) suggests that the bacterium may migrate systemically from twigs to leaves. Pruning operations will also transmit the disease (Goodman and Hattingh, 1988).

Following foliage infection, summer cankers develop in the green tissue of the shoot, but they usually become sealed off by a periderm layer, and as cankers tend to dry out during the course of summer, the viability of bacteria therein is largely reduced; thus, except in certain localities, summer cankers are of no importance as overwintering sites for the bacterium, or in initiating infections the following spring. In general, it is the late infections of shoots, occurring during rains just before and during leaf fall in the autumn, when the host resistance mechanism of producing a periderm barrier is reduced, which constitute the primary inoculum source for the following spring.

A warm, moderate season with temperatures of 19°C–28°C and with light, frequent rains accompanied by fairly heavy winds and heavy dews is most favorable for severe infection. The disease tends to appear and spread in the spring, and then makes little progress through the summer, but late infections occur in the autumn. In culture, bacteria have survived ice-box conditions of –2°C to +2°C for 5 months. The disease is not usually found in arid regions.

Strain differences have mostly not been noted in *X. arboricola* pv. *pruni*, but Du Plessis (1988) has found differential virulence to peach, plum, and apricot cultivars.

Detection and Inspection Methods

X. arboricola pv. *pruni* can be detected by a detached-leaf bioassay (Randhawa and Civerolo, 1985) and by isolation (Gitaitis et al., 1988). Although serological techniques have been developed for other *X. campestris* pathovars on fruit crops, it seems that none is yet available for *X. arboricola* pv. *pruni*.

Means of Movement and Dispersal

X. arboricola pv. *pruni* has a limited capacity for local dispersal by rain splash in orchards. In international trade, it is likely to be carried on plants for planting (except seeds) of host species, including budwood. The bacterium may also be found on fruits.

Control Measures

Resistant cultivars are available, and *Prunus* breeding programs in North America attach considerable importance to *X. arboricola* pv. *pruni* resistance. No direct control methods are suggested, but bactericides have been evaluated (Du Plessis, 1983). Care should be taken to ensure that budwood is obtained from disease-free trees, preferably grown in arid regions. Orchard management practices in New Zealand have been reviewed by Young (1987).

Phytosanitary Risk

X. arboricola pv. *pruni* is listed as an A2 quarantine pest by EPPO (OEPP/EPPO, 1978) and is of quarantine significance for IAPSC. The disease is rated as of little economic value by the EPPO countries where it currently occurs, but it is absent in several major countries producing *Prunus*. Its behavior elsewhere in the world suggests that it would be likely to establish more widely in the EPPO region, although, in general, the bacterium would not present a threat to arid regions.

Phytosanitary Measures

The EPPO-specific quarantine requirements recommend that consignments of plants for planting (except seeds and tissue cultures), and fruits, of *Prunus* should come from a field found free from the disease by growing-season inspection (OEPP/EPPO, 1990).

7.3 SWEET CHERRY

7.3.1 BACTERIAL CANKER OF SWEET CHERRY

Pathogen: *Pseudomonas syringae* pv. *syringae*

Bacterial canker is the number-one killer of young sweet cherry trees. This disease is particularly destructive in higher rainfall districts of Oregon such as the Hood River and Willamette Valleys. It is still an important disease in drier climates such as the Dalles, the Mosier area, and Central Washington. Where conditions have been favorable for disease development, tree losses of 75% have been observed in young orchards. Under normal conditions, losses between 10% and 20% are not uncommon. In the Mid-Columbia district, sweet cherry trees have been planted beyond areas that experience the weather-moderating effects of the Columbia River, thus increasing the risk of these orchards to spring frosts and cold winters that promote bacterial canker.

Symptoms

The main symptom on young sweet cherry trees is a canker that appears slightly sunken and darker than other areas on the trunk or larger scaffold limbs. The inner tissue of cankers is orange to brown, and narrow, brown streaks extend into healthy tissue above and below the canker.

A clear-to-amber ooze that may be gummy or sticky may or may not be associated with the cankers. Cankers may eventually girdle limbs or trunks. Leaves droop, turn light green to yellow, and limbs or entire trees may die during hot weather. Infected dormant buds often are killed in the Willamette Valley; this symptom is known as dead bud.

If blossom infection occurs, cankers subsequently form on twigs and spurs, and the dead flowers remain attached on fruit spurs (Photo 7.11). Leaf and fruit infections are rare in Oregon's major fresh sweet cherry production regions. Leaf spots are at first water-soaked but then become dry and brown. Fruit infections appear as small, dark, sunken lesions with distinct margins.

Disease Cycle

P. syringae colonizes the surfaces of many plants, including weeds and grasses commonly found in orchards. It has also been found in healthy-appearing wood and buds of sweet cherry. In the Willamette Valley, bud infections (dead bud) are common. These infections originate at the base of outside bud scales between November and February, but a small percentage takes place through leaf scars formed when the leaves drop in the fall.

Infection of young trees occurs mainly in tissue damaged by spring frost, severe winter freezes, pruning cuts, insect injuries, or at leaf scars. Tree death is generally greatest in the first 2 years after

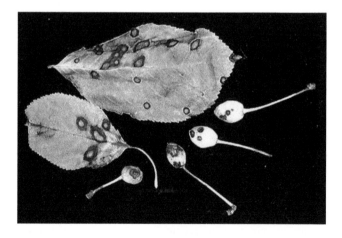

PHOTO 7.11 Bacterial canker symptoms on leaves and fruits of sweet cherry. (Courtesy of MacSwann, I., pnwhandbooks.org.)

planting, and trees are seldom killed 8 years or more after planting. Severe infections and a high rate of tree death are related to heading cuts.

P. syringae populations on sweet cherry trees increase 10- to 100-fold during bloom in mid-April, the time many orchards are planted. *P. syringae* populations are low during summer, but with the return of autumn rains and cooler temperatures, *P. syringae* is detected at high levels prior to and during leaf fall in sweet cherry orchards. Infection at leaf scars can be high at this time and results in tree death. Infection of sweet cherry also increases when tissue is exposed to severe temperatures during early and midwinter. Gummosis is associated with many cankers and is a common source of bacterial cells that can be spread by rain and insects.

Disease Management

Pruning
P. syringae can infect the woody tissue of sweet cherry at wound and injury sites. However, severe infections and a high rate of tree death are related to heading cuts. Since bacteria often are spread by rain, heading cuts should be made during dry weather. Heading cuts become resistant to infection after 1 week in summer but may take 3 weeks in winter. If rain occurs before cut tissues become resistant to infection, protective measures must be taken.

Scoring and summer pruning cuts are potential sites for infection. Scoring usually is done before the populations of *P. syringae* increase in the spring. Summer pruning is done during dry, hot weather when *P. syringae* populations are lowest. Since these wounds are potential sites for infection, growers should not make these cuts if summer rain occurs or if over-tree sprinklers are used for irrigation or evaporative cooling. Bacterial canker can be spread in summer or winter by cutting through active cankers and then immediately using the same pruning tool to make heading cuts on healthy trees. Leaf scars can become naturally infected and are another area that needs protection.

Cultivar Susceptibility
The cultivars Rainier and Regina appear more resistant than Sweetheart and Bing. In a Willamette Valley survey, Sandra Rose and Regina had less dead buds and blossom infections than Sweetheart, Staccato, or Solamente.

Rootstocks
Rootstock significantly affects cultivar susceptibility to bacterial canker. Death of trees on Mazzard was 30% but 77% when trees were on Gisela 6 rootstock. While no Bing on Colt rootstock died, mortality of Bing on Gisela 6 was 90% in one study. Trees on Gisela rootstocks have shown increased susceptibility in field observations. Bing on Krymsk 5 had smaller heading cut cankers than trees on Mazzard or Gisela 6, and 43% of trees died on Krymsk 5 compared to 50% on Mazzard.

Integrated Disease Management
An integrated approach is necessary for successful management of bacterial canker of sweet cherry. Because all natural and manmade injuries are susceptible to infection by *P. syringae*, injuries should be avoided whenever possible. Infection sites should also be protected until healing occurs. Bacterial canker is reduced when tree trunks are painted white to reduce sun scald–related winter injury.

Growers are encouraged to make heading, pruning, and scoring cuts only during dry weather. For this reason, pruning in the summer can help to keep infection low, especially if the disease is already present within a block. However, in large cherry orchards or where trees are planted on productive rootstocks such as Gisela, it may not be possible or advantageous to prune all trees in summer. In this case, colder, dry periods of midwinter when temperatures are around freezing or slightly above can provide a reasonable alternative to summer pruning.

Heading cuts and leaf scars are particularly susceptible and require additional protection. Copper-based products provide poor control of bacterial canker and in some cases have resulted in more bacterial canker than in nonsprayed trees. None of the cultivar/rootstock combinations consistently showed high resistance to bacterial canker, but Bing on Gisela 6, which had a mortality of 90%, should not be planted in areas where bacterial canker is a problem.

In addition, it is important to plant cherry trees in well-drained soils and keep trees moderately vigorous through proper irrigation and nutrition. Field observations indicate that stressed trees show higher levels of infection than those with a moderate amount of vigor.

7.4 WALNUT

7.4.1 BACTERIAL BLIGHT OF WALNUT

Pathogen: *Xanthomonas campestris* pv. *juglandis*

Carpathian walnut is susceptible to a destructive bacterial blight caused by *X. campestris* pv. *juglandis*. Severe leaf infections can defoliate trees by early August, which reduces the health of the tree and development of the crop. The most economically important damage occurs when the developing nut is infected.

It is a major bacterial disease of walnuts worldwide. The pathogen infects flowers, shoots, leaves, buds, and fruits. The premature fruits drop; that is, up to 100% occurs with total crop loss. Staining caused due to infection reduces the quality of in-shell product, the walnut.

Symptoms

On leaves, infection appears first as reddish-brown spots and on the stems as black, slightly depressed spots often girdling the shoots (Photo 7.12). Young, infected leaves and catkin buds turn dark brown or black and soon die. The disease is serious on nuts, where it causes black slimy spots of varying sizes (Photo 7.13). The organism penetrates the husk, the shell, and occasionally the edible meat. Late-season infection produces black rings on husks.

PHOTO 7.12   Brown necrosis on green twig of walnut. (Courtesy of Gardan, L., French National Institute for Agricultural Research [INRA], Paris, France.)

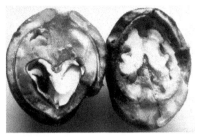

PHOTO 7.13 Bacterial blight of walnut on fruits. (Courtesy of Gardan, L., French National Institute for Agricultural Research [INRA], Paris, France.)

PHOTO 7.14 Canker on twig of walnut caused by *Xanthomonas*. (Courtesy of Gardan, L., French National Institute for Agricultural Research [INRA], Paris, France.)

In walnut blight, one to several black lesions may appear on catkins. Infected nuts develop black, slightly sunken lesions at the flower end (end blight) when young; more lesions develop on the sides of the nut as it matures (side blight). Shoots develop black lesions and leaves show irregular lesions on blade (Photo 7.14).

One to several black lesions may appear on male catkin flowers in spring. Infected young nuts develop black, slightly sunken lesions at the flower end. Additional lesions develop on the sides of the nut as it matures. Shoots develop black lesions along the stem. Leaves show irregular-shaped black lesions on the blade (Photo 7.15). Leaf infections can lead to defoliation when severe.

Disease Cycle

The bacterium overwinters in infected buds and catkins. Buds with the highest bacterial populations are the ones most likely to develop blight. During early spring growth, bacteria spread along developing shoots and nuts. There seems to be very little secondary spread to other shoots and trees by rain drip. This results in local infection centers within a tree or orchard. A frequent prolonged rain,

PHOTO 7.15   Bacterial blight of walnut on leaves. (Courtesy of Scortichini, M., atlasplantpathogenicbacteria.it.)

just before and during bloom and for about 2 weeks after, results in severe blight outbreaks within these local infection centers.

Walnut blight bacteria overwinter primarily inside infected dormant buds. Infection can occur as soon as buds break and growth begins in spring. All green tissue is sensitive to infection. Rain is important for the spread of bacteria to green tissue and causes new infection. In Ontario, this disease appears to be more prevalent in seasons with rain and high humidity and is active throughout spring and summer.

All green tissues are sensitive to walnut blight infections. Economic damage occurs when the developing nut is infected. Early leafing varieties are most severely affected, and the disease tends to be more severe in northern California.

Environmental Factors Involved in Disease Development
Rainfall and free moisture cause infection and dissemination of the pathogen. Duration of surface wetness is critical for infection. As little as 5 min is required for susceptible fruits to get infected. Relative humidity increasing within the tree canopy is implicated with increased incidence of walnut blight.

Forecasting
A model, Xanthocast, can determine blight risk. Temperatures during periods of leaf wetness are measured and a daily index is calculated. A sliding 7-day accumulation of the daily index is calculated.

Disease Management
 1. Prune out diseased twigs and branches, if practical.
 2. Management of this disease depends on the application of protective sprays to buds, flowers, and developing nuts. In orchards with histories of walnut blight damage, protective treatments at 7- to 10-day intervals during prolonged wet springs are necessary for adequate disease control. In areas or years with less intensive rainfall, spray intervals can be stretched and weather forecasts can help with spray timing.
 3. In years with high rainfall during catkin flowering, treatments may be applied when 30%–40% of the catkins emerge. (Note: This is usually 7–10 days before pistillate flowers emerge.) In most years, the first application can be delayed and should be applied when 30%–40% of the buds reach the "prayer" stage (when terminal leaves of pistillate flower

buds first unfold and appear like hands in a prayer position). A second spray should be done 7–10 days later to effectively treat the pistillate flowers that weren't sufficiently open during the initial application. Additional treatments can be timed using weather predictions. A spray prediction model (Xanthocast) is available at www.irrigate.net to help determine the need for additional treatment.

4. Applications of copper spray in June and July, in periods of prolonged wetting, at 7- to 10-day intervals, can help reduce the incidence of walnut blight infection. As a precautionary note, resistance of bacterial blight to copper spray has become a serious problem in other countries growing Carpathian walnuts. Use copper sprays only when necessary to control infections.

5. Spray at early prebloom (when catkins begin to enlarge), late prebloom (when shoots begin to expand), and early postbloom. In California, initial treatments are recommended when the terminal buds break. Resistance to copper products has been seen widely in California and may be a problem in the Pacific Northwest. However, copper-resistant bacteria do not seem to cause as much disease as ones that are copper-sensitive. Most spray adjuvants have not improved blight control; however, silicone-based adjuvants show promise.

Biological Control

Serenade MAX at 2–4 lb/A. Active ingredient is a protein from *Bacillus subtilis* strain QST 713. Effectiveness as a commercial treatment is unknown in the Pacific Northwest. 4 h reentry.

7.4.2 Crown Gall of Walnut

Pathogen: *Agrobacterium tumefaciens*

The United States is the largest producer of walnuts in the world, with nearly all the walnut acreage concentrated in California. In California, walnuts (*Juglans regia*, the "English" walnut) are commonly grafted onto Paradox rootstocks, which have a *J. regia* male parent and a black walnut female parent, generally *Juglans hindsii* (Potter et al., 2001). While Paradox is vigorous and performs well in a variety of conditions, this rootstock is extremely susceptible to crown gall disease (McKenna and Epstein, 2003).

Symptoms

Crown gall appears as rough, abnormal galls at or below the soil surface on roots or trunk. Live galls are not hard but soft and spongy. The centers of older galls decay. Young trees become stunted. Older trees often develop secondary wood rots.

Young galls are somewhat soft and spongy, not hard, and lack annual growth rings.

Crown galls (CG) and root galls (RG) are caused by *Agrobacterium* spp. Root galls generally have little impact, but crown galls may affect walnut tree growth and yield.

Galls sometimes are visible at ground level; most galls are below ground on the crown (the juncture between the main roots and the trunk) or scattered along the roots. In contrast to galls on the crown, galls on the roots are generally smaller and appear to have little impact on most trees.

Pathogen

Crown gall on walnuts is caused by two species of bacteria: *A. tumefaciens* and *A. rhizogenes* (Young et al., 2001). *A. tumefaciens* is also known as *A. tumefaciens* biovar 1, *A. radiobacter*, and *Rhizobium radiobacter*, while *A. rhizogenes* is also known as *A. tumefaciens* biovar 2 and *Rhizobium rhizogenes* in California. *A. tumefaciens* is a more common causal agent of crown gall in Paradox than *A. rhizogenes* (Kaur and Epstein, unpublished), although the reverse may be true in Oregon or other locales (Moore et al., 1996).

Crown gall affects nut and fruit trees in both nurseries and orchards. Based on observations of stone fruit trees, Moore (1976a, 1976b) and Alconero (1980) suggested that some infections are latent, occurring in the nursery, but only developing galls after transplantation.

The transplanted Paradox trees that had been next to trees with galls were significantly more likely to have galls at the crown than transplanted trees that had been next to gall-free trees in the nursery (14% vs. 3%). The location history of a tree in the nursery did not have a significant effect on the incidence of galls on the roots but did have a significant effect at $P < 0.05$ level on the incidence of galls overall. The location history of the tree in the nursery can have a significant effect on the incidence of galls in an orchard, demonstrating that the pathogen can be acquired in the nursery, but a visible gall is not produced until after transplantation.

The wounding treatment had no effect on the incidence of galls on the crown, but did have a slight effect on the incidence of galls overall because of a slightly higher incidence of galls on wounded than unwounded roots. McKenna and Epstein (2003) demonstrated that galls are associated with wounds that penetrate into the cambium or perhaps the phloem. In the crown, the cambium is the thin layer between the bark and the hardwood, and the phloem is a thin layer on the inside of the bark.

Wounds incurred during the normal course of harvesting and transplanting trees from commercial nurseries are rarely sufficiently deep to induce a gall, particularly in the crown region. To date, root-pruned transplants from a nursery were planted and dug again after 2 years (McKenna and Epstein, 2003; Kaur and Epstein, unpublished), where galls were rarely present in locations with root-pruned transplants.

The bacterium *A. tumefaciens* survives in soil and gall tissue. Bacteria enter primarily through wounds. Crown gall is most damaging to trees that are 1–8 years old. Paradox rootstock is especially susceptible. Gall decreases growth and yield.

Disease Management

The incidence of crown gall can be reduced by planting noninfected, "clean" trees. Before planting, make sure trees stay moist and the roots do not dry out. It is also important to carefully handle trees to avoid injury as much as possible, both at planting and during the life of the tree in the orchard. Although preplant, preventive dips or sprays with a biological control agent are available, their effectiveness can be variable on walnut trees.

Look for and treat crown gall during the growing season when the orchard is not wet because moisture favors the bacterium. When established orchard trees are infected with crown gall, you can use a combination of surgery, flaming, or a bactericide to treat the tumors. The best time to treat is during the spring because with rapid tree growth occurring, new callous tissue is formed relatively quickly.

The treatment is most effective for small galls on young trees. The procedure, however, can be expensive and difficult to carry out, depending on the size and location of the galls. If trees less than 4 years old are severely affected with galls, it is more economical to remove the trees and replant. If galls develop on trees after they are 7 or 8 years old and the trees appear healthy otherwise, then treatment probably is not necessary. For trees between 4 and 7 years old, the decision whether to treat galls or remove trees depends on the severity of galling and the cost of treatment relative to the cost of replacing trees.

To treat crown galls, first remove soil away from the crown and roots to completely expose the gall. Soil can be safely removed using pneumatic equipment such as air compressors or hydraulic excavation with fire suppression equipment or water washers. To flame the gall, use a propane cylinder or bottle and slowly move the torch tip around the margin of the gall, creating a red-hot zone that is about 1 in. wide. It may be advantageous to surgically remove the gall in order to gain access to all parts of the gall margin. If surgery is used, be sure to sterilize the tools with heat before advancing to the next tree. Leave the treated areas uncovered for a year and retreat if galls begin to regrow. Treatment success is about 80%.

As an alternative to flaming, galls can be treated with a bactericide such as Gallex, but treatment success depends on the complete removal of the gall first.

When replanting a previously affected site, remove as many of the old tree roots as possible, grow a grass rotation crop to help degrade leftover host material and reduce pathogen levels, and offset the new trees from the previous tree spacing to minimize contact of healthy new roots with any infested roots and soil that may remain. Before placing the tree in the planting hole, line the hole with clean soil and then fill the hole with clean soil at planting. Keeping the crown area dry may help reduce disease severity.

Extrapolating from these results, if walnut growers remove soil a year after planting and see crown galls in a high proportion of the trees, they can then predict the reduction in nut yield in the first 4 years of production. Nut yield will be reduced by up to 17% if gall affects less than one-quarter of the crown, by 18%–36% if gall affects between one-quarter and half of the crown, and by 36%–54% if gall affects between half and three-quarters of the crown.

Consequently, 1- to 4-year-old trees with severe gall should be replaced. Surgical removal of crown galls is generally recommended on less severely infected trees that are 1–4 years old. More detailed recommendations and methods are described in the UC IPM Management Guidelines for Crown Gall on Walnut (UC IPM Online, 2007).

7.4.3 Deep Bark Canker of Walnut

Pathogen: *Brennaria (=Erwinia) rubrifaciens*

During the summer of 1995 and subsequent years, bark cankers were observed in walnut trees (cv. Hartley grafted on *J. hindsii*) imported from California in 1978 and growing in Badajoz, Spain. Two foci were found in an orchard of 200 ha where 80 walnut trees were affected. Cankers were observed on trunks and branches, and dark exudates staining the bark appeared mainly in summer. Isolations were performed from affected tissue using King's B medium, which yielded *Brenneria* (*Erwinia*)-like colonies (Hauben et al., 1998).

Symptoms
The symptoms that characterize deep bark canker are the deep cracks running down scaffolds and trunks. A reddish-brown to dark brown substance oozes from these cankers from late spring through early fall, giving them a "bleeding" appearance (Photo 7.16). Internally, dark brown to

PHOTO 7.16 Deep bark canker symptoms on walnut trunk. (Courtesy of Scortichini, M., atlasplantpathogenicbacteria.it.)

black streaks of varying width extend through the inner bark and may run for many feet up and down the scaffolds and trunks of affected trees. Because these streaks occur deep in the bark in the region of the phloem, the disease was named phloem canker or deep bark canker to distinguish it from shallow bark canker, which is restricted to the outer layers of bark.

Another typical internal symptom of deep bark canker is the numerous small, round, dark spots that extend into the wood beneath the cankered areas.

Deep bark canker infections develop first on the trunk or lower scaffold; only one or two scaffold limbs are affected. As the disease progresses upward, the branch weakens slowly over time. After many years, most branches are affected, and the tree becomes less productive. The symptoms do not extend into the rootstock.

Geographical Distribution
Deep bark cankers are geographically found in the United States and Europe.

Disease Cycle
Deep bark canker occurs in all walnut-growing areas of the Central Valley but is rarely a problem in coastal growing areas. Deep bark canker is most common and most severe on the Hartley cultivar. The canker does not kill trees, but it may further debilitate trees already weakened by other factors, including inadequate irrigation, poor water infiltration, restrictive soils, as well as insect and disease pests.

The deep bark canker pathogen is most commonly transmitted in symptomless graft wood used to develop new trees. The disease may also be spread when the bacterial pathogen is introduced into a deep wound that exposes the phloem, such as wounds caused by a shaker and possibly woodpeckers. Shallow wounds and pruning cuts are not infected.

The pathogen survives the winter in cankers or dried exudate on the tree surface. The bacteria become active in late spring and begin to ooze from the cankers, together with plant sap. At this time bacteria may be spread by wind-blown rain to wounds on uninfected trees. Trees are susceptible to infection from April to October and almost completely resistant in winter. Cankers may lengthen by about 1 ft in spring and about 2 ft in summer. The bacteria spread within the tree through nonconducting parts of the phloem tissue. Movement of nutrients in the phloem is impaired, slowly weakening the affected branches.

Because high temperatures favor the development of the disease, deep bark canker is more prevalent in the Central Valley than in coastal areas. The disease sometimes develops on trees that have not been injured. It is thought that these are latent infections that were introduced when the tree was grafted.

Disease Management
To avoid predisposing trees to infection by deep bark canker, keep them healthy by practicing good water management, fertilization, pruning, and pest control. These management practices also reduce canker development in infected trees and keep them in production in most cases. Only trees growing on poor soils or under particularly adverse conditions may never fully recover. Infected trees may remain free of symptoms unless stressed.

Deep bark canker cannot be cured by any known chemical means. Cutting away the cankered areas both with and without applying copper or sodium hypochlorite has not proven successful and is also harmful to the trees. Attempts to halt the disease by injecting antibiotics have also failed.

7.4.4 SHALLOW BARK CANKER OF WALNUT

Pathogen: *Brennaria (=Erwinia) nigrifluens*

Symptoms
The most characteristic symptoms of shallow bark canker are brownish-to-black round spots or areas, usually several in a group, on the trunk or lower scaffolds. Newly infected areas have a

PHOTO 7.17 Shallow bark canker symptoms on walnut trunk. (Courtesy of Giorcelli, A. and Gennaro, M., Italian Society of Silviculture and Forest Ecology, Viterbo, Italy.)

margin of water-soaked bark and a central spot of black ooze that later dries, leaving a tarlike black spot (Photo 7.17). Just under the surface, dark brown areas of dying tissue are formed in the outer bark. These superficial cankers can be extensive, but they seem to cause little damage to the tree. Shallow bark canker rarely extends into the inner bark, as does deep bark canker.

Shallow bark canker affects many commercial walnut cultivars, but the full range of susceptible hosts has not been determined. Because the damage is superficial and does not result in economic loss, shallow bark canker is not considered a major disease of walnut. It is not known how the pathogen infects walnut trees or how it develops and spreads.

Disease Management
As with deep bark canker, shallow bark canker is often severe in stressed trees. Improving tree vigor may help contain the disease. Cankers can be cut away, but this practice is not recommended because the disease is not serious enough to warrant possible wound damage and other infections.

7.5 ALMOND

7.5.1 BACTERIAL SPOT OF ALMOND

Pathogen: *Xanthomonas arboricola* pv. *pruni* (Xap)

Synonyms: *X. campestris* pv. *pruni*
Bacterial spot is a relatively new almond disease in California. It has currently been found predominantly on the cultivar Fritz in Colusa, Merced, Stanislaus, and San Joaquin counties.

Symptoms
Symptoms can be observed on leaves as well as on fruit nuts.

Symptoms first appear in mid-April to early May. Amber-colored gum exudes from nuts with reddish lesions on the hull surface similar to anthracnose and leaf-footed plant bug feeding are found.

In bacterial spot, leaves become spotted especially where water collects (e.g., along leaf margins), turn yellow, and drop prematurely. Green twigs (less than a year old) can have visible lesions or cankers.

Bacterial spot is a common problem in Australia and growers there have been forced to abandon the two most severely affected varieties, Fritz and Ne Plus, due to extensive crop loss.

On close inspection, severely affected trees may have a majority of nuts and leaves with characteristic lesions. The symptoms of bacterial spot become more obvious as temperatures increase during the late spring and summer. The infection period, however, is predominantly in the spring and during periods of high moisture and mild-warm temperatures.

On Leaves

Both young and old infected leaves exhibit leaf spot and tatter symptoms, but it is believed that most infection is initiated in the leaves while young. The lesions are generally clustered in the area of the leaf that stays wet longer, that is, leaf tips, sheltered parts of the leaf blades, along the midrib. The lesions are circular, angular, or irregularly shaped and reddish to dark in color. As the lesions dry out, shot holes and leaf tatter result. These bacterial spot symptoms are easily confused with those caused by the fungal disease "shot hole" and also by copper phytotoxicity. Affected leaves, in each of these cases, may fall prematurely.

On Nuts

Infected nuts develop corky lesions from which a lightly colored-tan ooze and gum may stream or clump. The gum contains bacteria. Small lesions on nuts may be confused with insect injury sites. The larger bacterial spot lesions that are sunken and surrounded by a gray yellow area are also reminiscent of those caused by fungal pathogens. Infected nuts are often clustered within the canopy; some may fall prematurely. Others may remain as "stick tights." These mummies harbor viable bacteria and serve as a source of inoculum thereafter.

On Twigs

Twig lesions have not been observed as commonly as the leaf and nut symptoms. The twig lesions on current season's wood are dark and elongated along the length of the twig, slightly depressed and often have a shiny, greasy appearance with a water-soaked margin. If the lesion expands, it may girdle the twig and dieback will occur. In stone fruit, open cankers develop from these lesions on older wood but these are yet to be found on almonds.

Epidemiological Aspects

Bacterial spot needs wetness to spread; bacteria are spread from cankers or mummies by dripping dew and splashing or wind-blown rain to newly emerged leaves. It overwinters on mummies and possibly in twig cankers.

Disease Cycle

It is most likely that bacterial spot will be introduced to an orchard in budwood or nursery trees, via wind-blown rain from neighboring infected stone fruit or almonds, or on equipment.

Once present within an orchard, insects, birds, equipment, and irrigation (especially if overhead) may play a role in tree-to-tree spread of the bacteria. Infected leaves, nuts, and mummies within the tree canopy or fallen are sources from which bacteria may be splash distributed during rains and overhead irrigation.

Once present on susceptible tissue, the bacteria enter wounds or natural entry points, in a moisture film. These may be injury sites, pruning wounds, growth or frost cracks, leaf scars, axils, microscopic sites of wind, dust or sand abrasion, or injury sites caused by hail or chemical sprays.

The rates of infection and disease development are dependent on environmental conditions. Moist conditions and warm temperatures promote disease development and proliferation of the bacteria. While late spring/summer rains frequently present perfect conditions for infection and disease development, the required humidity may also be induced by heavy dew, fogs, and irrigation. In the absence of conditions conducive to infection, the bacteria have the capacity to survive extended periods in protected areas on the trees. During cold periods and dormancy, it is known that the bacteria may persist in mummies and in protected sites like buds, axils, and twig lesions. The overwintering or survival sites are not necessarily symptomatic.

Disease Management

Practice prevention measures as is the case with other bacterial diseases. Research in other crops suggests that the following preliminary management guidelines may reduce bacterial spot. Research is needed to determine the benefits of in-season bactericide treatments.

- During the season, blow fallen nuts into the center between rows and grind them up into small pieces that easily degrade.
- Practice sanitation when moving between orchards by brushing off shoes; sweep out all nuts from trailers or hoppers.
- If possible, harvest before fall rains.
- After harvest, defoliate trees to reduce inoculum and improve the visibility of mummies and coverage of dormant sprays.
- Remove and destroy mummy nuts; pole or shake and then disc or mow.
- Copper plus oil applications before winter rain may help prevent disease.
- A delayed-dormant copper plus oil application may help prevent disease.

Avoidance:
- Check the source of budwood used by nursery, for symptoms of disease.
- Check nursery growing conditions and treatment program for bacterial spot.
- Purchase nursery stock grown in less bacterial spot–prone areas and in heavier soil.
- Check all incoming trees of susceptible varieties for cankers and lesions.
- Minimize plantings of NePlus and Fritz.
- Avoid overhead irrigation of susceptible cultivars. Overhead irrigation provides both a means of spread for the bacteria and induces conditions favorable for disease development. Orchards using high-frequency drip irrigation may also see higher levels of disease incidence as a result of higher humidity associated with constantly moist soil.
- Avoid planting susceptible cultivars in exposed or windy sites, especially if soil is light.
- Avoid planting stone fruit alongside almonds, or almonds alongside an existing stone fruit orchard.

Protection:
- Establish wind breaks.
- Practice excellent orchard hygiene.
  - Remove fallen fruit, prunings
  - Remove mummies during dormancy
  - Harvest least susceptible trees first
  - Clean all equipment
- Practice good frost control.
- Avoid tree, leaf, and nut injuries.
- Do not prune or tree train during wet weather.
- Prune such that air flow through canopies is enhanced.
- Do not cross-use equipment for both almonds and infected stone fruit.

7.5.2 ALMOND LEAF SCORCH

Pathogen: *Xylella fastidiosa*

Symptoms

Leaf tissue dies when xylem plugging due to bacteria results in insufficient water arriving at the leaf margins. Almond leaf scorch (ALS) symptoms first appear on individual leaves in early June to mid-July. The leaf tips or margins initially turn light gray-green. The scorching occurs with the onset of hot weather.

By late July, symptoms are fully developed and are most noticeable. It is important to identify and mark affected trees, while the scorched areas of the tree canopy contrast clearly with healthy green leaves. Once harvest begins, mites, dust, and drought stress combined with tree shaking often make signs of ALS difficult to detect.

Scorch symptoms first appear in a branch, scaffold, or portion of the tree, but will subsequently spread to affect the entire tree. The rate of symptom spread from when first visible to infecting the entire tree can occur slowly over several years, or relatively fast, infecting the entire tree from one season to the next. Regardless, the disease will infect more of the tree with each succeeding year. It may be easily overlooked when only a few leaves on one branch are affected. ALS is also known as Golden Death because of the striking yellow color of a fully infected tree's canopy. Within few years, affected trees lose vigor, become unproductive, and may eventually die.

In early summer (late June), leaves appear with marginal leaf scorch (brown, necrotic [dead] leaf tissue). Usually, a narrow band of yellow (chlorotic) tissue is inward from the dead tissue, but the sudden appearance of leaf scorch symptoms prompted by hot weather may not allow the narrow chlorotic band to develop for several days. As the disease progresses, affected twigs on limbs die back from the tip (Mircetich et al. 1976). The most characteristic symptom of ALS is leaf scorching (Photo 7.18) followed by decreased productivity and general decline. Even very susceptible varieties take many years to die completely, but nut production is severely reduced within a few years in most varieties. In the first summer after infection by insect vectors, there will be only a few scorched leaves. During the winter, many of these infections fail to survive, depending on the severity of winter. By the second summer after infection, growers may be able to see new infections and prune

PHOTO 7.18 Leaf scorch symptoms on almond leaves. (Courtesy of Golino, D.A., University of California, Davis, CA, atlasplantpathogenicbacteria.it.)

them off about 3 ft below the lowest symptoms; however, even by the second year, symptoms may not be prominent enough to see in most cases.

ALS appears as a marginal scorching of leaves that begins as early as June and continues to develop during summer. A golden yellow band develops between the brown necrotic edge and the inner green tissues of the leaf. Disease symptoms may appear first on one branch or a portion of one scaffold. As years go by, more and more trees are affected until the whole canopy is involved. Infected trees bloom and leaf out later than healthy trees, are stunted, are less productive, and have reduced terminal growth. Trees with ALS usually survive for many years.

In Sacramento Valley, almond orchards with symptoms similar to those of ALS by the bacterium *Xy. fastidiosa* were reported.

ALS symptoms spread more rapidly and affect a larger number of trees in some varieties than in others. The varieties Peerless and Long IXL are extremely sensitive. Nonpareil is also sensitive. Carmel and Mission are fairly resistant or tolerant.

Geographical Distribution

California trees demonstrating signs of ALS (Purcell, 1980; Almeida and Purcell, 2003) were first observed in the mid-1930s near Riverside. ALS was also noted in a few randomly scattered almond trees in Los Angeles and Contra Costa Counties in the early 1950s. ALS has since spread to commercial orchards throughout California. The disease is still (2003) rare in the San Joaquin Valley, even in orchards next to vineyards with Pierce's disease (PD); the reasons for this are not known. Strains of the causal bacterium, *Xy. fastidiosa*, that cause PD cause ALS, but not all strains from almond with ALS can cause PD (Hendson et al., 2001; Almeida and Purcell, 2003); this can explain how ALS can occur at high levels in almond without PD in adjacent vineyards, as in the Delta region of Contra Costa County. Reports of ALS in India (Jindal and Sharma, 1987) have not confirmed the isolation of the causal bacterium.

Disease Spread

The bacterium (*Xy. fastidiosa*) that causes ALS also causes Pierce's disease of grapevines and alfalfa dwarf disease. Many common weeds and riparian plant species, including Bermuda grass, rye, fescue grasses, watergrass, blackberry, elderberry, cocklebur, and nettle, are hosts and serve as reservoirs of inoculum. Common annual orchard weeds such as annual bluegrass, burclover, cheeseweed, chickweed, filaree, London rocket, and shepherd's purse have also been found to be infected. Weed-to-tree or tree-to-weed spread is also a possibility. The bacteria are vectored by xylem-feeding insects such as leafhoppers and spittlebugs. The most probable vectors for almond are the red-headed and green sharpshooters. However, it is not believed that currently identified vectors can spread the disease from an infected tree to a healthy one. The disease may become more important if the glassy-winged sharpshooter (GWSS) (*Homalodisca vitripennis* [=*Homalodisca coagulata*]), which can spread the disease, becomes established in almond-growing areas.

Pathogen and Vectors

ALS and PD of grapevines are caused by the same bacterium, *Xy. fastidiosa*, although there are almond and grape strains of the disease that can both be found in almond. The disease presents a potentially serious threat to California almond orchards since it is spread by sharpshooter leafhoppers that feed on the water-conducting xylem. These insects carry the pathogen from plant to plant, and common annual weeds in the orchard can be sources of infection. At this time, there is no evidence that infected trees are the source of bacteria to infect other trees, but sharpshooters feeding on infected trees can infect annual weeds, and sharpshooters feeding on the infected annual weeds can infect additional trees in the orchard. Irrigated pasture, weedy grasses, alfalfa, and permanent cover crops are the most common habitat for sharpshooters in almond-growing regions of California. The green sharpshooter *Draeculacephala minerva* and the red-headed sharpshooter *Carneocephala fulgida* are the most common vectors of the pathogen found in almond orchards. The GWSS

H. coagulata could also successfully spread the pathogen in almond orchards because, unlike other known vectors, it prefers to feed on trees. The small, green potato leafhopper (*Empoasca* sp.), prune leafhopper (*Edwardsiana prunicola*), and the white apple leaf hopper (*Typhlocyba pomaria*) that commonly feed on almond leaves are not vectors of this disease.

Effects of Glassy-Winged Sharpshooter
Some increase in ALS is expected in California almond orchards that are near breeding habitats (such as citrus groves) of the GWSS, but as yet it is hard to predict how severe the problem with GWSS will be for ALS. Remember that most new infections are not visible until the second or third year after infection, and this may take even longer with tolerant varieties of almond. GWSS has been observed feeding but not reproducing on almonds from just after leaves appear in February until March and early April. Insecticides in orchards to control vectors that have entered almond orchards will probably not be very effective because the vectors may continue to enter the orchards from outside sources. So far there are no protocols for controlling GWSS in almond.

Disease Management
First-year symptoms may be confined to a few inches from the infection site and not be noticed for 2 or 3 years. Movement within a tree may be slow and it requires several years to infect the entire tree. In other cases, spread throughout the tree appears to have occurred within a year.

If discovered early and only in one branch, the infection may be removed by pruning off a primary scaffold 5–10 ft below visible symptoms. If this is attempted, flag the pruned tree and observe it in subsequent years for indications of the disease.

If the orchard is young (5–10 years old), the best course of action may be to remove infected trees. In older orchards (16–20 years old), it may be more cost-effective to keep infected trees, because the entire orchard is normally removed between 22 and 25 years of age and infections will probably not significantly impact yields before then. The most difficult part is to decide on what has to be done when infected trees are found in orchards 11–15 years old. The answer may depend on whether there are other young orchards nearby, how long the orchard is expected to last before it is likely to be removed, and, once mapped, whether the number of infected trees increases rapidly.

Sharpshooter populations increase slowly and the insects disperse slowly. Grass-feeding sharpshooters require year-round access to plants on which they can feed and reproduce. Clean cultivation of almond orchards for a 6-week period at any time of the year (like during harvest) should prevent the establishment of in-orchard vector populations. Thus, cover crops in almond orchards should not pose a threat. The most common habitats for sharpshooters in the Central Valley are irrigated pastures, alfalfa fields with grass weeds, and permanent cover crops.

No chemical or nutritional treatments control ALS. In addition, no research or practical experience suggests that disease occurrence is reduced by controlling the vectors with insecticides. If you suspect that a tree may be infected, first test its leaf tissue for excess salts, particularly chloride and sodium.

Salt injury, especially chloride burn, may be mistaken for ALS. Sometimes the two are indistinguishable. With salt damage, there is usually just healthy, green tissue and dead, brown tissue without the yellow margin between the healthy and dead tissue.

Salt injury may occur at any time but often worsens as the growing season progresses. Ordinarily, it is a result of excess salinity in soil or water. Unlike ALS, salt injury affects numerous trees in one concentrated area rather than individual trees widely scattered throughout an orchard.

If sodium and chloride levels are normal, and salinity has been eliminated as a possible cause of the problem, you may wish to have the tree tested for leaf scorch. The best time to test for ALS is July through September.

The same testing practices and labs that can detect *Xy. fastidiosa* in grapes can be used for testing almond leaves for ALS. The cost per sample runs from approximately $150 to $400 depending on the lab and the number of tests run.

In young orchards less than 10 years old, early identification and removal of diseased trees may minimize the problem and reduce further tree losses. If your orchard is 18–20 years old, you may want to simply live with some infected trees until the orchard is removed and replaced. Orchards between 10 and 18 years old are the difficult call—do you remove the infected trees or live with the problem? The answer to this question may depend on what is around the infected block. If infected trees jeopardize a newly planted nearby orchard, then removing the trees might be the best course.

Varietal Susceptibility

Varieties that appear more susceptible in the field include Peerless, Sonora, Winters, Livingston, and Wood Colony. Nonpareil is also susceptible and can be significantly affected. The disease is rare in Carmel and Butte and is seen less often in other varieties.

7.5.3 Bacterial Canker of Almond

Pathogen: *Pseudomonas syringae* pv. *morsprunorum*

Bacterial canker is a serious but common disease caused by two related species of bacteria (*P. syringae* pv. *morsprunorum* and *P. syringae* pv.) and affects all stone fruits (plums, gages, damsons, cherries, peaches, nectarines, apricots, and sweet almonds). Trees are particularly susceptible during their early years before cropping. Infection is most likely to occur during the autumn and winter when bacterial populations are carried in wet and windy weather infecting leaf scars and pruning wounds. The first signs of bacterial canker, however, are not normally noticed until spring/early summer.

Symptoms

Sunken areas with elongated, flattened cankers that ooze sticky gum appear on the bark and affected branches die back. Leaves on these branches either do not develop or are small and yellow and die soon. If cankers occur on a main stem of the tree, the entire tree will probably die. During the summer months, the organism infects the leaves and sometimes the shoots. Round brown spots appear on the leaves and subsequently develop into holes, giving them a "shot-holed" appearance.

Bacterial canker can easily be confused with gummosis, a disorder of stone fruits that usually occurs after freezing weather. The patches of gum appear on the surface of branches and trunks, but with gummosis, the gum arises from healthy wood, whereas with bacterial canker, the gum oozes from diseased tissue and has a sour smell.

Control Measures

Avoid pruning stone fruits during the winter since the wounds formed may allow the entry of the disease. Any necessary pruning or cutting back should be carried out during the summer months between May and the end of August.

Orchard Management

Bacterial canker must be tackled quickly in order to save the tree. Remove and burn cankered branches by pruning them back into healthy wood at least 20–30 cm from any visible damage. Wounds should be painted with a protective paint such as Growing Success "Prune & Seal" or Arbrex "Seal & Heal," which are both suitable for use by organic gardeners. It is very important that the pruning tools are properly disinfected by immersing the blades in boiling water for 5–10 min before using them on other trees. For nonorganic gardeners, dipping the blades in methylated spirit will also do. If canker occurs on the main stem, the entire tree will die. In such a case, it is necessary to uproot and burn the tree immediately to prevent the canker from spreading to other trees.

Chemical Control

Trees affected by canker should also be sprayed with copper-based fungicides such as Bayer Garden "Fruit and Vegetable Disease Control" or Vitax "Bordeaux mixture," making three applications from late summer to mid-autumn following the manufacturer's guide.

7.5.4 HYPERPLASTIC CANKER OF ALMOND

Pathogen: *Pseudomonas amygdali*

Symptoms

Infection affects branches and twigs of the host plant and results in tree decline. Short-distance spread of the disease occurs from active cankers during wind and rain. Symptoms of hyperplastic canker infection can be detected from leaf emergence. Infected trees have swollen bark canker around leaf scars and wounds. As the symptoms progress, a longitudinal crack develops in the swollen bark (Photo 7.19). These cankers look like as the bark has been peeled back to show discolored wood beneath. Longitudinal crack with peeled bark symptoms with soft, rough tissue appears at the margins of cankers. If there are multiple cankers, they can girdle shoots. The buds in cankers will not develop, so affected trees generally have little foliage. There are no specific leaf symptoms; however, infected trees decline because of the lack of new growth. The diseases progresses by *P. amygdali* on almond is correlated with the amount of indole-3-acetic acid (IAA) and cytokinins, respectively, which are related to the length of the incubation period and to the final length of the cankers. Nonpareil shows some tolerance to this pest, so detection inspections should focus on other cultivars.

Host

Primary hosts of this bacterium are almonds and peach.

Geographical Distribution

Hyperplastic canker is currently found only in Greece, Afghanistan, and Turkey.

PHOTO 7.19 Symptoms of hyperplastic canker on almond twig. (Courtesy of Psallidas, P.G., Benaki Phytopathological Institute, Kifissia-Athens, Greece, atlasplantpathogenicbacteria.it.)

Control Measures
Use clean planting material from a known source that utilizes high health and hygiene strategies. Additional protection can be obtained by visually checking for cankers in all planting materials before use. Protect scars and wounds on tree stems and leaves to reduce the chance of infection. Check your orchard frequently for the presence of new pests and unusual symptoms.

7.5.5   CROWN GALL OF ALMOND

Pathogen: *Agrobacterium tumefaciens*

Symptoms
Rough, abnormal galls appear on roots or trunk. Galls are soft and spongy. The centers of older galls decay. Young trees become stunted; older trees often develop secondary wood rot.

Disease Spread
The bacteria survive in gall tissue and in soil. They enter only through wounds. Crown gall is most damaging to young trees either in the nursery or in new orchard plantings. Peach-almond hybrid rootstocks are more susceptible to crown gall than Nemaguard rootstocks.

Disease Management
Crown gall is best prevented by purchase of trees from a reputable nursery accompanied by careful handling to avoid injury as much as possible, both during planting and during the life of the tree in the orchard. Preplant treatment is for prevention only. Galltrol is a preparation of the biological control agent *A. radiobacter-84*. It is effective only as a preventive treatment and is used as a root dip or spray before heeling in or planting. It does not eradicate existing galls. Chlorine bleach root dips or sprays are not effective as a crown gall protectant.

Strains of *A. tumefaciens* resistant to Galltrol and Norbac have been reported. Their occurrence is not widespread, but failure to control crown gall with these materials should be reported. Eradication involves removal of existing galls and topical application of Gallex.

7.6   PLUM

7.6.1   BACTERIAL CANKER OF PLUM

Pathogen: *Pseudomonas syringae*

Symptoms
Symptoms are most obvious in spring and include limb dieback with rough cankers and amber-colored gum. There may also be leaf spot or blast of flowers and young shoots. The sour sap phase of decline may not show gum and cankers, but the inner bark can be brown, fermented, and sour smelling. Flecks and pockets of bacterial invasion in bark occur outside canker margins. Frequently, infected trees sucker from near-ground level; cankers do not extend belowground.

The bacteria can invade all the above-ground parts of a plant. Symptoms vary widely between hosts and different climatic conditions but are more commonly initiated on green foliage. Although bacterial canker may affect trees of any age, it is more often associated with death of trees in the first or second year after planting. The tree flowers and begins its growth normally but limbs or whole trees collapse and die within a matter of weeks in spring. Leaves often remain attached to the tree. Death usually occurs in the same season and young trees seldom, if ever, struggle on dies in the next season. Once the top dies, many shoots arise from just below-ground level because the disease does not invade the roots.

In mature trees, under the right climatic conditions, infection can spread quickly, killing large branches in a matter of weeks.

Areas of infection, particularly under bark of mature branches, are usually sunken. These cankers often exude large amounts of gum and often give off a characteristic vinegar (sour) smell.

Slicing bark off the affected area reveals a brown or tan discoloration in the bark. The outer layers of wood are often stained brown color in contrast to the normal white, but the inner wood (xylem) is not discolored except in the more severe cases.

Symptoms on leaves vary greatly, but generally first appear as water-soaked spots. These spots eventually become dry and brittle and fall out of the leaf, giving it a shot-hole appearance. Infected fruits often develop depressed spots with dark centers and sometimes have underlying gum pockets.

Host
Apricots and plums especially, but all stone fruit are susceptible.

Predisposing Factors
Bacterial canker enters the tree through cracks, wounds, leaf scars, and bud scales during cool wet weather in autumn and winter, though the symptoms will not be obvious until the next spring.

P. syringae survives in or on plant surfaces, is spread by splashing rain, and is favored by high moisture and low temperatures in spring. The disease is worse in low or sandy spots in the orchard. Vigorous trees are less susceptible to bacterial canker. Young trees, 2–8 years old, are most affected. The disease rarely occurs in the first year of planting and is uncommon in nurseries.

Disease Management
Firstly, when purchasing trees from nurseries, ensure they are free from bacterial canker. Badly infected trees are a source of infection for adjacent trees and are generally stunted and often die. Remove all diseased trees in young (less than 4 years old) plantings and replace with healthy trees. The bacteria will not infect a tree via its root system, so you can replant in the same spot.

In older plantings, careful pruning is an alternative. Limbs with developing cankers can be cut off and the wounds sealed to prevent further damage, but make sure the limb is cut back to clean wood. Do this only after leaf growth has started and while the tree is actively growing. Dip all pruning tools in an effective disinfectant (e.g., methylated spirits) after each cut and paint all wounds with a copper-based paint or Bordeaux paste.

In difficult situations where the disease is hard to control and particularly in young trees, a cover spray program according to which copper is sprayed during two springs, two autumns, and one winter is suggested. They are applied at early budswell, 7–10 days later, 25%–50% and 90%–100% of leaf fall, and midwinter. Copper oxychloride, plus oil to act as a sticker, is an acceptable spray to use. Bordeaux mixture is an alternative. Applications of copper-based fungicides cause shotholing of the leaves and can damage fruit. For this reason, copper sprays should not be applied between pink bud and the commencement of leaf fall. Where the severity of disease is such that a copper spray at 25%–50% leaf fall is warranted, a half-strength spray should be used at this timing.

Under normal circumstances, a minimum program of at least one early budswell and one autumn leaf fall spray should be sufficient.

Planting trees that are budded or grafted about 32 in. above the root crown can help suppress bacterial canker infections. Bacterial canker tends to mostly affect weak trees, so any management practice that improves tree vigor (e.g., lighter, more frequent irrigation, improved tree nutrition, and nematode management) will help to reduce the incidence of this disease. Trees on Lovell peach rootstock are more resistant than others; those on plum rootstocks are most susceptible. Delayed pruning may help.

In light sandy soils and some heavy soils, successful control has been achieved with preplant fumigation for nematodes. Nematodes stress trees and predispose them to bacterial canker. Preplant fumigation for nematode control reduces the severity of bacterial canker in newly planted orchards. The benefits of preplant soil fumigation for control of bacterial canker usually last only a few years; in some areas, only limited improvements in disease control occur following soil fumigation.

Bactericide applications have no reliable effect on bacterial canker and their use is not recommended. Similarly application of copper during dormancy has not been shown to protect against bacterial canker.

7.6.2 Bacterial Spot of Plum

Pathogen: *Xanthomonas campestris* pv. *pruni*

Other names for the disease are bacteriosis, shot hole, and black spot. The causal bacteria can attack fruit, leaves, and twigs. Fruit loss on some varieties can be very high. Early and severe defoliation can affect fruit size and the winter hardiness of buds and wood.

Symptoms

The symptoms of bacterial spot are quite different from other diseases of stone fruits. They may be confused with nitrogen deficiency and spray injury. The disease first appears as small, water-soaked, grayish areas on the undersides of leaves. Later the spots become angular and purple, black, or brown in color. The mature spots remain angular and are most numerous at the tip ends and along the midribs of leaves (Photo 7.20). The infected areas may drop out, giving the infected leaves a shot-holed, tattered appearance. On plum, the shot-hole effect is more pronounced than on other stone fruits. Infected leaves eventually turn yellow and drop. Severe defoliation often results in reduced fruit size, increased sunburn, and fruit cracking. As a result, tree vigor and winter hardiness are also reduced. Other leaf spot diseases and spots due to spray injury tend to be much more circular in outline. Often, these are not confined by veins in the leaf as is bacterial spot. Leaf spots due to nitrogen deficiency are normally red in color.

A fruit infected early in the season develops unsightly blemishes and may exhibit gumming. Since the infected areas cannot expand with increased fruit size, the spots crack. Pits or cracks on the fruit surface extend into the flesh and create large, brown-to-black depressed areas on the fruit surface. Lesions that develop during the preharvest period are usually superficial and give the fruit a mottled appearance. On plum, the fruit symptoms are likely to be quite different in that large, black, sunken areas are most common. On a few varieties, small pit-like spots occur.

The cankers on plums eventually appear quite different. On susceptible varieties, the bacteria may survive for 2 or 3 years, slowly enlarging and deepening the cankered area. The results are deep-seated cankers deforming the small branches, so they have a knotty appearance. Some of these branches may be killed or they may break from the weight of the fruit.

Disease Cycle

The bacteria overwinter in the twigs, buds, and symptomless plant tissue. In the spring, the bacteria are spread by rain to leaves, shoots, and fruit. Spring infections can occur anytime after the leaves begin to unfold. Temperatures above 17°C and warm rains are needed for the bacteria to multiply,

PHOTO 7.20 Bacterial spot on leaves and plum fruit. (Courtesy of Scortichini, M., atlasplantpathogenicbacteria.it.)

become exposed, and be disseminated. After these first infections, which are rarely noticed but do initiate the disease each year, the severity of the secondary infections depends entirely on the weather. A moderately warm season with light, frequent rains accompanied by heavy winds favors severe outbreaks of bacterial spot. Any recent injury to the leaves or fruit, such as wind-blown soil particles and hail, may result in severe outbreaks.

Secondary spread of the bacteria can occur from oozing summer cankers and leaf and fruit lesions during warm, wet weather. The systemic movement of the bacteria from leaves and shoots contributes to the formation of cankers. These cankers can be spread by budding to healthy nursery trees.

Disease Management
Maximum use of resistant varieties is the most effective control measure. Resistance in plums is not as common. Nurserymen are well aware of the degree of susceptibility of the varieties they sell and they can provide good information for specific areas. Since trees in poor vigor are more susceptible, orchard management programs should be designed to maintain good vigor. Major outbreaks of bacterial canker in young orchards are often attributed to poor cultural practices.

No completely successful spray programs exist for control of bacterial spot. Chemical sprays can help reduce the amount of fruit and leaf infection but must be applied before symptoms occur. In seasons when disease incidence is light, special programs do help. In those when infections are numerous, spray programs can reduce the number of infections, but not enough to prevent defoliation and fruit infection. Chemical applications suppress the development of disease but do not eliminate it. Because chemical control is uncertain, the use of resistant varieties appears to be the best control strategy.

7.7 CHESTNUT

7.7.1 Horse Chestnut Bleeding Canker

Pathogen: *Pseudomonas syringae* pv. *aesculi*

Over the past few years, *Pseudomonas* bleeding canker caused by the bacterial pathogen *P. syringae* pv. *aesculi* (Pae) has become a serious and widespread problem of horse chestnut trees (*Aesculus hippocastanum* L.) in the United Kingdom.

Bleeding canker is an infection of the bark of horse chestnut by the bacterium *P. syringae* pv. *aesculi* and several species of the fungus *Phytophthora*, which causes the affected bark to bleed a dark sticky fluid. It is specific to horse chestnuts and both the white-flowered *Ae. hippocastanum* and the red-flowered *Aesculus carnea* are affected. Cankers can be seen at any time of year.

Several species of the fungus *Phytophthora* have long been known to cause cankers (bark infections) in horse chestnut, though cases were relatively uncommon. But in recent years, there has been an upsurge in cases where *Phytophthora* could not be detected. Recent work in the United Kingdom and the Netherlands has established that the bacterium *P. syringae* pv. *aesculi* is the cause of these new cases.

Research on the bacterium is still in progress. It may require wounds to infect (which may include naturally occurring lenticels, or pores, in the bark) or might exist on plant surfaces and be spread by wind-blown rain.

Symptoms
The bleeding fluid is produced by the tree in response to the infection, which kills the inner bark, cambium, and outer layers of wood, causing disruption to water and nutrient transport. If the canker girdles the stem, the stem dies.

- Dark or reddish-brown sticky liquid oozes from cracks in the bark where the infections occur. In dry weather, this dries out to form a black deposit (Photo 7.21).

PHOTO 7.21   Symptoms of bleeding canker on chestnut main trunk. (Courtesy of Lamiot, F., en.wikipedia.org.)

- Cutting away the outer bark over infected areas will reveal a brown or purple discolored area of inner bark, with a diffuse edge if the infection is still spreading and a sharply defined edge if it is stable. Healthy inner bark is white or pinkish in color.
- On older cankers, the dead bark may fall away to expose the wood.

The effect on the tree is variable. Some infections last for years, more or less stable, and with little effect on the crown. Others spread rapidly and cause crown thinning, die-back, and even death of part or whole of the tree.

Disease Management
- Some infections stabilize, so do not immediately resort to removal but keep affected trees under observation.
- If the tree's condition continues to deteriorate, consider removing it because it may be a source of infection for other horse chestnuts.
- Burn the affected material if possible, rather than composting or chipping.
- Disinfect tools with a product such as Jeyes Fluid after use.
- Consider raising horse chestnuts *in situ* from seed to avoid the risk of importing the disease on infected plants.
- There is no chemical control currently available.

7.8   HAZELNUT

7.8.1   Bacterial Canker or Decline of Hazelnut

Pathogen: *Pseudomonas avellanae* Janse et al. (1997)

Synonyms: *P. syringae* pv. *avellanae*

Common names: Moria in Central Italy, stem dieback of hazelnut.

Symptoms

Bacterial Decline

Initial symptoms of bacterial decline may develop in winter during the blossoming of the male inflorescences. Infected catkins release only sparse amounts of viable pollen and often completely wilt. The dead catkins tend to remain firmly attached to the twig during the winter. In February and March, female inflorescences on infected trees may fail to enlarge properly and become necrotic. In spring, diseased trees may exhibit a delayed bud break and leaf emergence. Emerging leaves may wilt and die rapidly on individual branches.

In other cases, trees without any previous symptoms of decline may exhibit pale green foliage in early spring. Such trees often wilt and die during summer. The most striking symptoms occur in summer when leaves on one to all branches rapidly wilt. Wilted foliage and immature fruit remain attached to the twigs for many weeks and sometimes into winter. Necrotic spots do not develop on leaves or nuts.

Bacterial Canker

In autumn, cankers develop on branches and the trunk (Photo 7.22). Diseased bark turns reddish brown, and a brownish discoloration of the sapwood is apparent if the bark is stripped from the branch.

The first symptoms are rapid wilting of twigs, branches, and trees during spring and/or summer. Leaves usually remain attached to the twigs after withering. The bacterium infects mainly in autumn through leaf scars and, from there, the pathogen can move systemically within the plant, with the potential to reach the roots and cause cankers on the trunk and branches. Root necrosis may also occur. Infected trees that survive the winter often die the following summer. In severe infections, trees are completely killed in one season. Acidic soils and spring frost probably are important for predisposing trees. Not much is known about the mode of transmission. Very similar symptoms, but with a less aggressive disease development, are caused by strains that are not *P. avellanae* but closely related to *P. syringae* pv. *syringae* (Scortichini et al., 2002) and recently described as *P. syringae* pv. *coryli* (Scortichini et al., 2005).

Host

Hazelnut (*Corylus avellana*).

Geographical Distribution

The disease was first reported in Greece in 1976 (Psallidas and Panagopoulos, 1979) and the causal organism was first described by Psallidas in 1984. The disease has been observed in Greece

PHOTO 7.22   Bacterial canker on trunk and decline of hazelnut tree. (Courtesy of Scortichini, M., atlasplant-pathogenicbacteria.it.)

and Italy, where *P. avellanae* has caused substantial losses in N Greece and also in central Italy (Scortichini et al., 2001).

Disease Cycle

P. avellanae primarily infects through leaf scars in early autumn. Leaf scars during this period are not fully suberized and can be infected by rain-splashed or wind-disseminated bacteria. Application of 10 μL of a 1×10^5 CFU/mL suspension (i.e., around 1000 cells) of *P. avellanae* on a single leaf scar in early autumn resulted in 100% infection. Once inside the twig, the bacterium overwinters in bark. In spring, the bacterium moves systemically from the infected twig to other branches and even roots. An adult tree can be killed in 7 months following inoculation of 25 leaf scars scattered on 1-year-old twigs randomly distributed through the canopy. Two-year-old trees can be killed by inoculating as few as two or three leaf scars. After spring frosts, *P. avellanae* may also colonize cracks on multiple branches on the same tree; mortality may occur later that spring or early summer. Presently, there is no evidence for the presence of epiphytic populations of *P. avellanae* on leaf or bark surfaces. The bacterium has also not been isolated from nuts taken from infected trees. Bacterial infection may result in the formation of longitudinal cankers on branches and the trunk during summer and autumn. The bacterium survives adverse weather in the bark of twigs and roots. Insects may play a role in disseminating *P. avellanae* from an infected tree to a nearby healthy tree. Several scolytid beetles, including *Xyleborus* (*Anisandrus*) *dispar* (L.) and *Xyleborus saxesenii* (Ratz.), are attracted by the terpenes released by diseased trees. Adult insects may come into contact with the bacterium during oviposition, and larvae may be contaminated during tunneling. *P. avellanae* has been isolated from both larvae and adult scolytid beetles, but it has not been conclusively demonstrated that these insects can transmit the disease. The disease cycle of bacterial canker or decline of hazelnut is illustrated in Figure 7.3.

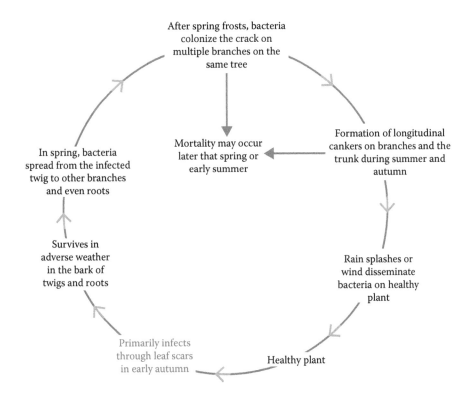

FIGURE 7.3   Disease cycle of bacterial canker or decline of hazelnut caused by *P. syringe* pv. *avellance*.

Economic Impact

Bacterial canker was devastating in young plantations of cultivar Palaz in northern Greece in 1976. Since it was first discovered in the Latium area in central Italy in the late 1990s, bacterial canker and decline have resulted in the mortality of more than 40,000 trees. It continued to damage trees on approximately 1000 ha in this area. Losses are estimated to be $1.5 million on an annual basis, and the disease is considered a serious problem. Currently, the inoculum pressure of the pathogen seems to be reduced.

Disease Management

Effective control of bacterial canker and decline in hazelnut is difficult. The production of disease-free nursery plant material is very important. The best way to avoid the disease is to prevent the introduction of latently infected plants. Monitoring of orchards for early disease symptoms is essential. Inspections should preferably be made during spring and summer with the aim of detecting wilting twigs and branches. Infected plant parts must be removed from the orchard and should be burned. In case of completely wilted trees, the roots and suckers also must be removed. Pruning and/or sucker removal should be avoided during humid periods. After a branch is cut, it is advisable to seal the wound with wax or Bordeaux mixture. It may take several years to eradicate the pathogen in severely damaged orchards. Taking into consideration that *P. avellanae* mainly lives inside the plant, the efficacy of copper-based compounds, the only chemicals allowed in Italy for controlling bacterial disease, is rather low. However, a combination of agronomic techniques and applications of copper-based materials at key times can partially manage the disease. As a general rule, farmers apply copper-based chemical immediately after pruning, spring frost, hail, windy storms in early autumn, and at the beginning and middle of leaf drop to reduce the possibility of wound colonization by the bacterium. Orchards having very acidic soils require lime application to increase the soil pH. It is also very important to control Scolytidae by using chromotropic traps. Recent progress in the control of bacterial decline has been achieved by artificial induction of systemic acquired resistance (SAR). The use of acibenzolar-*S*-methyl (CGA 245704, by Syngenta Crop Protection), registered in Europe as Bion (Actigard in the United States) was found effective. In 2001, after 5 years of field trials, Bion was registered in Italy for bacterial canker and decline of hazelnut. The minimum concentration to attain good control without inducing any phytotoxic effect is 5 g a.i./hL (25 g a.i./ha). Five applications of Bion once a month from late April to July were required to reduce the number of dead trees and branches. A final application in September after harvest was also important. In orchards sprayed with Bion, in which wilted trees and branches were annually removed, the mean number of dead trees and braches was 25% lower than in the orchards treated only with copper oxychloride at leaf fall. However, the efficacy of the product was lower in orchards where wilted twigs and/or branches were not removed. Therefore, an integrated approach including inoculums destruction (wilted twigs, branches, and trees), copper treatments at critical times, insect control, and application of acibenzolar-*S*-methyl may result in satisfactory management of bacterial decline.

7.8.2   Bacterial Blight in Hazelnut

Pathogen: *Xanthomonas campestris* pv. *corylina*

Hazelnut (*Co. avellana* L.) represents an economically important nut crop of Italy. Italy is the second-largest producer of hazelnut worldwide after Turkey. In Italy, the production of this crop is concentrated in Campania, Latium, Sicily, and Piedmont regions, in order of importance. The province of Viterbo has 90% of the hazelnut cultivations of Latium region, where Tonda Gentile Romana is the predominant cultivar in over 85% of the orchards.

Bacterial blight of hazelnut is caused by *X. arboricola* pv. *corylina* (*Xac*). The disease first occurred in the United States on *Corylus maxima* and further spread in other continents. Recent reports of bacterial blight disease on hazelnut are from countries like Iran, Germany, Poland, and

Chile, explaining the movement of the pathogen between the countries via propagation materials. However, this disease is not widespread in Europe and, as such, the European and Mediterranean Plant Protection Organizations included this pathogen in the A2 list of quarantine pathogen. The damage caused by this bacterial pathogen is mainly on young hazelnut plants (1–4 years old) in orchards killing up to 10% of plants. The losses are even more severe in nurseries, where suckering is widely practiced on the mother plants. However, devastating damage can also occur on older (7–8 years old) plants.

In Italy, *X. arboricola* pv. *corylina* was first reported in Latium and successively in Campania regions. During the early 1990s, endemic presence of the pathogen was described in central Italy. Recently, hazelnut plants infected by *X. arboricola* pv. *corylina* were also noticed from the Italian islands. Nonetheless, no economically important loss, associated with hazelnut bacterial blight, was reported previously from central Italy.

The current status of the bacterial blight has been changing drastically, for some years now. Frequent occurrence of this disease with severe damage, found across the orchards, is a matter of serious concern for growers. A prime example could be the severe canker symptoms caused by bacterial blight on cv. Tonda di Giffoni, the only Italian cultivar that did not bear the canker symptoms in the field for a century.

Host
The main host of *X. arboricola* pv. *corylina* is hazelnuts (*Co. avellana*), but *Corylus colurna*, *Co. maxima*, and *Corylus pontica* are also susceptible.

Geographical Distribution
Bacterial blight of hazelnut was described for the first time in the western United States (Oregon) in 1913 (Barss, 1913). After this initial report, it was found (possibly through introduction) in Yugoslavia (Sutic, 1956), Italy (Noviello, 1969), Turkey (Alay et al., 1973), France (Luisetti et al., 1976; Prunier et al., 1976), Russia (Koval, 1978), the United Kingdom (Locke and Barnes, 1979), Australia (Wimalajeewa and Washington, 1980), Algeria (Gardan, 1982), and more recently in Chile (Guerrero and Lobos, 1987). For more information, see Gardan and Deveaux (1987).

EPPO Region: Algeria, Denmark, France, Italy, the Netherlands, Russia, Slovenia, Spain, Switzerland, Turkey, the United Kingdom, Yugoslavia.
Asia: Turkey.
Africa: Algeria.
North America: Canada (British Columbia; IMI, 1996), the United States (Oregon, Washington).
South America: Chile; Oceania: Australia (Western Australia).
EU: Present.

Symptoms
Small, angular or round, yellowish green, water-soaked spots develop on leaves, later turning reddish brown. Buds may turn brown and fail to leaf out. Shoots may develop but are infected. The first infections on current-season stems consist of dark green water-soaked areas on bark, which turn reddish brown. Lesions may girdle stems and kill them. Dead leaves often cling to girdled stems for a long time. Lesions or cankers may extend into the main scaffold or trunk. Use a pocket knife to determine the extent of canker development. Lesions may be wet and ooze in spring.

Leaf buds and pistillate-flower-bearing buds are also susceptible. The outer bud scales are infected first and then the bacteria move into the bud itself. Buds may be completely killed or only partially damaged. Shoots emerging from buds generally become infected from infected bud scales.

Bacterial blight can cause lesions that encircle the trunk of young trees and cause them to die (Photo 7.23). These lesions can be difficult to detect, but close examination shows the bark to be

PHOTO 7.23 Symptoms of bacterial blight on leaf and fruits. (Courtesy of Morone, C., Regione Piemonte, Servizio Fitosanitario Regionale, Torino, Italy, atlasplantpathogenicbacteria.it.)

slightly sunken and reddish-purple in coloration. If you remove the bark at the crown with a knife, the tissue beneath is brown. A sticky liquid containing many bacterial cells may ooze out of the lesions during periods of high humidity, and dead leaves will generally cling to the girdled trunks for some time. Bacterial blight infections are easiest to see in late spring. Leaves develop small, angular spots that are less than 3 mm in diameter, reddish brown, and surrounded by a yellowish-green circle of tissue. Leaf lesions eventually coalesce at the tip of the leaf. Infections on developing nut husks are less common but can appear as dark brown or black spots.

Detection and Identification
Symptoms are different in orchards and nurseries. In orchards, hazelnut is grown like other fruit trees, suckers being pruned away every year. In nurseries, on the other hand, suckering is encouraged on the mother plants to produce shoots for layering. The mother plants bear densely crowded long young shoots, on which the disease can spread very readily.

In Nurseries
Shoots more than 1 year old show bud dieback, necrosis of the shoot tips, and spots on leaves in spring after bud-burst. Shoots may dry out entirely as the bacterium spreads downward, either girdling the base, causing dieback of the distal portion, or causing cankers 10–25 cm long with longitudinal surface cracks. Brownish-black necrosis may appear on the convex side of the layered shoots. This necrosis can spread to the stump and girdle the shoot, resulting in its complete dieback.

The new growth shows oily lesions starting at the tip and spreading rapidly back: buds form limited necrotic cankers, at which the shoots are liable to break. Buds below the necrotic zone then develop abnormally, giving a characteristic bushy appearance. Leaves show numerous oily polygonal lesions that may run together to cause a general chlorosis of the lamina and premature leaf fall.

In Orchards
While bud cankers, necrosis, and dieback of new lateral shoots and cankers are seen, leaf symptoms are rare. Fruits show "black heel" symptom with browning of the shell and corresponding part of the involucre, which are covered with bacterial slime. Oily lesions 3–7 mm long are sometimes seen on the involucre and shell before lignification.

One of the most characteristic symptoms is necrosis of emerging growth from buds in late spring (Locke and Barnes, 1979; Wimalajeewa and Washington, 1980; Gardan and Devaux, 1987; Guerrero and Lobos, 1987). New growth exhibits oily lesions that start at tips and progress down emerging stems. Diseased shoots become necrotized and dry. Shoots may dry out entirely as the bacterium spreads downward, girdling the base and causing dieback of the distal portion. Necrosis can spread to the stump and girdle the shoot, resulting in complete dieback. Black spot and streak may be found on young stems, and cankers may also be found on twigs and branches (Lelliot and Stead, 1987). The pathogen rarely causes dieback of branches or stems older than 3–4 years old (Schuster, 1924; Noviello, 1969). The necrosis is only on the superficial bark and does not involve the xylem. Small, black, necrotic spot lesions are superficially present on fruits and cupules (Noviello, 1969; Gardan and Devaux, 1987). Leaves show numerous polygonal water-soaked yellowish-green to dark-green lesions, which may merge together causing a general chlorosis of the lamina and premature leaf fall (Gardan, 1986). The glucoprotein secretion factor of 0.18 is necessary for expression of water-soaking reaction within 3 days during *Xanthomonas corylina*–noisetier interaction (Borkar, 1989a). The plants are more susceptible when young and succulent (Miller, 1949). Leaf symptoms are rare in orchards, while bud cankers, dieback of new lateral shoots, and cankers are frequently observed. On fruit, oily lesions are sometimes seen on the involucres and shells before lignification (Gardan, 1986). The disease may be particularly severe, causing dieback of seedlings either in the nursery or in a young orchard (Scortichini, 1995).

Latent infections as well as low population level of *X. corylina* on noisetier buds/leaves were detected under field condition in France using specific immunoglobulin under immunofluorescence (Borkar, 1989b). Buds of noisetier plant were found to harbor more population of *Xanthomonas coylina* as latent infection than leaves and twigs. Specific immunoglobulins had an advantage over nonspecific immunoglobulins to detect *X. corylina* among epiphytic bacterial flora of noisetier.

Pathogen

X. arboricola pv. *corylina* is a Gram-negative rod with a single polar flagellum. It is strictly aerobic. Like other *Xanthomonas* bacteria, it produces a yellow carotenoid pigment in the culture medium.

X. arboricola pv. *corylina* (formerly *X. campestris* pv. *corylina*) attacks buds, leaves, branches, and trunks. Occasionally, it attacks nuts but seldom invades roots. Tree mortality due to this disease is commonly found in orchards the first few years after planting. The most serious phase of the disease is trunk girdling and killing of trees up to 8 or 10 years old; 1- and 2-year-old twigs are attacked and killed. Rain splash or the movement of infected nursery stock spreads the bacteria. Seedborne transmission has been reported, but no data are presented to support the claim. Infections are indirect, through wounds or by bacterial invasion of blighted buds and shoots of current-season growth. The disease has been associated with high stress sites.

In Italy, higher disease incidence was found in regions with higher rainfall and a greater difference between high and low temperatures during frost events. Higher disease incidence for soils with higher nitrogen levels was also found.

Disease Cycle

The bacterial pathogen enters through open stomata (on leaves) and wounds on the plant. It survives from one season to another in cankers and infected buds, surviving better in the large branch and trunk lesions than in the smaller twig lesions (less than 8 mm in diameter). It generally does not attack and kill branches that are more than 3 years old. Trunk lesions develop from pruning wounds or migration of the bacteria from adjacent infected buds or shoots. Pruning and suckering young trees with unsterilized pruners will spread the disease.

Bacterial blight overwinters in cankers on branches and inside infected buds. New bacteria ooze from cankers during the growing season. Bacteria are spread by rain to new infection sites. Bacteria enter through open stomata on the leaves and tender branches and through wounds or cracks on older branches. The bacteria first infect the outer bud scales and then move into the inner bud. Buds may be completely killed or partially damaged. Shoots that emerge from infected buds generally become infected from the bud scales as they grow past.

Temperatures above 20°C favor infection, although infection can occur at lower temperatures if the wetting period is long enough. Moisture must be present on the plant tissue for infection to occur. Infection time can be as short as 1 h where temperatures are warm. Infection by bacterial blight is greater after a winter where freezing injury has occurred.

Buds are susceptible to bacterial infection from summer, when they are three-quarters grown, to when they open the following spring. Twigs and branches are susceptible in summer and fall. Leaves are susceptible until they have reached their maximum size. The bacterium does not infect root systems.

Susceptibility

Buds appear to be the most susceptible from when they are three-quarters grown, until they open in the spring.

Twigs and branches are also most susceptible during the same time period. Twigs become tolerant to stomatal infections after they have stopped growing and become woody. Any infections that occur after this stage are from wounds or infections from adjacent buds.

It is not known if the disease infects the plant systemically but the bacterium does not seem to move into the roots.

Weather Factors

Wet weather is an important factor in the spread of bacterial blight. The pathogen can be picked up from cankers and carried by water droplets onto branches below the cankered branch. Moisture must be present on the plant tissue for infection to occur, but the infection time can be as short as 1 h for leaf infections if the leaf is wet during that period.

Disease incidence also seems to increase following freezing weather. This may be because the trees are weakened or because there may be more entry sites through wounds.

Spatial patterns of pedoclimatic data, analyzed by geostatistics, showed a strong positive correlation of disease incidence with higher values of rainfall, thermal shock, and soil nitrogen—a weak positive correlation with soil aluminum content and a strong negative correlation with the values of Mg/K ratio. No correlation of the disease incidence was found with soil pH. Disease incidence ranged from very low (<1%) to very high (almost 75%) across the orchards. Young plants (4 years old) were the most affected by the disease, confirming a weak negative correlation of the disease incidence with plant's age. Plant cultivars did not show any difference in susceptibility to the pathogen.

The need to carry out a detailed epidemiological study was emphasized following the outbreaks of this disease across central Italy. Apparently, distribution and incidence of the disease were heterogeneous across the Viterbo province, which suggested the possible role of pedoclimatic factors in disease occurrence. Recent finding of the disease in other European countries and the outbreaks in central Italy could in part be associated with the possible effect of climate change. The role of the latter on crop–disease interaction has been a matter of serious concern for many authors. Regarding the soil, its physical and chemical properties play an important role in the plant health and the consequent disease occurrence and spread. In addition, crop management practices influence significantly the occurrence and control of hazelnut bacterial blight.

Other Factors

Researchers could not isolate the pathogen from soil under severely infected trees in Oregon trials. This suggests that the pathogen does not survive in soil. It does not appear that insects have an important role in disease spread.

Means of Movement and Dispersal

The main mode of spread is through infected planting material. The potential for natural spread can be considered as relatively low. Seeds from fruits picked on infected trees can produce infected seedlings.

Economic Impact

Four countries account for the bulk of world's hazelnut production: Turkey, Italy, Spain, and the United States (in decreasing order). In the United States, bacterial canker is considered the most serious disease of hazel and of more economic value than all other hazel diseases put together.

In the United States, bacterial canker is considered to be the most serious disease of hazelnut in terms of economic value. The greatest yield losses are seen in 1- to 4-year-old orchards, in which up to 10% mortality has been recorded (Gardan, 1986). Losses are due to reduced yield and reduction in the development of young, nonbearing trees. Older plants are rarely killed. Economic losses depended on the disruption of buds and fruiting shoot caused by the bacterium (Noviello, 1969). Losses in yield varied from 1% to 10% (Miller, 1949). In France, over 250,000 young plants have been destroyed since 1975. In 1983, 1300 ha suffered the loss of about fifty 7- to 8-year-old trees and 2 ha of 4-year-old trees were killed (Gardan, 1986).

Disease Management

Although bacterial blight is widespread in the Pacific Northwest, it can be managed. Identification is the first step in controlling this disease. Sometimes bacterial blight can be confused with other diseases such as sunscald and winter damage, but lab tests can confirm the presence of the bacterial pathogen. It is easiest to test for the bacterium during the spring.

Buds can be infected but may not show symptoms for over 200 days. This means that healthy-looking trees can be infected. All young trees (planting stock) should be handled as though they were infected.

Healthy trees are less susceptible than weakened trees, so growers should encourage good growing conditions.

Trials conducted in Oregon showed that the removal of infected plant material helped reduce the spread of disease but did not eliminate it. Sprays of Bordeaux mixtures (6-3-100) in late summer (August) were sufficient to control the disease. However, in exceptionally rainy years, three sprays—in late summer, late fall when leaves were about three-quarters off the tree, and early spring when the buds were opening—were necessary to control the disease.

The Oregon trials showed that fixed copper with a spreader sticker was also an effective deterrent. Guardsman copper oxychloride 50% (PCP No. 13245) and UAP Copper Spray (copper oxychloride 50%, PCP No. 19146) are registered in Canada.

Use 3–9 kg of copper oxychloride per hectare. Make the first application in August/September before fall rains, the second application when three-fourths of leaves have fallen, and the third in early spring before bud set. Use low rate on small trees and high rate on large trees. Apply in 1000 L water/ha by ground spray only. There should be a maximum of three applications per year.

Disinfect pruners between cuts (Table 7.2).

Infected branches should be pruned well below the cankers in late winter to reduce the source of new inoculum. All prunings should be removed from the orchard and burned or buried. Copper sprays are registered to help manage bacterial blight. Up to three applications can be applied per year. The first application of copper spray is to be given in late August or early September and the second application in the fall when three-quarters of the leaves have fallen. For severe infections, a third application of copper spray can be made the next spring just before bud break occurs.

TABLE 7.2
Disinfectant Treatments for Cutting Knives

Best Disinfectants	Treatment Time
5% Virkon	Quick dip
10% Bleach[a]	Quick dip
DCD Floralife (16 mL/L)	Quick dip
Ethanol 70%	20 s

[a] Household strength (5.25% sodium hypochlorite). Prune out infected branches, making cuts 60–100 cm below the infected branches. Burn the bran.

Cultural Control

- Plant pathogen-free nursery stocks in early winter. Do not let roots dry out.
- Prune out infected twigs and branches. Make cuts 2–3 ft below affected branches.
- Sterilize pruning tools between cuts with shellac thinner (70% ethyl alcohol).
- Control sun scald during summer using a shield or white paint on trunks.
- Mulch around the base of newly planted trees with chipped, composed tree debris to reduce moisture stress.
- Irrigate during the first three summers after planting to reduce moisture stress.
- Apply nitrogen fertilizer in the spring based on leaf tissue analysis the prior August.

Chemical Control

Apply sprays in late August or early September before the first heavy rains. If fall rains are heavy, apply another spray when 75% of the leaves have dropped. Note that most copper products do not allow to use until after harvest.

- Bordeaux 6-3-100.
- Champ Dry Prill at 10.67–16 lb/A with 1 pint superior-type oil/100 gal water; can be used only after harvest; 48 h reentry.
- C-O-C-S WDG at 11.6 lb/A plus 1 pint superior-type oil/100 gal water; can be used only after harvest; 48 h reentry.
- Copper-Count-N at 8–12 quarts/A; can be used only after harvest; 48 h reentry.
- Cuprofix Ultra 40 Disperss at 10–15 lb/A; *Oregon and Washington only*; 48 h reentry.
- Kocide 3000 at 7–10.5 lb/A plus 1 pint superior-type oil/100 gal water; can be used only after harvest; 48 h reentry.
- Monterey Liqui-Cop at 1–2 Tbsp/gal water.
- Nordox 75 WG at 8–13 lb/A; can be used only after harvest; 12 h reentry.
- Nu-Cop 50 DF at 8–12 lb/A with 1 pint superior-type oil/100 gal water; can be used only after harvest; 48 h reentry.

Prevention

Once established, this serious pathogen cannot be eradicated except by removal of all *Corylus* (hazelnut) plants. Therefore, the most important element of control is to introduce only disease-free planting material into an orchard (Gardan, 1986; Gardan and Devaux, 1987). Standard hygienic practices in affected orchards such as removing and destroying affected shoots and disinfecting pruning tools may reduce the impact of the pathogen. The application of protective copper-based sprays such as copper oxychloride may be efficacious but probably must be applied annually as a prophylactic routine; when the disease is seen to be severe, it is unlikely that spray applications will be effective. Treatments with copper compounds must be done in the spring, beginning at bud break, and repeated in the autumn, at leaf fall (Gardan and Devaux, 1987; Scortichini, 1995). Spring sprays are most efficacious. The number of applications varies depending on the length of the rainy season (Noviello, 1969).

The use of resistant cultivars (cultivars Négret, Gunslebert, Segorbe, Logued'Espagne, Merveille de Bollwiller) or the species *Co. pontica* (Gardan, 1986) is recommended when planting a new orchard. The Graham filbert was reported as a *Corylus* species with a degree of resistance to blight (Brooks and Olmo, 1958). Different cultivars of hazelnut are characterized by a different degree of susceptibility, but none are immune (Noviello, 1969).

Less vigorous plants due to inappropriate cultural techniques are more susceptible to the disease. It is, therefore, important to avoid plantations in soil with poor drainage (Noviello, 1969).

7.9 CASHEW

7.9.1 BACTERIAL LEAF AND FRUIT SPOT OF CASHEW NUT

Pathogen: *Xanthomonas campestris* pv. *mangiferaeindicae*
In 2003 and 2004, leaves and young fruits of cashew nut plants showing an undescribed disease symptom were observed on plants of an early-dwarf clone in a commercial orchard in Ceará and Piauí states in northeastern Brazil.

Geographical Distribution
Brazil, India.

Symptoms
Initial symptoms consist of angular, water-soaked, dark-to-black spots on the leaf and at the midrib vein surrounding the leaf veins. Eventually, lesions also extended from the midrib to the secondary veins, delineating the vein system of the leaf. In young, green fruits, symptoms were large, dark, oily spots surrounded by conspicuous water-soaked areas.

Pathogen
A yellow-pigmented bacterial colony was consistently recovered from the disease lesions on nutrient yeast-extract dextrose agar medium (3 g of meat extract, 5 g of peptone, 10 g of dextrose, 5 g of yeast extract, and 18 g of agar per liter). Physiological tests revealed the bacterium belonged to the genus *Xanthomonas*.

Polymerase chain reaction (PCR) amplification of bacterial DNA using RST2 (Leite et al., 1994) and Xcv3R (Trindade et al., 2007) primers resulted in identical band patterns to mango isolates *X. campestris* pv. *mangiferaeindicae*.

Restriction fragment length polymorphism (RFLP) analysis of PCR-amplified products of six isolates of *X. campestris* pv. *mangiferaeindicae* was conducted with *Hae*III, which showed different profile patterns on agarose gel, indicating genetic variability among these isolates.

Pathogenicity was demonstrated by gently piercing and misting cashew leaves with a bacterial suspension adjusted to 10^6 CFU/mL. After 8 days, foliar symptoms similar to those observed in the field developed on all inoculated plants, and reisolated bacteria were characterized and found to be *X. campestris* pv. *mangiferaeindicae*.

This is the first description of commercially grown cashew plants as host to *X. campestris* pv. *mangiferaeindicae* in Brazil.

The disease cycle has not been worked out for this disease.

This disease may pose a serious problem to the cashew-growing industry in Brazil.

Control Measures
Not worked out.

7.10 GRAPE

7.10.1 CROWN GALL OF GRAPEVINE

Pathogen: *Agrobacterium vitis* Ophel and Kerr (1990)

Synonyms: *Rhizobium vitis* (Ophel and Kerr, 1990; Young et al., 2001)
        A. tumefaciens biovar 3
        Agrobacterium biovar 3

Crown gall is an important disease of wine grapes when they are grown in cold climates. Incidence may range from a few vines in a vineyard to 100% of the vines. Gall development may result in girdling of vines and reduced vigor and yield.

Crown gall on grape, caused by the bacterium *A. vitis*, occurs throughout the world where grapes are grown. In Pennsylvania and New York, crown gall can lead to vine decline and mortality in vineyards (Stewart and Wenner, 2004). Grapevines grown in areas subject to freezing winter temperatures are especially vulnerable to crown gall because freeze injuries provide a wound where the disease can initiate. The crown gall pathogen survives systemically within the grapevine and causes disease at wounded areas on the vine (Lehoczky, 1968). There are both tumorigenic and nontumorigenic strains of *A. vitis* found on grape (Burr et al., 1998). Tumorigenic strains are detected only in soils that have been used to grow grapes and can induce infected vines to develop large galls at above and/or below the soil line. Nontumorigenic strains can be found on the roots of wild vines and have not been associated with gall formation. Tumorigenic strains of *A. vitis* are initially introduced into new vineyard soils by planting infected nursery stock. Studies have shown that *A. vitis* can survive in small pieces of dead grape debris remaining in the soil for at least 2 years after the removal or death of a vine (Burr et al., 1995), while other researchers have found that grape roots may remain viable in the soil for at least 5 years after the removal of a vine (Raski et al., 1983). Vines that are free of crown gall can be infected when planted in soil with debris remaining from a vine that was infected with *A. vitis* (Burr et al., 1995). Therefore, the best way to prevent crown gall in the vineyard is to prevent the site from being contaminated with infected plants from the beginning. Infected vines may remain symptomless until the vine is injured. Injury that leads to expression of crown gall can result from freezing, pruning, grafting, training vines, and from other mechanical devices employed in maintaining the vineyard. As the gall forms, the tissues that normally function to conduct water from the roots to the shoots and photosynthetic products from the leaves to the roots become highly disorganized and lose their ability to function. Large galls that girdle the vine result in significant vine decline, and may lead to vine death. If the gall is smaller and localized, the vine can survive with a reduced, but functional, vascular system, resulting in a reduction in vine vigor and productivity.

Symptoms

Gall formation on the aerial part of the vines is the most common symptom associated with this disease. The bacterium that causes crown gall may be present in plants that do not show any symptoms. Galls are usually noticed as swellings near the base of the vine and up the trunk (Photo 7.24). Galls on roots of grape are not typical; however, the bacteria can induce a localized necrosis of roots. Young galls are soft, creamy to greenish in color, with no bark or covering. As they age, the tissue darkens to brown. The surface becomes open and the texture becomes moderately hard and very rough. The surface tissue of the galls turns black as it dies, but the bacterium remains alive in the vine.

PHOTO 7.24   Crown gall on grapevine. (Courtesy of Brown Jr., W.M., Bugwood.org.)

Newly formed galls are often first noticed in June–July and appear as pale-colored, fleshy, convoluted tissue immediately beneath the bark layer. Crown gall may develop on any freshly wounded, woody portion of the vine. It usually shows up low on the trunk or at the graft union. Galls have been observed below the soil surface, which can lead to underestimating the incidence of crown gall in a block of vines (Stewart and Wenner, 2004). *A. vitis* also causes areas of root necrosis on infected vines (Burr et al., 1987).

Host
A. vitis has a host range limited to grape (*Vitis* species), whereas *A. tumefaciens* has a wide host range of more than 600 plant species in over 90 families. In Oklahoma, crown gall of grape is usually caused by *A. vitis*.

How *Agrobacterium vitis* Causes Crown Gall
When a grapevine cell is wounded, it releases a compound that attracts *A. vitis* to the cell. The bacterium then transfers a piece of its DNA (called T-DNA) into the plant cell DNA. When integrated into the plant DNA, the bacterial T-DNA directs the cell to produce an overabundance of the plant hormones auxin and cytokinin that lead to the uncontrolled division and growth of the infected plant cells, resulting in gall formation. The T-DNA also instructs the infected plant cells to produce compounds called opines that serve as a carbon and nitrogen source for *A. vitis*, giving them an advantage over other bacteria.

Geographic Distribution
These pathogens are distributed worldwide, specifically in the United States, Europe, and Australia. Severe losses may occur when countries aberrantly apply quarantine regulations for this ubiquitous organism.

Disease Cycle
The pathogen can live in soil for short periods and enters plants through root system. Once within the vine, the pathogen colonizes the xylem tissue. The vine usually remains symptomless. An injury is required for initiation of gall formation. Such an injury could arise from freeze damage. Plant cells will die with freeze but bacteria will survive. Bacteria in the xylem system invade the damaged cells and initiate infection. Therefore, aerial galls can occur on grapevines. Mechanical injuries can also lead to formation of galls in the crown area. Galls tend to form on the lower trunk and sometimes at graft unions inhibiting graft-take. The formations of galls inhibit the growth of vascular tissue cutting off nutrient flow. Grape crown gall can lead to vine decline or death. In grapevines, certain strains can cause extensive necrosis. The disease cycle of crown gall of grapevine is illustrated in Figure 7.4.

As indicated, crown gall develops at wound sites, primarily during wound healing in the cambium, where dividing cells are susceptible to infection. Infected cells overproduce plant growth hormones, leading to gall formation and enlargement.

Economic Impact of Crown Gall
Burr et al. (1995) have put together some hypothetical numbers and made a few assumptions to illustrate the economic impact of crown gall on a vineyard and winery operation. The numbers can vary widely, but the economic and psychological impact of crown gall on growers and wine makers is not to be underestimated.

Vineyard management costs:

- Cost of vines: $3.50 × 2 (original and replacement) = $7.00
- Labor to plant: $2.50 × 2 (original and replacement) = $5.00
- Labor to train: $3.00 × 6 years = $18.00

Cost per vine = $30.00

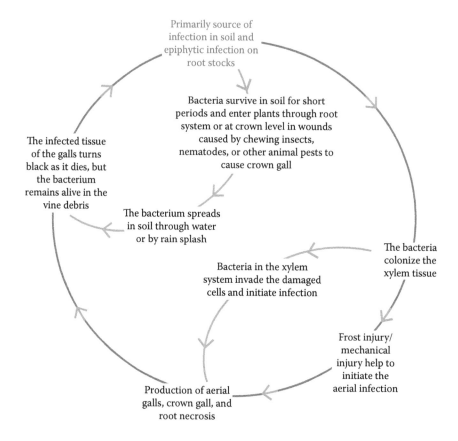

FIGURE 7.4 Disease cycle of crown gall of grapevine caused by *A. vitis*.

One-third acre or approx. 200 vines replaced due to crown gall: 200 × $30.00 = $6000

Loss of wine sales:

- 1 vine produces 10–15 lb grapes/year
- 1/3 acre (200 vines) = approx. 1 ton grapes
- 1 ton grapes = approx. 750 bottles of wine
  750 bottles wine × $9.00 = $6750 lost in wine sales lost/year
  $6,750 × 6 years = $40,500 lost in wine sales over 6 years

Potential economic loss in replant and wine sales over a 6-year period = $46,500

Control Measures
Control of crown gall can be achieved at best when followed in totality.

Before Planting
Losses of grape plants due to crown gall may be minimized with some considerations before vineyard sites are selected and before vineyards are planted. Criteria to consider include the following:

- Select sites with good air and water drainage.
- Avoid planting in frost-prone areas.
- Select rootstocks that are resistant to crown gall. Certain rootstocks such as Courderc 3309, 101-14 Mgt, and Riparia Gloire are resistant, whereas Teleki 5C and 110 Richter

are susceptible. Resistant rootstocks can reduce the amount of crown gall that appears on susceptible scions.

- Select hardy varieties where possible. In general, *Vitis vinifera* cultivars are all highly susceptible.
- Prepare sites with nutrients and lime before planting, if necessary, to avoid vine stress due to poor nutrition or low pH.
- Consider pest control programs for nematodes and phylloxera before planting through the use of soil fumigation or rootstocks. Feeding damage by these pests provides sites for the entry of crown gall bacteria.
- Plant old vineyard areas where crown gall was present only after grapevines have been removed for at least 2 years. This is important because crown gall bacteria can survive in the remnants of the old grape plants until the debris decomposes. Success in reducing crown gall from the soil by leaving soil fallow or rotating to other crops may vary depending on the amount of vine debris that is left in the soil and how fast it breaks down.
- Do not propagate wood taken from galled vines.
- Specify that plant material you purchase be propagated from crown gall–free plants. Producers must promote the production of certified indexed crown gall–free vines.
- Hot water treatment of vines is effective in reducing crown gall infection levels in planting materials.

After Planting
There is little that can be done to control this disease once it is established in the vineyard.

Avoid injury to vines (winter, mechanical, and human). Remove suckers when shoots are small (3–6 cm) to reduce trunk damage and promote rapid healing of wounds. Removing larger shoots before they harden will result in clean and small scars.

- Hill young vines with 30 cm or more of soil or other material to protect them from cold temperatures. Remove this material carefully to avoid mechanical damage.
- Prune to delay bud break (late pruning) on varieties prone to early bud break.
- In France, application of K_2O instead of nitrogen fertilizers is used to improve the resistance of the vines to cold.
- Biological control of grape crown gall has been effective in trials and is a promising management tool for the future.

Although biological control of crown gall in other plants through *A. radiobacter* K-84 is highly effective, this strain does not prevent the disease in grapes. However, an alternative biological control bacterium, *A. vitis* strain F2/5, shows promising disease control and is under further investigation. Strain F2/5 is nontumorigenic and is effective in experiments when applied to grape wounds before they are inoculated with tumor-causing bacteria. F2/5 is not commercially available, and research is being done to determine its efficacy in field trials.

Prevention and Management Strategies
The most important step in preventing crown gall is to avoid planting infected material in the vineyard. When choosing the types of grapes to grow, keep in mind that *V. vinifera* cultivars are especially susceptible to crown gall and phylloxera. One way to avoid problems with these diseases is to graft *V. vinifera* on resistant rootstocks. Studies have shown that some rootstocks such as 3309C, 101-14 Mgt, and Riparia Gloire that have been used for phylloxera resistance may also provide resistance to crown gall (Burr et al., 1998; Sule and Burr, 1998). In one study, crown gall–susceptible *V. vinifera* vines grafted on Gloire rootstocks were infected with *A. vitis* and still developed symptoms after 3–4 years, but they were able to recover compared to nongrafted *V. vinifera* (Sule and Burr, 1998). In some instances, the infected trunk had died, but a new trunk from below

the gall survived without any galling. From that research it was concluded that "rootstocks can greatly affect the severity of crown gall infection of grapevine" (Sule and Burr, 1998). When choosing a vineyard site, avoid frost-prone areas and wet, heavy soils to prevent freeze injury to potentially infected grapevines. Before planting, be sure to control soilborne nematodes. Studies have shown that the incidence of crown gall is higher on nematode-infected vines (Sule et al., 1995). After planting, it is important to avoid external mechanical wounding of trunks and canes as well as freeze injury. Ways to avoid freeze injury include the following: planting in sites that are not prone to cold winter temperatures and selecting cultivars suitable to the site; not planting cold-sensitive *V. vinifera* cultivars in a frost-prone area; hilling-up young vines to protect graft unions and the lower trunk in young, own rooted vines; balance pruning and not overcropping; applying K_2O (potash) instead of nitrogen fertilizers to improve the resistance of vines to cold; and avoiding vigorous late-season growth, so vines will harden off for winter. Vines infected with crown gall that are not girdled can often be pruned and retrained to regain productivity. Galled and girdled trunks can be removed and new trunks brought up for training. Partially girdled trunks may outgrow the gall. Success of this process is determined by the susceptibility of the cultivar and the environmental conditions that trigger gall development. Successive generations of galls can form on a single vine resulting from freeze damage and other forms of mechanical damage to plant cells. When managing crown gall–affected vines, keep the following things in mind. Maintain two trunks on crown gall–infected vines and be prepared to bring up new shoots for trunk renewal. Remove galled prunings from the vineyard and burn them if possible. When replacing vines killed by crown gall, remove the dead vine and as much of the root mass as possible because *A. vitis* can live for years on decaying roots. Copper compounds applied for downy mildew may help manage crown gall. The crown gall pathogen can survive in the debris of roots, trunks, and canes in the soil, so it is important to remove any *A. vitis*–infected plant material from the vineyard and burn if possible. This may help to reduce levels of the pathogen in the soil and help prevent the infection of replants. The cost of replanting, and loss of production and profit, can be high for severely infected vineyards.

7.10.2 BACTERIAL BLIGHT OF GRAPEVINE

Pathogen: *Xylophilus ampelinus*

Synonyms: *Xanthomonas ampelina* Panagopoulos
Xyl. ampelinus (Panagopoulos)

Common names: Bacterial blight of grapevine
Tsilik marasi (in Greece)
Maladie d'Oleron
Nécrose bactérienne de la vigne (in France)
Vlamsiekte (in South Africa)
Mal nero (in Italy)

Bacterial blight of grapevine is a serious, chronic, and destructive vascular disease of grapevine, affecting commercially important cultivars. It is widespread in the Mediterranean region and South Africa and may occur in other regions. The pathogen survives in the vascular tissues of infected plants.

Symptoms
Bacterial necrosis of grapevines is characterized by typical symptoms such as cankers on stems and petioles, by necrotic foliar spots, and by bud death.

In early spring, buds on infected spurs fail to open or make stunted growth and eventually dies. Affected spurs often appear slightly swollen because of hyperplasia of the cambial tissue. Cracks appear along such spurs and become deeper and longer, forming cankers. Young shoots may develop pale yellowish-green spots on the lowest internodes. These expand upward on the shoot, darken, crack, and develop into cankers. Cracks, and later cankers, also form on more woody branches later in spring. In summer, cankers are often seen on the sides of petioles, causing a characteristic one-sided necrosis of the leaf. They may also appear on main and secondary flowers and fruit stalks. Leaf spots and marginal necrosis occur sometimes. Gum formation is not necessarily a symptom.

Infection usually occurs on the lower two to three nodes of shoots that are 12–30 cm long and spreads slowly upward. Initially, linear reddish-brown streaks appear, extending from the base to the shoot tip. Lens-shaped cankers then develop. Shoots subsequently wilt, droop, and dry up. Discoloration is less common on very young shoots, but the whole shoot dies back. Where there is severe infection, a large number of adventitious buds develop, but these quickly die back. Infected shoots are shorter, giving the vine a stunted appearance. Tissue browning is revealed in stem cross-sections. Infected grape bunch stalks show symptoms similar to the infected shoots (Photo 7.25).

Leaves may be penetrated via the petiole and then the veins, in which case the whole leaf dies. Alternatively, leaves are penetrated directly via the stomata, resulting in the development of angular, reddish-brown lesions. Infection through the hydathodes results in reddish-brown discolorations in the leaf tips. Pale yellow bacterial ooze may be seen on infected leaves when humidity is high.

Immature flowers turn black and die back.

Roots may also be attacked, resulting in retardation of shoot growth, regardless of grafted or natural rootstock.

Pathogen

Xyl. ampelinus is a Gram-negative, nonspore-forming rod with one polar flagellum, 0.5–0.8 × 1.6–3.2 μm. Occasionally isolates have some filamentous cells 8–10 times longer than usual. They occur singly or in pairs. In culture, they exhibit the following characteristics: At 25°C, growth is slow, appear as nonmucoid, smooth, yellow, round entire colonies, measure 0.4–0.8 mm in diameter, and develop in 6–10 days on yeast-glucose-chalk agar. On nutrient agar with 5% sucrose and on sucrose-peptone-salts agar, there is more growth and a deeper shade of yellow.

Growth on artificial medium is very slow and a heavy inoculum is advisable as growth may fail if streaks are made from dilute suspensions. On nutrient agar, colonies are barely visible (0.2–0.3 mm diameter) in 6 days at 26°C increasing to 0.6–0.8 mm in 15 days, and they appear circular, entire, glistening, translucent, pale yellow. On nutrient agar with 5% sucrose or on GYCA (containing 1% glucose, 0.5% yeast extract, 3% calcium carbonate and 2% agar), growth is faster and deeper yellow. Best growth is obtained on 2% galactose, 1% yeast extract, 2% calcium carbonate, and 2% agar, on

PHOTO 7.25 Bacterial blight infection causes scorching on leaf, cracks, and canker on vine. (Courtesy of Lopez, M.M., Instituto Valenciano de Investigaciones Agrarias, Valencia, Spain, atlasplantpathogenicbacteria.it.)

which it is deep yellow and produces a diffusible brown pigment. In minimal synthetic medium, 0.1% glutamate is required for growth but not L-methionine. No fluorescent pigment is produced on King's medium B or any similar medium.

No strain hydrolyzes aesculin, arbutin, casein, gelatin, sodium hippurate, or starch, shows pectolytic activity on potato tissue, and reduces nitrate to nitrite, produces ammonia, indole, arginine dihydrolase, ornithine and lysine decarboxylases, or phenylalanine deaminase. H_2S is produced from cysteine and thiosulfate by all strains, but from peptone by some strains only. The malonate and gluconate tests are negative, and hence tyrosinase evidently varies. Lipolysis is positive on Tween 80 but negative or weak on tributyrin. Urease is strongly positive, catalase-positive, and oxidase-negative.

Acid is produced without gas in 6–8 days from L-arabinose and D-galactose, but not from glucose in Hugh and Leifson's medium with peptone reduced to 0.1%. Metabolism is strictly aerobic. In Dye's Medium C with bromothymol blue indicator and 0.5% carbon source, acid is produced from L-arabinose and D-galactose and slowly by some isolates from glycerol, but not from adonitol, arbutin, cellobiose, dextrin, dulcitol, D-fructose, D-glucose, glycogen, *meso-inositol*, inulin, lactose, maltose, D-mannitol, D-mannose, D-melibiose, amethyl-D-glucoside, raffinose, L-rhamnose, D-ribose, salicin, D-sorbitol, L-sorbose, sucrose, trehalose, or D-xylose.

Some strains may produce acid from glucose and Oleron strains are reported to produce acid from maltose. Citrate, fumarate, malate, and tartrate are used as sole carbon sources. Gluconate, maleate, malonate, oxalate, benzoate, formate, and propionate are not used. The last three inhibit the small amount of growth seen on the minimal medium in the absence of a carbon source. Lactate is used by some strains. Growth fails, or is very poor, in presence of 0.02% triphenyltetrazolium chloride. Maximum tolerance of NaCl is 1%. Minimum temperature for growth is 6°C, while the maximum is 30°C.

The G + C content of the DNA is 68 to 69 tool% by thermal denaturation. Type strain NCPPB 2217; ICMP 4298; ATCC 33914 (Bradbury, 1991).

The bacterium differs from *Xanthomonas vitis-carnosae*, *Xanthomonas vitis-trifoliae*, and most other members of the genus due to its very slow growth, low maximum temperature, low salt tolerance, inability to produce acid from numerous sugars (including glucose), and failure to hydrolyze gelatine, casein, starch, and aesculin. It most nearly resembles *Xanthomonas albilineans*, but differs from this and other xanthomonads due its strong urease activity and ability to use tartaric acid salts as carbon source.

Host Range
The only known host of bacterial blight is *V. vinifera* (grapevine).

Distribution
Bacterial blight is a chronic, systemic disease of significant economic value.

Bacterial blight is known to occur in Asia, South Africa, Greece, France, Spain, Italy, Turkey, Portugal, Russia, South America, and the Canary Islands.

Disease Cycle
The life cycle of *Xyl. ampelinus* has not been completely elucidated. Primary infections occur mainly on shoots of 1- or 2-year-olds, via leaves, blossoms, and grapes. The pathogen is also readily transmitted with pruning tools (Ridé et al., 1977) and enters healthy tissues mainly through pruning wounds, especially in wet and windy weather. The bacteria then spread to other shoots in the early summer.

Epidemiology of bacterial blight indicates that no insect vector of importance has been found. The major sources of infection are apparently infected propagating material and epiphytic bacteria that enter through wounds.

Bacteria overwinter in the vines and emerge probably in spring and are carried to healthy shoots, most probably in wind and rain. Wounds may facilitate entry but are not needed for primary infection.

The disease is associated with warm moist conditions, and spread is favored by overhead sprinkler irrigation. From initial disease foci, local spread in vineyards tends to occur along rows.

Dispersal

The disease can be transmitted by propagating material, grafting, and pruning knives. Bleeding sap appears to be an important source of contamination. Illegally imported plants pose the greatest risk and, if such material is infected, the disease is likely to become established.

Bacteria are spread by moisture to plants where infection may occur through wounds, leaf scars, and other sites. Infection may also occur without wounds (Serfontein et al., 1997).

The spread is favored by overhead sprinkler irrigation. From initial disease foci, local spread in vineyards tends to occur along rows.

Survival

The bacterium overwinters in the vines. Shoots are susceptible to infection during autumn and winter and nonsusceptible during spring and summer. The bacterium is able to survive in wood, and thus may be transmitted between locations in infected cuttings.

The ability of *Xyl. ampelinus* to survive for several years inside plants without inducing symptoms may result in a latency period.

Identification Keys

X. ampelina is an anomalous member of the genus *Xanthomonas*, which is readily distinguished on the basis of a few tests. The pathogen may be rapidly detected in grapevine sap using indirect immunofluorescence technique.

Economic Impact

Losses arise from reduced productivity and shortened life of diseased vines. Some cultivars are more susceptible than others and no control measures are known.

Severe infection of susceptible cultivars can lead to serious harvest losses. In 1940, Du Plessis observed harvest losses of 70% and more in South Africa. Vines infected in the previous year deteriorated and died back in subsequent years. Since 1956, however, the disease has only appeared sporadically in South Africa and, where controlled by copper sprays, is of no economic value.

In France, since 1968, serious damage has been reported, particularly on Alicante-Bouschet and Ugni Blanc vines in Charente, and on Grenache and Macabeu in Languedoc. Vines growing on their own roots in the irrigated areas around Narbonne are most severely affected. The disease is also of increasing importance in Spain (Lopez et al., 1980).

In Greece, the disease is widespread in Crete, especially in Iraklion county, where it occurs mainly on the very susceptible cultivar Sultanine. It has recently spread to some other Aegean islands. On the mainland, where it was previously limited to the Kynegos area in the South Peloponnesos on cv. Corinthe noir, it has recently appeared in two of the best grape-growing country in West Peloponnesos, where large areas of this economically important cultivar are threatened.

Control Measures

Control can be obtained only through viticultural practices. Chemicals have failed to control the disease (Panagopoulos, 1987). Infected shoots should be destroyed. Pruning should be carried out in dry weather and as late as possible. All pruning tools should be thoroughly disinfected during the operation. Overhead sprinkler irrigation should not be used.

The varieties Sultana and Barlinka are very susceptible. Hermitage is moderately susceptible and Green Grape and Riesling are moderately resistant.

7.10.3 PIERCE'S DISEASE OF GRAPE

Pathogen: *Xylella fastidiosa* Wells et al. (1987)

Synonyms: *Xy. fastidiosa* subsp. *fastidiosa*
          Xy. fastidiosa subsp. *pierceli*

Common names: Pierce's disease, California vine disease, Anaheim disease of grapevine

PHOTO 7.26   Leaf scorching symptoms of pierce disease on leaves of grape. (Courtesy of Scortichini, M., atlasplantpathogenicbacteria.it.)

Primary host: Grapevine, peach, Japanese plum, fox grape, almond.

Symptoms

On grapevine, the early sign of infection is leaf scorch (Photo 7.26) or sudden drying of a green leaf that turns brown, and adjacent tissues turn yellow or red. This may spread to the whole leaf that will shrivel and fall off, leaving the petiole attached. Diseased stems have patches of brown and green tissues. Mature infected plants may produce stunted, chlorotic shoots. Highly susceptible cultivars may not survive more than 2–3 years. More tolerant cultivars may survive more than 5 years. Young vines are more susceptible.

There are numerous symptoms expressed by susceptible cultivars after infection. The first symptom is usually uneven marginal leaf necrosis that often appears near the point of infection. Since the disease inhibits water movement in the vine, symptoms often appear during heat stress or near veraison (color change) in the cluster. The clusters of heavily infected vines may actually collapse during this time of high water and carbohydrate movement.

Another diagnostic symptom of PD is the abscission of leaf blades from shoots with retention of leaf petioles. In addition, as winter approaches, new shoots become woody and develop periderm on 1-year-old shoots. This periderm formation usually begins at the basal portion of a shoot and progresses toward the growing tip. In infected grapevines, periderm formation is not uniform, usually resulting in green "islands" at the nodal area while the internodal portion of the stem becomes brown.

While each of these symptoms can be confused with one or more other non-related factors, the occurrence of several symptoms together provides strong suspicion of infection in a susceptible host.

In vines that are infected in spring, symptoms of Pierce's disease first appear as water stress in midsummer, caused by blockage of the water-conducting system by the bacteria. The occurrence of the following four symptoms in mid- to late summer indicates the presence of Pierce's disease: (Almeida et al., 2005) leaves become slightly yellow or red along margins in white and red varieties, respectively, and eventually leaf margins dry or die in concentric zones; (Almeida et al., 2008) fruit clusters shrivel up like a raisin; (Almeida et al., 2008) dried leaves fall leaving the petiole (leaf stem) attached to the cane; and (California Winery Advisor, 2009) wood on new canes matures irregularly, producing patches of green, surrounded by mature brown bark. Delayed and stunted shoot growth occurs in spring following infection even in vines that did not have obvious symptoms the preceding year.

Leaf symptoms vary among grape varieties. Pinot Noir and Cabernet Sauvignon have highly regular zones of progressive marginal discoloration and drying on blades. In Thompson Seedless, Sylvaner, and Chenin Blanc, the discoloration and scorching may occur in sectors of the leaf rather than along the margins.

Usually only one or two canes will show Pierce's disease symptoms late in the first season of infection, and these may be difficult to notice. Symptoms gradually spread along the cane from the point of infection out toward the end and more slowly toward the base. By midseason some or all fruit clusters on the infected cane of susceptible varieties may wilt and dry. Tips of canes may die back; roots may also die back. Vines of susceptible varieties deteriorate rapidly after symptoms appear. Shoot growth of infected plants becomes progressively weaker as symptoms become more pronounced.

Climatic differences between regions can affect the timing and severity of symptoms, but not the type of symptoms. Hot climates accelerate symptoms because moisture stress is more severe even with adequate soil moisture.

A year after the vines are infected, some canes or spurs may fail to bud out and shoot growth is stunted. New leaves become chlorotic (yellow) between leaf veins and scorching appears on older leaves. From late April through summer, infected vines may grow at a normal rate, but the total new growth is less than that of healthy vines. In late summer, leaf burning symptoms reappear.

Host

Major host: *Citrus sinensis, Prunus persica, V. vinifera.*

Minor host: *Citrofortunella microcarpa, Poncirus trifoliate, Pr. armeniaca, Prunus domestica, Prunus dulcis, Pr. salicina, Vitis labrusca, Vitis riparia, Citroncirus, Citrus, Fortunella, Vitis.*

Incidental: *Acer rubrum, Morus rubra, Platanus occidentalis, Prunus angustifolia, Quercus rubra, Ulmus Americana, Vinca minor.*

Wild/weed: *Sorghum halepense,* Cyperaceae, Lilaeaceae, Poaceae, and various other woody plants.

Geographical Distribution

Asia: India, Taiwan.

North America: Mexico, the United States (Alabama, Arizona, California, Florida, Georgia, Louisiana, Mississippi, Missouri, Montana, North Carolina, South Carolina, Texas, Kentucky, New York, and West Virginia).

Central America and Caribbean: Costa Rica.

South America: Argentina, Brazil, Venezuela.

Life Cycle

Xy. fastidiosa proliferates only in xylem vessels, roots, stems, and leaves. The bacteria may spread rapidly through the plant. The vessels are ultimately blocked by bacterial aggregates (Photo 7.27) and by tyloses and gums formed by the plant. Pierce's disease strains of the bacterium are acquired by vector insects, with no latent period, and persist in infective adult insects indefinitely. All sucking insects that feed on xylem fluids are potential vectors. Leafhoppers (Cicadellidae) and froghoppers (Cercopidae) are the most important in North America. Such insects live in permanent pastures beside vineyards that contain weed hosts of the bacteria. Persistence of the disease is determined by mild winter temperatures. Freezing temperatures eliminate the bacteria in plant tissue. Survival of insect vectors is a determining factor in reinfection by insect vector blue-green sharpshooter (BGSS).

PHOTO 7.27   Leaf scorching symptoms of Pierce's disease on branches of grape. (Courtesy of Scortichini, M., atlasplantpathogenicbacteria.it.)

Grapevines become infected when a sharpshooter that carries the bacterium feeds on tender tissue. These insect vectors are very efficient at transferring the bacterium during feeding and infection is likely.

Once a grapevine is infected, the bacteria multiply and colonize the xylem, or water-conducting tissue of the plant. This vascular constriction inhibits the movement of water through the grapevine and often results in first visible symptoms noted during periods of heat or drought stress.

It is important to distinguish between two groups of grapevines: susceptible cultivars and tolerant cultivars. Once a susceptible cultivar is infected, there is no known, approved method of treating the infection and the disease will most probably be fatal to the vine. Cultivars vary in the length of time it takes the pathogen to cause vine death. Tolerant cultivars appear to have internal mechanisms of suppressing bacterial numbers to the point that the vine can live and be productive even in the presence of the bacteria. There is preliminary evidence that some nonsusceptible cultivars may in fact be resistant to infection. All native Texas species of *Vitis* are believed to be tolerant of PD, which potentially makes them carriers of the bacterium. As a consequence, removal of adjacent, wild grapevines is imperative to disease management.

Pathogen

Xy. fastidiosa is a fastidious Gram-negative, xylem-limited, rod-shaped bacterium.

The bacterium that causes PD is limited to the xylem (water-conducting elements of plants). Insects with piercing/sucking mouthparts that feed on xylem sap transmit the bacteria from diseased to healthy plants. Vines develop symptoms when the bacteria block the water-conducting system and reduce the flow of water to affected leaves. Water stress begins in midsummer and increases through fall.

One of the key features of grape strains of *Xy. fastidiosa* is that grape strains infect a wide range of plant species (Freitag, 1951). Recent studies (Hill and Purcell, 1995) have shown that the fate of *Xy. fastidiosa* can vary greatly from one plant species to another. The bacteria can move internally from cell to cell in blackberry as in grape, but not within California mugwort. All of these plants are important breeding plants of the BGSS in Northern California.

Xy. fastidiosa (*Xf*) is associated with a large number of diseases, including PD, phony peach disease, plum leaf scald, citrus variegated chlorosis, scorch of almond, oleander leaf scorch, and several other diseases (Hopkins and Purcell, 2002; Costa, 2004; Almeida et al., 2008;

Montero-Astua et al., 2008; Singh et al., 2010). This bacterium has a very wide host range and has been reported to cause diseases in over 100 plant species of both monocots and dicots (Hopkins and Purcell, 2002). Different subspecies or strains of *Xy. fastidiosa* can cause disease in different plants, and, in some cases, the bacterium can inhabit the plant without causing symptoms. The fastidious bacterium resides in the plant xylem tissue or water-conducting vessels and is spread exclusively by xylem-feeding insects known as sharpshooter leafhoppers (Hill and Purcell, 1997). PD is known to be prevalent within the United States from Florida to California, and in other countries throughout Central and South America.

In 1892, Newton Pierce was the first plant pathologist to describe the "mysterious vine disease" eventually known as PD. Since the 1880s, *Xy. fastidiosa* has been reported to cause considerable losses in Southern California, destroying more than 35,000 acres of vineyards. As a result, nearly all viticulture was forced to move northward due to the inability to produce a quality grape in this area due to PD.

The term "fastidiosa" indicates that *Xy. fastidiosa* is difficult to recover from host plant xylem, primarily because it does not grow on common bacterial media (Wells et al., 1987). However, there has been some recent success in isolation and culturing of certain *Xy. fastidiosa* strains. Although *Xy. fastidiosa* is classified as a single species, there is a complex and poorly defined relationship between hosts and strains of this unique pathogen (Hopkins, 1989). Currently there are several *Xy. fastidiosa* strains that, based on the 16S–23S intergenic spacer region, were grouped under several subspecies (Schaad et al., 2004). *Xy. fastidiosa* subsp. multiplex causes disease in peach, plum, almond, elm, pigeon grape, sycamore, and other trees. *Xy. fastidiosa* subsp. *pauca* causes disease in citrus and coffee, while *Xy. fastidiosa* subsp. *fastidiosa* causes disease in grape, alfalfa, almond, and maple (Schaad et al., 2004). Using the same basis for classification, Hernandez-Martinez et al. (2007) reclassified *Xy. fastidiosa* into four subspecies with the addition of *Xy. fastidiosa* subsp. *sandyi*, which causes disease in oleander, daylily, jacaranda, and magnolia. They also proposed that strains isolated from mulberry with mulberry leaf scorch (MLS), which belong to *Xy. fastidiosa* subsp. *fastidiosa* (Chen et al., 2002), be grouped into a separate subspecies because of their genetic differences from *Xy. fastidiosa* subsp. *fastidiosa* and because MLS strains did not infect grapes or oleander. On the other hand, a classification scheme using fluorescent amplified-fragment length polymorphism (fAFLP) revealed a strain that infects mulberry, grapes, and temecula, which belongs to the *Xy. fastidiosa* subsp. *fastidiosa* group (Kishi et al., 2008). Thus, different methods of classification revealed grape-infecting strains in different *Xy. fastidiosa* subspecies groups.

Vectors

Virtually all sucking insects that feed predominantly on xylem fluid are potential vectors (Purcell, 1989). Leafhoppers (Cicadellidae) in the subfamily Cicadellinae (sharpshooters) and spittle bugs or froghoppers (Cercopidae) are by far the most common species of known vectors within the natural range of *Xy. fastidiosa* in North America. *Cicadella viridis* (Cicadellinae) and the meadow spittle bug, *Philaenus spumarius* (Cercopidae), are common and widespread in central and southern Europe. In California (the United States), all tested members of the subfamily Cicadellinae (including *Ca. fulgida*, *D. minerva*, and *Graphocephala atropunctata*) were vectors of the grapevine strain. *H. coagulata*, *Homalodisca insolita*, *Oncometopia orbona*, *Graphocephala versuta*, and *Cuerna costalis* are reported vectors of the peach strain (Turner and Pollard, 1959; Yonce, 1983). All these are xylem-feeding suctorial insects that acquire the bacterium rapidly on feeding (less than 2 h). The bacterium adheres to the mouthparts and is released directly from them when the insect feeds again (Purcell et al., 1979). It multiplies in the vector but does not circulate in its hemolymph, nor does it require a latent period before transmission (unlike a phytoplasma). Transmission is usually from wild, generally symptomless, hosts to cultivated hosts (grapevines, peaches) rather than between cultivated hosts, though the latter can occur. Feeding preferences of *G. atropunctata* for different grapevine cultivars have been noted (Purcell, 1981).

In California (the United States), species such as *D. minerva* and *Ca. fulgida* inhabit permanent pastures alongside vineyards, or live on weeds within them. Irrigation and weed control

2 mm

5502051

PHOTO 7.28   The glassy-winged sharpshooter leafhopper vector of *Xy. fastidiosa*. (Courtesy of Pest and Diseases Image Library, Bugwood.org.)

practices that focus on preferred host plants, including *Cynodon dactylon* and *Echinochloa crusgalli*, increase vector populations and the spread of the bacterium (Purcell and Frazier, 1985). Other species, such as *G. atropunctata*, multiply on grapevine but overwinter on a variety of other wild hosts.

In coastal California, the principal vector of PD was *G. atropunctata*, known as the BGSS (Simpson et al., 2000). BGSS overwinters in riparian vegetation but readily feeds and multiplies on grapevines. These insects transmit *Xy. fastidiosa* in a persistent manner, with a short or even no latent period required between acquisition and transmission (Purcell and Finlay, 1979). In the southeastern United States, another sharpshooter, *H. vitripennis* Germar (Turner and Pollard, 1959), was identified as a vector of *Xy. fastidiosa, H. vitripennis* (formerly *H. coagulate*, Takiya et al., 2006), popularly known as the GWSS (Photo 7.28), which was introduced to California in the 1990s. BGSS and GWSS have similar feeding habits. However, BGSS prefers to feed on young leaves and tissues, while GWSS, although it can exist on these tissues, is also found on woody parts of the plants (Hopkins, 1981). At present, GWSS is considered to be the most efficient vector of PD because it displays greater mobility, is capable of feeding on older woody stems, and has a wider host range than the native sharpshooters (Larsen, 2000). Its establishment in California threatens the overall wine grape, table grape, and almond industries (Redak et al., 2004).

Pathogen–Vector–Host Interaction

Three events are essential for the successful transmission of the pathogen to the plant. First is the acquisition of the pathogen by a vector from a diseased host plant. Acquisition efficiency is directly proportional to the population of the bacteria from the infected hosts (Hill and Purcell, 1997; Redak et al., 2004). Similarly, efficiency of acquisition also depends upon the sharpshooter feeding site whether it be leaves, stems, or older woody branches; however, no report suggests a preferred feeding site by the vector (Hill and Purcell, 1997). The second event is the attachment or retention of the acquired bacteria into the vector foregut. De la Fuente et al. (2007) suggested that type-I and type-IV pili are involved in the attachment to the cuticle within an environment described as a "fast fluid flow" condition (De la Fuente et al., 2007). Microscopy studies revealed that the bacteria form a biofilm in the foregut of the sharpshooters where they are polarly attached to the cuticle, with the cuticle being shed at each molt (Newman et al., 2004). Lastly, the pathogen must be detached and introduced into a new host during vector feeding. The lack of a latent period suggests that the population of bacteria within the vector is not a significant factor affecting the transmission efficiency (Almeida et al., 2005).

Chatterjee et al. (2008) used green fluorescent protein–tagging of *Xy. fastidiosa* to monitor the correlation between the colonized vessels and symptom expression. They found a fivefold increase in colonized vessels from symptomatic leaves compared to symptomless leaves (Newman et al., 2003; Chatterjee et al., 2008). It was also observed that vessels in symptomatic leaves contained large bacterial colonies while asymptomatic leaves from infected plants had vessels that contained very small colonies of bacterial cells. This finding suggested that vessel blockage determined symptom expression.

Like many plant pathogenic bacteria, *Xy. fastidiosa* harbors a cell-to-cell signaling system that plays an important role in colonization, pathogenesis, and biofilm formation (Bodman et al., 2003). A set of genes known for the regulation of pathogenic factor (rpf) is required for the synthesis of diffusible signaling factor, involved in the formation of biofilm in the foregut of the vector (Chatterjee et al., 2008). Rpf was also involved in colonization of both the vector and the host (Newman et al., 2004) and pathogenesis in the host (Chatterjee et al., 2008). Genomic characterization of *Xy. fastidiosa* also revealed its richness in adhesins and hemaglutinin-encoding genes that are involved in the interaction with vectors and hosts, and cell-to-cell attachment in biofilm formation (Chen et al., 2005). Whether these virulence factors are secreted via type-I or type-II secretion systems is not yet clear, but sequence analysis reveals that *Xy. fastidiosa* possesses genes for both type-I and type-II secretion systems for the export of exoenzymes involved in plant cell wall degradation and allow bacterial colonization of plant xylem (Simpson et al., 2000; Jha et al., 2005).

Disease Detection

Preliminary PD diagnosis is commonly done by symptom identification. However, visual diagnosis is not definitive due to the similarities of PD symptoms to those of other grape diseases, with the possible exception of the "matchstick" symptom indicator of *Xy. fastidiosa*–infected grapevines (Krell et al., 2006). Leaf scorching, the typical PD symptom, is absent during the early stages of infection (Hopkins, 1981), and there are inconsistencies in the correlation between symptom expression and bacterial populations in the leaf (Gambetta et al., 2007). Thus, limitations of visual diagnosis can lead to incorrect disease management decisions and may increase the likelihood of disease spread and dispersal. Reliable diagnosis is dependent upon the use of diagnostic techniques such as bacterial culturing, enzyme-linked immunosorbent assay (ELISA), and PCR. However, each method has its own limitations. Culturing the pathogen is a difficult task, especially during the early season, presumably due to sampling and low population numbers. ELISA is very useful for late-season identification of PD but the test is not reliable for early-season diagnosis (Smart et al., 1998), presumably for the same reason of low population numbers. ELISA, however, is still commonly used because of its capability of large-scale detection at a lower cost. PCR is an efficient technique for *Xy. fastidiosa* detection due to the availability of PCR primers that target different specific components of the genome, particularly the 16S rRNA region (Chen et al., 2005). Different PCR-based techniques are being evaluated to determine the most economic, sensitive, fast, and reliable method of early PD detection. Currently, the most sensitive, cost-effective method is the use of real-time PCR with the use of sap flow during bud break. This technique gets rid of the intricacy of DNA extraction and the inhibitory compounds that limit sensitive detection. Clearly, the reliability of the various diagnostic techniques depends on the sample, timing, and plant tissue tested.

Risk of Pierce's Disease Applicable to the Midwest

The primary factor to consider with respect to the probability of PD development in the Midwest is climate. In the years since the discovery of PD, reports of its presence have been restricted to areas with a subtropical climate. In the United States, its distribution has been mostly limited to the southernmost grape growing states and West Coast, with the greatest distribution and greatest disease severity occurring in Florida and California. The correlation between disease incidence and the concentration of vineyards may be coincidental. However, this pattern of disease seclusion to subtropical climates appears to be a result of the intolerance of *Xy. fastidiosa* and its respective vectors to cold temperatures. Even though *Xy. fastidiosa* is favored by a subtropical climate, it also is sensitive to elevated temperatures, and growth above 34°C does not occur (Feil and Purcell, 2001).

There are no reports of any of the known insect vectors of *Xy. fastidiosa* being present in the Midwest, including *H. vitripennis* (syn. *coagulata*), *D. minerva*, *Xyphon fulgida*, *Graphocephala coccinea*, *G. versuta*, and *G. atropunctata* (IFAS, Global Invasive Species Database). However, the Midwest is home to many related xylem-feeding, hemipteran insects. Many of the known vectors of *Xy. fastidiosa* increase their numbers and overwinter on lush grasslands, riparian buffer strips, and alfalfa fields, which are also prevalent in the Midwest. Of further concern is the observation that alfalfa may be an alternate host for *Xy. fastidiosa* (Sisterson et al., 2008).

Many xylem-feeding insects found throughout the Midwest are known to feed on the younger, greener tissues of their respective host plants. Theoretically, if *Xy. fastidiosa* and a suitable vector were introduced into the Midwest, the insect would likely also feed on the younger tissues. If *Xy. fastidiosa* is introduced to these parts of the grapevine, disease severity is expected to be low and overwintering inside the plant hindered because of vine pruning practices. The risk of disease in the Midwest, aside from the effect of temperature, relies on two important parameters: the introduction of *Xy. fastidiosa* and the association between *Xy. fastidiosa* and a xylem-feeding insect pest of grapes.

Little work has been done to correlate the effects of severe cold weather, such as seen in the Midwest during a typical winter, and its effect on the survival of either *H. vitripennis* or *Xy. fastidiosa*. However, there is a relationship between low temperatures (5°C) and the corresponding growth rate of *Xy. fastidiosa*. Henneberger et al. (2004) isolated *Xy. fastidiosa* from the roots and shoots of sycamore trees (a natural alternate host) at various times throughout the year at varying temperatures. They constructed a plot correlating the frequency of *Xy. fastidiosa* isolation with the respective temperature when the isolation was attempted, which revealed a decrease in isolation success as the temperature approached 0°C. These data suggest that survival of *Xy. fastidiosa* is minimal at temperatures below freezing. In a similar study, Feil and Purcell (2001) found a similar correlation between temperature and the growth rate of *Xy. fastidiosa*. They were able to show that, *in vitro*, the optimal growth conditions for this bacterium ranged from 25°C to 32°C and that growth declined and was even suppressed when temperatures fell below 12°C.

These findings conflict with the discovery of *Xy. fastidiosa* strain that causes MLS (Chen et al., 2002). Based on restriction RFLP and random-amplified polymorphic DNA (RAPD) molecular genome assays, it was found that this strain is closely related to the subspecies that causes PD. Furthermore, the mulberry trees from which the bacterium was isolated were growing naturally in two different locations, namely, Massachusetts and Nebraska, suggesting that this particular strain has adapted to the harsh temperature fluctuations of the Midwest and Northeast.

More recently, a report from Oklahoma revealed that the table grape cultivar "Concord" displayed typical symptoms of PD. Real-time PCR analysis of DNA isolated from vine samples using available specific primers coupled with ELISA confirmed the presence of *Xy. fastidiosa* (Smith et al., 2009). This finding in Oklahoma has caused concern among Midwest wineries because of its geographical proximity. The recent identification of *Xy. fastidiosa* in grapes suggests the possibility that the strain of *Xy. fastidiosa* infecting woody trees was able to overcome previous grape cultivar resistance. Currently, there are no studies to investigate the ability of this strain to infect any of the widely grown grape cultivars throughout the Midwest. Until this is done, it will not be possible to extrapolate the possibility of PD on the Midwestern grape and wine industry.

Climate Change Adaptations

Over the past 100 years, average temperatures across the northern Midwest have increased 2°C, while average temperatures have decreased in the southern Midwest by 0.5°C. These trends are projected to continue (National Assessment Synthesis Team, 2000). A shift to warmer temperatures throughout the Midwest may enable establishment of organisms that would otherwise be unable to exist in these areas. An increase in the minimum temperature by 12°C might shift a particular region's climate such that freezing would be avoided altogether. The pathogen and vector may also be able to adapt in the presence or absence of climate change. Rising temperatures would also lengthen the growing season and potentially increase drought conditions throughout the Midwest. In combination, these elements could enable an increased proliferation of insect pests and pathogens. Drought stress not only increases the probability of PD incidence but may also increase the expression of disease symptoms on stressed plants due to compromised defense system. In fact, McElrone et al. (2003) revealed that occlusion of the xylem vessels by *Xy. fastidiosa* was exaggerated when host plants experienced drought stress (McElrone et al., 2003; Garrett et al., 2006). Symptoms appear when a significant amount of xylem becomes blocked by the growth of the bacteria. (This bacterium is also responsible for alfalfa dwarf disease and ALS in California.) Insect vectors for PD belong to the sharpshooter (Cicadellidae) and spittlebug (Cercopidae) families. The BGSS (*G. atropunctata*) is the most important vector in coastal areas. The green sharpshooter (*D. minerva*) and the red-headed sharpshooter (*Ca. fulgida*) are also present in coastal areas but are more important as vectors of this disease in the Central Valley. Other sucking insects, such as grape leafhoppers, are not vectors.

A new PD vector, the GWSS, has recently become established in California. This vector is a serious threat to California vineyards because it moves faster and flies greater distances into vineyards than the other species of sharpshooters. The GWSS occurs in unusually high numbers in citrus and avocado groves and on some woody ornamentals. Until now, these plants have not been sources of PD vectors.

Since the early 1990s, the GWSS has been seen in high numbers in citrus along the coast of southern California. It subsequently has become locally abundant further inland in Riverside and San Diego Counties. In 1998 and 1999, high populations of GWSS on citrus and adjacent vineyards were seen in southern Kern County, and in 2001, hundreds of vines had PD. The GWSS is expected to spread north and eventually has become a permanent resident of various habitats throughout northern California.

GWSS feeds and reproduces on a wide variety of trees, woody ornamentals, and annuals in its region of origin, the southeastern United States. Crepe myrtle and sumac are especially preferred. It reproduces on eucalyptus, coast live oaks, and a wide range of trees in southern California. But because GWSS is a relatively new arrival to the state, it is not clear yet in which regions and habitats it will become permanently established.

The principal breeding habitat for the BGSS is riparian (riverbank) vegetation, although ornamental landscape plants may also harbor breeding populations. As the season progresses, these insects shift their feeding preference, always preferring to feed on plants with succulent growth. In the Central Valley, irrigated pastures, hay fields, or grasses on ditch backs are the principal breeding and feeding habitats for the green and red-headed sharpshooters. These two grass-feeding sharpshooters also occur along ditches, streams, or roadsides where grasses and sedges provide suitable breeding habitat.

Some vines recover from PD the first winter following infection. The probability of recovery depends on the date of infection. Infections that occur until June have the greatest probability of surviving until the following year. Recovery rates also depend on grape variety; recovery is higher in Chenin Blanc, Sylvaner, Ruby Cabernet, and White Riesling, compared to Barbara, Chardonnay, Mission, Fiesta, and Pinot Noir. Thompson Seedless, Cabernet Sauvignon, Gray Riesling, Merlot, Napa Gamay, Petite Sirah, and Sauvignon Blanc are intermediate in their susceptibility to this disease and in their probability of recovery. In tolerant cultivars, the bacteria spread more slowly within the plant than in more susceptible cultivars. Once the vine has been infected for over a year (i.e., bacteria survive the first winter), recovery is much less likely.

Young vines are more susceptible than mature vines. Rootstock species and hybrids vary greatly in susceptibility. Many rootstock species are resistant to PD, but the rootstock does not confer resistance to susceptible *Vinifera* varieties grafted on to it. Finally, the date of infection strongly influences the likelihood of recovery: late infections (after June) by BGSS, green sharpshooters, and red-headed sharpshooters are least likely to persist the following growing season. This may not be the case with GWSS, however, because it feeds on leaves near the base of the cane, as well as on 2-year-old dormant wood.

Most, if not all, sharpshooter species go through five larval, or instar, stages in which they apparently lose the ability to transmit the disease with each molt. In areas of rampant infection, it is assumed that alternate sources of the bacterium are widely available. Keeping vineyards and adjacent areas free of potential alternate hosts is essential for long-term management of PD. Monitoring insect populations, especially after habitat disturbance such as cutting of adjacent hay fields, can greatly assist growers in the judicious use of insecticides.

Disease Management

Insecticide treatments aimed at controlling the vector in areas adjacent to the vineyard have reduced the incidence of PD by reducing the numbers of sharpshooters immigrating into the vineyards in early spring. The degree of control, however, is not effective for very susceptible varieties such as Chardonnay and Pinot Noir or for vines less than 3 years old. If a vineyard is near an area with a history of PD, then the plant varieties are less susceptible to this disease. Monitor and treat for insect vectors.

During the dormant season, remove vines that have had PD symptoms for more than 1 year; they may be chronically infected and are unlikely to recover or continue to produce a significant crop. Also, remove vines with extensive foliar symptoms on most canes and with tip dieback of canes even if it is the first year that symptoms have been evident. From summer through harvest, mark slightly symptomatic vines; reexamine for symptoms the following spring through late summer or fall and remove vines that have symptoms for a second year. Pruning a few inches above the graft union of vines with moderate foliar symptoms (some canes on entire cordons without symptoms or no symptoms at the bases of most canes) may eliminate PD and allow vigorous regrowth the following year, but symptoms will reappear in many (30%–40%) or most of these severely pruned vines the second year.

For table grapes, examine vines for poor bud break in spring. Later in the season, look for pests and damage.

Because the GWSS feeds much lower on the cane than other sharpshooters in California, late-season (after May–June) infections and infections occurring during dormancy made by the GWSS can survive the winter to cause chronic PD. This enables vine-to-vine spread of PD, which has not been the case in California. Vine-to-vine spread can be expected to increase the incidence of PD exponentially rather than linearly over time, as has been normal for California vineyards affected by PD. Insecticide treatments of adjacent breeding habitats, such as citrus groves, have been the most effective approach.

Removing diseased vines as soon as possible when PD first appears in a vineyard is also critical to help reduce the infection rate. Early and vigilant disease detection and vine removal are recommended for any vineyards that experience influxes of the GWSS.

Long-term studies are being conducted on the effect of riparian vegetation management in reducing disease incidence and severity in North Coast vineyards. Riparian vegetation management has proven to be effective in reducing the damaging spring populations of BGSS. Because these areas are ecologically sensitive and regulated by federal, state, and local legislation, unauthorized removal of vegetation is prohibited or restricted. Vegetation management of these areas must be acceptable or beneficial for wildlife and water quality and to maintain the integrity of the riparian habitat.

Because there is no known control for PD, the act of planting susceptible cultivars in areas where PD is known to exist assumes an inherent risk.

Remove Wild Grapevines

In Texas, wild hosts of the grape pathogen have not been identified. In other states, grape strains of *Xy. fastidiosa* have been isolated from wild grape, ragweed, alfalfa, and almond trees. As a precaution, it is recommended that wild grapes be removed from around the vineyard.

Remove Diseased Vines

Based on foliage and cane symptoms confirmed by laboratory diagnosis, diseased plants should be immediately destroyed. Regardless of varietal tolerance, any vine with symptoms of this disease should be pulled up or cut off at the ground and removed from the vineyard. Since observations indicate that the disease can spread from vine to vine within the vineyard, removal of diseased vines reduces the potential sources of inoculum that could be transmitted by insect vectors.

Vector Management

The disease is vectored by certain kinds of xylem-feeding insects, mainly the leafhopper group known as sharpshooters. All of the insect species responsible for vectoring PD in Texas are not known at this time. There are species of leafhoppers that inhabit Texas vineyards and adjacent wild hosts that look like sharpshooters that are not known to vector PD. Sharpshooters tend to be significantly larger than other species of leafhoppers found in and adjacent to vineyards.

The difficulty of vector management as a means to manage PD is the inability to identify all potential vectors within and adjacent to the vineyard, so chemical control of vectors is tenuous at best. Nonetheless, the current thinking in California is that vector transmission occurs primarily from host plants adjacent to the vineyard, so California growers practice vector control in areas adjacent to the vineyard. Growers should use caution when choosing insecticides to ensure that specific pesticide labels permit such use.

The pattern of PD spread in Texas more closely parallels that observed in Florida, where significant vine-to-vine spread of the disease occurs. This would indicate that insecticidal control of vectors within the vineyard may also be needed.

Based on the best information available, the following vector control recommendations are suggested:

Establish and maintain a 150 ft buffer (minimum) around the vineyard through mechanical or chemical mowing or cultivation.

The California experience would indicate that the greatest danger from transmission of PD through sharpshooter vectors is shortly after bud break and decreases as the season progresses. Starting at bud break and continuing for 6 weeks, sample the vegetation in the area outside and adjacent to the buffer, or, in the absence of a buffer, sample the vegetation adjacent to the vineyard. Sampling consists of using a standard sweep net and taking a minimum of eight 25-sweep samples at least twice a week. If adult sharpshooter numbers exceed an average of 1 per 25-sweep sample, insecticidal treatment may be justified.

Treat a 65 ft band adjacent to the buffer or a 130 ft band adjacent to the vineyard in the absence of a mowed buffer. If it is not possible to treat adjacent vegetation, it might be appropriate to treat the vineyard itself. The problem with this approach is that if the alternate host reservoir for the sharpshooter vectors is large and the buffer is small or absent, then within vineyard, treatments may be ineffectual in keeping sharpshooters out. Twice-a-week spraying for 4–6 weeks following bud break may be necessary, but only if sweep samples indicate that a threshold population has been reached.

Insecticides should be used judiciously. Unfortunately, the greater the number of sprays, the more likely secondary pest outbreaks will be created, especially with spider mites.

Use an insecticide registered for use for the target area. In most cases (and for all sites external to the vineyard), sharpshooters are not listed as a target pest on the label. Specific use restrictions for grapes and alternate hosts will be found on the label.

Vineyard Floor Management

Because there is limited information as to other species serving as a source of the PD bacterium, many growers utilize clean cultivation to eliminate any possible inoculum source within the vineyard. Weed growth under the trellis can be controlled with cultivation, or herbicides, but management of the vineyard floor between the rows has become problematic. Clean cultivation can have serious drawbacks such as the potential for serious soil loss due to erosion. The use of cover crops in vineyard row centers has several advantages over cultivation, including increased equipment mobility, the preservation of soil structure within the vineyard, and erosion control.

Because at this point we do not know what plant species constitute propagative alternate hosts of PD, the decision on what plant species growers should plant or encourage on the vineyard floor is still only a guess. In light of these considerations, it may be wise to plant (drill or no-till seed) cool season, annual cover crops such as annual rye grass or oats in October and encourage cover crop growth during the months that grapevines are dormant. These annual plants have a low probability of contracting the causal bacterium and would be growing during a period when transmission to grapevines is not believed to occur.

Cover crop height can be managed by mowing and is easily controlled during the spring with low-rate glyphosate applications. This practice keeps cover crop roots in place to support equipment traffic, helps reduce erosion, and establishes an organic material layer that inhibits the germination of indigenous weed species. When annual rye grass is used for this purpose, additional suppression of weed seed germination may be observed due to the allelopathic properties of rye. Additional applications of glyphosate or glufosinate can be used throughout the growing season to keep developing weed populations in check. Preemergence herbicides can also be incorporated into a vineyard floor management program.

There are several disease management strategies used with some success in areas with severe PD. These include host resistance, vector control, and changes in cultural practices. Used in combination as an integrated pest-management strategy, they should prove to be fairly successful in preventing the infection and spread of *Xy. fastidiosa* in Midwestern vineyards.

The most popular grape cultivars of European and American descent are susceptible to PD to some degree. However, there are a few native species found in the southeastern part of the United States that have shown resistance to *Xy. fastidiosa*. Muscadine grape (*Vitis rotundifolia*) cultivars have varying degrees of resistance to *Xy. fastidiosa* but are not immune to the pathogen. Although some muscadine grape cultivars may be highly resistant to PD, they lack the fruit quality that most popular cultivars offer. Also, *V. vinifera* cultivars have shown differences in PD resistance in California, with Petit Sirah, Chenin blanc, and Sylvaner among the most resistant cultivars (Hopkins and Purcell, 2002). These cultivars may be useful for resistance gene introduction into popular wine and table grape cultivars.

Potential transgenic grapes resistant to PD are highly controversial and are not currently considered within the grape industry (National Research Council, 2004). Wine-making, an "Old World" industry based on historical European wine regions, carries with it original wine grape names representing different cultivars, tradition, and prestige. The general opposition to genetically altering the classic cultivars is that they will not be the same cultivar anymore, hence losing their status among the top wine-making grapes. Plant breeders and the skeptical wine consumers are at a standstill, involving regulatory and marketing issues, with the growers left in the middle, scrambling for a better solution to PD. If resistance genes can be identified and introduced into a desirable but susceptible grape genome, new cultivars may be produced having varied levels of resistance to PD that may or may not be accepted in the trade.

With management of any mobile pest, such as the GWSS, timing is crucial. Targeting the first generation of GWSS is most effective in controlling *Xy. fastidiosa* transmission (Hopkins and Purcell, 2002). These studies have shown that an early insecticide application before the appearance of the first generation of GWSS helps to slow PD spread. Insecticides may also need to be applied to surrounding vegetation containing hosts of GWSS and other insect vectors of *Xy. fastidiosa*. Although effective for shorter annual growth periods, it is less effective in areas with longer growing seasons and warmer climates, such as those experienced in Florida (Hopkins and Purcell, 2002). Because systemic insecticides last a few months inside plants, a second application may be needed in areas where the growing season is extended.

Several insecticides (i.e., foliar-applied acetamiprid and soil-applied imidacloprid and thiamethoxam) have been tested for control of GWSS and other insects transmitting *Xy. fastidiosa*

(Bethke et al., 2001). The most effective insecticides for insect control, inhibiting bacterial transmission, and PD spread are the neonicotinoids (Almeida et al., 2005). These insecticides are systemic and aid in disrupting the *H. vitripennis–Xy. fastidiosa* interface.

7.10.4 BACTERIAL LEAF SPOT OF GRAPEVINE

Pathogen: *Xanthomonas campestris* pv. *viticola*

Symptoms

Under favorable atmospheric condition, the bacterial disease causes several minute water-soaked lesions on the lower surface of the leaves by 6 days of infection. These spots rapidly expand up to 2 mm in next 3 days and remain water-soaked, which gradually turns brown. However, under normal weather condition, faint water-soaked spots develop on the lower surface of the leaves by 10th day of infection and increase up to 1 mm in diameter by 15th day and thereafter turn gradually brown. Thus, the leaf spots developed are dark brown, irregular, 1–2 mm in diameter, and coalescing sometimes to form larger spots (Photo 7.29). Leaves are particularly susceptible to the disease when these are 60–75 days old. Susceptibility also varies with leaf position (Chand et al., 1991). Mechanically damaged leaves readily get infected. Vein infection is also common. Petiole and stem show elongated, 1–2 cm cankers spots. The infected leaves start drying after 2 months. The infected leaves after drying are found firmly attached to the stem. Slight pressure, however, results in complete disintegration of the dried leaves. Kochenko (1993) reported development of canker tumors on grape root due to the bacterium from Southern Ukraine. Isolation from the leaf spots, infected vein, cankers spots on the cane and dry infected leaves yield *Xanthomonas*. The varieties Thompson Seedless, tas-A-Ganesh, Sonaka, and Manik Chaman are widely attacked by the bacterium (Kore et al., 1992).

Geographical Distribution

The disease was first reported from India in October 1969 on a grape variety, Anab-e-Shahi, at Agricultural College Farm, Tirupati, and since then its sporadic occurrence has been reported from other grape-growing areas. In Maharashtra, the disease was first reported in 1972; however, it was not serious until 1982. As the acreage of Thompson Seedless increased after 1972 in Maharashtra, the disease assumed significant importance after 1982 in all the grape-growing areas of the state. Chand et al. (1991) reported 85%–100% infection of the bacterium on the vineyard in different locations of the country except from Sangli and Nashik in Maharashtra where infection level was less than other parts of the country. Around 1992, a severe outbreak of the disease was observed on commonly grown grape varieties in all the grape-growing areas in Punjab (Soni and Parmar, 1993).

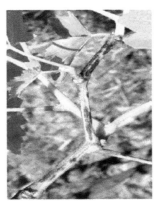

PHOTO 7.29 Symptoms of *Xanthomonas* on leaf, twig, and fruit. (Courtesy of S.G. Borkar and Kadam, S., PPAM, MPKV, Rahuri, India.)

Pathogen

The pathogen that causes this disease is *X. campestris* pv. *viticola* (Nayudu, 1972; Dye, 1978). Typical *Xanthomonas* strains of pv. *viticola* seldom occur in nature; however, albino strains of *X. campestris* pv. *viticola* are widely prevalent in nature (Borkar, 1991, personal observation) as obtained by Nayudu as a type strain. The bacterium grows very fast at 28°C–30°C producing much mucoid growth and survives for 15–20 days under laboratory condition. Long stored cultures of the bacterium (4 months at 4°C) lose their ability to grow and their pathogenicity. However, in infected leaves, the bacteria are found to survive for up to 2 years or more (Borkar, 1991, unpublished data), which served as a primary source of inoculums for fresh infection. *X. campestris* pv. *viticola* isolates vary in producing disease and resistance reaction on various grapevine cultivars. At least five cultivars—Bhokari, Arkashyam, Kalishebi, Dogris America, and Karolina black rose—varied in their reaction to one or other isolates of *X. campestris* pv. *viticola*. Cultivar Haithi was uniformly susceptible to all the isolates, whereas cultivar Berlain Riperia was uniformly resistant to all the isolates. Based on the reactions obtained, a scheme for classification of races of *X. campestris* pv. *viticola* was derived by using six grapevine differentials. On the basis of susceptible or resistant reaction on these differentials, at least 32 races of *X. campestris* pv. *viticola* can be identified. By using this scheme of race identification at least five races—race 1, 9, 19, 20, and 22—of *X. campestris* pv. *viticola* were identified, which are prevalent in grape vineyards of Nashik region (Borkar, 2002).

The bacterium *X. campestris* pv. *viticola* isolated from different vineyards of various locations in Nashik district of Maharashtra did not infect the neem plant (*Azadirachta indica*) as reported by Nayadu (1972); however, these induce hypersensitive reaction on neem. Under artificial infection, the bacterium infects mango plant (Chand and Kishun, 1990). The bacterium also produces hypersensitive reaction on leaves of pangara plant (*Erythrina indica*), which is quite common on borders of vineyard in Maharashtra and was also used earlier as supporter for climbing grapevine.

Disease Cycle

Old, dry infected leaves attached to the grapevine and stem cankers act as primary source of inoculums for fresh infection. *Phyllanthus maderaspatensis* may act as additional source of inoculums particularly in new areas where grape is introduced for first time as these are being commonly infected by the bacterium. In infected leaves, the bacterium was found to survive for 2 years and act as a primary source of inoculums. Observations during the rainy period indicated fresh infection 14–16 days following the rain, suggesting rain as one mode of dissemination of the pathogen. Rainy periods are favorable for heavy incidence of the disease.

Epidemiological Parameters

A temperature of 28°C with 90%–92% relative humidity coupled with rain flashes or showers of 8–14 mm was found favorable to cause the infection of the bacterium on the vine. In Nashik region, the fresh infection of vine by *X. campestris* pv. *viticola* occurs during July to September and even on September-pruned vines in Karnataka (Ramesh Chand et al., 1991) when an atmospheric condition with a temperature range of 29°C–33°C and rain showers almost a week or alternatively prevails. In Punjab, the disease appears during third to fourth weeks of March and rapid development takes place if outbreak is followed by rain showers. The disease becomes severe in rainy season (July–September). The progress of infection decreases as the temperature increases and no further infection occurs as temperature reaches 35°C or more. After October pruning, the disease was not found to appear in Maharashtra and Karnataka (Ramesh Chand et al., 1991) because of unfavorable condition for infection of fresh vine growth, though the bacterium remains in previously infected disease left over in the field.

Control Measures

Chemical Control

Spraying of copper fungicide (0.2%) with streptomycin sulfate (500 ppm) as soon as the infection or probability of infection appears is recommended, and the spraying should be continued at an interval of 10–12 days until favorable weather parameters prevail for disease incidence. Copper oxychloride and Bordeaux mixture reduce disease intensity but are much less effective in areas with high and frequent rainfall. Prophylactic sprays of these chemicals every 7 days from pruning until the onset of dry weather are recommended for low-rainfall areas (Ramesh Chand et al., 1992). In other areas, chemical control should be combined with late pruning. Antibiotics fail to give significant control. The pathogen may develop resistance to certain formulation when these are used rigorously and repeatedly. Kadam (1990) reported a high incidence of bacterium in Pune and Solapur districts of Maharashtra that could not be controlled with repeated sprays of Agrimycin 100 (75 g) + copper oxychloride (250 g) in 50 L of water. Ramesh Chand et al. (1991) also reported that bactericides like copper oxychloride, streptomycin sulfate, tetracycline, and Bacterionol-100 are not effective in controlling this disease. The bacterium showed widespread resistance to copper and streptomycin. The isolates differed in their resistance to these bactericides (Chand, 1992). *In vitro* adaptation to higher dose of copper and streptomycin is also observed in the isolates. Resistance to streptomycin increases several fold in mutant. Antibiotics like eviprin, tetracycline, and streptomycin sulfate (600 ppm concentration) were found effective on laboratory testing. The bacterium is sensitive to streptocycline; however, resistant strains develop with increasing concentration of streptocycline during subsequent transfer. Addition of vitamin B-12 (91 ppm) facilitates the development of resistant mutant. However, aureofungin (10 ppm) removes the resistant factor and makes these strains susceptible to streptocycline (Gaikwad and Naik, 1992). The minimum inhibitory concentration (MIC) of streptocycline for the bacterium *in vitro* is 25 ppm. Streptomycin sulfate, teramycin, and bactericin, although inhibited the bacterium at higher concentration, were totally ineffective at lower concentration. Concentration of the antibiotic/fungicide just below the MIC level induces the resistance in the bacterium to the chemical, thereby forming the mutant colonies resistance to the pesticide. The MIC for streptomycin sulfate is 500 ppm, terramycin 100 ppm, bactericin 100 ppm, copper oxychloride 0.05%, dithane M-45 0.01%, dithane Z-78 0.1% and Bordeaux mixture 0.2%. MIC for combination of copper oxychloride + streptocycline is 0.05% + 100 ppm; for Bordeaux mixture + streptocycline is 0.05% + 75 ppm and for copper oxychloride + streptomycin sulfate is 0.2% + 500 ppm respectively. Pesticides, particularly streptomycin sulfate, bactericine, dithane Z-78, Bordeaux mixture, and combination of copper oxychloride/Bordeaux mixture + streptomycin, induce the mutation into the bacterium. The mutation frequency of these pesticides varies from 1×10^{-5} to 7×10^{-5}. The pesticide-resistant mutants are able to grow on higher concentration of the same pesticide. For example, Dithane Z-78–resistant mutants (0.05%) were able to grow on 0.2% concentration, Bordeaux mixture–resistant mutants (0.2%) on 0.5% concentration, and copper oxychloride–resistant mutants (0.05%) on 1.0% concentration of the same pesticide. Cross-resistance pattern to pesticide is also recorded in the bacterium. The mutant of streptomycin sulfate, bactericin, Dithane Z-78, and Bordeaux mixture show resistance toward copper oxychloride (0.05%). Similarly, mutants of streptomycin sulfate, bactericin, and Bordeaux mixture are resistant toward Dithane Z-78 (0.1%). Copper oxychloride-resistant mutant shows resistance toward Dithane Z-78 and Bordeaux mixture (0.2%). These resistant mutants were found pathogenic on grapevine and cause typical leaf spot symptoms (Borkar, 1997). Similarly, application of Bavisitin at 0.1% to control anthracnose was found to increase the intensity of bacterial leaf spot; however, application of Bordeaux mixture drastically reduces the bacterial leaf spot infection (Mohan et al., 2001) and is therefore recommended as an alternative.

Disease Resistance
V. vinifera is highly susceptible while other Vitaceae genera and some *Vitis* species are highly resistant. Seedless *V. vinifera* cultivars are more susceptible than seeded cultivars. Among seedless cultivars, colored ones were more susceptible than white cultivars (Chand, 1992). Grapevine varieties Sahebi Ali, Berlan riperia, Gordo blanca, and Kismis red are immune to the bacterium (Borkar, 2005) and can be used in the breeding program for disease resistance to bacterial leaf spot pathogen. Grapevine cultivars Bhokari, Arkashyam, Kali Sahebi, PS III-12-4, PS III-11-1, PS III-11-4,

Kismis red, Karoline Black rose, Musket, and Dogrij-America showed resistance (hypersensitive reaction) to some isolates of the pathogen, thereby indicating variation in the virulence of the bacterium and probably presence of races in *X. campestris* pv. *viticola* (Borkar, 2002). Chand (1991) also found variation in the isolates that fell into three groups based on the fatty acid profiles. Grapevine varieties Anab-eSahi, Black round, Kismis churnim Rubi red, Thompson Seedless, Tas-a-Ganesh, Sonaka, Manikchaman, Katta Karkan, Raosahebi, Malaga, Black cornichon, Rose of Peru, and Pandhari Sahebi are susceptible to *X. campestris* pv. *viticola* (Borkar, 2005) and, therefore, their plantation should be avoided in disease-prone areas.

Cultural Practices
Pruning of infected parts of vine and its destruction.

7.10.5 BLACK KNOT OF GRAPEVINE

Pathogen: *Agrobacterium tumefaciens* (Smith and Townsend) Conn.

Symptoms
The bacterium very frequently produces the galls on the aerial parts. Cane galls may occur on the grape, often up to 3–5 ft above the ground level in the form of isolated or confluent excrescences that are often more or less elongated parallel to the length of the cane. In the beginning, these appear like a wound tissue or callus, but later in the season, they become dark brown or almost black due to weathering and decay of the outer tissues; this phase of the disease on grape is commonly known as black knot in the United States (Hedgcock, 1910).

Geographical Distribution
The pathogen is distributed in the United States and Sporadic.

Disease Cycle
The bacteria that survive in black knot are brought to the fresh infection site on minor bark cracks of freshly grown vine through rain splashes or by contaminated pruning tool or by insect carriers, which cause a fresh infection.

Control
1. Sanitary practices
2. Spraying of prune portion with copper fungicide + antibiotic formulations
3. Prevention of wounding
4. Surgery

7.11 MANGO

7.11.1 BACTERIAL BLIGHT AND CANKER OF MANGO

Pathogen: *Xanthomonas campestris* pv. *mangiferaeindcae*

Synonyms: *Erwinia mangiferae* var. *indicae*
Phytobacterium mangiferaeindicae
Pseudomonas mangiferaeindicae
X. campestris mangiferaeindicae
X. campestris pv. *mangiferaeindicae*
X. campestris subsp. *mangiferaeindicae*
Xanthomonas mangiferaeindicae

Common names: Bacterial black spot, bacterial blight, and canker of mango

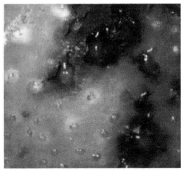

PHOTO 7.30 Dark-colored cankerous bacterial spots on mango leaves and on fruits with bacterial oozing. (Courtesy of Nongmaithem Prabeena Agriculture Research Centre, Reliance Foundation, Mouda, India.)

It is one of the serious diseases and is reported from many countries. In India, it is reported from Maharashtra, Delhi, Uttar Pradesh, Tamil Nadu, and Karnataka.

Symptoms

The disease affects all aerial parts of the plant, including fruits. Leaf and fruits symptoms are more common. In severe infections, twig and branch cankers are found. Groups of minute, water-soaked lesions, delimited by veins, appear toward the tip of the leaf. They increase in size to about 1–4 mm, become raised, and turn brown to black in color (Photo 7.30). Sometimes these spots are surrounded by a chlorotic halo. Large necrotic patches may be formed by coalescing of several spots. The patches sometimes dry up with a decrease in atmospheric humidity. The leaf symptoms are very conspicuous on young newly formed leaves. Brown spots are seen on such leaves when there is heavy infection in warm and very humid weather, but when there are no heavy rains, drops of amber-colored exudation can be seen on those leaves. The spots are often rough and raised. The young leaves invariably shed and can be seen scattered on the orchard floor. When a major portion of the lamina surface of older leaves are affected, they also fall down; otherwise, spotted leaves remain on the tree. The incubation period on leaves varies from 2 to 14 days (Patel et al., 1948). In leaves, the bacteria induce hypertrophy of tissues by enlargement of intercellular spaces of spongy parenchyma early in the infection process. Epidermis is ruptured and bacterial cells ooze out.

Fruit symptoms appear as small water-soaked spots on lenticels. These later become star-shaped, black, erumpent, and exude infectious gum. Cracks may appear in the skin of the fruit. Often, a "tear-stain" is observed. Badly affected fruits drop prematurely. Fruits with cracked skin invite other secondary pathogens, and postharvest rot of fruits invariably occurs. Even on the tree, such fruits with cracked skin invite ants and insects, and half-mature fruits undergo rotting and shed. Xylem plugging and phloem distortion also occur. Twig cankers are potential source of inoculum and reduce the resistance of the branches and twigs to high winds.

Different strains cause different predominant symptoms. The leaf spot isolates and fruit or twig canker isolates are different.

Host

Cashew (*Anacardium occidentale*), Brazilian pepper (*Schinus terebinthefolius*), ambarella (*Spondias cytherea* or *Spondias dulcis*), and other members of the plant family Anacardiaceae.

Disease Cycle

In the orchards, the bacteria are phylloplane residents throughout the year. They have been reported as epiphytes on buds where their population is favored by high relative humidity and temperature of 15°C–20°C. They also survive in cankers on twigs and smaller branches, stone of the fruit, and as phylloplane microflora on weeds. The primary inoculum is brought by contaminated nursery stock, wind-driven rains, and grove maintenance operations. The disease spreads rapidly during rains.

Bacterial cells are rain-splashed from the source of survival to uninfected parts. Long-distance spread is caused by infected planting stock. The bacteria enter the fruits through bruises and other types of injuries. Disease development is favored by high humidity (RH above 90%) and a temperature range of 25°C–30°C. Maximum infection of fruits occurs when minimum and maximum (night and day) temperatures are 22°C and 25°C. Rainfall is a major weather factor affecting fruit infection. High wind velocity is also a favorable factor for the disease. Many insects such as ashy weevil (*Myllocerus discolor* var. *variegata*), leaf webber (*Orthega vadrusalia*), and bugs (*Canthecona furcellata*) mechanically transmit the bacteria via their legs and mouth parts (Kishun, 1986). Some other members of Anacardiaceae such as cashew (*An. occidentale*) are hosts of the bacteria.

In India, the disease remains dormant during November to March due to low temperature (11.8°C–22°C) and the leaf infection is considerably reduced by the fall of the infected leaves. Kent mangoes showed close relationship between rainy season and incidence of bacterial leaf spot. One day of rainfall did not affect the disease appearance, whereas 4 days of rainfall led to 36.9% disease incidence.

Planting the orchard in a humid region, use of spray irrigation systems, planting very sensible mango cultivars, and so on increase the disease incidence during the rainy season.

High disease prevalence was observed in Ghana, indicating the suitability of environmental conditions in this region for the development of mango bacterial canker. The budwood for these blocks was imported from Burkina Faso in 2002 and symptoms were observed in these blocks shortly after establishment.

Geographical Distribution

India, Australia, the Comoros Islands, Japan, Kenya, Malaysia, Mauritius, New Caledonia, Pakistan, the Philippines, Réunion, Taiwan, Thailand, the United Arab Emirates (Gagnevin and Pruvost, 2001), Hawaii, Egypt, and Ghana.

Economic Importance

The disease incidence and disease severity have varied from 0.52% to 42.0% and from 15% to 90%, respectively, in different states. Up to 80% loss may occur in areas with high winds and heavy rains.

Disease Management

Chemical Control

Twig canker persists and initiates fruit infection. The disease spread is rapid during the rains and becomes severe in July–August. Five application of Bordeaux mixture (4:4:50) with spreader was recommended by Wager (1937), whereas four sprays of 6:6:50 Bordeaux mixture was found to be effective by Marloth (1947). Sprays with fungicides such as copper-containing materials and Agrimycin (Viljoen and Kotze, 1972) have also been advocated. Shekhawat and Patel (1975) recommended orchard sanitation by way of removal of infected materials and seedling treatment as preventive measures. Agrimycin-100 proved best in arresting the disease development. Plantomycin (200 ppm) followed by Agrimycin + Bavistin (1000 ppm) was reported to be the best against this disease. Streptomycin sulfate followed by aureofungin has been recommended by Prakash and Raooff (1985) to control bacterial disease of mango.

Some workers have claimed control of the disease by streptocycline spray. In North India, two sprays of 300 ppm streptocycline with 0.3% copper oxychloride in May at 10-day interval were recommended.

Three sprays of streptocycline (0.01%) or Agrimycin-100 (0.01%) after first visual symptom at 10-day intervals and monthly sprays of carbendazim (Bavistin 0.1%) or copper oxychloride (0.3%) are effective in controlling the disease.

Varietal Resistance

A large number of mango varieties were screened for resistance to the disease, and none was found to be resistant. Langra, Dashelin, Chausa, Bombal Zardalu, Sunder Langra, Gulabkhas, Kesar, and

Mankurad were moderately resistant, Bombai Green, Hernasagar, Fazali, Swamrekha, and Anupam were susceptible. In South India, severe natural infection in the varieties such as Mulgao, Alphanso, Neelam, Rumani, Bangalore, and Baneshan were reported. In the Tarai area of Uttaranchal (India), Amrapali and Dashehri have been found highly susceptible to leaf blight but not to fruit cankers. Bombai green was less severely affected while Langra had leaf infection only in traces, but at times, heavy fruit infection and loss are reported. Mishra (1995) screened 212 mango varieties and found 95 to be free from fruit canker.

7.12 BANANA

7.12.1 BANANA XANTHOMONAS WILT

Pathogen: *Xanthomonas campestris* pv. *musacearum*
More than 70 million people in 15 countries in sub-Saharan Africa depend on banana for their livelihood and food supply. Their food security is being threatened by the arrival and spread of two devastating banana diseases—banana bunchy top disease and banana *Xanthomonas* wilt (BXW).

Symptoms
BXW symptoms can be sorted into two domains: symptoms on the inflorescence and symptoms on the fruit. Symptoms on the fruit are usually used to distinguish BXW from alternative banana diseases. A bacterial ooze is excreted from the plant organs and this is a mandatory sign that BXW may be present. Common symptoms on the fruit include internal discoloration and premature ripening of the fruit. A cross-section of the BXW-infected banana is characterized by the yellow-orange discoloration of the vascular bundles and dark brown tissue scaring. Symptoms on the inflorescence include a gradual wilting and yellowing of the leaves plus wilting of the bracts and shriveling of the male buds (Photo 7.31).

PHOTO 7.31 Shrivelled male bud and uneven ripening of fruit. (Courtesy of Blomme, G., Bioversity International, Maccarese, Italy, www.musarama.org.)

PHOTO 7.32 Wilting and yellowing of leaves. (Courtesy of Blomme, G., Bioversity International, Maccarese, Italy, www.musarama.org.)

The disease causes loss both through death of the plant and rotting of the fruit. The leaves gradually turn yellow and start looking lifeless as if they were melting under intense heat (Photo 7.32). They eventually turn brown and die.

In flowering plants, the first symptoms of insect transmission are a drying rot and blackening of the male bud that start with the outer bracts and eventually extend to the rachis. The fruits ripen unevenly and prematurely, turning from green to yellow and black rapidly. The pulp of the rotting fruits shows rusty brown stains.

Internal symptoms revealed through a cross-section of an infected pseudostem are yellow-orange streaking of the vascular tissues and the presence of a yellow bacterial ooze that can also be seen from any other infected plant part (Photo 7.33).

PHOTO 7.33 Cream to pale yellow bacterial ooze appears on the cutting of pseudostem. (Courtesy of Blomme, G., Bioversity International, Maccarese, Italy, www.musarama.org.)

In the absence of other symptoms, the leaf symptoms of *Xanthomonas* wilt can be confused with those of *Fusarium* wilt. In plants affected by *Fusarium*, yellowing and wilting of the leaves typically progress from the older to the younger leaves. The wilted leaves may also snap at the petiole and hang down the pseudostem. In plants affected by *Xanthomonas*, the wilting can begin with any leaf and the infected leaves tend to snap along the leaf blade.

Transmission

Through Soil
Soil is one of the main sources for *X. campestris* pv. *musacearum* inoculum.

X. campestris pv. *musacearum* may contaminate the soil for 4 months and more. BXW aware-ness campaigns have helped reduce the numbers of farmers growing bananas on contaminated plantain soil aiding in the control of the disease overall.

Through Air
It is widely thought that *X. campestris* pv. *musacearum* bacteria is transmitted by airborne vectors to exposed male flowers (see plant reproductive morphology). *X. campestris* pv. *musacearum* bacteria has been isolated from the ooze and nectar excreted from openings of fallen male flowers. Insects such as stingless bees (Apidae), fruit flies (Drosophilidae), and grass flies (Chloropidae) transmit the disease from banana to banana after being drawn to the infected nectar. If the disease has been transmitted by insects, the symptoms tend to first appear on the male buds of the banana plant.

Through Insect
Fresh wounds offer bacteria a point through which they can enter or exit the plant. Bacteria have been isolated from the sap and ooze collected from the cushions to which the male flowers were attached and the scars made by the fallen bracts, and to a lesser extent from the nectar of flowers.

An insect visiting the male bud of an infected plant can get bacteria on its body through a wound that exudes bacteria-laden ooze. The bacteria on the insect's body can then infect healthy plants when the insect visits healthy plants that have similar wounds.

Even though female flowers are more visited than male flowers and the loss of their bract also leaves a scar, experiments suggest that infection only occurs through the cushions of male flowers as no flower infection was found to occur when the male bud was removed. It could be because the female bract scars are less numerous and less accessible than the male bract scars.

In Uganda, the bacterium was isolated from stingless bees (*Plebeina denoiti* and undetermined species of the Apidae family), honeybees (*Apis mellifera*), fruit flies (Drosophilidae family), and grass flies (Chloropidae family).

Cultivars that have persistent bracts and flowers are less likely to contract the disease from flying insects visiting the male bud, although some cultivars whose bracts and flowers do fall off also seem to escape infection. In general, East African highland bananas (EAHB) seem less prone to insect transmission, maybe because their inflorescence is less attractive to insects. The cultivar Kayinja, on the other hand, is very prone to floral infection. Insect transmission is less frequent at high alti-tudes; probably the lower temperatures are not favorable to insect vectors. Male bud infections were rare at altitudes of 1700 m and above in Ethiopia and the Masisi district (1700 m) of DR Congo, where the disease was first observed.

Through Tools
Cutting tools used by farmers, such as a machete, can be a key mechanism by which the disease is spread in intensively managed cropping systems. Transmission occurs when a farmer uses a cutting tool on an infected plant, where it comes in contact with the bacteria in the sap or ooze, and then on a healthy plant. The regular harvesting of leaves for sale or other domestic use, which is common in Uganda, can contribute to the spread of the disease through contaminated tools.

Knife (panga) is used almost universally in African agriculture. Use of contaminated knives was a common route for disease spread when the disease first originated, but increased knowledge of

BXW transmission has led to increased numbers of knives being disinfected after use. Herbicides are now advised as a more economical and effective way of destroying infected banana crop.

Through Infected Plant Material

BXW infects all parts of the plant. Disease spread has been primarily linked with the transport of planting material for replanting. Other parts of the plant such as the male buds (used in banana beer production) and mulch (banana waste material) can also serve as novel material for the spread of the disease.

The spread, over short and long distances, has been associated with the movement of symptomless but infected suckers for replanting. Latent infections could explain the spread of *Xanthomonas* wilt in previously disease-free areas by farmers who unwittingly exchanged what they thought were healthy suckers. Banana plant residues can also spread the disease.

Distribution and Spread of the Disease

Xanthomonas wilt is currently found only in eastern Africa and the northeastern corner of the Democratic Republic of Congo. It was first reported on enset and banana in Ethiopia in 1968, although earlier records report a disease consistent with these symptoms as present in the 1930s. Spread beyond Ethiopia was not reported until the disease was found in Uganda in 2001. Subsequent spread to other countries in eastern Africa has proceeded rapidly. In many instances, however, the exact time of introduction is not known.

Although it is said to have been first observed in 2001 by farmers in the North Kivu Province, the disease was not confirmed in the Democratic Republic of Congo until 2004. In Rwanda, *Xanthomonas* wilt was first observed in the northern part of the country in 2005. It was also reported to have spread to Tanzania in 2005 and to Kenya (Mbaka et al., 2009) and Burundi in 2006. An analysis of data collected between 2001 and 2006 in Uganda and in six countries of the Great Lakes region in 2007 suggests that the number of newly affected areas declined between 2004 and 2005 and that the spread of the disease was changing from being more or less continuous to more isolated outbreaks.

Leaf wetness has also been implicated in disease establishment. Inoculated leaves of 3-month-old tissue-culture plants that were kept wet for 72 h developed symptoms within 14 days, contrary to the control plants whose leaves were allowed to dry (Mwangi et al., 2006). The results may explain why a higher percentage of nonflowering Pisang Awak was infected in the high-altitude region of Masisi in DR Congo (26%) than in the midaltitude region of Central Uganda (6.1%–9.3%) (Mwangi et al., 2006). In the high altitudes, rainfall is evenly distributed throughout the year, as opposed to two rainy seasons interspersed with a dry season at midaltitudes. High plant densities could also contribute to disease spread when inoculum splashes down from the taller infected plants to the suckers growing below. Goats and other livestock can carry the bacteria in their mouth and as such spread the disease to healthy plants. Pests such as weevils and nematodes can also help the bacteria in the soil gain entry into the plant through the injuries they make on the root system. The disease cycle of BXW is illustrated in Figure 7.5.

Alternative Hosts

Cereals, especially maize, may serve as a reservoir to the bacteria (Anonymous, 2006).

Economic Impact

BXW or banana bacterial wilt first reached epidemic levels in Uganda in 2001. Despite control efforts, it has now spread to many locations in Uganda and other countries of the Great Lakes region, including DRC, Kenya, Rwanda, and Tanzania. Analysis from Uganda indicated that, if unchecked, the disease could result in national cumulative losses of $4 billion over a 5-year period.

BXW causes wilting of leaves, premature ripening of bunches and rotting of fruit, and death of the plant, causing damage of up to 100%. Contaminated tools, insects, and even birds spread the bacteria, but long-distance transmission is often induced by human through the movement of planting material carrying latent infections. All banana cultivars are susceptible to the disease, and BXW has to be controlled through eradication or field management practices that reduce pathogen spread.

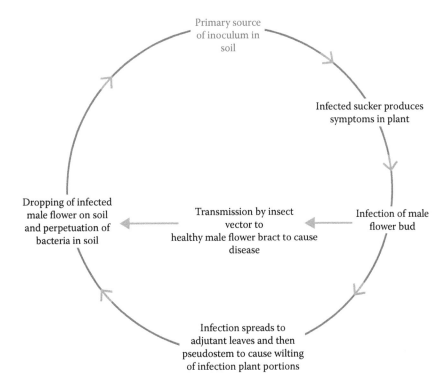

FIGURE 7.5 Disease cycle of banana xanthomonas wilt caused by *X. campestris* pv. *musacearum.*

In Uganda, participatory management campaigns were initially successful in reducing infections, and in pilot villages where Farmers' Field Schools were introduced by FAO in 2006, the disease was practically eliminated. Recently, however, the disease is showing resurgences in areas where it was previously controlled, probably resulting from a form of fatigue and the lack of sustained support system and incentives to the farming communities, including availability of clean planting material. Control practices, including destruction of infected plants, removal of male buds, and use of clean tools, are physically onerous. Besides, more information on the disease epidemiology and spread is needed.

BXW, or banana bacterial wilt, or enset wilt, after being originally identified on a close relative of banana, *Ensete ventricosum*, in Ethiopia in the 1960s (Yirgou and Bradbury, 1968), emanated in Uganda in 2001 affecting all types of banana cultivars. Since then, BXW has been diagnosed in Central and East Africa, including banana-growing regions of Rwanda, Democratic Republic of Congo, Tanzania, Kenya, Burundi, and Uganda.

Of the numerous diseases infecting bananas, BXW alongside banana bunchy top virus has been the most devastating in recent years. Global concern arose over the livelihoods of African banana farmers and the millions relying on bananas as a staple food when the disease was at its worst between the years 2001 and 2005. It was estimated that in Central Uganda from 2001 to 2004, there was a 30%–52% decrease in banana yield due to BXW infection (Castellani, 1939). Although extensive management of the disease outbreaks has helped reduce the impact of BXW, even today BXW continues to a pose a real problem to the banana farmers of Central and East Africa. There is a saying: a journey of 1000 miles begins with one step. This is the story of the long walk toward the elimination of BXW disease that is threatening to wipe out the valued banana crop in East Africa. Bananas and plantains are important sources of food for over 100 million people in sub-Saharan Africa as well as income for over 50 million smallholder farmers. Kalyebara et al. (2006) reported that if BXW is not controlled, Uganda stands to lose an estimated $295 million worth of banana output valued at farm gate prices. Since the disease was discovered, banana farmers have used several cultural methods such as sanitary measures to try and control its spread without much success. As the saying goes, necessity is the mother of invention, and scientists have had no option but to explore a lasting solution to this malady. This led to the birth of the Banana Bacterial Wilt project; this project develops bacterial wilt–resistant bananas for use by smallholder farmers in the Great Lakes region. The BXW project is a public–private partnership that brings together the African Agricultural Technology Foundation (AATF), the International Institute of Tropical Agriculture (IITA), and the National Agricultural Research Organisation (NARO) of Uganda. In 2004, IITA and NARO scientists started developing transgenic bananas resistant to BXW in a joint project funded by Gatsby Charitable Foundation. AATF joined the project in 2005 and successfully negotiated for two genes (*Pflp* and *Hrap*) from Academia Sinica, Taiwan, for use in the project free of royalty. The three organizations agreed to work together and, therefore, signed a tripartite agreement in 2009 to guide the partnership especially on roles and responsibilities. IITA and NARO were responsible for developing, transforming, and evaluating transgenic bananas that are resistant to BXW, while AATF was mandated to coordinate the partnership and provide expertise in the management of intellectual property and regulatory issues. The project then commenced with funding from AATF with additional support from Gatsby Charitable Foundation and USAID. The developmental research work is now being carried out both in Uganda and in Kenya by Dr. Leena Tripathi of IITA and Dr. Wilberforce Tushemereirwe of NARO. The project uses modern tools of biotechnology to develop BXW-resistant banana varieties through genetic modification of banana varieties preferred by farmers of East Africa.

The progress made so far by this project is positive following the successful transformational work in the laboratories and subsequent conduct of the first confined field trial in NARO's Kawanda field station in October 2010. The scientists are looking forward to a victorious war against the BXW once the product complies with the necessary regulatory requirements in the project countries.

There is hope that most countries in sub-Saharan Africa will put in place enabling legislations that allow the development and growing of genetically modified agricultural products to curb food production constraints in Africa. It is estimated that the first BXW-resistant varieties from these efforts could get to farmers in 6–7 years depending on research results and regulatory approvals in each of the project countries.

Control Measures
Control of BXW is based upon a variety of methods to help prevent the spread of the disease. Vigilance and the quick removal of infected plants remain critical to minimizing the spread of the disease.

Experience and research, as well as an understanding of how the bacteria move through the plant, have shown that the impact of *Xanthomonas* wilt can be mitigated by the adoption of crop-management strategies that have proved successful against other banana bacterial wilts, in particular early removal of the male bud to prevent transmission by insects and strict sanitation on the farm to avoid transmission through contaminated tools. In areas where *Xanthomonas* wilt is present, the practice of using banana plant material as mulch is also discouraged as it may contribute to the spread of the disease.

Debudding
The risk of infection through the inflorescence has been shown to be markedly reduced by the timely removal of the male bud by means of a forked stick (not cutting it off with a knife). In disease-free and infected areas, the recommendation is to remove male bud of healthy plants as soon as the last hand of fruit has formed. Using a forked stick not only avoids the risk of moving bacteria around on knives if the plant is infected but also enables farmers to remove out-of-reach male buds.

Disinfecting Tools
Once the disease has been detected in a field, disinfecting tools (pangas, knives, leaf removers, hose, etc.) before using them on other plants would help prevent the spread of the bacteria to healthy plants.

The tools can be sterilized by flaming until just too hot to touch (not red hot) or by soaking in a sodium hypochlorite solution (one cup of household bleach in five cups of water).

Removing Infected Plants
From the early days of the epidemics, farmers were advised to uproot diseased mats and to dispose the plant debris, before replanting using clean planting material. (Studies on the survival of bacteria in the soil suggest that it should be safe to replant bananas after 6 months.) The practice, however, was not widely adopted, especially among resource-poor households. Since one person can remove only about two mats per day, removing a large number of diseased mats turned out to be particularly laborious. Moreover, even after destruction of the mat, suckers might emerge, requiring farmers to frequently return to the field to ensure complete destruction of the mats. A 2010 household survey targeting Uganda's main production systems, the intensively managed EAHB production systems, and the less intensively managed Kayinja ones, showed that 53% of EAHB farmers and 30% of Kayinja farmers uprooted the entire mat.

Mats can also be destroyed by injecting an herbicide into the pseudostem of the mother plant. Notwithstanding, health and environmental issues (the most effective herbicide turns out to be more persistent in the environment), the adoption by farmers has been poor. Some of the possible reasons that have been cited are inaccessibility of herbicides in rural areas, perceived high cost of herbicides, and reluctance to inject infected mother plants as asymptomatic suckers will also die.

The elimination of an infected mat can also be done gradually by cutting down all the plants at soil level and destroying emerging suckers. The rhizomes are left to rot, a process that can take up to 24 months if no measure is taken to accelerate decomposition, during which time the soil will remain infective for bananas. Uganda farmers have used the practice to clear heavily infected banana plots for planting annual crops.

Single Diseased Stem Removal
An alternative to uprooting the entire mat is cutting off the pseudostem at soil level. In the early days of the epidemic, studies on the systemicity of the bacteria suggested that the practice would be effective in preventing the entry of the bacteria in the rhizome, but only if the diseased plant was at the first stage of a flower infection (shrivelled bract) and only for "Kayinja." However, long-term studies have since shown that the bacteria do not go on to systematically colonize all the suckers attached to the rhizome. Moreover, even when the bacteria are present in a sucker, the disease may not develop or develop much later (latent infections), making the single diseased stem removal technique a recommended alternative to uprooting the entire mat.

Beyond what individual farmers can do on their plots, a number of measures, such as intensive surveillance and reporting of new outbreaks (with prompt action to investigate them and take action) and strict control of the movement of plant material from infected areas to unaffected ones, have been proposed. However, in plots of Kayinja, which are frequently neglected and for which ownership and responsibility for maintenance are often obscure, the likelihood of stopping the progress of the disease through coordinated actions is judged low.

BXW-Resistant Banana
No banana cultivars in Central and Eastern Africa have shown any resistance to BXW despite some varieties, such as those in the Pisang Awak region, which show increased susceptibility. Scientists have recently transferred two genes from sweet green pepper to bananas in order to confer resistance to BXW. This is a promising step forward in circumventing the time-consuming and expensive practices of disease management such as "debudding."

Pflp and *Hrap* genes encoding the proteins plant ferredoxin–like amphipathic protein (*Pflp*) and hypersensitive response (HR)-assisting protein (*Hrap*) were isolated from sweet pepper and introduced into the genome of East African bananas by genetic engineering. The two proteins induced an HR and SAR within the banana plant after being exposed to the bacterial pathogen. It was reported that over half of the transgenic bananas were resistant to BXW.

So far, no resistant cultivars have yet been identified, and field observations in areas where the bacterium is present suggest that all commonly grown cultivars succumb to the disease. Laboratory experiments on seven cultivars and a wild species showed differences in susceptibility, with the cultivar Kayinja being the most susceptible, whereas 10% of the potted *Musa balbisiana* plants that had showed early signs of necrosis eventually recovered and did not wilt. Similarly, all the 3-month plants from 42 genotypes tested in a pot trial eventually developed the disease, except for the *M. balbisiana* plants, from which no bacterial cells were recovered. In later screenhouse and field trials, *M. balbisiana* plants were symptom-free 6 weeks after a single inoculation at dosages that caused disease on the Pisang Awak controls (Kumakech et al., 2013). Disease symptoms were observed 6 weeks after a second inoculation, but only at the highest dosages.

Some cultivars, however, possess characteristics (such as persistent male bracts and flowers or no male bud) that make it harder for the bacteria to infect the plants under natural conditions. While these cultivars escape transmission by insect vectors, they are still vulnerable to transmission through cutting tools.

Other Measures

Early on in the outbreak, affected and threatened countries were urged to develop national and regional strategies. In Uganda, the Ministry of Agriculture, Animal Industry and Fisheries (MAAIF) and the NARO embarked on an intensive program to raise awareness of the problem and enable farmers to identify the symptoms of the disease and to implement control measures. The country was divided into endemic areas (where the disease was present), frontline (bordering the endemic areas), and threatened (disease-free) in the hope of isolating, and eventually eradicating, the disease from the endemic areas.

A number of approaches were used to reach out to farmers through the extension system, media (local radio, television, newspapers, and brochures); civic, cultural, and local leaders; and a participatory development communication approach that consisted in organizing, at the local level, farmers, local leaders, extension staff, researchers, and communication specialists. An analysis of different approaches used in Uganda to mobilize farmers suggests that farmer field schools have been more effective in reducing the incidence of *Xanthomonas* wilt than the traditional approach of using extension services and mass media.

In 2006, a regional Crop Crisis Control project (C3P) was initiated to reduce the spread and impact of the disease through education and training in East and Central Africa. The project adopted a cascade training model; that is, those who participated in a regional training were expected to conduct further training in their country. The people trained in-country would then take the training to the community level. In less than a year, the regional trainings had reached more than 1,000 extension and research staff, while more than 30,000 farmers had been trained at the community level. However, considerable differences in capacities and institutional structures between countries affected implementation.

Coupling data on the incidence of the disease with other types of information, such as the share of bananas in daily food intake and the level food insecurity, was suggested to identify areas that should be targeted in priority.

Scientists have also mentioned the lack of incentive for farmers living in threatened, but not yet affected, areas to adopt preventive measures.

7.12.2 Bacterial Wilt or Moko Disease of Banana

Pathogen: *Ralstonia solanacearum* Phylotype II

Common name: Moko disease

Symptoms

Bacterial wilt due to *R. solanacearum* is very destructive, especially during hot and wet seasons. Plants wilt and die suddenly.

PHOTO 7.34 Symptoms of moko disease of banana. (Courtesy of Borkar, S.G. and Gaikwad, R.T., PPAM, MPKV, Rahuri, India.)

The young plants are affected severely. In the initial stages, the bacterial wilt is characterized by the yellowish discoloration of the inner leaf lamina close to the petiole. The leaf collapses near the junction of the lamina with the petiole. Within a week, most of the leaves exhibit wilting symptoms. The presence of yellow fingers in an otherwise green stem often indicates the presence of moko disease. The most characteristic symptoms appear on the young suckers that have been cut once and begin regrowth. These are blackened and stunted. The tender leaves from the suckers turn yellow and necrotic.

An initial symptom of an infected banana is that one of the youngest three leaves turns pale-green or yellow in color and breaks down at the junction of the petiole and the pseudostem (Photo 7.34). All the other leaves follow to collapse around the pseudostem after 1 week. An infected finger or fruit shows dry and rotted pulp that is colored brown or black and the presence of bacterial discharges.

Moko disease affects ornamental *Heliconia* spp. It also affects commonly grown dessert bananas (*Musa*), plantains (*Musa paradisiaca*), and cooking bananas, especially Bluggoe (ABB).

Host Plant
Potato, tomato, tobacco, eggplant, banana, and plantain are the major hosts, but peanut, bell pepper, cotton, sweet potato, cassava, castor bean, ginger, and other solanaceous weeds are also affected by the pathogen.

Geographical Distribution
The Philippines, Australia, Hawaii, and Northern Queensland.

Central and South America: Belize, Brazil, Colombia, Costa Rica, Ecuador, El Salvador, Grenada, Guatemala, Guyana, Honduras, Mexico, Nicaragua, and Panama, Peru, Surinam, Trinidad, and Venezuela.

Economic Impact
Moko is one of the most serious diseases of banana and plantains. It has caused severe losses in peasant crops of bananas and plantains in South and Central America and the Caribbean. Moko has been recorded as a continuing problem in Mexico and Belize. Reductions in yield due to moko of up to 74% have been reported in Guyana. A single localized root invasion by moko can soon become an epidemic if flowers are being produced at that time and if the bacteria begin to ooze from the male bud.

Predisposing Factors for Disease Development
1. Crop residues left in the field that were infected by *R. solanacearum*
2. Injured roots caused by farm tools or by soil pests
3. High soil pH
4. Poor and infertile soil
5. Nematodes present in the soil

Control Measures

Early detection and destruction of the suspected plants may help in preventing the spread of the disease. All the tools used for pruning and cutting should be disinfected with formaldehyde. Removal of the male flowers as soon as the last female hand emerge help in minimizing the spread of the disease as the insects can carry the disease-causing bacterium on the male flowers. An integrated approach to control the disease should be followed, including the following:

1. Remove and destroy all infected plants immediately.
2. Control nematodes.
3. Rotate crops other than solanaceous crops. Rice, corn, beans, cabbage, and sugarcane are found to be resistant to bacterial wilt.
4. Since the bacteria can be transmitted through farm tools, wash or expose them to heat before using in another field.
5. Remove and chop the plants surrounding the infected mat or within the radius of 6 m from the infected banana plant to prevent further spread of the disease.
6. The primary means of moko control is exclusion from areas where the disease does not occur.
7. Good cultural practices involving removal of infected plant material and disinfection of all tools and machinery reduce the risk of spread of this disease. Among edible bananas, there is no banana resistant to moko.

7.12.3 BANANA BUGTOK DISEASE

Pathogen: *Pseudomonas solanacearum*

Of all the bacterial diseases that continuously inflict severe damages to banana plantations worldwide, the Bugtok disease is claimed as the most "stubborn." The disease is caused by a bacterium called *P. solanacearum*, which is transmitted by sucking insects. Banana farmers in tropical regions have reported that bugtok is really difficult to deal with. Though several control strategies had been developed to minimize the infestation, the disease continues to spread mainly due to poor cultural management.

In the Philippines, the main targets of bugtok are the cardaba and saba, the varieties best used in making quality chips, flour, and ketchup. The bugtok outbreak hits one of the country's top suppliers of banana, particularly the municipality of Cabadbaran, Agusan del Norte, and the overall production of banana tremendously declined.

This alarming incidence prompted the officials and researchers of the Caraga Regional Integrated Agricultural Research Center in Region XIII to conduct a technology demonstration project in an effort to teach banana farmers the most effective way to combat this "stubborn" disease.

Symptoms

The most discernible symptom of the disease is the discoloration of the fruit pulp that is most intense at the core. In fruit with a light infection, the discolored parts are interspersed with soft fruit pulp. All fruits within a bunch can be discolored in severe infections, but the distribution of discolored fruits within a bunch is random in plants that are less severely infected.

Unlike moko disease, bugtok-infected plants appear outwardly normal to the untrained eye. The leaves remain green and fruit seems to develop normally. However, the bracts of the male inflorescence, if left in the fruit bunch, fail to dehisce. This gives the male inflorescence a dry and loose appearance. This character is the only external symptom that can differentiate healthy from infected plants.

Internally, brown vascular streaks can be observed in the fruit peduncle, the fruit stem, and the pseudostem. Browning is less intense at the base of the pseudostem but discoloration sometimes extends to the corm of the plant. There is convincing evidence that infection occurs via the inflorescence and that bugtok disease is transmitted by insects, probably thrips. Bagging the young inflorescence as it emerges from the crown produces bugtok-free fruit, an indication that insect vectors play a role in the spread of the disease. Sucker transmission is unlikely.

Experts who worked with bugtok disease project (1998–2001) reported that the most evident symptom of bugtok is the red and black discoloration running from the core toward the whole pulp of the fruit. The fruit pedicel or stem shows yellowish-brown discoloration. Plants infected with bugtok do not wilt or die, but since the disease is restricted to the fruits, these become hard even when ripe or cooked. Experts attributed this condition to the entry of pathogens inside the fruit, which later multiply and invade the fruit's vascular system. The bracts (modified leaves) of the male bud (puso) do not detach or fall off from the stem even if they are already dried.

Economic Impact
Banana plants highly infested with bugtok fail to produce healthy and marketable fruits, thus cutting the income of farmers down to zero level.

Control Measures
Cultural farm practices employed are desuckering, stem and mat sanitation, leaf pruning, and bagging and debudding.

Desuckering is usually done to maintain the health and vigor of the plants. This is also practiced to prevent competition for nutrient and water and to avoid overcrowding. Extension personnel advise the farmer-cooperators that the ideal number of suckers for a banana plant should not exceed three.

Since weeds serve as refuge and breeding place for insect pests and diseases, stem and mat sanitation should be done at least twice a month. Banana farmers should strip the dry portion of the leaf sheaths, cut them into small pieces, and file along the middle row of the plants' rows and base to prevent the growth of weeds.

Likewise, banana leaves should be regularly pruned to eliminate the inoculum source of the disease. Leaves with 50% dryness must be trimmed and burned right away.

On the other hand, bagging and debudding involving wrapping the male bud or "heart" with plastic bag, cement bag, sack, or any material with an open end should be followed. The farmer-cooperators were advised to remove the bag 14–15 days after shooting.

Of all the cultural methods employed in the management of this disease, the researchers concluded that bagging is the most effective technique that completely eradicates bugtok. They explained that bags serve as protective covering against insects that may carry the bacterium responsible for bugtok. They also observed that banana plants are most susceptible to bugtok during their flowering stage.

In 3 years, the project yielded positive results, as farmer-cooperators were able to harvest a total of 2909 bugtok-free bunches. Farmers who strictly followed the bagging technique and other cultural management practices in the techno-demo site had augmented their income, ranging from P8, 400.00 to P14, 810/ha per harvesting season.

7.12.4 BANANA BLOOD DISEASE

Pathogen: *Ralstonia solanacearum* Phylotype IV

Synonyms: *Pseudomonas celebensis*

Banana blood disease is a wilt disease caused by a bacterium that invades the vascular tissues. The name "blood disease" was originally adopted because droplets of a thick milky white, yellow or red-brown liquid often ooze out of the vascular tissues of infected plants at cut surfaces. The disease affects cultivars of both AAA and ABB genomic groups.

Blood disease was first reported about 80 years ago from southern Sulawesi (formerly Celebes), Indonesia, where it caused the abandonment of dessert banana plantations being developed on the adjacent Salayar (Salieren) islands. The disease was the subject of extensive investigations by Ernst Gäumann in the early 1920s, who showed that it was caused by a Gram-negative bacterium that he named *P. celebensis*. Gäumann found the disease widely distributed throughout southern Sulawesi and observed symptoms in wild *Heliconia* and *Musa* spp. in which it was apparently endemic. It was not found in Java or other islands. A quarantine order restricting movement of banana fruit and vegetation from Sulawesi was probably instrumental in limiting further spread of the disease until 1987, when an outbreak was confirmed in West Java. There are recent, unconfirmed reports from Kalimantan and the northern Maluku islands.

Symptoms

Blood disease is commonly seen in the Pisang Kepok cultivar (ABB/BBB; "Saba") but other groups are also affected. Symptoms are similar to those of moko disease in Latin America and vary according to the growth stage of the plant and the route of infection.

Fully expanded leaves of plants of all ages show a conspicuous transient yellowing, followed by loss of turgor, desiccation, and necrosis. In mature plants, the base of the petiole collapses, causing wilted leaves to hang down around the pseudostem. The youngest leaves cease emerging and develop whitish and later necrotic panels in the lamina. Daughter suckers may show general wilting, but infection is not always systemic, and healthy suckers are sometimes produced. Internally, vascular bundles exhibit a reddish-brown discoloration (Photo 7.35) which, depending on the mode of infection, may extend throughout the plant or may be confined to the central fruit stem. If kept moist, cut vascular tissues exude droplets of bacterial ooze that can vary in color from white to reddish brown or black.

The inoculated banana seedling with the bacterial ooze caused wilting of the seedling after 5 weeks of inoculation. Blackening and shriveling of male flowers are frequently found in mature plants, and vascular discoloration can be traced into the peduncle and down the fruit stem. Blackening sometimes extends into the lower fruit bunches, but these often appear outwardly healthy (Photo 7.36). Internally, fruits in all bunches are usually uniformly discolored reddish-brown and rotten.

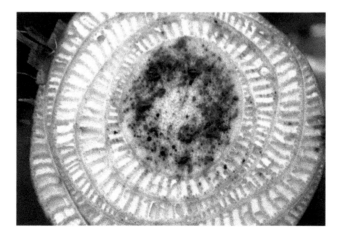

PHOTO 7.35   Internal discoloration, also in the central pseudostem. (Courtesy of Eden-Green, S., Natural Resources International [NRI], U.K.)

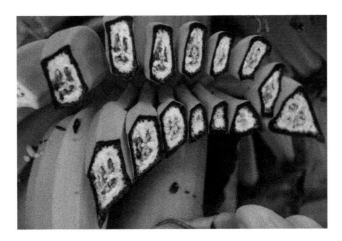

PHOTO 7.36   Uniform, internal discoloration of the fruits. (Courtesy of Schmidt, S.M., University of Amsterdam, Amsterdam, the Netherlands.)

Pathogen

Firstly described in 1920s found in Celebes (Sulawesi Island) as *P. celebensis*. As the member of *R. solanacearum* species complex in Phylotype IV (moko pathogen in Phylotype II). Slow growing on agar medium with a diameter of 2–3 mm in 5 days on TZC medium. Do not infect *Solanaceous* plants. Insect transmitted (by pollinators, decay visiting insects, banana pests). In the field, cooking bananas (kepok, saba, pisang awak, etc.) were found to be highly susceptible to blood disease bacterium (BDB) infection compared to dessert banana.

Although there are many similarities in the symptoms and epidemiology of blood and moko diseases, the causal bacteria show distinctive phenotypic and genetic differences. The first isolation of commonly used culture media showed the following distinctive features of the BDB: production of nonfluidal colonies, which are smaller and develop more slowly than those of *P. solanacearum*; utilization and production of acid from galactose and glycerol but not from glucose and other carbohydrates used to differentiate strains of *P. solanacearum*; and failure to reduce nitrate. Motility has not been observed in any of the isolates examined so far, and attempts to demonstrate intracellular accumulation of poly b-hydroxybutyric acid (a key diagnostic feature of *P. solanacearum*) have given equivocal results.

Unlike moko strains, blood disease isolates are not pathogenic to solanaceous hosts, including tomato, tobacco, and *Capsicum*, but symptoms are readily reproduced by mechanical inoculation to the corm or pseudostem of banana plants of all ages. Genetic analyses, by whole genome RFLP groupings, comparison of partial 16S ribosomal DNA sequences, and analysis of tRNA consensus primer amplification products, indicate that the BDB is closely related to, but distinct from, strains of *P. solanacearum*. The precise taxonomic status of the pathogen requires further study but, as the specific epithet *P. celebensis* is no longer valid in the international bacteriological nomenclature, it is recommended that the trivial name "blood disease bacterium" be used at present to distinguish the pathogen from moko strains (*P. solanacearum* race 2) that also occur in South East Asia.

Phylogenetically, BDB is confirmed as a close relative of the *R. solanacearum* species complex and is included as a member in the phylotype IV within this species complex, which is the most diverse phylotype as it contains two other taxa, *Ralstonia syzygii* and *R. solanacearum*.

In the literature, the members of phylotypes I, II, III, and IV of the *R. solanacearum* species complex are shown to be phenotypically diverse; however, they are closely related on the basis of sequencing of 16S rRNA, 16S–23S intergenic transcribed spacer (ITS) region, endoglucanase (egl), and polygalacturonase genes. It had been previously suggested that high DNA homology values were also found among the *R. solanacearum* biovars 1, 2, and 3 strains (currently assigned as phylotypes I, II, and III). Thus, the exact taxonomic position of BDB and the other members of the *R. solanacearum* species complex remains unresolved.

Data from the sequencing of 16S rRNA, 16S-23S rRNA ITS region, and DNA–DNA hybridization (DDH) demonstrate that the *R. solanacearum* species complex should be separated into three genomic species. Phenotypic data including classical phenotypic, Biolog GN2 MicroPlate metabolic fingerprinting, and chemotaxonomic methods also support this new insight. The first genospecies contains only *R. solanacearum* phylotype-II strains including the type strain. The second genospecies is the merger of the three taxa included in phylotype-IV strains to be a single genomic species, *R. syzygii*. The last genospecies includes *R. solanacearum* strains belonging to phylotypes I and III.

Several phenotypic and chemotaxonomic characteristics are proven to distinguish the three taxa in the genospecies *R. syzygii* (phylotype IV); therefore, three subspecies are proposed to represent *R. syzygii*, BDB, and *R. solanacearum* within the four phylotypes IV. Phenotypic methods including physiological and biochemical analyses, Biolog GN2 MicroPlate metabolic fingerprinting test, and chemotaxonomic markers successfully identified several differentiating characteristics of the four phylotypes. The sequencing of 16S rRNA and 16S–23S ITS spacer region supports the separation of the *R. solanacearum* into three genomic species. The three novel species cluster together in the phylogenetic tree derived from the 16S rRNA sequence analysis. In the ITS dendrogram, phylotypes I and III strains cluster separately but their branches are next to each other, whereas all phylotypes II and IV strains form a single cluster each. However, the sequence similarity of the members of the *R. solanacearum* species complex is more than 97%, which cannot resolve their taxonomic issues. Sequencing of *egl* gene showed that BDB strains were clearly separable from *R. syzygii* strains but not from *R. solanacearum* phylotype-IV strains.

Repetitive PCR and multilocus sequence typing (MLST) were selected for determining the genetic diversity of 25 and 4 BDB strains, respectively. BDB strains showed high degree of internal diversity on the basis of their fingerprint patterns. However, MLST data showed that among the four housekeeping genes (*adk*, *gapA*, *gdhA*, and *ppsA*) and one megaplasmid (*hrpB*), the four BDB strains showed their high level of similarity.

Based on the data obtained, the following taxonomic proposals are made: emendation of the descriptions of *R. solanacearum* and *R. syzygii* and descriptions of *R. syzygii* subsp. *syzygii* comb. nov. (R 001T = LMG 10661T = NCPPB 3446T) for the current *R. syzygii* strains, *R. syzygii* subsp. *indonesiensis* subsp. nov. (R-46900T) for the current *R. solanacearum* phylotype-IV strains, *R. syzygii* subsp. *celebensis* subsp. nov. (R-46908T) for the BDB strains and *Ralstonia pseudosolanacearum* sp. nov. (LMG 9673T = NCPPB 1029T) for the *R. solanacearum* phylotype-I and -III strains.

Disease Cycle
Gäumann (1921, 1923) found that the bacteria can survive for over a year in soil infested by decaying diseased plant tissues and can infect the banana plant through wounds on suckers, pseudostem, and fruits. The sequence of symptoms depends on the route of infection and the growth stage of the plant. There is evidence that this disease is probably transmitted by insects visiting the male flowers. The cultivar Pisang Kepok (ABB group) in Indonesia is thought to be highly susceptible because the male flower nectar has high sugar content, making it particularly attractive to insects that spread the bacterium from male bud to male bud. Following this route of infection, blackening and shriveling of male flowers are frequently found. Then, the bacteria move into the fruit and cause a reddish dry rot of the pulp. Afterward the bacteria move down into the pseudostem toward the suckers. As the disease progresses, all leaves become gradually yellow and necrotic

(Stover and Espinoza, 1992), then wilt, collapse, and hang down. Red-to-brown necrotic marks are seen toward the center of the pseudostem and/or peduncle when cut transversely. The male bud below the fruit may ooze droplets, especially from flower and bract scars in those genotypes that shed flowers and bracts. This is thought to be one reason why certain ABB cultivars are much more susceptible to infection than other bananas (Davis et al., 2001). An additional symptom may occur on ABB cultivars.

Insect transmission would account for the rapid spread of the disease (over 25 km per annum in some areas) since it was first noted in Java. Spread can also occur in infected planting material and the pathogen can probably persist in soil or plant debris. Fruits from infected plants may also be a source of infection. Because affected bunches can appear normal, these may be marketed and subsequently discarded by the consumer near back-garden banana plants.

Impact

It was seen that blood bacterial wilt disease caused by BDB was involved in the case of low production of bananas in Indonesia (Supriadi, 2005). The national loss of banana production due to blood bacterial wilt disease was estimated to be around 36% in 1991 (Muharom and Subijanto, 1991). The damage of banana mats was extremely serious in certain districts where ABB genomic groups were planted such as Bondowoso and Lombok; in such case the disease incidence could reach over 80% (Mulyadi, 2002; Supeno, 2002; Supriadi, 2005). Now, the pathogen has been distributed in 90% of provinces in Indonesia with various disease incidences from ten thousands to millions of banana clusters (Subandiyah et al., 2006). Blood bacterial wilt disease remains difficult to control due to poor fundamental knowledge about the ecology and epidemiology of the disease. How long does the pathogen survive in soil? Does the pathogen associate with root systems of nonhost plants? How widespread is the problem in the naturally occurring *Helliconia* and *Musa* spp.? It is obvious that in-depth studies on the ecology and epidemiology of BDB are urgently required (Fegan, 2005).

Control Measures

Until recently, the distribution of blood disease was very limited, but it is now spreading rapidly in Java and poses a serious threat to neighboring islands. Plant quarantine regulations, including controls on the movement of fruits, need to be strictly enforced to limit further spread. Within affected areas, sanitation measures developed for moko are likely to be effective against blood disease, particularly disinfection of cutting tools, field sanitation, and selection of disease-free planting materials (including, where possible, avoidance of cultivars with dehiscent male flower bracts, which are considered particularly vulnerable to infection via newly exposed bract scars). Removal of male flower buds (denavelling) may be effective, but it should be noted that Gäumann reported evidence for infection via both male and female flowers.

7.12.5 Rhizome Rot and Tip Over of Banana

Pathogen: *Erwinia chrysanthemi* pv. *paradisiaca*

Synonyms: *Erwinia carotovora*

Rhizome rot and tip over of banana caused by *E. carotovora* was reported to damage cultivar Gros Michel in Central America and Honduras (Stover, 1972). The disease has been reported on Nendran and Dwarf Cavendish banana in Kerala, Tamil Nadu, and Gujarat (Singh, 1990). Chattopadhyay and Mukherjee (1986) reported the pseudostem rot of banana caused by *E. chrysanthemi* pv. *paradisiaca* on Giant Governor in West Bengal. Survey conducted in Karnataka during 1999–2000 revealed that the rhizome rot was noticed at Gokak, Raibagh, and Bijapur districts (Anonymous, 2002). It was observed in Jalgaon district of Maharashtra in 2009 (Nagarale, 2009) and is known as rhizome rot or collar rot of banana.

PHOTO 7.37   Symptoms of rhizome rot and tip over of banana. (Courtesy of Department of Agriculture, Government of Kerala, Kerala, India, www.kissankerala.net.)

Symptoms

The symptoms include wilting of newly planted rhizome, yellowing of plants, pockets of dark brown area on pseudostem of infected rhizomes, splitting of the pseudostem, and infected matured plants. Internal symptom of vascular discoloration of the pseudostem (Photo 7.37) is observed.

Rotting of newly planted rhizomes with failure to sprout, stunting and yellowing of young plants, and toppling over of mature fruited plants are the major symptoms noticed. The infected rhizomes show pockets of dark brown or yellow water-soaked areas with dark peripheral rings in abundance in the cortex region in the beginning and then throughout the rhizomes producing decay leading to cavity formation. In advanced stage of infection, splitting of the pseudostem is the prominent symptom. When such infected rhizomes are examined by cross-sectioning, yellow or dark brown discolored patches in the vascular region are prominently observed. The infected young plants become stunted and they could be easily pulled up. The mature plants do not show yellowing and stunting but are easily blown or toppled over, and remnants of rhizomes remain attached with the pseudostem. Due to this symptom, the disease is referred to as tip-over or snap-off disease.

Disease Transmission

Nagrale (2011) reported that the nematode *Radopholus similis* was found to carry the bacterium externally as well as internally and was responsible for the transmission of the pathogen in the field.

Disease Management

Nagrale (2011, 2013) proposed the following management practices for the control of rhizome rot/collar rot of banana.

Chemical Control

Streptocycline (1000 ppm), tetracycline (1000 ppm) streptocycline + copper oxychloride (1000 ppm + 0.1%), and Bordeaux mixture (0.6%) showed 100% control of the disease.

Biological Control

P. fluorescens, *Bacillus thermophilus* (MPKV strain), and *Trichoderma harzianum* were found effective in controlling the disease.

Use of Organic Amendments

Addition of neem seed cakes and castor seed cakes was effective in reducing the pathogens population in soil, thereby helping to control the disease.

Crop Rotation

Crop rotations with groundnut, bitter gourd, pigeon pea, soybean, and chickpea should be followed. These crops are found resistant to *E. chrysanthemi* pv. *paradisiacal*.

Resistance

Different banana cultivars differ in their reaction to collar rot bacterium. The cultivars (cvs) Grandnain, NRCB-03, and NRCB-01 were highly susceptible, while straight finger, Mahalaxmi, Safed velchi, MAS, Basrai, and Udhyam were susceptible. The varieties Lalkel, Red banana, and Sarkar chonya were immune to the disease pathogen.

7.13 POMEGRANATE

7.13.1 BACTERIAL BLIGHT OF POMEGRANATE

Pathogen: *Xanthomonas axonopodis* pv. *punicae*
Bacterial blight is one of the most devastating diseases of pomegranate occurring in major pomegranate-growing states of India. The disease has not been reported from other parts of the world except south Africa. Bacterial blight has been observed damaging the pomegranate crop in moderate to severe proportion, resulting in enormous losses in the states of Maharashtra, Karnataka, and Andhra Pradesh. In Maharashtra, major districts affected by the disease are Solapur, Sangli, Osmanabad, Pune, Nasik, Latur, Aurangabad, Jalna, and Ahmednagar. In Karnataka, Bagalkot, Gadag, Koppal, Bellary, Chitradurga, and Bijapur are main blight-affected districts.

Bacterial blight was first reported in India from Delhi (Hingorani and Mehta, 1952). Subsequently, it was reported from Karnataka (Hingorani and Singh, 1959), Himachal Pradesh (Sochi et al., 1964), Haryana (Kanwar, 1976), and Maharashtra (Dhandar et al., 2004; Sharma et al., 2008). The disease was prevalent only in India until 2010, when it was also reported from South Africa in 2010.

Symptoms

One to several small water-soaked, dark-colored irregular spots on leaves resulting in premature defoliation under severe cases appear. The pathogen also infects stem and branches causing girdling and cracking symptoms. Spots on fruits are dark brown and irregular in shape, slightly raised with oily appearance, which split open with L-shaped cracks under severe cases (Photo 7.38).

Pathogen

Variation in pathogen *X. axonopodis* pv. *punicae* does exist (Raghuwanshi et al., 2013) in Maharashtra. The bacterial pathogen isolated from different locations in Maharashtra state and their ISSR analysis confirmed the uniqueness of all four isolates as they formed separate clusters. Akkalkot–Solapur isolate was the most divergent, while Deola–Nashik and Sangamner–Ahmednagar isolates were most similar. No correlation was observed between the virulence of isolates and banding pattern. Mondal and Mani (2009) reported the relationship between ERICPCR-generated fingerprints with

PHOTO 7.38 Symptoms of bacterial blight of pomegranate on leaves, fruits, and twigs. (Courtesy of Borkar, S.G., Raghuwanshi, K.S., and Pokharel, R., PPAM, MPKV, Rahuri, India.)

pathogenic variability in *X. campestris* pv. *punicae* (i.e., *X. axonopodis* pv. *punicae*). High genetic variability was observed among the strains of *X. axonopodis* pv. *punicae* on RAPD analysis (Giri et al., 2011).

Disease Cycle

Primary source of inoculum is infected cuttings. Secondary source of inoculum is wind-splashed rains. Continuous/intermittent rainfall for a longer period, congenial maximum (29.4°C–35.6°C) and minimum temperature (19.5°C–27.3°C), and relative humidity (63%–87%) were found favorable for the development and spread of the disease.

Bacterial blight of pomegranate-affected leaves, twigs, and fruits are potential sources of primary inoculum. The secondary spread of bacterium is mainly through rain and spray splashes, irrigation water, pruning tools, humans, and insect vectors. Entry is through wounds and natural openings. Bacterial ooze serves as secondary inoculums for spread of the disease in field under favorable conditions. Severity increases during June and July and reaches a maximum in September and October and then declines. Bacterial cells are capable of surviving in soil for >120 days and also survive in fallen leaves during the off-season. High temperatures and low humidity or both favor disease development. Optimal temperature for growth of bacterium is 30°C; thermal death point is about 52°C. The disease cycle of bacterial blight of pomegranate is illustrated in Figure 7.6.

Impact of Disease

Ramesh et al. (1991) reported 60%–80% yield loss due to blight in Karnataka. Bacterial blight of pomegranate has assumed epidemic proportions (40%–85% severity) in northern Karnataka (Bijapur, Bellary, Bagalkot) seriously threatening its cultivation.

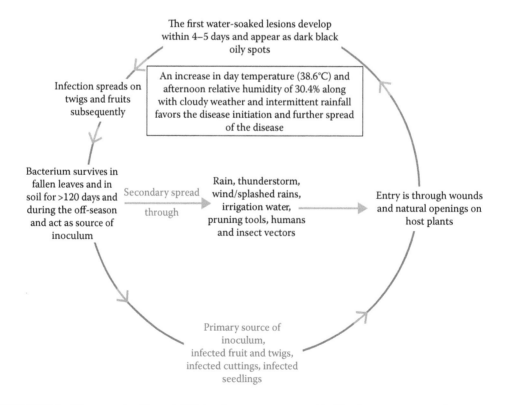

FIGURE 7.6   Disease cycle of bacterial blight of pomegranate caused by *Xanthomonas axonopodis* pv. *punicae*.

Kumar et al. (2006) revealed 20%–90% disease severity in Bijapur and Bagalkot districts. Similarly 71.14% severity was reported in Bellary district (Yenjerappa et al., 2004). The survey of 82 pomegranate orchards in Maharashtra revealed that bacterial blight was observed up to 100% severity in some orchards (Anonymous, 2007). Bacterial blight of pomegranate, of late, has become a major constraint in important pomegranate-producing states of Maharashtra, Karnataka, and Andhra Pradesh of India. In view of enormous losses that may extend up to 60%–80% in unmanaged orchards under epidemic conditions, surveys of important pomegranate-growing states conducted during 2006–2008 revealed disease prevalence in different districts of Maharashtra— Solapur, Sangli, Pune, Nashik, Osmanabad, Aurangabad, Latur, and Jalna—in mild to severe form. Ahmednagar district, which was free from blight until 2007, also revealed severe blight infections in a few orchards of Sangamner taluka during July 2008. Satara and Dhule districts of the state, however, were found free from bacterial blight. In Andhra Pradesh, disease was prevalent in mild to severe form in most pomegranate-growing areas of Ananthpur district, whereas in Karnataka, disease was prevalent in mild to moderate form in Bijapur, Gadag, Koppal, and Bagalkot districts. All pomegranate cultivars grown in the region—Bhagwa, Ganesh, Mridula, and Arakta—were susceptible to the blight pathogen.

Epidemiological Parameter

Studies on disease epidemiology at Solapur revealed that though disease existed throughout the year at a temperature range of 9.0°C–43.0°C and varying humidity (30.0% to >80.0%), disease development was rapid during the rainy months from July to September due to free water and high humidity. The Mrig bahar crop regulated in June–July (summer rainy season) revealed more disease pressure due to higher infection rate (0.2/unit/day) compared to the Hasta bahar crop regulated in September–October (autumn season), which had an infection rate of 0.08/unit/day, thereby indicating the brisk spread of the disease in the rainy season. Studies on pathogen's survivability revealed that *X. axonopodis* pv. *punicae* was able to survive in naturally infected leaves collected from the diseased orchard for 1 year under laboratory conditions (at 25.0°C–38.0°C).

Control Measures

Kumar et al. (2006) evaluated sprays with streptocycline (500 ppm) + copper oxychloride (2000 ppm) (33.3%) and compared these with control (78.5%) after eighth spray. A maximum mean yield of 9.3 tons/ha was recorded in streptocycline (500 ppm) + copper oxychloride (2000 ppm) followed by 8.50 tons/ha in bromopal (500 ppm) + copper oxychloride (2000 ppm), while the untreated check yielded 2.95 tons/ha. Manjula et al. (2003) reported that paushamycin (500 ppm), streptocycline (500 ppm), and K-cycline (500 ppm) were very effective in controlling the bacterial blight of pomegranate.

On the Maharashtra isolates of the pathogen, six chemical treatments showed complete control under *in vitro* conditions, while the rest varied in their response to each isolate. Streptocycline 250 ppm showed least effectiveness against isolates from Nashik, Sangamner, and Akkalkot regions and was effective against Pandharpur isolates, while Captan was little less effective against isolate from Pandharpur region but it was very effective against the remaining three isolates. The study indicated that selection of fungicide and antibiotics for mitigating the disease is very important because the efficacy of chemicals varies from location to location. Besides streptocycline, other chemicals like Bordeaux mixture, Captan, and bromopol were observed effective against the bacterium alone or in combination with *Copper hydroxide* or copper oxychloride. Alternative use of these chemicals will help to avoid the development of resistance in bacterium. Pesticide resistance is developed at a particular location and hence it is suggested to use alternative pesticides. Earlier, Ravikumar et al. (2011) reported that spray of streptocycline 500 ppm + copper oxychloride 200 ppm was effective against the bacterial blight of pomegranate.

Bacterial blight is becoming very severe, which is responsible for killing most of the pomegranate trees in the country. It is an airborne bacterium and also spreads through the use of planting material. It is difficult to root out the disease unless and until mass eradication measures are taken up throughout the country with the help of department, university, and growers. Pomegranate crop should be taken only 3–4 years after planting; pruning should be done in August–September in order to get lesser severity and with less number of sprays.

A field trial for the management of bacterial blight was conducted in an orchard having 6-year-old Bhagwa during 2007–2008. Adoption of an integrated disease management schedule that included practices like orchard sanitation, avoidance of rainy season crop (Mrig bahar), and regulation of the Hasta bahar crop, and judicious sprays of antibiotic streptocycline (500 ppm) in combination with fungicides like carbendazim (0.15%)/mancozeb (0.2%)/copper oxychloride (0.25%) at 15-day interval resulted in 82.2% disease control and gave yield of 16 tons/ha. In separate trials, antibiotic Bactronol (2-bromo, 2-nitropropane, 1–3 diol) at 750–1000 ppm concentrations was also found effective in managing bacterial blight.

The pathogen in Nashik district has developed resistance to antibiotics like streptocycline and streptomycin sulfate. A package of practices developed by Borkar and Raghuwanshi of Mahatma Phule Agricultural University, known as MPKV protocol, is followed in the state with cent percent disease control.

MPKV Protocol

MPKV protocol for eradicating and combating oily spot bacterium in pomegranate

1. Spray Bacti-Nashak (2-bromo-2-nitrol propen-1,3,diol) (500 ppm) at 5 g/10 L of water after total harvest of earlier bahar fruits.
2. A resting period of 3 months is must. During resting period, the infected leaves or plant should be sprayed at 10-day interval. If no infection is observed, then spray at 30-day interval.
3. Defoliation with ethereal at 15 mL/10 L of water.
4. Pruning of infected twigs, and collection and burning of infected twigs, fallen leaves, and fruits.
5. Spraying of defoliated plant with Bacti-Nashak (500 ppm) at 5 g + Captan (0.5%) at 50 g/10 L of water.
6. Pasting of trunk with neem oil followed by application of Bacti-Nashak (500 ppm) at 5 g + Captan (0.5%) at 50 g/10 L of water on trunks.
7. Dusting of bleaching powder at 60 kg/ha on the soil surface and plant basin.
8. Spraying of new growth with 250 ppm Bacti-Nashak at 2.5 g/10 L of water followed by Bordeaux mixture 1% at 100 g copper sulfate + 100 g lime/10 L and Captan 0.25% at 25 g/10 L of water at 10-day interval.
9. Note 1: This spray sequence should be continued at an interval of 10 days during rains, high humidity, and cloudiness. If these conditions do not prevail, then spray at 20-day interval, and if no infestation is observed, then spray at 30-day interval. Use sticker at 0.1% during rainy season.

While management practices at other places include the following.

1. New orchards should be established with certified disease-free planting material or tissue culture–raised saplings.
2. Spray Bordeaux mixture (1.0%) during dormancy.
3. Spray with streptocycline (0.025%) in combination with copper oxychloride (0.25%) or carbendazim (0.15%) at 15-day interval for five to six times starting from leaf initiation stage during rainy season and postrainy season.
4. If possible, cut ends should be pasted with Bordeaux (10%) paste after every pruning.
5. Follow orchard sanitation measures strictly. Fallen twigs, leaves, and fruits should be destroyed outside the orchard premises.

6. Apply copper formulations + streptocycline or carbendazim + streptocycline 0.05% and other bactericides if disease pressure is high and weather conditions are favorable.

7. In bacterial blight–prone areas only Hasta bahar or late Hasta bahar crop is recommended.

7.14 PINEAPPLE

7.14.1 BACTERIAL HEART ROT OF PINEAPPLE

Pathogen: *Erwinia chrysanthemi*

The first reported outbreak of bacterial heart rot of pineapple (*Ananas comosus* var. *comosus*) in Hawaii occurred in December 2003. Of immediate concern was the differentiation of heart rot caused by *E. chrysanthemi* from a soft rot caused by *E. carotovora* subsp. *carotovora* because of regulatory issues. Presumptive identifications of the isolated bacteria were made using bacteriological tests (including reactivity with an *Erwinia*-specific monoclonal antibody, E2) and by comparisons with identifications obtained by two general methods: carbon source utilization profiling (Biolog) and 16S rDNA sequence analysis. The panel of bacteriological tests consistently differentiated *E. chrysanthemi* from *E. carotovora* subsp. *carotovora* and other nonquarantine organisms. BOX-PCR fingerprint patterns further differentiated the pineapple-isolated *E. chrysanthemi* strains from those obtained from other plants and irrigation water. Pineapple leaf inoculations revealed that only *E. chrysanthemi* from pineapple produced water-soaking and rot similar to that observed on the original symptomatic plants, thus identifying these strains as the causal agents of the outbreak.

Symptoms

The following symptoms appear: water-soaked lesions on the white basal sections of leaves in the central whorl, which may spread to all leaves in the central whorl; midportions of leaves become olive green in color with a bloated appearance; infected fruits exude juices and the shell become olive green; cavities form within the fruit.

Disease is thought to be spread from the juices of infected fruits. Bacteria in the juice can enter leaves through wounds, and ants act as vectors for the bacteria.

Pathogen

The bacterial pathogen from Hawaii, originally called *E. chrysanthemi*, was thoroughly characterized by biochemical tests, molecular assays, and DDH data and now is classified as *Dickeya zeae*. Pineapple strains are unique, however, and they were placed into a separate subgroup from *Di. zeae* strains previously found in water, soil, and ornamentals. Hawaiian pineapple strains further formed two clearly separate groups, each with distinct DNA fingerprints and different reaction patterns with *Dickeya*-specific monoclonal antibodies. Strains associated with pattern A were initially less virulent than strains associated with patterns B and C, and the A strains were not encountered in later surveys. The A, B, and C types were originally associated only with the planting stocks received from Central America. In subsequent years, C types were also found in drainage ditches bordering affected fields and in a reservoir used for irrigation. New strains (D and E types) later appeared in pineapple fields and irrigation water.

Di. zeae was identified in an outbreak of bacterial stalk rot of corn (maize) on Oahu, and the corn strains were genetically similar to the C-type pineapple strains. Two pineapple strains from Malaysia and four strains from the Philippines were similar to C-type strains found in Hawaii. For the first time, it is now possible to identify the pineapple pathogen with a few genetic markers and to determine its closest relatives in a worldwide collection of *Dickeya* strains. The specific monoclonal antibody will distinguish the pineapple pathogen from contaminants and other *Dickeya* species.

Impact

New knowledge generated about disease spread from infected sites in Hawaii has impacted plant quarantine regulations as well as industry practices. No pineapple heart rot disease was found in Maui following four extensive surveys of entire plantations. As a result, commercial plantations on

Maui were declared disease-free. Movement of planting stocks from Oahu to Maui was prohibited and the disease outbreak was contained on Oahu. The results eventually had impact on the pineapple industries in the Philippines where the disease is prevalent, and measures to control the disease are enhanced by the ability to rapidly identify and destroy the major sources of infected materials. The discovery that *Dickeya* strains genetically similar to the pineapple pathogen were found on ornamentals, corn, and taro on Oahu could have major impact on taro distribution to other islands and could impact the future expansion of the corn seed industry in Hawaii. The impact of pineapple heart rot disease on local pineapple production was immense because of the regulatory issues that followed the first discovery of the disease in Hawaii.

Disease Management
Remove and destroy infected fruits; avoid the use of infected crowns for seed material to prevent the spread of the disease; planting to avoid flowering when adjacent field is fruiting can reduce disease development; use of miticides and control of ants can significantly reduce disease incidence.

7.15   GUAVA

7.15.1   BACTERIOSIS OF GUAVA

Pathogen: *Erwinia psidii*

Symptoms
The bacterium affects mainly branches and twigs of guava trees, producing a collapse of vascular tissues and dieback. Leaves, blossoms, and green fruits are also affected.

Infection of guava twig and principal vein of leaf spreads through the blade in the form of irregular lesions.

Host
Guava (*Psidium guajava*)

Pathogen
The species *E. psidii* was described in Brazil, in 1987, causing bacteriosis on guava.

Control Measures
Recommended control methods include pathogen-free propagating material and cultural practices such as pruning infected branches and providing protection against wind and trickle irrigation.

7.16   PAPAYA

7.16.1   BACTERIAL CANKER AND DECLINE OF PAPAYA

Pathogen: *Erwinia* spp.
Papaya (*Carica papaya*) is an economically important tropical cash crop for export, as well as a significant vegetable crop in small farming systems worldwide. Early in the twentieth century, a disease caused by *Bacillus papaya* in Java (Von Rant, 1931) was identified; this bacterium was further typed to the genus *Erwinia* by Magrou (1937). Nelson and Alvarez (1980) described a purple stain of *Car. papaya*, due to *Erwinia herbicola*. More recently, Trujillo and Schroth (1982) reported a bacterial decline of papaya caused by two *Erwinia* species in the Mariana Islands. Finally, reports from Webb (1985) in the U.S. Virgin Islands, from Frossard et al. (1985) and Prior et al. (1985) in the French West Indies, and from Guevara et al. (1993) in Venezuela, all consistently described this disease, which was named "bacterial stem canker."

Symptoms
Typical symptoms of bacterial stem cankers on Solo-type cultivars include greasy, water-soaked lesions and spots on leaves, which evolve into foliar, angular lesions. Firm, water-soaked cankers

PHOTO 7.39   Water-soaked areas on the base of leaf stalks and crown of papaya infected with *E. papayae*. (Courtesy of Maktar, N.H., Kamis, S., Mohd Yusof, F.Z., and Hussain, N.H., New Disease Reports, www. ndrs.org.uk.)

develop on the stem (Photo 7.39) and fruits (Photo 7.40) and sometimes lead to the destruction of papaya trees (Photo 7.41) (Prior et al., 1985; Webb, 1985).

Preliminary identification of this novel bacterium by classical phenotypic tests led Trujillo and Schroth (1982), Frossard et al. (1985), Prior et al. (1985), and Webb (1985) to place this bacterium in the genus *Erwinia* and in the *Amylovora* group according to Dye (1968). An *Erwinia* sp. was the cause of blackish, water-soaked, mushy cankers that occurred near or in leaf axils of upper portions of papaya stems in the northern Mariana Islands. This disease was prevalent on wild and commercial papaya trees after near-typhoon storms. The pathogen gained entrance through wounds caused by wind damage to the foliage. A different *Erwinia* sp. was the cause of a systemic blight of papaya resulting in rapid decline of commercial orchards. Symptoms of this disease are dark, angular, greasy, water-soaked lesions on the underside of the leaf causing chlorosis and necrosis of the foliage. Water-soaked lesions on petioles and upper portions of the stems are followed by systemic invasion and rot of the tree top and finally tree death in <6 weeks.

The African snail *Achatina fulica*, a vector of the decline of *Erwinia* sp., spread it from diseased to healthy trees in the feeding process. The declined *Erwinia* sp. was recovered from the fresh excreta of snails collected from diseased trees. High resistance to the pathogen was observed in wild papaya trees of the northern Marianas. Commercial varieties tested showed high to low susceptibility. "Kapoho Solo," one of the most important commercial lines in Hawaii, was highly susceptible. The two *Erwinia* pathogens do not correspond with any of the defined *Erwinia* species and subspecies. The diseases are referred to as the *Erwinia* mushy canker disease of papaya and the *Erwinia* decline disease of papaya.

The term "St. Croix papaya decline" applies to a variety of factors, acting alone or in concert, that affect the normal development of papaya. In St. Croix, papayas respond to a number of adverse conditions by premature abscission of leaves and cessation of apical growth, which result in a symptom called "pencil point." This condition has been observed in trees suffering from viral infections, root rot caused by *Pythium butleri*, bacterial canker, water stress, and nutrient

PHOTO 7.40   Fruits infected by *E. papayae*—flesh with water-soaked areas. (Courtesy of Maktar, N.H., Kamis, S., Mohd Yusof, F.Z., and Hussain, N.H., New Disease Reports, www.ndrs.org.uk.)

PHOTO 7.41   Advanced stage of papaya dieback caused by *E. papayae*. Bending of water-soaked leaf stalks giving a skirting effect' leading to dieback and death of trees. (Courtesy of Maktar, N.H., Kamis, S., Mohd Yusof, F.Z., and Hussain, N.H., New Disease Reports, www.ndrs.org.uk.)

deficiencies caused by low nitrogen levels and/or high soil pH. Greasy spots on papaya stems caused by viral infections are often misdiagnosed by growers as bacterial cankers. Leaf spots caused by *Corynespora cassicola* and the canker bacterium are very different but may be found on the same leaf and are frequently misdiagnosed. *C. cassicola* is also a common secondary invader of cankers caused by the bacterium.

Pathogen

Bacterial canker of papaya is caused by an *Erwinia* sp. similar to the bacterium causing the Mariana and Java papaya diseases. The St. Croix isolates, however, possess a combination of physiological and biochemical characteristics that distinguish them from the Mariana isolates as well as other *Erwinia* groupings. Most notable is the absence of flagellae that are similar to those in *Erwinia stewartii*; this makes this bacterium unique in the genus. As mentioned by Trujillo and Schroth (1982), classification of these papaya pathogens by the pathovar system is impractical because it does not take into consideration bacteria that do not fit the characteristics of presently defined species. Species ranking of the St. Croix pathogen will be made by DNA homology matrix typing (Schroth and Hildebrand, 1983) after further study.

Gram-negative bacteria isolated from water-soaked lesions on papaya were typically slow-growing, giving colonies of 2–3 mm diameter after 3–4 days of incubation. Colonies were hyaline on YBGA (0.7% yeast extract, 0.7% bactopeptone, 0.7% glucose, and 1.5% agar; pH 7.2) or on King's medium B at 25°C, with a mucoid and white-to-creamy white color after 2–3 days. A non-diffusible blue pigment was observed on King's medium B.

Bacterial canker of papaya (*Car. papaya*) emerged during the 1980s in different islands of the Caribbean. Nineteen strains of Gram-negative, rod-shaped, nonspore-forming bacteria isolated from papaya were compared to 38 reference and type strains of phytopathogenic Enterobacteriaceae and related bacteria. Phylogenetic analysis of 16S rRNA gene sequences showed that the papaya strains belonged to the genus *Erwinia*. The DNA G+C content of strain CFBP 5189^T, 52.5 mol%, is in the range of the genus *Erwinia*. The 19 papaya strains were all pathogenic to papaya and were differentiated clearly from type or reference strains of phytopathogenic Enterobacteria and related bacteria by phenotypic tests. The papaya strains constituted a discrete DNA hybridization group, indicating that they belonged to a unique genomic species. Thus, strains pathogenic to papaya belong to a novel species for which the name *Erwinia papaya* sp. nov. is proposed, with the type strain CFBP 5189^T(=NCPPB 4294^T).

Further, *Erwinia mallotivora* was isolated from papaya infected with dieback disease showing the typical symptoms of greasy, water-soaked lesions and spots on leaves. Phylogenetic analysis of 16S rRNA gene sequences showed that the strain belonged to the genus *Erwinia* and was united in a monophyletic group with *E. mallotivora* DSM 4565 (AJ233414). Earlier studies had indicated that the causal agent for this disease was *E. papayae*. However, current studies, through Koch's postulate, have confirmed that papaya dieback disease is caused by *E. mallotivora*. This is the first new report that emphasized *E. mallotivora* as a causal agent of papaya dieback disease in Peninsular Malaysia. Previous reports have suggested that *E. mallotivora* causes leaf spot in *Mallotus japonicus*. However, present research confirms it to be pathogenic also to *Car. papaya*. This study included a selection of 19 strains from a collection of 72 strains that were isolated in 1982–1991 from *Car. papaya* in different islands of the Caribbean and 38 type or reference strains of phytopathogenic Enterobacteriaceae that belong to the genera *Erwinia*, *Enterobacter*, *Brenneria*, *Pantoea*, *Pectobacterium*, and *Samsonia*.

Predisposing Factors

Observations of losses occurring from disease outbreaks show that this disease is most destructive on St. Croix during the short rainy seasons. Symptoms, however, may be observed throughout the entire year. Although rainfall disseminates the pathogen, free moisture following deposition to a susceptible site on the host does not increase pathogen survival or symptom severity. In fact, the severity of symptoms appears to be decreased by 72 h of free moisture. These findings indicate that the pathogen is well adapted to the semiarid climate of St. Croix and may provide an answer, in part, why this disease is not a significant problem on surrounding, wetter islands.

Under St. Croix conditions, the pathogen did not survive well in association with papaya roots or in decaying diseased plant material, indicating that it is a transient soil inhabitant (Buddenhagen, 1970). The bacterium does survive for indefinite periods in the cankers and leaf infections of affected papaya trees and on the leaves of tomato and cantaloupe. Unlike the "Mariana decline" pathogen, the St. Croix bacterium has not been associated with the African snail (*Ac. fulica*) (Trujillo and Schroth, 1982) or insect vectors.

Control Measures

Attempts to control this disease with bactericides, antibiotics, and the locally isolated antagonistic pseudomonad described were not effective. Current trials using a tank mix of copper hydroxide and mancozeb (Dithane M-45) (Conover and Gerhold, 1981) also have shown this treatment inadequate for controlling this disease.

At present, the most effective control strategy for the canker disease is the use of resistant cultivars. The most promising to date is Barbados Dwarf, which in field trials is often the only cultivar that remains standing after the first year of production.

Another control strategy under investigation is the use of suitable "barrier crops" that do not support epiphytic populations of the pathogen, that is, cassava, banana, and pigeon pea. Observations of small local farms where papayas are commonly intercropped with a wide variety of fruit and vegetable crops have shown a lower incidence of the canker disease than that found in monocultures.

7.16.2 INTERNAL YELLOWING OF PAPAYA

Pathogen: *Enterobacter cloacae*

Internal yellowing of Hawaii-grown papayas was first observed in 1984 in fruits that were treated with a two-stage, hot-water immersion that was developed to disinfest papaya of fruit fly eggs and larvae.

Symptoms

Internal yellowing disease is characterized by yellow discolored tissue with diffuse margins around the seed cavity of infected fruit. Most infections are present around the blossom end and the middle of the fruit and often included a portion of the seed cavity. A distinctly rotten odor is consistently present. No external symptoms are observed in naturally or artificially infected fruit.

Damage

Internal yellowing affects the internal flesh of ripening fruit without displaying any external symptoms. The disease is characterized by soft, yellow-discolored flesh with diffuse, spreading margins and an offensive, rotting odor in ripening papaya fruit. Infected tissue, which are observed only when fruit are cut open, occur in areas around and including the portions of the seed cavity, usually near the calyx and middle sections of the fruit. Occasionally, vascular tissue near the stem end becomes infected and appears yellow to yellowish tan in color.

Fruit ripeness is a factor in the incidence of internal yellowing with ripe and overripe fruit having the highest incidences of infection. Symptom expression also varies with fruit ripeness. In mature but not fully ripe fruit, infected tissues appear as "pockets" of yellow discolored flesh, while in ripe fruit, infected areas do not have distinct margins and appear "diffuse."

During the period 1984–1986, disease incidence in Hawaii of the hot-water-treated fruit was as high as 9.7%. Since then, recent surveys indicate that this level is often surpassed, depending on the time of year and the packinghouse.

Pathogen

En. cloacae is a Gram-negative, rod-shaped bacterium that has peritrichous flagella, measures $0.3–0.6 \times 0.8–2.0$ μm, is oxidase-negative, catalase-positive, and is facultatively anaerobic. The bacterium is positive for beta-galactosidase, arginine dihydrolase, ornithine decarboxylase, citrate utilization, nitrate reduction, and Voges-Proskauer reaction. Production of lysine decarboxylase, hydrogen sulfide, urease, tryptophan deaminase, and indole is all negative. Acid is produced from many carbon sources. Reactions to many of these biochemical tests may be obtained using API 20E strips (Analytab Products). In addition, *E. cloacae* is negative for phenylalanine deaminase and pectate degradation.

En. cloacae grows well on standard bacteriological media on which yellow pigment and purple stain are not produced. Colonies are creamy tan on yeast extract–dextrose–calcium carbonate agar, and dark pink to burgundy with translucent margins on tetrazolium chloride agar. Miller-Schroth medium is useful for initial screenings because of the positive reaction (orange colonies) of *E. cloacae* on this medium.

The bacterium isolated from diseased papaya fruits and demonstrated to cause the internal yellowing disease was identified as *En. cloacae* on the basis of biochemical and physiological tests and by comparison with the type strain of *E. cloacae*, ATCC 13047. This pathogen differed from two strains (PP-1 and PP-2) of *E. herbicola* isolated from fruits with the purple-stain disease. Strains of *E. cloacae*, but not *E. herbicola*, were recovered from hot-water treatment tanks, papaya flowers, and the oriental fruit fly. The internal yellowing disease was found to be more prevalent and widespread than the purple stain disease. Purple stain disease, however, was observed occasionally during surveys and strains isolated from diseased fruit conformed to physiological and biochemical characteristics of *E. herbicola*, confirming a previous report (Nelson and Alvarez, 1980).

There are five other reports of bacterial diseases of papaya caused by *Erwinia* spp. (Von Rant, 1931; Leu et al., 1980; Trujillo and Schroth, 1982; Frossard et al., 1985; Webb, 1985). *B. papaya* was reported to cause a disease of papaya in Java (Von Rant, 1931), but Magrou (1937) placed this bacterium in the genus *Erwinia*. Leu et al. (1980) reported that *Erwinia cypripedii* caused a black rot of papaya trees that also affected fruit in Taiwan. Trujillo and Schroth (1982) reported two bacterial diseases of papaya trees caused by *Erwinia* spp. Webb (1985) reported that an *Erwinia* sp. caused a bacterial canker of papaya trees in St. Croix, U.S. Virgin Islands, whereas Frossard et al. (1985) reported that an *Erwinia* sp., probably in the group *E. amylovora*, caused a decline in papaya trees in the French Antilles.

The isolation of plant pathogenic strains of *E. cloacae* is rare but not unique. *Erwinia dissolvens* (originally *Pseudomonas dissolvens*) and *Erwinia nimipressuralis*, both placed in *E. cloacae* (Richard, 1984), were reported to cause stalk rot of corn (Rosen, 1922) and wetwood of elm trees (Carter, 1945), respectively.

Disease Cycle

En. cloacae has been isolated from papaya flowers, homogenates of papaya seeds, and the crop and midgut of the oriental fruit fly (*Dacus dorsalis* Hendel), and recent studies claiming an apparent attractive nature of *Dacus dorsalis* to *E. cloacae* suggest that fruit flies may possibly be involved in the transmission of the bacterium to papaya. It is suspected that after *E. cloacae* is transmitted to papaya flowers by fruit flies or other insects, the pathogen remains quiescent during fruit development until symptoms are expressed when the fruit are fully ripe.

Although *E. cloacae* was recovered from hot-water treatment tanks, it is believed that these tanks are not a major source of infection in processed papaya fruit because of the relatively rare occurrence of internal yellowing symptoms in the vascular tissue near the stem-end of the fruit.

A report on *E. cloacae* isolated from homogenates of papaya seeds in 1972 suggests that this organism may have been present in a nonpathogenic form for many years. Monthly samplings from five papaya packinghouses, which process fruit from different areas on the island of Hawaii, indicate that the incidence of internal yellowing is sporadic and may be affected by environmental factors.

The susceptibility of papaya fruits to internal yellowing increased as the fruits ripened. When fruits at different stages of ripeness were artificially inoculated with strain PV-5, the ripest fruits inoculated were most susceptible to disease. Similarly, when fruits were inoculated at color-break stage and checked for internal yellowing at different stages of ripeness, disease incidence increased with fruit ripeness.

The incidence of internal yellowing was lower among hot-water-treated fruits than among untreated fruits. Hot water treatment of papaya has been used to control postharvest fungal diseases of papaya (Akamine and Arisumi, 1953), and its apparent activity against a bacterial disease of papaya enhances the usefulness of this method of control.

Based on the observation of internal yellowing symptoms occurring primarily at the blossom end and middle of the fruit, and the recovery of *E. cloacae* from papaya flowers, it is assumed that the flower may have been the site of infection. The pink disease of pineapple (Rohrbach and Pfeiffer, 1975) is another bacterial disease in which the flower is the site of entry for the pathogen. Because *E. cloacae* also was recovered from fruit files, it is suspected that fruit files, and possibly other insects, may aid in the dispersal of the pathogen. If the pathogen enters the fruit through the flowers, it remains dormant during flowering to fruit development. Physiological changes that accompany fruit ripening then permit full development of the bacterial disease.

Disease Management
Control of *E. cloacae* is currently limited to postharvest hot-water quarantine treatments that effectively reduce the incidence of internal yellowing in papaya fruit. Without the hot-water treatment, incidence of internal yellowing may be as high as 43%, depending on the packing house and the time of year.

The hot-water treatments may be inducing resistance in the fruit to the bacterial pathogen, but the biochemical and physiological mechanisms involved are not known.

7.16.3 PURPLE STAIN OF PAPAYA

Pathogen: *Erwinia herbicola*

Symptoms
The diagnostic characteristic of the disease is a reddish-purple coloration of the entire vascular and latex tissue. Immediately after fruits are cut open, the flesh is speckled bright purple, but the pigment faded after several hours, leaving a greenish-yellow color in vessels, ducts, and surrounding parenchymal tissue, uncharacteristic of papaya fruit. The tissues became soft and translucent and rotted as the fruit ripened. Microscopic examination of the diseased vascular tissue and latex ducts showed bacteria but no fungi.

Pathogen
One predominant colony type was consistently isolated from fruits with purple stain symptoms. Yellow brown mucoid colonies formed on Yeast Extract Dextrose Calcium Carbonate Agar (YDC), and white mucoid colonies with red centers formed on Triphenyltetrazolium Chloride (TZC) medium. The purple pigment appeared 7 h after the cultures were streaked on TZC or YDC.

The pathogen is a Gram-negative, single rod, 0.6–1 μm × 1–2 μm (average, 0.7 μm × 1.5 μm) in diameter with peritrichous flagella, and has the following characteristics of the genus *Erwinia*: facultative anaerobe, negative for oxidase and indole production, and positive for catalase and acetoin production. Acid was produced from arabinose, ribose, glucose, rhamnose, sucrose, maltose, mannitol, inositol, α-methyl glucoside, and dextrin but not from xylose, raffinose, lactose, melibiose, cellobiose, or dulcitol. Gas was not produced from glucose, sucrose, or rhamnose.

All strains of *Erwinia* that produced purple pigment in culture also caused a purple coloration of the vascular tissue in papaya fruit. Fruit inoculated in the laboratory developed purple coloration in the vascular tissue of the stem end and seed cavity, but pigmentation throughout the latex ducts was less striking than that observed in naturally infected fruits. Nonpigment-producing strains of *E. herbicola* isolated from apple and papaya produced no symptoms in papaya fruit; neither did other species of *Erwinia*. The papaya pathogen did not cause symptoms on papaya seedlings and did not affect the other inoculated plants.

Pigment production was greatest when bacteria were grown at 25°C or 31°C in liquid shake culture containing 0.5% sucrose and 1% peptone and buffered at pH 5.5 but the pigment was unstable, and the liquid shake culture solution was colorless after 24 h. Neither growth nor pigment production occurred at 47°C. Only slight growth and no pigment production occurred at 10°C. Bacteria grew well at 37°C but did not produce the pigment; However when cultures were restreaked and incubated at 25°C or 31°C, pigment was produced. The absorbance peaks of the pigment in water, 95% ethanol, and pyridine were 595, 575, and 580 nm, respectively.

The unusual aspect of this disease is its causal agent, which has bacteriological characteristics of *E. herbicola*, a species not normally considered a plant pathogen. Unlike *E. herbicola*, the papaya pathogen produces a purple to dark-blue pigment within papaya fruits and in culture. However, pigment production alone would not warrant the exclusion of the papaya strains from the *E. herbicola* spp. *E. chrysanthemi* is the only *Erwinia* reported to produce a dark-blue, water-insoluble pigment, indigoidine (Starr et al., 1966). The pigment produced by the papaya pathogen differs from indigoidine in its solubility properties and absorbance spectra in pyridine (Kuhn et al., 1965). The difference in color between the two pigments is readily recognized when *E. chrysanthemi* and the papaya strains are grown on YDC.

Although purple stain disease occurs sporadically, it is potentially important during winter months. Because the disease causes no external symptoms, the incidence is particularly difficult to assess, and infected fruits could easily reach the consumer. Fruits with off-colored greenish-yellow flesh, in the market before fully ripe, may reflect the occurrence of the disease, even though the purple pigment is not detected.

Control Measures
At present no control measures are known.

7.17 CUSTARD APPLE

7.17.1 BACTERIAL WILT OF CUSTARD APPLE

Pathogen: *Ralstonia solanacearum*

Common name: crown rot in custard apple

Symptoms
Most leaves exhibit pale or yellow color. The disease is diagnosed by examining the trunk at ground level for discoloration of wood under the bark. The bark around the crown at or just below the ground level decays. A slice of bark removed from above the affected area will show a dark discoloration of the water-conducting tissue. The disease often occurs on trees that have just started fruiting. Wilting is most common in late summer.

Young trees may rapidly wilt and decline, often with severe defoliation. Leaves that stay on the tree are dull green and hang almost vertically. In older trees, a slow decline occurs over about 2 years, generally with little or no yellowing of the leaves. Affected trees have a dark discoloration of the water-conducting tissues in the basal trunk and large roots.

Host
Potato, tomato, eggfruit, capsicum, and custard apple.

Pathogen
P. solanacearum biovar 3 was identified as the causal agent of two disease syndromes in custard apples (*Annona* spp.), namely, sudden death of young trees and the decline of mature trees. Cross inoculation tests using cultures of *P. solanacearum* from custard apple and tomato on *Annona squamosa*, *Annona squamosa* × *Annona cherimola* hybrids, *Capsicum annuum*, three cultivars of *Lycopersicon esculentum*, and *Solanum tuberosum* failed to distinguish differences in virulence

among these isolates. Bacterial wilt resistant rootstock clones, particularly of *Annona cherimola*, constitute the most effective control measure available.

Disease Spread

The bacterium is common in soil and is carried over in crop residues and weed hosts. It spreads in irrigation and rain water, particularly downhill, and may spread by root contact.

Control Measures

No treatment is available. Mulching and reducing crop load may help to prolong the life of affected trees. Avoid planting in areas that have grown tomatoes, potatoes, eggfruit, or capsicums within the past 2 years and had the bacterial wilt problem. Do not plant in poorly drained sites, and improve drainage by mounding. Use cherimoya rootstocks.

7.18 CITRUS

7.18.1 CITRUS CANKER

Pathogen: *Xanthomonas axonopodis* pv. *citri* (Hasse) Vauterin et al. (1995)

Synonyms: *X. campestris* pv. *citri* (Hasse) Dye (1978)
           Pseudomonas citri Hasse
           Xanthomonas citri (Hasse) Dowson
           Xanthomonas citri f. sp. *aurantifolia* Namekata and Oliveira
           Xanthomonas citri (ex Hasse) nom. rev. Gabriel et al., *X. campestris* pv. *aurantifolii* Gabriel et al.

Common names: Asiatic canker (A strains), bacterial canker of citrus, cancrosis A (A strains), cancrosis B (B strains), citrus bacteriosis (D strains), false canker (B strains), Mexican lime cancrosis (C strains), South American canker (B strains)

Symptoms

Citrus canker can be a serious disease where rainfall and warm temperatures are frequent during periods of shoot emergence and early fruit development. This is especially the case where tropical storms are prevalent. Citrus canker is mostly a leaf-spotting and fruit rind-blemishing disease (Photo 7.42), but when conditions are highly favorable for infection, infections cause defoliation, shoot dieback, and fruit drop. The disease symptoms appear as cankerous leaf lesions, stem bark canker, and cankerous lesions on fruits.

PHOTO 7.42 Symptoms of bacterial infection on citrus leaves, twigs, and fruits. (Courtesy of Borkar, S.G. and Yumlembam, R.A., PPAM, MPKV, Rahuri, India.)

Leaf Lesions

Citrus canker lesions start as pinpoint spots and attain a maximum size of 2–10 mm in diameter. The eventual size of the lesions depends mainly on the age of the host tissue at the time of infection and on the citrus cultivar. Lesions become visible about 7–10 days after infection on the underside of leaves and soon thereafter on the upper surface. The young lesions are raised or "pustular" on both surfaces of the leaf, but particularly on the lower leaf surface. The pustules eventually become corky and crateriform with a raised margin and sunken center. A characteristic symptom of the disease on leaves is the yellow halo that surrounds the lesions. A more reliable diagnostic symptom of citrus canker is the water-soaked margin that develops around the necrotic tissue, which is easily detected with transmitted light.

Fruit and Stem Lesions

Citrus canker lesions on fruit and stems extend to 1 mm in depth and are superficially similar to those on leaves. On fruit, the lesions can vary in size because the rind is susceptible for a longer time than for leaves, and more than one infection cycle can occur. Infection of fruit may cause premature fruit drop, but the fruits that remain on the tree until maturity have reduced fresh fruit marketability. Usually the internal quality of fruit is not affected, but occasionally individual lesions penetrate the rind deeply enough to expose the interior of the fruit to secondary infection by decay organisms. On stems, lesions can remain viable for several seasons. Thus, stem lesions can support long-term survival of the bacteria.

Leafminer Interaction

The Asian leafminer (*Phyllocnistis citrella*) can infest leaves, stems, and fruit, and greatly increase the number of individual lesions that quickly coalesce and form large irregular-shaped lesions that follow the outlines of the feeding galleries. Leafminers feed on the epidermis just below the leaf cuticle. Numerous cracks occur in the cuticle covering leafminer galleries, providing means for bacteria to penetrate directly into the palisade parenchyma and spongy mesophyll that are highly susceptible to infection. Citrus foliar wounds normally callus within 1–2 days; however, the extensive wounds composed of the entire leafminer feeding galleries do not callus for 10–12 days, greatly extending the period of susceptibility of galleries to infection. Leafminer infestations can be very prevalent and severe, producing hundreds, if not thousands, of potential infection courts on individual trees. When bacterial dispersal events occur in the presence of the leafminer, not only is inoculum production greatly exacerbated but so is the potential for infection over the entire dispersal range.

Host

Hosts include numerous species, cultivars, and hybrids of citrus and citrus relatives, including orange, grapefruit, pummelo, mandarin, lemon, lime, tangerine, tangelo, sour orange, rough lemon, calamondin, trifoliate orange, and kumquat.

Geographical Distribution

Asia: Indigenous to and widespread as A strain throughout Asia, occurring in Afghanistan, Bangladesh, Cambodia, China (Fujian, Guangdong, Guangxi, Guizhou, Hubei, Hunan, Jiangsu, Jiangxi, Sichuan, Zhejiang), Hong Kong, India (Andaman Islands, Andhra Pradesh, Assam, Haryana, Karnataka, Maharashtra, Punjab, Tamil Nadu), Indonesia (Java), Iran, Iraq (Ibrahim and Bayaa, 1989), Japan (Honshu, Kyushu, Ryukyu Archipelago, Shikoku), Korea Democratic People's Republic, Korea Republic, Lao, Malaysia (peninsular, Sabah), Maldives, Myanmar, Nepal, Oman, Pakistan, the Philippines, Saudi Arabia, Singapore, Sri Lanka, Taiwan, Thailand, United Arab Emirates, Viet Nam, Yemen.

Africa: Present as A strain in Comoros, Côte d'Ivoire, Gabon, Madagascar, Mauritius, Mozambique (eradicated), Réunion, Seychelles, South Africa (eradicated), and Zaire.

North America: Mexico (D strain only; declared no longer to occur, the original disease being attributed to *Alternaria limicola* by Palm and Civerolo, 1994); the United States (introduced into Florida in 1912 and spread to Alabama, Georgia, Louisiana, South Carolina and Texas; eradicated in Florida by 1933 and from all the United States by 1947; the A strain reappeared in Florida in 1986 [Whiteside et al., 1988] and an eradication program is currently being conducted to eliminate the disease and preclude establishment and dissemination of the pathogen; after a period when eradication was thought to have been successful, the disease appeared again in private gardens in the Miami area in 1995).

Central America and Caribbean: The previous version of this data sheet (EPPO/CABI, 1992) mentioned unconfirmed reports in several countries of this region. After checking, all have been found to be erroneous.

South America: Argentina (A strain along the coast, B strain only in small isolated foci on lemons in southern Entre Rios); Brazil (A and C strains; São Paulo, in the region of Presidente Prudente; Paraná, northeast, north and west central; Mato Grosso do Sul, east, southeast and south; Santa Catarina; unconfirmed reports in Mato Grosso, Minas Gerais, Rio Grande do Sul); Paraguay (A, B, and C strains; east and west [Chaco central]); Uruguay (A strain under eradication; Salto, on north bank of River Uruguay; Paysandu, north; B strain eradicated since 1985).

Oceania: Australia, at one time on Thursday Island, Queensland (Jones et al., 1984), but now eradicated there (Catley, 1988); since the late 1980s, at one location in Northern Territory. Christmas Island (Shivas, 1987), Cocos Islands, Fiji, Guam, Northern Mariana Islands, Micronesia, New Zealand (eradicated), Palau, Papua New Guinea.

Pathogen

The bacterium *X. axonopodis* pv. *citri* is rod-shaped, Gram-negative, and has a single polar flagellum. Colonies on laboratory media are usually yellow due to "xanthomonadin" pigment production. When glucose or other sugars are added to the culture medium, colonies become very mucoid due to the production of an exopolysaccharide slime. A semi-selective medium can be prepared by adding an antibiotic, kasugamycin, which inhibits many contaminants but not xanthomonads. The maximum and optimum temperature ranges for growth are 39°C and 28°C–30°C, respectively.

Serology, bacteriophage typing, fatty acid profiles, PCR, and DNA analysis are useful for the identification and classification of bacterial isolates into pathovars. However, when such techniques are unavailable, strains of *X. axonopodis* pv. *citri* can be distinguished from other pathovars by a panel of susceptible and resistant citrus hosts. Bioassays can also be run on detached leaves or leaf disks. On detached leaves, lesions will be apparent for 2–4 days after inoculation, and different isolates can be tested side by side on the same leaf for comparison.

Host Susceptibility

In general, in field plantations, grapefruit, Mexican limes, and trifoliate orange are highly susceptible to canker; sour orange, lemon, and sweet orange are moderately susceptible; and mandarins are moderately resistant. Within orange cultivars, early maturing cultivars are more susceptible than midseason cultivars, which are in turn more susceptible than late-season cultivars. However, when plant tissues are disrupted by wounds or by the feeding galleries of the Asian leafminer, internal leaf tissues (mesophyll) are exposed. When this occurs, all cultivars and most citrus relatives that express some level of field resistance can become infected.

Pathogen Diversity and Distribution

There are distinct forms of citrus canker disease caused by various pathovars and variants of the bacterium *X. axonopodis*. Because symptoms are generally similar, separation of these forms from each other is based on host range, cultural and physiological characteristics, bacteriophage sensitivity, serology, DNA–DNA homology, and by PCR analysis of genomic DNA. The latter assays demonstrate that these forms are genetically unique.

The Asiatic form of canker (canker A), caused by *X. axonopodis* pv. *citri* (syn. *Xanthomonas citri*, *X. campestris* pv. *citri*), is by far the most widespread and severe form of the disease.

Cancrosis B, caused by *X. axonopodis* pv. *aurantifolii*, is a disease of lemons in Argentina, Paraguay, and Uruguay, but Mexican lime, sour orange, and pummelo are also susceptible. Cancrosis B isolates can be differentiated serologically from canker A, but not from cancrosis C.

Cancrosis C, also caused by *X. axonopodis* pv. *aurantifolii*, was isolated from Mexican lime in Brazil. The only other known host for this bacterium is sour orange.

Other forms of canker bacteria have at times been reported. For example, canker D, sometimes called citrus bacteriosis, was reported on Mexican lime in Mexico in 1981, but its existence remains controversial. Isolates were discovered in Oman, Saudi Arabia, Iran, and India that produce canker A-like lesions only on Mexican lime (termed A* strains) and appear to be distinct serologically from all other known forms. Another atypical form of canker A bacterium has been described from Réunion and surrounding islands in the Indian Ocean that has high levels of resistance to a number of antibiotics. Recently, a variant of canker A, known as canker A^w, was discovered on the southeast coastal areas of Florida. The canker A^w isolate is distinct via PCR primer reaction and has a host range apparently restricted to Mexican lime and Alemow (*Citrus macrophylla*). Other variants not described, yet probably exist.

Disease Cycle

Bacteria propagate in lesions in leaves, stems, and fruit. When there is free moisture on the lesions, the bacteria ooze out and can be dispersed to infect new growth. Wind-driven rain is the main dispersal agent and wind ≥ 8 m/s (18 mph) aids in the penetration of bacteria through the stomatal pores or wounds made by thorns, insects (leafminer), and blowing sand. Pruning causes severe wounding and can lead to infection. Multiplication of bacteria occurs mostly while the lesions are still expanding, and numbers of bacteria produced per lesion are related to general host susceptibility.

The bacteria remain alive in the margins of the lesions in leaves and fruit until they abscise and fall to the ground. Bacteria have also been reported to survive in lesions on woody branches up to a few years of age. Bacteria that ooze onto plant surfaces do not survive and begin to die upon exposure to rapid drying. Death of bacteria is also accelerated by exposure to direct sunlight. Survival of exposed bacteria is limited to a few days in soil and to a few months in plant refuse that is incorporated into soil. On the other hand, the bacteria can survive for years in infected plant tissues that have been kept dry and free of soil.

Epidemiology

Infection

Leaves, stems, and fruits become resistant to infection as they mature unless they are wounded. Almost all infections occur on leaves and stems within the first 6 weeks after initiation of growth. The most critical period for fruit rind infection is during the first 90 days after petal fall. Any infection that occurs after this time results in the formation of only small and inconspicuous pustules. Because the fruits are susceptible over longer periods compared to leaves, infections can result from more than one dispersal event, resulting in lesions of different age on the same fruit. Fruit can also act as an indicator of time of infection. Lesion age can be estimated on fruit and help determine when infection occurred, and this information can be related to meteorological events, such as storms, that occurred at that time.

Pathogen Dispersal

Cankers by wind and rain are mostly spread only for short distances, that is, within trees or to neighboring trees. Canker develops more severely on the side of the tree exposed to wind-driven rain. Spread over longer distances, up to several miles, can result from severe meteorological events such as tropical storms, hurricanes, and tornadoes. A recent study determined that 99% of infections that occur within a 30-day period are located within 594 m (1950 ft) of prior infected trees during normal weather conditions, that is, when normal rain storms occur but not when tropical storms and hurricanes occur. Hurricanes and tropical storms greatly increase citrus canker infection and can spread the bacteria over many miles. During 2004, Florida was subjected to three hurricanes that crossed and affected the majority of the commercial citrus industry. Bacterial dispersal gradients of up to 53 km were recorded and hundreds of new outbreaks were subsequently discovered. However, long-distance spread more often occurs with the movement of diseased propagating material, such as budwood, rootstock seedlings, or budded trees. There is no record of seed transmission. Commercial shipments of diseased fruit are potentially a means of long-distance spread, but there is no authenticated record of this ever having happened. Nursery workers can carry bacteria from one nursery to another unless hands, clothes, and equipment are disinfected. Such spread can also result from contaminated budwood or contaminated budding equipment. Pruning, hedging, and spray equipment have been demonstrated to spread the disease within and among plantings. Wooden harvesting boxes that contained diseased fruit and leaves and are later taken to disease-free orchards have also been implicated in long-distance spread.

Disease Management

Exclusion
The first line of defense against citrus canker is exclusion. Citrus canker still does not exist in some countries or regions of countries where climatic conditions are favorable for pathogen establishment, which is probably because of rigid restrictions on the importation of propagating material and fruit from areas with canker. Unfortunately, with increased international travel and trade, the likelihood of *X. axonopodis* pv. *citri* introduction is on the rise as it is with many exotic pests and pathogens. Documentations of six separate introductions of citrus canker into Florida have occurred since 1985, demonstrating that even with eradication, reintroduction is a continual process and problem.

Sanitation
Numerous cases of new infections of citrus canker are linked to human and mechanical transmission. Humans can carry bacteria on their skin, clothing, gloves, hand tools, picking sacks, ladders, etc. Vehicles can become contaminated by brushing wet foliage or coming in contact with plant material. Machineries such as tractors, implements, sprayers, and hedgers can similarly become contaminated and even inadvertently transport plant parts. In areas where citrus canker is resident, it is necessary to construct decontamination stations for personnel, vehicles, and machinery that are sprayed with bactericidal compounds.

Eradication
Once introduced into an area, elimination of inoculum by removal and destruction of infected and exposed trees is the most accepted form of eradication. To accomplish this, trees may be uprooted and burned, or in urban areas, they may be cut down and chipped and the refuse disposed of in a landfill. In Florida, state law requires that all citrus trees within 579 m (1900 ft) of infected trees be removed in both residential and commercial situations.

Other Practices
In countries where the disease is well established and severe, only the more resistant types of citrus, such as Valencia oranges and mandarins, may be profitable. In regions where canker is endemic, certain cultural practices are used to reduce the severity of the disease. It is imperative to avoid working in infected orchards when the trees are wet from dew or rain. The reduction of wind is another primary concern. Wind speeds are reduced by deployment of windbreaks on the perimeter of the orchard or between the rows. A reduction in wind speed lowers the probability of direct penetration of stomates by bacteria as well as the entry of wind-induced injuries on foliage and fruit.

Chemical Control

Where canker is a major problem, control requires integration of appropriate cultural practices including sanitation, windbreaks, and leafminer control with frequent applications of copper sprays. Copper sprays have been shown to reduce infection somewhat. Because the fruit is susceptible to canker during the first 90 days after petal fall, it is important to maintain a protective coating of a copper material on the fruit surface during this period. Two or three treatments may be needed for this purpose, depending on rainfall and cultivar susceptibility. Windbreaks can greatly reduce spread and severity of disease and increase the efficacy of copper sprays. Leafminer control is particularly important on young trees and certain cultivars that have a high proportion and greater frequency of vegetative growth flushes.

Significance

History

Citrus canker probably originated in Southeast Asia, which is the ancestral home of citrus, but the disease continues to increase its geographic range in spite of the heightened regulations imposed by many countries to prevent introduction. Citrus canker presently occurs in over 30 countries in Asia, the Pacific and Indian Ocean islands, South America, and the United States.

Citrus canker was first described after it was discovered in the United States in the Gulf States in 1915. However, canker lesions were noted on herbarium specimens collected in India as early as 1827. The Gulf States' outbreak, which included seven southern states, resulted from a shipment of infected trifoliate orange nursery stock from Japan. The disease also appeared earlier this century in South Africa and Australia. However, it was reportedly eliminated in these countries as well as the Gulf States through nursery and orchard inspections, quarantines, and the on-site burning of infected trees. Subsequent epidemics have occurred in Australia, Argentina, Uruguay, Brazil, Oman, Saudi Arabia, and Réunion Island. In some locations, eradication efforts have been attempted but failed. In others, active eradication campaigns continue (Australia, Brazil, Florida).

Citrus canker was found in Florida in 13 locations from 1985 to 1992. Through extensive inspection and tree removal, eradication was believed to have been achieved. However, the disease re-emerged in commercial plantations in Manatee County, Florida, in June 1997, where eradication efforts had previously taken place. The age of the oldest lesions found indicated that the disease had been in the area for about 1–1.5 years. This outbreak has largely been suppressed by the destruction of several hundred acres of infected commercial citrus plantations.

A new and extensive outbreak was discovered in urban Miami, Florida, in 1995. The original Miami outbreak consisted of approximately 14 square miles of infected residential properties when first discovered in September 1995, but had expanded to over 202 square miles as of December 1998. The oldest lesions in the Miami area indicated that the disease had existed in that area for about 2–3 years prior to discovery. Severe tropical weather patterns have affected Miami in the past several years, including hurricanes, tropical storms, tornadoes, and numerous rainstorms associated with high winds. These along with human movement have spread the infestation and greatly exacerbated the epidemic. Nearly all Florida counties within the commercial citrus range have had one or more outbreaks and about 10% of the commercial acreage has been removed in attempts to eradicate the disease.

Genomic analyses of bacterial isolates from the Miami and Manatee County areas have demonstrated significant differences between these isolates that suggest that the bacteria in the two areas were the result of two independent introductions. In July 1998, another outbreak was discovered in commercial citrus in Southwest Florida. Genomic analysis indicated that a bacterial isolate from this outbreak was identical to that found in Miami and, therefore, likely resulted from the transport of inoculum or infected plant material from the Miami area.

Impact

Citrus canker is an extremely costly disease. Worldwide, millions of dollars are spent annually on prevention, quarantines, eradication programs, and disease control. Undoubtedly, the most serious consequence of citrus canker infestations is the impact on commerce resulting from restrictions to interstate and international transport and sale of fruit originating from infested areas. The USDA, Animal and Plant Health Inspection Service, in collaboration with the Florida Department of Agriculture and Consumer Services, Division of Plant Industry, has formed a joint state/federal eradication campaign to eliminate the disease. An average of over $50 million per year and over 600 personnel are presently dedicated to this program. In Florida alone, costs of running an eradication program from 1995 through 2005 plus compensation to commercial growers and homeowners for residential citrus destroyed approached $1 billion.

7.18.2 CITRUS BACTERIAL BLAST

Pathogen: *Pseudomonas syringae*

Symptoms

Infections caused by *P. syringae* usually start as black lesions in the leaf petiole and progress into the leaf axil. Once the petiole is girdled, leaves wither, curl, and eventually drop. Entire twigs may die back. The damage is most severe on the south side of the tree, which is exposed to the prevailing winter winds. Diseased areas are covered with a reddish-brown scab. Infections result in small black spots on the fruits (Photo 7.43).

Geographical Distribution

It occurs in regions possessing a cool wet season particularly in parts of Australia, Japan, South Africa, certain Mediterranean countries, the Soviet Union, and the United States (California but not Florida).

Disease Cycle

P. syringae pv. *syringae* is a normal inhabitant of citrus leaves and becomes even more abundant on leaf surfaces during prolonged periods of wetness caused by rain or fog and relatively low temperatures. Infection occurs when bacteria enter injuries in shoots or fruit caused by wind, heavy rain, insect feeding, hail, or storms. Young, succulent shoots and leaves are most susceptible to infection. The disease rarely progresses when the temperature is above 20°C or below 8°C.

Predisposing Factors

Bacterial blast, also known as citrus blast or black pit, is restricted mainly to citrus-growing areas in the Sacramento Valley, where wet, cool, and windy conditions during winter and spring favor the development and spread of the blast bacterium. Leaves and twigs of oranges and grapefruit and the fruit of lemon are most susceptible to infection. The bacterium infects small injuries caused by thorn punctures, wind abrasions, or insect feeding.

 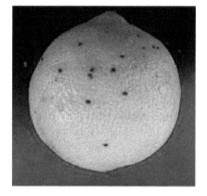

PHOTO 7.43 Symptoms of bacterial blast on citrus stem and fruit. (Courtesy of Catara, A., Università di Catania, Catania, Italy.)

Disease Management

Preventive treatment against bacterial blast alone is generally not economical, but sprays against brown rot or Septoria may provide some protection against bacterial blast. Certain cultural practices can reduce the incidence of bacterial blast.

Planting windbreaks and using bushy cultivars with relatively few thorns help prevent wind injury; pruning out dead or diseased twigs in spring after the rainy period reduces the spread of the disease; and scheduling fertilization and pruning during spring or early summer prevents excessive new fall growth, which is particularly susceptible to blast infection.

Copper and Bordeaux sprays are acceptable for use in organically managed citrus groves.

Treatment Decisions

In the Sacramento Valley where blast is an annual problem, apply treatments each year at the onset of cool, wet periods.

7.18.3   Citrus Bacterial Spot

Pathogen: *Xanthomonas alfalfae* subsp. *citrumelonis* (synonym: *X. axonopodis* pv. *citrumelo*)

Citrus bacterial spot is known to occur only under nursery conditions. There are three groups of isolates based on laboratory assays: aggressive, moderately aggressive, and weakly aggressive. Only aggressive isolates were historically spread in field nurseries naturally by wind-blown rain and overhead irrigation. All strains can be mechanically moved from tree to tree under normal nursery operation. When infected trees are transplanted into groves, the bacterium rapidly dies off and the disease becomes undetectable in a few months. The activity of the citrus leaf miner exacerbates citrus bacterial spot.

Symptoms

The leaf symptoms of citrus bacterial spot are very similar to those of citrus canker, but the lesions are flat and not raised. Foliar lesions show necrotic centers that often can crack or drop out (Photo 7.44) and are surrounded by water-soaked margins. Lesions produced by aggressive strains have more pronounced water-soaked margins than those of citrus canker (Photo 7.45). Bud graft and growth on infected rootstocks, especially *Swingle*, can be infected. Fruit infection is rare and has been reported only on flying-dragon trifoliate.

PHOTO 7.44   Symptoms of citrus bacterial spot on citrus leaf. (Courtesy of Serrano, D. et al. Citrus Diseases. USDA/APHIS/PPQ Center for Plant Health Science and Technology, [09–14], 2010.)

PHOTO 7.45 Symptoms of citrus bacterial spot on citrus leaf. (Courtesy of Graham, J.H., UF/IFAS Citrus Extension, Lecanto, FL.)

Citrus bacterial spot can affect common scion and rootstock species and is most severe on *Swingle citrumelo*, trifoliate orange rootstock, and grapefruit scions.

The disease is known to occur only in Florida.

Control Measures

Not yet worked out but copper sprays can reduce the disease.

7.19   ORANGE

7.19.1   Citrus Blast Disease of Orange and Mandarin

Pathogen: *Pseudomonas syringae* pv. *syringae*

In the spring of 2004, severe outbreaks of a disease resembling citrus blast (Whiteside et al., 1988) were observed on trees of orange (*Ci. sinensis* cv. Washington) and mandarin (*Citrus reticulate* cv. Marisol) in the Turkish Mediterranean regions of Adana and Antalya. The damage was serious in a 50 ha citrus orchard in Antalya, with a disease incidence of nearly 100%.

Symptoms

Characteristic disease symptoms are first seen on leaves as water-soaked lesions and black areas on the petiole wings. Later lesions extend to the midvein of leaves and to the twigs surrounding the base of the petiole. Finally, the leaves dry and roll, while still firmly attached, before eventually dropping without petioles. The necrotic areas on twigs further enlarge and the twigs are eventually killed within 20–30 days.

Pathogen

Twelve isolates of a bacterium, consistently isolated from infected leaves, petioles, and twigs, which formed fluorescent colonies on King's medium B, were purified and used for further studies. All isolates were Gram-negative and tested negative with oxidase, pectolytic activity, starch hydrolase, arginine dihydrolase and nitrate reduction negative; and tested positive with levan, gelatin hydrolase. They produced hypersensitive reactions (positive) on tobacco leaves (*Nicotiana tabacum* cv. Samsun N.). All produced acid from glucose, glycerine, arabinose, mannitol, sorbitol, sucrose, and xylose but not from lactose and maltose. The test results conformed to the characteristics of *P. syringae* pv. *syringae* (Braun-Kiewnick and Sands, 2001) as the causal organism of citrus blast and were similar to those of reference strain NCPPB 2307 of *P. syringae* pv. *syringae* used in the study. Fatty acid analysis (Atatürk University, Erzurum, Turkey) confirmed the bacterial strains as *P. syringae* pv. *syringae* with similarity indices of 81%–94%.

Epidemiological Parameters

Injury to tissues, such as those that occur during wind, driving rain, sandblasting, and frost, facilitates entry of the bacteria into the tissue. Several days of rain appear to be required for infection to take place.

Disease Management

As in citrus blast.

7.19.2 CITRUS GREENING DISEASE OF ORANGES

Pathogen: *Candidatus liberibacter asiaticus*
Candidatus liberibacter africanus
Candidatus liberibacter americanus

Common name: *Huanglongbing*, HLB

Citrus huanglongbing (HLB), previously called citrus greening disease, is one of the most destructive diseases of citrus worldwide. Originally thought to be caused by a virus, it is now known to be caused by unculturable phloem-limited bacteria. There are three forms of greening that have been described. The African form produces symptoms only under cool conditions and is transmitted by the African citrus psyllid *Trioza erytreae*, while the Asian form prefers warmer conditions and is transmitted by the Asian citrus psyllid *Diaphorina citri*. Recently, a third American form transmitted by the Asian citrus psyllid was discovered in Brazil. This American form of the disease apparently originated in China. In North America, the psyllid vector, *Dia. citri*, of HLB is found in Florida, Louisiana, Georgia, South Carolina, Texas, and Hawaii, and recently arrived in Southern California from Mexico. HLB is known to occur in Florida, Lousiana, South Carolina, Georgia, Cuba, Belize, and the Eastern Yucatan of Mexico. A federal quarantine restricts all movement of citrus and other plants in the family Rutaceae from Asian Citrus Psyllid or HLB-infested areas into California in order to prevent introduction of the disease.

Primary Host

Citrus sp. and some related plants, box thorn or Chinese box orange (*Severinia buxifolia*), wood apple (*Limonia acidissima*), white ironwood (*Vepris lanceolata*), and mock orange or orange jasmine (*Murraya paniculata*).

Symptoms

Symptoms are many and variable: yellow shoots, twig dieback, leaf drop, leaves with blotchy yellow/green coloration (Photo 7.46) similar to the symptoms of zinc nutritional deficiency, enlarged veins that appear corky, excessive fruit drop, small and misshapen fruit, fruit that remains green

PHOTO 7.46 Citrus greening disease symptoms on leaves and fruits. (Courtesy of Lotz, J.W., Florida Department of Agriculture and Consumer Services, Tallahassee, FL, Bugwood.org.)

at one end (the stylar end) after maturity, fruit with mottled yellow/green coloration, small, dark, aborted seed inside fruit, discolored vascular bundles in the pithy center of the fruit, bitter tasting fruit, and silver spots left on fruits that are firmly pressed.

The time from infection to the appearance of symptoms is variable, depending on the time of year, environmental conditions, tree age, host species/cultivar, and horticultural health ranging from less than 1 year to several years.

Damage

The HLB bacteria can infect most citrus cultivars, species, and hybrids and even some citrus relatives. Leaves of newly infected trees develop a blotchy mottle appearance. On chronically infected trees, the leaves are small and exhibit asymmetrical blotchy mottling (in contrast to Zinc deficiency that causes symmetrical blotching). Fruit from HLB-infected trees are small, lopsided, poorly colored, and contain aborted seeds. The juice from affected fruit is low in soluble solids, high in acids, and abnormally bitter. The fruit retains its green color at the navel end when mature, which is the reason for the common name "citrus greening disease." This fruit is of no value because of poor size and quality. There is no cure for the disease and rapid tree removal is critical for prevention of spread.

Pathogen and Disease Cycle

Candidatus liberibacter are Gram-negative bacteria with a double-membrane cell envelope. *Candidatus liberibacter asiaticus*, *africanus*, and *americanus* are found in plants only in the phloem cells. The bacteria are transmitted by psyllids, a type of insect, as they feed. *Candidatus liberibacter asiaticus* and *Candidatus liberibacter americanus* are transmitted by the adults of the citrus psyllid *Dia. citri* Kuwayama. *Candidatus liberibacter africanus* is transmitted by the adult psyllid *T. erytreae* Del Guercio. The bacteria can be acquired by the insects in the nymphal stages and the bacteria may be transmitted throughout the lifespan of the psyllid. Eggs are laid on newly emerging leaves and hatch in 2–4 days. Five nymphal instars complete development in 11–15 days. The entire life cycle takes 15–47 days, depending upon temperature, and adults may live several months with females laying up to 800 eggs in a lifetime.

In an orchard, diseased trees are clustered together, with secondary infections produced 25–50 m away. *Candidatus liberibacter africanus* is found at elevations greater than 700 m and is less heat tolerant than *Candidatus liberibacter asiaticus*. *Candidatus liberibacter americanus* resembles *Candidatus liberibacter africanus* in being less heat tolerant. Infections of *Candidatus liberibacter asiaticus* and *Candidatus liberibacter americanus* are more severe than *Candidatus liberibacter africanus* and can lead to the death of trees. The three disease agents (*Candidatus liberibacter* spp.) are not distinguishable from each other based on the symptoms produced.

Current Geographic Distribution

Candidatus liberibacter africanus is found in eastern, central, and southern Africa. *Candidatus liberibacter americanus* is found in Sao Paulo State, Brazil. *Candidatus liberibacter asiaticus* is found in Asia from Japan to South China, SE Asia, and the Indian subcontinent to Pakistan, the Arabian Peninsula (not including Iran), Brazil, Cuba (2009), Dominican Republic (2009), Mexico (2009) and Florida (2005), and Louisiana (June 2008) in the United States. The vector *Dia. citri* is more widely spread in south and central America, including Mexico (at least since 2004), and in the United States in Texas (2001), Louisiana (May 2008), Alabama (August 2008), Georgia (August 2008), Mississippi (August 2008), South Carolina (August 2008), and California (September 2008), posing a threat to the citrus industry in these areas.

Economic Impact

HLB is one of the most devastating diseases of citrus and since its discovery in Florida in 2005, citrus acreage in that state has declined significantly. If the disease were to establish in California, the nursery industry would be required to move all of their production under screenhouses, and pesticide treatments for the vector would be instituted resulting in greatly increased pesticide costs (three to six treatments per year) and indirect costs due to pesticide-induced disruption of integrated pest-management programs for other citrus pests. An expensive eradication program would need to be instituted to remove infected trees in order to protect the citrus industry.

Control Measures

To date, control of the disease is based on planting HLB-free citrus germplasm, eradication of infected citrus plants, and control of the vector with systemic insecticides. Countries with HLB learn to manage the disease so that they can still produce citrus. In California, the best strategy is to keep this disease out. This goal is supported by both federal and state quarantine regulations and the University of California's Citrus Clonal Protection Program, which provide a mechanism for the safe introduction of citrus germplasm into California.

7.20 RASPBERRY

7.20.1 FIRE BLIGHT OF RASPBERRY AND BLACK BERRIES

Pathogen: *Erwinia amylovora*

While fire blight is most common in pears and apples, it also affects raspberries and blackberries (*Rubus* spp.). Summer red raspberries cultivars K81-6 and Boyne are particularly susceptible. Losses result from berry necrosis and from tip dieback of primocanes. Fruit losses of 65% or more have been reported on thornless blackberries in Illinois. Although still relatively rare on raspberries, this disease has become increasingly common on certain red raspberry cultivars in Wisconsin.

Symptoms

The most obvious and striking symptoms are blackened cane tips, which bend over and die, resulting in a "shepherd's crook" appearance (Photo 7.47). Infections may proceed down the cane for up to 8.0 in., can girdle canes, and may produce cream-colored bacterial ooze under high moisture conditions. As the disease progresses down the cane (Photo 7.48), the veins of leaf and portions of the leaf surrounding the midvein turn black, and lesions become water soaked. Carmel-colored bacterial ooze comes out of lesions in beads during humid periods. Diseased plant parts become purplish black. Entire leaves may wither and die. Typically, discoloration and dieback are limited

PHOTO 7.47 Shepherd's crook—a classic symptom of fire blight on raspberry apical stem. (Courtesy of Olson, B., Oklahoma State University, Stillwater, OK, Bugwood.org.)

PHOTO 7.48 Symptoms of fire blight on raspberry plant showing discoloration and death of leaf tissue along the veins due to the spread of the bacteria within the plant. (Courtesy of Olson, B., Oklahoma State University, Stillwater, OK, Bugwood.org.)

to succulent young growth. In addition, the disease can affect fruit clusters. Infected peduncles (the stalks of fruit clusters) turn black and the young developing berries do not mature, become brown, dry, and very hard, and remain on pedicel. Entire fruit clusters may be infected, but generally a few berries in each cluster remain healthy.

Pathogen
Fire blight is caused by the bacterium *E. amylovora*. Although this is the same organism that causes fire blight on pear and apple, it is a different strain. Thus the strain that attacks raspberries and blackberries will not infect apple or pear and vice versa. However, it has been found that "Boyne" raspberries can be infected by the apple strain, but this is an exception.

Disease Cycle
The bacterial spread from plant to plant is by insects, wind, and splashing rain water. Rain, high humidity, and warm temperatures favor disease development. It is not known how and where the bacteria overwinter, although they likely survive in cankers on infected canes.

Probably apple strain attacks raspberry, but not vice versa. Bacteria present from overwintering cankers on diseased plant material serve as source of inoculums. Warm temperatures (18°C–25°C) and light rain favor infections; prolonged host flowering due to wet cool springs is a predisposing factor for prolonged infection. Flowers, fruit, cane tips that are succulent are more susceptible to infection. Bacteria enter the plant, through flower or mechanical damage.

Disease Management

Cultural Methods
Cultural methods are the most important means of managing this disease. The following practices offer effective methods for limiting the spread of the disease.

1. Purchase and plant only certified, pathogen-free plants obtained from reliable nurseries.
2. Practice good sanitation. Pruning is the best method of control. Remove and destroy diseased canes from the planting as soon as you see them. Disinfest pruning shears in a 10% household bleach solution (containing one part bleach and nine parts water) between each cut to avoid transmitting bacteria to healthy canes. Isopropyl alcohol (70%) or quaternary ammonia may also be used, but the bleach solution is more effective.

3. Manage insect pests to avoid a possible means of moving the bacteria from plant to plant.
4. Do not overfertilize. Vigorous, succulent growth appears to be most susceptible.
5. Orient rows and prune and thin plants to maximize air circulation around the plants. This will help lower the relative humidity within the plant canopy.
6. Destroy wild or abandoned brambles growing nearby. These plants may serve as inoculum sources for fire blight.

Cultivar Resistance

Fire blight affects both red and black raspberries and blackberries. The effect of fire blight on purple raspberries is not known. To date, there has been no thorough study of resistance to fire blight among commercially available cultivars of raspberry. However, red raspberry cultivars do vary somewhat in resistance to the diseases. Of the more popular cultivars grown in Wisconsin, Fallgold, Latham, and Boyne are known to be susceptible. Plant-resistant cultivars are such as "Chilcotin," "Newburgh," or "Nova."

Chemical Control

Because fire blight has occurred only infrequently on raspberry over the years, no chemical controls have been developed. Copper sprays act as protectants, not control. Sprays must be directed into open flowers to be effective.

Bordeaux mixture (8-8-100). Apply one spray at the delayed dormant/bud-bursting stage. Apply a second spray in the fall before rains start. Thoroughly wet the canes.

Or use the following:

Copper spray fungicide or copper oxychloride 50 (50% copper oxychloride) at 2.0 kg in 1000 L of water per ha (0.8 kg in 400 L of water/acre). Begin protection at the bud-bursting stage. Apply in fast drying conditions to minimize the risk of plant damage. Repeat at 14-day intervals until three sprays have been applied. Thoroughly wet canes at each treatment. Do not apply within 1 day of harvest.

Caution: Do not use bluestone or copper sulfate alone as it washes off readily and may cause plant injury. Copper-tolerant strains of this bacterium have been detected from blueberry.

Fall–winter infection: No satisfactory chemical control for the fall–winter infection period has been determined. The following may be beneficial:

Bordeaux Mixture (8-8-100). Apply spray before fall rains start (about October). Thoroughly wet the canes and apply in fast drying conditions.

Or, use the following:

Copper spray fungicide or copper oxychloride 50 (50% copper oxychloride) at 2.5 kg/1000 L of water/ha (1 kg/acre in 400 L of water); or

Serenade Max (14.6% *B. subtilis*) at 1.0–3.0 kg/ha (0.4–1.2 kg/acre). Apply before fall rains and again during dormancy before spring. Serenade may be applied up to and including the day of harvest.

Note: Serenade is a bacterial-based biofungicide. It is approved for organic production.

Applications of following chemical before fall rains and as a delayed application after rainfall to control the disease is recommended.

- Bordeaux 12-12-100 or
- Champ WG at 4 lb/A plus crop oil at 1 quart/A; 48 h reentry or
- Cuprofix Ultra 40 Disperss at 2.5–3 lb/A; 48 h reentry or

- Kocide 3000 at 1.75 lb/A plus 1 quart/A crop oil; 48 h reentry or
- Monterey Liqui-Cop at 4 teaspoons/gal water or
- Nu-Cop 50 DF at 4 lb/A with 1 quart/A crop oil; 48 h reentry or
- Phyton 27 AG at 20–40 fl oz/100 gal water; 48 h reentry

7.20.2 Bacterial Blight of Raspberry

Pathogen: *Pseudomonas syringae* pv. *syringae*

The other important bacterial disease reported on raspberry is *Pseudomonas* blight, expressed as collapse of floricane laterals, followed by wilting and dieback of entire floricanes, and were recorded on raspberry plants "Willamette" in western Serbia in 2002–2004. Isolated bacterial strains were identified as *P. syringae*.

Symptoms

Brown water-soaked spots develop on leaves and petioles of developing laterals or young shoots (Photo 7.49). In wet weather, spots enlarge rapidly, killing leaves and shoots. Brown streaks extend from these dead tissues down under the bark. If the infection does not fully girdle the stem, affected laterals show a characteristic downward bend. If infected laterals are not killed, they are often stunted.

Bacterial blight can occur at two periods during the year. In the spring, blight symptoms appear as a sudden wilting and blackening of new shoots, cane tips, laterals, and leaves. Affected laterals have a distinct "crooking" or downward bend. This type of damage is often associated with temperatures just above 0°C and is usually not a problem after mid-May.

The most serious phase of the disease is believed to occur in the fall in fields that are actively growing later than normal. These fields seem to be susceptible to infection that shows as dead buds and black streaking of the cambium layer under the bark. This damage is usually not noticed until spring and can be confused with injury due to spur blight or winter injury.

PHOTO 7.49 Symptoms of bacterial blight on raspberry. (Courtesy of Plant Clinic Collection, 2011. Oregon State University, pnwhandbooks.org)

The symptoms are very similar to fireblight symptoms, causing blackening or browning of tissue. The disease is slow in spreading unlike fireblight. The disease can also be confused with several other disorders like winter injury, cane borer, or some herbicides injury.

Damage

This bacterial disease is seldom a problem but can occasionally cause severe losses.

Pathogen

Strains of the bacterium *P. syringae* pv. *syringae* are found on many plants, including blueberry, cherry, apple, and pear.

Predisposing Factors

Outbreaks usually are associated with cold, 0°C–0.5°C, and moist weather. Succulent growth due to high nitrogen is more susceptible when conditions favor infection. The disease on raspberry occurs occasionally in the Pacific Northwest.

Cool and moist weather favor the disease. The bacteria overwinter on the canes and buds. The bacteria enter the host through frost damage or mechanical damage. Bacteria are systemic in the plant, and infection spreads as long as weather is cool and wet; once the weather begins to warm up, disease subsides until fall.

Disease Cycle

The bacterium survives on leaf surfaces, in healthy buds, and on weeds. It may be spread by splashing rain, wind, insects, and infected planting stock.

Monitoring

Check developing laterals and young shoots for symptoms. Where fall conditions may have promoted the development of blight, inspect buds for damage. Look for black streaking under the bark near the buds.

Control Measures

No specific control measures have been developed because of the sporadic nature of the disease. However, follow the disease management practices as described for fire blight of raspberry for the management of bacterial blight.

7.20.3 CROWN GALL AND CANE GALL OF RASPBERRY

Pathogen: *Agrobacterium tumefacience*

Symptoms

Soil-borne bacterial pathogen that infects raspberry roots, crowns, and lower stems through wounds, causes crown gall. The bacteria induce the plant tissue to grow abnormally, which results in the production of spongy wart-like galls on infected crowns and roots (Photo 7.50). The galls eventually become hard and woody as they age. Raspberry plants with numerous galls become stunted and weak. The leaves turn yellow and dry up at the edges. The galls disintegrate over the winter and release the disease-causing bacteria into the soil. New galls often form in the area of the old galls the next spring.

The first symptoms are usually woody swellings or galls on the crowns or canes at ground level. These galls range from the size of a pea to the size of a tennis ball. Root infections may go undetected until galls are so numerous that the vigor of the plant is affected. In some plantings, where the disease has become established, the fruiting canes produce short, weak laterals. The leaves turn yellow and dry at the edges and curl up with the onset of warm weather. Root systems from these dying plants resemble a string of beads because of the frequency of galls.

Damage

Crown gall poses a serious threat to the production of susceptible raspberry varieties. If infected planting stock is used, yield can be significantly reduced.

PHOTO 7.50   Symptoms of crown gall on raspberry canes and roots. (Courtesy of Plant Clinic Collection, 2011.)

Disease Cycle

The crown gall bacterium present in same fields serves as primary inoculums for infection. It can also be introduced on infected planting stock. Once introduced into the field, the bacteria survive almost indefinitely in decaying root galls or in alternate hosts. Wounds resulting from insect injury and cultivation or mechanical harvester damage encourage new infections.

Disease Management

Use certified raspberry plants. Never use plants from sources where crown gall has been reported. Do not use plants containing visible galls.

Where only a few plants in a field are infected, entire plants (including the complete root system) should be removed carefully and burned. Take care when removing canes and pruning because the bacteria can be spread on the pruning shears. Disinfect pruning shears by dipping in 5% Virkon, Chemprocide or CleanGrow, or a 1:10 dilution of household bleach. (*Caution*: Bleach is corrosive to metal blades.) Minimize root and cane injury by controlling root weevils and nematodes, avoiding close cultivation and making sure that catch plates on mechanical harvesters are working properly.

Field experience has shown that the variety Meeker does not develop galls while Saanich and Chemainus are susceptible.

Do not plant new canes in infested soil. Maintain good soil fertility. Avoid injury to crown and roots.

Biological Control

Dygall is a formulation of a naturally occurring bacterium that is antagonistic (i.e., kills) to the crown gall bacterium. It is applied to cuttings or plant roots before planting in infested soils. It is to be used by trained nursery personnel only.

Chemical Control

None.

7.21   STRAWBERRY

7.21.1   ANGULAR LEAF SPOT OF STRAWBERRY

Pathogen: *Xanthomonas fragariae*, Kennedy and King

Common names: Angular leaf spot (English)
Taches angulaires (French)
Blattfleckenkrankheit (German)
Maculatura angolare delle foglie (Italian)
Mancha angular da folha (Portuguese)

Angular leaf spot (ALS) is a bacterial disease caused by *X. fragariae*, a pathogen highly specific to both the wild and the cultivated strawberry, Fragaria x ananassa.

ALS is an important disease of winter-produced strawberries worldwide, and it is considered a quarantine disease by the European Union. The disease was reported first in the United States in Minnesota in 1960; since then, it has been found in almost all cultivated strawberry areas in the United States and is reported throughout North America and in Africa, Australia, Europe, New Zealand, and South America. The disease has been introduced into new areas on infected plants and is a major economic concern for nurseries, especially those involved in international trade. International plant health organizations have listed ALS as a disease of increasing concern and have implemented quarantine measures to curtail further spread of *X. fragariae*. The economic consequence of ALS to growers has not been well documented. Severely infected leaves might lead to decreased plant vigor and yield. Direct yield loss can occur if peduncles (fruit stems) and calyxes (leaf-like hulls on fruit) become infected.

In California, the largest U.S. producer of fresh strawberries, ALS is a minor disease that occurs especially during rainy weather or when overhead sprinkler irrigation is used. Although the disease is considered minor for fruit production fields in California, it is considered a problem for nurseries that export plants to Europe and other countries with quarantine restrictions.

In Florida, ALS is the only strawberry disease caused by a bacterium. The disease can be especially problematic during seasons when overhead irrigation is used many times for freeze protection.

Symptoms
Leaf spots first appear on the lower surfaces of leaves as tiny, water-soaked lesions that are delimited by veins (Photo 7.51). The angular spots appear yellow to pale green and translucent when held up to light, but dark green and opaque when viewed from above (Photo 7.52). Under wet conditions, a slimy white film often oozes from the spots. Upon drying, this film becomes scaly, and white flakes can be easily scraped from the leaves. As the disease progresses, spots become more numerous, merge, and become visible on the upper surfaces of leaves as reddish-brown dead areas. The edges of leaves may appear ragged as dead tissue breaks off. At advanced stages, ALS is difficult to distinguish from fungal leaf-spotting diseases such as common leaf spot (*Mycosphaerella fragariae*) and leaf scorch (*Diplocarpon earliana*).

While leaf lesions are the most common symptom of ALS, the bacteria can also infect the calyx. Calyx infection causes direct losses as it makes fruit unmarketable. Infected calyx tissue first appears dark green and water-soaked but later turns black. Fruit tissue nearest the calyx becomes water-soaked (Photo 7.53). *X. fragariae* can move into the roots, crowns, and stolons without showing obvious symptoms. This type of infection can cause sudden collapse and death of plants.

X. fragariae can cause vascular collapse, although this is uncommon in California. This symptom initially appears as a water-soaked area at the base of newly emerged leaves. Shortly after, the whole plant suddenly dies, much like plants infected with crown rot. *X. fragariae* is also associated with strawberry blossom blight in California.

PHOTO 7.51   Symptoms of angular leaf spot on leaves. (Courtesy of Ferrin, D., Louisiana State University Agricultural Center, Baton Rouge, LA, Bugwood.org.)

PHOTO 7.52   Symptoms of angular leaf spot on leaves. (Courtesy of Florida Division of Plant Industry, Florida Department of Agriculture and Consumer Services, Tallahassee, FL, Bugwood.org.)

Hosts

Fragaria ananassa, the predominant cultivated strawberry, whose progenitors derive from hybridization between *Fragaria chiloensis* and *Fragaria virginiana*, is the main host, but its numerous cultivars vary a great deal in susceptibility. *F. virginiana*, *Fragaria vesca*, *Potentilla fruticosa*, and *Potentilla glandulosa* have been infected following experimental inoculation. Among *Fragaria* spp., only *Fragaria moschata* is immune (Kennedy and King, 1962b; Kennedy, 1965; Maas, 1984). Cultivated strawberries are the host of concern throughout the EPPO region.

Geographical Distribution

X. fragariae was first described in 1962 in North America. It probably spread from there, with planting material, to other continents but this is only a presumption.

PHOTO 7.53 Symptoms of angular leaf spot on leaves and fruits. (Courtesy of Sikora, E., Auburn University, Auburn, AL, Bugwood.org.)

EPPO region: Locally present in France (Rat, 1974), Greece (Panagopoulos et al., 1978), Israel (found but not established), Italy (including Sicily; Mazzucchi et al., 1973), Portugal (Fernández and Pinto-Ganhao, 1981), Romania (Severin et al., 1985), Spain (Lopez et al., 1985) and Switzerland (Grimm et al., 1993).

Asia: Israel (found but not established), Taiwan, India (detected in 2012 and reported eradicated [Borkar and Pokharel, 2012].

Africa: Ethiopia, Réunion (detected in 1986 and reported eradicated; Pruvost et al., 1988).

North America: The United States (California, Florida, Kentucky, Minnesota, Wisconsin) (Kennedy and King, 1962a).

South America: Argentina (Alippi et al., 1989), Brazil (Minas Gerais, Rio Grande do Sul, São Paulo), Chile (found in 1992 and eradicated), Ecuador (found but not established), Paraguay, Uruguay, Venezuela, Mexico (Fernández-Pavía et al., 2014).

Oceania: Found in the past but eradicated in Australia (New South Wales, Victoria) and New Zealand.

EU: Present.

Disease Cycle

Residues of infected leaves and crown infections on runners used for planting are sources of inoculum for primary infections. In the residues of infected leaves, in or on soil, the bacterium survives from one crop to the next. In dried infected leaves, kept in the laboratory, the bacterium may survive for at least 2.5 years. Bacterial cells are transferred from residues to young leaves at the beginning of the growing season. From crown infection pockets, the bacterium causes lesions along the veins at the base of the youngest leaves, which develop in the apical crown region. The bacterium exudes from primary lesions, and bacterial cells are spread by aerosols, caused by rain and sprinkler irrigation, and transported by wind to healthy leaves. Penetration occurs through the stomata. Infections of the crowns occur through local wounds or downward from the affected leaves. During the growing season, several cycles of secondary infections may occur. The bacterium may attack flowers, but not fruits. During epidemics, when environmental conditions favor exudation and spread, the bacterium may cause systemic infections associated with crown pockets. Systemic infections may arise under damp nursery conditions. The conditions favoring infection are moderate to cool daytime temperatures (about 20°C), low nighttime temperatures, and high humidity (Maas, 1984). For more information, see Kennedy and King (1962a), Hildebrand et al. (1967), Maas (1984).

The pathogen overwintering in leaf debris in or on top of the soil apparently infects new leaves as they emerge in the spring. The pathogen can also persist in the crowns of infected plants and systemically invade emerging tissue. Disease development is favored by cool to moderate daytime temperatures of 18°C–21°C, cold night time temperatures (near freezing), high relative humidity, and wet conditions brought on by rain, irrigation, or dew. The white, slimy film on leaf lesions contains the pathogen and is spread within a planting by splashing rain and overhead irrigation water. Young, vigorously growing tissue is highly susceptible to infection. Bacterial populations and symptoms subside during hot, dry weather, but then rebound when temperatures become cool in the fall. *X. fragariae* apparently does not infect plants other than strawberry.

Disease Development and Spread

Little is known regarding the epidemiology of ALS; however, the development of the disease seems to be favored by warm days (20°C) and cold nights (3.5°C–5°C).

The primary source of inoculum in a new field is contaminated transplants. Secondary inoculum comes from bacteria that exude from lesions under high moisture conditions. Bacteria can survive on dry infested leaves and tissue buried in the soil for up to 1 year.

The pathogen can be spread easily by harvesting operations when wet and cool conditions favor the production of bacterial exudate. The pathogen also can be dispersed by rain and overhead sprinkler irrigation.

Means of Movement and Dispersal

The bacterium is spread locally by splash dispersal. Commercial strawberry runners used for planting may spread the bacterium over short and long distances. They may still bear old, whole or torn, infected leaves or have crown infection pockets. Moreover, almost invisible fragments of infected leaves may be hidden in the apical crown region or between the roots (Kennedy and King, 1962b).

This bacterium is not free living in soil. It can, however, overwinter in soil on previously infected plant material. Transmission is by splashing water. It is host-specific and highly resistant to degradation and can persist in the soil for long periods of time. It is killed by methyl bromide/chloropicrin mixture used as a preplant fumigant, so it is very likely that most initial infections in fields that have been fumigated originate from contaminated plants. Lesions on the leaf surface serve as a source for secondary inoculum.

Pathogen

X. fragariae is an aerobic, Gram-negative, nonspore-forming, noncapsulate rod, its size averaging 0.4 × 1.3 μm. Most cells are nonmotile, but some have a single polar flagellum. On beef-extract-peptone agar, or similar medium without added carbohydrate, colonies are circular, entire, convex, glistening, and translucent to pale-yellow (Bradbury, 1977).

Detection and Inspection Methods

The presence of the bacterium in affected plants may be confirmed by direct isolation or indirect immunofluorescence antibody staining (IFAS) of suspensions obtained by macerating in a mortar some water-soaked spots in a small volume of distilled water (Mazzucchi et al., 1973). *X. fragariae* in direct isolation is difficult because the growth of the bacterium is very slow and its colonies are easily overgrown by those of secondary organisms. Isolation may be successful if the suspension is streaked on yeast dextrose chalk agar plates and incubated at 27°C at high humidity. Colonies develop more frequently where a mass of cells is transferred in close association. The first colonies are visible after 4–5 days. The colonies are circular, 0.5–1.0 mm in diameter, yellow-pigmented, dome-shaped, with entire edges. The pure culture is distinguishable from other phytopathogenic xanthomonads by at least seven characteristics (no growth at 33°C; no hydrolysis of aesculin; no acid from arabinose, galactose, trehalose, cellobiose; 0.5%–1.0% maximum NaCl tolerance, etc.) (Kennedy and King, 1962a; Bradbury, 1984).

IFAS can be used successfully on the concentrated suspension or on its 10-fold dilution. In positive cases, millions of small, rounded fluorescent bacteria can be seen. When using differential dilutions of antiserum, no interference has yet been reported due to the existence of cross-reactions with other bacteria. Detection is quite difficult on symptomless runners (Mazzucchi and Calzolari, 1987). Any crown-infection pockets can be detected only by histological examination of single runners, which is difficult to apply to large lots. Moreover, the runners may be so well cleaned that any small residues of old infected leaves are almost invisible. Recently a sampling detection method for symptomless runners was studied. Cleaned crowns of the sample are cut in quarters and homogenized. The bacteria are concentrated from the thick suspension by centrifuging. IFAS is applied to the final pellet. Although the method is not very sensitive, preliminary applications were quite encouraging. ELISA has also been tried out as a detection method (Lopez et al., 1987).

Economic Impact
Like other strawberry leaf blights, *X. fragariae* causes a certain reduction in yield, but generally the disease is not destructive. However, heavy losses may occur with frequent overhead sprinkler irrigation.

Control Measures
ALS is a threat if *X. fragariae* is present in a strawberry planting and the environment is favorable for infection and disease development. Thus, the best strategy for controlling this disease is avoidance of the pathogen. Plants should be purchased from a reputable nursery and closely inspected for symptoms. However, be aware that plants can carry the pathogen without showing symptoms. DNA-based detection methods may permit nurseries to screen stock more rigorously in the future and reduce the risk of selling contaminated plants.

Because *X. fragariae* survives in leaf debris, new plantings should not be established in soil containing infected leaves or near infected plants.

Fixed copper compounds applied as directed on product labels may reduce bacterial populations and protect tissue from infection. However, chemical control of ALS has been inconsistent. Copper is active on the surfaces of plants but will not eliminate infections and probably does not affect systemic movement of the pathogen. Resistance of commercial cultivars of strawberry to ALS is not known.

The best way to control ALS is to use pathogen-free transplants. Since this is not always possible, growers should avoid harvesting and moving equipment through infested fields when the plants are wet. Minimizing the use of overhead sprinklers during plant establishment and for freeze protection also reduces the spread of the disease. The use of surfactant-type spray adjuvants should also be avoided when ALS is a threat since these products often help bacteria penetrate through the stomata and may enhance disease development.

Copper-based products can provide effective control of the disease in some instances, but low rates of copper should be used since phytotoxicity (reddening of older leaves, slow plant growth, and yield decrease) has been documented with repeated sprays. A number of copper products are labeled for ALS control on strawberry, such as copper hydroxide, copper oxychloride, basic copper sulfate, cuprous oxide, and various other copper compounds. All of these active ingredients suppress ALS, but it is important to apply the correct amount. Trial results have shown that preventive, weekly applications of copper fungicides at 0.3 lb of metallic copper per acre were effective in reducing disease symptoms without causing phytotoxicity on the plants. However, trial results have also shown that when disease pressure is low to moderate, the use of copper sprays did not significantly increase yield. Copper products can increase yield and decrease the possibility of fruit rejection only when environmental conditions are highly favorable for infection and spread.

Many other products have been tested over the years in the search for an alternative to copper. Actigard, a plant-resistant activator, manufactured by Syngenta, has been shown to suppress ALS. Actigard is used to control bacterial spot disease on tomatoes in Florida, but it is not currently approved for use on strawberry. However, registration materials have been submitted, and this product may be available for use during the strawberry season.

Use of healthy planting material and avoidance of conditions favoring disease are the main control methods. Treatments with copper-containing products have some effectiveness, but have to be applied very intensively, with a risk of phytotoxicity. Resistance to *X. fragariae* exists in breeding material, but not yet in commercial cultivars (Maas, 1984).

Phytosanitary Risk
EPPO considers *X. fragariae* as an A2 quarantine organism (OEPP/EPPO, 1986b), while IAPSC also considers it as of quarantine significance. The disease is absent from most strawberry-growing countries in Europe but probably has the potential to establish there. For *X. fragariae* indeed, the general climatic conditions that are said to favor disease in North America tend to occur in Central and Northern Europe rather than in the Mediterranean countries, where *X. fragariae* has been recorded until now. However, the specific influence of overhead sprinkler irrigation may be more characteristic of the Mediterranean countries. *X. fragariae* is certainly sufficiently damaging to deserve its quarantine status.

Phytosanitary Measures
EPPO recommends, in its specific quarantine requirements (OEPP/EPPO, 1990), that strawberry-planting material from infested countries should be derived from mother plants kept free from *X. fragariae* as part of a certification scheme (in preparation by EPPO), and in addition that the place of production should have been found free from the disease during the last five growing seasons. In addition, visual inspections during the dormant period can be useful. Inspectors should look for typical angular spots on old leaves or on their remains still attached to the runners. Samples from lots kept in cold storage must be inspected immediately after the runners are taken out and thawed. The spots can no longer be seen after only 1 day at room temperature.

7.21.2 BACTERIAL WILT OF STRAWBERRY

Pathogen: *Pseudomonas solanacearum* E.F. Smith.

Bacterial wilt of strawberry caused by *P. solanacearum* has been observed in nurseries in some areas of Shizuoka Prefecture, killing around 30% seedlings at most every year.

Symptoms
The earliest field symptoms are dropping of the older leaves of youngest plants. Infected plants may show wilting during the hottest period of the day and recover during cooler periods. Infected plants die within a few days after the appearance of the first symptoms. Older plants appear resistant to the bacterium.

    P. solanacearum invades only a limited number of xylem trachea cells even in young plants. Parenchymatous tissues are attacked preferentially and large lysigenous cavities are formed and become filled with bacteria. The symptoms found on the cultivar HOKO WASE under the natural conditions were the acute wilting of the younger leaves, resulting soon in total death of the plants.

Host
Not known whether the isolates of *P. solanacearum* from strawberry are specific for strawberry, or if isolates of *P. solanacearum* from other hosts are also pathogenic to strawberry.

Geographical Distribution
Japan, Taiwan.

Pathogen
The cultures isolated from the diseased plants were identified as pathovars 3 (biovar III) and 4 (biovar IV). In the inoculation tests, however, the disease development was rather mild and slow regardless of the inoculation techniques such as the prick method or the immersion method in which the root systems or the cut-end of runners were dipped into the bacterial suspensions for 10 min.

The pathogen could survive in the tissues of some inoculated plants until next spring. The reisolation trials made 8–10 months after inoculation revealed that the rates of the carrier plants were about 10% in the cultivar FUKUBA and 5% in HOKO WASE. In most cases, these plants grew as vigorous as the noninoculated healthy plants and set fruits normally. In the diseased plants, either under the natural conditions or artificially inoculated, plugging by bacterial cells was usually restricted to only a small number of tracheary elements in a xylem tissue, where the cell walls were often broken down forming bacterial pockets. It was remarkable that the bacterial fissures were often observed in the parenchymatous tissues of the infected stems even when the invasion of the vascular tissues was slight.

Control Measures
Not worked out.

7.21.3 MARGINAL CHLOROSIS OF STRAWBERRY

Pathogen
An elongated bacterium has been uniquely associated with this disease.

Symptoms
Dwarfing of plants, cupping, and yellowing of leaves, especially at leaflet margins occurs.

Host Range
Cultivated strawberry.

Geographical Distribution
It is widespread and severe in France and Spain in nurseries and in fruiting fields.

Transmission
Marginal chlorosis spreads in the field by unknown means.

Control
Not worked out yet.

7.22 MULBERRY

7.22.1 BACTERIAL BLIGHT OF MULBERRY

Pathogen: *Pseudomonas syringae* pv. *mori*

Symptoms
Leaf spots are small, irregular, and brown-black, usually with yellow haloes. Infected leaves are often distorted. Infected buds may become disfigured as they swell. Young shoots may show rapid necrosis and dieback. Occasional stem cankers occur with brown streaks in the wood, which may exude ooze. Infected bushes and trees often appear stunted. Infected plants show blossom and tip dieback with lesions on flowers and fruits.

In the spring, the first sign of infection is seen as young leaves emerge from buds. The expanding buds develop large blackened areas. On the blade, midrib, and veins of the young leaves, small irregularly shaped brown-to-black spots develop (Photo 7.54). These spots are surrounded by a yellow halo. The rapidly expanding leaves may become curled or distorted. Long ragged cankers develop on the infected young shoots, which often die.

Predisposing Factors
Bacterial blight is most prevalent in prolonged, rainy springs, with the wettest areas showing the most extensive symptoms and infections.

Cool, wet weather in spring favors *P. syringae*. Young leaves are more susceptible to infection. Bacteria appear to gain entry into the leaf by colonizing and destroying epidermal idioblasts.

PHOTO 7.54   Symptoms of bacterial blight on mulberry leaves. (Courtesy of Brown Jr., W.M., Bugwood.org.)

Young idioblasts of mulberry do not have calcium deposits, while older leaves have these concretions (cystoliths). New infections start each spring from bacteria that have lived through the winter in infected shoots from the previous season.

Disease Spread
Bacterial leaf spot is spread by bacteria oozing from cankers during wet weather and is distributed by rain and wind. Wet weather favors the spread of the disease.

Control Measures
This disease is difficult to control once it has become well established in a tree. Control efforts should include various methods.

Hygiene
In autumn, prune out and burn all dead shoots to eliminate as much as possible of the bacterial carryover from season to season.

In spring, prune out and burn blighted shoots as soon as they are detected.

Cultural Control
- Minimize wounds to limbs and new shoots.
- Prune out and destroy infected shoots and branches during the late dormant season.
- Space plantings to provide good air circulation.

Chemical Control
The following chemicals are registered on weeping mulberry; do not use on edible types.

- Copper-Count-N at 1 quart/100 gal water; 48 h reentry or
- CuPRO 2005 T/N/O at 0.75–3 lb/A (or 1–3 Tbsp/1000 ft^2) dormant or at 0.75–2 lb/A when new growth is present; 24 h reentry or
- Junction at 1.5 lb/100 gal water; 24 h reentry or
- Monterey Liqui-Cop at 3 Tbsp/gal water or
- Nu-Cop 50 DF at 1 lb/100 gal water; 48 h reentry

Spraying does not reliably control bacterial blight and the best defense is to remove affected branches. Prune out these affected branches during dry weather; wet, rainy weather will further spread the disease. Do not overfertilize affected plants, and water only at soil level. Mulberry trees with bacterial blight damage that is confined to the leaves and small twigs will usually recover, providing affected branches are pruned off. If cankers appear on the mulberry's trunk, it will most likely not recover and you should remove the tree.

7.22.2 Bacterial Wilt of Mulberry

Pathogen: *Enterobacter cloacae*

In August of 2006, a new bacterial disease was noted in Hangzhou mulberry orchards of Zhejiang Province, China, where bacterial wilt of mulberry caused by *R. solanacearum* was previously reported (Xu et al., 2007). In the summer, this new disease caused severe wilt, especially on 1- or 2-year-old mulberry plants that resulted in premature plant death.

Symptoms

Leaf wilt symptoms generally started on older leaves at the bottom of the plant and spread to the younger leaves. The leaves of infected plants became withered and dry, turned dark brown, and eventually the plants became defoliated. The root xylem of infected plants was moist and discolored with brown stripes. The phloem was asymptomatic; however, in severe infections, the phloem was decayed. The observation of wilting proceeding from the bottom of the plant to the top distinguishes this disease from bacterial wilt caused by *R. solanacearum*.

Pathogen

Five bacterial strains isolated from infected mulberry plants showed characteristics similar to those of the standard reference strain of *En. cloacae* subsp. *cloacae* IBJ0611 from China, but differed from *R. solanacearum* IBJ35, *Enterobacter cancerogenus* LMG2693^T, and *En. cloacae* subsp. *dissolvens* LMG2683^T from the University of Gent, Belgium, in phenotypic tests, including the Biolog Identification System version 4.2 (Biolog Inc., Hayward, CA), pathogenicity tests, transmission electron microscopy (TEM, KYKY-1000B, Japan) observation, and gas chromatographic analysis of fatty acid methyl esters (FAMEs) using the Microbial Identification System (MIDI Company, Newark, DE) with the aerobic bacterial library (TABA50).

Isolates were Gram negative, facultative anaerobic, and rod-shaped, measuring 0.3–1.0 × 1.0–3.0 µm with peritrichous flagella. Colonies on nutrient agar were light yellow, smooth, circular, entire, and convex with no green fluorescent diffusible pigment on King's medium B (Xu et al., 2007). Weak hypersensitive reaction was observed on tobacco 3 days after inoculation. All five strains were identified as *En. cloacae* with Biolog similarity of 0.662–0.863 and FAMEs similarity of 0.632–0.701. Inoculation of 10 6-month-old intact mulberry plants of cv Husang with cell suspensions containing 10^9 CFU/mL by pinprick at the base of the stem reproduced symptoms observed in natural infections.

En. cloacae has been reported from the United States as the cause of internal yellowing of papaya fruits (Nishijima et al., 1987) and rhizome rot of edible ginger (Nishijima et al., 2004). This was the first report of mulberry wilt caused by *En. cloacae* in China.

Control Measures
Not yet worked out.

7.23 AVOCADO

7.23.1 Bacterial Canker of Avocado

Pathogen: *Pseudomonas syringae* or *Xanthomonas campestris*

In 1980, cankerous lesions were observed on avocado trees in South Africa (Myburgh and Kotzé, 1982). The causal agent was identified as *P. syringae* (Korsten and Kotzé, 1985). Prior to this, cankers were observed on trunks and branches of avocado in southern California, but the occurrence has generally been infrequent and the cause not determined (Cooksey et al., 1993). Bacterial canker occurs sporadically in all major avocado-growing areas of South Africa (Korsten, 1984; Korsten and Kotzé, 1987), several counties of California (Ohr and Korsten, 1990; Cooksey et al., 1993), and in Australia (Scholefield and Sedgley, 1983). At present, it is of little economic impact and the percentage of trees showing canker symptoms in groves is usually low (Cooksey et al., 1993). However, severely infected trees show retarded growth, defoliation, and low fruit yields (Korsten, 1984). The disease occurs predominantly on the cultivars Hass and Edranol and to a lesser extent on Fuerte (Korsten and Kotzé, 1987). Although symptoms are similar in all countries, the causal agent appears to differ.

P. syringae has been isolated from the periphery of discolored streaks in South Africa and Australia (Korsten, 1984), and X. campestris in California (Cooksey et al., 1993). Isolates from South Africa and California were pathogenic to avocado seedlings in artificial inoculation studies, but not those from Australia. However, Australian researchers consider their P. syringae isolate to be a secondary invader and attribute cankerous symptoms to a boron deficiency. Furthermore, although X. campestris was isolated from almost half the samples screened in California, the possibility exists that some of the samples could have had black streak (Cooksey et al., 1993). X. campestris has also been isolated in South Africa but failed to induce symptoms upon artificial inoculation (Korsten, 1984). Various analyses suggest that P. syringae from South Africa and X. campestris from California represent new pathovars (Cooksey et al., 1993). Neither pathogen appears to be particularly aggressive or destructive in pathogenicity tests.

In South Africa, the cankers were associated with P. syringae, but similar symptoms in California are associated with X. campestris and not with P. syringae.

Symptoms

Symptoms Caused by *Pseudomonas syringae*
Cankerous lesions appear first as slightly sunken and darker areas on the bark, with a necrotic, watery pocket underneath the bark (Korsten and Kotzé, 1987). In more advanced cankers, the bark splits at the edge of the lesion, allowing fluid to ooze out. As the fluid dries, it leaves a white powdery residue at the periphery. During spring and autumn, the white residues surrounding the lesions can easily be spotted on trunks or branches. Cankers are 2–10 cm in diameter, and usually appear first at the base of the tree from where they spread upward, mostly in a straight line. Reddish-brown necrotic tissue normally is present in the cortex underneath the canker, with similarly colored streaks extending for up to 30 cm above and below the pocket. Necrotic streaks between cankers are usually in the xylem, sometimes toward the center of branches or trunks (Cooksey et al., 1993). No leaf or fruit symptoms are associated with the disease (Korsten and Kotzé, 1987).

Symptoms Caused by *Xanthomonas campestris*
Bacterial cankers appear as slightly sunken, dark areas on the bark and vary in size from about 1–4 in. in diameter. Bark around cankers may crack. Fluid often oozes and dries, leaving a white powder around or over the lesion. Usually cankers appear and spread upward in a line on one side of the trunk or branch. Cutting under the bark surface reveals a decayed, reddish-brown necrotic pocket, which may contain liquid. Dark streaks in the wood radiate out both above and below from the lesions. Severely affected trees may have pale, sparse foliage and low yields on one branch or on the entire tree, but this is rare.

Although symptoms of the canker disease in California and South Africa are similar, the disease in California was associated with X. campestris rather than P. syringae. Both pathogens caused a similar spreading necrosis in stems of inoculated avocado plants. Cankers resembling early symptoms on mature trees developed after inoculation of young plants with the X. campestris strains from avocado in California, but mature trees have not yet been inoculated to observe the development of large cankers.

Both pathogens have been inconsistently isolated from canker samples (Korsten and Kotze, 1987). This is not too surprising; however, the high numbers of saprophytic bacteria that invade the necrotic tissues were associated with the cankers. In addition, several of the samples from California probably had black streak disease (Jordan, 1981; Jordan et al., 1983) instead of bacterial canker.

Neither the pathogen appears to be particularly aggressive or destructive in pathogenicity tests with avocado plants nor have any symptoms been produced by inoculation of leaves or fruit. Although widespread geographically, the percentage of trees showing canker symptoms in groves is usually low. Some heavily infected trees have shown poor growth and yields, however, suggesting that further studies on variation in aggressiveness among strains and the source and spread of these pathogens should be pursued.

Pathogen

X. campestris has now been isolated from 25 avocado samples with bacterial canker symptoms from 14 groves in San Diego, Orange, Los Angeles, Ventura, and Santa Barbara Counties. No *P. syringae* was recovered from any of the diseased samples. A combination of traditional physiological tests, isozyme analyses, RFLP analyses, DNA fingerprinting with rare-cutting enzymes, and the computerized Biolog system was used to compare the different isolates of the California pathogen. The strains of *X. campestris* from each site are related to each other but are distinct from the strains of *X. campestris* from other host plants, indicating that they represent a previously undescribed pathovar of this species.

Inoculations of Hass avocado plants with five of these strains resulted in vascular and pith necrosis that spread above and below the inoculation point over time to a distance of 15 cm or more. The bacterium spreads at a rate of 0.5 cm/day above and below the inoculation point in the stem of 2-year-old plants. Spread below the inoculation point stopped when it reached the graft union, however. In addition to the internal, spreading necrosis produced by the inoculation of nearly all of these strains, some plants have developed cankers that look like young versions of the large cankers seen on naturally infected trees.

Disease Spread

Bacterial canker presumably spreads through nursery practices, since young cankerous lesions occur on newly planted trees (Korsten, 1984). The disease appears to be systemic, considering the necrotic tissue connecting cankerous pockets and the phenomenon that young cankers usually appear acropetally to old lesions in infected trees. The epiphytic phase of the pathogen has been detected in South Africa on leaves and twigs by means of monoclonal antibodies raised against *P. syringae* (Korsten and Kotzé, 1987). Insects apparently are not involved in disease transmission (Korsten, 1984).

Control Measures

Chemical Control

Cutting out cankerous pockets and applying copper sulfate to the wounds have been attempted in California. Trees injected with 3% streptomycin or 1% chloramphenicol had fewer cankerous lesions than untreated trees (Korsten, 1984).

Cultural Control

Normally the disease is a minor problem. Usually no control is necessary on established trees.

If the disease is severe and yield is affected, remove the tree. Keep trees healthy and provide good cultural care. Provide appropriate amounts and frequency of irrigation and good uniformity of water distribution among trees. Use certified, disease-free nursery stock if available. Regularly inspect young trees, and remove and dispose young trees if they are infected. Nurseries should use stringent sanitation, regularly screen stock for disease, and dispose affected trees so they are not planted.

7.23.2 Bacterial Blast of Avocado

Pathogen: *Pseudomonas syringae* van Hall

In August 1983, a blast disease of avocado pear occurred on four different plantations in the southern parts of the state of Mexico at about 2000 m elevation and in an area of about 50 ha, where many avocados (*Persea americana* Miller) are grown. Frost may occasionally occur early in the spring or late in the autumn when the fruit is setting or maturing. The role of frost was not ascertained but may predispose to disease. The cultivar Fuerte was particularly susceptible.

Symptoms

The disease was characterized by anthe appearance of dark brown irregular areas that merged and covered a significant part of the maturing fruit. Affected areas took on a variety of forms. In some cases, they were either raised above the surface of the fruit or frequently located in the central or distal part. In other cases, lesions were not raised and were frequently depressed. Later lesions cracked, exposing the mesocarp tissue. As these dark areas enlarged, the mesocarp tissue softened and became sunken. Infected avocado pears were randomly distributed throughout the crown of 10-year-old trees. In advanced stages, the avocado pear produced brown exudates from the lenticels. Dried exudates had a whitish appearance and dissolved rapidly with rain. A delimitation zone between diseased and healthy tissues was formed at some part of the fruit and the decayed section frequently fell off. Entire fruit drop was a common occurrence with heavy infection. The vascular bundles of many fruit turned black. When this disease was first observed in August, some trees lost as much as 10% of the fruit, and as the disease advanced, up to 20% of the fruit was lost within 2 months.

Symptoms similar to these have been described previously as a bacterial avocado blast in Israel (Volcani, 1946) and in California (Smith, 1926a,b). Both reports indicate that the disease was caused by *P. syringae* van Hall.

Control Measures

Not yet worked out.

7.23.3 Crown Gall of Avocado

Pathogen: *Agrobacterium tumefaciens*

In the fall of 1962, a Fuerte tree with a large swelling or gall at the ground level was observed on an avocado property in the San Luis Rey Heights section of Fallbrook.

Symptoms

A gnarled, irregularly cracked, and swollen growth on the lower trunk at and just below the ground line is observed on an avocado tree; the tree had been wounded in that area about a year previous to the appearance of the gall.

Descriptions of this disease have appeared in the literature on other plant for over 100 years. The bacterium causes galls on a wide variety of plants, including rose, apple, pear, peach, cherry, other deciduous fruit trees, tomato, grape, willow, poplar, sugarbeet, walnut, Pelargonium, and many others. However, crown gall has not been mentioned in any of the publications on avocado to date. E. F. Smith did report in 1920, however, that he was able to produce "with difficulty" some galls on "Persea," which was probably *Pe. americana*, the avocado (Smith, 1920). These inoculations were apparently made on seedling plants in his greenhouse, along with inoculations on a number of other plants. In 1945, G.R. Hoerner also reported obtaining galls on *Pe. americana* by inoculating stems of seedling plants (Hoerner, 1945). He obtained only two galls from six inoculations, however, and the galls took 72 days to develop. C.O. Smith, working at Riverside, stated that he had obtained negative results in his attempts to inoculate avocado with the crown gall bacterium (Smith, 1912). No other attempts have apparently been made to inoculate avocado plants, and the bacterium has not been isolated from naturally occurring galls on avocado prior to this report.

Thus the observation of 1962 was the first report of actual occurrence of the bacterial disease, crown gall, caused by *A. tumefaciens* on avocado under natural conditions. Occasional swellings have been seen in the past on avocado trunks, but the condition is not common, and this was the first instance of isolation of the bacterium in culture and reproduction of the disease on avocado and tomato seedlings in the greenhouse. It is probable that avocado is quite resistant to crown gall, based on the difficulty in producing galls on plants receiving heavy inoculation. In addition, the early report by E.F. Smith of gall production on avocado by artificial inoculation noted that galls were produced with difficulty. It is not anticipated, therefore, that this will be a significant problem on avocado trees.

Control Measures
Currently, no definite control measures are recommended other than avoiding wounds on the lower trunk of the trees, which may facilitate the entrance of the bacterium. Nurserymen should be alert for any occurrence of the disease in the nursery; crown gall could become a problem in nurseries, as it has in the case of other types of nursery stock. If it does appear in planting stock, nurserymen should use measures that have been developed for control of the problem on rose and other woody plants, involving removal of diseased material and fumigation of soil and planting with disinfected pathogen-free planting stock.

7.24 KIWI

7.24.1 BACTERIAL CANKER OF KIWI FRUIT

Pathogen: *Pseudomonas syringae* pv. *actinidiae*

P. syringae pv. *actinidiae* was first identified in Japan in the 1980s where the disease caused damage in kiwifruit orchards. It has since also been identified in Korea.

The disease was noticed in Northern Italy in 1992 where it remained sporadic and with a low incidence for around 15 years. In 2007–2008, economic losses started to be observed particularly in the Lazio region and the possible spread of the disease, to other kiwifruit-producing regions in Italy, began to cause concern. *P. syringae* pv. *actinidiae* is continuing to emerge in the Mediterranean region where the secretariat of European and Mediterranean Plant Protection Organisation (EPPO) has included the disease to the EPPO Alert List.

P. syringae pv. *actinidiae* was detected in New Zealand in the Bay of Plenty (Te Puke) in November 2010. Since then the disease has spread widely throughout the Bay of Plenty and is now also present (October 11, 2012) in Waikato, Franklin District, and Coromandel. The number of *P. syringae* pv. *actinidiae* positive orchards is now over 1600 and 57% of the hectares on kiwifruit orchards now contain at least some infected vines.

Symptoms
Symptoms include angular-shaped spots, often associated with a halo, although not all leaf spots clearly exhibit the halo, brownish discoloration of buds and, in advanced stages of infection, the leakage of red-rusty gum. Not all symptoms appear at the same time.

P. syringae pv. *actinidiae* infection can result in the death of kiwifruit vines. *P. syringae* pv. *actinidiae* carries no risk associated with human or animal health and does not affect plants other than kiwifruit vines.

Growth of the bacteria outside/inside the vines can result in leaf spotting, cane/leader dieback and, in extreme cases, vine death accompanied by the production of exudates (a rusty red liquid discharge).

Disease Spread and Predisposing Condition
Symptoms are usually expressed during spring and autumn when climatic conditions are favorable, that is, cool temperatures, persistent rains, and high humidity. *P. syringae* pv. *actinidiae* is temperature-sensitive and active between 10°C and 20°C but limited by temperatures over 25°C.

The disease can be spread via windborne pollen, strong winds, and heavy rainfalls. It is also believed to be spread by footwear, vehicles and orchard tools, and animals and humans. The bacterium infects the plant through natural openings (stomata and leaf axis) and wounds.

The bacterial disease has had a serious economic impact on kiwifruit production. The virulent form of *P. syringae* pv. *actinidiae*, known as *P. syringae* pv. *actinidiae* V, is known to have seriously affected Gold kiwifruit orchards in Italy and to have infected orchards across other countries in Europe (Spain, France, Switzerland, Portugal). The haplotype, or form, found in New Zealand is genetically similar to the European haplotype, but genetically distinct from the form responsible for damage in Japan and Korea.

The bacterial disease that has blighted New Zealand kiwifruit orchards originated in China and was also responsible for an outbreak in Italy and Chile according to new researches.

Kiwifruit canker, or *P. syringae* pv. *actinidiae*, has spread to more than 1000 New Zealand orchards since it was discovered in the Bay of Plenty region in November 2010.

In the early stages of the disease outbreak, the government contributed $25 million, matched dollar for dollar by industry, for the management of *P. syringae* pv. *actinidiae*.

Combating the disease was expected to cost between $310 million and $410 million over the next 4 years with the long-term bill rising to between $740 million and $885 million.

Losses and Impact
New Zealand kiwifruit production particularly gold kiwifruit has been rocked by the news of bacterial disease due to *P. syringae* pv. *actinidiae* and that was the sobering news for the world's number-two kiwifruit producer after Italy is coming into sharper focus. Ministry of Agriculture and Forestry (MAF) has released a half-year update that shows loss of vines since the virulent strain stuck in November 2010 could take production of Gold kiwifruit from 30 million trays in 2011 to 20 or even 10 million trays in the 2012 season.

P. syringae pv. *actinidiae* has hit hardest around Te Puke in the Bay of Plenty where 41% of the total planted area is located. At this stage, just over a quarter of kiwifruit orchards in New Zealand are known to have the bacterium present. Overall, up to 20% of Gold kiwifruit in Te Puke could still be harvested and up to 80% in the wider Bay of Plenty. This would yield an export volume of around 16 million trays of Gold kiwifruit. *P. syringae* pv. *actinidiae* affects Green kiwifruit orchards at a slower rate, but orchard infection numbers have ramped up during the 2011 spring. The impact of *P. syringae* pv. *actinidiae* on green orchards in the medium term remains uncertain. Overall, export volumes are expected to fall 21% to 89 million trays and export returns are expected to fall 18% to $862 million, for the year ending March 31, 2013.

This is a direct result of the production problems associated with the *P. syringae* pv. *actinidiae* (v) outbreak which will see up to 50% of the older Gold Kiwifruit variety (Hort16a) perish after contracting the disease in the 2010/2011 growing season. Total production of kiwifruit in 2012 is down by 21% and is estimated to be 343,300 tons.

The success of the sector over the past 10 years initially began with the total control of flow of fruit on to world markets (excluding Australia) by Zespri on behalf of the growers. After a shaky start due to fruit loss problems, Gold kiwifruit has been a raging success for Zespri and the growers who purchased licenses to grow the limited supply variety. Nowhere is this more evident than with exports to Japan. In 2010, Japan took 17% of the total volume but returned 31% of the total value of all exports of kiwifruit.

The volume of kiwifruit was kept at between 55,000 and 59,000 tons from 2004 to 2009, and only in 2010 did the volume rise to 61,000 tons. Now with *P. syringae* pv. *actinidiae* (v) ravaging the Gold crop, Zespri may find marketing green kiwifruit more difficult than ever to maintain the price premiums it has enjoyed in the past, even though the crop will be smaller. The prognosis for the industry is not encouraging over the next 2–3 years with the effects of *P. syringae* pv. *actinidiae* (v) likely to further reduce the export volume in 2014 and onward.

There will be rationalization in the postharvest sector over the next 2 years. However, a serious, well-funded, and coordinated response to the incursion is being mounted.

Control Measures
In the long run, plant breeding is likely to be the key to living with the disease. Already more toler-ant cultivars are being identified and it is likely that replanting/grafting over to one or two of these cultivars have started recently.

7.25 WATERMELON

7.25.1 BACTERIAL FRUIT BLOTCH OF WATERMELON

Pathogen: *Acidovorax avenae* subsp. *citrulli*

Bacterial fruit blotch (BFB) has great potential to cause significant economic losses to cucurbit production and has been responsible for up to 90% losses of marketable yield in some watermelon fields.

BFB is a serious disease infecting watermelons. The causal organism, *Aci. avenae* subsp. *citrulli*, was first identified in the United States, in Florida in 1989 and in Missouri in 1994. The disease was subsequently detected in South Carolina, North Carolina, Maryland, Delaware, and Indiana as the growing season progressed that year. It is suspected that the cause for the initial outbreak was seed infested with the bacterium. Since then, fruit blotch has been detected every year to varying degrees of significance in melon-producing areas nationwide.

In April 2001, BFB was detected on the watermelon variety Carousel growing in a Florida trans-plant house. Infested seed from Syngenta, the producers of Carousel, was the suspected source of the outbreak.

This pathogen can cause significant yield losses depending on the stage of growth in which it infects the crop. Infections that develop early can destroy up to 100% of the marketable watermelon crop.

Symptoms
Symptoms of BFB can be observed on the seedling transplants, mature leaves, and fruits.

Seedling Transplants
Water-soaked, "oily" areas on the underside of the cotyledons or seedling leaves often paral-leling the veins with a yellow halo are characteristic symptoms. The infected areas dry up and become elongated, angled, black, necrotic patches. Some seedlings will collapse and die imme-diately from infection. Others can retain the bacterial infection and will not exhibit symptoms until fruit set.

Leaves
Because the vines do not drop their leaves when infected with *Aci. avenae*, the lesions are difficult to see. The lesions tend to be small, dark, and angled. Leaf lesions are significant reservoirs of bac-teria for fruit infection. Leaf lesions will usually be observed at temperatures above 32°C. Stems, petioles, and roots are not infected and thus do not show symptoms.

Fruits
Bacteria from leaf lesions can spread and infect developing fruit. Infection of fruit occurs at flow-ering and early fruit set. Two- to three-week-old developing fruit that has not formed a wax layer is most susceptible to *Aci. avenae* infection. Once the fruit matures and develops a wax layer, it is more difficult for the bacteria to invade the fruit. The diagnostic symptom of BFB is a dark green stain or blotch on the upper surface of the developing fruit. The blotch may be 0.5 in. in diameter at first, but will rapidly expand to cover the entire fruit surface (Photo 7.55) within a week if environ-mental conditions are favorable. The bacterial infection does not extend into the meat of the melon, but will cause the rind to rupture (Photo 7.56), enabling infection by secondary pathogens that cause the fruit to rot.

PHOTO 7.55   Fruit with small, water-soaked spots. (Courtesy of Holmes, G., California Polytechnic State University, San Luis Obispo, CA, Bugwood.org.)

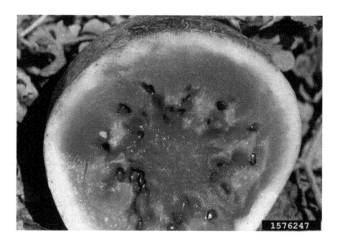

PHOTO 7.56   Cross-section through mature fruit shows external and internal damage. (Courtesy of Holmes, G., California Polytechnic State University, San Luis Obispo, CA, Bugwood.org.)

Pathogen

Aci. avenae subsp. *citrulli* is a Gram-negative, rod-shaped bacterium with average dimensions of 0.5×1.7 µm and is motile by a single polar flagellum. The bacterium is strictly aerobic, positive for oxidase activity, and grows at 41°C but not at 4°C. *Aci. avenae* subsp. *citrulli* grows on many general bacterial growth media and produces smooth, round, cream-colored, nonfluorescent colonies after 48 h on King's medium B. Based on physiological characteristics, the pathogen was initially classified as a member of the group-III pseudomonads and identified as *Pseudomonas pseudoalcaligenes* subsp. *citrulli*. Based on nucleic acid hybridization data and analysis of whole-cell protein and carbon substrate utilization profiles, the pathogen was reclassified as *Aci. avenae* subsp. *citrulli*.

Originally *Aci. avenae* subsp. *citrulli* was reported to cause watermelon seedling lesions and blight, but not fruit rot. Additionally, the original strain recovered from plant introductions at the Georgia (GA) Agricultural Experiment Station did not induce a HR on tobacco leaves. In contrast, strains recovered from the first BFB outbreaks in commercial watermelon fields in 1989 were HR-positive and caused severe fruit rot. This discrepancy led to confusion about the identity of the causal agent of BFB. However, based on fatty acid analysis, DNA fingerprinting data, and carbon substrate utilization, it was shown that at least two distinct subgroups of the pathogen existed. In general, group-II strains are highly aggressive on watermelon, mildly aggressive on other cucurbits, and include strains that caused the original BFB outbreaks in commercial fields. In contrast, group-I strains are mildly to moderately aggressive on a wide range of cucurbit hosts and include the strains that caused watermelon seedling blight on plant introductions at the GA Agricultural Experiment Station. This strain might have mutated to reduced aggressiveness, or it might represent a genetically distinct subpopulation. At present, there is still much that is unknown about the biology of *Aci. avenae* subsp. *citrulli.*

Disease Cycle

A pathogen is introduced with a contaminated watermelon seed. Contaminated seed results in infected seedlings with characteristic lesions on cotyledons. Overhead irrigation disperses the pathogen throughout transplant facilities. A few infected seedlings threaten all others. Infected seedlings with inconspicuous lesions are transplanted to the field.

Under field conditions, high relative humidity and high temperature favor BFB development. *Aci. avenae* subsp. *citrulli* is disseminated by wind-driven rain and irrigation water and causes foliar lesions and blight symptoms. Infected vines are usually not killed by infection, and in most cases, symptoms on mature foliage are difficult to recognize. Foliar lesions and epiphytic populations serve as reservoirs for *Aci. avenae* subsp. *citrulli* inoculum and contribute to secondary infections. The bacteria penetrate leaves via stomata and wounds and remain in the apoplast (intercellular spaces) of infected tissue. There is no evidence of systemic bacterial migration through the plant. For fruit infection, bacteria deposited on the rind of fruit 2–3 weeks after anthesis (flowering) swim through open stomata to initiate infections. After this period, stomata become blocked by the deposition of waxes on the fruit surface that prevent further invasion. Hence, mature fruits are generally not susceptible to natural infection by *Aci. avenae* subsp. *citrulli* unless wounding occurs.

During the early stages of fruit development, BFB symptoms are inconspicuous or absent, but appear suddenly at, or just prior to, harvest maturity. Surface lesions expand rapidly, and fruits may eventually rot in the field. *Aci. avenae* subsp. *citrulli* can overwinter in decaying infested rind tissues or in infested seeds that can produce volunteer plants in subsequent crops.

Disease Development

The fruit blotch bacterium can be introduced into watermelon fields by infested seed, infected transplants, natural spread from alternate hosts (wild cucurbits), or volunteer watermelon. The bacterium can be a surface contaminant of seed harvested from infected watermelon. Infected transplants represent the most important means of disease transmission because watermelon fruit blotch can spread throughout the transplant operation, can be asymptomatic on transplants, and can lead to high numbers of infected transplants entering field plantings.

BFB disease development is favored by warm wet weather such as those that exist in Florida in May–June during the spring watermelon season and August–September during the fall watermelon season. The disease can develop quickly in these weather conditions. It can cause 100% infection from just a few primary infection sites within an infested field.

Bacteria within leaf lesions of infected transplants can cause infections on susceptible (2–3 weeks old) watermelon fruit. The naturally occurring waxy layer that develops on maturing watermelon hinders bacterial invasion and infection unless the fruit rind becomes injured and the protective wax layer is compromised.

In Missouri, the majority of watermelon acreage is field established by transplants. The infected seed is sown in the greenhouse in soilless potting media and plastic trays. The greenhouse is an excellent environment for the spread of BFB. High temperatures and humidity are requisites for the development of BFB. Many transplant production greenhouses top-water transplants, which aids in splash dispersal of the pathogen to other plants. Bottom watering of transplants will help control BFB. Movement of infected plants from the greenhouse to the field can have serious consequences. Once in the field, high temperature, humidity, and wet weather favor the spread of the disease.

Diseased fruit decays and infected seeds and cucurbit weeds (e.g., citron) are hosts for the BFB pathogen. BFB cannot survive for more than a few weeks during the summer without a plant host.

Disease Management

Since there are no resistant commercial cultivars, successful BFB management depends on the exclusion of primary inoculum, cultural practices, and the use of copper-based bactericides. Currently, no single tactic is effective for managing BFB as the efficacy of chemical control depends on environmental conditions. Additionally, effective disease management must be practiced in seed and transplant production.

Bacterial Fruit Blotch Management in Seed Production Fields

The most important management strategy for BFB is the exclusion of *Aci. avenae* subsp. *citrulli* by using pathogen-free seeds and seedlings. It is impossible to guarantee pathogen-free seed, but seed health testing reduces the risk of outbreaks. Currently, the standard seed test for BFB is the seedling grow-out assay in which 10,000–30,000 seed/lot are planted under conditions conducive for BFB development. Seedlings are visually inspected after 18 days for BFB symptoms.

To reduce the risk of seed infestation, cucurbit seeds are produced in dry, cool climates in regions of countries with no history of BFB, and 3- to 5-year crop rotations with noncucurbit hosts are routinely practiced. Seed fields are visually inspected and only seeds from BFB-free fields are used.

Currently, no chemical or physical seed treatments are 100% effective at eradicating *Aci. avenae* subsp. *citrulli*. While seed treatments including thermotherapy, NaOCl, fermentation, HCl, and peroxyacetic acid significantly reduce BFB seedling transmission, they can adversely affect seed physiology. Two factors that influence seed treatment efficacy are (1) the inability of seed treatments to penetrate the seed coat and (2) the location of bacteria on/in the seed. Because the risk of BFB development in transplant houses is high for seed with low levels of infestation, seed treatment alone cannot control BFB.

Moreover, seed treatments used to reduce bacterial contamination of watermelon seeds can inhibit germination of triploid seedless melons.

BFB has been reported to infect cantaloupe and tomato, but the preferred host is watermelon.

Bacterial Fruit Blotch Management in Transplant Production Facilities

To avoid introducing *Aci. avenae* subsp. *citrulli* into transplant houses, only tested, pathogen-free seeds are planted. To minimize secondary pathogen spread, physical contact with seedlings is kept to a minimum and sterilized transplant flats are used for each crop to limit the risk of carry-over contamination. Adequate fertilization and pest control programs are implemented to promote healthy seedling growth.

To prevent pathogen spread, traffic and the movement of equipment between houses are minimized. Ideally, seed from different lots should be planted in separate greenhouses; however, this is usually not feasible and instead 60 cm (24 in.)-high plastic barriers should be established between plantings. If possible, ebb and flow irrigation should be used instead of overhead irrigation to reduce splash dispersal of *Aci. avenae* subsp. *citrulli*. However, this is cost-prohibitive and rarely implemented. With overhead irrigation, watering should be done at midday to facilitate rapid drying of plant surfaces, and the water delivery pressure should be low to limit aerosol generation.

To facilitate early detection of BFB outbreaks, visual inspection of seedlings is routinely conducted. Since it is possible for *Aci. avenae* subsp. *citrulli* to survive epiphytically on asymptomatic seedlings, all seedlings produced in transplant houses with BFB should be discarded. This is an expensive option since planting will be delayed and targeted market windows will be missed. Hence, in cases where it is economically unfeasible to discard exposed seedlings, symptomatic plants and those in the immediate vicinity should be discarded. The remaining seedlings should be treated with copper-based bactericides. Additionally, relative humidity in the transplant house should be reduced by increasing air flow, and aggressive copper bactericide–based BFB management programs should be implemented in the field. After each seedling production cycle, transplant houses should be completely disinfested and left empty for at least 2–3 weeks before planting cucurbits.

Bacterial Fruit Blotch Management in Fruit Production Fields

Though several watermelon plant introductions are resistant, all commercial cultivars appear to be susceptible to the disease. To exclude *Aci. avenae* subsp. *citrulli* from commercial fruit production fields, tested seed and BFB-free seedlings are used to plant fields. Three-year rotations with non-cucurbit hosts are routinely employed. Debris, including watermelon culls from fields with BFB, is deep plowed at the end of each growing season. Subsequent cucurbit crops are established in new fields physically separate from the outbreak site. If possible, alternative hosts such as wild and volunteer cucurbits should be eliminated from regions around production fields and transplant houses.

For preventive management, biweekly applications of copper-based bactericides, for example, cupric hydroxide, copper hydroxosulfate, or copper oxychloride, are employed at the recommended rate or weekly at half the recommended rate. Preventive sprays are initiated at or before anthesis (flowering) and continued until fruits are mature. If BFB symptoms develop, weekly applications of the full recommended rate of copper-based bactericides are employed. To prevent spread, work is not conducted in fields with BFB when the foliage is wet, and field equipment is decontaminated before moving between fields.

Significance

BFB has great potential to cause significant economic losses to cucurbit production and has been responsible for up to 90% losses of marketable yield in some watermelon fields. BFB was first observed in 1965 on watermelon at the USDA plant introduction station, Griffin, Georgia, where it caused seedling blight symptoms. The disease was confined to seedlings at the station, suggesting that it was due to the introductions of plant seeds. Furthermore, no outbreaks were reported in commercial watermelon fields during this period. By 1988, BFB outbreaks were observed in commercial watermelon fields in the Mariana Islands (in the North Pacific Ocean). In 1989, the first outbreak in the continental United States was observed in Florida. It is clear now that the source of inoculum in the commercial watermelons was independent of the seedling blight outbreak at the USDA Plant Introduction Station in Georgia. Since 1989, BFB outbreaks have occurred sporadically in the major cucurbit-producing states in the United States, with significant economic impact. Most notably, BFB caused significant losses in 1994, when the seedborne nature of the pathogen, along with the lack of adequate seed health assays, resulted in widespread outbreaks in Florida, Georgia, Indiana, South Carolina, and Texas. These outbreaks brought national attention to BFB and highlighted the general threat posed by seedborne diseases. The repercussions of these outbreaks included direct economic losses, as well as costly lawsuits against seed and transplant producers. The economic magnitude of these lawsuits forced some seed companies out of business, while others suspended watermelon seed sales in certain "high-risk" states. Even today, some seed producers do not guarantee the performance of their seed in South Carolina for fear of litigation. Fortunately, with the implementation of routine seed health testing, many companies resumed the sale of watermelon seed with reduced risk of BFB transmission. Despite the implementation of routine seed health testing, sporadic BFB outbreaks continue to occur, and in 2000 and 2001, significant widespread outbreaks occurred across the United States. Additionally, outbreaks have occurred on a wide range of cucurbit hosts in Australia, Brazil, China, Costa Rica, Israel, Japan, Nicaragua, Taiwan, and Thailand.

This wide geographical distribution of BFB indicates that seed is still an important primary inoculum source for BFB and suggests a need for improved management strategies.

7.25.2 Bacterial Rind Necrosis of Watermelon

Pathogen: *Erwinia* sp.

This disease has been reported from Hawaii, Texas, Florida, and California in watermelon. The disease has also been reported in melon in Texas. Bacterial rind necrosis (BRN) was first observed in Illinois in 2000.

Although BRN has not become a severe annual problem in North Carolina, it is important because it does show up sporadically and particularly because fruits cannot be identified readily as being infested before cutting.

Symptoms

Characteristic symptom of BRN of watermelon is a brown, corky, dry necrosis of interior of the rind (Photo 7.57), which rarely extends into the flesh (Photo 7.58). The affected area may vary from a single small spot (1/8 in. in diameter) to the entire rind. There are rarely any external symptoms on watermelon. In the case of severe internal necrosis, the fruit may be misshapen. No symptoms on the foliage have been reported.

Although surface roughening may occur in the area external to the diseased area, accurate identification of uncut fruits as diseased is usually difficult. Consequently, the consumer may purchase an otherwise acceptable fruit and find upon slicing an unattractive, diseased interior.

PHOTO 7.57   Bacterial rind necrosis of watermelon. (Courtesy of Brock, J., University of Georgia, Athens, GA, Bugwood.org.)

PHOTO 7.58   Bacterial rind necrosis of watermelon extends into the flesh. (Courtesy of Brock, J., University of Georgia, Athens, GA, Bugwood.org.)

Disease Cycle

Little is known about the disease cycle and epidemiology of BRN of watermelon. The studies indicate that the rind necrosis is incited by bacteria that are normally residents of the healthy host. Under some predisposing environmental conditions, these resident bacteria multiply to a population high enough to cause disease.

Disease Management

Growing cultivars that are less susceptible to BRN is the only known control. Watermelon varieties vary both in the incidence of disease and in the severity of symptoms.

The cultivars fall into three classes: resistant, intermediate, and susceptible. The varieties Charleston Gray, Sweet Princess, Grayhoma, and Crimson Sweet were found resistant, whereas Blue Ribbon was susceptible.

7.26 MUSKMELON

7.26.1 BACTERIAL WILT OF MUSKMELON

Pathogen: *Erwinia tracheiphila*

Bacterial wilt is a common disease of muskmelons, but it does not affect watermelons. The bacterial pathogen, *E. tracheiphila*, multiplies within the vascular system of infected plants, causing rapid wilt and collapse of vines. The pathogen is transmitted by cucumber beetle vectors. Losses due to bacterial wilt can range from 10% to 20% in a disease-favorable season.

Symptoms

Initial symptoms of bacterial wilt include flagging or wilting of leaves on one or more vines. Symptom development proceeds rapidly; entire plants may collapse and die within a few days. Most of the loss caused by bacterial wilt occurs up to 3 weeks before harvest begins and coincides with springtime increases in cucumber beetle populations. In some seasons, wilt and collapse continue to occur after harvest begins. A diagnostic test for bacterial wilt can be performed in the field by cutting a wilted vine close to the main stem, rejoining the cut surfaces, then slowly drawing the sections apart. Bacterial wilt can be positively diagnosed if a thin strand of slime extends between the two sections.

Disease Cycle

Polycyclic.

Source of Primary Inoculums

Adult cucumber beetles that emerge in midspring transmit the bacterial pathogen.

Secondary Inoculum and Spread

Bacteria produced in infected plants are transmitted from infected plants to healthy plants by cucumber beetles.

Control Measures

Disease Resistance

No muskmelon varieties are resistant to bacterial wilt. Watermelons are not affected by the disease.

Cultural Control

Cultural practices appear to have no effect on disease development.

Chemical Control

Bacterial wilt control is directly related to the control of cucumber beetles. Soil-applied insecticides and repeated applications of foliar-contact insecticides are necessary for adequate disease control.

7.27 PASSION FRUIT

7.27.1 BACTERIAL BLIGHT OF PASSION FRUIT

Pathogen: *Xanthomonas axonopodis* pv. *passiflorae*

Symptoms

The pathogen causes local and systemic infections on seedlings and adult plants. On leaves, there can be two types of local symptoms: one or few large necrotic lesions at the leaf margins, with a translucent halo, which enlarge and coalesce, so the leaves appear to be burned (or many, small, water-soaked, dark green, generally rounded lesions, well defined, translucent, sometimes with chlorotic halo, scattered on the leaf). High relative humidity favors the exudation of the bacteria on the lesion edges. Afterward the lesions become necrotic and dark brown. Individual lesions reach up to 1 cm in diameter and they can coalesce and cause intense defoliation.

From the leaves, the bacteria spread throughout the plant via the xylem vessels to the branches and even to the stems and roots. The infected branches suffer progressive drying and show darkened vascular bundles. Fresh transversal section cuts of the infected branches can produce bacterial exudation. As the disease progresses, there may be intense defoliation, general decay, and death of the plant. The occurrence on fruits has been uncommon. However, incidence of this disease greatly reduces fruit production. Fruits are presented with dark or brownish green, anasarcous circular or irregular lesions with well-defined edges. Bacterial exudates when dry form a hard crust over the lesions. These spots penetrate the pulp, causing fruits to fall before maturation or making fruits unmarketable (Fischer Ivan and Rezende Jorge, 2008).

Infection and Spread

Infection occurs through natural openings or wounds followed by colonization of the pathogens in the inter-cell spaces and vascular tissues. Disease severity increases with high temperatures and relative humidity. Local dissemination of the bacterium is favored by wind, rain, irrigation, and also through infected seedlings.

Disease Management

Only preventive measures can be adopted as there are no effective chemical control measures available.

Seeds and seedlings should be taken from healthy plants of disease-free areas. Seed thermal therapy at 50°C for 15 min is efficient to eliminate pathogen without affecting germination of seeds.

New plantings should be done in the areas free from pathogens for at least 2 years.

Use of wind breaks and adequate amount of fertilizers can keep the pathogens in distance. Avoid working on wet plants to prevent spreading of diseases.

Use of adequate amount of nitrogenous fertilizers especially stimulates new shootings and delays maturation, making plants more susceptible to bacterium.

The elimination of diseased parts of the plants and disinfection of pruning tools and hands with bactericide products, such as those using quaternary ammonium and alcohol, may reduce the spread of pathogen.

Copper oxychloride and its mixture with mancozeb at 8- to 15-day interval decrease the intensity of disease. However, under frequent rains and favorable environmental conditions of pathogens, cupric fungicides or *streptomycin* sulfatem used, which are highly soluble in water, is washed away by rain. If there is no rain or no sprinkler irrigation, the product shows effective protection (Fishcer Ivan and Rezende Jorge, 2008).

7.27.2 GREASE SPOT OF PASSION FRUIT

Pathogen: *Pseudomonas syringae* pv. *passiflorae*

This is one of the most serious diseases of passion fruit in New Zealand. It infects leaves, stems, and fruit, leading to severe crop losses and even death of vines.

Symptoms

On Leaves

It causes irregular olive-green to brown lesions, often surrounded by a light-yellow halo. If unchecked, severe defoliation can result.

On the Stems

On young stem growth, the first signs of infection are small, slightly sunken, dark-green, water-soaked spots. These develop into light-brown, markedly depressed areas.

On Older Wood

Symptoms range from small, slightly sunken, smooth, dark-green circular spots to large, dark-brown, cracked lesions, which may completely girdle shoots and kill vines.

On Fruit

Early signs of infection on the fruit are small, dark-green, oily spots. These develop into roughly circular, greasy, or water-soaked patches. Premature fruits drop and fruit decay results.

Grease spot is said to be most active in autumn and winter, between March and August. However, a condition known as hard grease spot, also caused by *Pseudomonas passiflorae*, has become prevalent on passion fruit and is active in summer. The symptoms are similar to ordinary grease spot except that the fruit infections dry out and cause a hard brown patch on the skin, instead of leading to decay. This results in a downgrading of fruit and loss of income.

Infection and Spread

Penetration of the bacterium occurs most frequently via stomata and hydrathodes. Injury also contributes to the infection process. Infection is favored by high relative humidity, a water film on the leaf surface, and frequent rainfall. Local dissemination of the pathogen is enhanced by wind-blown rain and irrigation, and by workers handling wet plants, whereas long-distance dispersal occurs via seedlings (Manicom et al., 2003).

Disease Management

Seeds and seedlings should be from healthy plants and, if possible, should be obtained from disease-free areas. Alternatively, seed should be treated at 50°C for 15–30 min. Other complementary measures that should be adopted include planting in areas that have not had the disease for the preceding 2 years, using wind breaks, avoiding work on plants when they are wet, disinfesting pruning tools and hands, and using fertilizers judiciously, especially with respect to nitrogen. Chemical control is based on the use of mixtures of cupric and carbamate fungicides, or products that contain *streptomycin* or *oxytetracycline*. These measures have shown variable effectiveness that may be due to crop management, the quality and frequency of applications, the level of infection and susceptibility of the host plant, and virulence of the pathogen (Manicom et al., 2003).

REFERENCES

Akamine, E.K. and T. Arisumi. 1953. Control of postharvest storage decay of fruits of papaya of (*Carica papaya* L.) with special reference to the effect of hot water. *Proc. Am. Soc. Hortic. Sci.*, 61, 270–274.

Alay, K., Altinyay, N., Hancioglu, O. et al. 1973. Studies on desiccation of hazelnut branches in the Black Sea region. *Bitiki Koruma Bulteni*, 13, 202–213.

Alconero, R. 1980. Crown gall of peaches from Maryland, South Carolina, and Tennessee and problems with biological control. *Plant Dis.*, 64, 8.

Alippi, A.M., Ronco, B.L., and M.R. Carranza. 1989. Angular leaf spot of strawberry, a new disease in Argentina. Comparative control with antibiotics and fungicides. *Adv. Hortic. Sci.*, 1, 3–6.

Almeida, R.P.P., Blua, M.J., and J.R.S. Lopes et al. 2005. Vector transmission of *Xylella fastidiosa*: Applying fundamental knowledge to generate disease management strategies. *Ann. Entomol. Soc. Am.*, 98, 775–786.

Almeida, R.P.P., Nascimento, F.E., Chau, J. et al. 2008. Genetic structure and biology of *Xylella fastidiosa* causing disease in citrus and coffee in Brazil. *Appl. Environ. Microbiol.*, 74, 3690–3701.

Almeida, R.P.P. and A.H. Purcell. 2003. Transmission of *Xylella fastidiosa* to grapevines by *Homalodisca coagulata* (Hemiptera: Cicadellidae). *J. Econ. Entomol.*, 96, 264–271.

Anderson, A.R. and L.W. Moore. 1979. Host specificity in the genus *Agrobacterium. Phytopathology*, 69, 320–323.

Anderson, J.A., Buchanan, D.W., Stall, R.E. et al. 1982. Frost injury of tender plants increased by *Pseudomonas syringae* van Hall. *J. Am. Soc. Hortic. Sci.*, 107, 123–125.

Anonymous. 2002. Research Report. Group discussion of the AICRP and ICAR Ad-hoc research schemes on tropical fruits, Tamil Nadu Agricultural University, Coimbatore, India, March 18–21, 2002, p. 258.

Anonymous. 2006. Banana bacterial wilt - refining the "road map" for control in the September 2006 issue of the New Agriculturalist.

Anonymous. 2007. NRCP at a glance published by National Research Center on pomegranate (ICAR), Kegaon, Solapur, India.

Anonymous. 2013. Blood disease bacterium (blood disease bacterium of banana). CABI Data Sheet, January 23, 2013.

Arny, D.C., Lindow, S.E., and C.D. Upper. 1976. Frost sensitivity of *Zea mays* increased by application of *Pseudomonas syringae. Nature*, 262, 282–284.

Aubert, B., Faivre Amiot, A., and J. Luisetti. 1982. Phytosanitary selection work in Reunion on some fruit trees with a low chilling requirement. *Fruits*, 37, 87–96.

Barker, B.T.P. and O. Grove. 1914. A bacterial disease of fruit blossom. *Ann. Appl. Biol.*, 1, 85–97.

Barss, H.P. 1913. A new filbert disease in Oregon. *Oregon Agricultural Experiment Station Biennial Crop Pest and Horticulture Rep.*, 14: 213–223.

Bazzi, C. 1983. Biological control of crown gall in Italy. In: Grimm, R. (ed.), *Proceedings of the International Workshop on Crown Gall.* Swiss Federal Research Station for Fruit-growing, Viticulture, and Horticulture, Wadenswil, Switzerland, pp. 1–15.

Bazzi, C. and U. Mazzucchi. 1984. Update on the most important bacterial diseases of fruit crops in the nursery. *L'informatore Agrario*, 34, 51–62 (in Italian).

Bethke, J.A., Blua, M.J., and R.A. Redak. 2001. Effect of selected insecticides on *Homalodisca coagulata* (Homoptera: Cicadellidae) and transmission of oleander leaf scorch in a greenhouse study. *J. Econ. Entomol.*, 94, 1031–1036.

Bonn, W.A. and P. Fisher. 1989. Blister spot of apple (*Pseudomonas syringae* pv. *papulans*). Plantwise knowledge Bank. Factsheet booklet. Ministry of Agriculture and Food, Guelph, Ontario, Canada.

Borkar, S.G. 1989a. Glycoprotein secretion factor and dynamics of glycoprotein secretion by *Xanthomonas corylina* clone. *Indian J. Plant Pathol.*, 7(1), 36–37.

Borkar, S.G. 1989b. Use of specific immunoglobulin under immunofluorescence to detect the latent infection of *Xanthomonas corylina* in the field samples of noisetier. *Indian J. Plant Pathol.*, 7(1), 23–26.

Borkar, S.G. 1997. Chemotheropeutic resistance status of bacterium causing leaf spot on grapevine in Western Maharashtra, India. In: *Proceedings, Indian Phytopathological Society Golden Jubilee International Conference, International Conference on Integrated Plant Disease Management for Substainable Agriculture*, November 10–15, New Delhi, India, p. 323.

Borkar, S.G. 2002. Scheme for classification of races of *X.c.* pv. *viticola*, a leaf blight pathogen of grapevine. *Indian J. Plant Pathol.*, 20, 67–69.

Borkar, S.G. 2005. Search for resistance in grapevine cultivars to five races of bacterium *Xanthomonas campestris* pv. *viticola* prevalent in Indian Peninsula and inciting bacterial leaf spot/blight. *Indian J. Plant Pathol.*, 23.

Borkar, S.G. and R. Pokharel. 2012. Interception of bacterial leaf spot pathogen of strawberry in Mahabaleshwar region of Maharashtra. In: *Indian Phytopathological Society (West Zone). Symposium on Microbial Consortium Approaches for Plant Health Management*, October 30–31, 2012, Akola, India, p.118.

Bradbury, J.F. 1977. *Xanthomonas fragariae*. CMI descriptions of pathogenic fungi and bacteria no. 558. CAB International, Wallingford, U.K.

Bradbury, J.F. 1984. *Xanthomonas*. In: Krieg, N.R. and J.G. Holt (eds.), *Bergey's Manual of Systematic Bacteriology*, Vol. 1. Williams & Wilkins, Baltimore, MD.

Bradbury, J.F. 1986. *Guide to Plant Pathogenic Bacteria*. CAB International Press, Slough, U.K.

Bradbury, J.F. 1991. *Xylophilus ampelinus*. IMI descriptions of fungi and bacteria no.1050. *Mycopathologia*, 115, 63–64.

Braun-Kiewnick, A. and D.C. Sands. 2001. *Pseudomonas*. In: Schaad, N.D. (ed.), *Laboratory Guide for Identification of Plant Pathogenic Bacteria*, 3rd ed. APS Press, St Paul, MN, pp. 84–120.

Brooks Reid, M. and H.P. Olmo. 1958. Register of new fruit and nut varieties. List 13. *Proc. Am. Soc. Hortic. Sci.*, 72, 541–591.

Buddenhagen, I.W. 1970. The relation of plant pathogenic bacteria to the soil. In: Baker, K.F. and W.C. Snyder (eds.), *Ecology of Soil Borne Plant Pathogens*. University of California Press, Berkeley, CA, pp. 269–284.

Burr, T.J. 1982. Blister spot of apples. *NY Food Life Sci. Bull.*, 95.

Burr, T.J., Bazzi, C., Sule, S. et al. 1998. Crown gall of grape: Biology of *Agrobacterium vitis* and the development of disease control strategies. *Plant Dis.*, 82, 1288–1297.

Burr, T.J., Bishop, A.L., Katz, B.H. et al. 1987. A root-specific decay of grapevine caused by *Agrobacterium tumefaciens* and *Agrobacterium radiobacter* biovar 3. *Phytopathology*, 77, 1424–1427.

Burr, T.J., Reid, C.L., Tagliatti, E. et al. 1995. Survival and tumorigenicity of *Agrobacterium vitis* in living and decaying grape roots and canes in soil. *Plant Dis.*, 79, 677–682.

California Winery Advisor. 2009. Wine grape varieties. Online. Winery Advisor LLC, San Luis Obispo, CA.

Cancino, L., Latorre, B., and W. Larach. 1974. Pear blast in Chile. *Plant Dis. Rep.*, 58, 568–570.

Canfield, M.L., Baca, S., and Moore, L.W. 1986. Isolation of Pseudomonas syringae from 40 cultivars of diseased woody plants with tip dieback in Pacific Northwest nurseries. *Plant Dis.*, 70, 647–650.

Canfield, M.L. and L.W. Moore. 1992. Control of crown gall in apple (*Malus*) rootstocks using Copac E and Terramycin. *Phytopathology*, 82, 1153 (Abst.).

Carter, J.C. 1945. Wetwood of elms. III. *Nat. Hist. Surv. Bull.*, 23, 407–448.

Castellani, E. 1939. Su un marciume dell' Ensete. *Agric. Colon.*, 33, 297–300.

Catley, A. 1988. Outbreaks and new records. Australia. Eradication of citrus canker from the Torres Strait. *FAO Plant Protect. Bull.*, 36, 184.

Cazelles, O., Epard, S., and J.L. Simon. 1991. The effect of disinfection with oxyquinoline sulfate of the Berl. x Rip. 5C rootstock on the expression of crown gall in grape propagation. *Revue Suisse de Viticulture, d'Arboriculture et d'Horticulture*, 23, 285–288.

Chand, R. 1992. Sources of resistance to grape bacterial canker disease. *Vitis*, 31, 83–86.

Chand, R. and R. Kishun. 1990. Outbreak of grapevine bacterial canker disease in India. *Vitis*, 28(3), 183–188.

Chand, R., Kishun, R., Patil, B.P. et al. 1991a. Studies on grapevine bacterial canker disease. Transmission and efficacy of bactericide. *Indian J. Plant Protect.*, 19(1), 97–100.

Chand, R., Patil, B.P., and R. Kishun. 1992. Efficacy of different chemicals against grapevine bacterial canker disease (*Xanthomonas campestris* pv. *viticola*). *Indian J. Plant Protect.*, 20, 108–110.

Chand, R., Patil, B.P., Kishun, R. et al. 1991b. Management of bacterial canker disease of grapevine by pruning. *Indian J. Agric. Sci.*, 61(3), 220–222.

Chandler, W.A. and J.W. Daniell. 1976. Relation of pruning time and inoculation with *Pseudomonas syringae* van Hall to short life of peach trees growing on old peach land. *Hort. Sci.*, 11, 103–104.

Chatterjee, S., Newman, K.L., and S.E. Lindow. 2008. Cell-to-cell signaling in *Xylella fastidiosa* suppresses movement and xylem vessel colonization in grape. *Mol. Plant-Microbe Interact.*, 21, 1309–1315.

Chattopadhyay, P.K. and N. Mukherjee. 1986. A pseudostem rot of banana due to *Erwinia chrysanthemi* pv. *paradisiaca*. *Curr. Sci.*, 55, 789–790.

Chen, J., Groves, E.L., Civerolo, M. et al. 2005. Two *Xylella fastidiosa* genotypes associated with almond leaf scorch disease on the same location in California. *Phytopathology*, 95, 708–714.

Chen, J., Hartung, J.S., Chang, C. et al. 2002. An evolutionary perspective of Pierce's disease of grapevine, citrus variegated chlorosis, and mulberry leaf scorch diseases. *Curr. Microbiol.*, 45, 423–428.

Conover, R.A. and N.R. Gerhold. 1981. Mixtures of copper and maneb or mancozeb for control of bacterial spot and their compatibility for control of fungus diseases. *Proc. Fla. State Hortic. Soc.*, 94, 154–156.

Cooksey, D.A., Ohr, H.D., Azad, H.R. et al. 1993. *Xanthomonas campestris* associated with avocado canker in California. *Plant Dis.*, 77, 95–99.

Costa, H.S. 2004. Incidence of *Xylella fastidiosa* in landscape plants. December 15, 2004, Turf and Landscape Institute, Ontario, CA.

Crosse, J.E. 1959. Bacterial canker of stone-fruits. IV. Investigation of a method for measuring the inoculum potential of cherry trees. *Ann. Appl. Biol.*, 47, 306–317.

Davis, R.I., Moore, N.Y., and M. Fegan. 2001. Blood disease and Panama disease: Two newly introduced and grave threats to banana production on the island of New Guinea. In: *Food Security for Papua New Guinea. Proceedings of the Papua New Guinea Food and Nutrition 2000 Conference*, June 26–30, 2000, PNG University of Technology, Lae, Papua New Guinea, pp. 816–821.

De La Fuente, L., Burr, T.J., and H.C. Hoch. 2007. Mutations in type I and type IV pilus biosynthetic genes affect twitching motility rates in *Xylella fastidiosa*. *J. Bacteriol.*, 189, 7507–7510.

Dhandar, D.G., Nallathambi, P., Rawal, R.D. et al. 2004. Bacterial leaf and fruit spot: A new threat to pome-granate orchards in Maharashtra state. In: A paper presented in *26th Annual Conference and Symposium ISMPP*, October 7–9, 2004, Goa University, Goa, India, pp. 39–40.

Dhanvantari, B.N. 1977. A taxonomic study of *Pseudomonas papulans* Rose 1917. *N. Z. J. Agric. Res.*, 20, 557–561.

Dhanvantari, B.N., Johnson, P.W., and V.A. Dirks. 1975. The role of nematodes (*Pratylenchus penetrans, Meloidogyne hapla, Meloidogyne incognita*) in crown gall infection (*Agrobacterium tumefaciens*) of peach in southwestern Ontario. *Plant Dis. Rep.*, 59, 109–112.

Du Plessis, H.J. 1940. Bacterial blight of vines (Vlamsiekte) in South Africa caused by *Erwinia vitivora* (Bacc.) Du P. *Sci. Bull. Dept. Agric. S. Afr.*, 214, 105.

Du Plessis, H.J. 1983. Chemical control of bacterial spot on plums: Preliminary evaluation of bactericides. *Deciduous Fruit Grower*, 33, 413–418.

Du Plessis, H.J. 1987. Canker development on plum shoots following systemic movement of *Xanthomonas campestris* pv. *pruni* from inoculated leaves. *Plant Dis.*, 71, 1078–1080.

Du Plessis, H.J. 1988. Differential virulence of *Xanthomonas campestris* pv. *pruni* to peach, plum and apricot cultivars. *Phytopathology*, 78, 1312–1315.

Durgapal, J.C. 1977. Evaluation of rootstocks of pome and stone fruits and related wild species for resistance to crown gall. *Curr. Sci.*, 46, 389–390.

Dye, D.W. 1956. Blast of pear. *N.Z. Orchardist*, 29, 5–7.

Dye, D.W. 1968. A taxonomic study of the genus *Erwinia*. I. The *amylovora*' group. *N.Z. J. Sci.*, 11, 590–607.

Dye, D.W. 1978. Genus IX Xanthomonas Dowson 1939. From: Young, J. M.; Dye, D. W.; Bradbury, J. F.; Panagopoulos, C. G.; Robbs, C. F.; 1978: A proposed nomenclature and classification for plant pathogenic bacteria. *N.Z. J. Agric. Res.*, 21, 563–582.

EPPO/CABI. 1992. *Xanthomonas campestris* pv. *citri*. In: Smith, I.M., McNamara, D.G., Scott, P.R., and K.M. Harris (eds.), *Quarantine Pests for Europe*. CAB International, Wallingford, U.K.

EPPO/CABI. 1997. *Xanthomonas arboricola* pv. *pruni*. In: *Quarantine Pests for Europe*, 2nd ed. CAB International, Wallingford, U.K., pp. 1096–1100.

Fegan, M. 2005. Bacterial wilt diseases of banana: Evolution and ecology. In: Allen, C., Prior, P., and A.C. Hayward (eds.), *Bacterial Wilt Disease and the Ralstonia solanacearum Species Complex*. APS Press, St. Paul, MN.

Feil, H. and A.H. Purcell. 2001. Temperature-dependent growth and survival of *Xylella fastidiosa* in vitro and in potted grapevines. *Plant Dis.*, 85, 1230–1234.

Fernández, A.M.M. and J.F. Pinto-Ganhao. 1981. [*Xanthomonas fragariae* Kennedy & King - a new bacterial disease in Portugal]. *Agros*, 64, 5–8.

Fernández-Pavía, S.P., Rodríguez-Alvarado, G., Garay-Serrano, E. et al. 2014. First report of *Xanthomonas fragariae* causing angular leaf spot on strawberry plants in México. *Plant Dis.*, 98, 682.

Fischer Ivan, H. and A.M. Rezende Jorge. 2008. Diseases of passion flower (*Passiflora* spp.). Global Science Books, East Sussex, U.K.

Freitag, J.H. 1951. Host range of Pierce's disease virus of grapes as determined by insect transmission. *Phytopathology*, 41, 920–934.

Frossard, P., Hugon, R., and Ch. Verniere. 1985. Un deperissement du papayer aux Antilles francaises associe a un *Erwinia* sp. Du groupe *amylovora*. *Fruits*, 40(9), 583–595.

Gagnevin, L. and O. Pruvost. 2001. Epidemiology and control of mango bacterial black spot. *Plant Dis.*, 85, 928–935.

Gaikwad, U.V. and S.R. Naik. 1992. *In vitro* studies of resistance development in *Xanthomonas* of grape by streptocycline and removal of its 'R' factor by aureofungin. *Indian Phytopathol.*, 43 and 44, p. CXVII.

Gambetta, G.A., Fei, J., Rost, T.L., and M.A. Matthews. 2007. Leaf scorch symptoms are not correlated with bacterial populations during Pierce's disease. *J. Exp. Bot.*, 58, 4037–4046.

Gardan, L. 1982. La bactériose du noisetier. In: *3èmes journées françaises d'études et d'informations sur les maladies des plantes*. ACTA, Paris, France, pp. 489–495.

Gardan, L. 1986. *Xanthomonas campestris* pv. *corylina*. EPPO Data sheets on quarantine organisms. *Bull. OEPP/EPPO Bull.*, 16, 13–16.

Gardan, L. and M. Deveaux. 1987. La bactériose du noisetier (*Xanthomonas campestris* pv. *corylina*): Biologie de la bactérie. *Bull. OEPP/EPPO Bull.*, 17, 241–250.

Garrett, C.M.E. 1987. The effect of crown gall on growth of cherry trees. *Plant Pathol.*, 36, 339–345.

Garrett, K.A., Dendy, S.P., Frank, E.E., Rouse, M.N., and S.E. Travers. 2006. Climate change effects on plant disease: Genomes to ecosystems. *Annu. Rev. Phytopathol.*, 44, 489–509.

Gäumann, E. 1921. Onderzoekeningen over de bloedziekte der bananen op Celebes I & II. *Mededeelingen van het Instituut voor Plantenziekten*, 50:55 pp.

Gäumann, E. 1923. Onderzoekeningen over de bloedziekte der bananen op Celebes I & II. *Mededeelingen van het Instituut voor Plantenziekten*, 59:47 pp.

Giri, M.S., Prasanthi, S., Kulkarni, S. et al. 2011. Biochemical and molecular variability among *Xanthomonas axonopodis* pv. *punicae* strains, the pathogen of pomegranate bacterial blight. *Indian Phytopathol.*, 64(1), 1–4.

Gitaitis, R.D., Hamm, J.D., and P.F. Bertrand. 1988. Differentiation of *Xanthomonas campestris* pv. *pruni* from other yellow-pigmented bacteria by the refractive quality of bacterial colonies on an agar medium. *Plant Dis.*, 72, 416–417.

Gloyer, W.O. 1934. Crown gall and hairy root of apples in nursery and orchard. *NY Agric. Exp. Stn. Bull.*, 638.

Goodman, C.A. and M.J. Hattingh. 1988. Mechanical transmission of *Xanthomonas campestris* pv. *pruni* in plum nursery trees. *Plant Dis.*, 72, 643.

Goodman, R.N., Grimm, R., and M. Frank. 1993. The influence of grape rootstocks on the crown gall infection process and on tumor development. *Am. J. Enol. Viticult.*, 44, 22–26.

Grimm, R., Lips, T., and J. Vogelsanger. 1993. [Angular leaf spot of strawberry]. *Schweizerische Zeitschrift für Obst- und Weinbau*, 128, 130–131.

Gross, D.C., Cody, Y.S., Proebsting, E.L. et al. 1983. Distribution, population dynamics, and characteristics of ice nucleation-active bacteria in deciduous fruit tree orchards. *Appl. Environ. Microbiol.*, 46, 1370–1379.

Gross, D.C., Cody, Y.S., Proebsting, E.J. et al. 1984. Ecotypes and pathogenicity of ice nucleation-active *Pseudomonas syringae* isolated from deciduous fruit tree orchards. *Phytopathology*, 74, 241–248.

Guerrero, C.J. and A.W. Lobos. 1987. *Xanthomonas campestris* pv. *corylina*, causal agent of bacterial blight of hazel in region IX, Chile. *Agricultura Tecnica Santiago*, 47, 422–426.

Guevara, Y., Rondon, A., Maselli, A. et al. 1993. Marchitez bacteriana del lechosero *Carica papaya* L. en Venezuela. *Agron. Trop. (Maracay)*, 43, 107–116 (in Spanish).

Hattingh, M.J., Roos, I.M.M., and E.L. Mansvelt. 1989. Infection and systemic invasion of deciduous fruit trees by *Pseudomonas syringae* in South Africa. *Plant Dis.*, 73, 784–789.

Hauben, L., Moore, E.R., Vauterin, L. et al. 1998. Phylogenetic position of phytopathogens within the Enterobacteriaceae. *Syst. Appl. Microbiol.*, 21(3), 384–397.

Hawkins, J.E. 1976. A cauterization method for the control of cankers caused by *Pseudomonas syringae* in stone fruit trees. *Plant Dis. Rep.*, 60, 60–61.

Hedgcock, G.G. 1910. Field studies of the crown gall of the grape. *U.S. Dept. Agric. Bur. Plant Ind. Bull.*, 183, 1–40.

Hendson, M., Purcell, A.H., Chen, D. et al. 2001. Genetic diversity of Pierce's disease strains and other pathotypes of *Xanthomonas fastidiosa*. *Appl. Environ. Microbiol.*, 67, 895–903.

Henneberger, T.S.M., Stevenson, K.L., Britton, K.O. et al. 2004. Distribution of *Xylella fastidiosa* in sycamore associated with low temperature and host resistance. *Plant Dis.*, 88, 951–958.

Hernendez-Martinez, R., de la Cerda, K.A., Costa, H.S. et al. 2007. Phylogenetic relationships of *Xylella fastidiosa* strains isolated from landscape ornamentals in southern California. *Phytopathology*, 97, 857–864.

Hewitt, W.B. 1939. A transmissible disease of grapevines. *Phytopathology*, 29, 10.

Hildebrand, D.C., Schroth, M.N., and S. Wilhelm. 1967. Systemic invasion of strawberry by *Xanthomonas fragariae* causing vascular collapse. *Phytopathology*, 57, 1260–1261.

Hill, B.L. and A.H. Purcell. 1995. Multiplication and movement of *Xylella-fastidiosa* within grapevine and 4 other plants. *Phytopathology*, 85, 1368–1372.

Hill, B.L. and A.H. Purcell. 1997. Populations of *Xylella fastidiosa* in plants required for transmission by an efficient vector. *Phytopathology*, 87, 1197–1201.

Hingorani, M.K. and P.P. Mehta. 1952. Bacterial leaf spot of pomegranate. *Indian Phytopathol.*, 5, 55–56.

Hingorani, M.K. and N.J. Singh. 1959. *Xanthomonas punicae* sp. Nov. on *Punica granatum* L. *Indian J. Agric. Sci.*, 29, 45–48.

Hoerner, G.R. 1945. Crown gall of hops. *Plant Dis. Rep.*, 29, 98–110.

Hopkins, D.L. 1981. Seasonal concentration of the Pierce's disease bacterium in grapevine stems, petioles, and leaf veins. *Phytopathology*, 71, 415–418.

Hopkins, D.L. 1989. *Xylella fastidiosa*: Xylem-limited bacterial pathogen of plants. *Annu. Rev. Phytopathol.*, 27, 271–290.

Hopkins, D.L. and A.H. Purcell. 2002. *Xylella fastidiosa*: Cause of Pierce's disease of grapevine and other emergent diseases. *Plant Dis.*, 86, 1056–1066.

Htay, K. and A. Kerr. 1974. Biological control of crown gall: Seed and root inoculation. *J. Appl. Bacteriol.*, 37, 525–530.

Ibrahim, G. and B. Bayaa. 1989. Fungal, bacterial and nematological problems of citrus, grape and stone fruits in Arab countries. *Arab. J. Plant Protect.*, 7, 190–197.

IMI. 1993. Distribution maps of plant diseases no. 520, 2nd ed. CAB International, Wallingford, U.K.

IMI. 1996. Distribution maps of plant diseases no. 699, 1st ed. CAB International, Wallingford, U.K.

Ishizawa, Y., Kyotani, H., Nishimura, K. et al. 1992. Methods for evaluating the degree of crown gall resistance and the varietal differences in peach. *Bull. Fruit Tree Res. Stn.*, 23, 37–46.

Jaburek, V. and J. Holub. 1987. Effect of the rootstock BD-SU-1 on the growth and productivity of selected peach cultivars. *Fruit Grow.*, 60, 192–195.

Janse, J.D., Rossi, P., Angelucci, L. et al. 1997. Validation of the publication of new names and new combinations previously effectively published outside the IJSB. List no. 61. *Int. J. Syst. Bacteriol.*, 47, 601–602.

Jha, G., Rajeshwari, R., and R.V. Sonti. 2005. Bacterial type two secretion system secreted proteins: Double-edged swords for plant pathogens. *Mol. Plant-Microbe Interact.*, 18, 891–898.

Jindal, K.K. and R.C. Sharma. 1987. Almond leaf scorch - A new disease from India. *FAO Plant Prot. Bull.*, 35, 64–65.

Jones, R., Moffett, M.L., and S.J. Navaratnam. 1984. Citrus canker on Thursday Island. *Aust. J. Plant Pathol.*, 13, 64–65.

Jordan, R.L. 1981. Avocado blackstreak disease; symptomatology, epidemiology, and attempts to identify the causal agent. PhD thesis, University of California, San Diego, CA.

Jordan, R.L., Ohr, H.D., and G.A. Zentmyer. 1983. Avocado blackstreak disease (ABS): A newly recognized major disease of avocado (*Persea americana* Miller) in California. (Abstr.). *Phytopathology*, 73, 960.

Kadam, V.C. 1990. Effect of different antibiotics on inhibition of *Xanthomonas* of grapevine. In: *Res. Rev. Sub-Committee Meeting in Plant Pathology*. Mahatma Phule Agricultural University, Rahuri, India, p. 35.

Kalyebara, M.R., Ragama, P.E., Kagezi, G.H. et al. 2006. Economic importance of the banana bacterial wilt in Uganda. *Afr. Crop Sci. J.*, 14(2), 93–103.

Kanwar, Z.S. 1976. A note on bacterial disease of pomegranate (*Punica granatum* L.) in Haryana. *Haryana J. Hortic. Sci.*, 5, 177–180.

Kennedy, B.W. 1965. Infection of potentilla by *Xanthomonas fragariae*. *Plant Dis. Rep.*, 49, 491–492.

Kennedy, B.W. and T.H. King. 1962a. Angular leaf spot of strawberry caused by *Xanthomonas fragariae* sp. nov. *Phytopathology*, 52, 873–875.

Kennedy, B.W. and T.H. King. 1962b. Studies on epidemiology of bacterial angular leaf spot on strawberry. *Plant Dis. Rep.*, 46, 360–363.

Kishi, L.T., Wickert, E., and E.G.D. Lemos. 2008. Evaluation of *Xylella fastidiosa* genetic diversity by Faflp markers. *Rev. Bras. Frutic. Jaboticabal - SP*, 30, 202–208.

Kishun, R. 1986. Role of insects in transmission and survival of *Xanthomonas campestris* pv. *mangiferaeindicae*. *Ind. Phytopathol.*, 39, 509–511.

Klement, Z. 1974. Apoplexy of apricots. Ill. Relationship of winter frost and bacterial canker and die-back of apricots. *Acta Phytopathol. Acad. Sci. Hungary*, 9, 35–45.

Klement, Z., Rozsnyay, D.S., Balo, E. et al. 1984. The effect of cold on development of bacterial canker in apricot trees infected with *Pseudomonas syringae* pv. *syringae*. *Physiol. Plant Pathol.*, 24, 237–246.

Kochenko, Z.I. 1993. Development of canker tumors on grape roots. *Zaschita Rastenii Moskva*, 7, 42–43.

Kore, S.S., Tekarle, J.R., and U.V. Gaikwad. 1992. Investigation on bacterial disease of grape incited by *X. C.* pv. *viticola*. *Indian Phytopathol.*, 43 and 44 Suppl., XXVI abstr.

Korsten, L. 1984. Bacteria associated with bark canker of avocado. MSc thesis, University of Pretoria, Pretoria, South Africa.

Korsten, L. and J.M. Kotzé. 1985. Bacterial canker of avocado. *S. Afr. Avocado Growers' Assoc. Yrbk.*, 8, 63–65.

Korsten, L. and J.M. Kotzé. 1987. Bark canker of avocado, a new disease presumably caused by *Pseudomonas syringae* in South Africa. *Plant Dis.*, 71, 850.

Koval, G.K. 1978. Diseases of hazel. *Zashchita Rastenii*, 8, 44–45.

Krell, R.K., Perring, T.M., Farrar, C.A. et al. 2006. Intraplant sampling of grapevines for Pierce's disease diagnosis. *Plant Dis.*, 90, 351–357.

Kuhn, R., Starr, M.P., Kuhn, D.A. et al. 1965. Indigoidine and other bacterial pigments related to 3,3 bipyridyl. *Arch. Mikrobiol.*, 51, 71–84.

Kumakech, A., Kiggundu, A., and P. Okori. 2013. Reaction of *Musa balbisiana* to banana bacterial wilt infection. *Afr. Crop Sci. J.*, 21(4), 337–346.

Kumar, R., Jahagirdar, S., Yenjereapp, S.T. et al. 2006. Epidemiology and management of bacterial blight of pomegranate caused by *Xanthomonas axonopodis* pv. *punicae*. In: Paper presented in the *First International Symposium on Pomegranate and Minor Mediterranean Fruits*, October 16–17, 2006, Adana, Turkey.

Larsen, D. 2000. Glassy-winged sharpshooter update—9/00. Online. GWSS features. *American Vineyard Magazine*, Malcolm Media, Clovis, CA.

Lehoczky, J. 1968. Spread of *Agrobacterium tumefaciens* in the vessels of the grapevine after natural infection. *J. Phytopathol.*, 63, 239–246.

Leite, R.P. Jr., Minsavage, G.V., Bonas, U. et al. 1994. Detection and identification of phytopathogenic *Xanthomonas* strains by amplification of DNA sequences related to the *hrp* genes of *Xanthomonas campestris* pv. *vesicatoria*. *Appl. Environ. Microbiol.*, 60(4), 1068–1077.

Lelliot, R.A. and D.E. Stead. 1987. Methods for the diagnosis of bacterial diseases of plants. In: Preece, T.F. (ed.), *Methods in Plant Pathology*, Vol. 2. Blackwell Scientific Press, London, U.K., 216pp.

Lemoine, J. and J.C. Michelesi. 1993. Agronomic behaviour of pears: Incidence of crown gall. *Arboriculture Fruitiere*, 465, 23–27.

Leu, L.S., Lee, C.C., and T.C. Huang. 1980. Papaya black rot caused by *Erwinia cypripedii*. *Plant Prot. Bull. Taiwan*, 22(4), 377–384.

Lindow, S.E. 1983. The role of bacterial ice nucleation in frost injury to plants. *Annu. Rev. Phytopathol.*, 21, 363–384.

Lindow, S.E., Hirano, S.S., Barchet, W.R. et al. 1982. Relationship between ice nucleation frequency of bacteria and frost injury. *Plant Physiol.*, 70, 1090–1093.

Locke, T. and D. Barnes. 1979. New or unusual records of plant diseases and pests. *Xanthomonas corylina* on cob-nuts and filberts. *Plant Pathol.*, 28, 53.

Lopez, M.M., Aramburu, J.M., Cambra, M. et al. 1985. [Detection and identification of *Xanthomonas fragariae* in Spain.] Anales del Instituto Nacional de Investigaciones Agrarias. *Serie Agricola*, 28, 245–259.

Lopez, M.M., Cambra, M., Aramburu, J.M. et al. 1987. Problems of detecting phytopathogenic bacteria by ELISA1. *Bull. OEPP/EPPO Bull.*, 17, 113–117.

Lopez, M.M., Gracia, M., and M. Sampayo. 1980. Studies on *Xanthomonas ampelina* Panagopoulos in Spain. In: *Proceedings of the Fifth Congress of the Mediterranean Phytopathological Union*, September 21–27, 1980, Patras, Greece, pp. 56–57.

Luisetti, J., Jailloux, F., Germain, E. et al. 1976. Caractérisation de *Xanthomonas corylina* responsable de la bactériose du noisetier récemment observée en France. *Comptes Rendus des Séances de l'Académie d'Agriculture de France*, 62, 845–849.

Maas, J.L. 1984. *Compendium of Strawberry Diseases*. American Phytopathology Society, St. Paul, MN.

Magrou, J. 1937. Genre *Erwinia*. In: Hauduroy, P., Ehringer, G., Urbain, A., Guillot, G., and J. Magrou (eds.), *Dictionnaire des bacteries pathogens*. Masson et Cie, Paris, France, pp. 214–215.

Maki, L.R., Galyan, E.L., Chang-chien, M.M. et al. 1974. Ice nucleation induced by *Pseudomonas syringae*. *Appl. Microbiol.*, 28, 456–459.

Manicom, B., Ruggiero, C., Ploetz, R.C. et al. 2003. Diseases of passion fruit. In: Plotez, R.C. (ed.), *Diseases of Tropical Fruit Crops*. CAB International, Wallingford, U.K., pp. 413–441.

Mansvelt, E.L. and M.J. Hattingh. 1986. Pear blossom blast in South Africa caused by *Pseudomonas syringae* pv. *syringae*. *Plant Pathol.*, 35, 337–343.

Mansvelt, E.L. and M.J. Hattingh. 1987. *Pseudomonas syringae* pv. *syringae* associated with apple and pear buds in South Africa. *Plant Dis.*, 71, 789–792.

Marloth, R.H. 1947. The mango in South Africa: Diseases and pests. *Fmg. S. Afr.*, 22, 615–619.

Mazzucchi, U., Alberghina, A., and A. Dalli. 1973. Occurrence of *Xanthomonas fragariae* Kennedy & King in Italy. *Phytopathologische Zeitschrift*, 76, 367–370.

Mazzucchi, U. and A. Calzolari. 1987. Detection of *Xanthomonas fragariae* in symptomless strawberry plants. *Acta Horticult.*, 265, 601–604.

Mbaka, J.N., Nakato, V.G., Auma, J. et al. 2009. Status of banana *Xanthomonas* wilt in western Kenya and factors enhancing its spread. *Afr. Crop Sci. Conf. Proc.*, 9, 673–676.

McElrone, A.J., Sherald, J.L., and I.N. Forseth. 2003. Interactive effects of water stress and xylem-limited bacterial infection on the water relations of a host vine. *J. Exp. Bot.*, 54, 419–430.

McKeen, W.E. 1955. Pear blast on Vancouver Island. *Phytopathology*, 45, 629–632.

McKenna, J.R. and L. Epstein. 2003. Relative susceptibility of *Juglans* species and interspecific hybrids to *Agrobacterium tumefaciens*. *Hort Sci.*, 38, 9.

Miller, P.W. 1949. Filbert bacteriosis and its control. *Oregon Agric. Exp. Stn. Tech. Bull.*, 6.

Mirow, H. 1985. Experiments on the control of crown gall on woody plants in the nursery. *Deutsche Baumschule*, 37, 300–301.

Mishra, A.K. 1995. Control of bacterial blight and canker of mango and suitable weather conditions. *Indian J. Mycol. Pathol.*, 25, 214–217.

Mohan, C., Thind, T.S., and P.S. Soni. 2001. Latrogenic effect of carbendazim on bacterial leaf spot of grape. *J. Mycol. Plant Pathol.*, 31(3), 353–354.

Mondal, K.K. and C. Mani. 2009. ERIC-PCR-generated genomic fingerprints and their relationship with pathogenic variability of *Xanthomonas campestris* pv. *punicae*, the incitant of bacterial blight of pomegranate. *Curr. Microbiol.*, 59(6), 616–620.

Montero-Astua, M., Saborio, R.G., Chacon-Diaz, C. et al. 2008. First report of *Xylella fastidiosa* in avocado in Costa Rica. *Plant Dis.*, 92, 175.

Moore, L.W. 1976a. Latent infections and seasonal variability of crown gall development in seedlings of three *Prunus* species. *Phytopathology*, 66, 1097–1101.

Moore, L.W. 1976b. Research findings of crown gall and its control. *Am. Nurseryman*, 144, 8–9.

Moore, L.W. and J. Allen. 1986. Controlled heating of root-pruned dormant *Prunus* seedlings before transplanting to prevent crown gall. *Plant Dis.*, 70, 532–536.

Moore, L.W., Canfield, M., and R. Hall. 1996. Biology of *Agrobacterium* and management of crown gall disease. In: Hall, R. (ed.), *Principles and Practice of Managing Soilborne Plant Pathogens*. APS Press, St. Paul, MN, p. 153.

Muharam and Subiyanto. 1991. Status of banana disease in Indonesia. Pp. 44–49 In: *Valmayor, R.V., Umali, B.E. and Bejosano., C.P. (eds). Banana diseases in Asia and the Pacific: Proceedings of a technical meeting on diseases affecting banana and plantain in Asia and the Pacific, Brisbane, Australia*, 15–18, April 1991. Montpellier, France: INIBAP, 1991.

Mulyadi, H.T. 2002. Blood disease intensity on banana caused by bacterium of *Pseudomonas solanacearum* in Bondowoso. In: *Proceedings of the 16th Congress and National Seminar of the Indonesian Phytopathological Society*, August 22–24, 2001, Bogor, Indonesia.

Mwangi, M., Tinzaara, W., Vigheri, N. et al. 2006. Comparative study of banana Xanthomonas wilt spread in mid and high altitudes of the Great Lakes region of Africa. In: *Conference on International Agricultural Research for Development, Tropentag 2006*, University of Bonn, Germany. October 11–13, 2006, http://elewa.org/pestdseagric.html.

Myburgh, L. and J.M. Kotze. 1982. Bacterial disease of avocado. *S. Afr. Avocado Growers' Assoc. Yrbk.*, 5, 105–106.

Nagrale, D.T. 2011. Studies of bacterial collar rot of banana. PhD thesis submitted to Mahatma Phule Agriculture University, Rahuri, India, p. 113.

Nagrale, D.T., Borkar S.G., Gawade, S.P. et al. 2013. Characterization of bacterial collar and rhizome rot of banana caused by strains of *Erwinia chrysanthemi* pv. *paradisiaca*. *J. Appl. Natl. Sci.*, 5(2), 435–441.

National Research Council. 2004. *California Agricultural Research Priorities: Pierce's Disease*. National Academics Press, Washington, DC.

Nayudu, M.V. 1972. *Pseudomonas viticola* sp. Nov., incitant of a new bacterial disease of grapevine. *J. Phytopathol.*, 73, 183–186.

Nelson, M.N. and A.M. Alvarez. 1980. Purple stain of *Carica papaya*. *Plant Dis.*, 64, 93–95.

Nesme, X., Beneddra, T., and E. Collin. 1990. Importance of crown gall in hybrids of *Populus tremula* L. x *P. alba* L. in forest tree nursery. *Agronomie*, 10, 581–588.

Newman, K.L., Almeida, R.P.P., Purcell, A.H. et al. 2003. Use of a green fluorescent strain for analysis of *Xylella fastidiosa* colonization of *Vitis vinifera*. *Appl. Environ. Microbiol.*, 69, 7319–7327.

Newman, K.L., Almeida, R.P.P., Purcell, A.H. et al. 2004. Cell-cell signaling controls *Xylella fastidiosa* interactions with both insects and plants. *Proc. Natl. Acad. Sci. USA*, 101, 1737–1742.

Nishijima, K.A., Alvarez, A.M., Hepperly, P.R. et al. 2004. Association of *Enterobacter cloacae* with rhizome rot of edible ginger in Hawaii. *Plant Dis.*, 88, 1318–1327.

Nishijima, K.A., Couey, H.M., and A.M. Alvarez. 1987. Internal yellowing, a bacterial disease of papaya fruits caused by *Enterobacter cloacae*. *Plant Dis.*, 71, 1029–1034.

Noviello, C. 1969. Infectious diseases of hazel. Annali della Facoltà di Scienze Agrarie della Università degli Studi di Napoli Portici, 3, 11–39.

OEPP/EPPO. 1978. Data sheets on quarantine organisms no. 62, *Xanthomonas pruni*. *Bull. OEPP/EPPO Bull.*, 8.

OEPP/EPPO. 1986b. Data sheets on quarantine organisms no. 135, *Xanthomonas fragariae*. *Bull. OEPP/ EPPO Bull.*, 16, 17–20.

OEPP/EPPO. 1990. Specific quarantine requirements. EPPO Technical documents no. 1008.

Ogawa, J.M., Zehr, E.I., Bird, G.W. et al. (eds.). 1995. *Compendium of Stone Fruit Diseases*. American Phytopathological Society, St. Paul, MN.

Ohr, H.D. and L. Korsten. 1990. Detecting bacterial canker. *California Grower*, 14, 22–27.

Ophel, K. and A. Kerr. 1990. Agrobacterium vitis sp. nov. for strains of Agrobacterium biovar 3 from grapevines. *Int. J. Syst. Evol. Microbiol.*, 40(3), 236–241.

Palm, M.E. and E.L. Civerolo. 1994. Isolation, pathogenicity, and partial host range of *Alternaria limicola*, causal agent of mancha foliar de los citricos in Mexico. *Plant Dis.*, 78, 879–883.

Panagopoulos, C.G. 1987. Recent research progress on *Xanthomonas ampelina*. *EPPO Bull.*, 17(2), 225–230.

Panagopoulos, C.G. and J.E. Crosse. 1964. Blossom blight and related symptoms caused by *Pseudomonas syringae* van Hall on pear trees. *Ann. Rep. E. Malling Res. Stn.*, 119–122.

Panagopoulos, C.G., Psallidas, P.G., and A.S. Alivizatos. 1978. A bacterial leaf spot of strawberry in Greece caused by *Xanthomonas fragariae*. *Phytopathologische Zeitschrift*, 91, 33–38.

Patel, M.K., Moniz, L., and Y.S. Kulkarni. 1948. A new bacterial disease of *Mangifera indica* L. *Curr. Sci.*, 17, 189.

Pierronnet, A. and J.P. Eyquard. 1993. *Prunus* rootstocks and crown gall. *Arboriculture Fruitiere*, 466, 37–41.

Potter, D., Gao, F., Baggett, S. et al. 2001. Defining the sources of Paradox: DNA sequence markers for North American walnut (*Juglans* L.) species and hybrids. *Scientia Hortic.*, 94, 70.

Prakash, O. and M.A. Raoof. 1985. Bacterial canker in mango (Abs.). In: *Second International Symposium on Mango*, Bangalore, India, 20–24 May, 1985. Acta Horticulture (ISHS) p. 59.

Prior, P., Béramis, M., and M.T. Rousseau. 1985. Le dépérissement bactérien du papayer aux Antilles françaises. *Agronomie*, 5, 877–885 (in French).

Proebsting, E.L. Jr. and H.H. Mills. 1978. Low temperature resistance of developing flower buds of six deciduous fruit species. *J. Am. Soc. Hortic. Sci.*, 103, 192–198.

Prunier, J.P., Luisetti, J., Gardan, L. et al. 1976. La bactériose du noisetier (*Xanthomonas corylina*). *Revue Horticole*, 170, 31, 40.

Pruvost, O., Fabrègue, C., and J. Luisetti. 1988. Mise en évidence de la maladie des taches angulaires du fraisier à l'île de la Réunion. *Fruits (France)*, 43, 369–373.

Psallidas, P.G. 1984. Bacterial canker of *Corylus avellanae*: The taxonomic position of the causal agent. In: *Proceedings of the Second Working Group of Pseudomonas syringae Pathovars*, April 24–28, 1984, Sounion, Greece, pp. 53–55.

Psallidas, P.G. and C.G. Panagopoulos. 1979. A bacterial canker of hazelnut in Greece. *Phytopatholologische Zeitschrift*, 94, 103–111.

Purcell, A.H. 1980. Environmental therapy for Pierce's disease of grapevines. *Plant Dis.*, 64, 388–390.

Purcell, A.H. 1981. Pierce's disease. Grape Pest Management. Publication no. 4102, Division of Agricultural Sciences, University of California, San Diego, CA, pp. 62–69.

Purcell, A.H. 1989. Homopteran transmission of xylem-inhibiting bacteria. In: Harris, K.F. (ed.), *Advances in Disease Vector Research*, Vol. 6. Springer, New York, pp. 243–266.

Purcell, A.H. and A.H. Finlay. 1979. Evidence for noncirculative transmission of Pierce's disease bacterium by sharpshooter leafhoppers. *Phytopathology*, 69, 393–395.

Purcell, A.H., Finlay, A.H., and D.L. McLean. 1979. Pierce's disease bacterium: Mechanism of transmission by leafhopper vectors. *Science*, 206(4420), 839–841.

Purcell, A.H. and N.W. Frazier. 1985. Habitats and dispersal of the principal leafhopper vectors of Pierce's disease bacterium in the San Joaquin Valley. *Hilgardia*, 53, 4.

Raghuwanshi, K.S., Hujare, B.A., and S.G. Borkar. 2013. Characterization of *Xanthomonas axonopodis* pv. *punicae* isolates from Western Maharashtra and their sensitivity to chemical treatments. *The Bioscan*, 8(3), 845–850.

Randhawa, P.S. and E.L. Civerolo. 1985. A detached-leaf bioassay for *Xanthomonas campestris* pv. *pruni*. *Phytopathology*, 75, 1060–1063.

Raski, D.J., Goheen, A.C., Lider, L.A., and C.P. Meridith. 1983. Strategies against grapevine fanleaf virus and its nematode vector. *Plant Dis.*, 67, 335–337.

Rat, B. 1974. Présence en France de la maladie des taches angulaires du fraisier. *Annales de Phytopathologie*, 6, 223.

Ravikumar, M.R., Wali, S.Y., Benagi, V.I. et al. 2011. Management of bacterial blight of pomegranate through chemicals/antibiotics. *Acta Hortic.*, 890, 481–482.

Redak, R.A., Purcell, A.H., Lopes, J.R.S. et al. 2004. The biology of xylem fluid-feeding insect vectors of *Xylella fastidiosa* and their relation to disease epidemiology. *Annu. Rev. Entomol.*, 49, 243–270.

Richard, C. 1984. Genus VI. *Enterobacter* Hormeche and Edwards 1960, nom. Cons. Opin. 28, Jud. Comm. 1963, 38: pp. 465–469. In: Krieg, N.R. and J.G. Holt (eds.), *Bergey's Manual of Systemic Bacteriology*, Vol. 1. Williams & Wilkins, Baltimore Co.

Ride, M., Ride, S., and D. Novoa. 1977. Donnes nouvelles sur la biologie de *Xanthomonas ampelina Panagopoulos*, agent de la necrose bacterenne de la vigne. *Annales de Phytopathologie*, 9, 87.

Rohrbach, K.G. and J.B. Pfeiffer. 1975. The field induction of bacterial pink disease in pineapple fruit. *Phytopathology*, 65, 803–805.

Rosen, H.R. 1922. The bacterial pathogen of corn stalk rot. *Phytopathology*, 12, 497–499.

Sands, D.C. and D.A. Kollas. 1974. Pear blast in Connecticut. *Plant Dis. Rep.*, 58, 40–41.

Schaad, N.W., Postnikova, E., Lacy, G. et al. 2004. *Xylella fastidiosa* subspecies: *X. fastidiosa* subsp. *piercei*, subsp. nov., *X. fastidiosa* subsp. multiplex subsp. nov., and *X. fastidiosa* subsp. pauca subsp. nov. *Syst. Appl. Microbiol.*, 27, 290–300.

Scholefield, P.B. and M. Sedgley. 1983. Avocado cultivars in the northern territory. In: *Proceedings of the Second Australian Avocado Research Workshop*, June 1983, Bogangar, New South Wales, Australia, p. 55.

Schroth, M.N. and D.C. Hildebrand. 1983. Toward a sensible taxonomy of bacterial plant pathogens. *Plant Dis.*, 67, 128.

Schuster, C.E. 1924. Filberts: 2. Experimental data on filbert pollination. *Oregon Agric. Exp. Stn. Bull.*, 208.

Scortichini, M. 1995. Le malattie batteriche delle colture agrarie e delle specie forestali. Edizione agricole, Bologna, Italy.

Scortichini, M., Rossi, M.P., Loreti, S. et al. 2005. *Pseudomonas syringae* pv. *coryli*, the causal agent of bacterial twig dieback of *Corylus avellana*. *Phytopathology*, 95, 1316–1324.

Scortichini, M., Rossi, M.P., and U. Marchesi. 2002. Genetic, phenotypic and pathogenic diversity of *Xanthomonas arboricola* pv. *corylina* strains question the representative nature of the type strain. *Plant Pathol.*, 51, 374–381.

Scortichini, M., Sbaraglia, M., Di Prospero, P. et al. 2001. Moria del nocciolo nel Viterbese e terreni acidi. *Inf Agrario*, 21, 85–88.

Scortichini, M. and A.M. Simeone. 1997. [Review of the bacterial diseases of apricot.] *Rivista di Frutticoltura e di Ortofloricoltura*, 59, 51–57 (in Italian).

Serfontein, S., Serfontein, J.J., and W.J. Botha. 1997. The isolation and characterisation of *Xylophilus ampelinus*. *Vitis*, 36(4), 209–210.

Serrano, D., Serrano, E., Dewdney, M., and Southwick, C. 2010. Citrus Diseases. USDA/APHIS/PPQ Center for Plant Health Science and Technology [09–14].

Severin, V., Stancescu, C., and E. Zambrowicz. 1985. [Angular leaf spot, a new bacterial disease of strawberry in Romania]. Buletinul de Protectia Plantelor, 1–2, 21–23.

Sharma, K.K., Sharma, J., Jadhav, V.T. et al. 2008. Bacterial blight of pomegranate and its management. *Indian Phytopathol.*, 61(3), 380–381.

Shekhawat, G.S. and P.N. Patel. 1975. Studies on bacterial canker of mango. *Zeitschrift for Pjlanzenkrankheiten und pflanzenschutz*, 82, 129–138.

Shivas, R.G. 1987. Citrus canker (*Xanthomonas campestris* pv. *citri*) and banana leaf rust (*Uredo musae*) at Christmas Island, Indian Ocean. *Aust. J. Plant Pathol.*, 16, 38–39.

Simpson, A.J.G., Reinach, F.C., Arruda, P. et al. 2000. The genome sequence of the plant pathogen *Xylella fastidiosa*. *Nature*, 406, 151–159.

Singh, H.P. 1990, Country paper. Report on banana and plantain, India. In: Valmayor, R.V. (ed.), *Banana and Plantation Research and Development in Asia and the Pacific*. International Network for the Improvement of Banana and Plantain, Montpellier, France, p. 189.

Singh, R., Ferrin, D.M., and Q. Huang. 2010. First report of *Xylella fastidiosa* associated with Oleander leaf scorch in Louisiana. *Plant Dis.*, 94, 274.

Sisterson, M.S., Thimmiraju, S.R., Daane, K. et al. 2008. Assessment of the role of alfalfa in the spread of *Xylella fastidiosa* in California. *Phytopathology*, 98, 147.

Smart, C.D., Hendson, M., Guilhabert, M.R. et al. 1998. Seasonal detection of *Xylella fastidiosa* in grapevines with culture, ELISA and PCR. *Phytopathology*, 88, 83.

Smith, C.O. 1912. Further proof of the cause and infectiousness of crown gall. *Calif. Agric. Exp. Stn. Bull.*, 235, 531–537.

Smith, C.O. 1926a. Similarity of bacterial disease of avocado, lilac and citrus in California. *Phytopathology*, 16, 235–236.

Smith, C.O. 1926b. Blast of avocado—A bacterial disease. *Calif. Citrograph.*, 11, 163.

Smith, D.L., Dominiak-Olson, J., and C.D. Sharber. 2009. First report of Pierce's disease of grape caused by *Xylella fastidiosa* in Oklahoma. *Plant Dis.*, 93, 762.

Smith, E.F. 1920. *An Introduction to Bacterial Diseases of Plants*. W.B. Saunders Co., St. Louis, MO, 688pp.

Smith, E.F. and C.O. Townsend. 1907. A plant-tumor of bacterial origin. *Science*, 25, 671–673.

Sochi, H.S., Jain, S.S., Sharma, S.L. et al. 1964. New records of plant diseases from H.P. *Indian Phytopathol.*, 17, 42–45.

Soni, P.S. and K.S. Parmar. 1993. Bacterial canker. A threat to grapevine in Punjab. *Indian Phytopathol.*, 46(3), 290 (Abstr.).

Starr, M.P., Cosens, G., and H.J. Knack-Muss. 1966. Formation of the blue pigment indigoidine by phytopathogenic *Erwinia*. *Appl. Microbiol.*, 14, 870–872.

Stefani, E., Bazzi, C., Mazzucchi, U. et al. 1989. [*Xanthomonas campestris* pv. *pruni* in Friuli peach orchards.]. *Informatore Fitopatologico*, 39(7–8), 60–63 (in Italian).

Stewart, E.L. and N.G. Wenner. 2004. Grapevine decline in Pennsylvania and New York. *Wine East*, 32(2), 12–21, 51.

Stover, R.H. 1972. *Banana, Plantain and Abaca Diseases*. Commonwealth Mycological Institute, Kew, England, p. 316.

Stover, R.H. and A. Espinoza. 1992. Blood disease of bananas in Sulawesi. *Fruits*, 47, 611–613.

Subandiyah, S., Hadiwiyono, Nur, E. et al. 2006. Survival of blood disease bacterium of banana in soil. In: *Proceedings of the 11th International Conference on Plant Pathogenic Bacteria*, July 10–14, 2006, Edinburgh, Scotland.

Sule, S. and T.J. Burr. 1998. The influence of rootstock resistance to crown gall (*Agrobacterium* spp.) on the susceptibility of scions in grape vine cultivars. *Plant Pathol.*, 47, 84–88.

Sule, S., Lehockzy, J., Jenser, G. et al. 1995. Infection of grapevine roots by *Agrobacterium vitis* and *Meloidogyne hapla*. *J. Phytopathol.*, 143, 169–171.

Supeno, B. 2002. Isolation and characterization of banana blood disease in Lombok. In: *Proceedings of the 16th Congress and National Seminar of the Indonesian Phytopathological Society*, August 22–24, 2001, Bogor, Indonesia.

Supriadi. 2005. Present status of blood disease in Indonesia. In: Allen, C., Prior, P., and A.C. Hayward (eds.), *Bacterial Wilt Disease and the* Ralstonia solanacearum *Species Complex* (pp. 395–404). APS Press, St. Paul, MN.

Sutic, D. 1956. [Bacterial canker of hazel]. *Zastita Bilja*, 37, 47, 53.

Takiya, D.M., McKamey, S.H., and R.R. Cavichioli. 2006. Validity of *Homalodisca* and of *H. vitripennis* as the name for glassy-winged sharpshooter (Hemiptera: Cicadellidae: Cicadellinae). *Annu. Entomol. Soc. Am.*, 99, 648–655.

Tawfik, A.E., Riad, F.W., and S. El Eraky. 1986. Field spread of crown gall and root-knot nematode infection to peach rootstocks in Wady-el-Mollake, Ismaelia. *Agric. Res. Rev.*, 61(2), 193–201.

Topp, B.L., Heaton, J.B., Russell, D.M. et al. 1989. Field susceptibility of Japanese-type plums to *Xanthomonas campestris* pv. *pruni*. *Aust. J. Exp. Agric.*, 29, 905–909.

Trindade, L.C., Eder, M., Biaggioni, L.D. et al. 2007. Development of a molecular method for detection and identification of *Xanthomonas campestris* pv. *viticola*. *Summa Phytopathol.*, 33, 16–23.

Trujillo, E.E. and M.N. Schroth. 1982. Two bacterial disease of papaya trees caused by *Erwinia* species in the Northern Mariana Islands. *Plant Dis.*, 66, 116–120.

Turner, W.F. and H.N. Pollard. 1959. Life histories and behavior of five insect vectors of phony peach disease. *Tech. Bull.*, 1188.

UC IPM. Pest Management Guidelines - Walnut Crown Gall. 2007. UC Statewide Integrated Pest Management Program. Retrieved June 12, 2015, from http://www.ipm.ucdavis.edu/PMG/r881100211.html (updated December 2007).

Vicedo, B., Peñalver, R., Asíns, M.J. et al. 1993. Biological control of *Agrobacterium tumefaciens*, colonization, and pAgK84 transfer with *Agrobacterium radiobacter* K84 and the Tra–Mutant strain K1026. *Appl. Environ. Microbiol.*, 59, 309–315.

Viljoen, N.M. and J.M. Kotze. 1972. Bacterial black spot of mango. *Phytoparasitica*, 4(3), 93–94.

Volcani, Z. 1946. Bacterial rot of avocado fruit. *Palestine J. Bot. Ser. R*, 5(2), 169–180.

Von Rant, A. 1931. Über eine Bakterienkrankheit bei dem Melonenbaume (*Carica papaya* L.) auf Java. *Zentbl Bakteriol Parasitenkd Infektkrankh Hyg*, 84, 481–487 (in German).

Vrain, T.C. and R.J. Copeman. 1987. Interactions between *Agrobacterium tumefaciens* and *Pratylenchus penetrans* in the roots of two red raspberry cultivars. *Can. J. Plant Pathol.*, 9, 236–240.

Wager, V.A. 1937. Mango diseases in South Africa. *Fmg. S. Afr.*, 12–14.

Waissbluth, M.E. and B.A. Latorre. 1978. Source and seasonal development of inoculum for pear blast in Chile. *Plant Dis. Rep.*, 62, 651–655.

Weaver, D.J. 1978. Interaction of *Pseudomonas syringae* and freezing in bacterial canker on excised peach twigs. *Phytopathology*, 68, 1460–1463.

Webb, R.R. 1985. Epidemiology and control of bacterial canker of papaya caused by an *Erwinia* sp. on St. Croix, U.S., Virgin Islands. *Plant Dis.*, 69, 305–309.

Wells, J.M., Raju, B.C., Hung, H.Y., Weisburg, W.G., Mandelco-Paul, L., and D.J. Brenner. 1987. *Xylella fastidiosa* gen. nov., sp. nov.: Gram-negative, xylem-limited fastidious plant bacteria related to *Xanthomonas* spp. *Int. J. Syst. Bacteriol.*, 37, 136–143.

Whiteside, J.O., Garnsey, S.M., and L.W. Timmer (eds.). 1988. *Compendium of Citrus Disease.* APS Press, St Paul, MN.

Whitesides, S.K. and R.A. Spotts. 1991. Induction of pear blossom blast caused by *Pseudomonas syringae* pv. *syringae. Plant Pathol.*, 40, 118–127.

Wimalajeewa, D.L.S. and W.S. Washington. 1980. Bacterial blight of hazel-nut. *Aust. Plant Pathol.*, 9, 113–114.

XU Li-hui, XU Fu-shou, LI Fang, Praphat Kawicha and XIE Guan-lin. 2007. Fungicidal activity of L-696, 474 and cytochalasin D from the ascomycete Daldinia concentrica against plant pathogenic fungi, 34(2): 141–146.

Yirgou, D. and J.F. Bradbury. 1968. Bacterial wilt of Enset (*Ensete ventricosa*) incited by *Xanthomonas musacearum. Phytopathology*, 58, 111–112.

Yonce, C.E. 1983. Geographical and seasonal occurrence, abundance, and distribution of phony peach disease vectors and vector response to age and condition of peach orchards and a disease host survey of Johnsongrass for rickettsia-like bacteria in the southeastern United States. *J. GA Entomol. Soc.*, 18, 410–418.

Young, J.M. 1977. *Xanthomonas pruni* in almond in New Zealand. *N.Z. J. Agric. Res.*, 20, 105–107.

Young, J.M. 1987. Orchard management and bacterial diseases of stone fruit. *N.Z. J. Exp. Agric.*, 15, 257–266.

Young, J.M., Kuykendall, L.D., Martínez-Romero, E. et al. 2001. A revision of Rhizobium Frank 1889, with an emended description of the genus, and the inclusion of all species of *Agrobacterium Conn* 1942 and *Allorhizobium undicola* de Lajudie et al. 1998 as new combinations: *Rhizobium radiobacter, R. rhizogenes, R. rubi, R. undicola* and *R. vitis. Int. J. Syst. Evol. Microbiol.*, 51(Pt 1), 89–103.

Zurowski, C.L., Copeman, R.J., and H.A. Daubeny. 1985. Relative susceptibility of red raspberry clones to crown gall. *Phytopathology*, 75, 1289.

SUGGESTED READING

Anonymous. 2008. Progress report of network project on mitigating the bacterial blight disease of pomegranate in Maharashtra, Karnataka and Andhra Pradesh. National Research Centre on Pomegranate (ICAR), Solapur, India.

Bradbury, J.F. 1987. *Xanthomonas campestris* pv. *corylina.* CMI descriptions of pathogenic fungi and bacteria no. 896. CAB International, Wallingford, U.K.

CMI. 1987. Distribution maps of plant diseases no. 340, 4th ed. CAB International, Wallingford, U.K.

Davis, R.I. and J.R. Liberato. 2006. Banana blood disease (blood disease bacterium). Updated on October 21, 2011, 9:31:13 AM. Retrieved from PaDIL, http://www.padil.gov.au.

Du Plessis, H.J. 1988. Bacterial spot disease of stone fruits; overview of findings. *Deciduous Fruit Grower*, 38, 128–132.

Hewitt, W.B., Frazier, N.W., and J.H. Freitag. 1949. Pierce's disease investigations. *Hilgardia*, 19, 207–264.

Johnson, K.B. 2000. Fire blight of apple and pear. *Plant Health Instruct.* doi: 10.1094/PHI-I-2000-0726-01. Updated 2005.

Kishun, R. and J.N. Chand. 1987. Studies on germplasm resistance and chemical control of citrus canker. *Indian J. Hortic.*, 44, 126–132.

Loreti, S., Gervasi, F., Gallelli, A., and M. Scortichini. 2008. Further molecular characterization of *Pseudomonas syringae* pv. *coryli. J. Plant Pathol.*, 90(1), 57–64.

Mircetich, S.M., Lowe, S.K., Moller, W.J. et al. 1976. Etiology of almond leaf scorch disease and transmission of the causal agent. *Phytopathology*, 66, 17–24.

Muhangi, J., Nankinga, C., Tushemereirwe, W. et al. 2006. Impact of awareness campaigns for banana bacterial wilt in Uganda. *Afr. Crop Sci. J.*, 14(2), 175–183.

OEPP/EPPO. 1986a. Data sheet on quarantine organisms, 134: *Xanthomonas campestris* pv. *corylina* (Miller et al. 1940) Dye 1978. *Bull. OEPP/EPPO Bull.*, 16, 13–16.

OEPP/EPPO. 2004a. Diagnosis protocols for regulated pests *Xanthomonas arboricola* pv. *corylina. Bull. OEPP/EPPO Bull.*, 34, 155–157.

OEPP/EPPO. 2004b. Diagnosis protocols for regulated pests *Xanthomonas arboricola* pv. *corylina*. *Bull. OEPP/EPPO Bull.*, 179, 179–181.

Ophel, K., Nicholas, P.R., Magarey, P.A. et al. 1990. Hot water treatment of dormant grape cuttings reduces crown gall incidence in a field nursery. *Am. J. Enol. Viticult.*, 41, 325–329.

Psallidas, D.G. and R. Vrijer. 1996. Reclassification of *Pseudomonas syringae* pv. *avellanae* as *Pseudomonas avellanae* (Spec. nov), the bacterium causing canker of hazelnut (*Corylus avellanae* L.). *Systemat. Appl. Microbiol.*, 19(4), 589–595.

Rohrbach, K.G. 1989. Unusual tropical fruit diseases with extended latent periods. *Plant Dis.*, 73, 607–609.

Ryder, M.H. and D.A. Jones. 1991. Biological control of crown gall using *Agrobacterium* strains K84 and K1026. *Aust. J. Plant Physiol.*, 18, 571–579.

Sasser, M. 1990. Identification of bacteria through fatty acid analysis. In: Klement, Z., Rudolph, K., and D. Sands (eds.), *Methods in Phytobacteriology*. Akademiai Kiado, Budapest, Hungary, pp. 199–204.

Scortichini, M. and M. Lazzari. 1996. Systemic migration of *Pseudomonas syringae* pv. *avellanae* in twigs and young trees of hazelnut and symptom development. *J. Phytopathol.*, 144, 215–219.

Scortichini, M., Dettori, M.T., Marchesi, U. et al. 1998. Differentiation of *Pseudomonas avellanae* strains from Greece and Italy by rep-PCR genomic fingerprinting. *J. Phytopathol.*, 146, 417–420.

Smith, E.F., Brown, N.A., and L. McCulloch. 1912. The structure and development of crown gall: A plant cancer. *USDA Bulletin* 255, Government Printing Office, Washington, DC.

Smith, J.J., Jones, D.R., Karamura, E. et al. 2008. An analysis of the risk from *Xanthomonas campestris* pv. *musacearum* to banana cultivation in Eastern, Central and Southern Africa. InfoMusa@ http://www. promusa.org/index.php?option=com_content&task=view&id=66.

Stewart, E.L., Wenner, N.G., Long, L., and Overton, B. (n.d.). Crown gall of grape: Understanding the disease, prevention and management. Retrieved August 8, 2015, from http://grape.cas.psu.edu/Diseases/Crown Gall/Crown gall of grape.pdf.

Tushemereirwe, W., Kangire, A., Ssekiwoko, F. et al. 2004. First report of *Xanthomonas campestris* pv. *musacearum* on banana in Uganda. *Plant Pathol.*, 53, 802.

Viana, F.M.P., Cardoso, J.E., and H.A.O. Saraiva. 2007. First report of a bacterial leaf and fruit spot of cashew nut (*Anacardium occidentale*) caused by *Xanthomonas campestris* pv. *mangiferaeindicae* in Brazil. *Plant Dis.*, 91(10), 361.

Von Bodman, S.B., Bauer, W.D., and D.L. Coplin. 2003. Quorum sensing in plant-pathogenic bacteria. *Annu. Rev. Phytopathol.*, 41, 455–482.

Walcott, R.R. 2005. Bacterial fruit blotch of cucurbits. *Plant Health Instruct.* doi: 10.1094/PHI-I-2005-1025-02.http://www.apsnet.org/edcenter/intropp/lessons/prokaryotes/Pages/BacterialBlotchport.aspx

Weaver, D.J. and E.J. Wehunt. 1975. Effect of soil pH on susceptibility of peach to *Pseudomonas syringae*. *Phytopathology*, 65, 984–989.

Whitesides, S.K. 1989. Epidemiology of disease caused by *Pseudomonas syringae* as an epiphyte and its frequency, distribution, and characteristics as an endophyte of pear. PhD dissertation, Oregon State University, Corvallis, OR.

Willems, A., Goor, M., Thielemans, S. et al. 1992. Transfer of several phytopathogenic *Pseudomonas* species to *Acidovorax* as *Acidovorax avenae* subsp. *avenae* subsp. nov., comb. nov., *Acidovorax avenae* subsp. *citrulli*, *Acidovorax avenae* subsp. *cattleyae*, and *Acidovorax konjaci*. *Int. J. Syst. Bacteriol.*, 42, 107–119.

Wilson, E.E., Zeitoun, F.M., and D.L. Fredrickson. 1967. Bacterial phloem canker, a new disease of Persian walnut trees. *Phytopathology*, 57, 618–621.

8 Bacterial Diseases of Vegetable Crops

8.1 POTATO

8.1.1 BACTERIAL WILT OF POTATO

Pathogen: *Ralstonia solanacearum*

Synonyms: *Pseudomonas solanacearum*

Common name: Brown rot, Southern wilt, sore eye or jammy eye

Bacterial wilt (BW) is one of the most destructive diseases of the potato, and the wilt pathogen has a very wide host range. It is a serious problem in many developing countries in the tropical and subtropical regions of the world. It is usually found between the latitudes 45°N and 45°S. It has been recorded in all Australian states except Tasmania.

Symptoms

The aboveground plant symptoms include wilting of one to two leaves on young plants during the heat of the day. Such plants tend to recover at night. On large-leafed plants, only the tissue on one side of the midvein may wilt. This is very characteristic for plants such as Nicotiana. Affected leaves turn yellow and remain wilted after a time. The area between leaf veins dies and turns brown. Usually, the main stem of the affected plants remains upright even though all the leaves may wilt and die (Photo 8.1).

Internal symptoms include light tan to yellow-brown discoloration of the vascular tissue. Long sections of infected stems reveal dark brown to black streaking in the vascular tissue as the disease progresses. As invasion proceeds, the pith and cortex of the stem become dark brown.

Typical symptoms include wilting, yellowing, and some stunting of the plants, which finally die right back. Wilting is first seen as a drooping of the tip of some of the lower leaves similar to that caused by a temporary shortage of water. At first, only one branch in a hill may show wilting. Affected leaves later become permanently wilted and roll upward and inward from the margins. The wilting then extends to leaves further up the stem and is followed by a yellowing of the leaves. This yellowing, wilting, and in-rolling of the leaves make diseased plants very obvious, especially when surrounded by healthy plants. The leaves finally turn brown and fall off, beginning at the base of the stem and continuing upward.

Symptoms in the tuber are very specific: brownish gray areas are seen on the outside, especially near the point of attachment of the stolon. Cut tubers may show pockets of white-to-brown pus or browning of the vascular tissues (Photo 8.2) which, if left standing, may exude dirty white globules of bacteria. As the disease progresses, bubbly globules of bacteria may exude through the eyes; soil will often adhere to the exuded bacteria, hence the name "sore eyes" or "jammy eyes."

Signs of the Pathogen

Slimy, sticky ooze forms tan-white to brownish beads where the vascular tissue is cut. When an infected stem is cut across and the cut ends held together for a few seconds, a thin thread of ooze can be seen as the cut ends are slowly separated. If one of the cut ends is suspended in a clear glass container of clean water, bacterial ooze will form a thread in the water.

PHOTO 8.1   Foliar symptoms of bacterial wilt on potato. (Photo courtesy of A.M. Varela, International Centre of Insect Physiology and Ecology (ICIPE), infonet-biovison.org.)

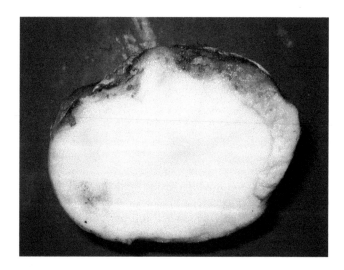

PHOTO 8.2   Cut tuber symptoms of bacterial wilt on potato. (Photo courtesy of A.M. Varela, International Centre of Insect Physiology and Ecology (ICIPE), infonet-biovison.org.)

Host

Bacteria *Ralstonia solanacearum* attack almost 200 plant species in 33 different plant families. These include economically important hosts such as tobacco, potato, tomato, eggplant, pepper, banana, peanut, and beans. Thorn apple and nightshade are two common weed hosts that are attacked by the disease pathogen. This constitutes one of the largest known host ranges for any plant pathogenic bacterium. Although the Solanaceae (potato family) contains the greatest number of susceptible species, many other dicot and a few monocot plants are also susceptible.

The common name for the diseases this organism causes varies with the host that is attacked. In tobacco, it is called BW or Granville wilt (for Granville County, North Carolina where it was observed as early as 1880) and moko disease in banana. It is sometimes called southern wilt or southern BW (in the northern hemisphere). This bacterium is noted for diseases caused outdoors in land areas bounded by 45°N and 45°S latitudes where rainfall averages above 100 cm/year, the average growing season exceeds 6 months, the average winter temperatures are not below 10°C, the average summer temperatures are not below 21°C, and the average yearly temperature does not exceed 23°C. It can be moved from such areas into the greenhouse industry in and on plants propagated in those regions and then sold to growers throughout the world.

Although the primary location of survival in the environment is in crop and weed hosts, it can also survive in soil. It can be readily spread through the movement of contaminated soil and infected vegetatively propagated plants, in contaminated irrigation water, on the surfaces of tools (cutting knives) and equipment used to work with the plants, and on soiled clothing.

Pathogen

The bacteria were first named *Bacillus solanacearum*. After several revisions, it was called, for many years, *Pseudomonas solanacearum*. The latest revision has settled on the name *Ralstonia solanacearum*. It is described as a nonspore-forming, Gram-negative, nitrate-reducing, ammonia-forming, aerobic, rod-shaped (0.5–1.5 μm) bacteria with one polar flagellum. Populations within this genus and species can be further divided into races and biovars based on differing host ranges, biochemical properties, susceptibility to bacteriophages, and serological reactions.

Bacteria *Ralstonia solanacearum* (earlier known as *Pseudomonas solanacearum*), based on the type of host plants it attacks, is divided into three races, and based on its biochemical properties, it is divided into four biovars. The most widespread strain in Australia is race 3/biovar 2. This strain is known to occur in New South Wales, Queensland, and it primarily attacks potato. Two other strains, which attack other hosts besides potato, are confined to the Northern Territory and Queensland.

Race 1 exist in the United States, where it attacks many floricultural and vegetable bedding plant crops including geraniums (all Pelargonium), Catharanthus, Impatiens, Ageratum, Chrysanthemum, Gerbera, Tagetes, Zinnia, Salvia, Capsicum, Lycopersicon, Nicotiana, Petunia, *Solanum melongena* (eggplant), Tropaeolum (nasturtium), and Verbena. Race 3 is tropical in distribution and does not occur naturally in North America. Race 3 biovar 2 (R3B2), inadvertently introduced to the United States in vegetatively propagated geraniums grown in Central America and Africa, is considered to be a major threat to agriculture in the United States because it causes brown rot in potato (Kim et al., 2003). Some of the other known hosts of R3B2 include Pelargonium, tomato, peppers, eggplant, bean, and beet. Weed hosts include black nightshade, climbing nightshade, horsenettle, Jimson weed, purslane, mustards, lambsquarters, and bittergourd. The bacteria can infect through roots and through any fresh wounds. The bacterium can be difficult to work with in the laboratory because it quickly loses pathogenicity and viability in artificial culture.

Disease Cycle

Potato wilt bacterium is a soil-borne organism primarily inhabiting the roots. It enters the root system at points of injury caused by farm implements, nematodes, and by other means. It is spread by irrigation water, flood waters, and contaminated soil.

BW of potato is generally favored by temperatures between 25°C and 37°C. It usually does not cause problems in areas where mean soil temperature is below 15°C.

Under conditions of optimum temperature, infection is favored by wetness of soil. However, once infection has occurred, severity of symptoms is increased with hot and dry conditions, which facilitate wilting.

The wilt bacterium is able to survive for periods up to 2–3 years in bare fallow soils and for longer periods in soils cropped to nonsolanaceous crops. Infected seed is an important method of dissemination, both locally and over considerable distances. Heavily infected tubers are not a problem since they generally rot away, but the contamination of the land in which they were grown poses a problem. However, slightly infected tubers, which show no visible symptoms, pose a serious threat of spreading the disease to new areas. Self-sown potatoes are extremely difficult to eradicate and, if a paddock is infected, the disease may remain in it for 5 or 6 years after the initial outbreak.

Bacteria can also be spread to clean tubers from an infected seed cutter. There is also a real danger of infection if second-hand bags are used or if half tonne bins have held infected potatoes. Growers should be aware of these risks and take precautionary measures.

Economic Importance

BW is responsible for causing considerable losses to the potato industry where the disease exists.

In the south east of Victoria, it has caused considerable losses in the past to the potatoes planted mainly in the swampy areas. However, the threat of the disease is potentially significant to the seed potato production industry. Some states and countries that import potato regard BW in the same light as black wart, ring rot, and potato cyst nematode and ban imports from areas known to be infected.

The disease can cause total loss of a crop and restrict the use of land for potato production for several years.

Disease Management

BW is difficult to control (or eradicate) because of the soil-borne nature of its causal organism. Therefore, following options should be considered in managing the disease.

Growing and propagating from pathogen-free plant material is the main way to avoid problems with *Ralstonia*, regardless of the race and biovar involved. Propagators must use pathogen-free potting soil or other media, establish stock plants that are tested and known to be free of the bacteria, train workers handling the stock plants in methods and procedures that prevent the pathogen from contaminating the potting soil or coming in contact with the stock plants, and then maintaining this system throughout the propagation phase of crop production.

There are no chemicals or biological agents that adequately control these bacteria. Infected plants *must* be discarded as soon as possible.

- Adopt rotations with pastures, cereals, and nonsolanaceous crops for periods exceeding 5 years.
- Use certified seeds from reliable sources. Exclusion of the disease may be exercised by quarantine or other legislative measures. For example, Tasmania, which so far has not recorded BW, is very careful to import only healthy seed. New Zealand and South Africa ban the importation of seed from areas known to have the disease.
- Planting in areas where BW has not occurred previously.
- Control self-sown potatoes.
- Control weed hosts such as nightshade, thorn apple, Narrawa burr around dames, along channels and in the paddocks after cropping potatoes.
- Avoid deep plowing—the organisms survive in the deep, cool layers of soil.
- Irrigation water should never be allowed to run freely over or below the soil surface. It should never be allowed to return to the dam or stream from which it is pumped nor to any other irrigation source.
- Regularly inspect crops for disease symptoms and remove and destroy diseased plants, tubers, and immediate neighbors.
- Use stock to clean up chats, discarded tubers, and crop debris, but do not allow the stock back onto clean paddocks.
- Do not return potato waste, for example, oversized, misshapen, and diseased tubers to paddocks, so as to minime the spread.
- Machinery taken onto a diseased paddock should be left on the paddock while it is being worked.
- Machinery removed from the paddock should then be washed clean with a disinfectant solution in a dedicated area where equipment are washed.
- Use high-pressure wash to clean machinery, sheds, and so on to remove soil adhering to any surfaces.

- Clothing and boots of people working in the paddock should be exchanged for clean items when leaving the paddock, or else boots should be washed in a suitable disinfectant.
- After harvest, all diseased and discarded tubers should be collected and buried at least 1 m underground.
- On no account should any of the produce from a diseased crop be kept as seed.
- Load and unload vehicles only in designated areas with sealed or hard ground or bare paddocks away from potato paddocks.
- Choose transport routes that minimize travel through potato paddocks and regions.
- If second-hand bags or half tonne bins have been used to hold potatoes, these should be thoroughly washed and disinfected before being used again. Bags should be disinfected or discarded.
- Ask visitors, contractors, and workers to wear overalls, gumboots, and overshoes on the property.

Action in the Event of Suspect Cases

Seed and ware growers, potato merchants, and importers are requested by the Department of Agriculture to examine potato tubers regularly for signs and symptoms of the disease. If the plants/tubers are suspected to be infected, please contact your local plant health inspector or contact the Department of Agriculture, Fisheries and Food.

8.1.2 Blackleg of Potato

Pathogen: *Erwinia carotovora* subsp. *atroseptica* (synonym: *Pectobacterium atrosepticum*)

Host: Potato (*Solanum tuberosum*)

The bacterium causing the blackleg disease of potato is a tuber-borne pathogen. The blackleg disease can cause severe economic losses to the potato crop. However, the occurrence of blackleg depends very much on the growing conditions, particularly temperature and rainfall after planting.

Symptoms

On Foliage

Blackleg disease sometimes develops early in the growing season soon after the plants emerge. This is referred to as an early blackleg and is characterized by stunted, yellowish foliage that has a stiff, upright habit. The lower part of the belowground stem of such plants is dark brown to black in color and extensively decayed. The pith region of the stem is particularly susceptible to decay, and in blackleg-infected plants, the decay may extend upward in the stem far beyond the tissue with externally visible symptoms. The typical blackening and decay of the lower stem portion is the origin of the "blackleg" designation for this disease. Young plants affected by blackleg fail to develop further and typically die.

In addition to early blackleg, the disease may also develop later during the potato-growing season. In more mature plants, blackleg appears as a black discoloration of previously healthy stems, accompanied by a rapid wilting, and sometimes yellowing, of the leaves. Black discoloration of the stems always starts below ground and moves up the stem, often until the entire stem is black and wilted. At the early stages of disease development in mature stems, the leaves may turn yellow and wilt, even before the black decay is evident. However, after the entire stem becomes diseased, it decays, becomes desiccated, and is often lost from view in the potato canopy.

Blackleg disease inevitably originates at the seed tuber from which the plant is grown. Bacterial decay that originates in broken or damaged stems is not to be confused with blackleg, although the symptoms have some similarity. This aerial stem rot is usually caused by *E. carotovora* subsp. *carotovora*, a close relative of the blackleg bacterium. Aerial stem rot is usually a lighter brown in color than blackleg and although the decay moves up the stem, it does not start below ground. A stem wet rot is much like blackleg in many aspects but is caused by yet another bacterium, *E. chrysanthemi*. Although stem wet rot is a significant disease in Europe, particularly in the Netherlands where it often cannot be distinguished from blackleg, it is not known to occur in North America.

On Tubers

There are two ways by which the blackleg bacterium may reach the progeny tubers produced on the potato plant. One important route of tuber infection is via the stolon by which the tuber is attached to the plant. Tubers with blackleg disease generally first become decayed at the stolon attachment site where the tuber tissue becomes blackened and soft. As the disease progresses, the entire tuber may decay or the rot may remain partially restricted to the inner perimedullary (or parenchymal) tissue, that is, the tissue inside the vascular ring.

An alternate route for the pathogen to attack progeny tubers is via the soil. As the blackleg disease causes the belowground stem and seed tuber to decay, the causal bacterium spreads from infected tissue into soil water and becomes distributed throughout the root zone in which the progeny tubers are growing. Bacterial cells enter lenticels of the progeny tubers and either become inactive, or, when conditions are favorable, initiate decay.

In a poorly managed potato storage environment, blackleg bacteria present on the surface of tubers can cause extensive decay. Sometimes when storage conditions are improved, decay lesions around tuber lenticels or mechanically damaged areas become arrested, resulting in a condition known as "hard rot." Hard rot is typified by slightly sunken, brownish-black, dry, necrotic lesions surrounding individual lenticels or damaged areas.

Once decay of potato tubers is incited by the blackleg bacterium, growth of secondary bacteria often contributes to the decay process and certainly modifies symptomatology of the disease. Hence, a general bacterial soft rot develops from the initial blackleg infection in tubers. Bacterial soft rot is characterized by total maceration of tuber tissue and seepage of a putrid, dark-colored liquid.

Pathogen

The causal agent of blackleg is *E. carotovora* subsp. *atroseptica*, a Gram-negative, rod-shaped bacterium closely related to enteric bacteria of importance as human and animal pathogens. Bacteria in the genus *Erwinia*, however, are not known to be harmful to humans or animals. When grown on a medium containing sodium polypectate, the blackleg bacterium develops pits or craters in the medium due to the excretion of pectolytic enzymes that liquefy the pectate. The pectolytic enzymes are, in fact, an important component of the pathogenicity factors of this and related bacteria. In recognition of the unique pectolytic activity of these bacteria, it has been proposed to place them in a separate genus, *Pectobacterium*. Furthermore, it has recently been suggested that on the basis of its genetic composition, the blackleg bacterium be considered a unique species, *P. atrosepticum*.

Several other subspecies of *E. carotovora* that cause disease in other crops have been described. These subspecies can cause decay of potato tuber slices but do not cause the blackleg disease. A new strain of *E. carotovora* has been described recently, which causes a blackleg-like disease of potato in Brazil. Preliminary results suggest that the *atroseptica* subspecies does not occur on potato where the newly described subspecies, tentatively named *brasiliensis*, occurs.

The most distinguishing feature of the blackleg bacterium is its pectolytic enzyme activity, but in contrast to many of the other pectolytic bacteria, it does not grow above 36°C. It is facultatively anaerobic, meaning that it can grow both with and without the presence of oxygen. It is motile with numerous peritrichous flagella. Most strains of *E. carotovora* subsp. *atroseptica* belong to serogroup I, although several other serogroups occur. Because of the relatively uniform serological type, serological methods such as enzyme linked immunosorbant assay and immunofluorescence can be used for its detection. Molecular methods including the polymerase chain reaction (PCR) are also available for the detection and identification of this bacterium.

The blackleg bacterium can be isolated on a selective medium such as the crystal violet pectate medium most efficiently at about 22°C. Colonies of pectolytic *Erwinia* can be readily identified on this medium with a dissecting microscope using oblique illumination, that is, by shining a light through the bottom of the Petri plate from an angle. Colonies transferred from Crystal Violet Pectate (CVP) grow on general bacteriological media such as nutrient agar. Pathogenicity of isolates can be easily determined on young (10–15 cm/4–6 in. high) potato plants by stabbing into the stem a toothpick smeared with bacterial cells. Symptoms of blackleg develop within 2 weeks.

Disease Cycle

The blackleg bacterium survives poorly in soil. Although other members of the pectolyic *Erwinia* survive in surface water and in the soil environment, all evidence suggests that the blackleg bacterium does not survive very well outside of association with host plant tissue. Hence, the seed tuber is the most important source of inoculum in the blackleg disease cycle. When a contaminated or infected seed potato is planted, one of three things may occur.

1. The blackleg bacteria may move via the vascular bundles directly into the growing plant and result in blackleg disease. If tuber contamination is confined to the lenticels, decay of the seed tuber occurs first, and when bacterial populations become great enough, invasion of the growing stem occurs. Both the processes of seed tuber decay and the spread of the pathogen into the stem are highly dependent on environmental conditions. Moist, cool conditions favor the disease.
2. When conditions are favorable for growth of the potato plant, no disease may occur even when the blackleg bacterium is present.
3. Decay of the seed piece may occur prior to the establishment of a plant, and this, too, is an important manifestation of blackleg, although other members of the pectolytic *Erwinia* can also cause seed piece decay.

The most frequently occurring situation that follows the planting of contaminated seed is the seed piece decays after a plant has been established, and no blackleg disease develops at all. In this case, the blackleg bacteria seeping from the decaying seed piece contaminate the entire root zone including developing progeny tubers. Surfaces of the progeny tubers become contaminated with the bacteria surviving particularly well in lenticels. In storage, the contaminated tubers may decay, develop hard rot symptoms, or remain symptomless. When symptomless but contaminated tubers are used for planting, which they often are, the cycle is repeated. The disease cycle of blackleg of potato is illustrated in Figure 8.1.

Contamination of potato tubers is exacerbated by harvesting and storage operations. A single tuber with blackleg decay may contaminate many additional tubers as they pass over conveyer belts on harvesters and bin pilers. During storage, decay and rotting of contaminated tubers is a common problem. Damaged tubers are particularly vulnerable to decay by the pectolytic *Erwinia*. The presence of moisture on stored tubers is also conducive to development of decay, since a film of water surrounding tubers causes them to become anaerobic. The lack of oxygen inhibits tuber metabolic activity and prevents them from staging a normal resistance reaction.

Disease Management

Preplanting

There was a time when almost all potato tubers were contaminated with the blackleg bacterium. That is no longer true today. The use of healthy tissue culture plantlets to initiate seed potato stocks has broken the cycle of carrying tuber contamination forward from year to year. Also by limiting the number of field generations to 5–7 years for production of individual seed lots after tissue culture, the buildup of tuber contamination is curtailed. Hence the incidence of blackleg is significantly lower than it was before the incorporation of tissue culture into seed potato production programs. Although disease reduction has been very significant in some geographic areas, the disease remains important in others where similar practices are used. The reason for the difference in disease

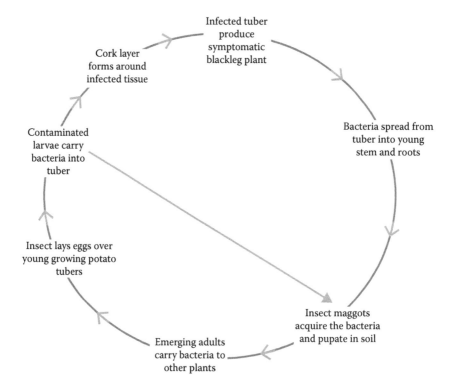

FIGURE 8.1   Diseased cycle of blackleg of potato caused by *Erwina caratovora* subsp. *atroseptica*.

incidence is unclear but is probably related to the rapidity by which new seed stocks are exposed to blackleg inoculum. The risk of exposing new seed stocks to inoculum will depend on specific agronomic practices and the ability of the bacterium to persist outside of potatoes in the prevailing climatic conditions of the different geographic locations.

At Planting
Planting limited generation seed in well-drained soil after soil temperature has increased above 10°C is recommended for avoiding the development of blackleg.

During the Growing Season
Roguing out blackleg-diseased plants including belowground portions reduces soil inoculum but is only a useful practice if precautionary measures are taken to prevent the contact of diseased tissue with other plants in the field.

At Harvest and during Storage
Avoiding injury to potato tubers during harvest is important to minimize decay in storage. Removal of decayed potatoes before they spread their contents over grading lines and bin pilers avoids spreading the bacterium to other tubers. Wound healing is important in the early phase of potato storage to prevent development and spread of rots. During storage, however, the potatoes should be kept at a low temperature with adequate aeration to provide a dry environment and to prevent condensation of moisture on tuber surfaces.

Significance
The different manifestations of potato blackleg as a disease of potato plants, seed piece decay, and storage rot all contribute to economic losses. Although the disease is now considered to be of minor importance in some potato-growing regions, it continues to be a major production factor in others. Control of the disease relies wholly on crop management practices and hence there are no chemical

control measures. Although cultivars vary in disease susceptibility, none is immune. Continued use of tissue culture–derived plantlets and minitubers (grown from plantlets in a protected environment) to initiate seed stocks coupled with limited generations of field planting are essential for minimizing the contamination of seed stocks and maintaining the level of control that has been achieved. In those areas where the disease is not adequately controlled by these measures, further research is required to determine the source from which *Erwinia*-free planting material becomes contaminated. Whether the bacterium survives in soil or irrigation water below the detection level threshold warrants investigation. Environmental spread of bacteria from late generation crops to new seed stocks via water, wind-driven rain, or insects also needs to be studied.

Molecular research on the pectolytic *Erwinia*, including the blackleg bacterium, has revealed many fascinating aspects concerning the genetics of pathogenicity in plant pathogenic bacteria. The complex genetic control mechanisms that modulate expression and excretion of pectolytic enzymes are now just beginning to be understood. The importance of biofilm formation and associated signaling mechanisms among bacterial cells and between bacteria and host is currently being investigated. Sequencing of the genome of the blackleg bacterium is underway and is expected to reveal even more about its ability to cause disease in potato and the mechanisms by which it does so.

8.1.3 Bacterial Soft Rot of Potato

Pathogen: *Pectobacterium carotovorum* sub sp. *carotovora*
Pectobacterium chrysanthemi

Symptoms
Soft rot on tubers first appears as small, tannish, water-soaked spots on the surface. These spots rapidly enlarge and the tissue decomposes in a soft, blister-like area on the surface of the tuber. Often, a slimy or watery substance oozes from breaks in the blister. The blister margin is darker than the tuber skin. Soft rot often follows bruising and is first white to cream colored (Photo 8.3). After exposure to air, it becomes brown to black. The boundary between the disintegrated and the sound tissue is sharp. It is nearly odorless at the stage. As secondary rot occurs, the rot becomes very foul smelling. The rot typically progresses to the point of a chalky white, foul-smelling mass.

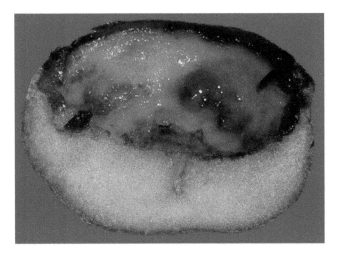

PHOTO 8.3 Symptoms of bacterial soft rot on potato tubers. (Photo courtesy of R. Yonzone, Uttar Banga Krishi Vishwavidyalaya, Cooch Behar, West Bengal, India.)

Soft rot symptoms on the foliage include weak, chlorotic (yellowed) plants with margins of leaflets curled upward. Stem lesions are usually light brown, and can be colorless, but not black. Stems will rot and become very mushy. Tuber rot will occur as point of infections often on an eye but can be generalized on the tuber.

Disease Cycle and Environment Favorable for Disease

Potato soft rot is caused by the bacterium *Erwinia carotovora* subsp. *carotovora*, a common soil resident.

Bacteria are present on all tubers and are associated with many kinds of plants. Infections in the field are favored by high soil moisture and high temperatures. Other factors include anaerobic conditions, enlarged lenticels, and invasion by other pathogens. Bacteria enter lenticels, growth cracks, or any injury. During and after harvest, soft rot is favored by immature tubers, adverse temperatures (pulp temperatures above 21°C at harvest), mechanical damage, and free water on tuber surfaces.

This bacterium can grow between the temperatures of 0°C and 32°C, with optimal growth between 21°C and 27°C. Bacterial soft rot occurs on a wide range of crops and is one of the most severe postharvest diseases of potatoes worldwide. Loss may occur during storage, transit, or marketing. All potato varieties are susceptible.

Contamination of potato tubers occurs anytime they come into contact with the bacterium, most commonly during harvest, handling, or washing. The bacterium invades the potato tuber chiefly through wounds. Most of the soft rot infections are in tissues that have been weakened, invaded, or killed by pathogens or by mechanical means. Soft rot in tubers is favored by immaturity, wounding, invasion by other pathogens, warm tuber and storage temperatures, free water, and low oxygen conditions. Tubers harvested at temperatures above 27°C can be predisposed to soft rot. Decay can be retarded by temperatures less than 10°C; the lower the temperature, the better. Immature tubers are susceptible to harvester-related injury and bacterial infection. Suberizing seed and treatment with fungicide is a tactic to reduce the risk of other seed infections that could lead to soft rot breakdown of the seed.

Disease Management

Although symptoms of bacterial soft rot do not begin in the field, control of bacterial soft rot does begin in the field. Other diseases that produce tuber lesions need to be controlled. Consider these suggestions:

- Delaying harvest until the skin has set reduces tuber injuries. This will reduce the entry points for the pathogen.
- At harvest, watch tuber-handling practices, and ensure good sanitary procedures to reduce spread of bacteria. Harvesting during wet, muddy conditions generally leads to an increase in bacterial soft rot in storage.
- Properly suberize potatoes to ensure wound healing and reduce the infection sites for the pathogen.
- Leave potatoes a minimum of 7 days after the vines are totally dead to encourage skin set and reduce bruising.
- Eliminate as much soil as possible before the tubers are stored, as soil can restrict air movement.
- If harvesting wet potatoes, ventilate continually until the potatoes are dry.
- Isolate and keep problematic potato lots in separate bin.
- Check the pile temperature at regular intervals. Early detection aids in control, thereby reducing loss. If elevated pile temperatures are detected, consider hot-spot fans. These 1/3- to 1/2-horsepower fans are about 16–18 in. in diameter. When run continuously for up to several weeks, these hot-spot fans can stop storage breakdown.
- Minimizing potato bruising, avoiding harvesting during wet conditions, and placing the potatoes into a disinfected storage are three easy, cheap, and effective control practices to reduce loss in storage from soft rot.

- Avoid washing seed or potatoes before storage unless absolutely necessary. This is a desperate salvage operation. Potatoes that have soft rot should be removed before storing and during packing. After washing, potatoes should be dry before shipping. Tubers need to be protected from bruises, excessive heat, or cold during harvest and transport.
- Use high-quality seed. Split applications of water-soluble calcium applied at 100–200 lb/acre during bulking have been shown to reduce infection and severity of soft rot. Harvest mature tubers with well-set skins and avoid mechanical injury.
- Avoid excessive soil moisture before harvest to reduce lenticel infection.
- Use clean water to wash potatoes and avoid water films on tuber surfaces during storage.
- Postharvest curing and storage temperatures can be critical components of soft rot management. Specific temperature recommendations vary depending on the level of decay evident at packing and the market destiny of the potatoes (i.e., processing, fresh market, or long-term storage).

8.1.4 BACTERIAL RING ROT OF POTATO

Pathogen: *Clavibacter michiganensis* subsp. *sepedonicus*

The disease is called "bacterial ring rot" because the rot appears in the vascular ring of the potato tuber.

Symptoms

In severely affected tubers, the vascular ring is brown to black in color, often with a cheesy or creamy ooze and many hollow spaces where the flesh has disintegrated. Dry cracks can usually be found on the surface of the tubers. In milder cases, the vascular ring (Photo 8.4) may show only broken, black lines or a yellowish discoloration. Leaves of infected plants may show interveinal yellowing (Photo 8.5), wilting, or no symptoms.

Survival and Spread

Once a crop or farm is infested with the bacterium, the disease will carry over from year to year and spread quickly. The bacteria can survive for 2–5 years in dried slime on the surface of machinery, crates, bins, or burlap sacking, even if frozen. Volunteer potato plants and plant debris, including infected cull tubers, will also carry the overwintering bacteria. The bacteria can be spread in rain and irrigation water and by insects, but wounds are needed for infection. The most important means of infection is cutting seed potatoes with contaminated knives.

PHOTO 8.4 Symptoms of bacterial ring rot pathogen on potato tuber. (Photo courtesy of William M. Brown Jr., Bugwood.org.)

PHOTO 8.5 Symptoms of bacterial ring rot pathogen on potato foliage. (Photo courtesy of William M. Brown Jr., Bugwood.org.)

Disease Management

Prevention
Plant only certified seed potatoes. Avoid using table stock for seed or importing seed from other disease-prone areas for planting. Disinfect knives frequently while cutting seed pieces to avoid spreading any disease that might be present but unnoticed. Bury cull piles and control volunteer potatoes. Clean and disinfect storage bins between crops and pressure-wash equipment to avoid spreading ring rot.

Regulations
Bacterial ring rot is a regulated disease in British Columbia. All potatoes grown in British Columbia are governed by Domestic Bacterial Ring Rot Regulation 93/59. Under this regulation, if symptoms of bacterial ring rot are found in any potato crop grown in British Columbia, the occurrence must be reported to the provincial Ministry of Agriculture and the crop must be detained until inspected. Bacterial Ring Rot Regulation 92/59 provides inspection of all potatoes imported to British Columbia.

8.1.5 POTATO SCAB

Pathogen: *Streptomyces scabies*

Streptomyces scabies or *Streptomyces scabiei* is a streptomycete bacterium species found in soils around the world (Lerat et al., 2009). Unlike most of the 500 or so *Streptomyces* species, it is a plant pathogen causing corky lesions to form on tuber and root crops as well as decreasing the growth of seedlings. Along with other closely related species, it causes the potato disease common scab, which is an economically important disease in many potato-growing areas.

Symptoms
Scab symptoms are quite variable and are localized on tubers (Photo 8.6) with no aboveground symptoms. Lesions are usually roughly circular, raised, tan to brown, corky in appearance, and between 5 and 10 mm in diameter (Hooker, 1981). The pathogen can induce erumpent, pitted, or superficial lesions on tubers, with the former being considered the most common (Loria et al., 1997). Although scab symptoms are not usually noticed until late in the growing season or at harvest, tubers are infected during the period of rapid tuber growth that commences when the tuber diameter reaches twice that of the stolon. Small, brown, water-soaked, circular lesions are visible

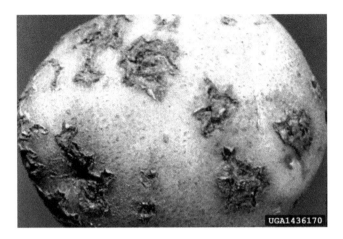

PHOTO 8.6 Symptom of common scab on potato tuber. (Photo courtesy of Clemson University—USDA Cooperative Extension Slide Series, Clemson, SC, Bugwood.org.)

on immature tubers associated with lenticels within a few weeks after infection (Lapwood, 1973). Mature tubers with a well-developed skin are no longer susceptible, but existing lesions will continue to expand as tubers enlarge.

The main route by which *Streptomyces scabies* is thought to enter into potato tubers is through the lenticels (pores for gas exchange) in the skin of potatoes. Other evidence suggests that they are also able to directly penetrate the skin of the potato causing infection.

Pathogen

It was first described in 1892, being classified as a fungus, before being renamed in 1914 and again in 1948. It can infect young seedlings as well as mature root of tuber crops, but it is most often associated with causing common scab of potato.

At least four other species of *Streptomyces* also cause diseases on potato tubers. The most widespread species other than *Streptomyces scabies* are *Streptomyces turgidiscabies* and *Streptomyces acidiscabies*, which can be distinguished based on their morphology, the way they utilize food sources, and their 16S RNA sequences. Unlike *S. scabies*, *Streptomyces acidiscabies* is predominantly seedborne rather than soilborne and be suppressed using insecticides and nematicides, suggesting that microfauna play a role in its transmission (Lambert et al., 2006). In 2003, three other species of *Streptomyces* that cause common scab symptoms were isolated in Korea and named *Streptomyces luridiscabiei*, *Streptomyces puniciscabiei*, and *Streptomyces niveiscabiei*. They differ from *Streptomyces scabies* by having spores that are of different colors. *S. ipomoea* causes a similar disease on sweet potato tubers.

There are also other species of *Streptomyces* found in scab lesions on potato tubers that do not cause disease. Sixteen distinct strains have been isolated from tubers, and based on a genetic analysis, they are found to be most similar to *Streptomyces griseoruber*, *Streptomyces violaceusniger*, *Streptomyces albidoflavus*, and *Streptomyces atroolivaceus*.

Streptomyces scabies is a streptomycete bacterium, which means it forms a mycelium made of hyphae, a growth form more usually associated with fungi. The hyphae of *Streptomyces* are much smaller than those of fungi (0.5–2.0 μm) and form a heavily branched mycelium. They are Gram-positive and have a high proportion of the DNA bases guanine and cytosine (Loria et al., 2003) (71%) in their genome. When cultured on agar, the hyphae develop aerial fragments, which bear chains of spores, giving the culture a fuzzy appearance. The chains of spores have the appearance of corkscrews and are gray in color (Loria et al., 2003). These chains allow it to be differentiated from other species that are virulent on potatoes. Each chain contains 20 or more spores that are 0.5 by 0.9–1.0 μm in size, smooth, and gray. Bacteria are often distinguished by their ability to grow on media containing different substances, which they either feed on or that inhibit their growth. Strains of *S. scabies* grow on the sugar raffinose, are unable to degrade xanthine, and when grown on media containing the amino acid tyrosine, they produce the pigment melanin, the same chemical that gives humans their skin color. This trait is often associated with their ability to cause disease, but it is not always present and is considered a secondary trait. They can be controlled by 10 μg of the antibiotic penicillin G/mL, 25 μg of oleandomycin/mL, 20 μg of streptomycin/mL, 10 μg of thallous acetate/mL, 0.5 μg of crystal violet/mL, and 1000 μg of phenol/mL. The lowest pH at which they will grow varies slightly between strains but is between 4 and 5.5.

Pathogenesis

Five toxins have been isolated from *S. scabies* that induce the formation of scabs on potato tubers. They are classed as 2,5-diketopiperazines (Borthwick 2012), with the most abundant having the chemical formula $C_{22}H_{22}N_4O_6$. The first two to be isolated in 1989 were thaxtomin A and thaxtomin B, of which thaxtomin A was the predominant compound. Thaxtoxin B has a hydrogen at C_{20} rather than a hydroxyl group, which differentiates it from thaxtoxin A (King et al. 1989). Three years later King et al. 1992. isolated several other toxins with similar structures to the first two they had isolated, which are thought to be precursors of thaxtomin A. Thaxtomin A is considered to be essential to produce symptoms and the pathogenicity of strains is correlated with the amount of thaxtomin A they produce. It is synthesized by a protein synthetase encoded by the *txtA* and *txtB* genes, forming a cyclic dipeptide, which is then hydroxylated by a cytochrome P450 monooxygenase encoded by *txtC*. The dipeptide is then nitrated by an enzyme similar to mammalian nitric oxide synthase at the fourth position on the tryptophan residue. All the genes required for thaxtomin biosynthesis are located on one part of the genome, termed the pathogenicity island, that is also found in *S. acidiscabies* and *S. turgidiscabies*, which is around 660 kb in length. The toxins are only produced once the bacteria have colonized a potato tuber, and it is thought they detect potatoes by sensing certain molecules present in their cell walls. Cellobiose, a subunit of cellulose, activates thaxtomin production in some strains, but suberin also acts as an activator, causing many changes to the proteome of the bacteria after it is detected.

The target of the toxins is unknown but there is evidence that they inhibit the growth of plant cell walls. They are neither organ nor plant specific, and if added to the leaves of various species, they cause them to die, indicating that the target is highly conserved (Joshi et al., 2010). Adding thaxtomin A to seedlings or suspended plant cell cultures causes them to increase in volume, and onion root tips treated with it are unable to form cell plates suggesting that it affects the synthesis of cellulose. Inhibiting the production of cell walls may aid *S. scabie* in penetrating plant cells, a key step in infection. The fact that scabs only form in regions of rapidly growing tissue is consistent with this hypothesis.

Host

Streptomyces scabies can infect many plants but is most commonly encountered causing disease on tuber and tap root crops. It causes common scab on potato (*Solanum tuberosum*), beet (*Beta vulgaris*), carrot (*Daucus carota*), parsnip (*Pastinaca sativa*), radish (*Raphanus sativus*), rutabaga (*Brassica napobrassica*), and turnip (*B. rapa*). It also inhibits the growth of the seedlings of both monocot and dicot plants. Potato varieties differ in their susceptibility to *S. scabies*. More resistant varieties tend to have fewer, tougher lenticels and a thicker skin.

Disease Management

Although complete control cannot be achieved, disease severity can be managed by cultural practices. The maintenance of high soil moisture can achieve adequate control of potato scab (Davis et al., 1974a, 1976; Adams et al., 1987; Loria et al., 1997) especially when applied early in the season (Barnes, 1972; Lapwood et al., 1973). Experiments conducted in the United Kingdom and the Netherlands showed that irrigation greatly reduced the damage caused by scab on carrot (Groves and Bailey, 1994; Schoneveld, 1994). Maintaining a low soil pH can also reduce scab through the application of elemental sulfur, gypsum (Davis et al., 1974b), or acid-forming nitrogen fertilizer, such as ammonium sulfate (Mizuno et al., 1995). Scab resistance is an objective of most breeding programs, though no completely resistant cultivars exist (McKee, 1958; Goth and Haynes, 1993), and this can be attributed to a poor understanding of possible resistance mechanisms (Loria et al., 1997).

Biological Control

Field studies on the biological control of potato scab using a nonantibiotic-producing actinomycete biofertilizer reduced scab severity from 10.7% to 2.3% (Nanri et al., 1992). Lesions were slight and potato production was three times higher in the biofertilizer subplot than in the control subplot. Viable counts of *S. scabiei* in the biofertilizer subplot decreased, while antagonistic fluorescent pseudomonads were found to increase. Further studies have demonstrated control of potato scab by two suppressive strains of Streptomyces (*S. diastatochromogenes* strain PonSSII and *S. scabiei* strain PonR) in a 4-year-field-pot experiment (Liu et al., 1995). Both strains significantly decreased scab on potato tubers of potato cv. Norchip compared with the nonamended control treatment. The suppressive strains did not affect tuber yield and were reisolated from the tubers over the 4 years of the experiment. Subsequently, the antibiotic produced by PonSSII was found to only inhibit pathogenic strains of *S. scabiei*, but not a range of other bacteria commonly found in soil (Eckwall and Schottel, 1997).

In further experiments, 93 *Streptomycetes* were isolated from lenticels of potato tubers grown in naturally disease-suppressive and disease-conducive soils (Liu et al., 1996). Of these, 22 strains showed greater antibiotic activity against virulent *S. scabiei* RB311 than PonR and PonSSII. These strains were nonpathogenic on leaf-bud tubers in greenhouse testing and significantly reduced scab without affecting tuber yield in field-pot tests. Other bioactive compounds have also been studied as possible control agents, particularly antibiotic substances from a red alga, *Laurencia okamurae* (Horikawa et al., 1996).

Impact

The disease is widespread in all the potato-growing regions of the world and reduces the marketability of table, processing, and seed potatoes and was ranked as the fourth most important disease in a 1991 survey of potato growers (Loria et al., 1997). As the disease affects marketability rather than yield, exact losses are difficult to quantify; however, some scattered reports exist in the literature. In Bangladesh in 1969, yield losses of potato tubers in cold storage at 22 locations amounted to about 2.2%–9.5% of the total. Losses were caused by dry rot (*Fusarium* spp.), bacterial soft rot (*Erwinia* spp.), common scab (*S. scabiei*), and physiological disorders (Khan et al., 1973); a similar story exists in Bosnia-Hercegovina where virus disease and scab are particularly destructive (Buturovic, 1972). In a study on the effect of radish scab, the disease badly affected the market quality and caused losses of up to 100% in some cases from farms in Berlin-Spandau, Germany (Koronowski and Massfeller, 1972).

8.1.6 *Dickeya solani* Rot of Potato

Pathogen: *Dickeya solani*

A new bacterial pathogen, *Dickeya solani*, has emerged as a major threat to potato production in Europe and Israel. This new pathogen causes blackleg-like symptoms, leaf wilts, and tuber soft rots. It has yet to be formally named but the name *Dickeya solani* has been proposed. It is a close relative of *D. dianthicola*, previously known as *E. chrysanthemi*, which has caused problems sporadically in potato crops across Europe since the 1970s and in England and Wales since 1990.

In recent years, the new, more aggressive species has established itself in a number of European countries (Belgium, Finland, France, Poland, the Netherlands, and Spain) and Israel. It was first confirmed on a potato crop in England and Wales in 2007 and found in 2009 in two Scottish ware crops, grown from non-Scottish seed. As yet *D. solani* has never been found on Scottish seed potatoes nor potatoes grown in Scotland from seeds of Scottish origin. It is evident from the experience of other European countries that the new species is more aggressive than other *Dickeya* species and *P. atrosepticum*, previously *Erwinia carotovora* subsp. *atroseptica*, the usual cause of potato blackleg in Northern Europe. Observations from other European countries indicate that, once established, *D. solani* will rapidly displace other species and take over as the principal cause of wilting and blackleg-like symptoms in potato crops. Production losses in Dutch seed potatoes reached €25M in 2007 due to downgrading and rejection of over 20% of stocks during certification, almost entirely due to this new pathogen.

Symptoms
Symptoms caused by *D. solani* on the growing plant closely resemble blackleg in many cases. Wilting can be rapid with black soft rotting extending internally up the vascular system of the stem from the infected seed tuber. Symptoms may vary depending on variety, and in some, wilting can occur with no obvious sign of blackleg. High incidences of wilting (as much as 20%) have been observed in England as a result of planting *Dickeya*-infected seed in the warm early season conditions of 2007 and 2009. The symptoms on the tuber include the soft rot with cream-colored putrid mass of tissues with a brown boundary to distinguish the rotted portion from healthy one.

Dickeya solani is adapted to warmer temperatures but can cause disease under Scottish conditions. Since it appears to be more aggressive at higher temperatures, there are implications for seed exports from Scotland to warmer countries.

Disease Prevention
It is clear that *D. solani* poses a significant threat to Scottish potatoes, both seed and ware. A recent Scottish Government consultation showed strong support from the industry for measures to tackle this new pathogen in order to maintain Scotland's high health status. A zero tolerance for all *Dickeya* species is therefore being introduced in the Scottish Seed Potato Classification Scheme (SPCS), from 2010 onward. Strengthened plant health measures will also enable inspectors to take action to protect both seed and ware crops from *Dickeya* infection.

Crops found to be infected with *Dickeya* will not be allowed to be used as seed. They may only be sold as ware for immediate consumption or processing, and conditions will be imposed to prevent any contamination of other crops.

Non-Scottish seed potatoes entering Scotland will also be tested for *Dickeya*, as well as tubers selected for the annual survey for brown rot and ring rot. Continued vigilance from the Scottish potato industry is also required, in addition to these official measures, to maintain Scotland's high plant health status. The Scottish Government would encourage all involved in the industry to comply with the statutory plant health controls and good practice, to protect their own business and prevent this new pathogen taking hold in Scotland.

8.1.7 ZEBRA CHIP DISEASE OF POTATO

Pathogen: *Candidatus liberibacter solanacearum*

Synonym: *Candidatus liberibacter psyllaurous'* (Hansen et al., 2008)

Common names: Zebra chip, zebra complex, psyllid yellows (English), punta morada, papa manchada, papa rayada (Spanish)

Symptoms
The characteristic aboveground plant symptoms of *Candidatus liberibacter solanacearum* infection in potato, tomato, and other solanaceous species resemble those caused by phytoplasmas and include stunting, erectness of new foliage, chlorosis and purpling of foliage with basal cupping of leaves, upward rolling of leaves throughout the plant, shortened and thickened terminal internodes resulting in plant rosetting, enlarged nodes, axillary branches or aerial tubers, leaf scorching, disruption of

fruit set, and production of numerous, small, misshapen, and poor-quality fruits (Munyaneza et al., 2007a,b; Liefting et al., 2009a; Secor et al., 2009; Crosslin et al., 2010; Munyaneza, 2010, 2012).

In potato, the belowground symptoms include collapsed stolons, browning of vascular tissue concomitant with necrotic flecking of internal tissues, and streaking of the medullary ray tissues, all of which can affect the entire tuber. Upon frying, these symptoms become more pronounced and crisps or chips processed from affected tubers show very dark blotches, stripes, or streaks, rendering them commercially unacceptable (Munyaneza et al., 2007a,b, 2008; Secor et al., 2009; Crosslin et al., 2010; Miles et al., 2010; Buchman et al., 2011a,b, 2012). The symptoms in potato tubers have led to the disease being named "zebra chip" (Munyaneza et al., 2007a,b; Secor et al., 2009; Crosslin et al., 2010; Munyaneza, 2010).

Host

Candidatus liberibacter solanacearum is known to primarily infect solanaceous species, including potato (*Solanum tuberosum*), tomato (*Solanum lycopersicum*), pepper (*Capsicum annuum*), eggplant (*Solanum melongena*), tomatillo (*Physalis peruviana*), tamarillo (*Solanum betaceum*), tobacco (*Nicotiana tabacum*), and several weeds in the family Solanaceae (Hansen et al., 2008; Liefting et al., 2008a,b, 2009a,c; Abad et al., 2009; Crosslin and Munyaneza, 2009; Lin et al., 2009; Munyaneza et al., 2009a,b,c; Secor et al., 2009; Wen et al., 2009; Brown et al., 2010; Crosslin et al., 2010; Munyaneza, 2010, 2012; Rehman et al., 2010; Sengoda et al., 2010). This *Liberibacter* species is transmitted to *Solanaceous* species by the potato/tomato psyllid, *Bactericera cockerelli* (Sulc). Recently, *Candidatus liberibacter solanacearum* has been detected in carrot plants (*Daucus carota*) in Finland and is vectored by the carrot psyllid, *Trioza apicalis* Forster (Munyaneza et al., 2010a,b, 2011b). Subsequently, this bacterium was also detected in carrots and *T. apicalis* in Sweden and Norway (Munyaneza et al., 2012b,c). Most recently, *Candidatus liberibacter solanacearum* has been reported in carrot, celery (*Apium graveolens*) and the psyllid *Bactericera trigonica* in the Canary Islands and mainland Spain (Alfaro Fernandez et al., 2012a,b). These recent findings on the association of *Candidatus liberibacter solanacearum* with non-solanaceous species and other psyllids suggest that it is likely to have more insect vectors and host plants than currently known.

Geographical Distribution

Candidatus liberibacter solanacearum has been detected in several European countries in carrot crops (and to a lesser extent in celery) in association with other psyllid species (*Bactericera trigonica and Trioza apicalis*). However, *Candidatus liberibacter solanacearum* has not been detected in potato or tomato crops in the EPPO region.

EPPO Region: Finland, France (transient; only in two carrot fields), Norway, Spain (mainland and Canary Islands), Sweden.

North America: Mexico, the United States (Arizona, California, Colorado, Idaho, Kansas, Montana, Nebraska, Nevada, New Mexico, Oregon, Texas, Washington, Wyoming)

Central America: Guatemala, Honduras, Nicaragua.

Oceania: New Zealand (recently introduced).

Pathogen

Candidatus liberibacter solanacearum is a phloem-limited, Gram-negative, unculturable bacterium that is spread from infected to healthy plants by psyllid insect vectors (Hansen et al., 2008; Munyaneza et al., 2008, 2010b; Secor et al., 2009; Munyaneza, 2012). It may be spread experimentally by grafting (Crosslin and Munyaneza, 2009; Secor et al., 2009). This *Liberibacter* species has also been shown to be transmitted both vertically (transovarially) and horizontally (from feeding on infected plant hosts) in *Bactericera cockerelli* (Hansen et al., 2008). No information is currently available on vertical transmission for *T. apicalis* and *Bactericera trigonica*. Although only limited experiments have been conducted on Liberibacter transmission, it appears that *Candidatus liberibacter solanacearum* is not transmitted through true seed from infected plants (Munyaneza, 2012). Four geographic haplotypes of *Candidatus liberibacter solanacearum* have recently been described (Nelson et al., 2011, 2013). Two haplotypes (LsoA and LsoB) are associated with diseases caused by this bacterium in potatoes and other solanaceous plants, whereas the other two (LsoC and LsoD) are associated with diseased carrots. Haplotype LsoA has been found primarily from Honduras and Guatemala through Western Mexico to Arizona, California, Oregon, Washington, Idaho, and New Zealand. Haplotype LsoB is currently found from Eastern Mexico and northward through Texas. These two haplotypes show same range as has been found in Finland, Sweden, and Norway (Nelson et al., 2011, 2013) and is associated with *T. apicalis*. LsoD was very recently described from infected carrots and the psyllid *B. trigonica* in Spain and the Canary Islands (Nelson et al., 2013). The four haplotypes are not yet known to elicit biological differences in the plant or insect hosts. These apparently stable haplotypes suggest separate longlasting populations of the bacterium. Interestingly, LsoA and LsoC were found to be genetically very close, despite their large geographic separation and the differences in both plant and insect hosts (Nelson et al., 2013). Furthermore, a study on the genetic diversity of Lso strains using simple sequence repeat markers identified two major lineages of this bacterium in the United States and only one lineage in Mexico (Lin et al., 2012). Similarly to other *Liberibacters*, scanning electron microscopy images of *Candidatus liberibacter solanacearum* in sieves of infected plants revealed that this bacterium has a rod-shaped morphology (Liefting et al., 2009a; Secor et al., 2009). The bacterium is about 0.2 μm wide and 4 μm long (Liefting et al., 2009a).

Effects of environmental conditions on *Candidatus liberibacter solanacearum* are not well known. However, temperature has a significant effect on the development of this bacterium. Compared to citrus-greening *Liberibacter* species, *Candidatus liberibacter solanacearum* appears heat sensitive as it seems not to tolerate temperatures above 32°C (Munyaneza et al., 2012a). *Bactericera cockerelli*, the insect vector of this bacterium in the Americas and New Zealand, appears to have similar sensitivity to heat.

Detection and Inspection Methods

The whole genome of *Candidatus liberibacter solanacearum* isolated from zebra chip–infected potatoes has recently been sequenced (Lin et al., 2011). Detection methods for *Candidatus liberibacter solanacearum* have been developed and include conventional and quantitative real-time PCR (Hansen et al., 2008; Crosslin and Munyaneza, 2009; Li et al., 2009; Liefting et al., 2009a; Lin et al., 2009; Wen et al., 2009; Crosslin et al., 2011; Munyaneza, 2012).

Uneven distribution and variation in the *Liberibacter* titer in different parts of infected plants has been observed, making detection of this bacterium by PCR sometimes inconsistent (Crosslin and Munyaneza, 2009; Li et al., 2009). Development of better and more accurate detection methods is currently underway. Visual symptom inspections in some infected plants such as potato tubers can be very reliable (see "Symptoms" section). Mixed infections of *Candidatus liberibacter solanacearum* and phytoplasmas have been reported in potato (Liefting et al., 2009b; Munyaneza, unpublished data).

Disease Spread and Movement

Candidatus liberibacter solanacearum can be moved by its vectors or in host plants meant for propagation. During international trade, infected planting material could carry the disease, or possibly also infective vectors (most likely as eggs). Seed potatoes infected with *Candidatus liberibacter solanacearum* generally do not germinate but, in rare cases, may produce infected plants (Henne et al., 2010; Pitman et al., 2011). However, these seed-borne infected plants are often weak and short-lived and do not significantly contribute to the disease spread (Munyaneza, 2012). Most importantly, the psyllids have to be present to spread the bacterium.

Economic Impact

Candidatus liberibacter solanacearum was first identified in 2008 (Hansen et al., 2008; Liefting et al., 2008a,b) and shown to be associated with zebra chip disease of potato, which has been observed since the 1990s with increasing impacts and linked to *Bactericera cockerelli* for the first time in 2007 (Munyaneza et al., 2007a,b). A comprehensive review of the history and association of this bacterium with zebra chip and other diseases of solanaceous crops was provided by Crosslin et al. (2010), Munyaneza (2010, 2012), and Munyaneza and Henne (2012). First reported in Mexico in the 1990s, zebra chip was documented causing serious economic damage in parts of Southern Texas in 2004–2005. The disease is now widespread in the South-Western, Central, and North-Western United States, Mexico, Central America, and New Zealand (Munyaneza et al., 2007a,b, 2009a; Liefting et al., 2009a; Secor et al., 2009; Crosslin et al., 2010, 2012a,b; Munyaneza, 2010, 2012; Rehman et al., 2010). This *Liberibacter* species also severely affects other important solanaceous crops, including tomato, pepper, eggplant, tamarillo (Liefting et al., 2009a; Munyaneza et al., 2009b,c; Brown et al., 2010), and tobacco (Munyaneza, unpublished data). It has been established that *Candidatus liberibacter solanacearum* is transmitted to solanaceous species by the psyllid *Bactericera cockerelli*. Recently, Munyaneza et al. (2010a,b) detected *Candidatus liberibacter solanacearum* in carrots affected by the psyllid *T. apicalis* in Finland, which constitutes the first report of *Liberibacter* in Europe and *Candidatus liberibacter solanacearum* in a non-solanaceous species. The bacterium was subsequently detected in carrots and *T. apicalis* in Sweden and Norway (Munyaneza et al., 2012a,b). Most recently, *Candidatus liberibacter solanacearum* has been detected in carrot, celery, and the psyllid *Bactericera trigonica* in the Canary Islands and mainland Spain (Alfaro-Fernández et al., 2012a,b; EPPO, 2012); symptoms in diseased plants had previously been attributed to phytoplasmas and spiroplasmas in this area.

The complex bacterium/vectors has caused serious damage to the potato and tomato industries in the Americas and New Zealand (Munyaneza et al., 2007a,b, 2008; Liefting et al., 2009a; Secor et al., 2009; Crosslin et al., 2010; Rehman et al., 2010; Guenthner et al., 2012) and to the carrot industry in Europe (Munyaneza, 2010; Munyaneza et al., 2010a,b, 2012b,c; Alfaro-Fernández et al., 2012a,b).

In the case of potato, plant growth is negatively affected; crisps or chips made from zebra chip–infected tubers show dark stripes that become markedly more visible upon frying, and hence are commercially unacceptable. Whole crops might be rejected because of high levels of the disease, occasionally leading to abandonment of entire potato fields. Potatoes from such fields are severely affected by zebra chip (Munyaneza et al., 2011a). Infected tubers usually do not sprout, and if they do, they produce hair sprouts or weak plants (Henne et al., 2010; Pitman et al., 2011). The bacterium is also associated with economically damaging diseases of other important solanaceous crops, including tomato, pepper, eggplant, tobacco, and tamarillo. In Europe, damage to carrots by *Liberibacter*-infected carrot psyllids can cause up to 100% crop loss (Munyaneza et al., 2010a,b, 2012b,c; Alfaro-Fernández et al., 2012a,b).

Control Measures
At present, applications of insecticides targeted against the potato and carrot psyllids are the only means to effectively manage diseases associated with *Candidatus liberibacter solanacearum* (Munyaneza, 2012). No plant resistance to the disease has yet been identified (Munyaneza et al., 2011a).

Phytosanitary Risk
Candidatus liberibacter solanacearum and its insect vector *Bactericera cockerelli* have been found to be serious and economically important pests of potatoes, tomatoes, and other solanacous crops in Western and Central United States, Mexico, Central America, and New Zealand. They would result in similar damage if introduced in the EPPO region. Quarantine considerations have already emerged in some regions where *Candidatus liberibacter solanacearum* has been documented. Some countries are now requiring specific testing for *Candidatus liberibacter solanacearum* prior to allowing import of potatoes (Crosslin et al., 2010; Munyaneza, 2012). Furthermore, Australia put in place additional quarantine requirements to import fresh tomato and pepper from New Zealand

after 2006, where growers need to ensure that crops for export have been produced in areas free of *Bactericera cockerelli* or the exported produce must be free of the psyllid.

Candidatus liberibacter solanacearum has already been documented in carrots in Finland, France, Sweden, and Norway, and it is suspected that this bacterium is more likely to be widespread in parts of Northern and Central Europe where its insect vector *T. apicalis* occurs or at least in regions where damage by this psyllid has been observed. Very recently, this bacterium was reported in carrots and the psyllid *Bactericera trigonica* in Spain and the Canary Islands. Nevertheless results of a preliminary study indicate that inoculation of *Candidatus liberibacter solanacearum* to potato by *T. apicalis* is not possible (Munyaneza, unpublished data). These observations suggest that the main pathway of introducing *Candidatus liberibacter solanacearum* into solanaceous species would be the introduction of infective *Bactericera cockerelli* into the EPPO region.

Phytosanitary Measures

EPPO recommends that vegetative material for propagation and produce (such as fruits) of Solanaceae should come from areas free of *Bactricera cockerelli* and *Candidatus liberibacter solanacearum*. Seed and ware potatoes should come from areas free of zebra chip. Alternatively high-grade seed potato may be imported under postentry quarantine, and ware potatoes may be imported only for industrial processing purposes.

8.2   TOMATO

8.2.1   BACTERIAL WILT AND CANKER OF TOMATO

Pathogen: *Clavibacter michiganensis* subsp. *michiganensis* (Smith) Davis et al.

Synonyms: *Corynebacterium michiganense* pv. *michiganense* (Smith) Dye & Kemp
Corynebacterium michiganense (Smith) Jensen

Common names: Bacterial canker, bird's eye (English)

Bacterial canker, although usually sporadic in its occurrence, is so destructive in nature that vigilance must be exercised in the selection and handling of seed stocks, the preparation and management of greenhouse soil beds or bags, and the selection and preparation of ground for field production. Bacterial canker is a vascular (systemic) and parenchymatal (superficial) disease with a wide array of symptoms resulting in loss of photosynthetic area, wilting and premature death, and the production of unmarketable fruit. Early recognition of the disease, especially in greenhouse crops, is essential if the disease is to be contained.

Symptoms

Bacterial canker, which may occur in tomato as a primary (systemic) or secondary (foliar) infection, shows a wide range of symptoms.

Primary infections originate from infected seed or from invasion of the vascular tissue of young seedlings. Symptoms, which may not show up until several weeks after infection, initially appear as wilting and downward turning of the lower leaves. The wilting generally progresses upward, unless the site of infection is in the upper part of the plant. Wilting is often seen only on one side of the leaf or one side of the plant (Photo 8.7). Plants may collapse and die, especially if infected at a very early stage. Generally, plants survive but are stunted, showing some or all of the symptoms described here, depending on their environment and stage of growth.

Tomato foliage infected with the canker organism has distinctive black leaf edges with no spotting on the interior of the leaves. Sometimes a thin yellow border is present between the dead leaf margins and healthy tissue.

If an infected stem is cut lengthwise, a light brown discoloration may be present in the vascular tissue (Photo 8.8), most noticeable at nodes and just above the soil line. As the disease progresses, this turns reddish-brown. Light-colored streaks are often visible on the outside of the stem.

PHOTO 8.7 Wilt symptoms on tomato plants caused by *Clavibacter michiganensis*. (Photo courtesy of Gerald Holmes, California Polytechnic State University, San Luis Obispo, CA, Bugwood.org.)

PHOTO 8.8 Close-up of cross-section through a diseased stem of tomato caused by *Clavibacter michiganensis* subsp. *michiganensis*. (Photo courtesy of Gerald Holmes, California Polytechnic State University, San Luis Obispo, CA, Bugwood.org.)

These may later darken and break open into cankers. With severe infections, a yellow ooze may exude from a cut stem when it is squeezed.

Fruit may develop relatively small spots with light brown centers, generally surrounded by a greasy white halo (Photo 8.9) (3–6 mm in diameter). These are known as bird's-eye spots. With bacterial canker lesions, this white halo generally remains as the fruit ripens, while in the case of bacterial spot, it disappears with time. Bacterial canker may also cause a darkening of the vascular tissues within the fruit. The fruit may show a black peppering at the vascular bundles under the calyx scar. Canker bacteria can grow in the vascular bundles within the fruit, all the way to the seed. This can result in visible yellowish strands from the stem to the seeds and internal infections in the seed.

With a secondary foliar infection, leaves develop brown-black margins with a thin, yellow (chlorotic) band. Leaflet edges may curl upward. Fruit may show bird's-eye spotting, as in a systemic infection. Secondary infections (no vascular system involvement) often have minimal impact on the crop, especially when initiated later in the season.

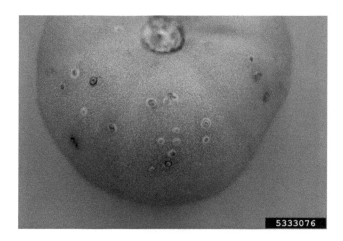

PHOTO 8.9   Symptoms of birds'-eye spot on tomato fruits caused by *Clavibacter michiganensis*. (Photo courtesy of Mary Ann Hansen, Virginia Polytechnic Institute and State University, Blacksburg, VA, Bugwood.org.)

Host

The main host of economic importance is tomatoes, but the pathogen has also been reported on other *Lycopersicon* spp. and on the wild plants *Solanum douglasii*, *Solanum nigrum*, and *Solanum triflorum*. A number of solanaceous plants are susceptible on artificial inoculation (for details see Thyr et al., 1975). Doubtful reports from other hosts include *Phaseolus* beans, peas, and maize. Recently, Stamova and Sotirova (1987) have also reported wheat, barley, rye, oats, sunflowers, watermelons, and cucumbers as hosts on artificial stem inoculation.

In the EPPO region, the main host is tomato, while some susceptible solanaceous weeds could be potential reservoirs of the pathogen.

Geographical Distribution

Clavibacter michiganensis subsp. *michiganensis* was first described in North America and presumably originated there.

EPPO region: Austria, Belarus, Belgium, Bulgaria, Czech Republic, Egypt, Finland (unconfirmed), France, Germany, Greece, Hungary, Ireland, Israel, Italy (including Sardinia and Sicily), Lebanon, Lithuania, Morocco, the Netherlands, Norway (eradicated), Poland, Portugal (eradicated), Romania, Russia (European, Siberia), Slovenia, Spain, Switzerland, Tunisia, Turkey, the United Kingdom (found in the past but not established), Ukraine, Yugoslavia.

Asia: Armenia, Azerbaijan, China (found in the past but not established), India (Madhya Pradesh), Iran, Israel, Japan, Lebanon, Turkey.

Africa: Egypt, Kenya, Madagascar, Morocco, South Africa, Tunisia, Togo, Uganda, Zambia, Zimbabwe.

North America: Widespread in Canada (British Columbia to Nova Scotia) and the United States (California, Florida, Georgia, Hawaii, Iowa, Illinois, Indiana, Michigan, North Dakota, Ohio, Wyoming), Mexico.

Central America and Caribbean: Costa Rica, Cuba, Dominica, Dominican Republic, Grenada, Guadeloupe, Martinique (unconfirmed), Panama.

South America: Argentina, Brazil (São Paulo), Chile, Colombia, Ecuador, Peru, Uruguay.

Oceania: Australia (New South Wales, Queensland, South Australia, Tasmania, Victoria, Western Australia), New Zealand, Tonga.

Diagnosing Bacterial Canker in the Greenhouse or the Field

On Seedlings
Marginal necrosis, tan-to-dark necrotic patches on the leaves and stems, and small, white, raised blisters on infected leaves are the symptoms of bacterial canker infection on young plants. Stunting, wilting, and stem splitting can also occur, especially in grafted seedlings. However, symptoms can take several to many weeks to develop following infection and, therefore, may not be visible at the seedling or transplant stage.

On Leaf and Plant
Leaf yellowing and necrosis around leaf margins called "firing" or "marginal necrosis" can indicate a foliar and/or systemic infection. When the stems or petioles are cut open, discoloration of the vascular tissues may be seen. In greenhouse-grown plants, symptoms appear as interveinal chlorotic to pale green patches that quickly become necrotic, giving a scorched appearance. Wilting in infected plants begins with the lower, older leaves, or leaves above the point of infection. Wilting may be asymmetric, appearing more on one side of the plant than the other. Infected leaves die, and light brown streaks or cankers, which may darken with age, develop on infected stems, calyx, and peduncles. Typical cankers can be common in the field but are rarely seen in the greenhouse. The vascular tissues become light brown to reddish brown and the pith appears mealy, brown, and dry. Older plants tend to be less susceptible to *Clavibacter michiganensis* subsp. *michiganensis* than younger ones and the disease tends to be more severe on plants infected early vs. late in their growth cycle.

On Fruit
Small dark spots on the fruit surrounded by a white halo or "bird's-eye" spots are characteristic of bacterial canker on field-grown fruit. Spots become raised and the centers turn brown with age. Infections, and the resulting spots, occur when *Clavibacter michiganensis* subsp. *michiganensis* bacteria are deposited on fruit by splashing water from rain or overhead irrigation, or mechanically during handling of the plants. When internally infected fruit are opened, yellowing or browning caused by the decay of the tissues may be seen.

In the greenhouse, bird's-eye spots are typically not observed, but fruits may appear netted or marbled, or they may remain symptomless. It is important to have an accurate diagnosis of any disease problem in tomatoes so that appropriate control measures can be taken.

Diagnostic kits have been developed for rapid, on-site identification of *Clavibacter michiganensis* subsp. *michiganensis*. However, it is advisable to submit a tissue sample to a reputable laboratory for confirmation of the diagnosis.

Disease Development and Spread

Sources of the Pathogen
The pathogen can survive in many environments including free living in infested soil for short periods, in over-seasoned plant debris in the soil, on weed hosts and volunteer plants, on contaminated stakes, and in association with seed. *Clavibacter michiganensis* subsp. *michiganensis* is a seed-borne pathogen, although rates of seed-borne infestation may be very low. Volunteer tomato plants from an earlier infected crop may harbor the pathogen, as can cull piles of diseased tomato plants. *Clavibacter michiganensis* subsp. *michiganensis* can infect or survive on some weeds such as nightshade, and several wild *Lycopersicon* species, and these can act as reservoirs of *Clavibacter michiganensis* subsp. *michiganensis* for new infections. Plant material (field tomatoes, weeds, cull piles) and soil infested with *Clavibacter michiganensis* subsp. *michiganensis* may be blown on wind and rain into the greenhouse. Recirculated water systems, such as ebb and flow irrigation, may also harbor the pathogen.

Pathogen Longevity
Clavibacter michiganensis subsp. *michiganensis* can survive for at least 5 years in infested tomato seed and still cause infection. The pathogen also survives for short (up to 3–4 weeks) periods of time

in soil and for up to 24 months or longer in infected plant debris. In the greenhouse, *Clavibacter michiganensis* subsp. *michiganensis* can survive for at least 1 month on surfaces such as cement and plastic and at least a year in plant material and rockwool. It survives better in cool and dry conditions than in hot, moist ones.

Environmental Conditions Favoring Disease Development
Development of bacterial canker is favored by warm (24°C–32°C) and moist conditions. In greenhouses, the disease tends to be more severe in the summer during long, hot days when plants are stressed. Bacterial canker is more likely to be found in wetter areas of the greenhouse (e.g., where water condenses and drips on plants) than in drier areas.

Transmission
Overhead irrigation during seedling production, movement through foliage by production workers, and rainfall on open fields favor the spread of *Clavibacter michiganensis* subsp. *michiganensis*, especially if plants have recently been staked or pruned. Once the disease appears in a field or greenhouse, the pathogen may spread to adjacent plants and infect them through pruning wounds and injury or through naturally occurring pores along the leaf surface (stomates) or leaf margins (hydathodes). The pathogen can also be moved quite easily by equipment during cultivation, especially with open field processor tomatoes. In the greenhouse, *Clavibacter michiganensis* subsp. *michiganensis* may move from infested to noninfested areas in recirculated water and from plant to plant in bags or trays. Equipment such as clippers, cutting blades, stakes, or plant ties may harbor the pathogen. Workers may transmit the bacteria via tools, hands, and clothing. During pruning and grafting, improperly disinfected tools may lead to disease spread. This spread can be particularly explosive following the grafting operation, when bacteria can be directly introduced into the vascular tissue of very young plants. Plants infected in this manner may develop more extensive symptoms at a much earlier growth stage compared to the natural infection process. *Clavibacter michiganensis* subsp. *michiganensis* can also infect roots, so it is important to sterilize the soil (if used in the production system) following a diseased crop.

Inoculum and Spread
Infected seed is probably the major source of primary (systemic) infections. The bacteria can be present on the surface of the seed as well as within the innermost layer of the seed coat. This makes the canker organism harder to eradicate with seed treatments than the spot and speck pathogens.

The organism can also be introduced from infected crop debris, weed hosts, or volunteer tomatoes, and contaminated equipment. Studies in the U.S. northern Midwest have shown that it can overwinter on infected debris. However, crop rotation and tillage before planting should reduce the risk of infection from this source.

The canker bacteria enter the plant through natural openings and wounds, including root wounds. Pruning or transplant clipping operations can introduce the bacteria directly into the vascular system, resulting in the more serious systemic infections.

Infections spread through splashing water, wind-driven rain, and fine water droplets or aerosols produced during storms. In the field, bacteria transfer by machinery or workers is probably not as significant as in the transplant greenhouse where plant density is high and growth conditions for the bacterium are optimal.

Disease Cycle
Infected tomato seeds give rise to contaminated seedlings. Where studied, not more than 1% seed transmission occurred (Grogan and Kendrick, 1953). Spread of the disease in the field or under glasshouse is favored by water (rainsplash, irrigation) and cultural practices (trimming, chemical sprays). The bacterium enters plant tissue through stomata and other natural openings, as well as wounds and roots.

Young plants have been shown to be more susceptible (Van Vaerenbergh and Chauveau, 1985). Nevertheless, under natural conditions, tomato plants seem to be susceptible throughout their life

(Rat et al., 1991). After infection, there is a long latent period before any symptoms appear (for details on biology and symptoms, see Strider, 1969).

The bacterium is located in the xylem vessels (Leyns and De Cleene, 1983) where it can cause lysigenous cavities. Infected vessels contain viscous granular deposits, tyloses, and bacterial masses (Marte, 1980). The pathogen also produces a toxic glycopeptide, which has biological activity (Miura et al., 1986). The bacterium survives for a long time in plant debris, soil, and on equipment and glasshouse structures. It probably does not survive long in soil per se. However, it remains viable for at least 8 months in seeds.

Economic Impact

Since the first report of the disease in the United States in 1910, *Clavibacter michiganensis* subsp. *michiganensis* has spread throughout the world and has caused serious losses to both glasshouse and field tomato crops, either by killing the young plants or disfiguring the fruits. In North Carolina (USA), a 70% reduction in yield has been recorded in some years. Recent experiments carried out in France have shown a yield loss of 20%–30% (Rat et al., 1991).

Disease Management

Only purchase seed that has been tested and found to have no evidence of *Clavibacter michiganensis* subsp. *michiganensis.*

Seed companies typically make every effort to produce seed in suitable production areas with low disease pressure or in greenhouses under strict sanitation to minimize the risk of infections.

Seed Health Issues

Ideally, seed from disease-free fields should be used to raise the tomato crops. In reality, one can never be certain that the seed is completely clean, so additional measures are taken to reduce the risk of disease.

Seed extraction alone, whether through fermentation or acid methods, is not a reliable method of eliminating bacteria on the seed. Extraction should be followed by a hot water, acid, or chlorine disinfection. All seed lots should be disinfected by one of these methods.

The hot-water method is the least desirable, as temperature and duration are limited by the need to keep seeds viable. On the other hand, hot-water treatment can be done on-farm, with equipment that is relatively easy to obtain and is better than any other treatment. Instructions must be followed carefully to minimize damage to the seed. Even so, significant losses of germination could occur in some seed lots.

Acid and chlorine disinfection are very difficult to do on-farm. Seed can be damaged if protocols are not followed exactly. Generally, growers must trust their seed supplier to perform the seed disinfection effectively.

The acid and chlorine treatments are very effective on spot and speck if done correctly. Some canker bacteria, however, may survive, as this organism has been found inside the seed coat.

Effective seed disinfection is critical to bacterial disease management in tomatoes.

One infected seed in 10,000 may be enough to cause a disease outbreak under the right conditions.

Hot-Water Treatment

Place seeds in a loosely woven cotton bag (such as cheesecloth). Leave lots of space in the bag for the seeds to move around. Prewarm the seeds for 10 min at 37°C in water. Place the prewarmed seeds in hot water at 50°C for 25 min, monitoring the temperature constantly. Cool immediately by placing the seeds in cold water for 5 min. Dry thoroughly. Expect to lose 5%–10% of viable seed.

Sanitation Procedures during Production

Seedling Production

Use only clean trays and flats for transplants. Use overhead watering sparingly to provide only enough moisture for seedling growth. Allow foliage to dry before the sun sets, as prolonged leaf wetness can lead to increased disease development, and use other strategies to minimize the time of leaf wetness. Inspect plants frequently for canker and other disease symptoms.

Grafting
Strict sanitation procedures must be in place during the grafting process. Seedlings should be thoroughly inspected for symptoms and suspected plants should be removed and tested. Ensure that seedling foliage is dry prior to grafting. Cutting tools should be disinfected regularly and workers should wear disposable gloves that are regularly changed or disinfected. Ethanol (70%–75%) and other disinfectants, such as Virkon, Chemprocide, and Kleengrow, are effective in disinfecting cutting tools and hands. Cutting blades should be soaked in a disinfectant solution long enough to ensure disinfection. Disinfectants should be changed regularly to ensure that the concentration is in the effective range. Disinfected tools and hands should be rinsed with clean water to avoid damage to the plant by the residual disinfectant. Workers should wear clean clothing or clean disposable coveralls.

Transplant Production
By following an effective disinfection protocol, the seed supplier does everything possible to ensure clean seed. Now the transplant grower must take steps to prevent the seed or seedlings from becoming infected in the greenhouse environment.

Transplant growers must maintain good sanitation practices. There is a wide range in types of transplant production facilities across the fresh-market and processing tomato industries, but in general, sanitation practices must include the following steps:

- Remove all plant material from the greenhouse before starting a new crop.
- Control weeds in and around the greenhouse.
- Start with sterile potting mix and trays.
- Use new, sterile trays if feasible, but if reusing trays, sanitize effectively by solarization or washing with disinfectant (for more information, see the *Transplant* section of OMAFRA Publication 363, *Vegetable Production Recommendations*).
- Disinfect racks, tools, equipment, and greenhouse surfaces before the growing season, and wooden racks must be soaked in the disinfecting solution for a minimum of 1 h.
- Avoid contact between seed lots, sanitize equipment and hands between lots; physically separate seed lots in the greenhouse.
- Avoid contact between tomato and pepper seed and plants, sanitize equipment and hands after handling, physically separate crops in the greenhouse (ideally grow in separate facilities).
- Minimize handling and human traffic in the greenhouse.

An important cultural practice for disease control during transplant production is minimizing leaf wetness. Attempt to reduce the number of hours of leaves wetness through timing of watering, control of relative humidity, ventilation, and heating.

Use low pressures when watering to minimize plant damage and splashing of water droplets that can contain bacterial cells.

Do not handle wet plants and ensure foliage is dry for shipping. Wet foliage and dripping water in plant trailers are very effective ways to spread disease and promote bacterial growth.

Field Production
Since all three bacterial pathogens can survive in crop debris, rotate tomatoes with nonhost crops. If tomato crop debris is remain buried into the top 15 cm (6 in.) of soil to speed decomposition, a 3-year rotation should be sufficient. Control weeds and volunteer tomatoes in and around the field, as they can act as reservoirs of disease.

Ensure good drainage and adequate fertility. After heavy rains, get excess water off the field as soon as possible and address problem areas with drop pipes or other measures.

Separate plant lots into different fields if possible. Consider using tall, barrier crops between plant lots and neighboring fields, but take care to maintain good air circulation within the field.

Try to avoid wet foliage during transplanting. It is difficult to dip the plug trays to wet the plugs without wetting the foliage, but it is beneficial if it can be achieved.

Ideally, growers would be able to keep workers out of the field when the foliage is wet. This, however, is not always practical. Keep in mind that earlier infections have more time to cause damage and that fruit lesions are initiated on young, green fruit. Late-season foliar symptoms are not a major concern.

Low-pressure systems are better if overhead irrigation is used, as they minimize splashing and plant damage. The potential for overhead watering to spread disease must be balanced with the potential benefits from irrigation if an overhead system is all that is available. Plants under stress are less able to withstand a disease outbreak.

Pruning, Harvesting, and Other Handling Operations
Greenhouses should be kept clean and thoroughly disinfected between crops. Hands and tools should be sanitized between plants or rows with a disinfectant solution to minimize pathogen spread. Clippers and knives that deliver a disinfectant to the cutting surface during each cut are commercially available, and using an effective disinfectant, such as Virkon or KleenGrow, can also help to prevent the spread of *Clavibacter michiganensis* subsp. *michiganensis*. Knives or clippers that do not deliver a disinfectant are not recommended for pruning, suckering, or harvesting. These operations may be carried out by hand and, with practice, the plant is not damaged and the wound is not touched. Hands should nonetheless be sanitized regularly. Any symptomatic plants should be removed immediately in sealed plastic bags or containers and buried in a pit away from greenhouses or field irrigation ditches. Be sure to avoid touching any healthy plants with hands, clothing, or the plastic bags during the plant removal process. Note that symptoms may take several weeks (2–7 weeks) to develop following infection, depending on conditions. Therefore, during an infection, there may be symptomless plants that serve as inoculum for spread of the disease. Thus, several neighboring plants in both directions down the row, as well as visibly infected ones, should be removed even if they do not show symptoms.

Tomato or pepper cull piles and solanaceous weeds must not be all owed in the vicinity of the greenhouse and production site. Recirculated water should be disinfected before being reused again as irrigation water.

Minimize the chance of disease in subsequent crops. Plant debris decomposes faster and the survival of *Clavibacter michiganensis* subsp. *michiganensis* decreases if the debris is plowed under and soil is moist.

Properly dispose vines and other plant residuals as far from the greenhouse and production fields as possible. Remove weeds in and around the greenhouse and fields. Thoroughly disinfect the greenhouse/screenhouse, machinery, tools, and crates between crops. In open field production, rotate tomatoes with a nonhost crop (peppers are hosts of *Clavibacter michiganensis* subsp. *michiganensis*) for 2–3 or more years. In greenhouses, where rotation is not possible, growing media such as rockwool and coconut dust, and ground covers should be changed to minimize the chance of carryover of *Clavibacter michiganensis* subsp. *michiganensis* to subsequent crops.

Genetic Resistance
Tomato cultivars with some resistance or tolerance to canker have been introduced, but there is little significant tolerance in commercial tomato varieties. Recent research has identified two genes that provide resistance toward bacterial canker. Further work is needed before cultivars with greatly improved resistance are available to tomato growers.

Chemical Control
Fixed copper bactericides are currently the only effective registered control products for bacterial disease on tomatoes. However, bacterial speck populations in Ontario have shown widespread

resistance to copper, and in some areas of the United States, bacterial spot has also developed resistance. Despite this, copper is still a useful tool for managing bacterial spot and canker in Ontario.

Research at Ridgetown College (University of Guelph), Agriculture and Agri-Food Canada, and elsewhere has shown that foliar-applied fixed copper sprays will reduce the number of bacterial cells on tomato foliage. Depending on the amount of bacteria present and the size and rate of growth of the plants, spray intervals of 7 days or less may be required. Copper sprays are less effective, to the point of ineffectiveness, when spray intervals are extended. Good spray coverage and the use of recommended rates are also very important.

The effect of the pH of the spray solution on the effectiveness of fixed copper formulations for bacterial disease control is unclear. The activity of fixed copper on the bacteria is due to free copper ions in the spray solution, the concentration of which changes with pH. Commercial pesticides, however, are generally formulated with buffers, surfactants, and other additives to ensure pesticide efficacy under the normal range of application conditions (including spray solution pH). Follow manufacturer's recommendations for mixing and application, as high concentrations of copper ions can damage plant tissue.

Bacteria reproduce very quickly. Although foliar sprays may clean the surface bacteria from a leaf, within a short period of time, the bacteria inside the leaf and those not controlled on the foliage (due to incomplete spray coverage) can build up population levels that can cause an outbreak.

In the past, copper spray programs may not have been applied at the right time to be fully effective. Bacterial disease does not affect each grower every year. It has been a common practice to begin an intensive copper spray program once lesions are present. We now know that starting a control program after symptoms have appeared is too late. It takes an incredibly high density of bacterial cells on the plant before symptoms are visible, and efforts to eradicate bacteria when they are at such high population levels are destined to fail.

The most current recommendations for copper spray programs for greenhouse transplant production and field production are found in OMAFRA Publication 363, Vegetable Production Recommendations. These recommendations outline a preventative program and must be followed closely to be effective. Making the decision to initiate a bacterial disease spray program when no symptoms are present can be difficult. However, with clean seed, the use of the nonchemical management practices described, and preventative copper applications in the transplant greenhouse and the field, relatively few sprays are required.

Many research trials across North America have shown that tank-mixing mancozeb with copper enhances bacterial disease control.

8.2.2   Bacterial Speck of Tomato

Pathogen: *Pseudomonas syringae* pv. *tomato* (Okabe) Young. Dye and Wilkie

Bacterial speck is caused by *Pseudomonas syringae* pv. *tomato*. Although this disease has been known since the early 1930s, it did not result in serious losses until the winter tomato crop of 1977–1978 in southern Florida and in 1978 in southern transplant fields, and in northern production areas where some infected transplants were shipped inadvertently. Cool, moist environmental conditions contributed to the development of the disease, which has now established itself as a major production problem in northern producing states.

Symptoms

Bacterial speck lesions may occur anywhere on the foliage, stems, or fruits. Symptoms are very difficult to visually distinguish from bacterial spot and can be confused with young, early blight lesions. On leaves, symptoms appear as black specks, usually no more than 2 mm in diameter, which are usually surrounded by a yellow halo (Photo 8.10). Speck lesions sometimes cause distortion of the leaf, as the infection restricts the expansion of leaf tissue. Lesions are often concentrated near leaf edges, and in some cases, leaf margin burn resembling bacterial canker. When numerous, lesions may coalesce, and entire leaflets may die. Severely infected seedlings may become stunted.

PHOTO 8.10 Disease lesions due to bacterial speck pathogen on tomato leaf. (Photo courtesy of Gerald Holmes, California Polytechnic State University at San Luis Obispo, Bugwood.org.)

Only green fruits less than 3 cm in diameter are susceptible to infection by the bacterial speck pathogen. Small (less than 1–3 mm), slightly raised black specks develop and are often surrounded by a narrow green to yellow halo (Photo 8.11). Lesions are usually superficial and can be scraped off with a fingernail. Red fruits are not susceptible to infection, likely due to a lack of entry points for bacteria; fruit hairs, which may break and allow bacteria to enter, are only present on young fruits. On fruits previously infected, black lesions remain after ripening.

Geographical Distribution
The United States, Israel, Australia.

Pathogen
Bacterial speck is caused by *Pseudomonas syringae* pv. *tomato*. Two races are present in Ontario, race 0 and race 1. The *Pto* gene, discovered by Ontario researchers, confers resistance to race 0.

This bacterium produces a number of compounds that help it infect and obtain nutrients from the tomato plant. One of these compounds is the plant-specific toxin coronatine, which is

PHOTO 8.11 Disease lesions due to bacterial speck pathogen on tomatofruit. (Photo courtesy of Gerald Holmes, California Polytechnic State University, San Luis Obispo, CA, Bugwood.org.)

responsible for the yellow halo surrounding leaf lesions and the stunting of young seedlings. The majority of bacterial speck strains that have been isolated from Ontario tomato fields are either resistant to or tolerant of copper-based bactericides.

Disease Cycle

The speck pathogen can survive from season to season in association with soil, debris from diseased plants, and seeds. Survival periods in soil may vary among geographic areas. In some northern areas of the United States, the pathogen can survive in fields for at least 1 year. In cool coastal areas of California, the pathogen appears to be ubiquitous and can be isolated from roots and foliage of many weed and crop plants, even from plants grown in soil with no history of tomato culture. The pathogen can be seedborne, as demonstrated in commercial seed lots in Australia and Israel; in fact, the pathogen persisted on dry seed for 20 years in Israel.

Infection occurs through wounds that result when leaf hairs and fruit trichomes are broken, an occurrence that appears to be natural on expanding fruit, and when other injuries are caused by wind, water, and plant movement associated with cultural practices. The bacteria occur in intact but dead trichomes on tomato leaves and can survive dry periods for at least 14 days. Long-distance spread can occur with seeds and transplants. Short-distance spread can occur on machinery and by workers during transplant clipping, transplanting, cultivation, and spraying and by strong winds and splashing rain or irrigation water. New lush foliage appears to be more susceptible than older hardened foliage. Only green fruits are infected; the pH of skin tissue on green fruit is about 6.3, whereas on ripe fruit, it is 5.2, which is too low to support bacterial growth.

A better understanding of the epidemiology of bacterial speck is essential for developing control measures. The question may be asked why this disease, which has been around for many years, has now moved to the forefront of tomato production. Research from different sections of the country offers several explanations. Circumstantial evidence suggests that the recent introduction of the bacterial speck organism into transplant production fields in southern Georgia originated on commercial seeds. If seeds are harvested by either the acetic acid extraction method or the fermentation process, the threat of seed-borne inoculum is greatly reduced. It is not known if the seeds planted in Georgia were extracted by the more controversial centrifugation method.

A cool (below 21°C) and moist (high relative humidity and prolonged period of free moisture) growing season in 1978 contributed to the outbreak of bacterial speck in southern transplants. Although these conditions are exceptional in the south, they represent typical growing conditions in the north, when young transplants are set out in spring. Research has shown that with appropriate temperature and leaf wetness, plants harboring a low resident bacterial population can show symptoms within as few as 3–5 days. In most cases in the field, symptoms can be expected in 6–10 days. This is significant in light of the finding that the bacterial speck pathogen can survive shipment and spread disease in the field, even though no disease symptoms were present on the plants during shipment or at planting time. Similarly, fruits that develop during cool, moist weather early in the season can be severely infected. Use of scanning electron microscopy has shown that bacteria can be associated with both glandular and nonglandular leaf hairs present on ovaries during anthesis (the period of flower expansion and stamen maturation). After anthesis, the leaf hairs are gradually lost, leaving openings in the young fruit epidermis. These sites may serve as areas for fruit infection. Uninjured fruits are most susceptible to infection from anthesis until the fruits reach a diameter of 1¼ in.

Another important factor in the epidemiology of bacterial speck is the survival of the organisms in either soil or host debris and on native weeds. In California, the leaves and roots of several weed and crop species maintained resident populations of bacterial speck pathogen. In Michigan, survival of the bacterium in buried tomato leaf tissue at various soil depths suggested that overwinter-diseased tissue was another source of primary inoculum.

Infection and disease development is promoted by cool moist weather. An increase in the resident population of the pathogen on tomato leaves is associated with a temperature of 16°C, even in the absence of a leaf-wetness period, and with a temperature of 16°C–26°C following a leaf-wetness period of 8–48 h. Moisture is necessary for infection to occur and for symptom development. Disease is enhanced by temperatures between 12°C and 26°C and relative humidity of about 80% or high enough to result in dew formation at night: when temperature or moisture conditions are unfavorable for 1 week, disease severity is limited on new foliage. New symptoms can appear from 3 to 5 days after inoculation when plants are wet for 24 h after inoculation. Masses of bacteria are extruded from cracks in the fruit-lesion surface.

Control Measures

1. Use disease-free, hot water–treated seeds.
2. If using southern grown transplants, strive to obtain disease-free transplants that have been produced with a good protective spray program (mancozeb plus fixed copper, with streptomycin as a replacement bactericide for copper in later sprayings if weather conditions favor speck development). *Note: Streptomycin can only be used on tomato plants before transplanting.*
3. Practice crop rotation because of the carryover of inoculum in plant debris and weeds.
4. Follow good weed control and sanitation programs before establishing the current season crop.
5. Practice a preventive copper + mancozeb spray program from anthesis until the first-formed fruits are one-third of their final size. After that point, the greatest risk of bacterial speck is passed; copper can be dropped from the program, and the full labeled rate of fungicide should be used to control foliar blights, especially early blight.
6. Resistance for bacterial speck has been identified in three tomato species and likely to be added to commercial varieties.
7. Regularly monitor tomato fields so that early outbreaks can be controlled. Obtain and plant high-quality seeds that do not have detectable, economically important levels of *P. syringae* pv. *tomato*. Use a hot-water seed treatment or treat seed with hydrochloric acid, calcium hypochloride, or other recommended materials. Hot-water treatments can reduce seed viability and germination percentages. Seed health testing and certification programs help regulate the availability and cleanliness of such seed. (Such seed tests usually involve the washing of a 10,000 seed sample and subsequent planting the liquid onto semi-selective medium.) Discard heavily infested seeds. Use resistant cultivars if such become available.
8. Inspect transplants and remove symptomatic plants and surrounding transplant trays. Sanitize benches that hold transplants, transplant trays, and equipment that comes into contact with plants.
9. Consider applying preventative spray applications (copper-based materials combined with maneb fungicides of newer chemistry such as acibenzolar-S-methyl) for protecting transplants.
10. Avoid using overhead sprinkler irrigation in the field.
11. With an appropriate disinfectant, periodically and regularly sanitize tools such as clippers and prunning shears.
12. Do not allow equipment or workers to pass through fields when foliage is wet.
13. Copper, mancozeb sprays provide some control.
14. Once the tomato crop is finished, incorporate the crop residues to enhance plant decomposition and the dissipation of bacteria.
15. Rotate to a nonhost crop before returning to tomato and do not allow volunteer tomato or weed hosts to survive.

8.2.3 Bacterial Spot of Tomato

Pathogen: *Xanthomonas vesicatoria* (ex Doidge) Vauterin et al.

Synonyms: *Xanthomonas campestris* pv. *vesicatoria* (Doidge) Dye

Common names: Bacterial spot, bacterial scab, black spot (English)

Bacterial spot is caused by *Xanthomonas campesiris* pv. *vesicatoria*. It is periodically a severe disease of tomatoes and sweet peppers in New York. Because bacterial spot and speck produce similar symptoms, they are often misdiagnosed.

Symptoms

The bacterial spot pathogen may produce lesions on all aboveground parts of the plant like leaves, stems, flowers, and fruits. It is difficult to reliably distinguish bacterial spot from bacterial speck based on visual symptoms, especially in the early stages.

Initial leaf symptoms are small, circular-to-irregular, dark lesions, which may be surrounded by a yellow halo. The lesions tend to concentrate on the leaf edges and tip and may increase in size to a diameter of 3–5 mm. Infected leaves may develop a scorched appearance. When spots are numerous, foliage turns yellow and eventually dies, leading to defoliation of the lower portion of the plant.

Fruit lesions are initiated only on green fruit, most likely because infection occurs through fruit hairs, which are present only on immature fruit. On fruit, the first symptoms are small, dark brown-to-black, raised spots (Photo 8.12). The lesions also may have a white halo, similar to the bird's-eye spot seen with bacterial canker. As the fruit ages, the white halos disappear. In contrast, bacterial canker fruit lesions retain their white halo. Bacterial spot lesions may increase in size to 4–6 mm in diameter and become brown, greasy-looking, and sometimes scabby.

Leaf symptoms are affected greatly by moisture. Under conditions of moderate moisture, leaf spots on mature leaves are small, black, and somewhat angular. The top surface may be greasy with a translucent center and a black margin. Center tissue soon becomes dry and thin and may crack. Spots are more numerous on young than on old leaves; sometimes spots occur only on one or two leaflets and on certain limited portions of a leaflet. Leaf lesions caused by the bacterial spot organism can be distinguished from those caused by the bacterial speck organism by use of a microscope (when a section of diseased tissue is placed in a drop of water, the spot bacteria ooze freely from the tissue, whereas the speck bacteria ooze out in a semi-gelatinous mass).

Under conditions of heavy moisture, such as those experienced in the wet full period in Florida (periods of continuous rainfall, heavy fog, or heavy dew), leaves take on a blighted or scorched appearance rather than retaining the typical small well-defined leaf spots. The blight symptoms begin as

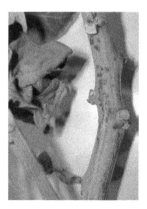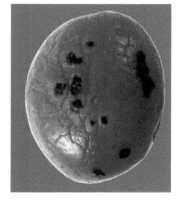

PHOTO 8.12 Disease lesions due to *Xanthomonas vesicatoria* on tomato leaves, stems, and fruits. (Photo courtesy of S. G. Borkar and Pratiksha Mote, PPAM, MPKV, Rahuri.)

yellowing along leaf margins and between veins. The yellowish tissue soon becomes water-soaked and greasy appearing. It finally dies, producing a leaf with symptoms suggestive of fertilizer burn.

Seedlings in seedbeds can become spotted severely enough to cause leaf yellowing and defoliation.

Host

Main hosts: Tomatoes and *Capsicum*. Various other Solanaceae.

Incidental hosts: *Datura* spp., *Hyoscyamus* spp., *Lycium* spp., *Nicotiana rustica*, *Physalis* spp., *Solanum* spp. (e.g., on fruits of potato).

Three pathotypes have been described (Cook and Stall, 1982), all attacking tomatoes: (1) tomato pathotype, giving hypersensitive resistance (HR) in all *Capsicum* cultivars; (2) *Capsicum* pathotype 1, giving susceptible reactions on *Capsicum* cvs. Florida VR2 and Early Calwonder; and (3) *Capsicum* pathotype 2, giving HR in Florida VR2 and susceptible reaction in Early Calwonder (*Capsicum* pathotype 2 was until recently reported from Florida (USA) only, but it has now been reported with the other two pathotypes in Taiwan (Hartman and Yang, 1990)).

Geographical Distribution

EPPO region: Widespread in Egypt, Greece, Hungary, Israel, Italy (including Sardinia and Sicily), Romania, Russia (European Siberia), Yugoslavia. Recorded in Austria, Belarus, Bulgaria, Czech Republic, France, Morocco, Poland, Slovakia, Slovenia, Spain, Switzerland (unconfirmed), Tunisia, Turkey. It probably occurs all over the Mediterranean area. Found in the past but not established in Azerbaijan, Germany (Griesbach et al., 1988), Kazakhstan.

Asia: At least in China (Jilin, Xinjiang), India (Andhra Pradesh, Delhi, Karnataka, Maharashtra, Rajasthan, Tamil Nadu), Israel, Japan (Honshu), Korea Democratic People's Republic, Korea Republic, Pakistan, the Philippines, Russia (Siberia), Taiwan, Thailand, Turkey. Found in the past but not established in Azerbaijan and Kazakhstan.

Africa: At least in Egypt, Ethiopia, Kenya, Malawi, Morocco, Mozambique, Niger, Nigeria, Réunion, Senegal, Seychelles, South Africa, Sudan, Togo, Tunisia, Zambia, Zimbabwe.

North America: Bermuda, Canada (British Columbia to Nova Scotia), Mexico, the United States (Arizona, California, Florida, Georgia, Hawaii, Iowa, Michigan, North Carolina, Ohio, Oklahoma). *Capsicum* pathotype 2 is restricted to Florida (USA).

Central America and Caribbean: At least in Barbados, Costa Rica, Cuba, Dominica, Dominican Republic, El Salvador, Guadeloupe, Guatemala, Honduras, Jamaica, Martinique, Nicaragua, Puerto Rico, St. Kitts and Nevis, St. Vincent and Grenadines, Trinidad and Tobago, the U.S. Virgin Islands.

South America: Argentina, Brazil (São Paulo), Chile, Colombia, Paraguay, Suriname, Uruguay, Venezuela.

Oceania: Australia (New South Wales, Queensland, Tasmania, Victoria, Western Australia), Fiji, Micronesia, New Zealand, Palau, Tonga.

Disease Cycle

The major sources of infection for these bacteria are thought to be seed and infected crop debris. Like the bacterial speck pathogen, they also may be present on volunteer tomato plants and on the surfaces of contaminated equipment (farm machinery, racks, greenhouse structures, tools). The bacteria are spread primarily by splashing water and wind-driven rain or mists produced during storms. In the field, spread by equipment or workers is probably of lesser importance than it is in the greenhouse, unless wounds are being opened up at the same time, as in a pruning operation or when plants are injured by a cultivator.

Bacteria enter the plant through natural openings (stomates and hydathodes) or wounds caused by wind-driven soil, insects, or mechanical damage (handling, wind whipping, high pressure sprayers).

The bacterial spot organism may be carried as a contaminant on tomato seed. This can occur during the seed extraction process. Entry of bacteria into plants occurs through natural plant openings (stomata and hydathodes) or through wounds created by wind-blown soil, insects, or cultural practices. Water soaking of the leaves, as caused by high-pressure sprays, greatly enhances bacterial spot infection. Moist weather and splashing rains are favorable for dissemination of bacteria. Bacterial spot may be present on tomato transplants produced in southern states, especially when frequent rains occur in these areas before plants are pulled. Once bacterial spot is introduced into the field, it can be difficult to control. The bacterial spot pathogen can also persist on infected plant debris in the soil for at least 1 year.

In the absence of living diseased plants, the pathogen has survived at least for 16.5 months and possibly for years on seed through northern winter periods, in the rhizosphere of dead tomato roots and in dead tomato stalks, especially when stalks are not plowed down in the fall and, remain relatively dry or in association with wheat roots during a winter period characteristic for Kentucky and in black nightshade (*Solanum nigrum*) and on ground cherry (*Physalis minima*) weeds in ditch bank during tomato-free periods in Florida.

Apparently, the bacteria cannot survive long in soil. Pathogen survival in fields appears to be associated with tomato plant debris, seeds from remaining tomato fruit, and diseased reservoir hosts such as weeds.

Disease can start from primary inoculum on seed, in soil-borne plant debris, or in reservoir hosts such as black nightshade. Some 1-year-old commercial seed lots appear to contain over 1% infected seed; about 1% of seedlings from such seed may have primary lesions of disease. Cotyledon leaves can become contaminated when they emerge from an infested seedcoat. Seedlings and transplants also can be infected by inoculum splashed or blown from soil, plant debris, and reservoir hosts. The bacteria get into plants through natural openings such as stomates in leaves and through wounds such as abrasions and broken hairs caused by wind-blown sand and the rubbing of plant parts against other plant parts, the soil surface, equipment, and workers. Bacterial spot bacteria grow and multiply in the leaf; these results in bacterial congestion, removal of air from within the leaf, and the water-soaked appearance. Symptoms can appear on tomato about 6 days after inoculation when conditions favor symptom development. The pattern of symptom development in plant beds suggests that considerable spread can occur in the beds. Long-distance spread and introduction of the disease into pathogen-free fields can occur on transplants. Secondary spread in plant beds and in production fields occurs primarily by splashing, blowing, and surface drainage of water that contains the bacteria. All plants in a field can become infected from an initial source present at only one edge of a field. Spread in plant beds and in production fields is enhanced by overhead irrigation, heavy and frequent rainfall, long dew periods, and high wind velocity. The direction of spread in fields is correlated with the direction of the prevailing winds. Fruit infection occurs through wounds such as abrasions, broken hairs, insect punctures, and growth cracks. Young leaves and fruit are more susceptible than older tissue. Infection is favored before and during inoculation by 100% relative humidity for periods of 24 h or longer, by long leaf-wetness periods that facilitate movement and introduction of bacteria, by conditions that produce wounds, and by driving rain. Disease development is favored by temperatures that fluctuate between 20°C and 35°C. High night temperatures of 24°C–27°C favor disease development, and low night temperatures of 17°C suppress disease development, regardless of day temperatures. High nitrogen levels are correlated with reduced disease severity.

Control Measures

Once present in tomato plantings, bacterial spot is difficult to control. Therefore, it should be prevented by eliminating or greatly reducing the initial inoculum if possible. Disease is prevented by using seed and transplants free of the spot organism and by planting in pathogen-free plant-bed and crop-production soil. Seed-borne inoculum is reduced or eliminated by treating tomato seeds with hot water or sodium hypochlorite solution. The hot-water treatment can be used on tomato seed that is less than 1-year old; seed is soaked at exactly 51°C for 25–30 min, then cooled, dried, and dusted with a fungicide. The sodium hypochlorite seed-soak treatment is used on seeds of any age—seeds are soaked in a 1.3% solution of sodium hypochlorite (one part of 5.25% clorox plus three parts of water) for 1 min and then are rinsed, dried, and dusted with a fungicide. The hot-water treatment controls the inoculum inside the seeds as well as on the seed surface; however, the germination percentage and rate for weak tomato seed (usually more than 1-year old) may be reduced by the treatment, especially if the treatment goes above 51°C. The sodium hypochlorite treatment controls inoculum only on the seed surface (at present, the treatment is registered for use in some areas). Following control practices should be adopted.

1. Use disease-free seed that has been produced in western states or seed that has been hot-water treated.
2. Purchase only certified disease-free transplants.
3. Rotate your tomato and pepper crops with nonhost crops.
4. Spray plants with streptomycin before transplanting. After transplanting, apply a mixture of mancozeb plus copper before the occurrence of disease. Protection is most needed during early flowering and fruit setting periods.
5. Tomato and pepper should be grown only where related plants, including weeds, have not been grown for 2 or 3 years. If possible, tomato vines should be removed or chopped up and hurled as soon as possible after harvest.
6. Try to avoid wheat in the rotation, since the bacteria have been found to survive near the surface of wheat.

8.2.4 BACTERIAL WILT AND SOUTHERN WILT OF TOMATO

Pathogen: *Ralstonia solanacearum*

Synonyms: *Pseudomonas solanacearum*

Ralstonia solanacearum (Smith) Yabuuchi et al. (formerly called *Pseudomonas solanacearum*), is a soil-borne bacterial pathogen that is a major limiting factor in the production of many crop plants around the world. This organism is the causal agent of brown rot of potato, BW or southern wilt of tomato, tobacco, eggplant, and some ornamentals, and moko disease of banana.

Symptoms

In tomato, BW first appears as flaccidity in the younger leaves. Under ideal environmental conditions, a complete and rapid wilt develops with advanced stages appearing within 2–3 days and plant death soon following (Photo 8.13). If environmental conditions are not optimal and the disease develops slowly, leaf epinasty may occur, and adventitious roots may appear on the stem.

When sectioned, the stem vascular system initially appears yellow or light brown. As the disease progresses, the stem becomes darker brown, and eventually the pith and cortex become brown. Water-soaked lesions may appear on the stem in the event of massive invasion of the cortex (Jones et al., 1991; Ji et al., 2004) (Photo 8.14).

The disease causes rapid wilting and death of the entire plant without any yellowing or spotting of leaves. All branches wilt at about the same time. When the stem of a wilted plant is cut across, the pith has a darkened, water-soaked appearance. A grayish slimy ooze is observed on pressing the stem. In later stages of the disease, decay of the pith may cause extensive hollowing of the stem. BW causes no spotting of the fruits. Affected roots decay, becoming dark brown to black in color. If the soil is moist, diseased roots become soft and slimy (Photo 8.15).

The wilting of the youngest leaves usually occurs during the hottest part of the day. This can easily go unnoticed because the leaves stay green but eventually the entire plant wilts and dies. These dramatic symptoms occur when the weather is hot (over 30°C), the humidity is high, and lots of rainfall has left the ground wet. It is also more common in soil with a high pH.

PHOTO 8.13   Field symptoms of southern wilt of tomato. (Photo courtesy of Don Ferrin, Louisiana State University Agricultural Center, Baton Rouge, LA, Bugwood.org.)

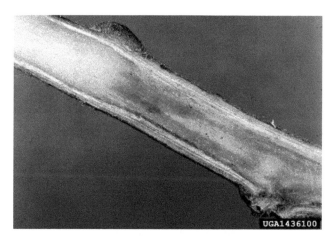

PHOTO 8.14   Browning of vascular system due to wilt pathogen. (Photo courtesy of Clemson University—USDA Cooperative Extension Slide Series, Clemson, SC, Bugwood.org.)

Mature, fruit-bearing plants are affected in midsummer with the wilting of a few leaves. This often goes unnoticed. Soon thereafter, the entire plant wilts suddenly and dies. Such dramatic symptoms occur when the weather is hot (30°C–35°C), and soil moisture is plentiful. Under less conducive conditions, wilt and decline will be slower, and numerous adventitious roots often form on the lower stems. In both cases, a brownish discoloration is present, first in the vascular system, and in advanced cases, spreading into the pith and cortex. The roots will exhibit varying degrees of decay.

Field Identification Test

You can diagnose BW by cutting the stem at the base of the plant. Look for discolored tissue. Suspend the stem in a glass of water. If it is infected, a white, slimy substance will ooze into the water within just a few minutes (Photo 8.16).

PHOTO 8.15 Root decay and basal canker of a tomato with southern bacterial wilt. (Photo courtesy of Don Ferrin, Louisiana State University Agricultural Center, Baton Rouge, LA, Bugwood.org.)

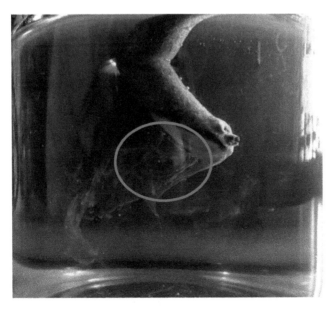

PHOTO 8.16 Oozing of bacterial wilt pathogen with formation of string in water as bacterial detection test. (Courtesy of R.A. Yumlembam and S.G. Borkar, PPAM, MPKV, Rahuri, Maharashtra, India.)

Host

Solanaceae species: Potato (*Solanum tuberosum*), tomato (*Lycopersicon esculentum*), eggplant (*Solanum melongena*), black nightshade (*Solanum nigrum*), bittersweet or climbing nightshade (*Solanum dulcamara*), and *Datura stramonium*.

Other nonsolanaceous hosts include *Portulaca oleracea, Brassica* spp., *Tropaeolum majus, Urtica dioica, Chenopodium album, Melampodium perfoliatum*, and *Geranium* (*Pelargonium hortorum*). *Musa* spp. (banana), *Musa paradisiacal* (plantain), *Nicotiana tabacum* (tobacco), *Arachis hypogeal* (peanut).

Geographical Distribution

In Africa it has been reported in Angola, Burkina Faso, Burundi, Congo, Ethiopia, Gabon, Gambia, Kenya, Madagascar, Malawi, Mauritius, Mozambique, Nigeria, Réunion, Rwanda, Senegal, Seychelles, Sierra Leone, Somalia, South Africa, Swaziland, Tanzania, Tunisia, Uganda, Zaire, Zambia, and Zimbabwe (CABI, 2005).

Asia, Australia (eastern), Central America (Costa Rica), Europe (western), South America. In North America there have been sporadic reports from Mexico and now the United States.

Pathogen

Ralstonia solanacearum is a Gram-negative motile rod. The organism grows aerobically and does not form endospores. Cells are 0.5–0.7 × 1.5–2.0 μm and are nonencapsulated. *Ralstonia solanacearum* is catalase positive, oxidase positive, and reduces nitrates. The pathogen does not hydrolyze starch and does not readily degrade gelatin. In broth culture, the organism is inhibited by concentrations of NaCl greater than 2% (Stevenson et al., 2001).

Kelman's tetrazolium chloride (TZC) agar or 2% sucrose peptone agar (SPA) should be used for isolation when *Ralstonia solanacearum* is suspected. After 2 days on TZC medium, virulent wild-type colonies appear as large, elevated, fluidal, and either entirely white or with a pale red center, while avirulent mutant colonies are butyrous, deep red often with a bluish border. On SPA, *Ralstonia solanacearum* colonies are white and fluidal with characteristic whorls (French et al., 1995).

Direct isolation of *Ralstonia solanacearum* can be obtained from plant ooze and exudates. The infected stems and/or petioles are cut using a sterile sharp knife or razor blade. If bacterial ooze does not actively appear, the plant material is squeezed between two fingers. A suspension of the ooze is prepared in sterile distilled water and then streaked onto either TZC or SPA plates. Pure cultures are usually easily isolated (French et al., 1995). *Ralstonia solanacearum* can also be isolated from water and soil using a modified Kelman's TZC medium (French et al., 1995).

Isolates of *Ralstonia solanacearum* rapidly lose virulence when maintained on laboratory media; however, the organism can easily be maintained for years in sterile distilled water or on agar slants covered with sterile mineral oil and stored at room temperature (Stevenson et al., 2001).

Races and Biovars

The pathogen species is subdivided into races based on host range. Currently, PCR is the primary means of definitive identification of pathogen race. Before PCR, a tobacco hypersensitivity test, developed by Lozano and Sequeira, was used to distinguish between races 1, 2, and 3—the most economically important races. The species is also subdivided into biovars based on the utilization of the disaccharides cellobiose, lactose, and maltose and oxidation of the hexose alcohols dulcitol, mannitol, and sorbitol (French et al., 1995).

Recently, it has been suggested that *Ralstonia solanacearum* should be considered a "species complex." Also, some consider the current race and biovar system inadequate since it is based on phenotypic measures and propose switching to a phylogenetically based hierarchical system. This scheme is based upon phylogenetic analysis of sequence information using RFLP typing, 16S rDNA phylogeny, and hrpB PCR-RFLP data. The system divides the species complex into phylotypes (phylogenetic grouping of strains), sequevars (group of strains with endoglucanase or *mutS* gene sequences diverging by <1%), and clones (group of strains exhibiting the same genomic fingerprint) (Fegan and Prior, 2004).

Disease Cycle

The bacteria are soil inhabitants and can survive in soil for long periods if the temperature does not drop too low. The bacteria can overwinter in diseased plant debris and weed hosts and they persist indefinitely in fallow soil that is moderately warm. The bacteria do not survive in soil in north temperate climates, but can overwinter in cold frames, hotbeds, and greenhouses. The bacteria can be spread by infected seeds, transplants, and cutting knives.

The bacteria enter plants through wounds made by cultivators, soil insects, broken roots on transplants, and through natural openings where secondary roots emerge. Several nematode species, *Meloidogyne hapla*, *M. incognita acrita*, and *Helicotylenchus nannus*, can increase the incidence and severity of wilt. Symptoms appear within 2–8 days after infection, depending on plant's age and susceptibility. Bacteria reach the large xylem vessels and spread throughout the plant. As they move, they escape into the intercellular spaces of the parenchyma cells in the cortex and pith, dissolve the cell wall, and make cavities filled with slimy masses of bacteria. If moisture is present on leaves, the bacteria can enter above the ground, but infection is more likely when inoculum is present just below the soil surface.

When diseased plant parts decay, bacteria are liberated in great numbers into the soil where they are disseminated in water, with particles of soil, and in association with diseased seedlings or tubers. The wilt organism is sensitive to high pH, low soil temperature, low soil moisture, and low fertility levels. Although the bacteria are able to reproduce and cause infection over a wide range of temperatures (15°C–37°C), the most favorable temperatures are 30°C–35°C. Pathogen isolates from a hot climate are favored by temperatures higher than those that favor pathogen isolates from a cooler climate.

High temperatures and high soil moisture generally favor *Ralstonia solanacearum*, the exception being certain race 3 strains that are pathogenic on potato and are able to grow well at lower temperatures (Stevenson et al., 2001). The pathogen is found in many different soil types and over a wide range of soil pH. The organism survives in infected plant material, vegetative propagative organs, wild host plants, and soil. Alternate hosts, especially latently infected weed species, are thought to play a major role in the overwintering ability of the organism in temperate regions (Stevenson et al., 2001; Ji et al., 2004). Bittersweet nightshade (*Solanum dulcamara*), a common aquatic weed, has been implicated as the source of the inoculum (Stevenson et al., 2001).

Sources of inoculum for agricultural fields and methods of spread include irrigation and surface water, aquatic weeds, infested soil and field weeds, contaminated planting material, latently infected vegetative propagative material, and contaminated farm tools and equipment. Once established in a field, plant-to-plant spread may occur when bacteria move from roots of infected plants to roots of healthy plants (Stevenson et al., 2001; Ji et al., 2004).

Ralstonia solanacearum enters the plant through wounds in the roots from cultivating equipment, nematodes, insects, and through cracks where secondary roots emerge (Agrios, 1997). The bacteria reach the large xylem elements and are spread into the plant, where they multiply. Once established in the xylem vessels, the bacteria are able to enter the intercellular spaces of the parenchyma cells in the cortex and pith in various areas of the plant. Here, *R. solanacearum* is able to dissolve the cell walls and create slimy pockets of bacteria and cell debris. Production of highly polymerized polysaccharides increases the viscosity of the xylem, which results in plugging (Shew and Lucas, 1991).

Predisposing Factors

- Injured plants, since the bacteria enters the roots through wounds caused by cultivation, improper planting, and nematodes or other root-feeding critters in the soil.
- Poorly draining, infertile, or heavy clay soil.
- Acidic soil.
- Hot, humid, or rainy conditions.
- Soil infected with the bacteria. BW can live for years in soil without a host plant present.

- Water runoff that spreads the bacteria.
- Weeds that can act as hosts to the bacteria without showing symptoms of BW.
- Infected tools, transplants, and imported soil.

Damage
Ralstonia solanacearum constitutes a serious obstacle to the cultivation of many solanaceous plants in both tropical and temperate regions. The greatest economic damage has been reported on potatoes, tobacco, and tomatoes. It can sometimes cause total crop losses. Disease severity mostly increases if *Ralstonia solanacearum* is found in association with root nematodes. Experiments conducted in India showed that the combined pathogenic effects of *Ralstonia solanacearum* and root-knot nematodes (*Meloidogyne javanica*) were greater than the independent effects of either (CABI, 2005).

The bacterium is a quarantine organism. The occurrence of different races and strains of the pathogen with varying virulence under different environmental conditions presents a serious danger to European and Mediterranean potato and tomato production. Absence of the bacterium is an important consideration for countries exporting seed potatoes.

Management of Bacterial Wilt in Tomato
Diseases are a major hurdle in tomato cultivation, both under controlled and field conditions. Among them, BW is most devastating on tomato causing yield losses of up to 90%. The disease can be managed with a holistic approach by adopting different control measures.

Cultural Management
- The disease can be kept under check if crop rotations are followed with crops such as maize, ragi, and okra, since they can reduce the pathogen population significantly.
- Soil solarization combined with fumigation reduces the pathogen considerably.
- Destruction of weeds, collateral hosts, and other off-season hosts will reduce the inoculum potential.
- Rotate away with tomatoes, peppers, eggplant, Irish potatoes, sunflowers, and cosmos for at least 3 years.
- Use plant varieties that are tolerant/resistant to BW. The varieties Neptune and Tropic Boy are partially resistant. In Kenya, the tomato varieties that have been claimed to be resistant to BW are Fortune Maker, Kentom, and Taiwan F1.
- Do not grow crops in soil where BW has occurred.
- Rogue out wilted plants from the field to reduce spread of the disease from plant to plant.
- Control root-knot nematodes since they could facilitate infection and spread of BW.
- Where feasible, extended flooding (for at least 6 months) of the infested fields can reduce disease levels in the soil.
- Use raised beds to improve soil drainage.
- Soil amendments (organic manures) can suppress BW pathogen in the soil.
- An alternative way to avoid the disease is to plant tomatoes in commercial potting soil in containers.

Application of Sawdust
Application of sawdust and peat moss are efficient in reducing the incidence.
Application of nitrite form of fertilizers is also capable of reducing the bacterial population.
Calcium concentration in soil should be increased to have better control of wilt disease.

Disease Avoidance
Since high soil temperatures and soil moisture enhance BW development, damages can be minimized by changing the date of planting, considering seasons that are less favorable for disease development.

Intercropping

In some developing countries, intercropping has been used as a means of reducing pathogen populations in the soil and root-to-root transmission.

In Burundi, a bean intercrop, which has a dense root system and grows quickly, was better than a crop such as maize, which develops slower and has a more dispersed root system.

Chemical Control

Soil treatment with chloropicrin, methyl bromide, or mixture of both was found effective in retarding the wilt development.

Application of bleaching powder (15 kg/ha) has also been found effective against this disease.

Seedling dip with streptocycline avoids early invasion and infection by the pathogen through wounds formed during transplanting.

Bacterinol-100 can be used as dry seed dresser, for nursery spray, and field application. Streptomycin sulfate or oxytetracycline when sprayed at 200 ppm at 7 days interval provides good control.

Tomato varieties such as Arka Abha, Sonali, DPT-38, and Arka Alok, which are resistant to BW can be used for cultivation.

Biological Management

Several biological control agents such as *Pseudomonas fluorescens*, *Bacillus licheniformis*, *Bacillus cereus*, *Bacillus subtilis*, and mycorrihiza are very effective in delaying and reducing the wilt development.

Good results have been encountered in Australia, the United States, and the Philippines by using "biofumigation" as soil treatment for BW.

8.2.5 Pith Necrosis of Tomato

Pathogen: *Pseudomonas corrugata*

Tomato pith necrosis, caused by *Pseudomonas corrugata* and other soil-borne species of *Pseudomonas*, has been observed sporadically throughout Louisiana since it was first observed here in 1983. The disease generally occurs on early planted tomatoes when the night temperatures are cool, the humidity is high, and plants are growing too rapidly because excessive nitrogen has been applied. Once the weather warms up, the plants tend to be able to outgrow the problem.

The disease occurs randomly within fields, and initial symptoms are usually observed at the time the first fruit clusters reach the mature green stage.

Symptoms

The symptoms generally consist of yellowing of young leaves followed by chlorosis and wilting of infected shoots in the upper part of the plant canopy (Photo 8.17). This wilting is usually associated with internal browning (or necrosis) of the basal portion of the stem. Dry, gray to brown to black lesions are also observed on petioles and stems, which may shrink, crack, or collapse. The pith of symptomatic stems is hollow or may appear to contain distinct chambers and often exhibits a dark discoloration (Photo 8.18). Additionally, adventitious shoots often develop profusely on the affected stems.

First symptoms of the disease appear during fructification. There are big oblong spots on leaves and leaf lobules roll upward; these symptoms are especially appreciable on sunny days. Diseased leaves look like "scalded" ones, though green in color. Necrotic dark green streaks (about 25–50 cm in length) develop on stricken stems. Strong maceration of the diseased tissues with destruction of a core is possible. Cracks are often formed later in the stems with inner emptiness and with brown core tissue. Dry necrotic lesions and adventitious root formation on tomato stem occur on pith necrosis–infected plants. Alternation of high (above 25°C) and low temperatures promotes bacteriosis development.

Geographical Distribution

Pith necrosis of stem of tomato occurs in England, France, and other countries. This disease is found in Leningrad, Kemerovo, Saratov, Volgograd, Moscow, Vologda Regions, in Tatarstan, and also in Armenia, Belarus, and Latvia.

PHOTO 8.17   Chlorosis and wilting of shoot associate with pith necrosis disease. (Photo courtesy of Gerald Holmes, California Polytechnic State University, San Luis Obispo, CA, Bugwood.org.)

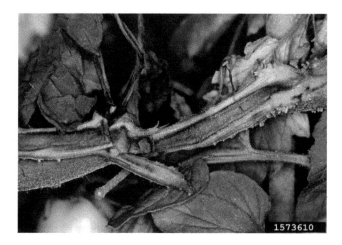

PHOTO 8.18   Pith necrosis symptoms due to *Pseudomonas corrugata* on tomato stem. (Photo courtesy of Gerald Holmes, California Polytechnic State University, San Luis Obispo, CA, Bugwood.org.)

Ecology

Pith necrosis of stem of tomato is distributed in greenhouses with the increased humidity (90%–95%) and temperature of both air (25°C–28°C) and ground. Drastic fluctuations of day and night temperatures promote development of this disease, as well as application of increased doses of nitric fertilizer that lacks potash and phosphoric, especially, boric fertilizers.

Life Cycle

The epidemiology of this disease is not well understood; it is possible that the bacteria are seedborne and most certainly survive in the soil in association with infected tomato debris

Economic Significance

Yield losses depend on a cultivated variety and on conditions of plant cultivation. In some greenhouses, the number of the struck plants reached 100% in Tatarstan, 8%–15% in Moscow Region, 58% in Kemerovo Region, and 35% in Latvia.

Control Measures

Cultural Control and Sanitary Methods
As *Pseudomonas corrugata* is a widespread opportunistic pathogen, cultural methods are most important for control.

Different preventive measures can be taken both in nursery and in the cultivation of tomato (Catara and Bella, 1996).

In nurseries, it is important to use pathogen-free seeds according to certification schemes. Furthermore, due to the ability of the bacterium to survive in soil, virgin soil should be used, but if not, the soil should be sterilized. However, when plantlets were transplanted in fumigated soil, a higher disease incidence was observed; it was suggested that natural antagonists against *Pseudomonas corrugata* were reduced by fumigation (Scortichini, 1989; Carroll et al., 1992). Thermal sensitivity of representative strains of the bacterium was assessed with the aim of developing control methods to reduce soil inoculum. Bacterial populations in soil were undetected only after a 30-min treatment at 60°C (Bella et al., 2003).

Excessive nitrogen should be avoided. Plants should be kept dry by rational irrigation and circulation and substitution of air. Avoiding low-night temperatures can prevent high humidity and free water on the plant surfaces.

Water used for irrigation should be checked for potential contamination with *Pseudomonas corrugate*. Sanitary methods such as disinfection of greenhouses and tools, regular disinfection of hands during cultural work, and avoiding wounding of plants are preventive measures against bacterial diseases in general (Naumann, 1980).

Although some authors recommend destroying plants as soon as they show symptoms in order to prevent dissemination of the disease (Naumann, 1980; Fiori, 1992), others advise growers not to remove them because spread among plants appears to be low. Nevertheless, removal of infected plantlets is very important in nurseries.

Preventive measures to minimize the occurrence of this disease in high tunnels include adequate ventilation to avoid high humidity levels (especially during cloudy weather), avoiding excessive nitrogen levels to prevent vigorous plant growth, incorporation of crop debris to speed decomposition of residue and associated bacteria, and crop rotation.

Chemical Controls
There is no effective treatment for this disease; however, affected plants may recover if environmental conditions improve (warm, sunny weather).

Regulatory Control
Only pathogen-free seeds, treated and/or tested by a reliable method, should be used. Transplants should be disease free.

Host-Plant Resistance
All 12 tomato cultivars tested for resistance proved to be susceptible or highly susceptible. Some wild species were resistant (Scortichini and Rossi, 1993).

Biological Control
In greenhouse tests, three different strains of *Agrobacterium radiobacter* (Agr-K84, Agr-K1026, and Agr-K84) reduced pith necrosis of tomato (Lopez et al., 1991).

8.2.6 Leaf Spot of Tomato

Pathogen: *Pseudomonas syringae* pv. *syringae*

In the spring of 2006 and 2007, grafted and nongrafted tomato plants (scion cv. Cuore di Bue, rootstock *Lycopersicon lycopersicum* × *L. hirsutum* cv. Beaufort) displaying stem and petiole necrosis were observed in many commercial greenhouses in the Piedmont of northern Italy.

Symptoms

Initial symptoms such as water-soaked circular lesions (2–3 mm in diameter) on stems and petioles develop 2–10 days after transplanting. These lesions eventually coalesce into brown-to-black areas as much as 1 cm in diameter. In some cases, necrotic areas progress from stem petioles to leaf tissues. Thereafter, plants wilt and die within a few days. In some greenhouses, more than 80% of young plants exhibit symptoms and production is severely reduced.

Pathogen

All colonies fluoresce under UV light when grown on King's B medium (King et al., 1954). Colonies are levan positive, oxidase negative, potato soft rot negative, arginine dihydrase negative, and tobacco hypersensitivity positive (LOPAT test; group Ia). In addition, all isolates are positive for arbutin and aesculin hydrolysis and utilize erythritol, but not adonitol, L (+)-tartrate, or DL-homoserine as a carbon source. The isolates also cause severe necrotic lesions on lemon fruits and lilac leaves (Scortichini et al., 2003). The bacterial colonies were identified as *Pseudomonas syringae* pv. *syringae* (Braun-Kiewnick and Sands, 2001). Also, repetitive-sequence PCR using the BOXA1R primer indicated that the isolates belong to pattern 4 of *Pseudomonas syringae* pv. *syringae* (Scortichini et al., 2003).

Geographical Distribution

Found in Italy as well as in Europe. A bacterial spot of tomato caused by *Pseudomonas syringae* pv. *syringae* has been reported in the United States (Jones et al., 1981).

Control Measures

Follow the cultural practices as in case of other bacterial tomato diseases. Copper sprays may reduce the incidence of the disease.

8.2.7 Soft Rot of Tomato

Pathogen: *Erwinia carotovora* subsp. *carotovora*

Symptoms

The symptoms of soft rot bacterium develop on stems and fruits.

On Stems

Soft rot develops in the inner stem, which becomes brown and slimy, then disintegrates, and becomes hollow. The damage may extend several centimeters above and below the site of infection. The affected plants wilt and die.

On Fruits

The infected fruit develops a thin outer skin and becomes filled with a slimy mass (Photo 8.19). The fruit collapses and dries into a shriveled mass when the skin is broken.

The affected stem and fruit have a rotten, foul-smelling odor. All stages of growth are affected, including harvested fruits.

Predisposing Factors

Soft rot bacteria are inhabitants of field soils. The bacteria survive in soil—longer in cool, wet conditions and spread to a shorter distance under dry, warm conditions. Survival of the bacteria in soil is linked mainly to soil temperature, but also to soil moisture and the presence of antagonistic soilborne microorganisms. The bacteria also can occur around the roots of many crops and weeds, in surface and underground waters, in rain, snow, aerosols, and on the mouthparts of insects. The bacteria are often transmitted through wounds created by workers pruning plants or harvesting fruit with contaminated tools or hands. Chewing larvae and insects, such as fruitworms, can also transmit the bacteria. Harvesting during rainy periods or washing contaminated fruits before shipment may disseminate the bacteria. Transmission of this bacterium may occur from the splashing of

PHOTO 8.19 Symptoms of soft rot on tomato. (Photo courtesy of Missouri Botanical Garden, St. Louis, MO, missouribotanicalgarden.org.)

infected soil into wounds of neighboring plants. Transmission also occurs when wounds are created using contaminated machinery and implements. Infected cull piles may attract air-borne insect vectors, leading to long-distance transmission.

The development of soft rot is favored by warm air temperatures (20°C–30°C) and high relative humidity. Infection of tomato tissue by *Pythium* spp. and *Phytophthora* spp. creates favorable conditions for the development of soft rot later.

Control Measures

Avoid injuring plants. Disinfect hands and tools when pruning tomato plants and wash contaminated clothing. Prevent the occurrence of insect wounds by controlling the pests. Avoid planting tomato crops after crops of potato or cabbage; instead, rotate with crops of bean, maize, and soybean. Plant in well-drained soil.

Remove affected plants and destroy them away from the field. Monitor neighboring plants for any symptom development.

Use deep well water if surface water is contaminated. Use chlorinated water to reduce the risk of infection during washing. This will not reduce soft rot development in fruit already infected with the organism. Harvest during dry weather and minimize fruit injury at harvest.

8.3 BRINJAL

8.3.1 BACTERIAL WILT OF BRINJAL

Pathogen: *Ralstonia solanacearum*

Synonyms: *Pseudomonas solanacearum*

The brinjal cultivation is mainly affected by BW and is a major production constraint.

This pathogen has a wide host range of more than 200 plant species, which makes its management difficult.

Symptoms

BW disease causes severe problem in brinjal cultivation. The characteristic symptoms of the disease are wilting of the foliage followed by collapse of the entire plant. The wilting is characterized by gradual, sometimes sudden, yellowing, withering, and drying of the entire plant or some of its branches.

Under unfavorable condition for disease development, the BW causes the leaf surface wilting, stunting, yellowing of the foliage, and finally collapse of the entire plant. Lower leaves may droop first before wilting occurs. The vascular system becomes brown. Bacterial ooze comes out from the affected parts. Plants show wilting symptoms at noon time and will recover at nights, but die soon.

Disease develops very rapidly in warm weather. Symptoms are very clear during morning or immediately after irrigation. Symptoms manifest initially as leaf drooping followed by wilting of entire plant within a few days. Recently wilted plants look green, a distinct symptom when compared to other vascular wilt diseases, which develops yellowing of the leaves. Vascular discoloration (brown) can also be seen in the wilted plant.

Diagnostic Test of Pathogen

A stream of milky white bacterial ooze can be noticed when the cut ends of the stem/root is kept undisturbed for few minutes in a clear glass container with water. This is a simple diagnostic method and could be used in the field to identify BW. In the laboratory, if the bacterial ooze is streaked in Tripheny tetrazolium chloride (TZC) medium, circular to oval shape, fluidal colonies with pink center will appear after 48 h of incubation which is unique to *Ralstonia solanacearum*. Molecular biology techniques like PCR can also be used in the detection of *Ralstonia solanacearum* from the soil and from the plant sap. The presence of bacteria can be seen by amplifying the specific region from its genome.

Pathogen

BW in brinjal is caused by *Ralstonia solanacearum*, a soil bacterium, formerly known as *Pseudomonas solanacearum*. The pathogen has five different races, each infecting different plant species. *Ralstonia solanacearum* strains are grouped into five biovars based on the biochemical tests. Brinjal BW is mostly caused by strains belonging to race 1 and biovars 3. Race 1 has wide host range including solanaceous vegetables like tomato and chili. Race 1 strains from Goa exhibit genetic variability and difference in virulence on brinjal.

Host Range and Distribution

Ralstonia solanacearum is a widely distributed pathogen found in tropical, subtropical and some temperate regions of the world. It has an unusually wide host range of 200 plant species belonging to more than 50 families. Majority of the hosts are dicots with the major exception being bananas and plantains. Most economically important host plants are found in Solanaceae family. Specific host range and distribution of *Ralstonia solanacearum* depends on the race and, to some degree, the biovars of the pathogen.

Survival and Spread

BW is a soil-borne disease and *Ralstonia solanacearum* is able to survive in the soil for long periods, which ranges from 1 to 10 years without a host plant. Moreover, it can colonize the nonhost plants including many weeds that serve as the symptomless carrier. The pathogen infects the plant through root injuries/wounds or at the site of secondary root emergence. Intercellular spaces of the root cortex and vascular parenchyma are subsequently colonized and cell walls are disrupted facilitating spread through vascular system. In xylem vessels, the bacteria multiply rapidly and finally block the translocation of water, which leads to wilting of the plant. After wilting and plant death, the bacteria were found to be released into the soil. The neighboring plants can be infected via root contact or spread of the pathogen through irrigation water. The pathogen can also enter through priming wounds, contaminated water sources, symptomless infected seedlings, as well as humans or machinery carrying infected soil.

Favorable Conditions
The disease occurs on all the types of soil, including sandy and clay soils. BW incidence is mostly prevalent in the acidic soils (soil pH < 7.0) and in the coastal humid areas. High temperature and moisture are favorable for disease incidence. Incidence of root-knot nematode predisposes the plant and will accelerate disease development.

Disease Management

Plant in a Disease-Free Field
Since the pathogen is soilborne, the disease will occur only if the pathogen is present in the field and environmental conditions are favorable for its development. Fields that have no previous history of BW and have not been planted with susceptible hosts during the previous season/year are less likely to have the pathogen. Regular rotation with other nonhost plants will reduce disease incidence. However, rotation with paddy for only one season doesn't make the field disease free.

Use Resistant Varieties and Pathogen-Free Seedlings
A simple and most promising way to control BW is to plant resistant varieties. However, commercial/preferred varieties with high and stable resistance to BW are not common. Kerala Agricultural University has recommended a few BW-resistant varieties—Surya, Swetha, and Haritha—which could be tried. Use resistant varieties like Pusa Bhairav, BB-7, BWR-12, Pant Rituraj, and H-8. The local preferred cultivars in Goa, that is, Agassaim and Taleigao, are highly susceptible to BW and the incidence ranges from 30% to 100% during *rabi*. When resistant varieties are not available, grafting with resistant root stocks could provide better option. Since the pathogens could spread through infected, symptomless seedlings, the nurseries should not be located in the fields with the history of BW. It is recommended to fumigate/solarize nursery soil for 15 days prior to sowing.

Seedling Treatment
- Dip seedlings into streptrocycline (1 g/40 L H_2O) for 30 min.

Prevent the Spread of the Pathogen
Infected plants should be removed and destroyed and also the infected portion of the field should be isolated if possible by preventing the waterflow into and from the field. Frequency of irrigation and quantity of water should be reduced. Flood irrigation in the infected field increases the disease incidence by twofold. Movement of people/machinery should also be limited. The field should be cleaned as soon as the disease is detected in the field, that is, the diseased fruits should be plucked and burnt. Other solanaceous crops are to be avoided in crop rotation.

Reduce the Pathogen Load in Infected Field
Once introduced into the field, pathogens like *Ralstonia solanacearum* are very difficult to eradicate, but the pathogen population/load can be reduced by the following methods, which in turn reduces the incidence of the disease:

1. Fumigation with chemicals like methyl bromide. However, it is highly toxic and not practical for small-scale farmers.
2. Crop rotation with paddy and other nonhost crops for three to four seasons.
3. Flooding the field for 1–3 weeks before planting will reduce BW.
4. Growing marigold (*Tagetes* spp.) in rotation or as intercrop will suppress the pathogen in addition to its anti-nematode effect. Growing *Brassica* spp. and incorporating the plants into soil at flowering stage will reduce BW incidence.
5. Application of organic manures like farmyard manure (FYM)/poultry manure in every year will increase the beneficial soil microflora and reduce the BW incidence.
6. Biological control.

Nursery Application

Treat the seeds with talc-based formulation of antagonistic *Pseudomonas fluorescens* (10 g/100 g of seeds) and make soil application of antagonistic *Pseudomonas fluorescens* (50 g mixed with 1 kg of soil and incorporated in the nursery bed).

Dip the seedlings in the antagonistic *Pseudomonas fluorescens/Bacillus subtilis* (at 25g talc formulation per liter of water) solution for 20–30 min just before transplanting. The leftover solution should be drenched around the root zones (50 mL/plant).

Treatment with biological control agents reduces the incidence of BW by more than 65% compared to control. Further, an increased yield (28%–54%) was observed for plants treated with biological control agents.

8.4 CAPSICUM

8.4.1 Bacterial Spot of Capsicum

Pathogen: *Xanthomonas euvesicatoria* and *Xanthomonas perforans* = [*Xanthomonas axonopodis* (syn. *campestris*) pv. *vesicatoria*], *Xanthomonas vesicatoria*, and *Xanthomonas gardneri*

Host: Pepper (*Capsicum* spp.) and tomato (*Solanum lycopersicum*) (economically important hosts)

Bacterial spot is one of the most devastating diseases of pepper grown in warm, moist environments. Once present in the crop, it is almost impossible to control the disease and prevent major fruit loss when environmental conditions remain favorable.

Worldwide, 16.5 million metric tons (36 million tons) of peppers are grown for fresh consumption and for use in condiments. For many people in the United States, the bell-shaped pepper fruit is the most familiar; however, nonbell-type peppers are more widely grown.

Symptoms

Bacteria attack the foliage, stems, and fruits of peppers to produce the disease symptoms.

On Fruits

On peppers, lesions may form on fruits, including the peduncle, but the major crop loss is due to the shedding of blossoms and young, developing fruit. Fruits that remain are usually nonmarketable because of poor quality. They may have lesions and are often misshapen and damaged due to excessive exposure to the sun as a result of defoliation. This can result in sunscald on the fruit. Diseased leaves drop prematurely resulting in extensive defoliation. As newly emerging leaves become infected and defoliation of older, diseased leaves continues, plants possess leaves mostly on their upper stems. Generally, defoliation is more common in peppers than tomatoes. Fruit spots are at first water-soaked and later become raised and scabby (Photo 8.20).

On Leaves

Because the most obvious symptoms occur on leaves, the disease is often referred to as "bacterial leaf spot." Symptoms begin as small, yellow-green lesions on young leaves that usually appear deformed and twisted, or as dark, water-soaked, greasy-appearing lesions on older foliage (Photo 8.21). Lesions develop rapidly to a size of 0.25–0.5 cm (0.1–0.2 in.) wide, and become tan to brownish red. Lesion shape is defined by leaf veinlets, so the shape is angular in shape rather than round that is more typical of fungal leaf spots or injury caused by some pesticides or other chemical sprays.

Lesions often are more numerous at the tip and margin of the leaf where moisture such as dew is retained. Under dry conditions, diseased leaves can develop a tattered appearance as the leaf margin and lesion centers become necrotic, dry up, and disintegrate. Lesion size is often larger and symptoms are more severe when extended periods (>12 h) of moisture-saturated tissue occur.

PHOTO 8.20   Symptoms of *Xanthomonas* pv. *vesicatoria* on capsicum fruit. (Photo courtesy of Howard F. Schwartz, Colorado State University, Fort Collins, CO, Bugwood.org.)

PHOTO 8.21   Symptoms of *Xanthomonas* pv. *vesicatoria* on capsicum leaves. (Photo courtesy of Howard F. Schwartz, Colorado State University, Fort Collins, CO, Bugwood.org.)

Host

The principal hosts are tomatoes and *Capsicum*. Various other Solanaceae, mainly weeds, have been recorded as incidental hosts particularly *Datura* spp., *Hyoscyamus* spp., *Lycium* spp., *Nicotiana rustica*, *Physalis* spp., and *Solanum* spp. (e.g., on fruits of potato). The forms of '*Xanthomonas vesicatoria*' reported many years ago (Hayward and Waterston, 1964) to attack Brassicaceae as well as Solanaceae are now referred to *Xanthomonas campestris* pv. *raphani*.

Three pathotypes have been described (Cook and Stall, 1982), all attacking tomatoes: (1) tomato pathotype, giving HR in all *Capsicum* cultivars; (2) *Capsicum* pathotype 1, giving susceptible reactions on *Capsicum* cvs Florida VR2 and Early Calwonder; and (3) *Capsicum* pathotype 2, giving HR in Florida VR2 and susceptible reaction in Early Calwonder (*Capsicum* pathotype 2 was until recently reported from Florida (USA) only, but it has now been reported with the other two pathotypes in Taiwan (Hartman and Yang, 1990)).

Geographical Distribution

EPPO region: Widespread in Egypt, Greece, Hungary, Israel, Italy (including Sardinia and Sicily), Romania, Russia (European, Siberia), and Yugoslavia. Recorded in Austria, Belarus, Bulgaria, Czech

Republic, France, Morocco, Poland, Slovakia, Slovenia, Spain, Switzerland (unconfirmed), Tunisia, and Turkey. It probably occurs all over the Mediterranean area. Found in the past but not established in Azerbaijan, Germany (Griesbach et al., 1988), Kazakhstan.

Asia: At least in China (Jilin, Xinjiang), India (Andhra Pradesh, Delhi, Karnataka, Maharashtra, Rajasthan, Tamil Nadu), Israel, Japan (Honshu), Korea Democratic People's Republic, Korea Republic, Pakistan, the Philippines, Russia (Siberia), Taiwan, Thailand, Turkey.

Africa: At least in Egypt, Ethiopia, Kenya, Malawi, Morocco, Mozambique, Niger, Nigeria, Réunion, Senegal, Seychelles, South Africa, Sudan, Togo, Tunisia, Zambia, Zimbabwe.

North America: Bermuda, Canada (British Columbia to Nova Scotia), Mexico, the United States (Arizona, California, Florida, Georgia, Hawaii, Iowa, Michigan, North Carolina, Ohio, Oklahoma). *Capsicum* pathotype 2 is restricted to Florida (USA).

Central America and Caribbean: At least in Barbados, Costa Rica, Cuba, Dominica, Dominican Republic, El Salvador, Guadeloupe, Guatemala, Honduras, Jamaica, Martinique, Nicaragua, Puerto Rico, St. Kitts and Nevis, St. Vincent and Grenadines, Trinidad and Tobago, the U.S. Virgin Islands.

South America: Argentina, Brazil (São Paulo), Chile, Colombia, Paraguay, Suriname, Uruguay, Venezuela.

Oceania: Australia (New South Wales, Queensland, Tasmania, Victoria, Western Australia), Fiji, Micronesia, New Zealand, Palau, Tonga.

Pathogen
Until the early 1990s, bacterial leaf spot of pepper and tomato was thought to be caused by a single bacterial species, *Xanthomonas campestris* pv. *vesicatoria*. In the early 1990s, two distinct genetic groups were shown to exist within strains of *Xanthomonas campestris* pv. *vesicatoria*. In 1995, Vauterin et al. (1995) restructured the species within the genus *Xanthomonas* and proposed *Xanthomonas vesicatoria* and *Xanthomonas axonopodis* pv. *vesicatoria*. Subsequently, four taxonomically distinct xanthomonads were identified and placed into four groups, designated A, B, C, and D. Jones et al. (2004) showed these four groups to be distinct enough to deserve species status: *Xanthomonas euvesicatoria* = *Xanthomonas campestris* (*axonopodis*) pv. *vesicatoria* (group A), *Xanthomonas vesicatoria* = *Xanthomonas vesicatoria* (group B), *Xanthomonas perforans* = group C strains, and *Xanthomonas gardneri* = group D strains. Pepper strains found within *Xanthomonas euvesicatoria* are the most widely distributed and cause the greatest economic loss in pepper. *Xanthomonas vesicatoria* and *Xanthomonas gardneri* are also known to cause bacterial leaf spot on pepper and can have a significant impact in regions where they are found. *Xanthomonas perforans* strains are not known to cause disease on pepper. Strains from all four species have been isolated from tomato.

Early work on bacterial leaf spot indicated that strains recovered from tomato and pepper were pathogenic on both plant species, and for many years, it was thought that cross infection could occur in the field. It was not until the 1970s that Cook (1973) demonstrated host specificity was associated with a hypersensitive reaction (HR) (Burkholder and Li, 1941; Cook, 1973; Canteros et al., 1991). Currently, three groups of strains are distinguished on the basis of virulence on tomato and pepper: tomato strains are virulent on tomato only, pepper strains are virulent on pepper only, and pepper–tomato strains are virulent on both crop species (Reifschneider et al., 1985). Within the pepper and pepper–tomato groups of strains, races of the pathogen can be distinguished by the reaction of various pepper lines.

Development of resistance to bacterial leaf spot of pepper began when Sowell (Sowell, 1960; Sowell and Dempsey, 1977) screened many plant introductions for resistance. Currently, four resistance genes that induce an HR have been identified within pepper (Burkholder and Li, 1941; Cook,1973; Canteros et al., 1991). These genes were identified from the following plant introductions: PI 163192 (*Bs1* gene), PI 260435 (*Bs2* gene), PI 271322 (*Bs3* gene), PI 235047 (*Bs4* gene).

The host range of the different strains can be determined by infiltrating a bacterial suspension into the leaf and observing the response. Within 18–36 h, a resistant response is indicated by a rapid collapse of the infiltrated area (HR). In a susceptible plant, the infiltrated area develops a chlorotic, water-soaked appearance, but not until 3–5 days after infiltration.

An HR is observed as a confluent necrosis when leaves are infiltrated with a concentrated bacterial suspension. Growth of the bacterial population is arrested during the development of an HR and disease symptoms are not observed (Stall and Cook, 1966; Hibberd et al., 1987). The HR is controlled according to the gene-for-gene model of resistance, that is, resistance is controlled by an avirulence gene in the pathogen and a resistance gene in the host (Flor, 1955; Ellingboe, 1984; Minsavage et al., 1990).

As sources of bacterial leaf-spot resistance have been identified, back-crossing these sources into the commercial, bacterial leaf spot–susceptible cultivar Early Cal Wonder was carried out for As sources of bacterial leaf spot resistance have been identified, back-crossing these sources into the commercial, bacterial leaf spot-susceptible cultivar Early Cal Wonder was carried out for *Bs1, Bs2 and Bs3, bs5, bs6 and Bs7*. Near isogenic lines were developed from Early Cal Wonder which became known as ECW10R, ECW20R, ECW30R, ECW12346R and ECW70R. These differential lines were used to identify races 0 to 5 of the pathogen. *Bs4*, which confers resistance to race 6, was identified in Capsicum pubescens PI 235047. Identification of *Bs4* also allowed for differentiation of four additional races, 7–10.

A host differential table (Table 8.1) was developed to identify races from pepper based on reactions on the ECW near-isogenic lines and PI 235047.

TABLE 8.1

Differentiation of Bacterial Spot Races Using Known Resistance Genes in Pepper

		Pepper Differential Lines				
Race	Functional Avirulence Gene	ECW No R Gene	ECW10R *Bs1* Gene	ECW20R *Bs2* Gene	ECW30R *Bs3* Gene	PI 235047 *Bs4* Gene
0	*avrBs1, avrBs2, avrBs3, avrBs4*	S	HRR	HRR	HRR	HRR
1	*avrBs2, avrBs3, avrBs4*	S	S	HRR	HRR	HRR
2	*avrBs1, avrBs2*	S	HRR	HRR	S	S
3	*avrBs2, avrBs4*	S	S	HRR	S	HRR
4	*avrBs3, avrBs4*	S	S	S	HRR	HRR
5	*avrBs1*	S	HRR	S	S	S
6	*avrBs4*	S	S	S	S	HRR
7	*avrBs2, avrBs3*	S	S	HRR	HRR	S
8	*avrBs2*	S	S	HRR	S	S
9	*avrBs3*	S	S	S	HRR	S
10	None	S	S	S	S	S

ECW, Early Cal Wonder; ECW10R, ECW20R, and ECW30R are near isogenic and differ by the presence of the *Bs1*, *Bs2*, and *Bs3* genes, respectively; S, susceptible reaction to *Xanthomonas campestris* pv. *vesicatoria*; HRR, hypersensitive-resistant reaction to *Xanthomonas campestris* pv. *vesicatoria* PI 234057 (*Capsicum pubescens*): *Bs4* gene confers hypersensitive resistance to *Xanthomonas campestris* pv. *vesicatoria* P6 and differentiates *Xanthomonas campestris* pv. *vesicatoria* P1 from *Xanthomonas campestris* pv. *vesicatoria* P7, *Xanthomonas campestris* pv. *vesicatoria* P3 from *Xanthomonas campestris* pv. *vesicatoria* P8, *Xanthomonas campestris* pv. *vesicatoria* P4 from *Xanthomonas campestris* pv. *vesicatoria* P9, and *Xanthomonas campestris* pv. *vesicatoria* P6 from *Xanthomonas campestris* pv. *vesicatoria* P10.

Sources of Infection

The disease is usually introduced first on the seed and is spread by infected seed and water droplets. Large numbers of bacteria occur in the spots and escape as soon as the surface becomes wet. The bacteria, carried in water droplets, can form new spots on the leaves, stems, or fruit where the droplet comes to rest. In wet weather, particularly if strong winds are blowing, the disease may spread rapidly through a crop from a few affected plants. Overhead irrigation acts in a similar manner to rain. Once the bacterium is established in the soil, it may persist for 2 or 3 years. The disease is favored by wet, windy weather and temperatures of 24°C–30°C.

Disease Cycle

Xanthomonas vesicatoria survives from one crop to another mainly on seed, but also in infected debris, for example, stalks. It may be able to survive in the soil to some extent, possibly in the rhizosphere of nonhost plants (Bashan et al., 1982a). Solanaceous weeds may act as alternate hosts.

In glasshouses, seed-borne infection is the only important consideration. Spread is primarily by rain splash or by overhead irrigation, but handling of young plants is also important (Goode and Sasser, 1980). Viable bacteria have been detected in aerosols over commercial fields, showing the possibility of aerial dispersal (McInnes et al., 1988). Leaves are infected through stomata and in fruits through small wounds, for example, abrasions and insect punctures. Only young fruits are infected. The bacterium can multiply epiphytically on young plants in the absence of symptoms. Thinning of directly seeded capsicum seedlings is reported to favor the spread of the disease, and it is recommended to thin in the afternoon when plants are dry and to use prophylactic hand washes (Pohronezny et al., 1990).

On *Capsicum*, it can multiply as a slime on the surface of young fruits, without symptoms, and cause shedding (Bashan and Okon, 1986). Disease is favored by heavy rainfall, high humidity (Diab et al., 1982a), and temperatures above 30°C, but not over 35°C (Diab et al., 1982b). The bacterium can survive on tomato and *Capsicum* seeds for periods of at least 10 years (Bashan et al., 1982b).

The bacteria have a very limited survival period of days to weeks in the soil, and thus their survival is almost always in association with debris from infected or diseased plants. The pathogens have been reported to persist in association with roots of wheat as well as a few weed species; weeds, however, are considered to play only a minor role in pathogen survival. Volunteer tomato plants and possibly pepper volunteers are potentially important sources of inoculum in some locations. In colder regions where vegetative material is killed, the bacteria survive very poorly, if at all. In these areas, reintroduction is primarily on contaminated seed or infected transplants.

The pathogen can survive in association with seed, either externally or internally. On externally infested seed, cotyledons may become infected upon contact with the seed coat and exhibit lesions soon after emerging from the soil. Bacteria are then readily splashed to new foliage and to other plants. Disease is a particular threat in transplant production. Transplant beds favor bacterial spot development because plants are irrigated frequently, crowded together, and humidity is typically high. In a 24-h period, the bacteria can multiply rapidly and produce millions of cells.

Although bacterial spot is a disease of warm, humid regions, it can develop in arid, irrigated regions. The bacteria can be spread by rain or by overhead irrigation. Bacteria also may be spread in water droplets when pesticides are applied with high-pressure sprayers.

Bacteria enter through stomata on the leaf surfaces and through wounds on the leaves and fruit, such as those caused by abrasion from sand particles and wind. Prolonged periods of high relative humidity favor infection and disease development. Symptom development is delayed or eliminated when relative humidity remains low for several days after infection.

Significance

Bacterial spot is one of the most devastating diseases of pepper. The disease occurs worldwide where pepper are grown in warm, moist areas. When it occurs soon after transplanting and when weather conditions remain favorable for disease development, the results are usually total crop loss. Current chemical control is limited to copper or copper combined with maneb sprays that provide only marginal success, thus making the disease very difficult to control once the epidemic is underway. When the disease occurs in commercial pepper fields early in the season, some farmers destroy the entire crop by disking because it is so difficult and economically costly to control once present in the field.

Disease Management

Prevention
The disease caused by the bacterium *Xanthamonas campestrus* pv. *vesicatoria* also occurs in tomatoes. There are three strains and the disease is mainly seedborne. A fraction of a percentage of infected seed can result in 100% infection in the land. As noted, bacterial spot is very hard to control. The best course of action is to ensure that transplanted seedlings are free of the pathogen.

When ordering seed, get an assurance from the seed company that the seed is either certified disease free or has been treated. This can take the form of immersion in hot water at 51°C for 30 min or soaking the seed in a 10% sodium hypochlorite (bleach) solution for the same duration of time. For chili and paprika farmers who keep their own seed, treatment is a worthwhile precaution, even if no symptoms are visible.

After treatment, sow the seed before it is fully dry. This will speed up germination. The disease can remain in undecomposed organic matter, but when a land is cultivated, decomposition is usually complete before a year is up. In this case, it is safe to replant peppers or tomatoes a year later.

Preventing Infection in Seedlings
Seedling nurseries should not buy infected seeds that can spread to other batches before going out to clients.

Seedling growers should always spray capsicums very thoroughly with a copper spray to which streptomycin at 200 ppm is added. Farmers often plant a number of varieties to spread the risk, evaluate new varieties, or produce different colors for specific markets.

The disease is usually not a problem with flood or drip irrigation, unless the climate is warm and moist. If you find any infected plants, use copper sprays immediately, wetting the underside of the leaves as well with drop arms on the sprayer or similar equipment.

Follow the sprayer when starting off to confirm that it is spraying underneath properly. Always use a sticker/spreader in the mixture. If you walk through the lands in the early stages and spot any infected plants, you can sometimes stop the spread by carefully removing these plants. There are varieties resistant to some or all of the three strain, but remember that resistance is not immunity.

Cultural Control
- Plant pathogen-free seeds for transplant production.
- Limit overhead irrigation or, better, produce peppers under furrow or drip irrigation.
- Do not pick or cultivate plants when they are wet.

Chemical Control
1. Treat seed with DryTec Calcium Hypochlorite Granular at 8 oz/gal water. Soak the seeds for 15 min with continuous agitation and then rinse them in potable water for 15 min. Dry the treated seeds to their normal storage moisture.
2. If disease is present, spray with fixed copper during wet periods.
   a. Champ formula 2 at 1.33–2 pints/A.
   b. Cueva at 0.5–2 gal/100 gal water on 7- to 10-day intervals. May be applied on the day of harvest.
   c. Cuprofix ultra 40D at 1.25–3 lb/A on 5- to 10-day intervals.

d. Firewall at 200 ppm beginning when seedlings are in two-leaf stage and continue on 4- to 5-day intervals until transplanted in the field.

e. Kocide 2000 at 1.5–2.25 lb/A or Kocide 3000 at 0.75–1.25 lb/A on 7- to 10-day intervals.

f. Liqui-Cop at 3–6 teaspoons/gal water.

g. Nu Cop 50 WP at 1.5 lb/A on 3- to 14-day intervals. Do not apply within 1 day of harvest.

Biological Control

Biological control options for bacterial spot are limited. However, a biological control method that uses bacterial viruses (bacteriophages) that specifically kill the bacterial pathogens is now available. Treatments with these bacteriophages, marketed as "AgriPhage," have been successful in reducing disease, especially in greenhouse transplant production.

8.4.2 BACTERIAL WILT OF CAPSICUM

Pathogen: *Ralstonia solanacearum*

Synonyms: *Pseudomonas solanacearum*

Bacterial Wilt (BW) caused by *Ralstonia solanacearum* is a serious disease in a wide range of crops.

BW of pepper is known in tropical, subtropical, and warm-temperate zones in Asia, South America, Oceania, and Africa (Elphinstone, 2005), and the isolates virulent to pepper were reported in North America in 2006 (Ji et al., 2007). The disease is also widely observed in Japan (Suzuki et al., 1964; Iwamoto et al., 1988;Osaki and Kimura, 1992; Monma et al., 1993; Horita and Tsuchiya, 2001). Damage has been observed even in the pepper cultivars Fushimi-amanaga and Manganji, which were previously known to be resistant (Matsunaga and Monma, 1999; Hashimoto et al., 2001; Tsuro et al., 2007). This suggests the occurrence of highly virulent strains in Japan. Hashimoto et al. made a survey of 81 isolates of the pathogen classified as biovars 3 or 4 in Kyoto, Japan, and found a highly virulent isolate, strain KP9547, from the pepper cultivar Fushimi-amanaga (Hashimoto et al., 2001).

Symptoms

Aboveground symptoms include wilting of one to two leaves in young plants during the heat of the day. Such plants tend to recover at night. On large-leafed plants, only the tissue on one side of the midvein may wilt. Affected leaves turn yellow and remain wilted after a particular time. The area between leaf veins dies and becomes brown. Usually the main stem of the affected plants remains upright, even though all the leaves may wilt and die (Photo 8.22).

Internal symptoms include light tan to yellow-brown discoloration of the vascular tissue. Long sections of infected stems reveal dark brown to black streaking in the vascular tissue as the disease progresses. As invasion proceeds, the pith and cortex of the stem become dark brown.

The disease occurs in scattered plants or groups of plants in the field. Wilting begins with the youngest leaves during warm or hot weather conditions during the day. The plants may recover, temporarily, in the evening under cooler temperatures. A few days later, a sudden, permanent wilt will occur. The wilted leaves maintain their green color and do not fall as the disease develops. The roots and lower part of the stem have a browning of the water-conducting portion (i.e., vascular system) of the plant. The invaded roots may rot due to infection from secondary bacteria.

Diseased roots or stems that are cut and placed in a small container of water will show a steady, yellowish or gray bacterial ooze coming from the cut end. This bacterial ooze is a key feature in diagnosing this disease. Such oozing is not found with Fusarium-infected plants, which die more gradually and have a drier, firmer stem rot than BW-infected plants. BW does not show an extensive darkening of the external part of the lower stem, which distinguishes it from Phytophthora blight.

PHOTO 8.22 Symptoms of bacterial wilt on capsicum. (Photo courtesy of A.M. Varela, International Centre of Insect Physiology and Ecology (ICIPE), infonet-biovison.org.)

Diagnosis of the Pathogen

Slimy, sticky ooze forms tan-white to brownish beads where the vascular tissue is cut. When an infected stem is cut across and the cut ends held together for a few seconds, a thin thread of ooze can be seen as the cut ends are slowly separated. If one of the cut ends is suspended in a clear glass container of clean water, bacterial ooze will form a thread in the water.

Disease Cycle

Soil is the primary source of disease. The bacterium can survive in soil for extended periods without a host plant.

The bacterium can also survive in diseased crop debris. The bacteria are released from the roots of the affected plant into the soil and can infect neighboring plants. Many weeds may harbor the bacteria in the roots and yet show no symptoms.

The bacterium enters pepper tissue through wounds on the roots arising from cultivation, natural wounds at emergence of lateral roots, insect feeding, and nematode feeding. When the diseased plant is removed from the field, the infected root pieces that remain in the soil provide bacteria for infection of new roots. The bacterium disperses through furrow irrigation or surface water, cultivation, transplanting, cutting/wounding, and pruning. Infested soil may be transported with seedlings, farm implements, or shoes of farm workers. Seed transmission in pepper is not considered important.

High temperatures (30°C–35°C) and high soil moisture favor disease development. High soil moisture increases the survival of *Ralstonia solanacearum* in soil, the rate of infection, the disease development after infection, and the number of bacterial cells released from the host into the soil. BW is a greater problem in heavy soils and in low-lying areas that can retain soil moisture for long periods.

Disease Management

Use an integrated approach to control this disease since the organism has many strains/races and a very wide host range (e.g., tomato, potato, tobacco, eggplant, banana, plantain, peanut, sweet potato, and many weeds). Avoid the use of contaminated water for irrigation. Do not irrigate a field with water containing runoff from other affected fields. Avoid contaminated land. If possible, plant during the cooler parts of the growing season.

Suitable rotations can only be determined through local experience because of the diversity of *Ralstonia solanacearum* strains and races, and the many agroclimatic zones where reports occur. Rotations of several years duration with maize, cotton, soybeans, grasses, and rice are used in various areas.

Eradicate weed hosts. Remove wilted plants, root debris, and volunteer hosts and burn them to reduce spread of the disease from plant to plant. Clean farm equipment after working in an infested field. Disinfest tools when used in an infested field. Wash with water or bleach or sterilize by flame. Wash the soles of shoes after working in an infested area. Work in the infested portion of a field after working in the noninfested areas.

For transplant production, use disease-free transplants, pasteurized soil medium, or fumigated plant beds. Avoid movement of infested soil or contaminated plant material into the nursery bed. Use proper sanitation measures for transplant production, and avoid damage to roots during transplanting.

Soil amendments may reduce BW in some locations. Consult with your local extension agent to determine possible treatments that may reduce the disease in your location.

Control root-knot nematodes and root-feeding insects since they may help the disease to establish and spread.

Growing and propagating from pathogen-free plant material is the main way to avoid problems with *Ralstonia*, regardless of the race and biovar involved. Propagators must use pathogen-free potting soil or other media, establish stock plants that are tested and known to be free of the bacteria, train workers handling the stock plants in methods and procedures that prevent the pathogen from contaminating the potting soil or coming in contact with the stock plants, and then maintain this system throughout the propagation phase of crop production.

There are no chemicals or biological agents that adequately control these bacteria. Infected plants must be discarded as soon as possible.

As is the case with all pathogens carried on vegetatively propagated crops, the purchaser of cuttings or prefinished plants must isolate all new, incoming plants as if the health of the plants were unknown, even if the plants have been certified as healthy. New plants must not be commingled or dispersed among other plants in the greenhouse from other sources. This procedure is crucial because keeping plants originating from one source together allows you to observe those plants as a group, detect any abnormalities within that group, and treat or discard those plants as a group without affecting or damaging plants from other sources. Keeping them together as a group in a defined area of the greenhouse also limits the area that may need to be quarantined, sanitized, or isolated, should a pathogen requiring a "stop sale" (such as *Ralstonia solanacearum*) be found.

Sources of resistance have been identified in sweet bell pepper. Variations in race and strain of *R. solanacearum* make it difficult to utilize these varieties in some regions.

Matos et al. found MC4 and MC5 to be resistant (Matos et al., 1990). MC4 was resistant to various isolates of biovars 1 and 3 and has been recommended for breeding programs in Brazil (Quezado-Soares and Lopes, 1995; Lopes et al., 2005). A resistant cultivar, Mie-Midori, and its progeny have been extensively used in the breeding of resistant cultivars (Matsunaga and Monma, 1999). In a preliminary report, a Malaysian accession, LS2341, was found to be resistant (Mimura et al., 2000). After screening 30 genotypes, two were found to be highly resistant to BW in India (Singh and Sood, 2004). However, most of these tests were done using only one or two isolates of the pathogen (Wang and Berke, 1997; Matsunaga and Monma, 1999; Fatima and Joseph, 2001; Singh and Sood, 2004). Because the pathogen of BW has a wide range of virulence, it is uncertain whether these resistant lines would be effective against various isolates collected from other regions.

8.4.3 BACTERIAL SOFT ROT OF CAPSICUM

Pathogen: *Pectobacterium carotovorum subsp. carotovorum*

Synonyms: *Erwinia carotovar* pv. *caratovara*

Other hosts: Wide range of plant species including many vegetable crops like tomato, chili, eggplant

Symptoms

This disease affects pepper stems and fruits. Internal discoloration appears on the stem, followed by hollowing-out of the pith and wilting. As lesions expand along the stem, branches break. Foliar chlorosis and necrosis may also develop. Symptoms of postharvest decay start as sunken, water-soaked areas around the edge of wounds or on the stem end next to the peduncle. These areas may be light or dark and become soft as they rapidly expand. Often, the epidermis splits open, releasing watery, macerated tissue.

Initial signs in capsicum leaves show darkened veins followed by leaf chlorosis and necrosis. The pith and vascular system within nearby stems may show dark-brown discoloration. As the disease progresses, dry, dark-brown or black stem cankers develop, often resulting in the breakage of branches. Bacterial ooze may be evident from diseased tissues, but this is not always the case. The affected plants wilt and die.

Soft rot frequently begins in the peduncle and calyx tissues of harvested fruit, but infection can occur through wounds anywhere on the fruit. Internal tissue near the site of infection softens and the lesion rapidly expands reducing the interior of the fruit to a watery mass in a few days. Fruits infected on the plant often collapse and hang on the plant like a water-filled bag (Photo 8.23). When the contents leak out, a dry shell of the pod remains.

The fleshy fruit peduncle is highly susceptible and most often is the point of infection. Both ripe and green fruit may be affected. Initially, the lesions on the fruit are light-to-dark colored with a water-soaked appearance, and somewhat sunken. The affected areas expand very rapidly, particularly under warm (25°C–30°C) and wet environments. In later stages, bacterial ooze may develop from affected areas, and secondary organisms often invade the rotted tissue. Affected fruits hang from the plant-like soft, water-filled bags and can give off a rotting odor.

Geographical Distribution

Worldwide.

PHOTO 8.23 Symptoms of bacterial soft rot on capsicum fruit. (Photo courtesy of Dipak Nagrale, NBAIM, Mau, Uttar Pradesh, India.)

Predisposing Factors

The bacterium is a common soil inhabitant and may survive on the surface of seed to act as a source of infection. The pathogen is spread by irrigation water and contact between fruit. The bacteria can enter through wounds caused by machinery, insect feeding, or natural openings. Insects such as flies can also spread bacteria.

Control Measures

In greenhouse operations, provide adequate air circulation to help reduce relative humidity. Avoid injuries to plants during the growing season and on fruits during harvest. Improved sanitation in the field and in packing houses is effective in reducing losses. All harvest equipment, the packing line, and packing boxes should be sanitized frequently. Dump tank water and packing line washers should maintain a minimum available chlorine concentration of 150 ppm at a pH of 6.0–7.5. Wet fruit should be dried promptly before packing and then cooled quickly to below 10°C.

8.4.4 BACTERIAL CANKER OF CAPSICUM

Pathogen: *Clavibacter michiganensis* subsp. *michiganesis*

Symptoms

Symptoms of bacterial canker in pepper include leaf and fruit spots and, less frequently, systemic wilt. In localized infections, symptoms first appear as small blisters or raised white spots on leaves and stems. Later, the centers of the leaf spots become brown and necrotic and develop a white halo. Stem lesions often develop a crusty appearance and elongate to form cankers. Symptoms on fruits first appear as very small, round, slightly raised spots. These spots gradually increase in size and may develop a brown center and a white halo. When these are numerous, spots merge and take on a crusty appearance (Photo 8.24). In systemic infections, a gradual wilting occurs followed by plant death.

Geographical Distribution

Australia, Brazil, China, Israel, South Korea, and the United States (California, Indiana, Michigan, and Ohio).

Predisposing Factors

The bacterium enters the plant via wounds and stomata. *Clavibacter* may be seedborne and may infest the seed externally or under the seed coat. High relative humidity and daytime temperatures between 25°C and 30°C generally favor the disease. Dense plant populations and overhead irrigation also provide an ideal environment to spread the bacterium. Insects, tools, and human contact may also aid the spread.

PHOTO 8.24   Field symptoms of bacterial canker on capsicum fruit and leaves. (Photo courtesy of semena.org.)

Control Measures

Sow only tested seed and certified transplants. Do not transplant peppers into ground used for tomatoes during the previous season. Clean cultivation equipment before entering a new field, avoid entering fields when foliage is wet, and incorporate plant debris immediately after harvest to help reduce losses. Never harvest fruits from symptomatic plants. Rogue all symptomatic and adjacent plants. Rotate to a nonhost for a minimum of 3 years if the disease is found in a field.

8.4.5 Syringae Seedling Blight and Leaf Spot of Capsicum

Pathogen: *Pseudomonas syringae* pv. *syringae*

Symptoms

Affected leaves or cotyledons develop irregular, water-soaked lesions that later become necrotic, turning dark-brown with a light center (Photo 8.25). Lesions may coalesce to form relatively large necrotic areas. Lesions with chlorotic halos are rare. However, under heavy disease pressure, large areas of the leaf may be affected and the whole leaf may turn yellow prematurely and drop. Infected fruit develop brownish-black, watery lesions that expand and rot. Symptoms of *Syringae* seedling blight can be confused with those caused by bacterial spot. However, the lower temperatures at which *Syringae* seedling blight occurs can help differentiate these two diseases.

Geographical Distribution

Southern and Southeastern Europe, Southern United States.

Predisposing Factor

Temperatures between 16°C and 24°C and high humidity favor *Syringae* seedling blight. Bacteria are spread by splashing water and enter the plant through natural openings or wounds.

Control Measures

Avoid low temperature and high humidity conditions in nurseries. Inspect seedlings for symptoms before transplanting to avoid introducing the disease to the field. Avoid overhead irrigation whenever possible.

8.5 CAULIFLOWER

8.5.1 Bacterial Fasciation of Cauliflower

Pathogen: *Rhodococcus fascians* (Tilford, 1936; Goodfellow, 1984)

Synonyms: *Corynebacterium fascians*

PHOTO 8.25 Symptoms of *Pseudomonas syringae* on capsicum leaves. (Photo courtesy of semena.org.)

Common names: Bacterial fasciation, leafy gall, cauliflower disease

Symptoms

Fasciation and formation of short, fleshy shoots and cauliflower-like galls under the influence of plant hormones (mainly cytokinins) are produced by *Rhodococcus fascians*. Galls and malformed shoots often start from the hypocotyls and are found close to soil/air or below the surface.

Host

Cabbage (*Brassica* spp.), carnation (*Dianthus* spp.), dahlia (*Dahlia* spp.), gladiolus (*Gladiolus* spp.), lily (*Lilium* spp.), mullein (*Verbascum* spp.), pelargonium (*Pelargonium* spp.), and sweet pea (*Lathyrus* spp.)

Geographical Distribution

Europe: Belgium, Czech Republic, Denmark, Estonia, France, Germany, Hungary, Italy, Latvia, the Netherlands, Norway, Russia, Slovakia, Sweden, Ukraine, the United Kingdom.

Asia: India, Iran.

Africa: Egypt.

America: Canada, Colombia, Mexico, the United States.

Oceania: Australia; New Zealand.

Disease Cycle

The bacterium lives outside or in the superficial tissues of the galls in small pockets. Wounds are not necessary for penetration. Transmission can take place via soil, infected planting material, or contaminated tools. The bacterium can survive in the soil, for up to several years, and can be present in seeds (Baker, 1950; Faivre-Amiot, 1967).

Economic Impact

The disease was first described from sweet pea by Brown in 1927 from the United States and the bacterium was described by Tilford in 1936. Damage for plants is usually low.

Control Measures

Healthy planting material, hygiene, and (steam) disinfection of growth substrate or soil are ways to control diseases caused by *Rhodococcus fascians*.

8.5.2   BACTERIAL LEAF SPOT OF CAULIFLOWER

Pathogen: *Pseudomonas maculicola (maculloch) stevans*

Synonyms: *Pseudomonas syringae* pv. *Machlicola*
　　　　　　Bacterium maculicolum McCulloch
　　　　　　B. maculicola (McCulloch) McCulloch
　　　　　　B. maccullochianum (McCulloch) Burgwitz
　　　　　　Phytomonas maculicola (McCulloch) Bergey et al.

Symptoms

Leaves of infected plants become covered with small irregularly shaped brown spots with yellow halos. This stage is difficult to see on the upper leaf surface, whereas the leaf underside clearly shows water-soaked brown spots. The small spots coalesce into large brown papery areas with yellow borders that tear, giving the leaves and plant a ragged appearance.

Bacteriosis attacks cotyledons, leaves, stems, peduncles, pods, and seeds. There are oily dark spots on cotyledons (they are better visible on the lower side of leaves). Adult plants have angular, slightly pressed, dark brown or violet (to black) spots with oily shade and semi-transparent limb. As a rule, leaves of strongly diseased plants are curtailed, turn yellow, dry up, and fall down. There are longitudinal black spots on the infected stems, sheet pedicles, and flowers. Symptoms of the disease are shown as small gray-brown spots on inflorescences. On infected heads of cauliflower, there are dark-brown spots, which may embrace the head completely at high air humidity in 2–3 days. Infected heads quickly decay. Black blight spots are observed on pod valves, both on external and internal surfaces of the diseased plants. The bacterial infection passes from pods to seeds, which become black.

Pathogen
Cells of *Pseudomonas syringae* pv. *maculicola* are straight bacilli, moving by one to five flagella, usually $0.9 \times 1.5–3.0$ μm in size, and Gram-negative and aerobic, forming a fluorescing pigment. On potato agar, colonies are round, brilliant, smooth, transparent in transmitted light, opalescent; later they become dirty-white with dense and raised center. Few characteristics of this pathogen are as follows: they do not curdle and peptonize milk; do not restore nitrates; do not form N_2S; produce indole; hydrolyze starch poorly; form NH_3; and utilize glucose and saccharose with production of acid, but do not utilize lactose and maltose. The optimum temperature for growth is 24°C–25°C—maximum is 29°C and minimum is lower than 0°C.

Geographical Distribution
Bacterial leaf spot of cauliflower is present in many countries (the United States, Denmark, Finland, Bulgaria, the Great Britain, Italy, China, Taiwan, etc.) where this crop is cultivated. Bacteriosis is also found in territories of the former Soviet Union.

Predisposing Factors
The infection/pathogen survives on vegetation residues and seeds. Secondary infection takes place into the plant through injuries.

Increased humidity (90% and higher) and temperature (17°C–20°C) play a significant role in bacteriosis development. Favorable weather conditions for pathogens of bacteriosis may provoke epiphytotics, which may destroy plantings of cauliflower. High severity of the disease is possible on seed shoots of late cauliflower planting, if the second half of vegetation period is excessively rainy.

Economic Significance
In addition to cauliflower, the pathogen attacks Brussels sprouts, and both white- and red-head cabbage. Yield losses depend on the cultivated variety and conditions of cultivation of plants. In Leningrad Region, cauliflower seeds affected by bacterial infection ranges from 2% to 35%, and infection of late-landed seed material by this pathogen can reach 100% in some vegetation seasons under favorable conditions. In Lithuania, the percentage of diseased plants is 17%–33%. Infection of some varieties of cauliflower reaches 64% in countries such as Ukraine that has favorable conditions; as a result, the yield and quality of marketable heads and seeds are reduced.

Control Measures
Control measures include optimal agriculture, maintenance of crop rotation, cultivation of relatively resistant varieties. Use seeds that are certified free from black rot bacteria, carefully remove plant residues, separate seeds from shrunken grains, treat the seeds with pesticides before sowing, and treat the plants with pesticides during vegetation period.

8.5.3 BACTERIAL SOFT ROT OF CAULIFLOWER

Pathogen: *Erwinia cartovora*

Symptoms
Water-soaked lesions on leaves and flower heads that expand to form a large rotted mass (Photo 8.26); surface of lesions usually crack and exude slimy liquid that turns tan, dark brown, or black on exposure to air.

PHOTO 8.26   Symptoms of bacterial soft rot on cauliflower. (Photo courtesy of Gerald Holmes, California Polytechnic State University, San Luis Obispo, CA, Bugwood.org.)

Predisposing Factors
Bacteria are easily spread on tools and by irrigation water; disease emergence favored by warm, moist conditions.

Disease Management
Chemical treatments are not available for bacterial soft rot and control relies on cultural practices; rotate crops; plant in well-draining soils or raised beds; only harvest heads when they are dry; avoid damaging heads during harvest.

8.6   CABBAGE

8.6.1   Black Rot of Cabbage

Pathogen: *Xanthomonas campestris* pv. *campestris*

Common names: Blight, black stem, black vein, stem rot, stump rot, and black rot.

Black rot, caused by the bacterium *Xanthomonas campestris* pv. *campestris*, is one of the most destructive diseases of cabbage and other crucifers. Cauliflower, cabbage, and kale are among the crucifers most susceptible to black rot.

Damage
In Kenya, black rot is endemic and causes much damage (Onsando, 1988, 1992). The disease is considered of intermediate economic importance in Mozambique (Plumb-Dhindsa and Mondjane, 1984). Black rot is widespread in Zimbabwe where it is considered the most important disease of Brassicas (Mguni, 1987, 1995).

Symptoms
Plants may be affected with black rot at any stage of growth. Seedling infection first appears as a blacking along the margin of the cotyledon (seed leaves), which later shrivels and drops off. Affected seedlings turn yellow to brown, wilt, and collapse. Leaf infections often result in a small, wilted, V-shaped infected area that extends inward from the leaf edge toward the midrib. Diseased areas enlarge and progress toward the base of the leaf, turn yellow to brown, and dry out. The veins of infected leaves, stems, and roots turn black as the pathogen multiplies (Photo 8.27).

In young plants raised from infected seeds, cotyledons show dark discoloration in the margins, which later turn black, shrivel, and drop off. In older seedlings, infection usually takes place

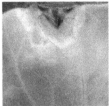

PHOTO 8.27   Symptoms of black rot bacterium on cabbage leaves and plant. (Photo courtesy of R.A. Yumlembam and S.G. Borkar, PPAM, MPKV, Rahuri, Maharashtra, India.)

from the lower leaves through the hydathodes. Large, yellow to dark-brown lesions form along the margins of the lower leaves. These lesions develop in the shape of a wedge or "V" around a hydathode. Initial lesions are surrounded by an indistinct pale-green, withered area. As disease advances, veins turn dark brown to black with a narrow yellow halo (Goto, 1992). Wilting of leaves and early defoliation and occasionally mishappen or deformed plants can be observed. Leaf spots may appear dry with a brown center. A network of black or brown veins may appear (depending on cultivar and bacteria strain) in the leaf spots and often in advance of yellow zones. The name black rot originates from the occurrence of a dark vein and black, rotting plants (Schaad and Thaveechai, 1983).The midribs when cut open show internal discoloration of the vascular tissues. Chlorotic spots or "pale mottle" (Cook et al., 1952; Alvarez et al., 1994) may appear on young leaves before the typical lesions, wilting, and blackening of veins without yellowing can also occur. Symptomless plants are common during the vegetative period until flowering. Pods from infected plants may show shriveling and darkening of irregular areas. Early invasion of the pods usually results in the abortion of all seeds (Cook et al., 1952). During periods of high temperature when the plant is near maturity, soft rot may occur. Soft rot may also be due to secondary invasion of *Pseudomonas* or *Pectobacterium* species.

The type strain and several other clones of *Xanthomonas campestris* pv. *campestris* are known to cause a distinctive blight symptom that later develops into black rot. Other strains of the pathogen cause typical black rot symptoms but not blight (Alvarez et al., 1994). Mguni et al. (1996) isolated three Biolog types (*X. campestris* pv. *campestris*, *X. campestris* pv. *campestris* A, and *X. campestris* pv. *campestris* B) from *Brassica* spp. cultivated in Zimbabwe. Type *X. campestris* pv. *campestris* A strains were normally associated with plants showing initial blight symptoms. Biolog type *X. campestris* pv. *campestris* B was mostly isolated from plants showing wilt and stem rot symptoms, while Biolog type *X. campestris* pv. *campestris* was associated with "V"-shaped lesions.

From the affected plant parts, bacterial ooze may also be seen. If the heads are already formed, soft rot of the head may set in.

Host Plants/Species Affected

Brassica juncea var. *juncea* (Indian mustard), *B. napus* var. *napobrassica* (swede), *B. napus* var. *napus* (rape), *B. nigra* (black mustard), *B. oleracea* var. *alboglabra* (Chinese kale), *B. oleracea* var. *botrytis* (cauliflower), *B. oleracea* var. *capitata* (cabbage), *B. oleracea* var. *gemmifera* (Brussels sprouts), *B. oleracea* var. *gongylodes* (kohlrabi), *B. oleracea* var. *sabauda* (Savoy cabbage), *B. oleracea* var. *viridis* (collards), *B. rapa* cultivar group *Mizuna*, *B. rapa* subsp. *chinensis* (Chinese cabbage), *B. rapa*subsp. *oleifera* (turnip rape), *B. rapa* subsp. *Pekinensis*, *B. rapa*subsp. *rapa* (turnip), *Capsella bursa-pastoris* (shepherd's purse), *Crambe*, *Erysimum cheiri* (wallflower), *Lepidium sativum* (garden cress), *Lepidium virginicum* (Virginian peppercress), *Matthiola incana* (stock), *Raphanus raphanistrum* (wild radish), *Raphanus sativus* (radish), and *Sinapis arvensis* (wild mustard).

Black rot is a pathogen of most cultivated *cruciferous* plants and weeds. Cauliflower and cabbage are the most readily affected hosts in the *crucifers*, although kale is almost equally susceptible.

Broccoli and Brussels sprouts have intermediate resistance and radish is quite resistant, but not to all strains. Kohlrabi, Chinese cabbage, rutabaga, turnip, collard, rape, jointed charlock (*Raphanus raphanistrum*), and mustard are also susceptible hosts.

Geographical Distribution

The pathogen that causes *black rot* is widely distributed in Africa, Asia, Australia and Oceania, Europe, North America, Central America, the West Indies, and South America. Black rot is endemic in Africa. It is the most important disease of Brassicas in Kenya, Zambia, and Zimbabwe (CABI, 2005).

It is also present in the Philippines, Norway, Australia, the United States (New York, Mississippi, Delaware, California, Florida), Mauritius, Bulgaria, Holland, South Africa, Rhodesia, Queensland, Mozambique, Greece, Israel, Yugoslavia, Hawaii, Thailand, Basilicata, and India.

Predisposing Factors

Symptoms of black rot vary considerably depending on the host, cultivar, plant age, and environmental conditions. The bacteria can enter plants through natural openings and wounds caused by mechanical injury on roots and leaves. Seed-borne bacteria infect the emerging seedlings through pores on the margin of the cotyledons and then spread systemically through the seedling. Infected seedlings grown in the greenhouse under cool conditions (below 15°C–18°C) frequently do not show any symptoms of the disease. When infected seedlings are transplanted to the field and temperatures rise to 25°C–35°C during periods of high relative humidity (80%–100%), they become stunted with dead spots on the cotyledons and may eventually wilt and die. In regions with temperate climates (where temperatures remain cool), disease symptoms on infected seedlings may not always be obvious or appear severe. Infected seedlings grown under cool conditions may ooze bacteria from pores and lesions, which then serve as a source of the pathogen for neighboring plants.

The bacteria produce a sticky polysaccharide called xanthan that eventually plugs the vascular tissue inside the veins causing them to collapse and turn black. During hot humid environmental conditions, the bacteria can move from the leaf into the stem through the xylem. Once inside the stem, the bacteria can move up or down to other parts of the plant including the roots. Systemically infected plants may produce chlorotic areas anywhere on the leaf. Severely infected leafy cole crops such as kale and cauliflower tend to shed their leaves from the bottom up leaving only a tuft of distorted leaves separated from the root system by a scarred barren stem.

Disease Cycle

The pathogen *Xanthomonas campestris* pv. *campestris* lives on the plant residue in the soil or in the seed. In a germinating seed, the bacterium enters through the stomata of the cotyledons and spreads first intercellularly in the substomatal region and then systemically into the vascular bundles. Bhide (1949) studied the mode of entrance of the bacterium in cabbage and reported that the organism invades the host normally through the hydathodes situated in the leaf margins.

Bacteria are spread within a crop primarily by wind-blown and splashing water and by workers, machinery, and occasionally insects. *Xanthomonas* campestris can survive on leaf surfaces for several days until dispersed to hydathodes or wounds, where infection can occur. Bacteria enter leaves through hydathodes when water exuded through these pores, at the leaf margin, during the night which is drawn back into the plant in the morning. Bacteria can enter leaves in 8–10 h, and wilt symptoms are visible as soon as 5–15 h later. Wounds, including those made by insects feeding on leaves and by mechanical injury to roots during transplanting, also provide entry sites. Wounds on roots are most important when transplants are dipped in water or the soil becomes saturated. Flea beetles can transmit *X. campestris* but were found to be ineffective vectors in New York. Bacterial movement into plants through hydathodes is restricted in resistant varieties; consequently, there are fewer infection sites and/or the affected area is much smaller in resistant varieties than susceptible varieties.

The secondary spread of the disease in the field is mainly through irrigation water, wind, implements, tools, and so on (Cook et al., 1952; Onsando, 1992). The disease cycle of black rot of cabbage is illustrated in Figure 8.2.

Disease Spread

Seed contaminated with black rot bacteria is considered the most important source of the pathogen and significantly contributes to the spread of this disease worldwide. As few as 3 infected seeds per 10,000 (0.03% infected seeds) can result in a black rot epidemic. Seeds should be tested and certified to be disease free with less than 1 in 30,000 infected seed.

The organism survives in infected crop tissue left on the soil until the crop tissue rots. However, the bacteria do not survive very long in soil as unprotected free living organisms. The black rot bacteria can also infect and survive on many crucifer weeds. This also contributes to the persistence and spread of the disease. It can grow and multiply on host tissue without infecting or causing disease.

Rain-splashed bacteria from contaminated plant residue left on the soil or from neighboring diseased plants is the primary method of disease spread throughout a field. The bacteria enter and exit through water-secreting glands called hydathodes located at the edges and tips of leaves. Hydathodes often produce a drop of water during periods of high humidity early in the morning. The pathogen spreads very quickly when rain droplets contaminated with bacteria splash onto healthy leaves and enter the hydathodes. The bacteria move into the leaf veins through hydathodes and begin to multiply, rot, and plug the veins. Contaminated water droplets that exude out of hydathodes of infected leaves can then be rain-splashed to other plants.

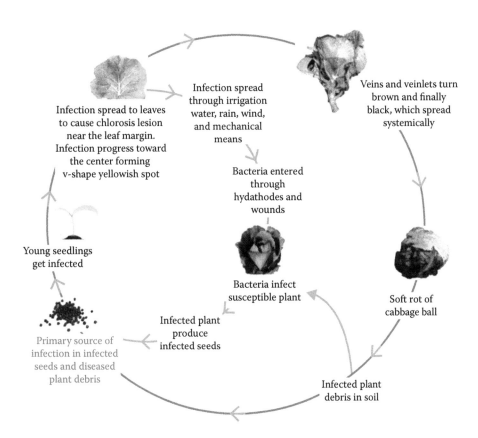

Infection spread through irrigation water, rain, wind, and mechanical means

Infection spread to leaves to cause chlorosis lesion near the leaf margin. Infection progress toward the center forming v-shape yellowish spot

Veins and veinlets turn brown and finally black, which spread systemically

Bacteria entered through hydathodes and wounds

Young seedlings get infected

Bacteria infect susceptible plant

Soft rot of cabbage ball

Primary source of infection in infected seeds and diseased plant debris

Infected plant produce infected seeds

Infected plant debris in soil

FIGURE 8.2 Disease cycle of black rot of cabbage caused by *Xanthomonas campestris* pv. *campestris*.

Black rot is more severe and widespread in fields that receive frequent early morning rains, particularly in May and June. Equipment, people, animals, and overhead irrigation can further spread the disease. Insects can also spread the bacteria; however, their contribution to the spread of black rot is limited.

Economic Impact

Black rot of cabbage caused by *X. campestris* pv. *campestris* cause huge losses of up to 50%–70%.

This disease is difficult for growers to manage and is considered the most serious disease of crucifer crops worldwide. The disease can cause significant yield losses when warm, humid conditions follow periods of rainy weather during early crop development. Late infections can provide a wound for other rot organisms to enter and cause significant damage during storage.

Black rot is considered the most important worldwide disease of crucifers. The disease is known to exist in the cool coastal climates of northern Europe and North America but was seldom a problem there until the 1990s. Its potential for crop damage is also considered low in New Zealand and parts of Australia. In many regions of Central and Eastern Europe, Central Asia (Kazakhstan), China, and tropical and subtropical regions of Asia, Africa, and South America where Brassicas are common and cultivated without crop rotation, black rot is always present. Seed production in those regions is commonly associated with high levels of seed-borne *X. campestris* pv. *campestris* (Williams, 1980). Even minor, visually undetectable development of black rot may considerably increase damage to plants by soft rot caused by *E. carotovora*, *Pseudomonas* spp., and other opportunistic pathogens (Dzhalilov et al., 1989).

For many years, the disease was considered of relatively minor importance to crucifer growers in the major northern production areas of the United States and Western Europe. Outbreaks of the disease were sporadic and limited. During the late 1960s, early 1970s, and 1990s, the frequency and severity of the disease increased. Approximately 70% of several million transplants from one single seedbed were systemically infected in the United States in 1973 (Williams, 1980). In 1976, losses of $US 1 million were estimated (Kennedy and Alcorn, 1980). In Florida, two cabbage crops are commonly grown every year. If temperatures remain cool in the late winter and early spring, black rot does not become a problem, but if temperatures turn warm, serious outbreaks often occur (Schaad, 1988). In Illinois and Virginia, black rot is one of the most important diseases of cabbage and cauliflower (Lambe and Lacy, 1982; Eastburn, 1989). In Canada, rutabaga (swede) producers lost up to 60% of their crop to black rot during the winter of 1979–1980 (McKeen, 1981).

Nemeth and Laszlo (1983) reported black rot as the cause of considerable damage in cabbage and cauliflower in Hungary. Radunovic and Balaz (2012) reported the presence of black rot in cabbage, kale, broccoli, and collard crops. In some regions of Russia, black rot caused 23%–57% losses on susceptible cabbage cultivars (Ignatov, 1992; Dzhalilov et al., 1989). Recurrent black rot epidemics have been reported from Italy during 1992–1994 and 1997 (Caponero and Iacobelis, 1994; Scortichini et al., 1994; Catara et al., 1999). In Israel, black rot causes major economic losses in cabbage, cauliflower, radish and kohlrabi, especially during the winter season (Kritzman and Ben-Yephet, 1990).

In Korea, black rot is considered an important disease of cabbage (Kim, 1986). Surveys conducted in 25 crucifer fields in eight provinces in Thailand, where black rot is known to cause severe losses, revealed the presence of the black rot organism from plants showing disease symptoms in 21 fields (Schaad and Thaveechai, 1983). Infected seed lots were reported from commercial seed lots in Japan (Shiomi, 1992) and in 1997–1998 black rot infected from 50% to 90% of plants of susceptible cabbage cultivars grown in three prefectures of Japan (Ignatov et al., 1997). During 1989–1992, *X. campestris* pv. *campestris* caused seed yield reductions in cauliflower in India (Shyam et al., 1994). In Himachal Pradesh, curd rot of cauliflower has been a menace to the seed crop and is the cause of huge losses to farmers in India (Shyam et al., 1994). Black rot appears annually in Manipur near the end of February. Its effects are more severe (up to 50% losses) in susceptible cultivars (Gupta, 1991). The widespread occurrence of black rot in Rajasthan, with a high incidence of seed infection, can be the cause of severe losses (Sharma et al., 1992).

In Kenya, black rot is endemic and the cause of much damage (Onsando, 1988, 1992). The disease is considered of intermediate economic importance in Mozambique (Plumb-Dhindsa and Mondjane, 1984). Black rot is widespread in Zimbabwe where it is considered the most important disease of Brassicas (Mguni, 1987, 1995). The pathogen was found in crucifer crops from the five agro-ecological regions of the country. Disease incidence was higher during 1994 (10%–80%) than during 1995 (10%–50%) (Mguni, 1996). In Mozambique, the disease was reported in the southern districts of Boane, Mahotas, and Chòkwé. In the Boane district, the highest incidence of black rot was recorded on variety Copenhagen Market (70%), Starke (67.9%), and Glory F1 (67.3%). In Chòwké, Tronchuda (Portuguese kale) was the least affected *Brassica* crop (Bila et al., 2009).

Disease Management

Black rot management begins with identifying potential disease sources and utilizing an Integrated Pest Management strategy including host resistance, planting disease-free seed, avoiding the spread of the disease, and sanitizing properly. Sanitation is the main method that reduces, excludes, or eliminates the initial sources of disease. General sanitation practices include crop rotation, disinfecting seed, rouging diseased plants, elimination of refuse piles, and eradication of alternative hosts.

Seed Treatment

Seed-borne inoculum significantly contributes to the spread of black rot bacteria. Growers should only plant tested certified seed <1 infected seed in 30,000 or 0.003% contamination. When the infection level of seeds is not known or disease-free seeds are not available, then the seeds should be treated to eliminate the bacteria. Growers who purchase transplants should request for proof that the seedlings were grown from disease-free or treated seed. During transplanting, diseased seedlings should not be planted in the field.

Seed treatments do not always eliminate 100% of the bacteria on or in the seed and may adversely affect seed germination and vigor. Soaking seeds in hot water at 50°C for 25–30 min is the most effective treatment for seed-borne black rot control. Weak seed, seed stored for several years, and seed of certain crucifer crops, such as cauliflower, kohlrabi, kale, rutabaga, and summer turnip, may be damaged by hot-water treatment; soak for 15 min at 50°C only.

The effect of the hot-water seed treatments on every variety of each individual crucifer crop has not been investigated. Growers are encouraged to treat a small portion of seed and plant in pots to determine the effect of the seed treatment on germination and vigor, prior to treating the entire seed lot.

Avoid Disease Spread

Use new seed trays each year to avoid contaminating the current year's crop with residual black rot bacteria from the previous year. If purchasing new trays each year is not economically feasible, used trays can be sterilized with steam, boiling water, or chemical disinfectants to eliminate potential contamination. Destroy infected seed trays immediately to prevent disease spread to other seedling trays.

Avoid soaking crates or bundles of transplant seedlings in tubs of water before transplanting. The black rot bacteria can spread from diseased to healthy seedlings by infecting leaf scars and wounds on roots when soaked in water.

Black rot bacteria can contaminate the surface of clothing, equipment, tools, and water sources. Reducing seeding rates and densities to promote good air circulation, facilitating the quick drying of plants, timing irrigation when plants will dry quickly, and restricting field activities until later in the day when fields are dry will help reduce disease spread. Working in diseased fields last will also avoid disease spread from infected to noninfected fields. Wash and disinfect equipment before moving from one field to another.

Field Selection

Field selection is very important due to the distance the pathogen can spread. Whenever possible, select fields as far away from fields grown to crucifer crops the previous year. Select fields that are well drained and that do not receive run-off water from areas or fields, where crucifers have been grown previously. Well-drained, light soils are best for crucifer production because they can be worked early in the season and facilitate earlier planting of transplants. Planting early can help avoid disease because environmental conditions are usually not conducive for the development and spread of black rot bacteria.

Crop Rotation

Planting disease-free, treated seeds or seedling transplants does not necessarily ensure a disease-free crop in the field. Crop rotation is also an important management tool. Black rot bacteria can survive in infected crop tissue in soil until the crop tissue breaks down and rots. The time required for crucifer crop debris to rot varies between regions depending on the temperature, amount of soil moisture, and soil type. For example, in the states of Georgia and Washington, which experience long, warm summers, it has been estimated that free-living bacteria can survive in infested soil for about 60 days and up to 615 days in infested host debris. The bacteria can survive longer in soil during cool, wet seasons than during hot, dry seasons. In Ontario, a 3-year rotation is recommended.

Weed Control

Black rot bacteria can infect and survive on many crucifer weeds including bird rape (*B. campestris*), Indian mustard (*B. juncea*), black mustard (*B. nigra*), shepherd's purse (*Capsella bursa-pastoris*), globe-podded hoary cress (*Cardaria pubescens*), pepper grass (*Lepidium densiflore*), and wild radish (*Raphanus raphanistrum*). Disease symptoms on weeds vary from small yellow V-shaped lesions on leaf margins to no visible symptoms. The pathogen can spread up to 30 m from infected plants (including weed hosts) to healthy plants. The pathogen not only infects and spreads from weeds to cruciferous crops but it can also survive on weed seeds and can grow and multiply on weed leaves without infecting or causing disease. Good weed control within fields will aid disease management; however, careful attention to weed control in ditches and along fence rows is also important.

Insect Control

The crucifer flea beetle (*Phyllotreta cruciferae*) can transmit black rot bacteria from infected plants to healthy ones; however, their importance in the spread of the disease is limited. Wounds caused by insects provide an entry point for the disease to infect plants during heavy dews or periods of rain. Insect control will help reduce the spread and severity of disease.

Cull Pile Management

Infected refuse or cull piles left in the field provide an excellent source of the black rot bacteria. Fresh cull piles left near fields can result in severe disease epidemics during the growing season. Prepare cole crops for market away from fields, and immediately chop and bury the diseased tissue cut from plants.

Resistant Varieties

The development of crop varieties with disease resistance or tolerance to black rot has been the focus of many cole crop breeding programs worldwide. Resistance to black rot was first identified in the Japanese cabbage cultivar, Early Fuji. Today, many crucifer hybrids with black rot tolerance are available for both fresh market and processing commercial production.

Crop Nutrition

The effect of plant nutrient management on the susceptibility of host crops to black rot infection is not fully understood. A balanced nutrient program may reduce the susceptibility of plants to disease infection. Excess nitrogen promotes lush vegetative growth and may increase plant susceptibility. Micronutrients may also be involved with the disease defense mechanisms of crucifer crops.

Chemicals

Seed treatment with 0.1% mercuric chloride for half an hour is effective in eradicating seed-borne infection. Hot-water treatment of the seed at 50°C for 30 min is also effective. Drenching the seed bed with 5% commercial formalin may help in checking the disease (Patel et al., 1949). In a small-scale field trial, Agrimycin 100 gave good control of the disease. Streptocycline was found to be effective in checking the *in vitro* growth of *Xanthomonas campestris* (Chakravarty et at., 1969).

Due to the variable regulations around (de-)registration of pesticides, we are for the moment not including any specific chemical control recommendations. For further information, we recommend you to visit the following resource: EU pesticides database (www.ec.europa.eu/sanco_pesticides/public/index.cfm).

Soil fumigation can significantly reduce black rot bacteria. Soil fumigation is expensive and alternative methods for managing plant pathogenic bacteria are needed. For more information on chemical control options, refer to OMAFRA Publication 363, Vegetable Production Recommendations.

8.6.2　Soft Rot of Cabbage

Pathogen: *Erwinia carotovora* subsp. *carotovora, atroseptica, and chrysanthemi*

Symptoms

Small, water-soaked areas appear and rapidly enlarge. Tissue becomes soft and mushy, and within a few days, the affected plant part may collapse. An offensive odor usually is present.

On cabbage and brussels sprouts, the first symptoms usually show as water-soaked or greasy spots on the leaves as blemishes or as white to grayish, soft areas at the base of the heads. Small lesions have no appreciable odor but extensive decay has a very putrid odor. The disease follows black rot with other diseases where a large proportion of the head is converted into a brownish-black slimy mass (Photo 8.28).

Soft rot disease causes a soft mushy breakdown on leaf stalks, heads, and storage roots. The decay is often foul smelling, but there is no mold associated with the rot. Postharvest rot by this organism is common where temperature of the harvested produce is allowed to rise and cool chain is not maintained.

Host

Cabbage, cauliflower.

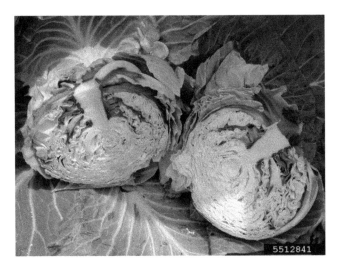

PHOTO 8.28 Bacterial soft rot on cabbage. (Photo courtesy of Gerald Holmes, California Polytechnic State University, San Luis Obispo, CA, Bugwood.org.)

Geographical Distribution

These pathogens are found worldwide. Bacterial soft rot occurs on wide range of vegetable all over the world and causes serious losses in field. In north India, especially in the valleys of Kashmir, Kullu, and Kalpa where cabbage and snowball crops are grown for seed production, the rotting of head and curds start just after chilling and melting of snow (Kapoor, 1987).

Predisposing Factors

The soft-rot bacterium, *E. carotovora* subsp. *carotovora* (syn. = *Pectobacterium carotovorum* subsp. *carotovorum*), enters through wounds caused by insects, other disease organisms, or by mechanical means. Under warm, humid conditions, uninjured tissue may become infected through natural openings. *Pseudomonas* spp. have been isolated from diseased heads and are thought to actively infect the host. Prolonged moisture from rain or irrigation and mild temperatures encourages disease development. Insects, tools, rain, clothing, or affected plant tissue can spread the bacteria. The bacteria survive in soil and plant debris.

Warm, wet weather favors the development and spread of the disease in the field, and warm, moist conditions favor its development and spread in transit. Temperatures between 20°C and 26°C favor the disease because there is less drying out of tissues. At 33°C, the organism does not grow as well as at 20°C, at 16°C it produces only one-half, and at 12°C only one-third as much rot as at 20°C, while at 8°C there is practically no invasion of tissues.

Disease Cycle

Soft rot bacteria persist in infected plant debris, in association with plant roots, in low numbers in the soil, and in association with several insects (Chakravarti and Hegde, 1972). Wounds such as leaf scars, insect injury, mechanical injury, and lesions caused by other pathogens are the primary avenues of soft rot bacterial invasion. Rainfall and high temperatures enhance infection in the field. Transit and storage infection may develop from bacterial contamination that occurred in the field or during postharvest from handling equipment and storage containers. Soft rot bacteria can grow over a temperature range of 5°C–37°C with an optimum temperature of about 22°C.

Control Measures

Disease management is based primarily on sanitation and cultural practices. Sufficient time should be allowed for crop residues to decompose before planting a second crop. Vegetable crops should be rotated with cereals or other nonsusceptible crops. Fields should be well drained to reduce soil surface moisture and plants should be spaced sufficiently to allow ventilation for rapid drying of foliage. Rain shelters to prevent soil splash and foliage wetting should also reduce soft rot incidence (Ren et al., 1995), especially during seed production.

Cultural Control
1. Set out plants in rows to allow good air drainage.
2. Cultivate carefully to minimize injuring plants.
3. Control frequency and source of irrigation water.
   a. Avoid frequent irrigation during head development.
   b. Time irrigation to allow the head to dry rapidly.
   c. Use well water, which generally is free of bacteria.
   d. Avoid stagnant water sources.
4. Clean and spray storage walls and floors with copper sulfate solution (1 lb/5 gal water). Bactericides such as Clorox, Lysol, and quaternary ammonium products also are effective.
5. Use a buffering material in storage, such as straw or paper to prevent injury to the heads.
6. Keep storage house humidity between 90% and 95% and the temperature between 0°C and 3°C.

Chemical Control
- Cueva at 0.5–2 gal/100 gal water on 7- to 10-day intervals.

Biological Control
Efficacy unknown in Oregon.
- Cease at 3–6 quarts in 100 gal water; for greenhouse plants only.

Preventive Measures
1. Use resistant cultivars.
2. Use seed that is certified free from black rot bacteria.

Cool Chain Handling
The control of bacterial soft rot in vegetables during transit and marketing involves careful trimming to eliminate other diseases, careful packing and handling to avoid unnecessary wounds, and good refrigeration in transit and while held in stores. Vegetables should be kept at 7°C or below to effectively check the development of bacterial soft rot.

8.6.3 BACTERIAL ZONATE SPOT OF CABBAGE

Pathogen: *Pseudomonas cichorii*
Bacterial zonate spot is primarily a disease of cabbage, although other members of the Cruciferae and Cucurbitaceae are susceptible.

It was first observed on Florida-grown cabbage in the Chicago market in 1953.

Symptoms
The disease is characterized by round-to-irregular lesions 1/16–1/2 in. in diameter, at first buff to light brown, becoming darker brown with age. The lesions are zoned or target-like and may occur on any part of the leaf. Superficially the lesions resemble those caused by *Alternaría brassicae* on cabbage. Infected areas are firm and pliable but only slightly soft. This is in contrast to the slimy type of decay produced by *Erwina carotovora*.

Predisposing Factors
The casual organism, *Pseudomonas cichorii*, enters cabbage through wounds. Lesions may occur on any part of the leaf but there is no evidence that the disease involves the vascular system. Best growth of the pathogen occurs at 23°C.

Control Measures
Control measures have not been developed, but sprays of copper containing bactericide may reduce the incidence of disease.

8.7 BROCCOLI AND BRUSSELS SPROUTS

8.7.1 BACTERIAL LEAF SPOT OF BRUSSELS SPROUTS AND BROCCOLI

Pathogen: *Pseudomonas syringae* pv. *maculicola*

A bacterium *Pseudomonas syringae* pv. *maculicola* causes bacterial leaf spot on broccoli and Brussels sprouts.

Symptoms
Tiny black to purplish spots appear on outer leaves. Yellow halos appear around the spots, and they eventually grow together to form light brown, papery areas. Symptoms may vary depending on the pathogen strain present in the field.

Predisposing Factors

Bacteria survive on infested seeds and crop residues as well as in soil. Cool, wet weather, which is common in the west of the Cascade Range, favors disease development before harvest in the fall.

Disease Management

Cultural Practices
- Plant pathogen-free seed.
- Avoid sprinkler irrigation in the seedbed once the crop has germinated and established. In the greenhouse, ebb and flow irrigation is preferred to overhead sprinkling.
- Shred and turn underdiseased crop refuse promptly after harvest to hasten the breakdown of the infected plant material.
- Do not plant cole crops the following year if the field has a significant level of infection.

Chemical Control
Cueva at 0.5–2 gal/100 gal water on 7- to 10-day intervals.

Biological Control
Efficacy unknown in Oregon.
   Cease at 3–6 quarts in 100 gal water; for greenhouse plants only.

8.7.2 BACTERIAL SOFT ROT OF BROCCOLI

Pathogen: *Pectobacterium carotovorum* subsp. *carotovorum*

Symptoms

Any plant part can be affected, but the most serious economic loss is from head infections. Plant tissue at first appears watersoaked and rapidly breaks down into a soft, mushy rot.

   This bacterial infection commonly affects the leaves. The disease is more likely where warm and wet conditions prevail. Leaves become yellow starting from the edges and moving toward the center. The infection causes the leaf veins and plant stems to turn black. As the infection increases, the broccoli head develops brown sunken areas of rotting matter, which rapidly extend (Photo 8.29). Much defoliation takes place especially for the more mature plants. Proper growth is inhibited. Plants remain small or die prematurely.

PHOTO 8.29   Bacterial soft rot on broccoli flower head. (Photo courtesy of C.H. Canaday, West Tennessee Experiment Station, University of Tennessee, Knoxville, TN.)

Predisposing Factors

The soft-rot bacterium, *Pectobacterium carotovorum* subsp. *carotovorum* (syn. *Erwinia carotovora* var. *carotovora*), enters through natural openings or wounds caused by insects and equipment. Prolonged moisture from rain or irrigation and mild temperatures encourage disease development. Bacteria survive in soil and plant debris.

Control Measures

Cultural Practices

1. Set out plant rows that will allow good air drainage.
2. Cultivate carefully to minimize injuring plants.
3. Control frequency and source of irrigation water.
   a. Avoid frequent irrigation during head development.
   b. Time irrigation to allow the head to dry rapidly.
   c. Use well water, which generally is free of bacteria.
   d. Avoid stagnant water sources.
4. Keep storage house humidity between 90% and 95% and the temperature between 0°C and 4°C.
5. It is best to plant disease-resistant varieties as a preventive measure. Also, crop rotation is helpful. Rotate crops every 2 years.
6. Warm, wet conditions favor soft rot development. Therefore, select fields with good drainage. It is best to irrigate when heads are already wet, such as at night, when dew is present. The next best choice would be during the day when heads are dry at the start and conditions will favor rapid drying afterward.
7. Avoid high nitrogen rates with susceptible varieties. Increasing nitrogen does not increase the amount of disease with resistant varieties. This is fortunate since 100–120 lb of actual nitrogen per acre is needed to obtain large, attractive heads. In an experiment conducted in Tennessee in 1989, incidence of heads with soft rot for the susceptible variety Premium Crop was 53% when fertilized with only 100 lb of ammonium nitrate (34-0-0) per acre but 84% when fertilized with 300 lb/acre. In contrast, soft rot incidence for Shogun was 8% and 5%, with 100 and 300 lb/acre, respectively.
8. Surfactants, which in most insecticides, also have been shown to increase soft rot severity. Therefore, get insects under control before soft rot begins to develop. If an insecticide must be used when soft rot is present, applications should be made when rain is not forecasted.
9. Finally, cut heads such that the stem stump is angled to permit water runoff. Water pooling on a flat stump will provide favorable conditions for soft rot. Bacteria in rotting stem stumps can be dispersed to heads not yet harvested.

Chemical Control

Cueva at 0.5–2 gal/100 gal water on 7- to 10-day intervals.

Biological Control

Efficacy unknown in Oregon.
Cease at 3–6 quarts in 100 gal water; for greenhouse plants only.

Varietal Resistance

Varieties, Shogun, Arcadia, and Marathon have performed well in several evaluations. For example, in 1994, soft rot was observed on only 4.5%–9% of the heads of these varieties, while 34%–64% of the heads were affected with Emperor, Green Comet, Green Valiant, Mariner, Packman, Paragon, Premium Crop, Southern Star, and Sultan. There were no significant differences among Shogun,

Arcadia, and Marathon in bacterial soft rot incidence, severity, number of heads harvested, or head diameter. Arcadia has a very attractive head with a purplish hue; it is a few days earlier than Shogun. Arcadia has become a popular variety in Maine. Shogun has a very large stem.

The varieties that are resistant to bacterial soft rot have dome-shaped, tight heads with very small beads. Other varieties with these characteristics are also likely to be resistant. Resistant varieties have blue green foliage; however, some susceptible varieties also have blue green foliage. Varieties with a flat head are more susceptible to soft rot because water will tend to sit on the head, providing favorable conditions for soft rot. Water will flow off a dome-shaped head.

8.8   BEANS

8.8.1   Common Blight of Beans

Pathogen: *Xanthomonas axonopodis* pv. *phaseoli* (Smith) Vauterin et al.

Synonyms: *Xanthomonas campestris* pv. *phaseoli* (Smith) Dye
Xanthomonas phaseoli (Smith) Dowson
Xanthomonas phaseoli var. *fuscans* (Burkholder) Starr & Burkholder
Xanthomonas phaseoli (Smith) Dowson var. *phaseoli*
Xanthomonas fuscans (Burkholder) Burkholder

Common names: Common blight, fuscous blight (English)

Common blight is a serious disease in many important snap- and dry bean–producing regions of the world. Although its effect on yield is difficult to estimate, workers have reported losses in the range 10%–45%.

Symptoms
Symptoms of common and halo blight diseases are very similar, and it is seldom possible from a superficial examination to be certain which one is present.

On seed
If the infection occurs when the pods are young, the seeds may rot or be variously wrinkled and shriveled. If the bacteria enter by way of the funiculus, only the hilum may be discolored, but this is difficult to detect on dark-seeded varieties. Strains producing the brown pigment (so-called *fuscans* strains) give more conspicuous seed discoloration.

On Seedlings
When grown from infected seeds, seedlings have injured or entirely destroyed growing tips. Angular, water-soaked areas frequently occur on the opposite sides of the primary leaves, indicating that the initial infection occurred while they were still folded together. Lesions on the stems of young seedlings begin as small water-soaked spots that gradually enlarge and sometimes become sunken. If these plants do not die, buds may arise in the axils of the cotyledons and produce dwarfed plants with few pods. Plants often exhibit a characteristic wilting during the heat of the day, with recovery of turgidity at night.

On Plants
Following infection in the field, small, water-soaked areas appear on the leaf, enlarge, and become encircled by a comparatively narrow zone of lemon-yellow tissue. These lesions turn brown, the leaf rapidly becomes necrotic, and defoliation may result. The diseased crop takes on a burned appearance, which distinguishes it from halo blight infections in which the crop appears generally more yellow. In systemic infections, reddish brown discoloration of the veins and watersoaking of adjacent tissues occurs. If leaf infection starts at the petiole, the main vein and its upper branches appear watersoaked at first and later take on a brick-red color. Lesions on the stems appear as reddish streaks, extending longitudinally. The stem surface often splits, releasing a yellow bacterial exudate (in halo blight infections, exudates are light cream or silver colored). Stem girdling occurs,

usually starting at the node above the cotyledonary attachment and is completed when the pods are half mature. These weakened stems often break at the node.

On Pods
Infections occur on any part as small, water-soaked spots that gradually enlarge and may be surrounded by a distinct zoning and narrow, reddish-brown or brick-red band of tissue. Infections may occur in the vascular elements of the sutures, causing water-soaking of the adjoining tissue. The infected tissue dries out and darkens, and droplets of yellow bacterial ooze may appear which, on drying, form a crust on the surface of older pod lesions. The whitefly *Trialeurodes abutiloneus* can cause similar leaf symptoms but, if present, nymphs will be found on the underside of the leaf at the center of each spot.

Host
The principal host is *Phaseolus vulgaris*, but other legume species are naturally infected, including *Phaseolus lunatus*, *Vigna aconitifolia*, and *Vulgaris radiata*. *Lablab purpureus* and *Mucuna deeringiana* are possibly natural hosts. *Phaseolus coccineus*, *Phaseolus acutifolius*, and *Lupinus polyphyllus* are hosts only by artificial inoculation (Bradbury, 1986).

In the EPPO region, only *P. vulgaris* is a significant host.

Geographical Distribution
EPPO region: Found in Egypt, Finland (unconfirmed), Lithuania, Moldova, Morocco (unconfirmed), Norway, Poland (unconfirmed), Sweden (unconfirmed). Widespread in Bulgaria, Hungary, Lebanon, and Spain; locally established in France, Germany, Greece, Italy, the Netherlands, Portugal (Madeira), Romania, Russia (European), Slovakia, Slovenia, Switzerland, Turkey, and Yugoslavia. Found in the past but not established in the Czech Republic, Israel.

Asia: Bangladesh, Brunei Darussalam, Cambodia, China (Heilongjiang, Henan, Hunan, Jilin, Jiangsu, Liaoning, Zhejiang), Cyprus, Georgia, Hong Kong, India (Delhi, Maharashtra, Rajasthan, Uttar Pradesh), Indonesia, Israel, Japan, Lebanon, Korea Democratic People's Republic, Korea Republic, Malaysia, Myanmar (Burma), Nepal, the Philippines, Sri Lanka, Taiwan, Thailand, Turkey, the United Arab Emirates (IMI, 1996), Viet Nam, Yemen.

Africa: Angola, Burundi, Central African Republic, Egypt, Ethiopia, Kenya, Lesotho, Madagascar, Malawi, Mauritius, Morocco, Mozambique, Nigeria, Rwanda, Somalia, South Africa, Sudan, Swaziland, Tanzania, Tunisia, Uganda, Zaire, Zambia, Zimbabwe.

North America: Bermuda, Canada (Ontario), Mexico, the United States (more prevalent east of the Rocky Mountains—Colorado, Hawaii, Michigan, Montana, Nebraska, New York, Texas, Wisconsin, Wyoming).

Central America and Caribbean: Widespread in Central America; Barbados, Costa Rica, Cuba, Dominica, Dominican Republic, El Salvador, Guatemala, Honduras, Jamaica, Martinique, Nicaragua, Panama, Puerto Rico, St. Vincent and Grenadines, Trinidad, and Tobago.

South America: Argentina, Brazil (widespread), Chile, Colombia, Ecuador, Paraguay, Uruguay, Venezuela.

Oceania: Australia (New South Wales, Queensland, Victoria, Western Australia), New Zealand, Samoa.

EU: Present.

Disease Cycle
The bacterial pathogen is commonly seedborne. As the seed germinates, bacteria contaminate the surface of the expanding cotyledon and spread to the leaves via natural openings and wounds, and eventually even to the vascular system. Throughout the season, bacteria can be spread to other plant parts by wind-driven rains or hail, insects, farm implements, or humans. Localized lesions on pods

and systemic invasion of pods lead to external and internal seed contamination. *Xanthomonas* can overwinter in seeds and infested bean straws and can survive in seed for over 15 years.

Environmental Factors
Common blight is favored by warm temperatures (28°C) and high humidity.

Natural movement occurs only over relatively short distances within or between fields. The only means of long-distance dispersal is by human transport of infected bean seed.

Economic Impact
It is not always possible to separate the losses caused by common blight and those due to halo blight, *Pseudomonas syringae* pv. *phaseolicola*, since they frequently occur together in the same field and even on the same plant. Moreover, the severity of blight varies from year to year depending on weather conditions. As early as 1918, 75% of the fields in New York State (USA) were affected and serious losses occurred. In the following years, losses of 20%–50% were recorded. In 1953, the disease was widespread in western Nebraska (USA) and the loss caused was estimated to be more than 1 million USD. In 1976, it was the most economically important bacterial disease of beans in the United States causing an estimated loss of 4 million USD (Kennedy and Alcorn, 1980). Losses for the field bean crop in Ontario (Canada) varied from a high of over 1,251,913 kg in 1970 to a low of 217,724 kg in 1972. A model for the assessment of yield losses from bacterial blight has been developed in Ontario, using aerial infrared photographic surveys to assess the disease levels (Wallen and Jackson, 1975).

In general, *Xanthomonas axonopodis* pv. *phaseoli* causes the most severe disease under fairly high temperature conditions (25°C–35°C) and also requires high rainfall and humidity. In the EPPO region, it is mainly present in eastern and southern areas, where it has a rather variable impact. In Romania, between 1962 and 1969, 45% of bacterial diseases of bean were caused by *Xanthomonas axonopodis* pv. *phaseoli*, while in Hungary in 1974 only 4% were caused by this pathogen.

Disease Management
Control methods (Severin, 1971) include planting disease-free seed, disease escape by suitable choice of planting date, crop rotation (an 85% reduction of attack was obtained by alternating bean and maize crops in Romania), sprays and dusts (e.g., with copper compounds or streptomycin), and resistant cultivars (Leakey, 1973). Numerous sources of tolerance have been identified, but breeding is complicated by the fact that different genetic systems control the reactions in pods and leaves (Coyne and Schuster, 1974).

Bacterial pathogens of dry and snap beans can be managed, but not eliminated, by carefully implementing integrated crop production strategies. These approaches reduce or delay disease severity during the critical periods of vegetative and reproductive plant growth.

Overwintering populations can effectively be reduced by employing a 3–4 year rotation with beans planted once every third or fourth year. Corn, small grains, and vegetables are recommended crop alternatives during two of these years. Proper sanitation of bean crop debris is important. It requires the complete incorporation of straw into the soil and the elimination of volunteer beans early the following year. This reduces diseased plants that provide the inoculums responsible for disease outbreaks in nearby bean fields. Bacterial pathogens can also survive in bean dust on contaminated harvest equipment, seed-cleaning equipment, and storage containers.

Always plant high-quality certified seed to minimize early season disease. Treat seeds with streptomycin, which effectively kills most external bacterial contamination.

Avoid cultivation practices in beans when plants are wet or when the stand is too tall to allow machinery to pass through without wounding the foliage. Thoroughly clean soil, weeds, and crop residues from equipment before moving to other fields. Reuse of irrigation run-off water is not recommended since pathogens are transmitted by it. Sprinkler irrigation can increase the spread and severity of bacterial foliar diseases by splashing bacteria from plant to plant and extending leaf wetness periods.

Planting resistant varieties is the best method to manage bacterial diseases of beans. Most older bean varieties, especially pintos and light red kidneys, are susceptible to these bacterial diseases.

Fortunately, blight resistant varieties are available in other market classes. Most navy and small white varieties, including Aurora, are resistant to halo blight.

The late-maturing dry beans "Great Northern Tara" and "jules" and the earlier-maturing "Valley" and "Great Northern Nebraska Sel. 27" have been reported resistant to common blight in temperate regions of the United States. Snap beans are somewhat tolerant. However, no cultivars widely grown in commercial fields in the United States are resistant.

In Colorado, copper-based bactericides have effectively reduced the populations of pathogenic pseudomonads on bean foliage and reduced the spread of these pathogens in infected crops. Common BWs are not controlled as effectively by copper spray programs, although some reduction in disease may be obtained with a preventive program. Applying these protectants early in the season (June–July) every 7–10 days during cool to warm, moist weather can decrease the establishment of bacterial pathogens. To ensure complete coverage, these products should be applied with at least 5 gal of water per acre. When pathogens have become established and disease symptoms are evident, copper sprays can help reduce their spread to healthy foliage and pods. But bacteria inside lesions are not affected. Commonly used cupric hydroxide bactericides include Kocide, NuCop, and Champ. Always read the product label carefully and consult extension and industry representatives for updated recommendations.

Closely inspect bean fields for symptoms of bacterial disease throughout the growing season, especially after prolonged periods of high humidity. Apply a bactericide every 7–10 days until plants are within 2–3 weeks from harvest; continue protection if infection is detected and weather conditions remain favorable for disease development. Hail-damaged beans should receive a preventive application of a copper bactericide to protect damaged plants during regrowth.

It must be emphasized that to minimize the impact of bacterial diseases on bean crops in the region, the producer must carefully integrate recommended strategies of crop rotation (2–3 years), sanitation, use of treated certified seed, deep plowing to eliminate infested bean debris in the field, varietal selection, stress and wound avoidance, and proper pesticide scheduling.

8.9   CHILI

8.9.1   BACTERIAL LEAF SPOT OF CHILI

Pathogen: *Xanthomonas campestris* pv. *vesicatoria*

It was first observed in the United States in 1912. At present, it is prevalent in many countries. It is the most common of the bacterial diseases of chili in India.

Symptoms
The leaves exhibit small circular or irregular, dark brown or black greasy spots. As the spots enlarge in size, the center becomes lighter surrounded by a dark band of tissue. The spot coalesces to form irregular lesions. Severely affected leaves become chlorotic and fall off. Petioles and stems are also affected. Stem infection leads to the formation of cankerous growth and wilting of branches.

On the fruits, round, raised water-soaked spots with a pale yellow border are produced. The spots turn brown developing a depression in the center, wherein shining droplets of bacterial ooze may be observed.

Disease Cycle
The disease is primarily seedborne. It spreads in the nursery and is further disseminated with infected transplants. Spattering rains are the chief means of dissemination. The bacterium subsists in infected debris.

The bacterium needs a temperature of 22°C–34°C with high humidity for maximum infection. Infection takes place at a wide range of temperatures from 15°C to 35°C but the optimum is 24°C–30°C. Plants that are 40- to 50-days old are most susceptible to this disease.

Disease Management
- Seed treatment with 0.1% mercuric chloride solution for 2–5 min is effective.
- Seedlings may be sprayed with Bordeaux mixture 1% or copper oxychloride 0.25%.
- Spraying with streptomycin should not be done after fruits begin to form.
- Field sanitation is important.
- Also seeds must be obtained from disease-free plants or certified seeds must be used.

8.10 PEA

8.10.1 BACTERIAL BLIGHT OF FIELD PEA

Pathogen: *Pseudomonas syringae* pv. *pisi* (Sacketi) Young, Dye & Wilkin

Synonyms: *Pseudomonas pisi* (Sackett)

On an average, severe epidemics occur once in 10 years and can cause some crops to fail completely. In the 1992 epidemic, yield losses from bacterial blights were estimated at 10% of the Australian field pea crop. The conditions favoring disease development are closely associated with wet weather and physical crop damage resulting from hail, wind, sand blasting, or vehicle/machinery/animal movement. The way these factors interact to produce an epidemic, however, is not fully understood.

Symptoms
Symptoms on peas may be found on all aerial plant parts, including stipules, leaflets, petioles, stems, tendrils, flower buds, and pods, but those on stems and stipules are most characteristic (Photo 8.30).

In dry weather, with occasional frost, symptoms usually appear on the stem near the soil as water-soaked and later olive-green to purple-brown spots. The infection extends upward to the stipules and leaflets, where veins turn brown to black and adjacent tissues become diseased in a fan-like pattern. The interveinal tissues may become watersoaked and then yellowish to brown, finally drying out and becoming papery.

In rainy weather, lesions on leaflets and pods begin as small, round, oval, or irregular dark-green water-soaked spots at first, and later enlarge and coalesce but are sharply defined by the veins. A cream-colored bacterial ooze may be found on the lesion surface, which, on drying, gives a glossy appearance. The leaflets later become yellowish and the spots brown and papery.

PHOTO 8.30 Field symptoms of bacterial blight on pea leaves and plant. (Courtesy of Mary Burrows, Montana State University, Bozeman, MT, Bugwood.org.)

Ripening pods become twisted and dry, and lesions on them sunken and greenish-brown. Lesions on the pod may be limited to a narrow band on the sutures. When pod invasion is mainly along the dorsal suture, the seed inside may be covered with bacterial slime. Infected seeds show a water-soaked spot near the hilum and/or are shriveled, with a brown-yellow discoloration.

Infection often takes place on sepals, spreading to the flowers, and flower buds may be killed before they open. If the infection spreads all over the plants, they may wither and die.

Under warm dry climatic conditions, however, the infection stops and the upper parts of the plants remain green and produce healthy flowers and pods. This may also be true for new axillary growth from the base of diseased plants.

Preemergence and postemergence damping-off may occur, and even advanced plants may be killed. Heavily infected seeds may be discolored, but light infection has no visible effect on seed.

The symptoms of bacterial blight caused by *Pseudomonas syringae* pv. *pisi* or *Pseudomonas syringae* pv. *syringae* are indistinguishable, from each other, on the pea plant.

Host
Major: *Pisum sativum*

Minor: *Vicia sativa*

Incidental: *Dolichos lablab, Lathyrus latifolius, Lathyrus odoratus, Vicia benghalensis, Vicia*

Artificial: *Medicago lupulina, Vigna unguiculata, Lathyrus*

Geographical Distribution
Europe: Armenia, Bulgaria, Croatia, Czechia, Denmark, France, Georgia, Greece, Hungary, Italy, Lithuania, Moldova, the Netherlands, Romania, Russia, Serbia, Slovakia, Ukraine, the United Kingdom

Asia: India, Indonesia, Israel, Japan, Kazakhstan, Kyrgyzstan, Lebanon, Nepal, Pakistan, Syria

Africa: Kenya, Malawi, North Africa, South Africa, Southern Africa, Tanzania, Zimbabwe

America: Argentina, Bermuda, Brazil, Canada, Colombia, Costa Rica, Mexico, the United States of America, Uruguay

Oceania: Australia, New Caledonia, New Zealand

Disease Cycle
Bacterial blight pathogens *Pseudomonas syringae* pv. *pisi* can survive on seed or pea trash, whilst *Pseudomonas syringae* pv. *syringae* can survive on a variety of host plants. The disease commonly becomes established within a field by sowing infected seed. During wet weather, bacteria spread from infected to healthy plants by rain splash, wind-borne water droplets, and plant to plant contact. Infection may occur at any stage of plant growth and is most prevalent following frosts.

Plants damaged by frosts or any other physical damage are more susceptible to infection. Rainfall, heavy dews, strong winds, and cold temperatures provide the most favorable conditions for spread of disease within crops. The optimum temperature for bacterial growth is 28°C, the minimum is 7°C, and maximum is 37°C.

Control Measures
Bacterial blight can be avoided by using an integrated approach to management that encompasses planting disease-free seed, crop rotation, variety selection, and avoiding early sowing.

Disease-Free Seed
Disease prevention can be obtained through production of disease-free seed by growing seed under semi-arid conditions and/or certification (Harris, 1964). Bacterial blight can be introduced by sowing contaminated seed. Do not use seeds from crops identified with bacterial blight during

field inspection. A field inspection should occur at mid-flowering to late pod fill. The field inspection will not control or eliminate bacterial blight but will help to identify very badly infected seed and enable growers to avoid sowing it in the following year. The bacteria remain viable on seed for at least 2 years. Choice of variety is not part of the current strategy to control bacterial blight, as current varieties do not vary greatly in their susceptibility. A seed test is available to identify infected seed. If purchasing seed, ask for a bacterial blight field inspection report. Seed dressings are not effective against bacterial blight.

Seed Test

A bacterial blight seed test is available and can be used to identify contaminated seed. Growers who wish to retain their seed in high-risk regions, that is, high rainfall, hail, strong wind, frost prone, with a history of bacterial blight, can utilize the seed test as an additional strategy to combat bacterial blight.

- The sensitivity of the test has been set at less than 0.05%, that is, 1 infected seed per 3000 seeds.
- This means low levels of bacteria can still remain undetected in seeds, and if conditions are conducive, the disease can still appear.
- The seed test does not guarantee a disease-free crop. Crops established with tested seed in which bacteria have not been detected, however, will produce less infected seed.
- Seeds with a positive test should not be sown in high-risk growing environments.

Host-Plant Resistance

Resistance to *Pseudomonas syringae* pv. *pisi* is present in peas, and this appears to have a gene-for-gene relationship with pathogenicity in *Pseudomonas. syringae* pv. *pisi*. Seven races have been identified by inoculation into a range of differential pea cultivars and race 2 is predominant. No resistance has been found to race 6 (Taylor et al., 1989).

Several pea varieties show resistance to one or more races of the pea blight pathogen. When planting these, races of pv. *pisi* may increase in the natural population. This was observed in the United Kingdom. Race 6 can mutate from race 2 (the race most commonly found and against which several UK varieties are resistant) following the loss of the avirulence gene A2 (Bevan et al., 1995; Reeves et al., 1996). Resistance breeding concentrates on the combination of six specific resistance genes and a race nonspecific resistance gene (from *Pisum abyssinicum*) in commercial peas (Taylor et al., 1994). Late varieties generally are less susceptible. Little genetic resistance has been reported other than in Partridgc-73, Vinco, and Line 3080 from New Zealand.

Paddock Selection

- A break of at least 4 years between field pea crops.
- Do not sow adjacent to field pea stubble, particularly downwind.
- Paddocks need to be suited to late sowing (good soil structure and drainage).
- Avoid paddocks prone to frost. Bacterial blight acts as an ice-nucleating agent, whereby plant tissues freeze sooner than normal. Bacterial blight–infected crops are more likely to suffer increased tissue damage from frost.
- Select paddocks with low weed pressure to minimize herbicide usage to preemergent or early postemergent spraying.

Sowing Time Adjustment

- Early sown crops are more vulnerable to bacterial blight infection than late sown crops; never sow earlier than recommended for your district.
- To reduce the likelihood of bacterial blight, sow at the later end of the recommended window for your district.

Avoidance of Crop Damage
- Bacterial blight is often associated with physical damage to the crop. Physical damage enables bacteria to enter plant tissue.
- Frost, hail, strong winds, sand blasting, and machinery can damage crops.
- Be aware that both domestic and feral animals can cause crop damage. Do not allow farm dogs to wander through crops.
- Bacterial blight severity can increase if plant tissue is damaged by herbicides, so minimize the use of postemergent sprays and avoid paddocks where sulfonylurea residues may be present.
- Avoid spraying or traveling in the paddock when rain is imminent as disease is more likely on freshly damaged and/or wet foliage.
- Controlled traffic will minimize damage by machinery. It will also allow a pathway for inspection (as long as pathways are wide enough to allow for field pea lodging).

Improved In-Crop Hygiene
- Bacterial blight can be spread by equipment.
- If bacterial blight infection is known, wash spraying equipment between paddocks with a high-pressure wash with disinfectant. This also applies to all other vehicles/machineries that enter field pea paddocks.
- Disinfectants should consist of 20% bleach or 70% methylated spirits.
- Crops should never be inspected when they are wet.
- Machine operators and farm workers should wear boots or waterproof trousers and wash them with disinfectant after leaving an infected paddock.
- Control volunteer peas in other crops.

Harvest Precautions
- Choose disease-free areas of your crop for seed and harvest this first.
- Dust originating from harvest, storage, handling, and grading of bacterial blight–infected crops can spread the disease to uninfected seed lots.

Stubble Destruction
- Stubble can be a potent source of inoculum.
- Bacterial blight can survive on stubble on the soil surface.
- Infected stubble and bacteria can be carried between paddocks by wind or grazing stock over summer.
- The survival time is significantly reduced by burying the stubble. In one field trial, survival of bacterial blight on stubble was reduced from 2 years to 11 months by burying pea trash 10 cm below the surface.
- Bury/destroy/bale/burn infected crop residues (stubble).

Rotations
To obtain a blight-free crop, peas should not be sown on land sown with peas in the previous year or adjacent to a pea stubble. Where possible, peas should not be grown on the same land more than once in 3 years. If disease occurs, the rotation should be extended to once in 4 years.

Stubble can be a significant source of inoculum. Destroy by burying, baling, or burning infected stubble. The survival time of inoculum is significantly reduced by burying pea trash 10 cm below the soil surface.

Chemical Control

Due to the variable regulations around (de-)registration of pesticides, for the moment, not any specific chemical control is recommended. For further information, visit the following resources:

- EU pesticides database (www.ec.europa.eu/sanco_pesticides/public/index.cfm)
- PAN pesticide database (www.pesticideinfo.org)
- Your national pesticide guide

Some copper-based, foliar products are registered on bacterial blight, but field trials have shown that their application after infection has not been successful.

Economic Impact

Damage to the plant is caused by necrotic spots and streaks that may occur on all aboveground parts and by infection of seeds. Necrosis results in a reduction of the assimilating surface, and when severe infections occur on peduncles, flowers, and pods, they wither and die. Pre- and postemergence damping-off has been observed.

Severe infections (reaching 100% disease incidence) and substantial crop losses have only recently been reported from winter-sown peas in southern France, New Zealand, and South Africa (Boelema, 1972; Taylor, 1972b), when hail, rainstorms, and frost prevail. It has been shown, however, that another pathogenic bacterium, *Pseudomonas syringae* pv. *syringae*, may also be involved in disease development under these conditions (Taylor, 1972b). Epidemic outbreaks (without severe crop losses) in Europe occur in pea cultivars sown in early spring under conditions of night frost and high humidity. A marked influence of these weather conditions was also established in Australia (Harris, 1964). Serious damage to *Vicia sativa* was reported from Greece in 1965 (Anon., 1965), and in Victoria, Australia, a 70% yield loss was observed (Forbes and Bretag, 1991). Individual pea plants showed a 24%–71% reduction in seed yield upon artificial inoculation (Roberts, 1993).

Economic losses arise from the reduced marketability of the crop, the presence of diseased pods and seeds, possible yield reduction, and indirectly through the application of phytosanitary restrictions. The disease as such, however, does not appear to be of great economic importance in the European area. In the state of New York, USA, the disease was observed in 1979 after an absence of 25 years (Hunter and Cigna, 1981). Bacterial blight is widespread in field peas in Victoria, but its severity varies greatly from crop to crop and between seasons. Severe epidemics can result in crop failure; however, losses are usually less than 20%.

8.10.2 Brown Spot of Pea

Pathogen: *Pseudomonas syringae* pv. *syringae*.

Symptoms

After initial water-soaked appearance, lesions on leaves resemble burning or physical injury; on stems, lesions are sunken and elongated, distorting stems, petiole and growing points (Photos 8.31 and 8.32).

Pathogen

The brown spot pathogen is seedborne, primarily as a colonizer of the seed surface (Hoitink et al., 1967). Six-month storage can significantly reduce contamination by the brown spot pathogen, although the pathogen could still be detected after storage for 1 year. The pathogen can survive in soil saprophytically for up to eight months.

Economic Impact

Brown spot of pea occurs in many parts of the world. Although it is not as serious as bacterial blight, it is very common in fall or winter-sown pea.

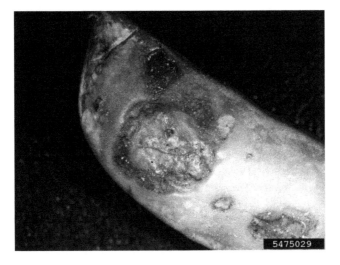

PHOTO 8.31   Field symptoms of brown spot pathogen on pea. (Courtesy of Mary Burrows, Montana State University, Bozeman, MT, Bugwood.org.)

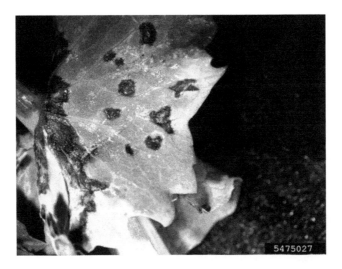

PHOTO 8.32   Field symptoms of brown spot pathogen on pea. (Courtesy of Mary Burrows, Montana State University, Bozeman, MT, Bugwood.org.)

Control Measures

Planting seeds free of the brown spot pathogen is an important means for control of the disease. Extra care should be taken during seed harvest to avoid seed contamination from infected pea fields. Infested seed could be used after storage for at least 1 year as the pathogen does not survive well on seeds. Because brown spot is very prevalent in peas physically injured by wind, frost, and hail, fall and winter planting should be avoided in areas with frequent brown spot epidemics (Butler and Fenwick. 1970).

8.11 CELERY

8.11.1 Bacterial Leaf Spot of Celery

Pathogen: *Pseudomonas syringae* pv. *apii*

Symptoms

Initial symptoms of bacterial leafspot are small, water-soaked spots that are visible from both sides of the leaf. The lesions usually are limited by leaf veins and thus have an angular, square, or rectangular appearance (Photo 8.33). These water-soaked lesions rapidly turn brown and with aging may dry out and become papery and tan. Lesions tend to be relatively small (less than 0.25 in. in diameter) and are restricted to leaves. On greenhouse transplants, bacterial blight lesions may develop extensively on the foliage. However, in the field, the disease usually is found only on the older leaves that are protected by the plant canopy, except where sprinkler irrigation is used. Under favorable conditions (free moisture), bacterial blight lesions may coalesce and cause considerable blighting of the foliage.

Disease Cycle

Pseudomonas syringae pv. *apii* is a seed-borne bacterium. Once introduced into transplant greenhouses, the pathogen can rapidly spread via splashing water. Disease development is favored by warm, moist conditions. Infected transplants carry the pathogen into production fields. In the field, widespread or severe symptoms generally do not develop unless the crop is sprinkler irrigated or subjected to a light frost during the production cycle. The pathogen survives in undecomposed celery residue.

Disease Management

Cultural Practices
- Use seeds that have been indexed free of *Pseudomonas syringae* pv. *apii*.
- Hot-water seed treatment (50°C for 25 min) will significantly reduce seed-borne inoculum, but may reduce seed germination.
- Using seed that is at least 2 years old can significantly reduce the incidence of this disease.
- Eliminate bacteria in seed by soaking at 45°C for 30 min.
- Obtain clean transplanting stock. Plants produced in ebb and flow greenhouses are less likely to be infected than those grown under sprinklers.

PHOTO 8.33 Symptoms of bacterial leaf spot on celery leaves. (Photo courtesy of Stacey Fischer, pnwhandbooks.org.)

- Do not mow or handle transplants when they are wet. If feasible, plant only in dry weather.
- Avoid mechanical operations in the field that can spread the bacteria, particularly when plants are wet with dew or irrigation water.

Disinfect transplant trays because bacteria may survive on dirty trays. In the greenhouse, lower the water pressure from overhead sprinklers because high pressures favor entry of the pathogen into celery leaves. In the field, avoid sprinkler irrigation. Excessive application of nitrogen fertilizers appears to favor disease development.

Chemical Control
Spray plants to prevent spread. Sprays usually are necessary only in cool, wet weather. Sprays are effective only if applied before blight appears to reduce bacterial populations on leaves. Later applications often are ineffective due to the bacteria's resistance.

- C-O-C-S WDG at 2–4 lb/A.
- Cueva at 0.5–2 gal/100 gal water on 7- to 10-day intervals.
- Kocide 2000 at 1.5 lb/A or Kocide 3000 at 0.75–1.5 lb/A on 5- to 7-day intervals.
- Nu Cop 50 WP at 1–2 lb/A on 3- to 7-day intervals. Do not apply within 1 day of harvest.

8.11.2 Bacterial Blight of Celery

Pathogen: *Pseudomonas cichorii* (Swingle) Stapp

Synonyms: *Pseudomonas endiviae* Kotte
Pseudomonas papaveris Lelliott et Wallace

Common names: Bacterial blight of celery, bacterial blight of endive, varnish spot of lettuce

Symptoms
The disease begins as small, water-soaked lesions on the leaf blade. Spots become dry and brown. The infected lesions coalesce to give blighted appearance to leaves. A second symptom, known as brown stem, may also develop (Pernezny et al., 1994). Brown stem results in a brown discoloration of the petiole, usually more evident on the petiole interior, closest to the crown. Browning is confined to the parenchyma surrounding vascular bundles, which appear healthy and green within diseased cortical and pith tissues.

Host
Cichorium endivia var. *crispum*, *Cichorium endivia* var. *latifolia*, *Lactuca sativa*

Geographical Distribution
Europe: France, Germany, Italy, Russia, the United Kingdom.

Asia: India, Japan, Taiwan.

Africa: South Africa, Tanzania.

America: Barbados, Brazil, Canada, Cuba, French West Indies, the United States.

Oceania: Australia, New Zealand

Disease Cycle
The *Pseudomonas* bacterium enters a leaf primarily through hydathodes and wounds. Severe symptoms can appear in as little as 3 days after infection, particularly if plants are misted before infection occurs. The number of lesions increases linearly as the misting period increases. Wounding is unnecessary for infection to occur on most hosts when moisture levels are high (such as misting).

Disease severity is enhanced at 28°C–29°C with little disease developing above 32°C. The organism is easily spread from leaf to leaf and plant to plant by splashing water, contaminated tools, insects, and by handling infected plants.

Control Measures
As in the case of leaf spot.

8.11.3   Soft Rot of Celery

Pathogen: *Pectobacterium carotovorum* subsp. *carotovorum* (syn. *Erwinia carotovora* subsp. *carotovora*)

Symptoms
The soft mushy rot of fleshy stems is characteristic in storage and transit. In the field, the celery heart may rot first. Small watery spots, becoming sunken and brown, also may appear at the base of stems. The disease cycle corresponds with bacterial soft rot of carrot.

Disease Management

Cultural Control
- Soft rot may follow bacterial blight, black heart, and mechanical injury, so avoiding those conditions will reduce infection.
- Long wet periods caused by a dense canopy and poorly timed irrigation encourage infection.
- Avoid using stagnant pond water for irrigation.
- Prompt hydrocooling and rinsing with chlorinated water reduces postharvest spread.

Chemical Control
Postharvest rinsing with chlorinated water. Use Agclor 310 at 1–1.1 gal/1250 gal water.

Note: Rotting celery contains high levels of toxins that react in sunlight to cause severe dermatitis in fair-skinned workers. Persons working in celery in sunny weather should wear gloves, long-sleeve shirts, and long pants.

8.12   LEEK

8.12.1   Bacterial Blight of Leek

Pathogen: *Pseudomonas syringae* pv. *porri*

In commercial settings, the bacterium affects primarily leek and has been reported from the United States, Engalnd, France, and New Zealand.

Onion, chives, and garlic have developed bacterial blight when inoculated with this pathogen in experimental situations.

Symptoms and Diagnostic Features
Young leaves show watersoaked, then yellow longitudinal lesions or stripes that later split and rot. The leaves can become curled and twisted as growth continues (Alvarez et al., 1978, Gitains et al., 1998). On older leaves, the pathogen cases yellow spots around wounds. Flowering stalks are very susceptible and develop deep water-soaked lesions that ooze bacterial exudates. Older stalk lesions are sunken, first yellow, then finally brown in color. Leek transplants can develop the disease while growing in green houses. Leaves of transplants develop yellow, and then brown, elongated lesions (Koike et al., 1999). Lesions usually involve the tips of the leaves.

Pathogen

Bacterial blight is caused by *Pseudomonas syringae* pv. *porri*. The designation pathovar (pv.) indicates this pathogen is host specific to the allium group and does not infect other crops, such as tomato, celery, and bean, which are susceptible to other *Pseudomonas syringae* pathogens. The pathogen is an aerobic, Gram-negative bacterium and can be isolated on standard microbiological media. It produces the cream-colored colonies typical of most pseudomonads. When cultured on Kings medium, the bacterium produces a diffusible pigment that fluoresces blue under ultraviolet light.

Disease Cycle

Pseudomonas syringae pv. *porri* is seedborne, so disease can be initiated in the field with direct seeded leek, or begin at the transplant stage in greenhouses. Spread of the pathogen at the seedling, transplant, or field stage is dependent on splashing water from rain or overhead sprinkler irrigation. The bacterium can survive in leek crop residues but will not persist in soil once residues are completely decomposed.

Control Measures

Use seed that does not have significant levels of the pathogen. Appropriate seed treatments can also contribute to the management of seed-borne inoculums. Rotate away from allium crops to reduce inoculums from crop residues. Irrigate using drip or Farrow Systems. Copper sprays may reduce spread. Avoid overfertilizing with high nitrogen materials.

8.13 LETTUCE

8.13.1 BACTERIAL BLIGHT OF LETTUCE

Pathogen: *Xanthomonas campestris* pv. *vitians*

Bacterial leaf spot of lettuce has been affecting coastal California crops for many years and has become a chronic problem. The disease was first noted in California in 1964 and became an economic concern in the 1990s. Bacterial leaf spot now occurs to some degree every season. In addition, it is possible that new strains of the pathogen may cause disease in previously resistant lettuce cultivars. For these reasons researchers are continuing to study the problem.

Symptoms

Early symptoms of bacterial leaf spot are small (1/8–1/4 in.), water-soaked spots that usually occur only on the older, outer leaves of the plant. Lesions are typically angular in shape because the pathogen does not penetrate or cross the veins in the leaf. Lesions quickly turn black; this is the diagnostic feature of this disease. If disease is severe, numerous lesions may coalesce, resulting in the collapse of the leaf. Older lesions dry up and become papery in texture, but retain the black color. Lesions rarely occur on newly developing leaves. If disease is severe, secondary decay organisms (bacteria, *Botrytis cinerea*) can colonize the leaves and result in a messy soft rot of the plant. Bacterial leaf spot can occur on all types of lettuce: iceberg, romaine, leaf, and butterhead.

Pathogen

Bacterial leaf spot is caused by *Xanthomonas campestris* pv. *vitians*. The taxonomy of this pathogen is unsettled and the name is likely to change in the next few years. This bacterium is a pathogen mostly limited to lettuce, though under greenhouse conditions, several weeds in the same plant family can develop bacterial leaf spot disease when inoculated; however, no natural infection on wheat by this pathogen showing leaf spot symptoms is observed in the field. Some researchers indicate that *Xanthomonas campestris* pv. *vitians* from lettuce can infect very different crops such as pepper and tomato when these plants are artificially inoculated; however, naturally infected pepper and tomato have never been found in California. Bacterial leaf spot disease of lettuce should not be confused with other *Xanthomonas* diseases. For example, bacterial spot disease of tomato and pepper is caused by a distinct pathovar (*Xanthomonas campestris* pv. *vesicatoria*); this pathogen will not infect lettuce. However, a related pathogen caused bacterial leaf spot on radicchio in California.

Differences in pathogen genotypes have been demonstrated and correlated to disease responses on resistant and susceptible lettuce cultivars. In California, the deployed lettuce germplasm is resistant to the strains of the pathogen collected many years ago in California. Therefore, samples of bacterial leaf spot disease should be analyzed so as to determine if novel, resistance-breaking strains are found in California. If this disease is encountered, samples can be submitted to the Cooperative Extension Diagnostic Lab in Salinas (1432 Abbott Street, Salinas).

Disease Cycle

The pathogen is highly dependent on wet, cool conditions for infection and disease development. Splashing water from overhead irrigation and rain disperses the pathogen in the field and enables the pathogen to infect significant numbers of plants. The pathogen can be seedborne, though the extent and frequency of seed-borne inoculum are not currently known. If lettuce transplants are grown from infested seeds, the pathogen may become established on plants during the greenhouse phase of growth. The bacterium can survive for up to five months in the soil. Therefore, infected lettuce plants and residues, once disked into the soil, can supply bacterial inoculum that can infect a subsequent lettuce planting. The bacterium has also been found surviving epiphytically on weed plants, though the significance of this factor is not known.

In terms of time of year, a very consistent pattern of bacterial leaf spot outbreaks is documented for the Salinas Valley. There is almost an annual pattern in which severe bacterial leaf spot occurs in August and September. Researchers have not clearly documented why the disease consistently occurs at severe levels in this late summer period.

Control Measures

Clearly the elimination or reduction of the use of overhead sprinkler irrigation will significantly curtail this disease in all situations, except when rains occur. Some resistant lettuce lines have been identified, though resistance is not widely available in currently used cultivars. Residual bacterial inoculum, left in the soil following an infected lettuce crop, will potentially cause problems for the next lettuce planting unless that planting is delayed for five months or longer. Therefore, crop rotation schemes need to be evaluated if bacterial leaf spot is a chronic problem in fields heavily planted with lettuce. Effective foliar sprays have not been identified for this disease. Lettuce seeds should be free of the pathogen.

The use of pathogen-free seeds is the first step in disease management. However, reliable seed assays and established threshold levels are not yet available. When possible, avoid sprinkler irrigation. Avoid planting back-to-back lettuce crops if the first crop was severely diseased and infected lettuce residue is present.

8.14 PARSLEY

8.14.1 Bacterial Leaf Spot of Parsley

Pathogen: *Pseudomonas syringae* pv. *apii*.

First seen in California in 2002, and continuing through 2010, a previously undescribed leaf spot disease has been affecting parsley throughout the central coast California and particularly in Monterey County.

Symptoms

Symptoms consist of small leaf spots that are usually less than ¼ in. in diameter. Spots are noticeably and consistently angular in shape, with the margins of the spot restricted by leaf veins. The color of the leaf spots can vary from light tan to brown to dark brown. These leaf spots penetrate the entire leaf, so that the spots will be visible from both the top and bottom sides of the infected tissue (in contrast to chemical damage or abrasion in which the symptom is usually only seen from the top side of the leaf). Spots rarely merged and coalesced, but in severe cases, the large number of spots resulted in blight-like symptoms.

Disease Development
The bacterium can be seedborne and can survive on tissue without causing symptoms (epiphytically) until conditions are conducive to a disease outbreak. The bacterium enters plants through wounds and natural openings. It can be transmitted from plant to plant by overhead irrigation, machinery, by insects, and by hands. The disease prefers warm temperatures, high humidity, and long hours of leaf wetness, at least 7 h/day over several days.

Pathogen
Researchers have documented that bacteria in the *Pseudomonas syringae* group are responsible for bacterial leaf spot of parsley. However, the cause (etiology) of this disease is complex and is due to two distinctive pathogens. A series of biochemical, physiological, and host range tests as well as extensive examination of pathogen DNA indicate that parsley bacterial leaf spot is caused by both *Pseudomonas syringae* pv. *apii* (also the pathogen causing bacterial leaf spot of celery) and *Pseudomonas syringae* pv. *coriandricola* (known as the pathogen-causing bacterial leaf spot of cilantro). It is interesting that in most cases parsley field was affected by either one or the other pathovar (pv.); in only a few cases were both pathovars found in the same planting. Therefore, this parsley disease is caused by two familiar pathogens that cause leaf spots on closely related species (celery, cilantro) in the Apiaceae plant family.

Of further interest are the experimental findings that these two pathovars are not restricted to a particular Apiaceae host plant. Parsley strains of pvs. *apii* and *coriandricola* both infected parsley, celery, and cilantro. Celery strains of pv. *apii* caused leaf spots on parsley, celery, and cilantro. Finally, cilantro strains of pv. *coriandricola* likewise caused disease on the same three crop species. Therefore, these two *Pseudomonas syringae* pathovars from Apiaceae hosts are not restricted to the original host but can cross-infect other Apiaceae crops. This cross infectivity could have important implications for growers and pest control advisors. For example, if bacterial leaf spot develops in a cilantro field, it would be possible for the pathogen to be splashed onto an adjacent celery or parsley crop and initiate disease in those plantings.

Managing Parsley Bacterial Leaf Spot
Control practices of this leaf spot disease of parsley will be similar to those used to manage the celery and cilantro problems.

- The initial source of inoculum for this parsley disease is not yet known. However, since *Pseduomonas syringae* pv. *apii* and *Pseduomonas syringae* pv. *coriandricola* are seedborne on celery and cilantro seed, respectively, contaminated parsley seed is a suspect source of the pathogen. Therefore, the use of pathogen-free seed or seed treatments may be appropriate.
- All three of these diseases depend on splashing water to disperse the bacteria and create favorable conditions for infection and disease development. Therefore, avoiding the use of overhead sprinkler irrigation is advisable where possible.
- Crop rotation with non-Apiaceae plants may be very important since these pathogens cross-infect other crops within this plant family.
- It is likely that diseased crop residues may still harbor viable bacteria, so back to back plantings of parsley, celery, and cilantro should be delayed until crop residues have dissipated. *Pseduomonas syringae* pathovars are not soil inhabitants, so once the host tissue has decayed and rotted away, the pathovars do not survive in the soil for long periods of time.

- Highly effective pesticides are not available for these bacterial leaf spot problems. While copper-based fungicides may provide some protection, these products are generally not sufficient to provide the high-quality produce demanded by the market.
- If seed contamination is suspected, soak seed in hot water at 50°C for 25 min.
- Irrigate when long periods of leaf wetness can be avoided, such as around sunrise, when dew is normally formed on leaves.
- Avoid fertilizers high in nitrogen as they stimulate lush growth that is very susceptible to bacterial leaf spot.

Cultural Practices
- Plant seeds under optimal conditions for quick emergence and vigorous growth.
- Use sterile soil or potting mix to get disease free seedlings. Also disinfect any tools and equipment that has to be used and which may contaminate the media.
- Avoid reusing pots or trays from a previous crop for propagation. If pots must be reused, then wash all the debris off and soak in a sanitizing solution or treat with aerated steam for 30 min.

Chemical Control
- Cueva at 0.5–2 gal/100 gal water on 7- to 10-day intervals.
- Cuprofix Ultra 40D on 10-day intervals at 2 lb/A is registered for parsley.
- Kocide 2000 at 1.5–2.25 lb/A or Kocide 3000 at 0.75–1.75 lb/A on 5- to 7-day intervals is registered for parsley.

Biological Control
- Double Nickel LC at 0.5–6 quarts/A on a 3- to 10-day interval.
- Serenade MAX at 1–3 lb/A on 2- to 10-day intervals is registered for parsley.

8.15 CUCUMBER

8.15.1 BACTERIAL WILT OF CUCUMBER

Pathogen: *Erwinia tracheiphila*

BW, caused by the bacterium *Erwinia tracheiphila*, is a common and destructive disease of cucurbits. The disease is most common on cucumber. Losses from BW vary from the premature death of occasional plants to as high as 75% of a crop.

Symptoms
On cucumber, BW first appears on leaves as dull green patches that rapidly increase in size. Within a day or two, the wilting symptoms spread to leaves, up and down the runner. In a short time, these and later affected leaves on the runner turn brown, wither, and die. The bacteria spread from the infected runner to the main stem and then to other runners. The entire plant soon wilts, shrivels, and dies (Photo 8.34).

BW is easily confused with other causes of wilting, including Fusarium or Verticillium wilts, root or stem rots, gummy stem blight, nematode damage, an excess or deficiency of moisture, damage by root- or stem-feeding insects, or an excess of fertilizer.

If a severely affected stem is cut across at the base and squeezed, a creamy white bacterial ooze may exude from the cut vascular tissue. After a minute or two, the droplets will adhere to a finger or knife blade and may be slowly pulled out into delicate, shiny threads about 1/4 in. long. This test is helpful in the diagnosis of the disease. The test works well for cucumber.

PHOTO 8.34   Cucumber plant infected with bacterial wilt. (Photo courtesy of Jason Brock, University of Georgia, Athens, GA, Bugwood.org.)

Disease Cycle

The bacteria overwinter in the digestive tracts of hibernating, adult striped, and spotted cucumber beetles (Photo 8.35). In the spring, the bacteria are introduced into the leaves of cucurbits by the beetles as they feed. The bacteria enter cucurbit tissue only through wounds, such as those produced by beetle feeding. When the beetles are absent, BW does not occur. The pathogen is not seed-transmitted and does not survive in soil.

Weather conditions have an indirect effect on the disease, per se the environmental conditions that favor the overwintering, feeding, and reproduction of the cucumber beetles subsequently affect the prevalence of BW. A good winter snow cover followed by warm temperatures in March and April can be expected to increase the number of overwintering beetles, and therefore, increase the early incidence of BW.

When a beetle feeds on a diseased leaf, its mouth parts become contaminated. The wilt bacteria are then introduced into the next several plants on which the beetle feeds. Beetles remain infectious for at least 3 weeks. Infection of cucurbit plants occur only when sufficient water from dew, rain, or irrigation is present to allow the bacteria to penetrate the inner leaf tissue via feeding wounds.

PHOTO 8.35   Cucumber beetles (spotted and striped), a vector of bacterial wilt pathogen. (Photo courtesy of Gerald Holmes, California Polytechnic State University, San Luis Obispo, CA, Bugwood.org.)

Wilting may be visible in several days, and the entire plant is invaded in 12–15 days. Field symptoms are usually apparent approximately 1 month after the beetles appear.

BW develops when daily temperatures average between 7°C and 22°C with an optimum of 11°C–15°C. The bacterium grows best at temperatures of 25°C–30°C, and under conditions of high relative humidity. Bacterial isolates differ in their virulence on each species of cucurbit. Some isolates that are highly virulent on cucumbers are also quite virulent on squashes, while others that are less virulent on cucumbers usually are not virulent on squashes. However, distinct cucumber and squash strains of the bacterium may exist.

The bacteria overwinter in the digestive system of the cucumber beetle. In the spring, BW is spread from plant to plant through both the striped and spotted cucumber beetles that feed on cucumbers and other relatives of this family. The bacteria are released through the insect excrement and move into host plants through the stomates and wounds, most likely the ones made when the insects feed. Insects ingest more bacteria as they feed on infected plants, and the cycle is repeated.

Epidemiology

The incubation period in the field from infection to symptom expression ranges from several days to several weeks and is influenced primarily by the age of the plant and the point of inoculation. The rate of BW in the field is most rapid while plants are young and succulent. Several factors contribute to a dramatic decrease in the rate of spread as plants mature: beetle populations naturally decline in midsummer; the older, more brittle leaves and stems are not preferred food sources; and bacteria introduced into new growth at vine tips must travel greater distances in the plant to induce wilting. Weather has little effect on wilt incidence, but it may influence the rate of symptom's expression. Ideal growing conditions for crops, including warm weather, adequate soil moisture, plentiful sunlight, and balanced nutrient concentrations, also appear to favor the development of the disease.

Disease Management Strategies

1. Protected plants with netting prevent cucumber beetles from feeding and infecting plants due to cover of netting or porous fabric.
2. Remove and destroy plant material when symptoms of wilting are first noticed. There are no cures for the disease. Beetles spread the bacterium from infected plants to healthy plants.
3. Grow susceptible crops on rotation every third year since beetles overwinter in the soil and carry the bacterium, the cycle can be disrupted only by planting the host in the area every third year.
4. Avoid planting cucurbits next to corn. Spotted cucumber larvae also feed on corn; avoiding close plantings of these two crops may help control the beetles on cucurbits.
5. Grow varieties that tolerate BW like butternut or a corn squash and Saladin or County Fair 83 cucumbers.
6. Dust plants with insecticide in the spring before the cucumber beetles have a chance to lay eggs (April–June). Apply pyrethrin or carbaryl (Sevin). Try an insecticide-bait combination such as Adios that has cucurbitacin, the beetle attractant, and a small amount of carbaryl (Sevin).

Start applications as the plants start to crack the soil, before the leaves appear, even if no beetles are evident. Frequent applications are necessary, especially in the seedling stage, to keep the foliage free of beetle-feeding wounds. Applications may be needed at 4–5 day intervals. Repeat after rains, especially if beetles are present. Early-season sprays or dusts are the most important step in controlling BW. Make treatment in late afternoon or evening to avoid damage to bees. The use of a systemic, soil-applied insecticide will provide moderate control for 5–6 weeks. Supplemental foliar insecticide treatments, however, will be necessary.

8.15.2 ANGULAR LEAF SPOT OF CUCUMBER

Pathogen: *Pseudomonas syringae* pv. *lachrymans*

Although the bacterium can attack a wide range of cucurbits, the disease mainly affects cucumbers in Virginia. Angular leaf spot (ALS) is a serious disease of the cucumber family that results in reduced yields and fruit of poor quality.

Symptoms

Leaves, stems, and fruit may be affected. The leaf spot is irregularly shaped with a water-soaked appearance; bounded by the leaf veins, the shapes of spots are angular. Bacteria may ooze from the spots in droplets, which dry to a white residue. The water-soaked area later turns gray and dies (Photo 8.36). Often, dead tissue is torn away from the healthy portion, leaving large, irregular holes. Water-soaked spots on fruits are smaller than on leaves, and they are circular. Lesions usually are superficial, but the injury may permit the entry of soft-rot organisms. Affected tissue becomes white and may crack open.

Disease Cycle

ALS bacterium is spread by wind, insects, machinery, and people who handle the plants and harvest the fruit while the vines are wet. The bacterium survives on plant debris and in the soil, and it can splash from the soil onto the plant by overhead watering and heavy rains. It also spreads through seeds collected from infected plants.

The bacterium can overwinter in seed and on diseased plant debris in the field. Seed-borne bacteria spread to the cotyledons when the seed germinates. Splashing rain spreads bacteria from the soil to plant parts and from plant to plant. The organism is easily spread in the field by cultivation equipment, harvesters, and by wind-blown rain. ALS is most active between 24°C and 28°C and is favored by high humidity.

Control Measures

Cultural Control

Use clean, bacteria-free seed. Practice a 2-year rotation out of cucurbits. Stay out of wet, infected fields. In the Hermiston area, do not irrigate with water draining from another cucumber field.

Good garden sanitation is an essential part of controlling the disease. Clean up plant debris regularly, and remove and destroy plant debris at the end of the season. The following control program is a combination of those used in other areas.

PHOTO 8.36 Angular leaf spot symptoms. (Photo courtesy of Gerald Holmes, California Polytechnic State University at San Luis Obispo, CA, Bugwood.org.)

Chemical Control
1. Applying fixed-copper products has reduced disease spread but reportedly has caused some stunting and leaf chlorosis in Oregon.
   a. Cueva at 0.5–2 gal/100 gal water on 7- to 10-day intervals.
   b. Cuprofix Ultra 40D at 1.25–2 lb/A on 5- to 7-day intervals.
   c. Kocide 2000 at 1–2.25 lb/A or Kocide 3000 at 0.5–1.25 lb/A on 5- to 7-day intervals.
   d. Liqui-Cop at 3–4 teaspoons/gal water.
   e. ManKocide at 2–3 lb/A on 7- to 10- day intervals. Under moderate to severe disease pressure, use the higher rate on 5- to 7-day intervals. Do not apply within 5 days of harvest.
   f. Nu Cop 50 WP at 1.5–3 lb/A. Do not apply within 1 day of harvest.
2. Seed treatments may help if affected seed is used.
3. Hot-water treatments with various chemicals (calcium propionate at 4.4 oz/gal water; acidic cupric acetate at 6.7 oz/gal water) for 20 min at 50°C reduced the frequency of disease in cucumber seedlings, but did not completely eliminate the pathogen. Hot water alone is not effective.

Resistant Varieties
Plant resistant cultivars. Pickling cultivars Regal, Royal, Pioneer, Express, Calypso, Cross Country, and Frontier have shown tolerance under Washington and Oregon conditions. Slicing cultivars Victory, Bel Aire, Raider, Encore, Poinsett 76, Slice Nice, Dasher II, Turbo, Quest, and Sprint-N are resistant.

8.16 CARROT

8.16.1 BACTERIAL LEAF BLIGHT OF CARROT

Pathogen: *Xanthomonas hortorum* pv. *carotae* (Kendrick) Vauterin et al.

Synonyms: *Xanthomonas campestris* pv. *carotae* (Kendnck) Dye
Xanthomonas carotae (Keridrick) Dowson

Common names: Bacterial leaf blight of carrot, root scab of carrot, carrot bacteriosis

Bacterial leaf blight of carrot is caused by *Xanthomonas campestris* pv. *carotae* and is a common disease wherever carrot is grown. It is particularly important in areas of frequent rainfall or extensive sprinkler irrigation.

Symptoms
Bacterial leaf blight incidence is often noticed in fields as brown areas about 3–4 ft in diameter. Leaf symptoms appear as irregular brown spots, often beginning on the leaf margins. Lesions initially have an irregular yellow halo and may appear watersoaked. Spots coalesce and cause a leaf blight and dark brown streaks develop on leaf petioles. Floral parts may also be blighted. A sticky amber-colored bacterial exudate, which is a diagnostic sign of the disease, may be present on leaves or observed flowing downward on petioles and flower stalks.

Host
Carrot (*Daucus carota* L. var. *sativa* DC).

Geographical Distribution
Europe: Italy, Poland, Russia.

Asia: Japan, Kazakhstan.

Africa: South Africa.

America: Brazil, Canada, the United States.

Oceania: Australia.

Disease Cycle

Xanthomonas campestris pv. *carota* is seedborne, survives on seed and is spread with carrot seed. The bacteria also survive in carrot debris but cannot survive in the soil in the absence of debris. Rain or sprinkler irrigation is required for optimum disease development. Warm weather favors infection and disease development. Optimum temperatures are between 26°C and 30°C; infection does not occur below 17°C. The pathogen is dispersed in splashing water. Plant-to-plant spread may occur under heavy dew conditions.

In most carrot-growing areas, bacterial blight does not warrant control. In a few areas, such as the Antelope Valley, severe outbreaks may occur.

Disease Management

Cultural Practices
- Start with certified, disease-free seed or treat seed with hot water (52°C for 25 min).
- Avoid overhead irrigation.
- Rotate out of carrots for 2–3 years.
- Plow infected crop debris after harvest under the soil to hasten decomposition.

Chemical Controls
- Application of copper bactericides can slow disease development, especially if applications begin when plants are young.

8.16.2 BACTERIAL SOFT ROT OF CARROT

Pathogen: *Erwinia carotovora* subsp. *atroseptica* (van Hall, 1902) Dye, 1969

Synonyms: *Erwinia carotovora* pv. *atroseptica* (van Hall, 1902) Dye, 1978

Bacterium atrosepticum (van Hall) Lehmann & Neumann, 1927
Bacterium carotovorum var. *atrosepticum (van Hall)* Hellmers & Dowson, 1953
Pectobacterium atrosepticum (van Hall) Patel & Kulkarni, 1951
Pectobacterium carotovorum var. *atrosepticum (van Hall)* Graham & Dowson, 1960
Bacterium carotovorum (Jones) Lehmann & Neumann/Burgwitz
Erwinia carotovora (Jones) Holl./Berg.
Pectobacterium carotovorum var. *carotovorum*
Erwinia atroseptica (van Hall) Jennison, 1923
Erwinia carotovora var. *atroseptica (van Hall)* Dye, 1969

Symptoms

Bacterial soft rot appears as a soft, watery, and slimy decay of the taproot. The decay rapidly consumes the core of the carrot, often leaving the epidermis intact. A foul odor may be associated with soft rot. Aboveground plant symptoms include general yellowing, wilting, and collapse of the foliage.

Under conditions of high soil temperatures and high moisture, soft rot bacteria can cause a pitting of carrot roots. Tissues partially decayed by *Pseduomonas carolovora* are invaded rapidly by fungi and other bacteria. *Rhizopus arrhizus* and *Rhizopus nigricans* frequently are found in storages on carrots partially rotted by *Erwinia corotovora*. In the field, *Botryris cinerea* and *Sclerotinia sclerotiorum* often permeate these decayed tissues. These secondary invaders make accurate diagnosis very difficult and, in certain instances, impossible.

Host

Allium (onions, garlic, leek, etc.), *Allium cepa* (onion), *Allium chinense* (rakkyo), *Allium tuberosum* (Oriental garlic), *Aloe arborescens*, *Apium graveolens* (celery), *Araceae*, *Beta vulgaris* var.

saccharifera (sugarbeet), *Brassica juncea*var. *napiformis* (Chinese mustard), *Brassica napus*var. *napobrassica* (rutabaga), *Brassica oleracea*var. *botrytis* (cauliflower), *Brassica oleracea*var. *capitata* (cabbage), *Brassica oleracea* var. *gongylodes* (kohlrabi), *Brassica rapa* ssp. *oleifera* (turnip rape), *Brassica rapa* subsp. *pekinensis* (Pe-tsai), *Brassicaceae* (cruciferous crops), *Capsicum* (peppers), *Capsicum annuum* (bell pepper), *Carthamus tinctorius* (safflower), Cichorium (chicory), *Cucumis melo* (melon), *Cucurbita*, Cucurbitaceae (cucurbits), *Cyclamen*, *Cymbidium*, *Daucus carota* (carrot), *Dioscorea* (yam), *Dracaena*, *Dracaena deremensis*, *Dracaena sanderiana*, *Foeniculum vulgare* (fennel), *Glycine max* (soyabean), *Helianthus annuus* (sunflower), *Ipomoea batatas* (sweet potato), *Iris* (irises), *Kalanchoe blossfeldiana* (Flaming katy), *Lactuca sativa* (lettuce), *Lycopersicon esculentum* (tomato), *Manihot esculenta* (cassava), *Musa* (banana), *Musa* × *paradisiaca* (plantain), *Nicotiana tabacum* (tobacco), *Oncidium*, *Onobrychis viciifolia* (sainfoin), *Opuntia ficus-indica* (prickly pear), *Orchidaceae* (orchids), *Oryza sativa* (rice), *Pandanus*, *Papaver somniferum* (Opium poppy), *Phalaenopsis amabilis*, *Phaseolus* (beans), *Pothos*, *Primula* (Primrose), *Primula malacoides* (Fairy primrose), *Primula obconica* (Top primrose), *Primula polyantha*, *Pyrus communis* (European pear), *Raphanus sativus* (radish), *Solanum tuberosum* (potato), *Sorghum bicolor* (sorghum), *Sorghum sudanense* (Sudan grass), *Strelitzia reginae* (Queens bird-of-paradise), *Triticum aestivum* (wheat), *Ullucus tuberosus* (Ulluco), *Xanthosoma sagittifolium* (yautia (yellow)), *Zantedeschia*, *Zantedeschia aethiopica* (Arum lily), *Zingiber officinale* (ginger).

Geographical Distribution
Europe: Bulgaria, Finland, Former Yugoslavia, France, Germany, Greece, Hungary, Italy, Lithuania, the Netherlands, Poland, Romania, Russian, Federation, Central Russia, Southern Russia, Western Siberia, San Marino, Slovenia, Spain, Sweden, Switzerland, Ukraine, the United Kingdom.

Asia: Bangladesh, China, Taiwan, India, (Haryana, Karnataka, Maharashtra, Rajasthan, Tamil Nadu, Uttar Pradesh), Indonesia, Iran, Iraq, Israel, Japan, (Hokkaido, Honshu, Kyushu), Jordan, Korea, the Philippines, Saudi Arabia, Singapore, Thailand, Turkey.

Africa: Algeria, Central African Republic, Congo, Egypt, Ethiopia, Libya, Malawi, Mauritius, Morocco, South Africa, Sudan, Zimbabwe, Central America and Caribbean, Costa Rica, Cuba, Honduras, Martinique, Panama, Puerto Rico, Saint Kitts, and Nevis.

North America: Canada—Alberta, British Columbia, Ontario, Prince Edward Island, Quebec; *The United States*: Arizona, California, Colorado, Florida, Georgia, Hawaii, Idaho, Louisiana, Maine, Minnesota, New Jersey, North Carolina, North Dakota, Ohio, Oregon, Tennessee, Texas, Washington, Wisconsin.

South America: Argentina, Bolivia, Brazil, Sao Paulo, Chile, Peru, Venezuela.

Oceania: American Samoa; Australia—Australian Northern Territory, New South Wales, Queensland, Tasmania, Victoria, Western Australia, New Zealand, Papua New Guinea.

Disease Cycle
Pectobacterium carotovora is a common soil-borne bacterium that attacks a wide range of fruits and vegetables. The bacterium enters carrots through various kinds of wounds. In the field, soft rot is most often associated with warm temperatures and standing water resulting from poor drainage, low areas, or leaky irrigation pipes. Carrots are most susceptible to infection when roots are mature and temperatures are warm.

Soil is the principal source of primary inoculum for stored carrots. Soil that contains debris from plants that were diseased the previous year is undoubtedly the most important inoculum source. The omnipresence of this bacterium on carrot roots seems to indicate that the pathogen probably lives and multiplies within the soil. If soft rot occurs on carrot roots in fields, the inoculum source sometimes can be traced back to carrot foliage from which it moved directly down to the roots. Although *E. carotovora* is a wound parasite, this presents no problem in gaining entrance to carrots. Harvest bruises, freezing injury, fungus invasion, and especially insect wounds offer penetration sites. Several fly species, especially *Hylemya cillicrura platura* (Meigen) and *H. brassicae* (Bouche), may carry the bacteria in their intestinal tracts. When they lay eggs near or on plants, the larvae emerge and become contaminated with the soft rot bacteria. When larvae bore into carrot roots, they introduce the bacteria into the carrots. There the bacteria multiply rapidly, produce an enzyme that softens the middle lamella between cells, and results in tissue softening associated with separation of cells from one another. By-products of the bacterial growth cause the cell contents to flow into the intercellular spaces where it serves as food for the bacteria. Such cell collapse results in the typical watery or slimy type of decay.

Disease Management
In the field, maintain good drainage and avoid practices that could wound roots. Avoid prolonged irrigation of mature carrots during warm months of the year. In the packinghouse, handle carrots carefully to avoid bruising and store them under cool conditions. Chlorine (Decco 240 and Seachlor 100) added to the wash water helps to eliminate the soft rot bacteria from carrot surfaces.

Cultural Practices
- For the earliest and latest seedlings of carrots, avoid fields subject to a high water table during wet conditions.
- Harvest carefully, particularly during warm weather.
- When soil temperatures are high, carrots harvested for immediate sale should be washed and cooled promptly and rinsed with clean, chlorinated water before being placed in a refrigerated holding area.
- Harvest crops intended for long-term storage after soil and air temperatures drop. Keep storage as close as possible to 0°C temperature and 85%–90% relative humidity.
- Thoroughly clean and disinfect bins between storage seasons.

Biological Control
- Double Nickel LC at 0.5–4.5 pints/A for soil application on 14- to 28-day intervals.

8.16.3 Scab of Carrot

Pathogen: *Streptomyces scabiei* corrig. (*ex* Thaxter, 1891) Lambert and Loria 1989, sp. nov., *nom. rev.*

Synonyms: *Streptomyces scabies* (Thaxter, 1891) Waksman and Henrici, 1948
Streptomyces acidiscabies

Common names: Scab

Symptoms
Symptoms on carrot are raised, brown, corky lesions on the taproot. Lesions measure 0.5–2 cm and are often elongated horizontally across the carrot root. Sunken lesions are also produced.

Host
Carrot, potato, radish.

Geographical Distribution
France, India.

Disease Cycle
Streptomyces scabies is a soil organism that invades the root through root openings such as sites of lateral root emergence or wounds. The pathogen causes localized death of cells and the formation

of corky tissue. This wound reaction results in the scab symptoms. Infection is favored by alkaline soils, dry conditions, and warm temperatures (optimum 20°C, with little activity below 11°C). Carrot is thought to be susceptible to infection for only 2–3 weeks during development, beginning when roots start to thicken (when root diameter is greater than 2 mm). This infection period is equivalent to 475°–625° days after planting. At this stage, the periderm grows through the epidermis, and the resulting small wounds are vulnerable to infection.

Control Measures

Rotate carrots with crops that are not susceptible to scab; however, crop rotations are not entirely successful because *Streptomyces* spp. survive in soil for very long periods. Irrigate to maintain adequate soil moisture, especially during the early stages of crop growth. There are indications of differences in cultivar susceptibility, so plant more tolerant varieties if these can be identified.

8.17   SWEET POTATO

8.17.1   Bacterial Stem and Root Rot of Sweet Potato

Pathogen: *Dickeya chrysanthemi* (Burkholder et al., 1953) Samson et al., 2005

Synonyms: *Erwinia chrysanthemi Burkholder* et al
             Erwinia carotovora (Jones) Bergey et al. f. sp. *parthenii Starr*
             Erwinia carotovora (Jones) Bergey et al. f. sp. *dianthicola (Hellmers) Bakker*
             Pectobacterium Parthenii (Starr) Hellmers
             Erwinia carotovora (Jones) var. *chrysanthemi Bergey* et al. (*Burkholder* et al.) *Dye*

Common names: Bacterial stem and root rot of sweet potato

Bacterial soft rot, also known as bacterial stem and root rot, is caused by the pathogen *Erwinia chrysanthemi*. Rotting can occur in the field as well as during shipment and storage.

Symptoms

This disease is more common in storage but may also affect plants in the field and in seedbeds.

The first symptom is the partial wilting of the plant; one or two branches may wilt, and eventually the entire plant may collapse and die. Discoloration of tissues inside the stem may also occur under some conditions (Photo 8.37).

PHOTO 8.37 Symptoms of bacterial stem and root rot of sweet potato on stem. (Photo courtesy of Charles Averre, North Carolina State University, Raleigh, NC, Bugwood.org.)

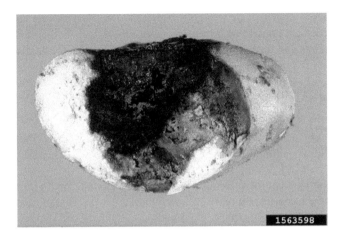

PHOTO 8.38   Symptoms of bacterial stem and root rot of sweet potato on storage root. (Photo courtesy of Charles Averre, North Carolina State University, Raleigh, NC, Bugwood.org.)

Water-soaked, sunken brown to black lesions are observed at the base of stems and on petioles. The disease initiates as small spots that grow overnight and form larger lesions.

On fibrous roots, localized lesions are observed, but the entire root system can be affected, showing the characteristic black, water-soaked appearance. There is also dark streaking in the vascular tissue of the roots.

On storage roots, small, sunken brown lesions with black margins can be observed on the surface (Photo 8.38), but more frequently the rotting is internal with no evidence outside. Affected tissue becomes watery. The disease is more common in the storage than in the field.

The pathogen enters host tissue through wounds. The characteristic evidence of this pathogen, as it occurs with some other bacterial diseases, is the peculiar smell produced in infected tissues, due to the invasion of saprophytes that live on decomposed organic matter.

Host
Dickeya chrysanthemi strains have been isolated from a number of different hosts in different countries (CABI, 2007). Chrysanthemum, daisy, tobacco, pepper, tomato, Irish potato, cabbage, eggplant, soybean, petunia, African violet, morning glory, and Cuscuta sp. have been reported susceptible to these bacteria (Schaad and Brenner, 1977).

Geographical Distribution
Erwinia chrysanthemi attacks several hosts in different regions of the world. However, the disease on sweet potato has only been reported from the United States. It is also suspected to occur in Central Luzon, the Philippines. It is common in Asia, Europe, Africa, North America, Central America, South America, and Oceania.

Disease Cycle and Epidemiology
The pathogen lives in plant debris and in the roots as well as on soils surrounding the roots of sweet potato and its other hosts. Dissemination is through infected planting material, irrigation water, tools, animal grazing, shoes of laborers, etc. Optimum disease development occurs at 30°C but the bacterium survives in a wide range of temperature below 27°C. No other factors, aside from temperature, have been reported to affect the development of the disease. Pathotypes of *Erwinia chrysantemii* have been reported to affect other plant species, but no reference was made on sweet potato. The bacteria can penetrate and enter the plant through wounded tissue on stem cuttings and can invade stems and the branches. The entire plant can be killed and roots can be rotted (Duarte and Clark, 1993).

Economic Importance

Bacterial stem and root rot caused by *Erwinia chrysanthemi* can be economically important because it destroys plants and fleshy roots. The amount of losses due to the disease has not been reported.

Disease Management

Cultural Practices
- Use of cuttings obtained from the uppermost portion or tips of vines. Proper postharvest handling to avoid wounding the storage roots. Crop rotation for at least 3 years with non-host crops.
- Handle sweet potatoes carefully during all stages of production. This is the most important control method for bacterial soft rot.
- Select mother roots from fields free of the disease.
- Cull roots infected during storage.
- Use vines cut above the soil's surface for transplanting.
- Use a handling system that does not involve the immersion of sweet potatoes in water.
- Application of soil amendments to reduce soil acidity.

Host-Plant Resistance

Differences in susceptibility have been observed. Assays on resistance should be tried with local varieties of sweet potato.

Chemical Control

No chemical control measures have been fully evaluated on sweet potato.

8.17.2 STREPTOMYCES ROT OF SWEET POTATO

Pathogen: *Streptomyces ipomoeae* (Person & Martin) Waksman & Henrici

Synonyms: *Actinomyces ipomoeae* Person & Martin

Common names: Soil rot, pox, pit, or ground rot of sweet potato, *Streptomyces* soil rot

Symptoms

All underground parts of the plant can be attacked. Dark brown to black spots of varying shapes and sizes are formed on roots, tubers, and underground parts of stems. In serious attacks, many of the fine feeding roots are either destroyed or more or less malformed. The aboveground parts of the plants then show poor growth and thin, pale green leaves. Yields are drastically reduced on such plants. In the early stages, lesions on the root-tubers are covered by the epidermis, but they crack and break up, leaving a hole or pit (Photo 8.39). These pits can be quite large and may girdle the sweet potato, preventing growth at that point. Continued growth on either side results in a dumbell-shaped or other misshapen sweet potato.

Host

Ipomoea batatas.

Geographical Distribution

Japan and the United States.

Disease Cycle

The disease is soilborne and most infection is thought to take place when plants are set out into already infested fields, but infection can occur in the nursery bed. The pathogen can be spread to

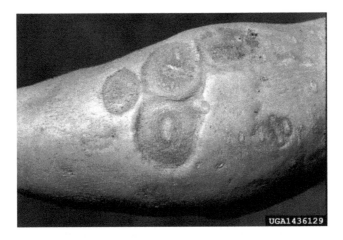

PHOTO 8.39   Streptomyces rot of sweet potato. (Photo courtesy of Clemson University—USDA Cooperative Extension Slide Series, Clemson, SC, Bugwood.org.)

new areas in a variety of ways. Infected planting material is probably the most frequent, but contaminated agricultural tools, workers' boots, feet of grazing animals, and even the wheels of vehicles may also play a part. Once established in a field, the organism can continue to live in the soil until the host is present and conditions are right for infection.

Economic Impact

Soil rot caused by *Streptomyces ipomoeae* is considered a minor disease, but under certain circumstances, it can produce a marked reduction in yield. In the past, this was a serious disease of sweet potato in the United States. Besides reducing yield, the disease also affects the quality of the storage root in size appearance and flavor. Affected storage roots are unmarketable in places demanding high quality.

Control Measures

Varietal Resistance
- Some cultivars, like Beauregard and Jasper, showed high levels of resistance, while some like Jewel and Centennial are susceptible.

Cultural Practices
- Crop rotation with 3–4 years between sweet potato crops.
- Avoid dryness of soil with timely irrigation.
- Careful selection of planting material.
- Avoid introduction from affected to unaffected areas, by restricting the movement of equipment, animals, manures, or planting material.
- If soil is mildly acidic, do not apply lime. Sulfur can be used to decrease soil pH, but may not be cost-effective.

Chemical Control
- Soil fumigation with chloropicrin.

8.17.3 Brown Rot or Bacterial Wilt of Sweet Potato

Pathogen: *Ralstonia solanacearum.*

Common names: Brown rot or bacterial wilt

Symptoms

In the field, this disease is usually observed on vegetable sweet potato, rather than on the tuber-producing type, causing yield losses of above 50%. Photo 8.40 shows the symptom development in sweet potato plants. Furthermore, bacteria ooze out from a wound site.

Sprouts from infected mother roots become wilty and the base of the stem shows progressive degeneration: initially the base of the stem becomes watersoaked before turning yellowish and brown. The vascular bundles turn brown, and this symptom extends upward. Infected sprouts usually fail to develop roots after transplanting.

Healthy sprouts or stem cuttings may become infected in the field. As for sprouts, the base of the stem becomes watersoaked, then yellowish brown. Brown streaks in the stem (in vascular tissue) may develop resembling Fusarium wilt. The plants may appear wilted, but may recover as their root system develops. They may appear stunted and hungry, with older leaves turning yellow.

Fibrous roots may have brown or water-soaked patches. Storage roots may show no symptoms, or may develop yellowish brown longitudinal streaks. In more severe infection, grayish water-soaked lesions may develop on the surface, and the storage root proceeds to decay with a distinctive odor.

Host

The sweet potato strain of *Ralstonia solanacearum* has been found to infect a number of other solanaceous crops, including tomato, potato, eggplant and capsicum, as well as peanut. Cereals are nonhosts and appropriate break crops in rotation with sweet potato.

Geographical Distribution

The causal organism, *Ralstonia solanacearum*, is globally distributed as a serious pathogen causing BW in a wide range of crops. However, the strain that causes BW of sweet potato has only been recorded in some parts of China. It is the subject of quarantine regulations restricting sweet potato and stock movement from the affected regions.

Seed potatoes are commonly exported from China to Vietnam, and they often carry BW. Any suspected case of BW in areas exposed to Chinese seed materials should be reported to plant protection authorities.

PHOTO 8.40 Symptoms of brown rot or bacterial wilt of sweet potato. (Photo courtesy of Gerald Holmes, California Polytechnic State University, San Luis Obispo, CA, Bugwood.org.)

Pathogen

The *Pseudomonads* are Gram-negative, flagellated, aerobic bacteria. Within *Ralstonia solona-cearum*, the sweet potato–infecting strain has been placed in race 1, group 2 on the basis of host range, being pathogenic on peanut and pepper but not on tobacco. It is grouped in biotype IV on the basis of physiological and biochemical tests. Strains in this group are unable to metabolize lactose, maltose, or cellobiose and are sensitive to the antibiotic oleandomycin (He et al., 1983).

Disease Cycle

The bacterium may be transferred either via soil or via plant material. A field may become infested through infested planting material, or via irrigation water or composts containing infested plants.

Development of the disease is favored in warm humid weather. It is also more severe on poorly drained clay loam soils than on sandy soils. However, it does not survive well in flooded conditions, such as in rice paddy. It prefers soils that are slightly acidic.

Economic Impact

BW can cause severe yield reduction, in the range 30%–80%. However, it is geographically restricted. Infected seed potato is an important factor in the distribution of this disease. Disease usually develops in localized areas associated with poor drainage. The bacterium attacks many different plant families, but most susceptible hosts are in the Solanaceae. The varieties Sebago and Green Mountain have tolerance to BW.

Control Measures

- Rotation with nonhost crops is the most important measure to reduce infestation on affected fields. The pathogen may survive for over 3 years in upland fields but is greatly reduced after 1 year in flooded paddy.
- Planting material should be sourced from disease-free crops.
- Cultivars differ in their susceptibility, although no immunity has been found.
- Soil amendments (lime) to reduce soil acidity may help reduce the severity of disease.
- Establishing the crop in the cooler months may reduce infection, which mostly occurs through wounded tissue immediately after transplanting. In China, quarantine regulations restrict the movement of sweet potato and livestock from infested areas.

8.18   RADDISH

8.18.1   BLACK ROOT ROT OF RADISH

Pathogen: *Acidovorax konjaci*

Also known as blight, stem rot, and black root rot, black rot is a serious bacterial disease that attacks only members of the crucifer family. Radishes (*Raphanus sativus*), an annual crop, are less suscep-tible to black rot than most other crucifers, but precautions should be taken to prevent the disease in radishes. Unfortunately, black rot cannot be cured after it infects a crop.

Symptoms

A range of symptoms are associated with black rot. One of the most common early symptoms is marginal leaf yellowing in the shape of a wedge or V. Leaf veins may turn black. Radish vegetables often develop irregularly shaped dark patches that spread and turn the entire root black. Cultivars that produce round radishes are less susceptible to the disease than long-rooted cultivars. Seedlings infected with black rot generally die quickly.

Geographical Distribution

Korea, Australia, the United States.

Disease Cycle

Black rot spreads primarily through infected seeds, transplants, and cruciferous weeds, such as wild mustards (*Brassica* spp.), pepper grasses (*Lepidium* spp.), and wild radish (*Raphanus raphanistrum*). Bacteria also may be spread by wind, splashing water, garden machinery, and even insects. It can survive in soil on infected plant debris until the debris decomposes completely, and then it can survive alone in soil for 40–60 days. Wet conditions and daytime temperatures of 24°C–35°C are optimal for black rot.

Disease Management

If black rot appears on a radish, remove and destroy the plant along with plants in a 3–5 ft radius around the affected plant. Remove all cruciferous weeds from the garden area because they could harbor the disease. Rotating crops and planting radishes in a new location each season are helpful. Do not plant radishes in a location where other crucifers were planted in the past 3 years.

Prevention

Treating seeds before planting them can reduce bacteria in the seeds, though it also reduces seed viability. Wrap radish seeds in a cheesecloth sack, and soak the sack containing the seeds in hot water for 20 min. Cool the seeds in cold water and spread them thinly to dry. Healthy radish plants have a stronger defense system than those that are weak due to poor cultural conditions. Radish plants grow best in a bright, sunny location with good air circulation and well-draining loamy or sandy soil. Water the plants regularly to keep their soil moist but not waterlogged.

8.18.2 BACTERIAL LEAF SPOT OF RADISH

Pathogen: *Xanthomonas campestris* pv. *raphani*

Geographical Distribution
Italy.

Symptoms
Discrete, water-soaked to greasy spots appear on leaves, with some spots surrounded by a narrow yellow halo.

The bacterium attacks the leaves and petioles causing small tan to white spots with narrow, yellowish, water-soaked zones on the leaves. The spots on the leaf petioles are black, sunken, and elongated. Severe infection results in defoliation and, in extremely severe cases, death may occur.

Disease Cycle
The causal bacterium is carried over in the crop residue and in infected seed. Once a plant is infected, further spread is by insects, rain, etc. During warm spring days, lesions are visible 4–5 days after infection. In cooler periods, development is slower. The bacterium will grow between temperatures of 5°C–35°C but is favored by temperatures between 27°C and 30°C.

Control Measures
Field sanitation is important in preventing infection. Rotation will also reduce the possibility of the disease becoming a problem. Use high-quality seed. Chemicals are not recommended.

8.19 CUCURBITS

Bittergourd, snakegourd, ridgegourd, pumpkin, roundgourd, bottlegourd, turnip.

8.19.1 ANGULAR LEAF SPOT OF CUCURBITS

Pathogen: *Pseudomonas syringae* pv. *lachrymans*

Symptoms
ALS of cucurbit pathogen *Pseudomons syringae* pv. *lachrymans* affects the leaves, petioles, stems, and fruit of cucurbits.

Small, circular spots appear first and soon after large, angular to irregular, water-soaked areas develop on the leaves. In wet weather, droplets of bacterial ooze exude from the spots on the lower leaf sides and dry into a whitish crust. Later, the infected areas turn gray, die, and shrink, often tearing away from the healthy tissue, falling off and leaving large, irregular holes in the leaves. On susceptible cultivars, the lesions often have yellow margins. On some resistant cultivars, the lesions are smaller and lack yellow margins.

Infected fruits show small, almost circular spots that are usually superficial, but when the affected tissues die, they turn white, crack open, and let soft rot fungi and bacteria enter and rot the whole fruit. Infection of young fruit may cause extensive fruit drop.

Under good growing conditions, the vines of indeterminate types of cucurbits, squashes, and melons grow away from older infected leaves, producing sufficient new growth to support normal fruit development. Occasionally in severe infestations, the growing tips of cucurbits vines are systemically invaded, become water-soaked and yellow, and stop growing.

ALS is the most important bacterial disease of pumpkin. Symptoms of ALS first appear as small, water-soaked (greasy) spots on the leaf surface. As lesions expand, they become delimited by leaf veins, thus developing their angular-shaped appearance. As lesions become aged and dry out, they may fall out leaving "shotholes" on infected leaves. Development of ALS is favored by warm, wet weather. Under good growing conditions, plants may outgrow the disease and produce a healthy fruit yield. In periods of heavy wind and rain, ALS may be splashed and carried to healthy fruit causing small, circular water-soaked lesions, which ruin aesthetic quality and predispose fruit to other pathogens.

Hosts

This pathogen not only can infect cucumber, zucchini squash, Summer Squash (*Cucurbita pepo* and *Cucurbita maxima*), and honeydew melon but also can infect muskmelon, cantaloupe, watermelon, other squashes, pumpkin, and various gourds. *Benincasa hispida* (wax gourd), *Citrullus lanatus* (watermelon), *Cucumis anguria* (gerkin), *Cucumis melo* (melon), *Cucumis sativus* (cucumber), *Cucurbita moschata* (pumpkin), *Cucurbita pepo* (ornamental gourd), *Luffa acutangula* (angled luffa), and *Sechium edule* (chayote).

Geographical Distribution
Worldwide.

Disease Cycle
The *Pseudomonas* bacterium is a seed-borne pathogen. In addition, the pathogen can overwinter in infested crop residues. This disease is widespread and particularly damaging in Illinois after extended and frequent summer rains, especially when temperatures are between 24°C and 28°C. Two weeks of dry weather will stop disease development.

The bacterium *Pseudomonas syringae* pv. *lachrymans* can enter the leaf through the stomates (breathing pores), hydathodes (water-excreting pores on the edge of the leaf), or wounds in the leaf tissue.

During wet weather, bacteria may ooze from the spots and later dry to a white residue.

ALS is favored by rainy weather and temperatures of 24°C–28°C. High nitrogen levels also favor the disease.

ALS thrives in warm humid conditions. The bacteria can infect all cucurbit crops and will infect all aboveground parts of the plant including leaves, fruit, and vines. When fruits are infected, the bacteria move deep into the fruits and infect the seeds.

The disease can be introduced into a field through contaminated seed. When humidity is high, a drop of clear to white sticky bacterial ooze forms on infections. These bacteria are moved from plant to plant via the hands and tools of workers, by insects, or by splashing water. The pathogen can survive in plant debris for over 2 years.

Disease Management

Use certified plants and pathogen-free seeds produced in arid locations. Do not grow cucurbits in the same field more than once every 3 or 4 years and avoid excessive nitrogen fertilization. Limit the use of overhead irrigation and avoid cultivating, harvesting, or otherwise handling plants when they are wet. Wherever feasible, cleanly plow under or collect and burn crop debris immediately after harvest. Apply a recommended bactericide at first sign of disease. Tank-mix the recommended bactericide with effective fungicides to protect the plants against fungal diseases.

Chemical Control

Copper can slow the spread of disease if the infection is caught early. Sprays are not effective once disease is severe. Sprays do not need to be continued if dry weather persists beyond 2 weeks.

Biological Control

The biological control agent Pentaphage (a lysate of the virulent strain of *Psedumonas syringae* pv. *syringae* by bacteriophages of five strains) was developed in Belarus. It was successfully used against *Pseudomonas syringae* pv. *lachrymans* on cucumbers under field condition. Pentaphage was most effective when applied at high RH (90%) in the morning and evening at intervals of 12–14 days (Korol and Bylinskii, 1994).

Spraying plants with a strain L33 of *Pseudomonas geniculata* (isolated from the rhizosphere of lucerne) can control the disease., incidence of *Pseudomonas syringae lachrymans*. The incidence was reduced to 5%–11% and disease intensity to 3%–5% compared with 26%–34% and 14%–18% respectively, in the control with a higher yield of 36–45 centners/ha (1 centner = 100 kg; Avagimov and Penteleev, 1984).

Systemic resistance to ALS was induced in cucumber plants by infection of first leaves with tobacco necrosis virus (Jenns et al., 1979).

Economic Impact

ALS is the most widespread bacterial disease of cucurbits. It has been reported in a wide range of cucurbits throughout the world. Early infection results in significant yield reduction in the number of fruits and fruit weight.

8.19.2 BACTERIAL WILT OF CUCURBITS

Pathogen: *Erwinia tracheiphila*

BW can cause severe losses in cucumbers and muskmelons, while squash and pumpkins are less severely affected. Watermelon is not affected. BW does not occur every year in Minnesota.

Symptoms

Leaves first appear dull green, wilt during the day, and recover at night. Leaves eventually turn yellow and brown at the margins, then completely wither, and finally die.

Wilt progression varies by crop. Cucumbers and melons wilt and die rapidly. Pumpkins take up to 2 weeks to completely wilt. Summer squash may continue to produce for several weeks even when infected.

Wilt progresses down the vine until entire vine is wilted or killed. Striped or spotted cucumber beetles in the garden are present, which are the vectors of the disease. If infected vines are cut close to the crown of the plant and the cross-sections pressed together, thread-like strands of bacterial ooze can be seen when the two halves are gently pulled apart again.

Disease Cycle

The bacteria overwinter in the gut of the striped and spotted cucumber beetles. Not all beetles carry the bacteria. Beetles that feed on infected plants pick up the bacteria. They then move to new plants, creating wounds through feeding. The bacteria are on the mouth parts or in the fecal matter of the beetle and enter the plant through the feeding wounds.

The bacteria multiply rapidly within the plant and plug the vascular tissue resulting in wilting of the vines. Once a plant is infected with BW, there is no way to control the disease. The bacteria cannot be transmitted in seed, do not survive in soil, and only survive in plant debris for a short period of time. It cannot overwinter in Minnesota in plant debris.

Disease Management
- Managing cucumber beetles provides the most effective control of BW.
- If disease appears in a few plants, rogue and bury these plants to prevent further spread of the disease.
- Pesticides will not help in managing a cucurbit plant infected with this bacterial disease.

8.19.3 Bacterial Spot of Cucurbits (Pumpkin)

Pathogen: *Xanthomonas cucurbitae*

In August 2012, leaves of pumpkin cvs. Gladiator, Aladdin, Apollo, and Super Hero in Kent County, Ontario, Canada, were observed to be infected with bacterial spot pathogen. Approximately 35 ha were affected with more than 50% of foliage and 60% of fruits damaged.

Symptoms
On the leaves, 1–4 mm irregular-shaped light brown to tan lesions, often with a chlorotic halo, appear. Mature fruits had 2–4 mm light brown to tan sunken lesions with dark borders (Photo 8.41) and later developed severe soft rot.

Pathogen
Isolations made from fruit yielded 95% of colonies with the following characteristics: opaque, light to bright yellow, glistening, circular, and flat, and Gram-negative.

Disease Cycle
Bacteria survive in association with infested crop residue. Bacteria are splash-dispersed to neighboring plants. Spread can be very rapid within fields. Long distance dispersal is with contaminated seed.

Bacteria infect fruits through natural openings in young, rapidly expanding fruit, prior to the development of a thick, waxy cuticle.

PHOTO 8.41 Bacterial infection of *Xanthomonas cucurbitae* on pumpkin fruit. (Photo courtesy of Gerald Holmes, California Polytechnic State University, San Luis Obispo, CA, Bugwood.org.)

Disease Management

Disease Resistance
All pumpkin varieties appear to be equally susceptible.

Cultural Practices
Normal rotations with noncucurbit crops will help prevent serious early season epidemics, unless inoculums are introduced via contaminated seed. Use commercially distributed seeds because seeds "saved" from previous crops are more likely to harbor bacteria.

Chemical Control
Copper sprays applied during early formation and expansion of fruit may result in substantially pumpkins with fewer symptoms.

8.20   SUGARBEET

8.20.1   BACTERIAL LEAF SPOTS OF SUGARBEET

Pathogen: *Pseudomonas syringae*

Symptoms
The symptoms appear as brown to black spots of irregular form and size mainly on the leaf edges or in hollows of the leaf blade. The dead tissue in the center of the spots often becomes brittle (Photo 8.42). The damage incurred can recover under dry and warm conditions.

Epidemiology
Wet and cool weather conditions over a longer period of time favor infection. The pathogen penetrates the leaf tissue through fissures or wounds (caused by hail or insects). Longer periods of rain and damage to the leaf increase the disease incidence and severity.

Economic Importance
Minor.

Control
No control measures have been developed.

PHOTO 8.42   Bacterial leaf spot on sugarbeet. (Photo courtesy of Ollie Martin, North Dakota State University Agriculture and University Extension, Fargo, ND.)

8.20.2 BACTERIAL VASCULAR NECROSIS AND ROT OF SUGARBEET

Pathogen: *Erwinia carotovora* subsp. *betavasculorum*

This pathogen is present in many native and cultivated soils. This pathogen can survive in some weedy hosts. Portions of the bacterium that cause potato blackleg can be pathogenic to sugarbeets. Plant wounding, excessive nitrogen or moisture, and warm temperatures (optimum is 27°C–28°C) favor disease development. The disease occasionally is severe in Idaho.

Symptoms

Black streaks are observed on petioles, and crowns may be blackened or produce froth. Vascular bundles are brown, and adjacent tissue turns pink when cut and exposed to air. Rot can become extensively soft or dry rot.

Cultural Practices

Most sugarbeet varieties have resistance, but losses still can occur.

- Maintain a 6–8 in. plant spacing, which helps to keep soil cooler.
- Minimize plant injury.
- Avoid excessive nitrogen.
- Avoid excessive irrigation.
- When hilling, avoid pushing soil into the crowns.

8.20.3 ERWINIA SOFT ROT OF SUGARBEET

Pathogen: *Erwinia carotovora* subspecies *carotovora*

Erwinia carotovora subspecies *carotovora* infects sugarbeet roots and causes *Erwinia* soft rot disease. The disease is difficult to detect aboveground, making it virtually unnoticeable until it has become severe. *Erwinia carotovora* subspecies *carotovora* moves from plant to plant via machinery, wind, and water to infect crops.

Symptoms

Pink to red-brown rot develops in root, which may be soft to dry rot. Roots may become hollow without dying and the plants show wilting symptoms. Occasionally brown, oozing lesions on petioles and crowns are observed.

Control Measures

- Use minimal fertilizer.
- Follow best practices for good soil structure.
- Plant early.
- Plant sugarbeets 6–8 in. apart.
- Select Hilleshög brand varieties with tolerance to *Erwinia*.

8.21 CLUSTER BEAN

8.21.1 BACTERIAL BLIGHT OF CLUSTER BEAN

Pathogen: *Xanthomonas cynopodium*

Symptoms

Water-soaked spots on leaves that enlarge and become necrotic; spots may be surrounded by a zone of yellow discoloration; lesions coalesce and give plant a burned appearance; leaves that die remain attached to plant; circular, sunken, red-brown lesion may be present on pods; pod lesions may ooze during humid conditions.

Disease Cycle
Disease can be introduced by contaminated seed; bacteria overwinters in crop debris; disease emergence favored by warm temperatures; spread is greatest during humid, wet weather conditions.

Disease Management
Plant-only certified seed; plant resistant varieties; treat seeds with an appropriate antibiotic prior to planting to kill off bacteria; spray plants with an appropriate protective copper-based fungicide before appearance of symptoms.

8.22 SPINACH

8.22.1 BACTERIAL LEAF SPOT OF SPINACH

Pathogen: *Pseudomonas syringae* pv. *spinaciae*

Symptoms
Initial symptoms of bacterial leaf spot consist of water-soaked, irregularly shaped spots that measure 0.12–0.25 in. in diameter. As the disease develops, these small spots enlarge to as much as 0.5–0.75 in. in diameter, are angular in shape, and turn dark brown in color. Occasionally, spots have black sections or edges, or they may have faint yellow halos. On leaves with numerous spots, the spots sometimes merge together, resulting in the death of large areas of the leaf. Spots are visible from both top and bottom sides of leaves. The disease occurs on both newly expanded and mature foliage.

Geographical Distribution
The United States, Italy, and Japan.

Disease Cycle
Bacterial leaf spot of spinach is always associated with overhead sprinkler irrigation and rainy springs. The pathogen appears to be host specific to spinach, and weed or other reservoir hosts have not been identified. The pathogen is likely to be seedborne.

Disease Management
Because the disease is uncommon, a management program is not necessary. If the disease occurs, avoid using overhead sprinklers.

8.23 CORIANDER

8.23.1 BACTERIAL LEAF SPOT OF CORIANDER

Pathogen: *Pseudomonas syringae* pv. *coriandricola*

Xanthomonas campestris pv. *coriandri* and *Xanthomonas hortorum* pv. *carotae* also have been reported to cause the disease. *Pseudomonas* sp. has been severe on cilantro in the Willamette Valley of Oregon and is transmitted by seed on coriander.

Symptoms
The disease is seen on young coriander plants as brown leaf spots 2–5 mm (⅛–¼ in.) in diameter, which may be surrounded by a water-soaked area. The leaf spots are often angular, being limited by the veins, and can be seen clearly on both leaf surfaces. Leaf spots may merge to cause a more extensive blight. On mature plants, there may be blackening of leaf veins, or black edges on leaves followed by leaf death. In severe infections, the bacterium can infect vein endings and spread down via the vascular system, resulting in dark longitudinal streaks on petioles and stems. If plants are left to bolt, then flowers and seed heads on infected plants shrivel, turn black, and collapse.

Leaves, petioles, and shoots develop brown necrotic lesions that look watersoaked. Severely infected plants may be stunted and yellowed. Flowers also may be attacked and show similar symptoms. Seeds decay can also occur.

Geographical Distribution
India, the United States.

Control Measures

Cultural Practices
- Plant seeds under optimal conditions for quick emergence and vigorous growth.
- Use sterile soil or potting mix to get disease free seedlings. Also disinfect any tools and equipment that has to be used and which may contaminate the media.
- Avoid reusing pots or trays from a previous crop for propagation. If pots must be reused, then wash all the debris off and soak in a sanitizing solution or treat with aerated steam for 30 min.

Chemical Control
There are no products approved specifically for the control of bacterial diseases on protected herbs, and it is unlikely that chemical control will be required under protection.

- Cueva at 0.5–2 gal/100 gal water on 7- to 10-day intervals.
- Cuprofix Ultra 40D on 10-day intervals at 2 lb/A is registered on parsley and at 1.3 lb/A for cilantro.
- Kocide 2000 at 1.5–2.25 lb/A or Kocide 3000 at 0.75–1.75 lb/A on 5- to 7-day intervals is registered for parsley.

8.24 COLOCASIA

Both the leaves and corms of colocasia are used as vegetable in Asian countries and south America. The crop is affected by two bacterial diseases.

8.24.1 BACTERIAL SOFT ROT OF COLOCASIA

Pathogen: *Erwinia carotovora*

Synonyms: *E. chrysanthemi*

Symptoms
The soft rot of corm is most prominent in the storage and during marketing. Bacterial soft rot–infected corms have a strong odor with watery soft rot ranging in color from white to dark blue (Photo 8.43).

Wounds and bruises caused by the feeding of insects and other animals and those inflicted at harvest are the most common infection courts for this disease. Abundant moisture is required for the invasion of the bacteria.

Disease Cycle
The bacterium survives in the soil and in the infected corm, which act as primary sources of infection. Secondary spread in the field is through irrigation water.

Control Measures
Use healthy corm for the planting. Avoid *Erwinia*-infected field for planting colocasia. Careful handling of corms to minimize injury at harvest, air drying of corms, and storage at low temperatures reduce the infection of bacteria.

PHOTO 8.43   Bacterial soft rot caused by *Erwinia* sp. (Photo courtesy of Yuan-Min Shen, Taichung District Agricultural Research and Extension Station, Taichung, Taiwan, Bugwood.org.)

8.24.2   Bacterial Leaf Spot of Colocasia

Pathogen: *Xanthomonas campestris* pv. *dieffenbachiae*

Symptoms

The bacterial leaf spot of taro or colocasia is reported from India and is also present in Hawaii. The symptoms are characterized by yellow or brown-necrotic marginal lesions of the leaf lamina with tan or pale yellow interveinal bleaching of the leaf (Photo 8.44). Under humid conditions, the infected lesions ooze the bacterium.

Disease Cycle

Warm and humid weather favors the disease initiation and development. The spread of the disease in the field is due to rain flashes and wind.

Control Measures

Practically no control measures are recommended.

PHOTO 8.44   Bacterial leaf spot on colocasia leaf. (Photo courtesy of Scot Nelson, flickr.com.)

REFERENCES

Abad, J.A., Bandla, M., French-Monar, R.D. et al. 2009. First report of the detection of 'Candidatus Liberibacter' species in zebra chip disease-infected potato plants in the United States. *Plant Dis.*, 93(1), 108–109.

Adams, M.J., Read, P.J., Lapwood, D.H. et al. 1987. The effects of irrigation on powdery scab and other tuber diseases of potatoes. *Ann. Appl. Biol.*, 110, 287–294.

Agrios, G.N. 1997. *Plant Pathology*, 4th ed. Academic Press, San Diego, CA.

Alfaro-Fernández, A., Cebrián, M.C., Villaescusa, F.J. et al. 2012a. First report of 'Candidatus Liberibacter solanacearum' in carrot in mainland Spain. *Plant Dis.*, 96, 582.

Alfaro-Fernández, A., Siverio, F., Cebrián, M.C. et al. 2012b. 'Candidatus Liberibacter solanacearum' associated with Bactericera trigonica-affected carrots in the Canary Islands. *Plant Dis.*, 96, 581.

Alvarez, A.M., Benedict, A.A., Mizumoto, C.Y. et al. 1994. Serological, pathological, and genetic diversity among strains of *Xanthomonas campestris* pv. *campestris* infecting crucifers. *Phytopathology*, 84, 1449–1457.

Alvarez, A.M., Buddenhagen, I.W., Buddenhagen, E.S. et al. 1978. Bacterial blight of onion, a new disease caused by *Xanthomonas* sp. *Phytopathology*, 68, 1132–1136.

Anon. 1965. The *Bacterium Pseudomonas pisi Sackett on Vetch and Peas in Greece*. Annual Report of the Phytopathological Station, Patras, Greece, pp. 31–34.

Avagimov, V.A. and A.A. Penteleev. 1984. Against angular spot. *Zashchita rastenii*, 7, 15 (in Russia).

Baker, K.F. 1950. Bacterial fasciation disease of ornamental plants in California. *Plant Dis. Rep.*, 34, 121–126.

Barnes, E.D. 1972. The effects of irrigation, manganese sulphate, and sulphur applications on common scab of the potato. *N. Irel. Minist. Rec. Agric. Res.*, 20, 35–44.

Bashan, Y., Diab, S., and Y. Okon. 1982a. Survival of *Xanthomonas campestris* pv. *vesicatoria* in pepper seeds and roots, in symptomless and dry leaves in non-host plants and in the soil. *Plant Soil*, 68, 161–170.

Bashan, Y. and Y. Okon. 1986. Internal and external infection of fruits and seeds of peppers by *Xanthomonas campestris* pv. *vesicatoria*. *Can. J. Bot.*, 64, 2865–2871.

Bashan, Y., Okon, Y., and Y. Henis. 1982b. Long-term survival of *Pseudomonas syringae* pv. tomato and *Xanthomonas campestris* pv. *vesicatoria* in tomato and pepper seeds. *Phytopathology*, 72, 1143–1144.

Bella, P., Greco, S., Polizzi, G. et al. 2003. Soil fitness and thermal sensitivity of *Pseudomonas corrugata* strains. *International ISHS Symposium on "Product and Process Innovation for Protected Cultivation in Mild Winter Climate." Acta Hortic.*, 614, 831–836.

Bevan, J.R., Taylor, J.D., Crute, I.R. et al. 1995. Genetics of specific resistance in pea (*Pisum sativum*) cultivars to seven races of *Pseudomonas syringae*pv*pisi*. *Plant Pathol.*, 44, 98–108.

Bhide, V.P. 1949. Stomatal invasion of cabbage by *Xanthomonas campestris* (Pammel) Dowson. *Indian Phytopathol.*, 2, 132–133.

Bila, J., Mortensen, C.N., Wulff, G.E. et al. 2009. Cabbage production and black rot disease management by smallholder farmers in southern Mozambique. *Afr. Crop Sci. Conf. Proc.*, 9, 667–671.

Boelema, B.H. 1972. Bacterial blight (*Pseudomonas pisi Sackett*) of peas in South Africa with special reference to frost as a predisposing factor. *Meded. Landbouwhogesch. Wageningen*, 72, 1–87.

Borthwick, A.D. 2012. "2,5-Diketopiperazines: Synthesis, Reactions, Medicinal Chemistry, and Bioactive Natural Products." *Chemical Reviews*, 112 (7): 3641–3716. doi:10.1021/cr200398y. PMID 22575049.

Bove, J.M. 2006. Huanglongbing: A destructive, newly-emerging, century-old disease of citrus. *J. Plant Pathol.*, 88, 7–37.

Bradbury, J.F. 1986. *Guide to the Plant Pathogenic Bacteria*. CAB International, Kew, U.K.

Braun-Kiewnick, A. and D.C. Sands. 2001 Pseudomonas. In: N.W. Schaad et al. (eds.), *Laboratory Guide for the Identification of Plant Pathogenic Bacteria*, 3rd ed. The American Phytopathological Society, St. Paul, MN, p. 84.

Brown, J.K., Rehman, M., Rogan, D. et al. 2010. First report of "Candidatus liberibacter psyllaurous" (synonym " Candidatus liberibacter. solanacearum") associated with 'tomato vein-greening' and 'tomato psyllid yellows' diseases in commercial greenhouses in Arizona. *Plant Dis.*, 94(3), 376.

Buchman, J.L., T.W. Fisher, V.G. Sengoda, and J.E. Munyaneza. 2012. Zebra chip progression: from inoculation of potato plants with liberibacter to development of disease symptoms in tubers. *American Journal of Potato Research*, 89: 159–168.

Buchman, J.L., Heilman, B.E., and J.E. Munyaneza. 2011a. Effects of Bactericera *Cockerelli* (Hemiptera: Triozidae) density on zebra chip potato disease incidence, potato yield, and tuber processing quality. *J. Econ. Entomol.*, 104, 1783–1792.

Buchman, J.L., Sengoda, V.G., and J.E. Munyaneza. 2011b. Vector transmission efficiency of liberibacter by Bactericera *Cockerelli* (Hemiptera: Triozidae) in zebra chip potato disease: Effects of psyllid life stage and inoculation access period. *J. Econ. Entomol.*, 104, 1486–1495.

Burkholder, W.C. and C.C. Li. 1941. Variation in *Phytomonas vesicatoria*. *Phytopathology*, 31, 753–755.

Burkholder, W.H., MacFadden, L.H., and A.H. Dimock. 1953. A bacterial blight of Chrysanthemums. *Phytopathology*, 43, 522–525.

Butler, L.D. and H.S. Fenwick. 1970. Austrian winter pea, a new host of *Pseudomonas syringae*. *Plant Dis. Rep.*, 54, 467–470.

Buturovic, D. 1972. Lista uzrocnika oboijenja bilja u Bosni i Hercegovini utvrtenih do 1971 godine. (A list of causal agents of diseases of plants in Bosnia and Hercegovina registered up to 1971.) Zbornik Radova, *Sarajevo*, 167–188.

CABI. 2005. *Crop Protection Compendium*, 2005 edn. CAB International Publishing, Wallingford, U.K. www.cabi.org.

CABI. 2007. *Crop Protection Compendium*. [Online] www.cabi.org/compendia/cpc/. Commonwealth Agricultural Bureau International (CABI), Wallingford, U.K.

Canteros, B.I., Minsavage, G.V., Bonas, U. et al. 1991. A gene from *Xanthomonas campestris* pv. *vesicatoria* that determines avirulence in tomato is related to avrBs3. *Mol. Plant-Microbe Interact.*, 4, 628–632.

Caponero, A. and N.S. Iacobellis. 1994. Foci of black rot on cauliflower in Basilicata. *Inf. Agric.*, 50, 67–68.

Carroll, N.B., Echandi, E., and P.B. Shomaker. 1992. Pith necrosis of tomato in western North Carolina. *NC Agric. Res. Serv. Tech. Bull.*, 300, 1–24.

Catara, V. and P. Bella. 1996. Pseudonjonas corrugata agente della necrosi del midollo del pomodoro. *Toecnica Agricola*, 4, 65–79.

Catara, V., Branca, F., and P. Bella. 1999. Outbreak of 'black rot' of *Brassicaceae* in Sicily. *Inf. Fitopatol.*, 49, 7–10.

Chakravarti, B.P. and S.V. Hegde. 1972. Bacterial blight of pea. *FAO Plant Prot. Bull.*, 20, 2122.

Chakravarty, S. 1969/2003. Black rot of cabbage caused by *Xanthomonas campestris* pv. *campestris*. In: R.S. Mehrotra (ed.), Palnt Disease. Tata McGraw-Hill Publishing Co., New Delhi. p. 676.

Cook, A.A. 1973. Characterization of hypersensitivity in *Capsicum annuum* induced by the tomato strain of *Xanthomonas vesicatoria*. *Phytopathology*, 63, 915–918.

Cook, A.A. and R.E. Stall. 1982. Distribution of races of *Xanthomonas vesicatoria* pathogenic on pepper. *Plant Dis.*, 66, 388–389.

Cook, A.A., Walker, J.C., and R.H. Larson. 1952. Studies on the disease cycle of black rot of crucifers. *Phytopathology*, 42, 162–167.

Coyne, D.P. and M.L. Schuster. 1974. Breeding and genetic studies of tolerance to several bean (*Phaseolus vulgaris* L.) bacterial pathogens. *Euphytica*, 23, 651–656.

Crosslin, J.M., Hamm, P.B., Eggers, J.E. et al. 2012a. First report of zebra chip disease and "*Candidatus Liberibacter solanacearum*" on potatoes in Oregon and Washington State. *Plant Dis.*, 96, 452.

Crosslin, J.M., Lin, H., and J.E. Munyaneza. 2011. Detection of '*Candidates Liberibacter solanacearum*' in the potato psyllid, Bactericera *cockerelli* (Sulc), by conventional and real-time PCR. *Southwestern Entomologist*, 36(2), 125–135.

Crosslin, J.M. and J.E. Munyaneza. 2009. Evidence that the zebra chip disease and the putative causal agent can be maintained in potatoes by grafting and *in vitro*. *Am. J. Potato Res.*, 86(3), 183–187.

Crosslin, J.M., Munyaneza, J.E., Brown, J.K., and L.W. Liefting. 2010. Potato zebra chip disease: A phyto-pathological tale. Online. *Plant Health Progress* doi:10.1094/PHP-2010-0317-01-RV.

Crosslin, J.M., Olsen, N., and P. Nolte. 2012b. First report of zebra chip disease and "*Candidatus Liberibacter solanacearum*" on potatoes in Idaho. *Plant Dis.*, 96, 453.

Davis, J.R., Garner, J.G., and R.H. Callihan. 1974a. Effects of gypsum, sulfur, terraclor and terraclor super-X for potato scab control. *Am. Potato J.*, 51, 35–43.

Davis, J.R., McMaster, G.M., Callihan, R.H. et al. 1974b. The relationship of irrigation timing and soil treatments to control potato scab. *Phytopathology*, 64, 1404–1410.

Davis, J.R., McMaster, G.M., Callihan, R.H. et al. 1976. Influence of soil moisture and fungicide treatments on common scab and mineral contents of potatoes. *Phytopathology*, 66, 228–233.

Diab, S., Bashan, Y., Okon, Y. et al. 1982a. Effects of relative humidity on bacterial scab caused by *Xanthomonas campestris* pv. *vesicatoria* on pepper. *Phytopathology*, 72, 1257–1260.

Diab, S., Bashan, Y., and Y. Okon. 1982b. Studies on infection with *Xanthomonas campestris* pv. *vesicatoria*, causal agent of bacterial scab of pepper in Israel. *Phytoparasitica*, 10, 183–191.

Duarte, V. and C.A. Clark. 1993. Interaction of *Erwinia chrysanthemi* and *Fusarium solani* on sweet potato. *Plant Dis.*, 77, 733–735.

Dye, D.W. 1969. A taxonomic study of the Genus *Erwinia*. II. "*Carotovra*" Group. *NZ J. Sci.*, 12, 81–97.

Dye, D.W. 1978. Genus V Erwinia Winslow, Broadhurst, Buchanan, Krumwiede, Rogers, and Smith, 1920. In Young et al. 1978. A proposed nomenclature and classification for plant pathogenic bacteria. *NZ J. Agric. Res.*, 21, 153–177.

Dzhalilov, F.S., Tivari, R.D., Andreeva, E.I. et al. 1989. Effectiveness of hydrothermal treatment and cabbage seed treatment against vascular bacteriosis. *Izvestiya Timiryazevskoi Sel'Skokhozyaistvennoi Akademii*, 5, 102–105.

Eastburn, D. 1989. Disease management of cabbage and broccoli-an IPM approach. *Trans. Illinois State Horticult. Soc.*, 123, 32–35.

Eckwall, E.C. and J.L. Schottel. 1997. Isolation and characterization of an antibiotic produced by the scab disease-suppressive *Streptomyces diastatochromogenes* strain PonSSII. *J. Ind. Microbiol. Biotechnol.*, 19, 220–225.

Ellingboe, A.H. 1984. Genetics of host-parasite relations: An essay. In: D.D. Ingram and P.H. Williams (eds.), *Advances in Plant Pathology*, Vol. 2. Academic Press, New York, pp. 131–151.

Elphinstone, J.G. 2005. The current bacterial wilt situation: A global view. In: C. Allen, P. Prior, and C. Hayward (eds.), *Bacterial Wilt Disease and the Ralstonia solanacearum Species Complex*. APS Press, St. Paul, MN, pp. 9–28.

EPPO. 2012. First report of 'Candidatus Liberibacter solanacearum' on carrots and celery in Spain, in association with *Bactericera trigonica*. EPPO Reporting Service. *Pests Dis.*, 6, 4–5.

Faivre-Amiot, A. 1967. Quelques observations sur la présence de *Corynebacterium fascians* (Tilford) Dowson dans les cultures maraichères et florales en France. *Phytiatrie-Phytopharamacie*, 16, 165–176.

Fatima, A.G. and S. Joseph. 2001. Reaction of different chili (*Capsicum annuum*) genotypes to bacterial wilt. *Capsicum Eggplant Newslett.*, 20, 82–85.

Fegan, M. and P. Prior. 2004. Recent developments in the phylogeny and classification of *Ralstonia solanacearum*. Presentation at *the First International Tomato Symposium*, Orlando, FL.

Fiori, M. 1992. A new bacterial disease of chrysanthemum: A stem rot by *Ps. corrugata* Roberts et Scarlett. *Phytopathol. Mediterr.*, 31, 110–114.

Flor, H.H. 1955. Host-parasite interactions in flax rust—Its genetics and other implications. *Phytopathology*, 45, 680–685.

Forbes, C.J. and T.W. Bretag. 1991. Efficacy of foliar applied *streptomycin* for the control of bacterial blight of peas. *Austr. Plant Pathol.*, 20(3), 115–118.

French, E.B., Gutarra, L., Aley, P. et al. 1995. Culture media for *Ralstonia solanacearum* isolation, identification, and maintenance. *Fitopatologia*, 30(3), 126–130.

Goode, M.J. and M. Sasser. 1980. Prevention the key to controlling bacterial spot and bacterial speck of tomato. *Plant Dis.*, 64, 831–834.

Goodfellow, M. 1984. Reclassification of *Corynebacterium fascians* (Tilford) Dowson in the genus *Rhodococcus*, as *Rhodococcus fascians comb. nov. Syst. Appl. Microbiol.*, 5, 225–229.

Goth, R.W. and K.G. Haynes 1993. Evaluation and characterization of advanced potato breeding clones for resistance to scab by cluster analysis. *Plant Dis.*, 77(9), 911–914.

Goth, R.W., Haynes KG, Wilson D.R., (1993). Evaluation and characterization of advanced potato breeding clones for resistance to scab by cluster analysis. *Plant Disease*, 77 (9): 911–914.

Goto, M. 1992. *Fundamentals of Bacterial Plant Pathology*. Academic Press, New York.

Griesbach, E., Lattauschke, G., Schmidt, A., and K. Naumann. 1988. On the occurance of *Xanthomonas campestris* pv. *Vesicrtoria* on pepper under glass and plastic. *Nachrichtenblatt fur den pflanzenschutz in der DDR*, 42, 176–178.

Grogan, R.G. and J.B. Kendrick. 1953. Seed transmission, mode of overwintering and spread of bacterial canker of tomato caused by *Corynebacterium michiganense. Phytopathol. Abstr.*, 43, 473.

Groves, S.J. and R.J. Bailey. 1994. The effect of irrigation upon the root yield and incidence of common scab of carrots. *Aspects Appl. Biol.*, 38, 217–221.

Guenthner, J., Bynum, E., Rush, C.M. et al. 2012. Zebra chip control cost based on psyllid population. *Potato J.*, 39(2), 197–201.

Gupta, D.K. (1991). Studies on black rot of cabbage in Manipur. *Indian J. Mycol. Plant Pathol.*, 21, 203–204.

Hansen, A.K., Trumble, J.J., Stouthamer, R. et al. 2008. A new huanglongbing species "*Candidatus liberibacter p syllaurous*" found to infect tomato and potato is vectored by the *Psyllid bactericera cockerelli. Appl. Environ. Microbiol.*, 74, 5862–5865.

Harris, D.E. 1964. Bacterial blight of pea. *J. Agric.*, 62, 276–280.

Hartman, G.L. and C.H. Yang. 1990. Occurrence of three races of *Xanthomonas campestris* pv. *vesicatoria* on pepper and tomato in Taiwan. *Plant Dis.*, 74, 252.

Hashimoto, N., Matsumoto, S., Oshikawa, M. et al. 2001. Varietal resistance among red pepper and sweet pepper cultivars to *Ralstonia solanacearum* isolated in Kyoto Prefecture [abstract in Japanese]. *Jpn. J. Phytopathol.*, 67, 201–202.

Hayward, A.C. and J.M. Waterston. 1964. *Corynebacterium michiganense. Descriptions of Pathogenic Fungi and Bacteria No. 19.* CAB International, Wallingford, U.K.

He, L.Y., Sequeira, L., and A. Kelman. 1983. Characteristics of strains of *Pseudomonas solanacearum* from China. *Plant Dis.*, 67, 1357–1361.

Hedges, F. 1926. Bacterial wilt of beans (*Bacterium flaccum facience*) including comparision with *Bacterium phaseoli. Phytopathology*, 16(1), 1–22.

Hellmers, E. and W.J. Dowson. 1953. Further investigation of potato black leg. *Acta Agric. Scand.*, 3, 103–112.

Henne, D.C., Workneh, F., Wen, A. et al. 2010. Characterization and epidemiological significance of potato plants grown from seed tubers affected by zebra chip disease. *Plant Dis.*, 94, 659–665.

Hibberd, A.M., Basset, M.J., and R.E. Stall. 1987. Allelism tests of three dominant genes for hypersensitive resistance to bacterial spot of pepper. *Phytopathology*, 77, 1304–1307.

Hooker, W.J. 1981. Common scab. In: W.J. Hooker (ed.), *Compendium of Potato Diseases.* American Phytopathological Society, St. Paul, MN, pp. 33–34.

Horikawa, M., Noro, T., and Y. Kamei. 1996. Comparison of antibacterial activity against causative agents of potato scab between antibacterial substances purified from a red alga, Laurencia okamurae, and commercial agricultural chemicals. *Marine Highland Biosci. Center Rep.*, 4, 17–21.

Horita, M. and K. Tsuchiya. 2001. Genetic diversity of Japanese strains of *Ralstonia solanacearum. Phytopathology* 91, 399–407.

Hunter, J.E. and J.A. Cigna. 1981. Bacterial blight of peas in New York State. *Plant Dis.*, 65, 612–613.

Ignatov, A., Vicente, J.G., Conway, J. et al. 1997. Identification of *Xanthomonas campestris* pv. *campestris* races and sources of resistance. In: *ISHS Symposium on Brassicas. 10th Crucifer Genetics Workshop*, Rennes, France, September 23–27, 1997, p. 215.

Ignatov, A.N. 1992. Resistance of head cabbage to black rot. PhD (Candidate of Sciences) thesis. TSKHA, Moscow, Russia, 25pp. (in Russian).

IMI. 1996. Distribution Mapis of Plant Diseases, No. 401, 3rd edn. CAB INTERNATIONAL, Wallingford, U.K.

Iwamoto, M., Ikeuchi, Y., Kobayashi, T. et al. 1988. Studies on the establishment of agronomical protection against soil-borne diseases of green peppers [in Japanese with English summary]. *Bull. Hyogo Pref. Agric. Inst.*, 36, 23–28.

Jennison, H.M. 1923. Potato blackleg with special reference to the etiological agent. *Reports from the Missowi Botanical Garden*, 10, 1–72.

Jenns, A.E., Caruso, F.L., and J. Kuc. 1979. Non-specific resistance to pathogens induced systemically by local infection of cucumber with tobacco necrosis virus, *Colletotrichum lagenarium* or *Pseudomonas lachrymans. Phytopathol. Mediterr.*, 18, 129–134.

Ji, P., Allen, C., Sanchez-Perez, A. et al. 2007. New diversity of *Ralstonia solanacearum* strains associated with vegetable and ornamental crops in Florida. *Plant Dis.*, 91, 195–203.

Ji, P., Momol, M.T., Olson, S.M. et al. 2005. New tactics for bacterial wilt management on tomatoes in the southern US. *Acta Hortic.*, 695, 153–160.

Jones, J.B., Jones, J.P., Stall, R.E. et al. (eds.) 1991. *Compendium of Tomato Diseases.* APS Press, St. Paul, MN.

Jones, J.B., Lacey, G.H., Bouzar, H. et al. 2004. Reclassification of the *Xanthomonads* associated with bacterial spot disease of tomato and pepper. *Syst. Appl. Microbiol.*, 27, 755–762.

Jones, J.B., McCarter, S.M., and R.O. Gitaitis. 1981. Association of *Pseudomonas syringae* pv. *syringae* with a leafspot disease oftomato transplants in southern Georgia. *Phytopathology*, 71, 1281–1285.

Joshi, M.V., Mann, S.G., and H. Antelmann. 2010. The twin arginine protein transport pathway exports multiple virulence proteins in the plant pathogen *Streptomyces scabies. Mol. Microbiol.*, 77(1), 252–271.

Kapoor, K.S. 1987. Bacterial soft rot (*Erwinia carotovora*) of cabbage. *Plant Dis.*, 71, 861.

Kennedy, B.W. and S.M. Alcorn. 1980. Estimates of US crop losses to procaryote plant pathogens. *Plant Dis.*, 64, 674–676.

Khan, A.A., Rahmans, S., and G. Kamaluddin. 1973. A preliminary survey of the diseases of potato in cold storage in Bangladesh. *J. Biol. Agric. Sci.*, 2, 15–21.

Kim, B.S. 1986. Testing for detection of *Xanthomonas campestris* pv. *campestris* in crucifer seeds and seed disinfection. *Kor. J. Plant Pathol.*, 2, 96–101.

Kim, S.H., Olson, T.N., Schaad, N.W. et al. 2003. *Ralstonia solanacearum* race 3, biovar 2, the causal agent of brown rot of potato, identified in geraniums in Pennsylvania, Delaware, and Connecticut. *Plant Dis.*, 87, 450.

King, R.R., Lawrence, C. H., and Calhoun, L. A. 1992. "Chemistry of phytotoxins associated with Streptomyces scabies the causal organism of potato common scab." *Journal of Agricultural and Food Chemistry*, 40: 834.

King, R.R., Lawrence, C.H., Clark, M.C., and Calhoun, L.A. 1989. "Isolation and characterization of phytotoxins associated with Streptomyces scabies." *Journal of the Chemical Society, Chemical Communications*, (13): 849.

King, E.O., Ward, M.K., and D.E. Raney. 1954. Two simple media for the demonstration of pyocyanin and fluorescin. *J. Lab. Clin. Med.*, 44(2), 301–307.

Koike, S.T., Cooperative, C., and J.D. Barak. 1999. Bacterial blight of leek: A new disease in California caused by *Pseudomonas syringae. Plant Dis.*, 83, 165–170.

Korol, A.L. and A.F. Bylinskii. 1994. Pentaphage against angular leaf spot. *Zashchita Rasteniĭ (Moskva)*, 4, 16–17.

Koronowski, P. and D. Massfeller. 1972. An actinomycosis of radish. *Nachrichtenblatt des Deutschen Pflanzenschutzdienstes*, 24, 152–154.

Kritzman, G. and Y. Ben-Yephet. 1990. Control by metham-sodium of *Xanthomonas campestris* pv. *campestris* and the pathogen's survival in soil. *Phytoparasitica*, 18(3), 217–227.

Lambe, R.C. and G.H. Lacy. 1982. Crown gall. *Am. Nurseryman*, 155, 113–114.

Lambert, D.H. and R. Loria. 1989. *Streptomyces scabies* sp. nov., nom. rev. *Int. J. Syst. Bacteriol.*, 39, 387–392.

Lambert, D.H., Reeves, A.F., Goth, R.W. et al. 2006. Relative susceptibility of potato varieties to *Streptomyces scabiei* and *S. acidiscabies. Am. J. Potato Res.*, 83, 67.

Lapwood, D.H., Wellings, L.W., and J.H. Hawkins. 1973. Irrigation as a practical means to control potato common scab (*Streptomyces scabies*): Final experiment and conclusions. *Plant Pathol.*, 22, 35.

Leakey, C.L.A. 1973. A note on xanthomonas blight of beans (*Phaseolus vulgaris*) and prospects for its control by breeding for tolerance. *Euphytica*, 22, 132–140.

Lehmann, K.B. and R. Neumann. 1927. *Bakteriologie insbesondere Bakteriologische Diagnostik. II. Allgemeine und spezielle Bakteriologie 7 Aufl.* J.F. Lehmann, Munchen, Germany.

Lerat, S., Simao-Beaunoir, A.M., and C. Beaulieu. 2009. Genetic and physiological determinants of *Streptomyces scabies* pathogenicity. *Mol. Plant Pathol.*, 10(5), 579–585.

Leyns, F. and M. De Cleene. 1983. Histopathology of the bacteriosis caused by inoculation of *Corynebacterium michiganense* and *Xanthomonas campestris* pv. *vesicatoria* in tomato stems. *Mededelingen van de Faculteit Landbouwwetenschappen, Rijksuniversiteit Gent*, 48(3), 663–670.

Li, W., Abad, J.A., French-Monar, R.D. et al. 2009. Multiplex real-time PCR for detection, identification and quantification of "Candidatus Liberibacter solanacearum" in potato plants with zebra chip. *J. Microbiol. Methods*, 78, 59–65.

Liefting, L.W., Perez-Egusquiza, Z.C., Clover, G.R.G. et al. 2008a. A new "Candidatus Liberibacter" species in *Solanum tuberosum* in New Zealand. *Plant Dis.*, 92, 1474.

Liefting, L.W., Sutherland, P.W., Ward, L.I. et al. 2009a. A new "Candidatus Liberibacter" species associated with diseases of solanaceous crops. *Plant Dis.*, 93, 208–214.

Liefting, L.W., Veerakone, S., Ward, L.I. et al. 2009b. First report of "Candidatus Phytoplasma australiense" in potato. *Plant Dis.*, 93, 969.

Liefting, L.W., Ward, L.I., Shiller, J.B. et al. 2008b. A new "Candidatus Liberibacter" species in *Solanum betaceum* (tamarillo) and *Physalis peruviana* (cape gooseberry) in New Zealand. *Plant Dis.*, 92, 1588.

Lin, H., Doddapaneni, H., Munyaneza, J.E. et al. 2009. Molecular characterization and phylogenetic analysis of 16S rRNA from a new species of "Candidatus Liberibacter" associated with zebra chip disease of potato (*Solanum tuberosum* L.) and the potato psyllid (*Bactericera cockerelli Sulc*). *J. Plant Pathol.*, 91, 215–219.

Lin, H., Islam, M.S., Bai, Y. et al. 2012. Genetic diversity of "Candidatus Liberibacter solanacearum" strains in the United States and Mexico revealed by simple sequence repeat markers. *Eur. J. Plant Pathol.*, 132, 297–308.

Lin, H., Lou, B., Glynn, J.M. et al. 2011. The complete genome sequence of "Candidatus Liberibacter solanacearum," the bacterium associated with potato zebra chip disease. *PloS One*, 6, e19135.

Liu, D., Anderson, N.A., and L.L. Kinkel. 1996. Selection and characterization of strains of *Streptomyces* suppressive to the potato scab pathogen. *Can. J. Microbiol.*, 42, 487–502.

Liu, D.Q., Anderson, N.A., and L.L. Kinkel. 1995. Biological control of potato scab in the field with antagonistic *Streptomyces scabies. Phytopathology*, 85(7), 827–831.

Lopes, C.A., Carvalho, S.I.C., and L.S. Boiteux. 2005. Search for resistance to bacterial wilt in a Brazilian capsicum germplasm collection. In: *Bacterial Wilt Disease and the Ralstonia solanacearum Species Complex*, ed. by C. Allen, P. Prior, and A.C. Hayward. APS press, St. Paul, MN, pp. 247–251.

Lopez-Lopez, G., Pascual, A., and J. Perea. 1991. Effect of plastic catheter material on bacterial adherence and viability. *J. Med. Microbiol.*, 34, 349–353.

Loria, R., Bukhalid, R.A., Fry, B.A. et al. 1997. Plant pathogenicity in the genus *Streptomyces*. *Plant Dis.*, 81, 836–846.

Loria, R., Coombs, J., Yoshida, M. et al. 2003. A paucity of bacterial root diseases: *Streptomyces* succeeds where others fail. *Physiol. Mol. Plant Pathol.*, 62(2), 65–60.

Marte, M. 1980. Histological and histochemical observations on tomato stems naturally infected by *Corynebacterium michiganense*. *Phytopathol. Z.*, 97, 252–271.

Matos, F.S.A., Lopes, C.A., and A. Takatsu. 1990. Identificação de fonts de resistência a *Pseudomonas solanacearum* en *Capsicum* spp. *Hortic. Brasil.*, 8, 22–23.

Matsunaga, H. and S. Monma. 1999. Sources of resistance to bacterial wilt in Capsicum. *J. Jpn. Soc. Hort. Sci.*, 68, 753–761.

McInnes, T.B., Gitaitis, R.D., McCarter, S.M. et al. 1988. Airborne dispersal of bacteria in tomato and pepper transplant fields. *Plant Dis.*, 72, 575–579.

McKee, R.K. 1958. Assessment of the resistance of potato varieties to common scab. *Eur. Potato J.*, 1, 65–80.

McKeen, W.E. 1981. Black rot of rutabaga in Ontario and its control. *Can. J. Plant Pathol.*, 3, 244–246.

Mguni, C.M. 1987. Diseases of crops in the semiarid areas of Zimbabwe: Can they be controlled economically. In: *Cropping in the Semiarid Areas of Zimbabwe. Workshop Proceedings, Agritex*, Harare, Zimbabwe, Vol. 2, pp. 417–433.

Mguni, C.M. 1995. Cabbage research in Zimbabwe. In: *Brassica Planning Workshop for East and South Africa Region*, Lilongwe, Malawi, May 15–18, 1995. GTZ/IPM Horticulture, Nairobi, Kenya, p. 31.

Mguni, C.M. 1996. Bacterial Black Rot (*Xanthomonas campestris* pv. *campestris*) of vegetable Brassicas in Zimbabwe. PhD thesis, Department of Plant Biology, The Royal Veterinary and Agricultural University, Copenhagen and Danish Government Institute for Developing Countries, Hellerup, Denmark.

Miles, G.P., Samuel, M.A., Chen, J. et al. 2010. Evidence that cell death is associated with zebra chip disease in potato tubers. *Am. J. Potato Res.*, 87, 337–349.

Mimura, Y., Matsunaga, H., Yoshida, T. et al. 2000. Pathogenicity of various isolates of *Ralstonia solanacearum* on tomato, eggplant and pepper varieties [abstract in Japanese]. *J. Jpn. Soc. Hort. Sci.*, 69(Suppl. 1), 231.

Minsavage, G.V., Dahlbeck, D., Whalen, M.C. et al. 1990. Gene-for-gene relationships specifying disease resistance in *Xanthomonas campestris* pv. *vesicatoria*—Pepper interactions. *Mol. Plant-Microbe Interact.*, 3, 41–47.

Miura, L., Romeiro, R., and J.C. Gomes. 1986. Production, purification and biological activity of an exotoxin produced *in vitro* by *Corynebacterium michiganense* pv. *michiganense*. *Fitopatol. Brasil.*, 11, 789–794.

Mizuno, N., Yoshida, H., and K. Yamamoto. 1995. Effect of ionic strength and fertilization method on the occurrence of potato scab. *Jpn. J. Soil Sci. Plant Nutr.*, 66, 639–645.

Monma, S., Narikawa, T., and Y. Sakata et al. 1993. Resistance to bacterial wilt in tomato cultivars [in Japanese with English summary]. *Bull. Natl. Res. Inst. Veg. Ornam. Plants Tea*, A6, 1–12.

Munyaneza, J.E. 2010. Psyllids as vectors of emerging bacterial diseases of annual crops. *Southwestern Entomol.*, 35, 417–477.

Munyaneza, J.E. 2012. Zebra chip disease of potato: Biology, epidemiology, and management. *Am. J. Potato Res.*, 89(5), 329–350.

Munyaneza, J.E., Buchman, J.L., and J.M. Crosslin. 2009a. Seasonal occurrence and abundance of the potato psyllid, Bactericera cockerelli, in south central Washington. *Am. J. Potato Res.*, 86, 513–518.

Munyaneza, J.E., Buchman, J.L., Goolsby, J.A. et al. 2010a. Impact of potato planting timing on zebra chip incidence in Texas. In: F. Workneh and C.M. Rush (eds.), *Proceedings of the 10th Annual Zebra Chip Reporting Session*, Dallas, TX, November 7–10, 2010, pp. 106–109.

Munyaneza, J.E., Buchman, J.L., Heilman, B.E. et al. 2011a. Update on potato variety screening trial for zebra chip under controlled field cage conditions. In: F. Workneh, A. Rashed, and C.M. Rush (eds.), *Proceedings of the 11th Annual Zebra Chip Reporting Session*, Dallas, TX, November 6–9, 2011, pp. 106–109.

Munyaneza, J.E., Buchman, J.L., Heilman, B.E. et al. 2011b. Effects of zebra chip and potato psyllid on potato seed quality, In: F. Workneh, A. Rashed, and C.M. Rush (eds.), *Proceedings of the 11th Annual Zebra Chip Reporting Session*, Dallas, TX, November 6–9, 2011, pp. 37–40.

Munyaneza, J.E., Buchman, J.L., Sengoda, V.G. et al. 2010b. Potato variety screening trial for zebra chip resistance under controlled field cage conditions. In: F. Workneh and C.M. Rush (eds.), *Proceedings of the 10th Annual Zebra Chip Reporting Session*, Dallas, TX, November 7–10, 2010, pp. 200–203.

Munyaneza, J.E., Buchman, J.L., Upton, J.E. et al. 2008. Impact of different potato psyllid populations on zebra chip disease incidence, severity, and potato yield. *Subtrop. Plant Sci.*, 60, 27–37.

Munyaneza, J.E., Crosslin, J.M., and J.E. Upton. 2007a. Association of *Bactericera cockerelli* (Homoptera: *Psyllidae*) with "zebra chip", a new potato disease in southwestern United States and Mexico. *J. Econ. Entomol.*, 100, 656–663.

Munyaneza, J.E., Goolsby, J.A., Crosslin, J.M. et al. 2007b. Further evidence that zebra chip potato disease in the lower Rio Grande Valley of Texas is associated with *Bactericera cockerelli*. *Subtrop. Plant Sci.*, 59, 30–37.

Munyaneza, J.E. and D.C. Henne. 2012. Leafhopper and psyllid pests of potato. In: P. Giordanengo, C. Vincent, and A. Alyokhin (eds.), *Insect Pests of Potato: Global Perspectives on Biology and Management*, Academic Press, New York, pp. 65–102.

Munyaneza, J.E., Sengoda, V.G., Crosslin, J.M. et al. 2009b. First report of *"Candidatus Liberibacter psyllaurous"* in potato tubers with zebra chip disease in Mexico. *Plant Dis.*, 93, 552.

Munyaneza, J.E., Sengoda, V.G., Crosslin, J.M. et al. 2009c. First report of *'Candidatus Liberibacter solanacearum'* in tomato plants in Mexico. *Plant Dis.*, 93, 1076.

Munyaneza, J.E., Sengoda, V.G., Buchman, J.L. et al. 2012a. Effects of temperature on *'Candidatus Liberibacter solanacearum'* and zebra chip potato disease symptom development. *Plant Dis.*, 96, 18–23.

Munyaneza, J.E., Sengoda V.G., Stegmark, R. et al. 2012b. First report of *"Candidatus Liberibacter solanacearum"* associated with psyllid-affected carrots in Sweden. *Plant Dis.*, 96, 453.

Munyaneza, J.E., Sengoda, V.G., Sundheim, L. et al. 2012c. First report of *"Candidatus liberibacter solanacearum"* associated with psyllid-affected carrots in Norway. *Plant Dis.*, 96, 454.

Nanri, N., Gohda, Y., Ohno, M. et al. 1992. Growth promotion of fluorescent pseudomonads and control of potato common scab in field soil with non-antibiotic actinomycete-biofertilizer. *Biosci. Biotech. Biochem.*, 56, 1289–1292.

Naumann, K. 1980. Pith necrosis of tomato. *Nachrichten blatt fur den pflanzenschutz in der. DDR*, 34, 157–159.

Nelson, W.R., Fisher, T.W., and J.E. Munyaneza. 2011. Haplotypes of *"Candidatus Liberibacter solanacearum"* suggest long-standing separation. *Eur. J. Plant Pathol.*, 130, 5–12.

Nelson, W.R., Sengoda, V.G., Alfaro-Fernandez, A.O. et al. 2013. A new haplotype of *"Candidatus liberibacter solanacearum"* identified in the Mediterranean region. *Eur. J. Plant Pathol.*, 135(4), 633–639.

Nemeth, J. and E.M. Laszlo. 1983. Bacterial black rot (*Xanthomonas campestris* (Pammel) Dowson 1939) of *Brassica* species. *Novenyvedelem*, 19, 391–397.

Onsando, J.M. 1988. Management of black rot of cabbage caused by *Xanthomonas campestris* pv. *campestris* in Kenya. *Acta Hortic.*, 218, 311–314.

Onsando, J.M. 1992. Black rot of crucifers. In: U.S. Singh, A.N. Mukhopadyay, and J. Kumar (eds.), *Plant Diseases of International Importance Chaube. Diseases of Vegetables and Oil Seed Crops*. Prentice Hall, Englewood Cliffs, NJ, pp. 243–252.

Osaki, K. and T. Kimura. 1992. Grouping of *Pseudomonas solanacearum* on the basis of pathogenicity to Solanum plants [in Japanese with English summary]. *Bull. Chugoku Natl. Agric. Exp. Stn.*, 10, 49–58.

Patel, M.K., Abhyankar, S.G., and Y.S. Kulkarni. 1949. Black rot of cabbage. *Indian J. Phytopathol.*, 2, 58–61.

Patel, M.K. and Y.S. Kulkarni. 1951. Nomenclature of bacterial plant pathogens. *Indian Phytopathol.*, 4, 74–84.

Pernezny, K., Datnoff, L., and M.L. Sommerfeld. 1994. Brown stem of celery caused by *Pseudomonas cichorii*. *Plant Dis.*, 78, 917–919.

Pitman, A.R., Drayton, G.M., Kraberger, S.J. et al. 2011. Tuber transmission of *"Candidatus Liberibacter solanacearum"* and its association with zebra chip on potato in New Zealand. *Eur. J. Plant Pathol.*, 129(3), 389–398.

Plumb-Dhindsa, P. and A. M. Mondjane. 1984. Index of plant diseases and associated organisms of Mozambique. *Trop. Pest Manage.*, 30, 407–429.

Pohronezny, K., Moss, M.A., Dankers, W. et al. 1990. Dispersal and management of *Xanthomonas campestris* pv. *vesicatoria* during thinning of direct-seeded tomato. *Plant Dis.*, 74, 800–805.

Quezado-Soares, A.M. and C.A. Lopes. 1995. Stability of the resistance to bacterial wilt of the sweet pepper 'MC-4' challenged with strains of *Pseudomonas solanacearum*. *Fitopatol. Bras.*, 20, 638–641.

Radunovic, D. and J. Balaz. 2012. Occurrence of *Xanthomonas campestris* pv. *campestris* (Pammel, 1895) Dowson 1939, on Brassicas in Montenegro. *Pestic. Phytomed.* (Belgrade), 27(2), 131–140.

Rat, B., Poissonnier, J., Goisque, M.J. et al. 1991. Le point sur le chancre bactérien. *Fruit Légumesi*, 86, 38–40.

Reeves, J.C., Hutchins, J.D., and S.A. Simpkins. 1996. The incidence of races of *Pseudomonas syringae* pv. *pisi* in UK pea (*Pisum sativum*) seed stocks, 1987–1994. *Plant Var. Seeds*, 9, 1–8.

Rehman, M., Melgar, J.C., Rivera, C.J. et al. 2010. First report of *"Candidatus liberibacter psyllaurous"* or *"Ca. liberibacter solanacearum"* associated with severe foliar chlorosis, curling, and necrosis and tuber discoloration of potato plants in Honduras. *Plant Dis.*, 94(3), 376–377.

Reifschneider, G.J.B., Bongiolo, N.A., and A. Takatsu. 1985. Reappraisal of *Xanthomonas campestris* pv. *vesicatoria* strains—Their terminology and distribution. *Fitopatologia Bras.* 10, 201–204.

Ren, J.P., Dickson, M.H., and R. Petzoldt. 1995. Improving the level of resistance to soft rot (*Erwinia carotovora*) by recurrent selection in *Brassica rapa*. *Eucarpia Crucferae Newslett.*, 17, 8687.

Roberts, S.J. 1993. Effect of bacterial blight (*Pseudomonas syringae* pv. *pisi*) on the growth and yield of single pea (*Pisum sativum*) plants under glasshouse condition. *Plant Pathol.*, 42, 568–576.

Samson, R., Legendre, J.B., Christen, R. et al. 2005. Transfer of *Pectobacterium chrysanthemi* (Burkholder et al. 1953) Brenner et al. 1973 and *Brenneria paradisiaca* to the genus Dickeya gen. *nov.* as *Dickeya chrysanthemi comb. nov.* and *Dickeya paradisiaca comb. nov.* and delineation of four novel species, *Dickeya dadantii* sp. *nov.*, *Dickeya dianthicola* sp. *nov.*, *Dickeya dieffenbachiae* sp. *nov.*, and *Dickeya zeae* sp. *nov. Int. J. Syst. Evol. Microbiol.*, 55, 1415–1427.

Schaad, N.W. (ed). 1988. Laboratory guide for identification of plant pathogenic bacteria. 2nd ed. The American Phytopathological Society, St. Paul, MN.

Schaad, N.W. and N. Thaveechai. 1983. Black rot of crucifers in Thailand. *Plant Dis.*, 67, 1231–1234.

Schaad, N.W. and D. Brenner. 1977. A bacterial wilt and root rot of sweetpotato caused by *Erwinia chrysanthemi*. *Phytopathology*, 67, 302–308.

Scheck, H.J. and J.W. Pscheidt. 1998. Effect of copper bactericides on copper-resistant and -sensitive strains of *Pseudomonas syringae* pv. *syringae*. *Plant Dis.*, 82, 397–406.

Schoneveld, J.A. 1994. Effect of irrigation on the prevention of scab in carrots. *Acta Hortic.*, 354, 135–144.

Scortichini, M. 1989. Occurrence in soil and primary infections of *Pseudomonas corrugata* Roberts and Scarlett. *J. Phytopathol.*, 125, 33–40.

Scortichini, M. 1994. Occurrence of *Pseudomonas syringae* pv. *actinidiae* on kiwifruit in Italy. *Plant Pathol.*, 43, 1035–1038.

Scortichini, M., Marchesi, U., Dettori, M.T. et al. 2003. Genetic diversity, presence of the syrB gene, host preference and virulence of *Pseudomonas syringae* pv. *syringe* strains from woody and herbaceous host plant. *Plant Pathol.*, 52, 277–286.

Scortichini, M. and M.P. Rossi. 1993. Response of some wild species of *Lycopersicon* and tomato cultivars to *Pseudomonas corrugata* Roberts et Scarlett. *Phytopathol. Mediterr.*, 32, 223–227.

Secor, G.A., Rivera-Varas, V., Abad, J.A. et al. 2009. Association of "*Candidatus* Liberibacter solanacearum" with zebra chip disease of potato established by graft and psyllid transmission, electron microscopy, and PCR. *Plant Dis.*, 93, 574–583.

Sengoda, V.G., Munyaneza, J.E., Crosslin, J.M. et al. 2010. Phenotypic and etiological differences between psyllid yellows and zebra chip diseases of potato. *Am. J. Potato Res.*, 87, 41–49.

Severin, V. 1971. Investigations on the prevention of the common blight of beans (*Xanthomonas phaseoli*). *Analele Institutului de Cercetari pentru Protectia Plantelor*, 7, 125–139.

Sharma, J., Agrawal, K., and D. Singh. 1992. Detection of *Xanthononas campestris* pv. *campestris* (Pammel) Dowson infection in rape and mustard seeds. *Seed Res.*, 20, 128–33.

Shew, H.D. and G.B. Lucas (eds.). 1991. *Compendium of Tobacco Diseases*. APS Press, St. Paul, MN.

Shiomi, T. 1992. Black rot of cabbage seeds and its disinfection under a hot-air treatment. *Jpn. Agric. Res. Q.*, 26, 13–18.

Shyam, K.R., Gupta, S.K., and R.K. Mandradia. 1994. Prevalence of different types of curd rots and extent of yield loss due to plant mortality in cauliflower seed crop. *Ind. J. Mycol. Plant. Path.*, 24(3), 172–175.

Sinclair, J.B. and P.A. Backman. 1989. *A Compendium of Soybean Diseases*, 3rd edn. American Phytopathological Society, St. Paul, MN.

Singh, Y. and S. Sood. 2004. Screening of sweet pepper germplasm for resistance to bacterial wilt (*Ralstonia solanacearum*). *Capsicum Eggplant Newslett.*, 23, 121–124.

Sowell, G., Jr. 1960. Bacterial spot resistance of introduced peppers. *Plant Dis. Rep.*, 44, 587–590.

Sowell, G. and A.H. Dempsey. 1977. Additional sources of resistance to bacterial spot of pepper. *Plant Dis. Rep.*, 61, 684–686.

Stall, R.E. and A.A. Cook. 1966. Multiplication of *Xanthomonas vesicatoria* and lesion development in resistant and susceptible pepper. *Phytopathology*, 56, 1152–1154.

Stamova, L. and V. Sotirova. 1987. Reaction of different crops to artificial inoculation with *Corynebacterium michiganense*. *Archiv für Phytopathologie und Pflanzenschutz*, 23, 211–216.

Stevenson, W.R., Loria, R., Franc, G.D. et al. (eds.) 2001. *Compendium of Potato Diseases*, 2nd ed. APS Press, St. Paul, MN.

Strider, D.L. 1969. Bacterial canker of tomato caused by *Corynebacterium michiganense*: A literature review and bibliography. *N.C. Agric. Exp. Stn. Tech. Bull*, 193: 1–110.

Suzuki, I., Sugahara, Y., Kotani, A. et al. 1964. Studies on breeding eggplants and tomatoes for resistance to bacterial wilt. I. Investigations on method of evaluating the resistance and on the source of resistance in eggplants and tomatoes [in Japanese with English summary]. *Bull. Hort. Res. Stn., Jpn. Ser. A*, 3, 77–106.

Taylor, J.D., Bevan, J.R., Crute, I.R. et al. 1989. Genetic relationship between races of *Pseudomonas syringae* pv. *pisi* and cultivars of *Pisum sativum*. *Plant Pathol.*, 38, 364–375.

Thaxter, R. 1891. The potato scab. *Conn. Agric. Expt. Stn. Rept.*, 1890, 81–95.

Thyr, B.D., Samuel, M.J., and P.G. Brown. 1975. New solanaceous host records for *Corynebacterium michiganense*. *Plant Dis. Rep.*, 59, 595–598.

Tilford, P.E. 1936. Fasciation of sweet peas caused by *Phytomonas fascians* n. sp. *J. Agric. Res.*, 53, 383–394.

Tsuro, M., Minamiyama, Y., and M. Hirai. 2007. QTL analysis for bacterial wilt resistance in Japanese pepper (*Capsicum annuum* L.) [in Japanese]. *Breed. Res.*, 9, 111–115.

van Hall, C.J.J. 1902. Bijdragen tot de kennis der Bakterieele Plantenzeikten, Cooperatieve Drukerijvereeniging "Plantijn". Amsterdam, Inaugural dissertation (Doctor's thesis), University of Amsterdam, Amsterdam, the Netherlands.

Van Vaerenbergh, J.P.C. and J.F. Chauveau. 1985. Host plant inoculation for detection of (latent) *Corynebacterium michiganense*. *Mededelingen van de Faculteit Landbouwwetenschappen, Rijksuniversiteit Gent*, 50(3a), 973–995.

Vauterin, L., Hoste, B., Kersters, K. et al. 1995. Reclassification of *Xanthomonas*. *Int. J. Syst. Bacteriol.*, 45, 472–489.

Waksman, S.A. and A.T. Henrici. 1948. Family III. Streptomycetaceae Waksman, S.A. and A.T. Henrici. In: Breed, Murray, and Hitchens (eds.), *Bergey's Manual of Determinative Bacteriology*, 6th ed. The Williams and Wilkins Co., Baltimore, MA, pp. 929–980.

Wallen, V.R. and H.R. Jackson. 1975. Model for yield loss determination of bacterial blight of field beans utilizing aerial infrared photography combined with field plot studies. *Phytopathology*, 65, 942–948.

Wang, J.F. and T. Berke. 1997. Sources of resistance to bacterial wilt in *Capsicum annuum*. *Capsicum Eggplant Newslett.*, 16, 91–93.

Wen, A., Mallik, I., and V.Y. Alvarado. 2009. Detection, distribution, and genetic variability of "*Candidatus liberibacter*" species associated with zebra complex disease of potato in North America. *Plant Dis.*, 93, 1102–1115.

Williams, P.H. 1980. Black rot: A continuing threat to world crucifers. *Plant Dis.*, 64, 736–742.

SUGGESTED READING

Brown, N.A. 1927. Sweet pea fasciation, a form of crown gall. *Phytopathology*, 17, 29–30.

Candidatus liberibacter solanacearum (zebra chip). (n.d.). Retrieved July 19, 2015, http://www.cabi.org/isc/datasheet/109434.

Catara, V. and P. Bella. 2009. Occurrence of tomato pith necrosis caused by *Pseudomonas marginalis* in Italy. *New Dis. Rep.*, 19, 58.

EPPO. 2011. *Curtobacterium flaccumfaciens* pv. *flaccumfaciens*. *OEPP/EPPO Bull.*, 41, 320–328.

EPPO/CABI. 1996. *Xanthomonas axonopodis* pv. *phaseoli*. In: I.M. Smith, D.G. McNamara, P.R. Scott, and M. Holderness (eds.), *Quarantine Pests for Europe*, 2nd edn. *CAB INTERNATIONAL*, Wallingford, U.K.

French-Monar, R.D., Patton, A.F. III, and J.M. Douglas. 2010. First report of "*Candidatus Liberibacter solanacearum*" on field tomatoes in the United States. *Plant Dis.*, 94(4), 481. http://apsjournals.apsnet.org/loi/pdis.

Garrett, K.A. and H.F. Schwartz. 1998. Epiphytic *Pseudomonas syringae* on dry beans treated with copper based bactericides. *Plant Dis.*, 82, 30–35.

Graham, D.C. and Dowson, W.J., 1960. The coliform bacteria associated with potato black-leg and other soft rots. Vol. I. Their pathogenicity in relation to temperature. *Ann. Appl. Biol.*, 48: 51–57.

Grogan, R.G. and K.A. Kimble. 1967. The role of seed contamination in the transmission of *Pseudomonas phaseolicola* in *Phaseolus vulgaris*. *Phytopathology*, 57, 28–31.

Hall, R. and C.B. Nasser. 1996. Practice and precept in cultural management of bean diseases. *Can. J. Plant Pathol.*, 18(2), 176–185.

Hall, R.J.B. 1991. *A Compendium of Bean Diseases*. American Phytopathological Society, St. Paul, MN. Paul, MN.

IMI. 1996. *Distribution Mapis of Plant Diseases*, No. 401, 3rd edn. CAB INTERNATIONAL, Wallingford, U.K.

Jones, C., Conn, K., and P. Himmel (eds.) 2006. Bacterial spot. In: *Pepper Eggplant Disease Guide.* Seminis Vegetable Seeds, Oxnard, CA, pp. 2–3.

Jones, J.B. and K. Pernezny. 2003. Bacterial spot. In: K. Pernezny, P.D. Roberts, J.F. Murphy, and N.P. Goldberg (eds.), *Compendium of Pepper Diseases.* American Phytopathological Society, St. Paul, MN, pp. 6–7.

Jones, J.B., R.E. Stall, and H. Bouzar. 1998. Diversity among *Xanthomonads* pathogenic on pepper and tomato. *Ann. Rev. Phytopathol.,* 36, 41–58.

Jones, R.K. 1993. Southern bacterial wilt. In: J.W. White (ed.), *Geranums IV.* Ball Publishing, Geneva, Switzerland.

Liefting, L.W., B.S. Weir, and S.R. Pennycook et al. 2009c. *"Candidatus Liberibacter solanacearum,"* associated with plants in the family Solanaceae. *Int. J. Syst. Evol. Microbiol.,* 59, 2274–2276.

Matsumoto, M. 1988. Studies on the occurrence of Gomasho of Chinese cabbage and its prevention (1) morphological and histochemical observations of Gomasho. *J. Jpn. Soc. Hortic. Sci.,* 57(2), 206–214.

McKenzie, C.L. and R.G. Shatters Jr. 2009. First report of *"Candidatus Liberibacter psyllaurous"* associated with psyllid yellows of tomatoin Colorado. *Plant Dis.,* 93, 1074.

Pith necrosis of tomato (*Pseudomonas corrugata*). (n.d.). Retrieved May 22, 2015, from http://www.plantwise. org/KnowledgeBank/Datasheet.aspx?dsid=44945.

Tabilio, M.R., Chiariotti, A., Di Prospero, P. et al. 1998. Hedgerows: A barrier against *Pseudomonas syringae* pv. *syringae* infections in an organic peach orchard. *Acta Hortic.,* 465, 703–708.

Taylor, J.D. 1972. Races of *Pseudomonas pisi* and sources of resistance in field and garden peas. *NZ J. Agric. Res.,* 15, 441–447.

Taylor, J.D. 1972. Specificity of bacteriophages and antiserum for *Pseudomonas pisi. NZ J. Agric. Res.,* 15, 421–431.

Zaumeyer, W.J. and H.R. Thomas. 1957. A monographic study of bean diseases and methods for their control. Technical Bulletin, USDA No. 865, 255pp.

Zaumeyer, W.J. 1932. Comparative pathological history of three bacterial diseases of bean. *J. Agric. Res.,* 44, 605–632.

Zaumeyer, W.J. 1930. The bacterial blight of beans caused by *Bacterium phaseoli. USDA Tech. Bull.,* 186, 36pp.

9 Bacterial Diseases in Spice Crops

9.1 ONION

9.1.1 BACTERIAL SOFT ROT IN ONION

Pathogen: *Erwinia carotovora* subsp. *carotovora, E. chrysanthemi,*
    Pseudomonas gladioli, and *Enterobacter cloacae*

Synonyms: *Pectobacterium carotovora* subsp. *carotovora, Dickeya chrysanthemi*

Bacterial decay, known as soft rot, is one of the most widespread and destructive storage diseases in onion. Soft rot generally starts in the field just before or during harvest. The disease causes serious losses to onions that have been wounded in the field or improperly handled and stored.

Bacterial soft rot pathogens have very broad host ranges and can attack many vegetables, such as carrot, potato, cabbage, and lettuce. Infection generally requires a wound caused by heavy, wind-driven rain, hail, insects, or cut necks during harvest. Splashing water, aerosols, contaminated equipment and workers, and insects spread soft rot bacteria. Bacterial soft rot pathogens are commonly found in and easily disseminated by irrigation water. *Erwinia* spp. survive between onion crops in soil, crop debris, and pathogenically on other crops.

Bacterial soft rot is a major postharvest disease in onion bulbs and causes serious problems for growers and packers in New Mexico. Although several types of bacteria are associated with postharvest decay through secondary infection, only *Pectobacterium carotovorum* and *Dickeya chrysanthemi* produce pectolytic enzymes and can enter a bulb without preexisting injury to cause the specific disease known as "soft rot." Bacterial soft rot is most prevalent during the rainy months of July and August or anytime when wet conditions prevail before or during harvest.

Symptoms

Symptoms of bacterial soft rot often appear as a soft, watery rot of individual scales that may advance and rot the entire bulb. A foul-smelling viscous fluid oozes from the neck when infected bulbs are squeezed. In the field, the youngest leaves or the entire foliage of affected plants appear bleached and wilted (Photo 9.1). Yield losses can be significant in the field and storage.

Bacterial soft rot is mainly a problem on mature bulbs. Affected scales first appear water-soaked and pale yellow to light brown when infected by *D. chrysanthemi* or bleached gray to white when infected with *Pectobacterium carotovorum* subsp. *carotovorum*. As the soft rot progresses, invaded fleshy scales become soft and sticky with the interior of the bulb breaking down. A watery, foul-smelling thick liquid can be squeezed from the neck of diseased bulbs.

Onion bulbs in the field may be affected before harvest, but infected bulbs usually go unnoticed until after harvest. Bacterial soft rot usually starts at the neck of the bulb, and progresses downward along one or more scales. Initially the tissue is water soaked and later it disintegrates into a soft, slimy mass. The decay does not spread readily from scale to scale. One or two scales may be completely rotten, while the remainder are fine. Eventually, the diseased bulbs can be detected by gently squeezing them, whereupon a watery fluid is exuded. An offensive sulfurous odor is usually associated with the liquid.

Although these bacteria can directly enter plant tissue, existing wounds may intensify the infection rate. The pathogen also requires moisture to infect plant tissues. Therefore, susceptibility increases if bulbs have mechanical injuries, sunscalds, or bruises, especially when stored under warm, humid conditions.

PHOTO 9.1 Symptoms on onion foliage, their collapse, and early season soft rotting in onion bulb due to soft rot bacteria. (Photo courtesy of Howard F. Schwartz, Colorado State University and Gerald Holmes, California Polytechnic State University, San Luis Obispo, CA, Bugwood.org.)

Geographical Distribution
Worldwide

Disease Cycle
Bacterial soft rots are a primary problem in onions, but not in garlic. Free water is essential for entry and spread of the bacteria. Wounds and senescent leaves are the means by which bacteria gain entrance into the bulb. The pathogens are soil borne and may spread via irrigation water.

The bacteria enter only through wounds created by insect feeding, or bruising, during harvest or packing. The disease spreads in storage if moisture is allowed to develop on bulb surfaces.

Soft rot bacteria commonly exist in the soil and plant refuse. They enter onions through wounds and aging tissue under moist conditions. Excessive irrigation during hot weather seems to favor a high incidence of soft rot.

Onion maggots are particularly effective in spreading the disease. They feed on bulbs, creating wounds for bacteria to enter. The bacteria may persist in the intestinal tract of the onion maggot larvae and adult flies and thus can be carried from one place to another. Other sources of wounds that serve as entry for bacteria include hail damage in leaves and developing bulbs, sunscald, and freezing. During harvest, the practice of cutting the onion tops creates openings for the bacteria. Also, bruising or other damage from mechanical activity leaves bulbs susceptible to infection, particularly if they are stored in warm, humid conditions. The disease cycle of soft rot of onion is illustrated in Figure 9.1.

Conditions for Disease Development
Bacterial soft rot is most common in onions in storage or transit; however, this disease can develop in onions in the field before harvest, after heavy rains, and also when leaves are drying. The main sources of inoculum are contaminated soil and crop residues. The bacteria is spread by splashing rain, irrigation water, and insects. This disease is favored by warm, humid conditions with an optimum temperature range of 20°C–30°C. However, soft rot can develop during storage or transit at temperatures above 3°C.

Disease Management

Cultural Practices
Avoid overhead irrigation once onions start to bulb (bulbing occurs about the time the bulb is twice the diameter of the neck) and control insect pests, such as onion maggots. Harvest only after onion tops are well matured. Provide for quick drying following topping, especially if temperatures are high.

Practice a 3-year or longer rotation to nonhosts such as small grains. Avoid reuse of irrigation tail water and overhead irrigation. Prevent bruising and wounding of plants and bulbs during field operations, harvest, and handling. Control onion maggots and other insects that can vector soft rot bacteria. Sever all roots during lifting to promote rapid drying of foliage and necks before topping.

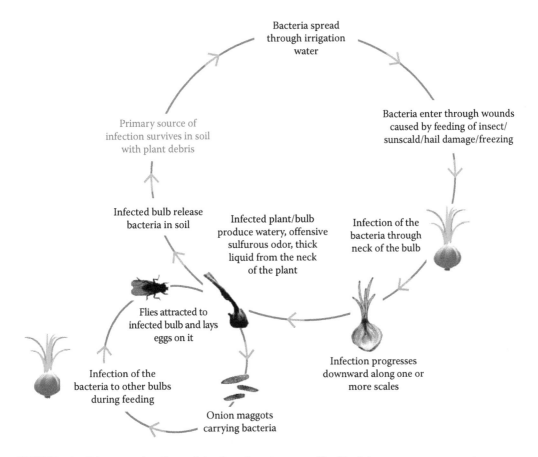

FIGURE 9.1 Disease cycle of bacterial soft rot in onion caused by *Erwinia carotovora* pv. *carotovora*.

Cure bulbs with abundant ambient air until necks are completely dry before storing bulbs at 0°C–3°C and 70% or less relative humidity.

Disease spread and infection may be reduced by copper-based bactericides. Allow onion tops to mature before harvesting and avoid damaging bulbs during harvest. Store onion bulbs only after they have been properly dried under appropriate temperature and humidity with good ventilation to prevent moisture condensation in the bulbs.

The first step in controlling soft rot is to control other diseases and avoid injury to the plants. This includes controlling insects such as onion maggots that wound plants and transmit the bacteria.

Allow the crop to mature completely before harvesting so that the neck areas are dried and they act as a barrier for bacterial movement from infected leaves to the bulbs. The tops should be dried out as much as possible after lifting and before topping. Mechanical toppers that cut the neck about ½ in. above the bulb should be designed to minimize bruising. Roller toppers should not be used because they rip and tear the neck leaves. At all stages of harvest and storage, onions should be handled carefully to avoid bruising.

Cure bulbs thoroughly so that the outer scales and neck tissues are completely dry. This is usually done with forced air in storage bins. Remove bulbs that show any sign of disease or injury before storage.

There are no soft rot–resistant onion varieties available.

Storms with hard rains, high winds, and hail cause wounds on leaves and bulbs that serve as entry points for bacteria. If these wounds, within hours after they occur, are protected with sprays containing copper-based materials, they will likely heal rapidly and infection will be reduced. Treatments after appearance of symptoms will not control bacterial soft rot.

Chemical Control

In Colorado, copper bactericides provide some bacterial control when applied before disease onset. Sprays should be initiated 2 weeks before bulb initiation and continued on a 5–10-day spray interval depending on weather conditions. Apply in a sufficient volume of water to ensure thorough coverage. Include a low rate of a nonionic surfactant to further improve coverage.

Copper-tolerant strains of the pathogens are common in the United States. Tank-mixing copper bactericides with a low rate of an Ethylenebisdithiocarbamates (EBDCs) fungicide such as maneb is essential for effective disease suppression. Tank-mixing coppers with zinc or iron can also enhance their activity.

9.1.2 Onion Sour Skin

Pathogen: *Pseudomonas (Burkholderia) cepacia*

Synonyms: *Burkholderia cepacia*

Sour skin, first described in 1950, has been reported from onion-growing areas all over the world. Losses often appear in stored onions, but infection usually begins in the field. The disease can be serious in individual fields, with yield losses of 5%–50%. Sour skin is primarily a disease in onions, but other *Allium* species are reported to be hosts.

Symptoms

Field symptoms often appear when one or two infected inner leaves turn light brown or yellow. A watery rot develops at the base of the leaves and proceeds to the neck, allowing the leaves to be easily pulled from the bulb. As the disease progresses, the outer bulb scales are infected. Infected scales develop a slimy pale yellow to light brown decay and may separate from adjacent scales, allowing the firm center scales to slide out when the bulb is squeezed. Infected bulbs often have an acrid, vinegar-like odor due to secondary invaders, especially yeasts, colonizing decaying bulbs.

Primary symptoms on onions include a slimy (but initially firm), pale yellow to light brown decay and breakdown of one or a few inner bulb scales. Adjacent outer scales and the center of the bulb may remain firm (Photo 9.2). Externally, bulbs appear healthy, but the neck region may soften after leaves have collapsed. In advanced stages, healthy scales can slip off during handling. Young leaves sometimes dieback, starting at the tips.

PHOTO 9.2 Yellowing of inner leaves and separation of bulb scales due to sour skin disease. (Photo courtesy of David B. Langston, University of Georgia, Athens, GA, Bugwood.org.)

A rot of bulb scales usually occurs at or near maturity and sometimes in storage. The bacterium does not appear to be strongly invasive, attacking plants that are damaged or weakened. Kawamoto and Lorbeer (1974) found that *Burkholderia cepacia* infected only wounded onion tissue in artificial inoculation tests. Inoculation with wounding resulted in lesion formation, but spray inoculations, without wounding, were unsuccessful. Lesions expanded very slowly in the absence of free moisture, and the rate of lesion development increased with surface leaf moisture and/or increasing temperature to 32°C.

In earlier experiments, a decrease in inoculum concentration from 100 million to 1000 cells/mL delayed initial symptom expression from 1 to 5 days in young onion leaf blades injected with *Burkholderia cepacia* (Kawamoto and Lorbeer, 1972b). As cell numbers increased to over 1000 million cells/mL, the first symptoms of water soaking and wilting became evident. Bacterial cells were found to have migrated through the intercellular spaces and formed large masses (Kawamoto and Lorbeer, 1972a). Maintenance of moisture allowed the formation of small groups of bacteria in the intercellular spaces of both blade and sheath. Host cells, except xylem vessels and epidermal cells, were compressed by the bacterial masses, and the large parenchyma cells in the blade were the first to collapse. Later, the more compact parenchyma near the periphery of the blade collapsed and tissues became macerated. In the sheath, bacteria were abundant among the loosely organized parenchyma adjacent to the adaxial epidermis. In the intercellular spaces of the closely packed parenchyma, beneath the abaxial epidermis, bacteria were usually restricted to dense, compact masses. Later it was found that bacterial numbers declined and none could be reisolated after leaves became dry and brittle. From these experiments, it was concluded that air-dried, diseased leaves were unlikely to be the source of inoculum in the field.

These findings were in part confirmed by Cother and Dowling (1985), and they concluded that bacteria were likely to gain entry through the neck or leaf blades as the foliage falls over, and the epidermis breaks, at maturity. Infections were thought to arise from contaminated irrigation water, and the expression of symptoms was heightened in hot weather.

Individual leaves affected by sour skin wilt and dieback. Internally, leaves develop a soft, watery rot. The fleshy scales associated with infected leaves rot to form a tan-colored slimy ring in the bulb. Adjacent rings may remain healthy. The neck of infected bulbs is soft when pressed.

Pathogen

Pseudomonas cepacia is obligately aerobic. The optimum growth temperature is 30°C–35°C. No growth occurs at 4°C, and most strains grow at 41°C. Denitrification is negative while nitrate is reduced to nitrite. It is oxidase positive and arginine dihydrolase negative and can liquefy gelatin.

Disease Cycle and Epidemiology

Apparently, onions are relatively resistant to *P. cepacia* prior to bulbing, or the environment does not become favorable for bacterial multiplication until after bulbing. Infection generally occurs through a wound when free water from rain, overhead irrigation, or flooding causes water congestion of the host tissue. The bacterium can gain entrance to the plant when onion tops are cut at harvest or through other wounds in the neck when the foliage falls over at maturity. Infection can also begin when water contaminated with bacterial cells strikes the younger upright leaves and flows down into the neck in the leaf blade *axil*. Young leaves are much more susceptible than mature leaves, which are usually symptomless. Infection can remain latent in the growing onion, and symptoms sometimes do not develop until the plant begins to bulb. Bacteria spread more rapidly in water-soaked tissue and when temperatures exceed 30°C. Infection advances into the bulb via the infected leaf and the corresponding scale. The infection does not move into adjacent scales.

Pseudomonas cepacia is commonly spread by heavy rains, overhead irrigation, and flooding, which splash the bacteria onto young or wounded foliage. Infection typically occurs through wounds, including those made when onions are cut at harvest. Infection can also occur when water lands on upright leaves and flows into leaf blade axils carrying the bacterium with it. Sour skin is favored by rainstorms and warm weather and develops rapidly at temperatures above 30°C. The disease cycle of onion sour skin is illustrated in Figure 9.2.

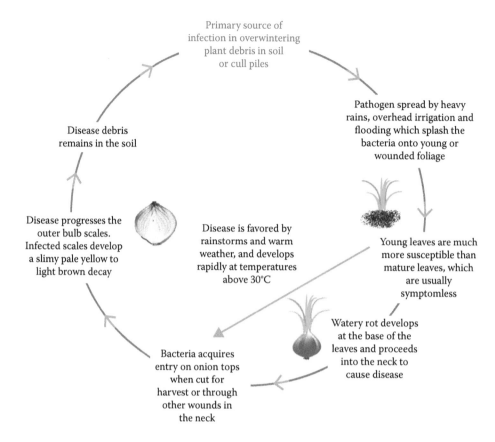

FIGURE 9.2   Disease cycle of onion sour skin caused by *Pseudomonas cepacia*.

Disease Management

Cultural Practices

Do not damage foliage prior to harvest or bulbs during harvest because *P. cepacia* enters the plant primarily through wounds. Onion crops should be harvested at maturity and the bulbs dried quickly. Storing onions at cool temperatures, 0°C, with adequate ventilation to prevent condensation on the bulbs will reduce storage losses resulting from this disease.

Since contaminated irrigation water has been implicated in the spread of the pathogen, the use of recycled or irrigation runoff water should be avoided. The method of irrigation has a substantial impact on the incidence of sour skin. Season-long overhead irrigation provides a favorable environment for infection by *P. cepacia*, whereas furrow irrigation results in almost complete absence of the disease. In experimental plots, the final four or five sprinkler irrigations were accompanied by increases in sour skin of 150%–300%. Where sour skin is a potential problem, changing from sprinkler to furrow irrigation, at least from bulbing to the end of the season, is advisable where feasible.

Studies from California, United States, have shown that *P. cepacia* can be controlled successfully by substituting furrow irrigation for overhead sprinkler irrigation (Teviotdale et al., 1990). Earlier work had shown that disease incidence was greater in plots irrigated by overhead sprinklers than in plots irrigated by furrow the entire season or irrigated by sprinklers until bulbing and then by furrow (Teviotdale et al., 1989). There was no correlation between the amount of water applied and incidence of disease. Similarly, there was no significant difference in the number of rotted onions between two planting densities.

Contaminated water is thought to be a major source of disease as disease incidence was found to increase when surface drainage water was used for irrigation as opposed to water drawn from deep wells (Irwin and Vaughan, 1972).

Economic Impact

In general, *P. cepacia* is not considered to be a serious pathogen of onion or other crops. While yield losses of 5%–50% have been noted (Schwartz and Mohan, 1995), the disease is sporadic, particularly in maturing bulbs or in storage (Cother and Dowling, 1985; Jaccoud Filho et al., 1987).

9.1.3 Bacterial Leaf Streak and Bulb Rot in Onion

Pathogen: *Pseudomonas viridiflava*

Symptoms

The first symptoms observed are oval, water-soaked leaf lesions, tipburn, and leaf streaking of varying lengths. Initially, leaf streaks are green but eventually darken to black. As infections become more severe and spread down the leaf, entire leaves collapse and dry. Leaf distortion and twisting may also occur (Photo 9.3).

Bulb infection is characterized by dark spots on outer scales and reddish brown discoloration of inner scales. Symptoms often develop in a ringlike pattern due to restriction of the rot by the scales.

Rotted scales are firmer than those caused by other bacteria, but infected bulbs can be colonized by other pathogens and secondary pathogens that can rot the bulb entirely. Yield losses of 20%–80% have been reported in Georgia, but the disease occurs infrequently in Colorado and the High Plains.

Life Cycle and Epidemiology

Bacterial streak and rot is caused by the bacterium *Pseudomonas viridiflava*. This bacterium efficiently colonizes the leaf surfaces of several weeds, crops, and onion. During cool, wet weather, the bacteria infect leaves and progress downward into the bulb where they cause a decay of inner scales. The pathogen survives between onion crops epiphytically and pathogenically on weeds and other crops.

PHOTO 9.3 Symptoms of *Pseudomonas viridiflava* on onion leaf and bulb. (Photo courtesy of Ronald D. Gitaitis, University of Georgia, Athens, GA, Bugwood.org.)

Conditions for Disease Development

This disease occurs particularly in winter and spring when temperatures are cool. Epidemics are associated with prolonged periods of rain, which favor progression of the disease. Excess fertilizer stimulates disease development. It is thought that frost damage may predispose onion plants to infection.

Disease Management

Cultural Control

Practice a 3-year or longer crop rotation to nonhosts such as small grains. Avoid rotations with pulse crops such as alfalfa, soybean, and dry bean, as these crops may serve as an alternative host of the bacterial leaf streak pathogen. Deeply bury crop residues after harvest to reduce pathogen survival. Eliminate weeds and volunteer onion in and around fields. If possible, avoid overhead irrigation and reuse of irrigation tail water. Drip irrigation may reduce disease severity compared to furrow irrigation. Apply adequate but not excessive nitrogen fertilizer, especially postbulbing or after storm damage. Avoid working in fields when the foliage is wet, which can easily spread bacteria throughout the field.

Plant wider rows in the direction of prevailing winds to increase air movement in the crop canopy and decrease periods of leaf wetness. Harvest and store onions only after thorough curing in the field and the packaging shed. Avoid wounding or bruising bulbs during harvest and storage operations. Store bulbs at 0°C–3°C with a relative humidity of 70% or less.

Chemical Control

Copper bactericides provide effective disease control in Colorado when applied before disease is observed. Sprays should be initiated 2 weeks before bulb initiation or when disease is first observed and continued on a 5–10-day spray interval depending on weather conditions. Apply in a sufficient volume of water, with a low rate of a nonionic surfactant, to ensure thorough coverage.

Copper-tolerant strains of the pathogen are common in the southeastern United States. If copper-tolerant strains are present, tank-mixing with a low rate of an EBDC fungicide such as maneb is essential for disease suppression. Tank-mixing coppers with zinc or iron can also enhance their activity.

Copper-Based Fungicides

Product List for Bacterial Streak and Bulb Rot

Pesticide	Product per Acre	Application Frequency (Days)	Remarks
Champ DP	1.33 lb	7–10	
Cuprofix	2.5–6 lb	7–10	Can be phytotoxic
Cuprofix MZ Disperss	3.5–5 lb	3–7	Maximum of 78.9 lb per season; 7-day pre-harvest interval (PHI)
Kocide 2000	1.5 lb	7–10	Can cause phytotoxicity to leaves
Kocide 3000	0.75 lb	7–10	Can cause phytotoxicity to leaves
ManKocide	2.5 lb	3–7	Maximum of 160 lb per season; 7-day PHI
Nordox	2–4 lb	7–10	
NuCop 50WP	2.0 lb	7–10	
Top Cop with S	2–3 qt	7–10	

9.1.4 Leaf Blight in Onion

Pathogen: *Xanthomonas axonopodis* pv. *allii*

Synonyms: *Xanthomonas campestris* pv. *allii* (Kadota et al.)

Xanthomonas leaf blight (also called onion bacterial blight or bacterial leaf blight) is caused by the bacterium *Xanthomonas axonopodis* pv. *allii*.

Xanthomonas leaf blight in onion has been reported from several parts of the world as an emerging disease. It was first observed in Barbados in 1971 and then spread to other continents (United States, Africa, and Asia).

The disease was first detected in the continental United States in Colorado in 1994; it has also been reported from Texas, California, Hawaii, Georgia, and several other tropical and subtropical onion-producing regions of the world. *Xanthomonas axonopodis* pv. *allii* is also pathogenic to other allium crops such as garlic, leek, chive, and Welsh onion. Strains of *X. axonopodis* pv. *allii* from Barbados are reportedly pathogenic to several pulse crops, including soybean and snap bean. However, strains from other onion-producing regions of the world are nonpathogenic or only weakly virulent to pulse crops.

In Colorado, *Xanthomonas* leaf blight (caused by *Xanthomonas axonopodis* pv. *allii*) is a common foliar, bacterial disease in onion. *Xanthomonas* leaf blight, also known as bacterial blight, in onion (*Allium cepa*) was first described in 1978 in Hawaii, but has rapidly spread to other onion and allium-producing regions in the East Caribbean, continental United States, South America, South Africa, Asia, and Réunion Island of France. A very similar disease was reported in onion in the Arkansas Valley in the 1940s and 1950s, but it was not reported again or investigated until 50 years later.

The disease reduces plant photosynthetic area, leading to stunting and a reduction in bulb size. Yield losses vary depending on infection timing, weather conditions, and cultivar susceptibility, but yield reductions of 20% or more are common.

Disease Symptoms

Xanthomonas leaf blight symptoms may appear at any stage of crop development on short-day onion cultivars, but generally develop during or after bulb initiation on long-day cultivars. Lesions initially appear as white flecks, pale spots, or lenticular lesions with water-soaked margins. Lesions quickly enlarge, become tan to brown in color, and cause extensive water soaking. Chlorotic streaks develop on some cultivars that may extend the entire length of leaves. When weather conditions become hot and dry, infected tissues or lesions dry out and become brittle, but retain their characteristic tan to brown color.

As the disease progresses, lesions fuse together and cause tip dieback, and extensive blighting of outer older leaves occurs. The loss of leaf area results in stunted plants and undersized bulbs. Under highly favorable disease conditions, all leaves may become completely blighted (Photo 9.4), resulting in premature plant death. Bulb rot is not known to occur. Symptoms are similar on chive (*A. schoenoprasum*), garlic (*A. sativum*), leek (*A. porrum*), shallot (*A. cepa* var. *ascalonicum*), and Welsh onion (*A. fistulosum*), but tend to be most severe on onion.

In case of severe outbreaks, premature plant death is observed. The disease is favored by temperatures higher than 27°C and severe outbreaks usually occur 7–10 days after a period of humid, rainy weather. In the United States, yield reductions of 20% or more are commonly observed in affected fields.

PHOTO 9.4 Field symptoms of *Xanthomonas* leaf blight on onion foliage. (Photo courtesy of Howard F. Schwartz, Colorado State University, Fort Collins, CO, Bugwood.org.)

Tip death and blighting of leaves reduces the plants' photosynthetic area, resulting in a reduction in bulb size. In some onion-producing regions, plants may also be killed before bulb initiation. Yield losses vary depending on the time of infection, but range from 20% to 100% under disease-favorable conditions. A bulb rot does not occur.

Predisposing Factor

Disease is favored by temperatures above 26°C. Frequent rains and high humidity promote disease development. Severe outbreaks are often associated with heavy rain, hail and windblown sand that damages foliage. Symptoms usually appear 7–10 days later. Spread of the pathogen within and between fields occurs with both overhead and furrow irrigation and movement of residual onion debris by field equipment. *Xanthomonas axonopodis* pv. *alli* is also seed transmitted. Frequent rains and overhead irrigation can initiate an epidemic from contaminated seed in semiarid environments. The bacterium survives on contaminated seed, in infested crop debris, and as an epiphyte or pathogen on volunteer onions, legumes, and weeds.

Disease Cycle

Xanthomonas axonopodis pv. *allii* survives between susceptible crops in association with contaminated seed and infested crop debris and epiphytically or pathogenically on volunteer onion, weed, and leguminous plants. Infection occurs when aerosols or splashing water deposit *X. axonopodis* pv. *allii* onto leaves. The bacterium then multiplies to form large populations in the presence of dew or other moisture, and infects hosts through natural openings, such as the stomata, or wounds. Leaf abrasion by wind and windblown sand appear to favor infection. The pathogen readily spreads within and between fields by surface irrigation water, and presumably by contaminated debris and exudates adhering to workers and equipment.

The bacterium persists and multiplies within and on common bean (*Phaseolus vulgaris*) and other leguminous plants, but generally causes mild or no disease symptoms on these hosts. However, some strains of the bacterium are aggressive on many leguminous hosts, such as lima bean (*Phaseolus lunatus*), soybean (*Glycine max*), winged bean (*Psophocarpus tetragonolobus*), moth bean (*Vigna aconitifolia*), and field pea (*Pisum sativum*), and causes disease symptoms on common bean identical to those by *X. axonopodis* pv. *phaseoli* (common bacterial blight).

Environmental, host, and cultural factors greatly affect the incidence and severity of bacterial diseases. *Xanthomonas* leaf blight is favored by high temperatures (>27°C) usually prevalent during July and August after bulb initiation. Symptoms usually appear on middle-aged to older leaves, even at low humidity.

Disease occurs when bacteria are blown or splashed onto leaves, where they multiply to form large populations (>10,000 colony forming units per leaf) in the presence of dew or other moisture, and infect through natural openings or wounds. Leaf abrasion by wind and windblown sand may favor infection. In Colorado, disease is associated with moderate to high temperatures and rainfall at bulb initiation and continuing through bulb development. Early season weather conditions do not appear to be significantly associated with disease appearance or severity in Colorado, but may be important when short-day onion varieties are grown under tropical or subtropical conditions. The pathogen can be disseminated between fields by irrigation water and contaminated workers and equipment. The pathogen survives between onion crops epiphytically and pathogenically on weeds, volunteer onion, leguminous hosts such as alfalfa, and contaminated seed and in crop debris.

Severe disease outbreaks often occur shortly (7–10 days) after a period of humid, rainy weather. Violent storms with hail and high winds cause plant wounding, which enable pathogens to enter and infect plant tissues. Wounds caused by damaging plants with farm machinery or field workers during cultivation can also spread pathogens, especially when the foliage is wet. Planting noncertified seed (which can be contaminated) or infected transplants and sets contribute to serious disease outbreaks due to the seed-borne nature of this pathogen.

Moderate to high temperatures and rainfall at bulb initiation and continuing through bulb development favor this disease. Frequent rains after bulb initiation, especially when driven by strong winds, favor severe disease epidemics. Overhead irrigation and humid, overcast conditions also appear to favor the disease. Distinct disease foci can develop and lead to secondary spread of the pathogen even when as little as 0.04% of seed is contaminated with the bacterium if abundant overhead irrigation is supplied to the crop. Frequent rains or overhead irrigation appear essential for seed contaminated with *X. axonopodis* pv. *allii* to produce an epidemic of *Xanthomonas* leaf blight in semiarid environments.

Many bacterial pathogens occur throughout the onion-growing areas of Colorado and surrounding states. Yield losses from bacterial pathogens, including bulb size and quality, range from a trace to 100%, especially when adverse environmental conditions persist after bulb initiation.

Disease Management

It must be emphasized that to minimize the impact of bacterial diseases on onion crops, the producer must carefully integrate recommended strategies of crop rotation, sanitation, use of clean seed and transplants, varietal selection, stress and wound avoidance, and proper pesticide selection and scheduling.

Plant seeds, transplants, and sets free of *X. axonopodis* pv. *allii*. Practice a 2-year or longer rotation to a nonsusceptible host such as winter wheat or corn. Avoid close rotations of onion and garlic with leguminous crops such as dry bean, soybean, or alfalfa. Varieties with complete resistance to *Xanthomonas* leaf blight are not commercially available, but varieties vary quantitatively in their reaction to the disease. Tolerant and moderately resistant varieties include white and red market-class varieties such as 'Cometa', 'Blanco Duro', and 'Redwing'. Yellow varieties such as 'X-202', 'Cannonball', and 'Vantage' are most susceptible.

Eliminate volunteer onion and weeds in and around fields. Prompt and thorough incorporation of crop debris into the soil after harvest reduces pathogen overwintering and survival between susceptible crops planted 2 or more years apart. *Xanthomonas* bacteria can survive between onion crops on weeds and crops such as dry bean, alfalfa, and other legumes; therefore, small grains, field corn, and other vegetables are preferred rotation crops. Avoid overhead irrigation and reuse of irrigation water. Avoiding excessive nitrogen fertilization can reduce *Xanthomonas* leaf blight severity.

Plant only pathogen-free seeds. Hot-water seed treatments may reduce seed contamination but can also reduce germination. Avoid rotations with pulse crops such as lentil, soybean, and dry bean, as these crops may serve as a reservoir of the *Xanthomonas* leaf blight pathogen. Deeply bury crop residues after harvest to reduce pathogen survival. Eliminate weeds and volunteer onion in and around fields. If possible, avoid overhead irrigation and reuse of irrigation tail water. Drip irrigation may reduce disease severity compared to furrow irrigation. Apply adequate but not excessive nitrogen fertilizer, especially postbulbing or after storm damage. Avoid working in fields when the foliage is wet, which can easily spread bacteria throughout the field. Plant wider rows in the direction of prevailing winds to increase air movement in the crop canopy and decrease periods of leaf wetness. Varieties vary widely in their reaction to *Xanthomonas* leaf blight, and highly susceptible varieties should be avoided. White and red market-class varieties tend to be less susceptible or more tolerant.

Chemical Control

Copper bactericides (e.g., cupric hydroxide formulations like Champ, Cuproxide, Kocide, NuCop) alone or tank-mixed with a low rate of an ethylenebis dithiocarbamate fungicide, such as maneb, provide effective disease control in semiarid production regions when initiated before disease development (1–2 weeks prebulbing). Spray intervals of 5–10 days are recommended in high gallonage to improve leaf coverage and canopy penetration. Work with disinfectants such as chlorine applied frequently to onion foliage with a ground rig provided little to no disease control.

Copper bactericides provide effective disease control in Colorado when applied before disease is observed. Sprays should be initiated 2 weeks before bulb initiation and continued on a 5–10-day spray interval depending on weather conditions. Apply in a sufficient volume of water, with a low rate of a nonionic surfactant, to ensure thorough coverage. Disease forecast models have been developed in Colorado that may improve spray timing and effectiveness.

Copper-tolerant strains of the pathogen have not been detected in the United States, but may appear with continued reliance upon copper bactericides. If copper-tolerant strains are present, tank mixing with a low rate of an EBDC fungicide, such as maneb, can enhance disease suppression. Tank-mixing coppers with zinc or iron can also enhance their activity.

Product List of Copper-Based Fungicides for *Xanthomonas* Leaf Blight

Pesticide	Product per Acre	Application Frequency (Days)	Remarks
Champ DP	1.33 lb	7–10	
Cuprofix	2.5–6 lb	7–10	Can be phytotoxic
Cuprofix MZ	5–7.25 lb	3–7	Maximum of 78 lb per season; 7-day PHI
Kocide 2000	1.5 lb	7–10	Can be phytotoxic to leaves
Kocide 3000	0.75 lb	7–10	Can be phytotoxic to leaves
ManKocide	2.5 lb	3–7	Maximum of 160 lb per season; 7-day PHI
Nordox	2–4 lb	7–10	
NuCop 50WP	2.0 lb	7–10	
Top Cop with S	2–3 qt	7–10	

Systemic acquired resistance induced by foliar applications of acibenzolar-*S*-methyl (e.g., Actigard) effectively manage *Xanthomonas* leaf blight in semiarid production environments. Biological control of *Xanthomonas* leaf blight with bacteriophage and commercially available antagonistic bacteria appear promising as well.

9.1.5 SLIPPERY SKIN OF ONION

Pathogen: *Burkholderia gladioli* pv. *alliicola*

Synonyms: *Burkholderia cepacia, Pseudomonas capaci*
Slippery and sour skin are caused by two different but related bacteria, *Burkholderia* (formerly *Pseudomonas*) *gladioli* subsp. *alliicola* and *B. cepacia*, respectively.

Symptoms
Intact onion bulbs with slippery skin can appear disease free until they are cut open. However, the inner scales of diseased bulbs are brown and water soaked, and the bulb turns soft and rotten (Photo 9.5). One indication of slippery skin is when the inner core slips out when pressure is applied to the base of the bulb. As with the sour skin disease, bacteria enter through the leaf tissue and the infection progresses down to the bulb tissue. Mature bulbs are highly susceptible to *B. gladioli* subsp. *alliicola* and can rot entirely after 10 days at room temperature.

Slippery skin (*B. gladioli* subsp. *alliicola*) can be differentiated from sour skin (*B. cepacia*) in that sour skin primarily attacks the outer fleshy scales. Both diseases are more prevalent in July and August after seasonal rains have begun.

Field Symptoms of Slippery Skin
- Field symptoms often appear as one or two wilted leaves in the center of the leaf cluster.
- These leaves eventually turn pale yellow and dieback from the tip, while older and younger leaves maintain a healthy green appearance. During the early stages of this disease, the bulbs may appear healthy except for a softening of the neck tissue.

PHOTO 9.5 Field symptoms of slippery skin in onion. (Photo courtesy of Howard F. Schwartz, Colorado State University, Fort Collins, CO, Bugwood.org.)

- In a longitudinal section, one or more inner scales will look watery or cooked.
- The disease progresses from the top of the infected scale to the base where it can then spread to other scales, rather than by spreading crosswise from scale to scale.
- Eventually, all the internal tissue will rot. Finally, the internal scales dry and the bulb shrivels.
- Squeezing the base of infected plants causes the rotted inner portion of the bulbs to slide out through the neck, hence the name "slippery skin."

Predisposing Factors

This bacterium requires moisture for infection and grows in the temperature range of 5°C–41°C. Severe disease can occur during periods of high rainfall combined with strong winds or hail. Heavy irrigation and persistent dews are also conducive to this disease. This bacterium is soil-borne and can be readily water splashed to the foliage and necks where it can enter through wounds. As the plant matures susceptibility of the plant also increases, as the mature plant being highly susceptible. In warm weather, approximately 30°C, infected bulbs can decay within 10 days. However, in storage, decay is gradual, for example, for a bulb to decay completely, it often requires 1–3 months.

Disease Management

Cultural Practices

The incorporation of green-manure crops the season before onions are grown, can reduce *B. cepacia* populations in soils.

Practice a 3–4-year crop rotation with small grains or sugarbeet. Avoid rotation with corn. Avoid reuse of irrigation water and overhead irrigation, especially after bulb initiation. Provide adequate but not excessive nitrogen fertilization, especially after bulb initiation or storm damage. Promptly harvest after bulbs have thoroughly cured in the field. Avoid wounding or bruising bulbs during harvest and storing operations. Promote rapid drying with forced air curing before long-term storage at 0°C–4°C with a relative humidity of 70% or less.

Chemical Approaches

Copper bactericides provide some disease control in Colorado when applied before disease is observed. Sprays should be initiated 2 weeks before bulb initiation and continued on a 5–10-day spray interval depending on weather conditions. Apply in a sufficient volume of water to ensure thorough coverage. Include a low rate of a nonionic surfactant to further improve coverage.

Copper-tolerant strains of the pathogens are common in the U.S. Tank-mixing copper bactericides with a low rate of an EBDC fungicide, such as maneb, is essential for effective disease suppression. Tank-mixing coppers with zinc or iron can also enhance their activity.

Copper-Based Fungicides

Pesticide	Product per Acre	Application Frequency (Days)	Remarks
Champ DP	1.33 lb	7–10	
Cuprofix	2.5–6 lb	7–10	Can be phytotoxic
Cuprofix MZ	5–7.25 lb	3–7	Maximum of 78.9 lb per season; 7-day PHI
Kocide 2000	1.5 lb	7–10	Can be phytotoxic to leaves
Kocide 3000	0.75 lb	7–10	Can be phytotoxic to leaves
ManKocide	2.5 lb	3–7	Maximum of 160 lb per season; 7-day PHI
Nordox	2–4 lb	7–10	
NuCop 50WP	2.0 lb	7–10	
Top Cop with S	2–3 qt	7–10	

Harvesting and Storage

Harvest onions when bulbs have reached full maturity. Do not store bulbs until they have been properly dried. Minimizing stem and bulb injury and avoiding overhead irrigation when the crop is approaching maturity can reduce losses from this disease. Bulbs should be stored at 0°C–2°C with adequate ventilation to prevent condensation from forming on the bulbs.

9.1.6 PANTOEA LEAF BLIGHT AND CENTER ROT IN ONION

Pathogen: *Pantoea ananatis*

The disease has been reported from Colorado, Georgia, and several other onion-producing regions of the world. Infection occurs when bacteria are deposited in or on leaves by wind, water splashing, or thrips. The disease is favored by moderate to warm temperatures and rainfall at bulb initiation, and continuing through bulb development. *Pantoea ananatis* can survive between onion crops epiphytically on weeds, on contaminated seed, and perhaps as a pathogen of melons.

Center rot in onion continues to be a problem in the production of Vidalia onions. The disease reduces both yield and quality of onion bulbs. Under certain conditions, center rot infections promote a soft rot of the bulb prior to harvest. Incidence and severity of center rot in Georgia in 2006 were the highest observed in the past 5 years. In some instances, entire fields were abandoned and the disease was responsible for hundreds of thousands, if not millions, of dollars in lost revenue. This was in spite of the fact that conditions during the 2006 onion-growing season, when center rot was the most devastating, were some of the driest experienced over the past 5 years. Center rot is a very unusual bacterial disease, as the vast majority of plant diseases caused by bacteria are favored by wet conditions.

Symptoms

Center rot can be a devastating disease when weather conditions favor the pathogen. Disease symptoms begin as necrotic, bleached areas on young leaves, which typically wilt (Photo 9.6).

Symptoms first appear as whitish to tan lesions with water-soaked margins, often on interior leaves. Foliar lesions can rapidly coalesce, progressing to wilt and dieback of affected leaves. The pathogen moves from the leaves into the neck and bulb causing yellowish to light brown discoloration. With severe infections, all leaves can be affected giving a bleached appearance to plants. This leaf blighting continues and the pathogen progresses downward into the bulb, causing a soft rot. Affected bulbs have a pale-yellow to yellow-orange decay and breakdown of one or more of the central inner bulb scales (Photo 9.7). Secondary bacteria can also invade the bulb, causing a more watery decay and breakdown of the entire bulb. However, the center rot pathogen

PHOTO 9.6 Wilt and dieback of onion leaves infected with *Pantoea*. (Photo courtesy of Howard F. Schwartz, Colorado State University, Fort Collins, CO, Bugwood.org.)

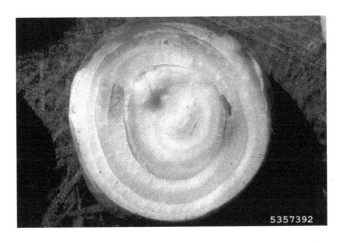

PHOTO 9.7 Bacterial decay of interior bulb tissue due to the center rot pathogen *Pantoea*. (Photo courtesy of Howard F. Schwartz, Colorado State University, Fort Collins, CO, Bugwood.org.)

alone can cause a breakdown of the entire bulb interior. Secondary bacterial infections rot the interior bulb tissue and produce a foul odor. Under conditions favorable to the disease, yield losses may approach 100%.

Predisposing Factors
The pathogen is seed borne and can survive on a few reported alternative hosts (corn, cotton, melon, pineapple, rice, and sugarcane). It may also survive epiphytically on weeds and crop debris. Spread can occur by wind, splashing water, and thrips. Infection is favored by moderate to warm temperatures and rainfall during bulb initiation.

Disease Management
Seed produced in high-risk areas should be tested for *Pantoea ananatis* and *P. agglomerans* before sowing. Some onion varieties are known to be more susceptible to this disease than others. Avoid planting these varieties where disease pressure is high. Control volunteer onions and thrips. Consider drip rather than sprinkler irrigation if possible, and avoid working in fields when foliage is wet. Avoid excessive nitrogen fertilization.

Plant seed free from the center rot pathogen. Eliminate weeds, volunteer onion, and cull piles that may be a source of *P. ananatis*. Deeply bury crop residues after harvest to reduce pathogen survival. Avoid reuse of irrigation water and overhead irrigation.

Disease tends to be more severe with furrow rather than drip irrigation. Avoid excess nitrogen, especially after bulb initiation and storm damage. Practice a 3-year or longer crop rotation to non-hosts such as small grains. Avoid working in fields when the foliage is wet, which can easily spread bacteria throughout the field. Plant wider rows in the direction of prevailing winds to increase air movement in the crop canopy and decrease periods of leaf wetness. Varieties vary in their reaction to center rot, and highly susceptible varieties should be avoided. Thrips and onion maggots are known to transmit the center rot pathogen, and should be controlled. Harvest and store onions only after thorough curing in the field and packaging shed. Avoid wounding or bruising bulbs during harvest and storing operations. Store bulbs at 0°C–4°C with a relative humidity of 70% or less.

Chemical Control
Copper bactericides provide some disease control in Colorado under low to moderate disease pressure when applied before disease is observed. Sprays should be initiated 2 weeks before bulb initiation and continued on a 5–7-day spray interval depending on weather conditions. Ensure thorough coverage by applying in a sufficient volume of water with a low rate of a nonionic surfactant. Disease forecast models have been developed in Colorado that may improve spray timing and effectiveness.

Copper-Based Fungicides

Pesticide	Product per Acre	Application Frequency (Days)	Remarks
Champ DP	1.33 lb	7–10	
Cuprofix	2.5–6 lb	7–10	Can be phytotoxic
Cuprofix MZ	5–7.25 lb	3–7	Maximum of 78 lb per season; 7-day PHI
Kocide 2000	1.5 lb	7–10	Can be phytotoxic to leaves
Kocide 3000	0.75 lb	7–10	Can be phytotoxic to leaves
ManKocide	2.5 lb	3–7	Maximum of 160 lb per season; 7-day PHI
Nordox	2.4lb	7–10	
NuCop 50WP	2.0 lb	7–10	

9.1.7 ENTEROBACTER BULB DECAY

Pathogen: *Enterobacter cloacae*
Bulb decay is caused by the bacterium *Enterobacter cloacae*. This disease was first reported in Colorado in 2000, but may occur in other onion-producing regions of the High Plains.

Symptoms
The exterior of the bulb remains asymptomatic while the inner scales show a brown to black discoloration and decay (Photo 9.8).

Bulb decay symptoms are generally absent in the field, but appear after 1–3 months in storage. Internal scales have a light to dark brown discoloration and decay; externally bulbs appear healthy. Yield losses of 1%–5% have been reported.

Predisposing Factors
This disease is observed in mature bulbs in the field after a period where air temperature reached 40°C–45°C. The bacterium is common in many environments and is considered to be an opportunistic pathogen on onions.

PHOTO 9.8 Symptoms of enterobacter bulb decay in onion. (Photo courtesy of Howard F. Schwartz, Colorado State University, Fort Collins, CO, Bugwood.org.)

Little is known about the life cycle of the pathogen, but disease appears to be favored by high temperatures (32°C–38°C or higher) near harvest. High manure application rates (>25 tons/acre) appear to increase the incidence of bulb decay in Colorado. The bacterium is a common soil inhabitant and is found in the digestive tracts of humans and animals.

Disease Management
Practices that reduce the incidence of other bacterial bulb rots may reduce bulb decay as well. The effect of copper bactericides or other pesticides on disease control is unknown.

9.1.8 BACTERIAL BROWN ROT IN ONION

Pathogen: *Pseudomonas aeruginosa*
It is a very serious disease in onions in storage. The infection occurs through wounds.

Symptoms
Dark brown discoloration in bulb scale is the characteristic feature of this disease. Browning of inner scale along with rotting is the main symptom of this disease. The rotting starts from the inner scale and spreads to outer scales. Apparently, the bulb seems to be healthy, but when pressed, a white oozing is noticed from the neck.

The rot begins at the neck of the bulbs, which later give out a foul smell through the neck when squeezed. In several cases, all of the bulbs in storage rot, giving out foul odor.

Disease Management
- Proper curing is required before storage.
- Use maleic hydrazide and isopropyl phenyl carbamate (IPC), 20 ml/L before 1 month of harvest.
- Neck cutting about 2.5–3.0 cm long above the bulb reduces bacterial infection.
- Light irrigation is required during the entire cropping period.
- Proper curing and rapid drying of the bulbs after harvesting is essential for controlling the disease.
- Affected bulbs should be discarded before storage.
- If rains occur during maturity, spraying streptocycline (0.02%) is recommended.

9.2 GARLIC

9.2.1 Bacterial Soft Rot of Garlic

Pathogen: *Erwinia carotovora* subsp. *carotovora*
    Erwinia chrysanthemi
    Pseudomonas gladioli
    Enterobacter cloacae

Symptoms

- Bacterial soft rot is mainly a problem on mature bulbs. Affected scales first appear water-soaked and pale yellow to light brown.
- As the soft rot progresses, invaded fleshy scales become soft and sticky with the interior of the bulb breaking down.
- A watery, foul-smelling thick liquid can be squeezed from the neck of diseased bulbs.
- Infection in the field during the bulb formation leads to withering and collapse of the foliage.

Survival and Spread

The pathogen is soil borne and survives in infected plant debris as well as in soil. It is also spread by splashing rain, irrigation water, and insects.

Favorable Conditions

This disease is favored by warm, humid conditions with an optimum temperature range of 20°C–30°C. However, during storage or transit, soft rot can develop when temperatures are above 3°C.

Disease Management

Follow the cultural practices and chemical control methods as described for soft rot of onion.

9.2.2 Bacterial Slippery Skin of Garlic

Pathogen: *Burkholderia gladioli* pv. *alliicola*

Symptoms

- Field symptoms often appear as one or two wilted leaves in the center of the leaf cluster. These leaves eventually turn pale yellow and dieback from the tip, while older and younger leaves maintain a healthy green appearance.
- During the early stages of this disease, the bulbs may appear healthy except for a softening of the neck tissue. In a longitudinal section, one or more inner scales were observed to be watery or cooked.
- The disease progresses from the top of the infected scale to the base.
- Eventually, all the internal tissue will rot. Finally, the internal scales dry and the bulb shrivels.
- Squeezing the base of infected plants causes the rotted inner portion of the bulbs to slide out through the neck, hence the name "slippery skin."

Survival and Spread

This bacterium is soil borne and can be readily water-splashed to the foliage and necks where it can enter through wounds.

Favorable Conditions

This bacterium requires moisture for infection and grows in the temperature range of 5°C–41°C. Heavy irrigation and persistent dews are also conducive to this disease.

Disease Management
Follow the cultural practices and chemical control methods as described for slippery disease of onion.

9.3 GINGER

9.3.1 BACTERIAL WILT OF GINGER

Pathogen: *Ralstonia solanacearum*
Synonyms: *Pseudomonas solanacearum, Burkholderia solanacearum*
Bacterial wilt is a significant disease of ginger root. The pathogen is widely distributed in areas where edible ginger has previously been cultivated. Historically, strains of *Ralstonia solanacearum* have been classified into five races based on their host ranges and into five biovars based on their differential ability to produce acid from a set of specific carbohydrates. *Ralstonia solanacearum* race 4 infects ginger in Hawaii and much of Asia. Bacterial wilt caused by *R. solanacearum* Biovar-3 is a soil- and seed-borne disease that occurs during southwest monsoon in India.

Symptoms
Water-soaked spots appear at the collar region of the pseudostem and progresses upward and downward. The first conspicuous symptom is mild drooping and curling of leaf margins of the lower leaves that spread upward. In the advanced stage, the plants exhibit severe yellowing and wilting symptoms (Photo 9.9). The vascular tissues of the affected pseudostems show dark streaks.

Ralstonia solanacearum colonizes the xylem, the water-conducting elements of a ginger plant's vascular system, and causes wilt. The affected pseudostem and rhizome start rotting (Photo 9.10) and when pressed gently extrude milky ooze from the vascular strands. Ultimately rhizomes rot emitting a foul smell.

PHOTO 9.9 Severe yellowing and wilting of ginger leaves due to bacterial wilt. (Photo courtesy of Scot Nelson, www.flickr.com.)

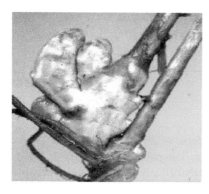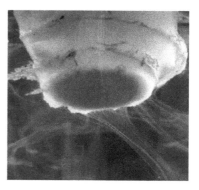

PHOTO 9.10   Symptoms of bacterial wilt pathogen as rotting of ginger rhizome. (Photo courtesy of Scot Nelson, www.flickr.com.)

Two types of symptoms can be observed in the field:

1. *Green wilt*: This is the diagnostic symptom of the disease. It occurs early in the disease cycle and precedes leaf yellowing. Infected green ginger leaves roll and curl due to water stress caused by bacteria blocking the water-conducting vascular system of the ginger stems.
2. *Leaf yellowing and necrosis*: Leaves of infected plants invariably turn yellow and then necrotic brown. The yellowing, however, should not be confused with another disease of ginger causing similar symptoms known as *Fusarium* yellows. Plants infected by the fungus *Fusarium oxysporum* f. sp. *zingiberi* do not wilt rapidly, as they do when affected by bacterial wilt. Instead, *Fusarium*-infected ginger plants are stunted and yellowed. The lower leaves dry out over an extended period of time.

Host

The host range of *R. solanacearum* race 4 is restricted to edible ginger (*Z. officinale*).

Disease Cycle

Ralstonia solanacearum spreads by infested soil adhering to hands, boots, tools, vehicle tires, and field equipment; in water from irrigation or rainfall; and within infected ginger rhizomes (Janse, 1996). This bacterium infects ginger roots and rhizomes through openings where lateral roots emerge or through wounds caused by handling parasitic insects or root-knot nematodes (Swanson et al., 2005). The pathogen survives in soils within infected plant debris and as free-living bacteria. Ginger crops can be completely lost to the disease in heavily infested soils.

Disease Management

The most efficient practices to manage bacterial wilt will integrate some or all of the following practices.

Site Selection

The cultivation site should be well drained and without recent history of ginger cultivation (at least 10 years). The best sites have rainfall dispersed evenly throughout the growing season, with a drier period near the end, before, and during harvest. Excessive rainfall and waterlogging favor bacterial wilt and other soilborne diseases. Do not plant downslope from another ginger field if runoff water can carry the pathogen into your field. Before planting, test the soil in your field for the pathogen by either PCR or a plant bioassay.

Planting Considerations

Avoid planting during very wet weather, as this promotes dispersal of the pathogen within fields on muddy boots, tools, and vehicles.

Site, Soil, and Bed Preparation

Use clean, pathogen-free tractors and equipment. Hose tractor down with clean water (not obtained from streams located below existing ginger fields) before entering the field to be planted with ginger. Preferably spray with 10% bleach solution, particularly on tractor wheels and blades. Prepare soil by plowing and harrowing so that the site and soils drain well after rainfall. Proper soil and bed preparation are essential for ginger production. Prior to planting, soils are typically plowed to a depth of 45–60 cm. Lime is incorporated to adjust the pH, and furrows are cut with a hand tiller to a depth of 30–45 cm, spaced about 150 cm apart. Fertilizer or organic amendments are placed at the bottom of the furrows and incorporated prior to seeding.

Pathogen-Free Seed

Growing disease-free ginger starts with planting pathogen-free seed. Bleach can be used to sterilize the surface of ginger seeds: dip them in a 10% bleach solution (one part commercial bleach to nine parts water) for 10 min. However, this may not eliminate bacterial infections within the rhizomes. A hot-water treatment consisting of exposing seeds to a constant 50°C temperature for 10 min is effective in controlling nematodes, but may not be effective for disease organisms such as *R. solanacearum* that are already present inside the rhizome.

Hilling, Cultivation, and Drainage

Hill the planting rows to promote aeration of the roots and allow adequate soil drainage. Hilling is often done by hand, even on commercial farms. The ginger crop, as it grows from the bottom of the furrows, is hilled three to five times during the growing season. This results in raised hilled beds that allow proper development of the rhizomes. Hilling of the ginger row, which is done at 6-week intervals, allows the rhizome to grow vertically. Ensure adequate soil drainage and diversion ditches to prevent the runoff of infected water sources into fields that are down slope from an infected field.

Limit Site Traffic and Pathogen Dispersal

Strictly limit nonfarm vehicles and visitors into ginger fields, as the pathogen can be in soils on truck tires or boots. Do not bring dirty tools to the farm from other locations. Disinfest boots by steeping into a 10% bleach solution before entering a ginger field. Erect fences to deter pigs and other animals from entering a ginger field.

Composting

Organic soil amendments (mulch, compost) promote microbial activity that may suppress *R. solanacearum* by competition, antibiosis, or both.

Crop Rotation and Intercropping

Rotate ginger with crops that are not hosts of the bacterial wilt pathogen, such as sweet potato and taro. Do not intercrop or rotate ginger with solanaceous crops, including tomatoes, peppers, and eggplant. The pathogen can reproduce in these crops and build up levels of bacterial wilt in the field. Ginger may also be rotated with grain crops such as corn or upland rice. Other possible crops for use in rotation or intercropping include green onion, soybean, sweet corn, cabbage, and Sun hemp.

Control of Other Pests

Control other pests that damage ginger plants, such as the lesser corn stalk borer (*Elasmopalpus lignosellus*) and the banana moth (*Opogona sacchari*), which may be controlled with applications of *Bacillus thurigiensis* (Bt). Injuries are infection courts for *R. solanacearum*, and some insects create wounds that favor the bacterial wilt pathogen.

Biofumigation with Green-Manure Crop

The use of volatile oils to kill or suppress a pathogen is known as biofumigation (Paret, 2010). These oils are a natural component of certain green-manure crops such as mint, palmarosa, and lemongrass. When these plants are turned or plowed into the soil several months before planting, they decompose and release essential oils. These oils are toxic to the bacterial wilt pathogen and can severely reduce its population in the field soil. However, the oils are very expensive if bought preprocessed, not naturally occurring as a component of green-manure crops, and in this case their use may not be economically feasible.

On-Time Harvest
Harvesting on time minimizes crop exposure to the pathogen.

Other Management Approaches
- The cultural practices adopted for managing soft rot are also to be adopted for bacterial wilt.
- Seed rhizomes may be treated with streptocycline 200 ppm for 30 min and shade-dried before planting.
- Once the disease is noticed in the field, all beds should be drenched with Bordeaux mixture 1% or copper oxychloride 0.2%.

9.4   TURMERIC

9.4.1   BACTERIAL WILT OF TURMERIC

The disease is reported from India and Japan.

Pathogen: *Ralstonia solanacearum*

Symptoms
- Rapid wilting and death of the entire plant without or with yellowing or spotting of leaves are the characteristic symptoms.
- All branches wilt at about the same time.
- When the stem of a wilted plant is cut across, the pith shows a darkened, water-soaked appearance.
- Grayish slimy ooze comes out on pressing the stem.
- In later stages of the disease, decay of the pith may cause extensive hollowing of the stem.

Favorable Conditions
The bacterium is especially destructive in moist soils at temperatures above 24°C.
  High soil temperature and moisture are favorable for the disease.

Disease Cycle
The disease rhizome as well as infected soils with *Ralstonia* serve as the primary sources of inoculums. The disease spreads in the field through irrigation water.

Control Measures
Use disease-free rhizomes for planting. Follow crop rotation for at least 3 years with a nonhost plant of *R. solanacearum*.

9.5   BLACK PEPPER

9.5.1   BACTERIAL LEAF SPOT IN BLACK PEPPER

Pathogen: *Xanthomonas beticola*
This bacterial disease was observed in black pepper plants growing adjacent to betel vine plants infected with bacterial leaf-spot at the College of Agriculture, Vellayani, Kerala. Later it was detected in farmers' fields in Trivandrum, Kerala.

Symptoms
This disease appears as minute, water-soaked lesions on the leaf lamina. As the lesions grow older, the center of the spot becomes black surrounded by a yellowish halo. Sometimes infection begins

from leaf margin also. Often, a dark pigmentation developing at the infection spot diffuses into sur-rounding parts of leaf lamina, eventually causing defoliation.

The pathogen isolated has been identified as *Xanthomonas beticola*, and the isolate gave positive results on pathogenicity tests.

Further details on this disease are lacking.

Control Measures
Not available in the literature.

9.6 SAFFRON

9.6.1 BACTERIAL SOFT ROT OF SAFFRON

Pathogen: *Burkholderia gladioli*

Symptoms
Field symptoms are observed in autumn and spring. They are characterized by rot on plants and spots on leaves and corms. In particular in autumn, during rainy, and mild periods just before blooming, the sheaths that wrap emerging shoots show brown lesions. Later the disease spreads to the leaves and flowers and they rot. Brown rounded marks, surrounded by widespread reddish brown halos, are observed on corms. At the last stage of the disease, under high humidity, the spots rot. Symptoms on leaves are observed in autumn and also in spring. Reddish brown spots, surrounded by widespread chlorotic halos, occur on foliar limbs. Subsequently, during wet periods, the veins and leaf edges are sometimes affected by the disease, and in some cases the leaves bend and distal parts wither.

9.7 CLOVE

9.7.1 DIEBACK OF CLOVE

Pathogen: *Ralstonia syzygii*

Common Name: Sumatra disease of clove

Symptoms
Dieback of trees begins in the crown and leads to tree death within 3 years of initial infection. The infected leaves turn chlorotic and drop from the tree or may wilt in the advanced stage of infection and remain attached; discoloration of vascular tissues is evident as gray-brown streaks are observed in the new wood; bacterial exudate may ooze out of tissue when cut.

Disease Cycle
Bacteria are limited to the water-carrying vessels in the tree (xylem); the disease is thought to be transmitted by *Hindola striata* and *Hindola fulva*, both sucking insect species.

Disease Management
An antibiotic oxytetracycline can be injected into the tree to slow the decline of infected trees but there is currently no known cure for the disease; several insecticides can control *Hindola* insect species, which are believed to spread the disease.

9.8 CARDAMOM

9.8.1 BACTERIAL CANKER OF CARDAMOM

Pathogen: *Xanthomonas* sp.
Bacterial canker (*Xanthomonas* sp.) on the capsules is common in Wynad area, Kerala, during monsoon.

PHOTO 9.11   Bacterial leaf spot infection on betel vine leaves. (Photo courtesy of S. G. Borkar and R. T. Gaikwad, PPAM, MPKV, Rahuri, Maharashtra, India.)

Symptoms

The disease starts as a small water-soaked lesion on pericarp of the capsule, which later develops into a yellow halo that is more conspicuous on drying.

Control Measures

Not worked out.

9.9   BETEL VINE

9.9.1   BACTERIAL LEAF SPOT OF BETEL VINE

Pathogen: *Xanthomonas campestris* pv. *beticola*

Bacterial leaf spot/leaf blight/stem blight is becoming serious in many of the betel vine-growing parts of India.

Symptoms

Brown leaf spots, surrounded by a translucent water-soaked area, appear on the lower surface, and a yellow halo appears on the corresponding upper surface (Photo 9.11). Elongated brown spots of variable length appear on the vine, resulting in stem canker. In severe infection, the stem cracks and a large area of leaf lamina is covered by the infection, causing leaf blight.

Control Measures

Application of 0.5% Bordeaux mixture or copper oxychloride 3 g/L effectively controls the disease.

REFERENCES

Cother, E.J. and V. Dowling. 1985. Association of *Pseudomonas cepacia* with internal breakdown of onion—A new record for Australia. *Aust. Plant Pathol.*, 14, 10–12.

Irwin, A.D. and E.K. Vaughan. 1972. Bacterial rot of onion and the relation of irrigation water to disease incidence. *Phytopathology*, 82, 1103.

Jaccoud Filho, D.de S., Romeiro, R.S., Kimura, O. et al. 1987. Bacterial rot of scale—A new disease of onion in Minas Gerais. *Fitopatol. Bras.*, 12, 395

Janse, J. 1996. Potato brown rot in Western Europe—History, present occurrence and some remarks on possible origin, epidemiology and control strategies. *Bull. OEPP/EPPO*, 26, 679–695.

Kawamoto, S.O. and J.V. Lorbeer. 1972a. Histology of onion leaves infected with *Pseudomonas cepacia*. *Phytopathology*, 62, 1266–1271.

Kawamoto, S.O. and J.W. Lorbeer. 1972b. Multiplication of *Pseudomonas cepacia* in onion leaves. *Phytopathology*, 62, 1263–1265.

Kawamoto, S.O. and J.W. Lorbeer. 1974. Infection of onion leaves by *Pseudomonas cepacia*. *Phytopathology*, 64, 1440–1445.

Paret, M.L., Cabos, R., Kratky, B.A. et al. 2010. Effect of plant essential oils on *Ralstonia solanacearum* race 4 and bacterial wilt of edible ginger. *Plant Dis.*, 94, 521–527.

Schwartz, H.F. and S.K. Mohan. 1995. *Compendium of Onion and Garlic Diseases*. APS Press, St. Paul, MN.

Swanson, J.K., Yao, J., Tans-Kersten, J.K. et al. 2005. Behaviour of *Ralstonia solanacearum* race 3 biovar 2 during latent and active infection of geranium. *Phytopathology*, 95, 136–143.

Teviotdale, B.L., Davis, R.M., Guerard, J.P. et al. 1989. Effect of irrigation on management of sour skin of onion. *Plant Dis.*, 73, 819–822.

Teviotdale, B.L., Davis, R.M., Guerard, J.P. et al. 1990. Method of irrigation affects sour skin rot of onion. *Calif. Agric.*, 44, 27–28.

SUGGESTED READING

Alvarez, A.M. 2005. Diversity and diagnosis of *Ralstonia solanacearum*. In: Allen, C., Prior, P., and A.C. Hayward (eds.), *Bacterial Wilt Disease and the Ralstonia solanacearum Species Complex*. American Phytopathological Society (APS Press), St. Paul, MN, pp. 437–447.

Alvarez, A.M., Trotter, K.J., and M.B. Swafford. 2005. Characterization and detection of *Ralstonia solanacearum* strains causing bacterial wilt of ginger in Hawaii. In: Allen, C., Prior, P., and A.C. Hayward (eds.), *Bacterial Wilt Disease and the Ralstonia solanacearum Complex*. American Phytopathological Society (APS Press), St Paul, MN, pp. 471–477.

Anonymous (2015) Sour skin of onion (*Burkholderia cepacia*). (n.d.). Retrieved June 22, 2015, from http://www.plantwise.org/KnowledgeBank/Datasheet.aspx?dsid=44940.

Bazzi, C. 1979. Identification of *Pseudomonas cepacia* on onion bulbs in Italy. *Phytopathol. Z.*, 95, 254–258.

Burkholder, W.H. 1950. Sour skin, a bacterial rot of onion bulbs. *Phytopathology*, 40, 115–117.

EPPO. 2009. *Xanthomonas axonopodis* pv. *allii* (an emerging disease of onion and garlic crops). European and Mediterranean Plant Protection Organization. Retrieved: October 21, 2010. https://www.eppo.int/MEETINGS/2008_meetings/PRA_xantho_allii.htm.

Gent, D.H. and H.F. Schwartz. 2005. Management of *Xanthomonas* leaf blight of onion with a plant activator, biological control agents, and copper bactericides. *Plant Dis.*, 89, 631–639.

Humeau, L., Roumagnac, P., and I. Soustrade. 2004. Unemaladieémergente de l'oignon à la Réunion. Le dépérissementbactériencausé par *Xanthomonas axonopodis* pv. *allii*. *Phytoma*, 573, 28–30.

Nelson, S. 2013. Bacterial wilt of edible ginger website. http://www.ctahr.hawaii.edu/gingerwilt/ (accessed September 23, 2013).

Roumagnac, P., Pruvost, O., Chiroleu, F. et al. 2004. Spatial and temporal analyses of bacterial blight of onion caused by *Xanthomonas axonopodis* pv. *allii*. *Phytopathology*, 94(2), 138–146.

Schwartz, H. and D. H. Gent. "*Xanthomonas* Leaf Blight of Onion - 2.951 - Colorado State University Extension." Colorado State University Extension. Accessed May 09, 2015. http://extension.colostate.edu/topic-areas/yard-garden/xanthomonas-leaf-blight-of-onion-2-951/.

Schwartz, H.F. and S.K. Mohan. 2008. *Compendium of Onion and Garlic Diseases and Pests*, 2nd ed. American Phytopathological Society, St. Paul, MN.

Shintaku, M., Kaneshiro, T., and C. Enriques. 1996. Detecting *Pseudomonas solanacearum* in edible ginger using the polymerase chain reaction. *J. Haw. Pac. Agric.*, 7, 11–19.

Trujillo, E.E. 1964. Diseases of ginger (*Zingiber officinale*) in Hawaii. http://www.ctahr.hawaii.edu/oc/freepubs/pdf/C2-62.pdf (accessed August 8, 2013).

10 Bacterial Diseases of Flowering Plants and Ornamentals

10.1 ROSE

10.1.1 BACTERIAL LEAF SPOT AND BLAST IN ROSE

Pathogen: *Pseudomonas syringae* pv. *morsprunorum*

Pseudomonas syringae pv. *morsprunorum*, on rose, was diagnosed only four times by the OSU Plant Clinic from 1962 to 1992. This is most commonly present in cool, wet spring weather.

Symptoms

Dark-brown, sunken spots appear on leaves, flower stalks, and calyx parts. Flower buds may die without opening. Black streaks appear on 1-year-old stems (Photo 10.1).

Disease Management

Cultural Practices
- Remove and destroy infected stems.
- Disinfect pruning shears before cutting more stems.

Chemical Control

Spray copper-based bactericide with antibiotics before fall rains begin and reapply when half the leaves have fallen. Repeat during spring to protect new growth.

10.1.2 CROWN GALL IN ROSE

Pathogen: *Agrobacterium tumefaciens*

Symptoms

Crown gall is characterized by the formation of outgrowths (galls) that vary in form and size. At first, the galls are very small with rounded outgrowths. They generally form just below the soil surface on the crown. Galls can also occur on roots and occasionally on aerial parts of rose plants (Photo 10.2). Development continues and the galls may become several inches in diameter. Galls have a rough, irregular surface and a diameter of ¼ to several inches. Young, developing galls are soft and white to light green, but they darken and harden with age. The surface of large woody galls often rots and sloughs off, while the internal tissues are white.

Roses with large, well-developed galls are stunted, have small yellow leaves, and produce few blooms.

Disease Cycle

The crown gall bacterium, *Agrobacterium tumefaciens*, is usually introduced into the landscape on diseased nursery stock. As the galls rot, bacteria are released into the soil where they survive for 2 to 3 years. Bacteria enter the plants through wounds or natural openings in the roots. Once inside the plant, the bacteria stimulate rapid cell division as well as an increase in cell size. Galls usually become visible several months after infection and may continue to grow depending on the vigor of the host plant and environmental conditions. Gall decay usually starts when there are not enough water and nutrients available for growth.

PHOTO 10.1   Symptom of bacterial leaf spot and blast on rose showing dark discoloration of the canes. (Photo courtesy of OSU Extension Plant Pathology Slide Set, pnwhandbooks.org.)

PHOTO 10.2   Formation of gall at crown level and on aerial stem of rose stem. (Photo courtesy of Jennifer Olson, Oklahoma State University, Stillwater, OK, Bugwood.org.)

Disease Management
- Transplant only disease-free plants.
- Avoid wounding during transplanting.
- Remove infected plants or plant parts as soon as galls are observed. Where possible, remove and discard all soil in and adjacent to the root system and replace with sterile soil to prevent reintroduction of the bacteria.

- Disinfect pruning and cutting tools frequently. Dipping in a 10% dilution of household bleach (one part bleach to nine parts water) for several minutes will effectively disinfect cutting tools. This should be done immediately after pruning out a gall or abnormal growth. Cut well below the galled area.
- During cultivation of roses, do not injure roots or crown area.
- If possible, wait for 1–2 years before replanting broad-leafed plants into the same location. If immediate replanting is desired, consider nonhosts that include most grassy plants.
- Inspect the roots and crown of roses for galls before planting. When planting or cultivating roses, avoid wounding the roots or crown of the plants.
- The root system of bare-root roses may be protected from crown gall with preplant root dip of the biological control agents, Galltrol-A and Norbac 84-C.
- Fumigation of new or renovated rose bends is suggested before planting.
- Destroy diseased roses and fumigate the area where the plants were removed.
- Where crown gall is a serious problem, consider replanting the area with woody ornamentals resistant to crown gall such as azalea, boxwood, crape myrtle, photinia, or wax myrtle.

10.2 CALLA LILY

10.2.1 BACTERIAL SOFT ROT IN CALLA LILY

Pathogen: *Pectobacterium carotovorum* subsp. *carotovorum*

Bacterial soft rot in calla lily (*Zantedeschia* spp.) is the main limiting factor of lily's production in many countries (Funnel, 1993; Wright, 1998; Snijder and van Tuyl, 2002; Wright et al., 2002; Cho et al., 2005; Krejzar et al., 2008; Mikicinski et al., 2010). Disease severity is related to the lack of effective control methods and rapid spread both in the field and during tuber storage (Vanneste, 1996; Wright et al., 2002).

In 1980, global losses in calla production caused by soft rot were assessed at US$100 million. In New Zealand alone, which is a big exporter of tubers and flowers, NZ$2 million is lost each year, which equals to 20% of the total income from Zantedeschia production (Vanneste, 1996; Wright et al., 2005).

Soft rot in calla lily is mainly caused by the *Pectobacterium carotovorum* subsp. *carotovorum* (formerly *Erwinia carotovora* subsp. *carotovora*) (Wright, 1998; Lee and Chen, 2002; Snijder and van Tuyl, 2002; Cho et al., 2005; Janse, 2006). However, bacteria identified as *Erwinia chrysanthemi* and *Xanthomonas campestris* pv. *Zantedeschiae* also proved to cause this disease in Taiwan (Lee and Chen, 2002; Lee et al., 2005). The pathogen mentioned earlier was also found in South Africa (Joubert and Truter, 1972). Recently, Krejzar et al. (2008) reported the first isolation of *Pseudomonas marginalis* and *Pseudomonas putida* from calla tubers with soft rot in the Czech Republic. However, their pathogenicity to the host plant was not proved.

Symptoms

Plant parts become soft and slimy. The disease initiation occurs at the base of leaves arising from a rhizome. Leaves may appear water soaked and start rotting and become slimy. The affected leaves turn dark green and may have spots on them and then start to wilt, turn yellow, and finally die. Typically, the foliage and flower stalks develop a brown, soft rot at the soil surface (Photo 10.3).

Predisposing Factors

The bacteria *P. carotovorum* enters plants through wounds. Wounds may be from physical injury to the rhizomes or from living organisms like insects or fungi. Soft rot may be the result of many other diseases such as those caused by *Pythium*. The disease is favored by moist conditions. Frost can increase the susceptibility and infection in the plant. The bacteria survive in old plant debris and infected rhizomes. Transplanting, irrigation water, or cutting and digging tools can spread the bacteria. This disease is considered to be the most important factor limiting the production of calla lily.

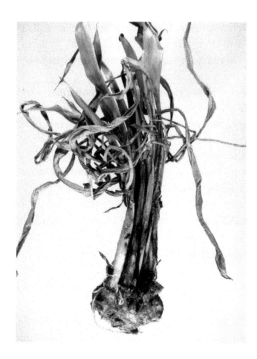

PHOTO 10.3 Symptom of bacterial soft rot on calla lily. (Photo courtesy of OSU Plant Clinic collection, 2012, pnwhandbooks.org.)

Control Measures

Cultural Practices
- When receiving rhizomes, unpack immediately and let it dry out under mild dry conditions to let abrasions heal.
- Discard any diseased rhizomes immediately. Wash hands and equipment well before handling more plant material.
- If soaking in gibberellic acid for increased flowering, ensure soak times are short and rhizomes dry quickly and thoroughly. Dip tanks should use fresh water and be cleaned frequently.
- Plant in well-drained soil or soilless media.
- Be sure to plant the rounded side down and the part where the roots and sprouts emerge at the top.
- Keep rooting media moist but not overly wet or dry. When irrigating, do not let water pool or splash. If an outbreak occurs, stop irrigating right away and let plants dry out before removing. Drip irrigation is better than overhead irrigation.
- Do not apply excessive nitrogen.
- It has been reported that yellow- and orange-colored cultivars are susceptible, while white- and cream-colored cultivars are not as susceptible.

Chemical Control
Fungicide drenches (a mix of two or more) are recommended to prevent injury from soil-borne fungal pathogens. Several bactericides are also used in the industry, but few are legal. Antibiotics are not labeled for this use. Disinfectants may be used but have a very short, if any, residual effect.

Mix Phyton-27® at 3 oz/10 gal water. Soak bulbs for 5 min and allow it to dry before planting.

PHOTO 10.4 Symptoms of bacterial leaf blight on calla lily. (Photo courtesy of Leslie Baker and Scot Nelson, https://www.flickr.com/photos/scotnelson/15274392763.)

10.2.2 BACTERIAL LEAF BLIGHT IN CALLA LILY

Pathogen: *Xanthomonas campestris* pv. *Zantedeschiae*

Bacterial leaf blight of white-flowered calla lily was first reported in South Africa, and the pathogen was described as *X. campestris* pv. *Zantedeschiae* (Joubert and Truter, 1972).

In Taiwan, leaf blight symptoms on white-flowered calla lily (*Zantedeschia aethiopica*) are generally observed during spring in some nurseries in Taipei, Taiwan.

Symptoms

The symptoms first appeared on the lower sides of leaves as small spots. The centers of the spots quickly turned brown and were surrounded by yellowish halos. The brown spots and halos enlarged rapidly and coalesced into irregular, dry, necrotic lesions (Photo 10.4). Some of these necrotic lesions became torn and partially detached or with a shot-hole effect.

Control Measures

The application of copper-based compounds at the proper time may help check the disease. The use of overhead or sprinkle irrigation should be avoided.

10.3 CARNATION

10.3.1 BACTERIAL SLOW WILT AND STUNT IN CARNATION

Pathogen: *Erwinia chrysanthemi*

Synonym: *Erwinia chrysanthemi* pv. *dianthicola* (*Hellmers*) *Dickey*

Bacterial slow wilt on carnation is a serious disease and is listed in EC plant health legislation.

Symptoms

Infected plants show wilting and stunting and often appear grayer than healthy ones. Symptoms may take a long time to appear after infection (hence the name "slow wilt") and are not always typical of a wilt disease. Wilting usually results from a slow rotting of the basal stem tissue around the vascular bundles. Leaves typically appear desiccated and become grayer or straw colored. Roots may rot away completely before shoot symptoms develop. Stunting is a characteristic symptom, and the disease has often been referred to as bacterial stunt. Infected plants may develop thickened basal stems, but these do not crack.

Hosts
Carnation and other species of Dianthus are found to be potential hosts of this disease. Almost all strains attacking *Dianthus* are *Erwinia chrysanthemi* biovar 1. Other biovars have different host ranges. *Erwinia chrysanthemi* from other ornamentals, such as chrysanthemum, dieffenbachia, and kalanchoe, are unlikely to infect carnation.

Geographical Distribution
This is widespread in Europe and also reported in northern United States and New Zealand.

Pathogen and Predisposing Factors
Erwinia chrysanthemi is a soft rot pathogen degrading succulent fleshy plant organs such as roots, tubers, stem cuttings, and thick leaves. It is also a vascular wilt pathogen, colonizing the xylem and becoming systemic within the plant. This latter aspect is the most alarming when vegetative propagation is involved. The pathogen can remain latent in stock plants (ornamentals, bananas) and can thus be spread in cuttings from them.

The bacterium is able to survive in the soil (on plant debris) so that infestation remains between two crops.

High humidity and free water favor spread and penetration of the bacteria. Disease development is dependent on high temperatures, generally 25°C–30°C.

Host specialization has not definitely been proved in *E. chrysanthemi*, except in pv. paradisiacal (Dickey and Victoria, 1980; Dickey, 1981). The pathogen is ubiquitous and isolates from maize and potato seem to be rather polyphagous, while philodendron and kalanchoe do appear to be differential hosts for temperate isolates (Janse and Ruissen, 1988).

Disease Cycle
As bacteria leak into the compost from the vascular tissues of infected plants, they spread to other cuttings. Infection can spread rapidly through a whole bed from only a few infected plants. Since infected plants occasionally remain symptomless for several years, the disease can also be spread unknowingly through cuttings taken from infected, symptomless plants and on contaminated cutting knives. Continued propagation from such plants can result in a rapid buildup of infection and eventually to a large-scale outbreak of disease. The disease is often a serious problem in mature plants as infections can remain symptomless for long periods.

Disease Management
These pathogens grow in raised beds pasteurized between crops. Use culture-indexed cuttings free of the pathogen. Destroy infected plants.

10.3.2 Bacterial Wilt in Carnation

Pathogen: *Burkholderia caryophylli* (*Burkholder*) (Yabuuchi et al.)

Synonyms: *Pseudomonas caryophylli* (Burkholder) Starr & Burkholder
Phytomonas caryophylli Burkholder

Common names: Bacterial wilt, bacterial stem crack (English)
Chancre bactérien de l'oeillet (French)
Welkekrankheit, Wurzelfäule der Nelken (German)
Chancro bacteriano del clavel (Spanish)

Bacterial wilt in carnations is caused by *Burkholderia caryophylli*, which was previously named as *Pseudomonas caryophylli*. The disease is characterized by abnormal colors of the leaves, soft rot of plant cortex, and gummosis or resinosis of the stem. Bacterial wilt in carnations causes devastating losses for areas in which carnations are mainly grown. Specifically, it is one of the major diseases to infect carnations in Japan, and it causes serious crop losses in carnations grown in warm districts.

Symptoms

Symptoms may take 2–3 years to manifest themselves, particularly when cuttings are mildly infected and maintained at relatively low temperatures. Foliage becomes grayish green, followed by yellowing and wilting and finally leading to death.

In stems, at soil temperatures below about 17°C, a rapid multiplication of cells leads to tension around the vessels and longitudinal, internodal stem cracks appear, usually at the base of the plant, and later develop into deep cankers. Initially, this cracking is very similar to the physiological cracking observed in certain cultivars. However, in pathogen-induced cracks, a brownish yellow bacterial slime is visible, often overgrown with saprophytic fungi such as *Cladosporium herbarum*. In some cases, the extrusions from the cankers leave the stems hollow.

At 20°C–25°C, cankers are rarer and wilting is the common symptom. Visual observation of peeled stems reveals sticky, brownish yellow, narrow or broad, longitudinal stripes in the vascular tissue; in cross section, these appear as irregular brownish spots with a water-soaked margin.

Roots of infected plants, once wilting occurs, are more or less rotten, the plants being easily pulled out of the soil, and upon cutting, the roots show discontinuous brown spots that distinguish the disease from that caused by *Phialophora cinerescens* that leaves the roots apparently symptomless (EPPO/CABI, 1996a).

Plants may survive about 1–2 months, but secondary invasion by fungi, such as *Fusarium* spp., accelerates death. Heavily infected cuttings wilt and die before roots are formed.

Leaves turn gray green and then yellow and eventually they die. Roots rot. Vascular tissue browns. Cracks develop in internode tissue. Slime oozes from these cracks when humidity is high.

Hosts

Carnations are the main host and are widely grown in the European and Mediterranean Plant Protection Organization (EPPO) region. However, *Dianthus barbatus* and *D. allwoodii* can be infected through artificial inoculation. In Florida (United States) and Japan, *Limonium sinuatum* is also reported to be infected (Jones and Engelhard, 1984; Nishiyama et al., 1988).

Geographical Distribution

This is widely distributed in the following regions:

EPPO region: Hungary (unconfirmed), Israel, Italy, Norway (unconfirmed), Poland (unconfirmed), Slovakia (unconfirmed), Sweden (unconfirmed), and Yugoslavia; found in the past but not established in Denmark, France, Germany, Ireland, the Netherlands, and the United Kingdom

Asia: China (Jilin; unconfirmed), India, Israel, Japan (Shikoku), and Taiwan

North America: The United States (Florida, Illinois, Indiana, Iowa, Massachusetts, Minnesota, New York, Pennsylvania, and Washington)

South America: Argentina, Brazil, and Uruguay

EU

Pathogen

Burkholderia caryophylli is a straight or slightly curved rod with rounded ends, occurring singly or in pairs; it is aerobic, nonsporing, and motile with one or several polar flagella; Gram negative; and sudanophilic with the dimensions 0.35–0.95 × 1.05–3.18 μm.

In PDA culture, colonies are round, smooth, and shining with regular margins: while cream colored at first, colonies darken with age. On nutrient agar, growth is slow and cells die rapidly; subculturing is not possible after about a week.

The carnation strain of *E. chrysanthemi* (EPPO/CABI, 1996b), which causes a similar disease, is readily distinguishable in nutrient agar culture by its rapid growth of grayish-white, lobate colonies. In addition, internal symptoms of the two diseases are different.

Disease Cycle

The natural spread of the pathogen is very slow and over extremely short distances. *Burkholderia caryophylli* can only enter the plant through natural openings or wounds. Considering that carnations are an ornamental plant, they are traditionally cut for sale. Carnation cutting is a large source by which infection takes place. Bacteria from an already infected plant can travel from plant to plant with each cut. Also, bacteria can pass from one cutting to another in the water of the propagating bed or if the cuttings are held in water. Wilting symptoms are found more reliably in crops grown at high temperatures (>30°C), whereas no symptoms or solely stem cracking symptoms are more common at lower temperatures (<20°C).

Protected environments (greenhouse, small gardens, etc.) are deemed more suitable for the pathogen success, opposed to unprotected environments, such as open fields. Turf and sphagnum peat have been reported to be a suitable environment for *B. caryophylli*.

The natural spread of the pathogen is very slow and over extremely short distances. The main path of distribution is by means of infected cuttings, which may be obtained from infected but symptomless mother plants.

Economic Impact

Burkholderia caryophylli has caused serious damage in the United States since its first report in 1940. Only minor losses occur in the EPPO region at present.

Control Measures

Cultural Practices

- Use culture-indexed plants free of the pathogen.
- Use sterilized soil or use a soilless potting mix.
- Remove diseased plants from production areas as soon as possible, including healthy-appearing plants within a radius of 1.5 ft from the diseased plants.
- Clean and wash hands and tools after handling plants.
- Thoroughly clean and sterilize the greenhouse between production cycles.
- Avoid overhead watering.
- Pot and propagate in pasteurized raised beds. Use clean, disinfested tools. Use culture-indexed cuttings free of the pathogen. Destroy infected plants.

There are no direct control measures. Disease-free mother plants should be used, and rooting beds and soil should be fumigated.

A testing method developed and in use in Denmark, "KPV-Metoden," enables diseased cuttings to be detected and destroyed at an early stage, thus preventing further dissemination of the disease and thus making it possible to eradicate the bacterium in 6–18 months.

Phytosanitary Risk

Burkholderia caryophylli is an EPPO A2 quarantine pest (OEPP/EPPO, 1978), in view of the limited number of EPPO countries in which it has been reported and the fact that it is readily carried on cuttings in international trade.

Phytosanitary Measures

In countries where the disease occurs, cuttings should be taken from separately grown mother plants derived from biologically tested, healthy cuttings. EPPO accordingly recommends that consignments should come from a place of production found free from *B. caryophylli* during the last growing season (OEPP/EPPO, 1990). However, the introduction of an EPPO-recommended certification scheme for carnation (OEPP/EPPO, 1991) provides a satisfactory alternative to such plant quarantine requirements.

Burkholderia caryophylli cannot be directly controlled by a chemical means. However, sanitary procedures can be done to prevent further infestation. Checking for signs of infected plant, which manifests itself as clogged and hyper-lignified vessel walls, before plant cutting and checking the soil to see that it does not have a presence of bacteria would ensure that further propagation would be prevented. However, in the instances of latent infections, these signs would not be present. Therefore, placing these cuttings under high temperature would accelerate the growth of bacteria, and the emergence of symptoms will allow for diagnosis. At its early stage, *KPV-Metoden* testing method can be applied, which allows for the infected cuttings to be detected and destroyed before it starts to infect other surrounding plants (Onozaki et al., 2002).

Although there are no commercial varieties of *Dianthus caryophyllus* that are resistant to *B. caryophylli*, there are some wild species that may have genes that are resistant to *B. caryophylli*. (Under a testing method that used cut-root soaking method to inoculate the different species of plants to *B. caryophylli*, *Dianthus capitatus* spp. *Andrzejowskianus* and *Dianthus henteri* were the two wild-type species that did not show any symptoms of *B. caryophylli*, and five other wild-type species along with *D. capitatus* spp. *Andrzejowskianus and D. henteri* were labeled as resistant due to a very low percentage of wilted plants.) It was because of the devastating effects and crop loss that breeding programs were initiated in 1988 by the National Institute of Floriculture Science (NIFS) in Japan.

10.4 GLADIOLUS

10.4.1 Bacterial Neck Rot and Scab in Gladiolus

Pathogens: *Pseudomonas marginata*
            Pseudomonas gladioli pv. *Gladioli*

Symptoms of Bacterial Neck Rot
Under wet conditions, and especially in heavy, poorly drained soils, numerous very tiny, reddish brown, slightly raised specks are seen first on the leaves (easily confused with thrip damage). These specks are numerous at the fleshy base of the leaf. In wet weather, elongated dead areas (soft and watery) occur when these specks or spots run together.

A shiny, varnish-like ooze is generally seen on leaves in humid weather. The neck rot causes the plant above the ground to fall over (Photo 10.5).

(a)

(b)

PHOTO 10.5 Field symptoms of (a) bacterial neck rot on gladiolus plant and (b) scab on gladiolus bulb. (Photo courtesy of T. Santha Das, santhagreens.com.)

Scab Symptoms

They are mainly seen on corms as irregular or round sunken brown spots with a shiny, brittle, varnish-like material (bacterial exudate) on the surface.

The scabby areas do not extend far into the flesh of the corm and may be removed leaving a depression. On the husks, elongated spots or lesions are dark brown to coal black. Holes are left with a rough coal black margin.

Disease Cycle

On corms and in soil refuse the pathogen survives for 2 years, favored by heavy, wet soils and warm weather and encouraged by heavy nitrogen fertilization.

Control Measures

- Rotate every 3 years. Control measures for other diseases usually take care of scab. Control chewing insects in the soil.
- Do not plant infected corms. Disinfest cutting knives frequently. Maintain good mite and insect control. Water in a manner that keeps leaf surfaces dry. Avoid working on wet plants.

10.4.2 Leaf Spot in Gladiolus

Pathogen: *Xanthomonas campestris* pv. *gummisudans* (McCulloch, 1924; Dye, 1978)

Synonym: *Xanthomonas gummisudans*

Common names: Leaf spot and scorch in gladiolus

Symptoms

Dark green, water-soaked leaf spots that are more or less rectangular are the first symptoms of the disease (Photo 10.6). These leaf spots later turn brown and necrotic. Bacterial ooze is present on the lesions under wet conditions. When there are many leaf spots, which can coalesce, plant and corm growth is retarded.

PHOTO 10.6 Leaf spot in gladiolus. (Photo courtesy of Melww at English Wikipedia; Photo courtesy of commons.wikimedia.org.)

Geographical Distribution

This is widely present in Australia, Canada, Finland, the Netherlands, South Africa, and the United States (McCulloch, 1924).

Disease Cycle

Bacteria enter the plant through stomata. The development of the disease is strongly dependent on free water (irrigation, wet weather).

Control Measures

Chemical control is not possible. If possible foliage should be kept dry.

10.5 ANTHURIUM

10.5.1 BACTERIAL BLIGHT IN ANTHURIUM

Pathogen: *Xanthomonas axonopodis* pv. *dieffenbachiae* (Mcculloch & Pirone) Vauterin et al.

Synonyms: *Xanthomonas campestris* pv. *dieffenbachiae* (Mcculloch et Pirone) Dye
Xanthomonas dieffenbachiae (Mcculloch et Pirone) Dowson

Common names: Bacterial blight in anthurium
Bacterial blight of aroids
Tipburn of philodendron

Bacterial blight disease caused by *X. campestris* pv. *dieffenbachiae* (*Xanthomonas axonopodis* pv. *dieffenbachiae*) affects a broad range of ornamental and edible aroids including anthurium, Colocasia (taro), *Aglaonema*, *Syngonium*, *Xanthosoma*, *Dieffenbachia*, *Epipremnum*, *Dracaena*, *Alocasia*, *Spathiphyllum*, *Rhaphidophora*, *Caladium*, and *Philodendron*.

Disease Symptoms

Early foliar symptoms start as water-soaked spots visible near the margins where hydathodes, filled with guttation fluid, serve as the most common port of entry. Tissues surrounding the infected areas turn yellow. Water-soaked spots coalesce, eventually forming large necrotic zones at leaf margins (Photo 10.7). The pathogen quickly moves into vascular tissues of petioles and stems, preventing the translocation of nutrients and water and producing symptoms of water stress that may resemble natural senescence. The main stem of systemically infected plants turns dark brown, and the growing point deteriorates, eventually leading to the death of the plant. When the spathe is infected, the disease is often called "flower blight." Less frequently, bacteria enter stomata, forming circular water-soaked lesions surrounded by chlorotic zones. Stomatal invasion often results in limited colonization of mesophyll tissues and does not necessarily lead to systemic infection.

(a)   (b)

PHOTO 10.7 Symptoms of bacterial blight on Anthurium (a) leaves and (b) flower. (Photo courtesy of D. Norman, UF/IFAS, edis.ifas.ufl.edu/pp292.)

The first visible symptoms are yellow (chlorotic), water-soaked lesions along the leaf margins that grow rapidly to form dead (necrotic) V-shaped lesions characteristic of this disease. *Xanthomonas* bacterial blight exhibits characteristic V-shaped, water-soaked lesions forming along the edges on *Anthurium* leaves.

Invading bacteria quickly spread throughout the plant. Leaves of systemically infected plants may exhibit a bronzed appearance. Floral quality may be reduced and/or flowers may become unmarketable. Eventually, plants wilt and die. *Xanthomonas* bacterial blight lesions can also appear on the flowers.

Geographical Distribution
This is widely distributed in the following regions:

Europe: Germany, Italy, the Netherlands, Romania, and Turkey
Asia: China, the Philippines, and Taiwan
Africa: Reunion Island in France and South Africa
America: Barbados, Bermuda, Brazil, Canada, Costa Rica, Dominica, Guadeloupe, Jamaica,
    Martinique, Puerto Rico, St. Vincent and the Grenadines, Trinidad and Tobago, the United
    States, and Venezuela
Oceania: Australia, French, and New Caledonia

Bacterial blight disease in anthurium was first reported in Brazil in 1960 and in Hawaii in 1971. In the humid tropics, high rainfall coupled with year-round high temperatures increases the severity and spread of bacterial blight disease making it a very important impediment to the cultivation of anthurium in the Caribbean.

The disease became a major problem in Hawaii in the 1980s and caused a major decline in the production of anthurium. In 1982, bacterial blight disease was accidentally introduced from Venezuela to Guadeloupe, causing severe decline of anthurium grown under traditional and greenhouse methods. This decline caused significant losses to commercial anthurium growers and was identified as an important limiting factor for the development of the crop in Martinique, Guadeloupe, and the rest of the Caribbean.

The disease was first reported in Jamaica in 1985 and along with the burrowing nematode, *Radopholus similis*, wiped out the anthurium industry in Jamaica. It was first identified in Trinidad in 1986 and has been responsible for the decline of the anthurium industry from 16 large anthurium farms in the early 1990s to only 5 large farms at present.

Bacterial blight disease has now been reported in most anthurium-producing countries including the mainland United States (Florida, New Jersey, and California), the Netherlands, and all other anthurium-producing Caribbean islands as well as South America, Tahiti, and Taiwan.

Host Range
The species of *Xanthomonas* that infects *Anthurium* has a very broad host range and is able to infect most aroid species; therefore, *Anthurium* plants may get blight when grown in close proximity to other aroids, such as *Dieffenbachia*, *Aglaonema*, and *Spathiphyllum*.

Major Hosts
Here are the following major hosts: *Anthurium andraeanum*, *Philodendron scandens, and Syngonium podophyllum*.

Minor Hosts
Aglaonema commutatum, Aglaonema crispum, Anthurium crystallinum, Anthurium Scherzerianum hybrids, Caladium bicolor hybrids, Colocasia esculenta, Philodendron selloum, Scindapsus aureus, Xanthosoma caracu, Xanthosoma sagittifolium, Anthurium, and *Araceae* are found to be minor hosts only.

Predisposing Factor

The bacteria infect *Anthurium* plants by entering pores (hydathodes) along the leaf margins. Bacteria may also enter if leaf tissues become torn through pruning or if leaf tissues are punctured by insects. When flowers are harvested, bacteria can enter via wounds.

Guttation droplets form at night when humidity is high and potting soil is warm and wet. Amino acids found in this guttation fluid are a source of food for the invading bacteria. Some infected plants are asymptomatic (do not show any disease symptoms) for months as the bacteria multiply. Bacteria in the guttation fluid from these asymptomatic infected plants can infect adjacent plants.

Infection and increase of the disease take place especially under warm (>25°C) and humid conditions. The bacterium can infect through wounds, hydathodes, and stomata. *Xanthomonas axonopodis* pv. *dieffenbachiae* may occur in low numbers on the leaf (epiphytically) or in the vascular system of the plant (latent infection). The bacterium can be spread by (latently) infected plants, splashing water (rain, irrigation), contaminated tools, wet clothes, infested soil, and possibly nematodes during planting, leaf pruning, and harvesting (Nishijima and Fujiyama, 1985).

There are at least three groups of *X. axonopodis* pv. *dieffenbachiae* affecting Araceae:

1. Strains from *Anthurium*, which are more virulent on *Anthurium* than other strains and have a broader host range
2. Certain strains from *Syngonium*, serologically closely related to *Anthurium* strains, also virulent on *Anthurium*, with a narrow host range
3. Strains from other Araceae, including strains from *Syngonium* other than those mentioned earlier, which are weakly virulent on *Anthurium* and have a narrower host range

Strains from *Syngonium*, first attributed to *Xanthomonas vitians*, were described as *X. campestris* pv. *syngonii* by Dickey and Zumoff in 1987, but the work of Chase et al. (1992) and Lipp et al. (1992) on a large number of *X. campestris* strains from aroids showed that there is little basis for a separate pathovar *syngonii*. The disease on *Anthurium* appears most damaging.

Bacteria can swim across wet surfaces; therefore, it is very important to keep the foliage dry.

Disease Impact

Foliar bacterial blight disease can severely decrease farm productivity and hence the profitability of anthurium cultivation. Infected leaves are removed, sometimes to a bare stem to reduce further disease spread. The reduced photosynthetic capacity as a result of leaf removal causes plant stunting. In cases where most of the leaves have been removed and the flowers are not produced at all, plants take longer time to mature or are unmarketable, being either deformed or small.

Systemic infection can occur independently of foliar infections. In such cases, the pathogen enters the plant via wounds or through the root system. The disease advances without apparent symptoms until the infected plant wilts and collapses. Symptoms of systemic infection start with yellowing of older leaves and petioles. Systemically infected leaves and flowers break off easily and may show dark-brown streaks at their base, which gradually enlarge. Cut petioles show yellow-brown vascular bundles. Systemic infection may also show water-soaked leaf spots when the bacteria enter the leaf parenchyma from the infected vascular bundles. Many farmers and researchers have observed that some cultivars resistant to foliar bacterial blight disease succumb rapidly to systemic infection and vice versa, indicating a two-phase differential basis for the disease.

The most serious disease problem to strike the Hawaii flower industry is bacterial blight caused by the pathogen *X. axonopodis* pv. *dieffenbachiae* (previously *X. campestris* pv. *dieffenbachiae*) (Nishijima, 1994). The disease was first reported in Kauai in 1972 (Hayward, 1972) but had little impact on the industry until 1981, when plants began to die in large numbers on farms in Hilo (Nishijima, 1988). The disease reached epidemic proportions in 1985–1989, destroying the production of approximately 200 small farms. Hawaii's production dropped from a record high of approximately 30 million stems in 1980 to 15.6 million stems in 1990 (Shehata, 1992). Following the implementation of an integrated disease management program, annual production losses were eventually reduced to 5% or less. However, due to the high cost of disease management, a few large farms now dominate the commercial markets.

In the production areas of *Anthurium* in North and South America, the disease is already a limiting factor. In Hawaii (United States), where the disease was first described by Hayward (1972), the total loss for the *Anthurium* industry in 1989 was US$15,782 per ha, that is, a total loss of US$2.74 million. Small farms have tended to reduce the area planted with this crop or to give up its cultivation (Shehata and Nishijima, 1989).

Once introduced into a new growing area, bacterial blight may result in 50%–100% loss of plants. This devastating disease has limited *Anthurium* production not only in Hawaii, but throughout the world where *Anthuriums* are produced. By 1992, it had been reported in the Philippines, Guam, Australia, Florida, Jamaica, Puerto Rico, Martinique, Venezuela, and Trinidad and has since been reported in India (Sathyanarayana et al., 1998) and the Netherlands (OEPP/EPPO, 2004). Bacterial blight affects most genera and species in the family Araceae (Lipp et al., 1992).

Control Measures

Bacterial blight is a contagious disease that is difficult to control. The pathogen spreads by excessive water splashing (rain or irrigation), movement of infected soil, and contact with infected plant material, tools (from leaf pruning and harvesting), and persons walking through infected plants. The bacterial blight pathogen could survive for many months in tissue-cultured plantlets without showing any symptoms. Furthermore, the pathogen can survive in leaf lamina, petiole, and root residues for as long as 4 months when tissues are left on the ground or buried 15 cm deep; the pathogen retains its pathogenicity during that time.

Chemical control of bacterial blight is ineffective, and to date, no chemicals have been described that can control or reduce the disease. Some chemicals used in the past have been shown to actually increase the severity of the disease.

1. Preventive Measures
    1. Use clean planting material.
    2. Install footbaths with disinfectant at entrances to greenhouses.
    3. Adjust plant density on beds to facilitate good aeration.
    4. Work from clean to diseased fields to minimize pathogen movement.
    5. Discard and burn infected plants.
    6. Disinfect tools and clothing regularly, change tools on different plots, and avoid exchange of materials between greenhouses.
    7. Ensure good drainage of the growing substrate.
    8. Use drip irrigation or microsprinklers.
    9. Avoid the presence of visitors not attached to the farm.
    10. Remove all debris and old media (or fumigate old media with methyl bromide) and let beds fallow for 2 months or fumigate with metam sodium and tarped in prior to replanting.
    11. Produce disease-free planting material from bud- or tissue-cultured plants away from production fields and ensure that the plants' nutritional needs are met to improve their vigor. Where the disease has been already established, sanitation practices should be performed, in addition to adhering to all the preventive measures mentioned earlier:
        a. Keep plants as dry as possible.
        b. Remove diseased leaves early and burn them.
        c. Avoid extreme climatic conditions particularly high temperatures.
        d. Adapt a fertilization regime to keep the production of glutamine in plants at a low level, as glutamine is a major food source for the bacteria. For instance, ammonium nitrate can be left out and desirable levels of potassium maintained.

2. Cultivar Resistance

The most effective control measure for bacterial blight disease is the use of resistant cultivars. The use of such cultivars will reduce the cost and risk associated with *Anthurium* cultivation. A rapid screening method for determining resistance to the disease at both the foliar and systemic phases has been developed at the University of the West Indies, which for the first time provides a practical means of identifying resistance to bacterial blight. With this two-stage screening method, several *Anthurium* cultivars with resistance/high tolerance to the disease combined with good horticultural attributes have been developed. The method is now being routinely used on a commercial farm in Trinidad with great success.

Resistant cultivars of *Anthurium antioquiense* may become infected with the pathogen but rarely develop systemic infection. The introduction of *A. antioquiense* in crosses with *A. andraeanum* has resulted in tolerant offspring (Kamemoto and Kuehnle, 1989; Kamemoto et al., 1990; Kuehnle et al., 1992, 1993). *A. andraeanum* and *A. antioquiense* cultivars also showed differential susceptibility to bacterial blight in foliar and systemic infection phases (Fukui et al., 1998).

Although some *Anthurium* are tolerant to *Xanthomonas axonopodis* pv. *dieffenbachiae*, natural genetic resistance to bacterial blight is not present in *Anthurium*, and breeding plants for tolerance through traditional means is time-consuming (Kamemoto and Kuehnle, 1996). Genetic engineering serves as a means of introducing resistance genes from nonplant origins into *Anthurium* plants. *Agrobacterium*-mediated gene transfer has been used to successfully transform *Anthurium* (Kuehnle et al., 1991). Genes that code the antibacterial peptides attacin and cecropin have been isolated from the cecropia moth (*Hyalophora cecropia*) and genetically engineered into *Anthuriums* (Kuehnle et al., 1992, 1993, 2004). Transgenic *Anthurium* plants expressing attacin were less susceptible to *X. axonopodis* pv. *dieffenbachiae* and had fewer numbers of bacteria present when compared to nontransgenic plants (Kamemoto and Kuehnle, 1996).

Kuehnle et al. (2004a,b) reported that two cultivars transformed to express the Shiva-1 lytic peptide (a synthetic analog of cecropin B) significantly resisted *Anthurium* blight. Comparisons were made between transgenic and nontransgenic lines of each cultivar as well as to a susceptible 'Rudolph' and tolerant 'Kalapana' lines used as controls. Resistance to infection was evaluated using a bioluminescent strain of *X. axonopodis* pv. *dieffenbachiae* to determine the amount of tissue colonized in wild-type and transgenic plants. Extensive bacterial colonization of major veins leading to the petiole was visualized by their bioluminescence in symptomless leaves. Other plants showed small sites of infection at leaf margins, whereas bioluminescence revealed a far greater level of tissue colonization. Disease progression in a transgenic line of 'Paradise Pink' was significantly reduced compared to the nontransgenic control, indicating increased tolerance. In contrast, one transgenic line of 'Tropic Flame' had increased susceptibility to blight, possibly due to the reduced transcription of the transgene. Other lines of *Tropic Flame* did not differ significantly from the controls. Additional testing of lines under field conditions is necessary before resistant transgenic cultivars are released.

3. Cultural and Chemical Controls

Components of an integrated management program for *Anthurium* blight include sanitation, disinfection of harvesting implements and containers, chemical sprays, modification of cultural practices, production of pathogen-free planting stocks *in vitro*, use of resistant cultivars, and biological control. Field sanitation, which involves removal of leaves showing early infections and elimination of systemically infected plants, was the principal method of reducing *Anthurium* blight in early years (Nishijima, 1988). Disinfection of cutting tools is important to prevent the spread of blight, since plant materials that show no symptoms have the potential for latent infection and the pathogen is spread during harvest. These two approaches, while useful, are insufficient for stopping disease spread.

Some chemicals and antibiotics were used with limited success at early stages of the blight outbreak in Hawaii (Alvarez et al., 1989, 1991; Nishijima and Chun, 1991), but these methods of control were later abandoned. Cultural practices, with a focus on nitrogen fertilization, were then examined (Mills, 1989). Chase (1988) suggested that lower fertilizer rates for potted *Anthuriums* could result in reduced disease and greater flower production. Sakai (1990) reported that higher levels of ammonium fertilizer led to higher amounts of amino compounds in guttation fluid when compared to nitrate fertilizers. He suggested that this promoted bacterial multiplication in the guttation fluid. The use of nitrate or inorganic fertilizers for plant growth was expected to reduce the amount of amino compounds in guttation fluid and thereby reduce blight incidence (Sakai, 1991; Sakai et al., 1992). Although greenhouse trials indicated a relationship between fertilizer treatments and blight susceptibility, field trials were inconclusive (Higaki et al., 1990, 1992).

Growing plants under plastic or glasshouses coupled with drip irrigation rather than overhead or sprinkler irrigation reduced the spread of the bacteria through aerosols and water splash and significantly reduced the incidence of blight in *Anthurium* seedling culture. Growing *Anthuriums* under cool and shaded conditions slows the progression of the disease. Inoculated plants exposed to temperatures greater than 31°C were more susceptible to disease than inoculated plants exposed to 26°C or lower temperatures. This difference in temperature can be regulated in commercial shade houses by strategically increasing airflow.

4. Tissue-Cultured *Anthuriums*

The best single means for blight disease management entails the use of plant material that is guaranteed to be pathogen-free. Tissue-cultured plants, although highly regarded and recommended to growers, have the potential of latent infection with *X. axonopodis* pv. *dieffenbachiae* (Norman et al., 1993; Norman and Alvarez, 1994). Microplants inoculated with a dilution series of the pathogen were killed when the pathogen was incorporated into the liquid medium at high populations, but microplants inoculated at very low population levels remained symptomless for 10 months. Furthermore, the pathogen was reisolated from the symptomless plants up to the time of deflasking a year later. Addition of coconut water to stimulate the growth of microplants provided a carbon source for quiescent bacteria, which then formed cloudy suspensions *in vitro* (Norman and Alvarez, 1994).

A triple indexing protocol developed for *Anthuriums* involves several steps as follows: (1) axillary buds are placed into culture to initiate the first generation of plantlets; (2) the bases of the plantlets are sectioned and tested for the presence of microorganisms using nutrient broth; (3) if the test is negative, the plantlets are cut into three nodal sections; (4) each nodal section is placed into liquid half-strength Murashige and Skoog medium with 15% coconut water; (5) this procedure is repeated two additional times if the plantlets test negative in nutrient broth. If the material tests positive at any point, the previously cultured material will be eliminated from further propagation (Tanabe et al., 1992). Triple indexing ensures that tissue-cultured plants do not serve as sources of inoculum.

Biological Control of Blight in *Anthurium*

The use of beneficial organisms, natural or modified, to control the effects of undesirable organisms was proposed to *Anthurium* growers during the initial stages of the bacterial blight epidemic (Cook, 1988). Cultures of microorganisms were soon isolated from internal petiole tissues of *Anthuriums* and examined as a means for biological control of bacterial blight (Fernandez et al., 1989). However, foliar applications of microorganisms antagonistic to *X. axonopodis* pv. *dieffenbachiae* resulted in inconsistent or insignificant control of the disease (Fernandez et al., 1990, 1991).

In later studies, Fukui et al. (1999) isolated bacteria from the guttation fluids of susceptible *Anthurium* cultivars ("Marian Seefurth" and UH1060) that did not succumb to infection by *X. axonopodis* pv. *dieffenbachiae* even under high inoculum pressure. Individually, these beneficial bacteria were not effective in suppressing the multiplication of *X. axonopodis* pv. *dieffenbachiae* in guttation fluids but were effective when applied in combination (Fukui et al., 1999). Foliar applications of the bacterial community reduced the infection of *Anthurium* leaves through hydathodes and wounds (Fukui et al., 1999). Using a combination of bacteriological tests, fatty acid analysis, and 16S rDNA sequence analysis, the strains were identified as *Sphingomonas chlorophenolica*, *Microbacterium testaceum*, *Brevundimonas vesicularis*, and *Herbaspirillum rubrisulbalbicans*. All four species survived on the surfaces of microplants up to 2 months and were effective in protecting microplants against infection (Alvarez and Mizumoto, 2001). Striking differences were observed when susceptible cultivar *Rudolph* was inoculated with *X. axonopodis* pv. *dieffenbachiae* and either treated (sprayed weekly) or not treated with the biocontrol agents. Nontreated plants developed typical blight symptoms after inoculation with the pathogen and died. In contrast, 75% of the treated plants survived and produced flowers after 22 weeks.

Fujii et al. (2002) demonstrated that biological control could be used simultaneously with genetic modification of *Anthurium* cultivars. The cultivars *Paradise Pink* and "Mauna Kea" were engineered to express the Shiva-1 lytic peptide, an antibacterial peptide, but did not inhibit the four species of beneficial bacteria when applied to these cultivars (Kuehnle et al., 2004a,b).

10.5.2 BACTERIAL WILT IN *ANTHURIUM*

Pathogen: *Ralstonia solanacearum* race 1 (Smith, 1896) Yabuuchi et al. (1996)

Synonyms: *Burkholderia solanacearum*
  Pseudomonas solanacearum

Symptoms

Leaf yellowing (chlorosis) is usually the first symptom observed. The disease spreads rapidly throughout the vascular system of the plant, turning veins in the leaves and stems a brown, bronze color (Photo 10.8). Bacterial ooze (brown slime) is present if cuts are made into the stems of highly infected plants. Plants exhibit wilt symptoms even though adequate soil moisture is available.

The disease causes necrosis of leaf laminae and petioles, spathes, and peduncles and eventually results in crown rot. Two months later, the plants begin producing new suckers from the rhizome, which may be free of symptoms. Preliminary reports suggest that the pathogen is a much greater problem for *Anthurium* plant cultivated in soil rather than in shade houses.

PHOTO 10.8   Yellowing (chlorosis) of *Anthurium* leaves due to bacterial wilt in *Ralstonia*. (Photo courtesy of D. Norman, UF/IFAS, edis.ifas.ufl.edu/pp292.)

Hosts

Major Hosts

Here are the following major hosts: *Lycopersicon esculentum, Nicotiana tabacum,* and *Solanum tuberosum.*

Minor Hosts

Arachis hypogaea, Capsicum annuum, Gossypium hirsutum, Hevea brasiliensis, Manihot esculenta, Ricinus communis, Solanum melongena, Zingiber officinale, and *Anthurium* are found to be minor hosts only.

Geographical Distribution

A new strain of *Ralstonia solanacearum* (Phylotype II-sequevar4NPB) was identified in the dry season of 2002 in a farm located on the northern range of Trinidad between Arima and Brasso Seco (on the way to Blanchisseuse) affecting local pink, local white, and local red cultivars.

Predisposing Factors

Cool greenhouse temperatures may temporarily mask symptoms and give bacteria time to spread. Symptoms appear rapidly during hot weather.

Bacterial wilt is spread via contaminated soil, water, tools, or worker contact.

Control Measures

A strict sanitation program is the most successful way to stop the spread of this pathogen and eventually eradicate it from a production facility. Fungicides that contain phosphorous acid have also been shown to be effective in preventing infection; however, they do not cure systemically infected plants (Norman et al., 2006).

Use disease-free propagation material. The bacterial wilt pathogen is easily spread via infected cuttings. Because the bacteria survive well in soil, both contaminated plant material and the supporting soil should be discarded. If pots and trays from contaminated infected plants are to be reused, they should be scrubbed free of adhering soil and then soaked in a disinfectant to kill the remaining bacteria. Knives and clippers should be sterilized between plants with a disinfectant containing a quaternary ammonium compound (Physan 20™, Green-Shield®) or diluted solution of bleach to prohibit spread.

10.5.3 BACTERIAL LEAF SPOT IN *ANTHURIUM*

Pathogen: *Acidovorax anthurii* (Gardan et al.)

Synonym: *Pseudomonas* sp.

Common name: Bacterial leaf spot disease in *Anthurium*

Symptoms

The disease produces both localized and systemic symptoms. Initially, the localized symptoms develop as angular lesions, oily in appearance and located either on the undersurface of the leaves and spathes, near the veins or along the margins. The lesions are surrounded by a chlorotic halo, which enlarges into black necrotic zones, darkening with age and resulting in deformed leaves. Some researchers have also observed white exudates oozing from the margins of the lesions when the humidity is high. On the spathes, the necrotic spots are surrounded by violet halos. Infection can progress in the veins and eventually lead to rotting and abscission of leaves and cut flowers.

Systemically infected plants show water-soaked areas at the junction between the petiole and leaf veins and general chlorosis of leaves. At the midrib, veins, and leaf margins, typical translucent water-soaked areas are evident, which eventually become necrotic. Many times, when the parent plant dies, multiple suckers, which may or may not show foliar symptoms, develop.

Hosts
Here are the commonly observed hosts: *Anthurium andreanum, Anthurium martinicense (Araceae)* by artificial inoculation, and *Dieffenbachia seguine (Araceae)*.

Geographical Distribution
This is widely present in Central America, Guadeloupe, Martinique, Trinidad, and Tobago.

Disease Cycle
The disease is highly contagious and the pathogen spreads by excessive water splashing (rain or irrigation), movement of infected soil, and contact with infected plant material, tools (from leaf pruning and harvesting), and persons walking through infected plants.

Control Measures
It can be controlled using cultural practices and applying chemicals. Cultural practices include the routine removal and burning of infected leaves, cut flowers, and plants, with regular disinfection of tools and clothes with 70% alcohol or 10% hypochlorite. One of the chemicals used to control the disease is Phyton-27 (a copper-based fungicide).

The bacterial leaf spot disease is especially devastating during the early stages of the crop and can result in decreased cut-flower production and death of plants. Proper management of this disease is both capital and labor-intensive, and the development of cultivars' resistance is important in reducing the risk and cost of cultivation.

10.6 TULIP

10.6.1 Yellow Pustule and Silvery Streaks in Tulip

Pathogen: *Curtobacterium flaccumfaciens* pv. *oortii*

Synonym: *Corynebacterium flaccumfaciens* pv. *oortii* (Saaltink and Mass Geestranus, 1969; Dye and Kemp, 1977)

Common names: Yellow pustule in tulip (bulb symptoms)
Hell fire (leaf symptoms) and silvery streak in tulip

Symptoms
The most conspicuous symptom on the bulb is a yellowing of the first white bulb scale that becomes visible when the brown skin is partly cracked or removed. Close inspection shows that this symptom is followed by the development of patches composed of many small white spots. These patches turn yellow and the tissue becomes raised, often with rupturing of the surface of the swollen area. On a transversely cut surface of a diseased bulb, the tissue of the infected scales shows yellow xylem bundles.

Severely diseased bulbs do not sprout after planting, but a certain percentage of moderately diseased ones develop into diseased plants. These plants are stunted, show a few silvery streaks along the leaf veins, and wither before flowering. The xylem in the stem of such plants is bright yellow. Microscopical examination of sections of discolored stem tissue shows numerous bacteria in the xylem vessels.

Plants with silvery streaks are noticed in greenhouses, mostly in the cultivar 'Paul Richter' and thereafter also in other cultivars. In the field, the symptoms are observed during and after flowering. On the upper surface of the leaf, silvery spots with a diameter of about 5 mm appear. The epidermis of these spots cracked very easily when touched, and the underlying parenchyma had a disorganized appearance. The leaves of some other plants showed no silvery areas, but they had a roughened appearance caused by a heavy cracking of the upper and lower epidermis. Such plants almost always showed a yellowish discoloration of the central part of the stem that could be followed into the first cell layers of the young growing bulb near the attachment to the stem.

The spread of the disease was considerable in the 1966–1967 season, but symptoms were seen only rarely in 1967–1968. The roughened appearance of the leaves has been seen for many years both in the Netherlands and elsewhere. The disease may have been endemic in the Netherlands for many years without serious consequences.

Hosts

Garden tulip (*Tulipa gesneriana*)

Geographical Distribution

This bacteria can be found in Europe (Denmark, the Netherlands, Romania, the United Kingdom, and England and Wales) and also in Asia (Japan, the Republic of Korea).

Pathogen

Identical bacteria were consistently isolated from the rough areas as well as from the discolored spots or streaks. These bacteria were mainly Gram positive. Isolates made on nutrient agar from yellow lesions on stored bulbs, however, were usually Gram negative but could be made Gram positive by transferring them from standard nutrient agar to protein-rich agar slopes (Eugon agar). This unstable Gram reaction may be related to the growth phase of the bulb and the medium in which the bacteria are cultivated. Both Gram-positive and Gram-negative isolates used for inoculation caused the same symptoms.

Disease Cycle

Curtobacterium flaccumfaciens pv. *oortii* is a systemic vascular pathogen and a wound parasite. Cold weather and wounds caused by rubbing of leaves, hail, and cultivation practices in spring favor disease development. Transmission is by infected bulbs and between plants in the field by wind-blown rain and sand and by contaminated implements used to cut the flowers.

Control Measures

Direct control of the disease is not possible. Preventive spray or dip in Captan-containing pesticides yields some protection. Removal of infected plants and bulbs and avoidance of working in the fields during humid weather conditions (especially when removing the flowers) reduces the risk of infection. Late planting (around November 30) to avoid critical night frost in spring may be beneficial.

10.6.2 BACTERIAL LEAF AND PEDUNCLE ROT IN TULIP

Pathogen: *Pectobacterium carotovorum*

The disease is reported from Turkey.

Symptoms

The bacterium produces lesions on leaf, bud neck, and bulb.

Control Measures

Resistance

Gander was the most susceptible variety in field, whereas salmon parrot exhibited the highest rate of bulb rot in storage.

10.7 ORCHIDS

10.7.1 BACTERIAL LEAF BLIGHT IN ORCHIDS

Pathogen: *Burkholderia gladioli*

The disease is reported from Japan and also continues to be a serious problem in Hawaii.

Symptoms

The bacterium infects the leaves. The bacterium is reported to be pathogenic to *odontroda* cv. in the dark with high-temperature stress and high humidity. The bacterium is also pathogenic to other orchids like *Calanthe*, *Habenaria*, and *Miltonia* sp.

Small to large leaf spots with or without water soaking or soft rot are observed in various orchid genera, including *Dendrobium*, *Oncidium*, and *Miltonia* sp. and hybrids.

Control Measures

Copper-containing fungicides with plant antibiotics may reduce the incidence of the disease.

10.7.2 LEAF SPOT IN ORCHIDS

Pathogen: *Acidovorax avenae* subsp. *cattleyae* (Pavarino, 1911; Willems et al., 1992)

Synonym: *Pseudomonas cattleyae*

Common names: Bacterial leaf spot and bud rot and brown spot in orchids

The disease was first described by Pavarino in Italy in 1911 and the causal organism in 1946 by Ark and Thomas in the United States.

Symptoms

Small water-soaked spots appear as the first symptoms caused by *Acidovorax avenae* pv. *cattleyae*. These spots enlarge rapidly under humid conditions and turn brown to black, often surrounded by a yellow halo. In advanced stages of the infection, spots are sunken and invaded by secondary organisms. Spots may coalesce, and whole leaves and plants are killed especially when the growing point becomes infected. Spots can be found not only on seedlings (where the infection can be devastating) but also on older plants (especially in *cattleya*). Infection occurs through wounds and stomata.

Overhead irrigation and manipulations during cultivation can cause rapid dispersal of the pathogen in greenhouses. Especially in seedlings the disease can cause heavy losses.

Hosts

The following are the hosts: *Cattleya* spp., *Phalaenopsis*, and *Paphiopedilum* (Venus' slipper) spp. and hybrids. Also, *Catasetum* spp., *Cypripedium* spp. (Lady's slipper), *Dendrobium* spp., *Doritaenopsis* sp., *Epidendrum* spp., *Epiphronitis veitchii*, *Ionopsis utricularioides*, *Miltonia* sp., *Oncidium* spp., *Ornithocephalus bicornis*, *Renanthera* sp., *Rodricidium*, *Rodriguezia*, *Rhynchostylis* spp., *Sophronitis cernua*, *Trichocentrum* sp., *Vanda* spp., *Vanilla* sp., and *Vuylstekeara* sp. have been reported as natural hosts.

Geographical Distribution

This is widely distributed in Australia (Stovold et al., 2001), Italy, the Netherlands, the Philippines, Portugal, Taiwan (Huang, 1990), the United States, and probably Venezuela (Trujillo and Hernández, 1999).

Control Measures

A combination of high temperature and high humidity should be avoided if possible. In an early stage, attempts can be made to cut out the infected spots, after which the cut surface should be dusted with an antibiotic (if permitted). When the infection is widespread in the greenhouse, spraying with a solution of natriphene or soaking of whole plants in orthophenylphenol, natriphene, or Physan (quaternary ammonium compound) for 1 h is often effective in reducing the disease. Avoidance of overhead irrigation, splash water, disinfection of tools, and isolation of infected plants are also helpful in control. Sprays with copper compounds or benzalkonium chloride (if permitted) can be applied as prevention.

10.7.3 BROWN ROT IN ORCHIDS

Pathogen: *Erwinia cypripedii* (Hori, 1911; Hammer and Huntoon, 1923)

Synonym: *Pectobacterium cypripedii*

PHOTO 10.9   Symptoms of brown rot in orchids caused by *Erwinia* spp. (Photo courtesy of Scot Nelson, flickr.com.)

Symptoms

This is first described in Japan by Hori in 1911. In orchids, *Erwinia cypripedii* initially causes water-soaked round to oval leaf spots that enlarge, turn dark brown, and become slightly sunken (Photo 10.9). When the leaf base and growing point are infected, plants rot totally and die.

It begins as a small brown spot in the leaf, and then it rapidly spreads throughout the rest of the plant. As soon as it reaches the basal portion of the plant, or its crown, the plant will die. The spread of this disease is so fast that it can destroy a multiple growth plant within 24 h. Often, the crown of the plant will completely rot and collapse, while the outer portions of the leaves still remain healthy looking.

Hosts

Primary hosts: *Cypripedium* and *Paphiopedilum* orchid spp. and their respective hybrids.
Secondary hosts: *Aerides japonicum*, *Phalaenopsis* spp., and *Carica papayae* (pawpaw) have also been reported as natural hosts.

Geographical Distribution

This kind of infection is evidently present in Africa, South Africa, Asia (Japan, Taiwan), Australia (South Australia), North America, and the United States (California, Florida).

Disease Cycle

The bacterium gains entry into the plant through injury, or possibly through the stomata.

The disease and organism may be spread by overhead irrigation, manipulations during cultivation and with diseased transplants.

Control Measures

Good ventilation and keeping roots aerated during cultivation and rapid drying of plants after watering are preventive measures to avoid rot caused by *E. cypripedii*. Compost should be dry before use and should not be too acidic. Avoidance of overhead irrigation, splash water, disinfection of tools, and isolation of infected plants are also helpful in control. Infected plants should be removed and destroyed. When the infection is widespread in the greenhouse, spraying with a solution of natriphene or soaking of whole plants in orthophenylphenol, natriphene, or Physan (quaternary ammonium compound) for 1 h is often effective in reducing the disease.

10.8 CHRYSANTHEMUM

10.8.1 Bacterial Leaf Spot in Chrysanthemum

Pathogen: *Pseudomonas cichorii*

The disease was reported from the subtropical experimental station Homestead in 1957.

Symptoms

The leaf spots are circular or elliptical and slightly sunken, up to 1 cm diameter, and coalesce into large necrotic areas. These spread upward from older leaves. Infection in flower buds may also occur.

Disease Management

It is controlled by weekly sprays of tribasic $CuSo_4$ at 4 lb/100 gal and Agri-Mycin 500 at 3 lb/100 gal without injury to leaves and flowers. Agri-Strep and Agri-Mycin 100 are slightly less effective.

10.8.2 Soft Rot and Wilt in Chrysanthemum

Pathogen: *Erwinia chrysanthemi* (Burkholder et al.)

Synonyms: *Erwinia carotovora* (Jones) Bergey et al. f.sp. *parthenii* Starr
Erwinia carotovora (Jones) Bergey et al. f.sp. *dianthicola* (Hellmers) Bakker
Pectobacterium parthenii (Starr) Hellmers
Erwinia carotovora (Jones) Bergey et al. var. *Chrysanthemi* (Burkholder et al.) Dye

Symptoms

Erwinia chrysanthemi causes soft rots and wilts, particularly stem rot in chrysanthemums. The infected plant rots at crown level when the infection occurs at soil level. The infected crown portion becomes soft and collapses, when the infection occurs beneath the soil; it invades the roots and clogs the xylem vessels that lead to wilting and eventually the death of the plant (Photo 10.10).

  Erwinia chrysanthemi is a soft rot pathogen degrading succulent fleshy plant organs such as roots, tubers, stem cuttings, and thick leaves. It is also a vascular wilt pathogen, colonizing the xylem and becoming systemic within the plant. This latter aspect is the most alarming when vegetative propagation is involved. The pathogen can remain latent in stock plants (ornamentals, bananas) and can thus be spread in cuttings from them.

PHOTO 10.10 Field symptoms of wilt in chrysanthemum plants. (Photo courtesy of Florida Division of Plant Industry, Florida Department of Agriculture and Consumer Services, Tallahassee, FL, Bugwood.org.)

Hosts

The following were found to be the hosts: *Dieffenbachia* spp., *Euphorbia pulcherrima*, *Kalanchoe blossfeldiana*, maize, *Philodendron* spp., potatoes, *Saintpaulia ionantha*, and *S. podophyllum*.

Erwinia chrysanthemi has also been naturally found to attack *Allium fistulosum*, *Brassica chinensis*, *Capsicum*, cardamoms, carrots, celery, chicory, *Colocasia esculenta*, Poaceae (such as *Brachiaria mutica*, *Brachiaria ruziziensis*, *Panicum maximum*, and *Pennisetum purpureum*), *Hyacinthus* sp., *Leucanthemum maximum*, lucerne, onions, pineapples, radishes, rice, *Sedum spectabile*, sugarcane, sorghum, sweet potatoes, tobacco, tomatoes, tulips, and glasshouse ornamentals (such as *Aechmea fasciata*, *Aglaonema pictum*, *Anemone* spp., *Begonia intermedia* cv. *Bertinii*, *Cyclamen* sp., *Dracaena marginata*, *Opuntia* sp., *Parthenium argentatum*, *Pelargonium capitatum*, *Phalaenopsis* sp., *Polyscias filicifolia*, and *Rhynchostylis gigantea*). *Primula obconica* has been erroneously mentioned as a host.

Geographical Distribution

Erwinia chrysanthemi has a worldwide distribution. Any kind of strain may occur in temperate countries, where outdoor and glasshouse plants are produced:

EPPO region: Algeria, Austria, Belarus, Belgium, Denmark, Egypt, Finland (found in the past but not established), France, Germany, Greece, Hungary, Israel, Italy, the Netherlands, Norway, Poland, Portugal, Romania, Russia (European), Spain, Sweden, Switzerland, the United Kingdom, and Yugoslavia

Asia: Bangladesh, China (Fujian, Hunan, Jiangsu, Jiangxi, Zhejiang), India (Bihar, Delhi, Karnataka, Uttar Pradesh, West Bengal), Iran, Israel, Japan (Hokkaido), Democratic People's Republic of Korea, the Republic of Korea, Malaysia (Peninsular), Nepal, the Philippines, Sri Lanka, and Taiwan

Africa: Algeria, Comoros, Côte d'Ivoire, Congo, Egypt, Réunion, Sudan, South Africa, and Zimbabwe

North America: the United States (California, Colorado, Connecticut, Florida, Georgia, Massachusetts, North Carolina, North Dakota, Nebraska, New York, Ohio, Pennsylvania, Texas, Virginia, and Wisconsin)

Central America and Caribbean: Aruba, Costa Rica, Cuba, Guadeloupe, Guatemala, Haiti, Honduras, Jamaica, Martinique, Panama, Puerto Rico, and St. Lucia

South America: Brazil (widespread), Colombia, Ecuador, French Guiana, Guyana, Peru, and Venezuela

Oceania: Australia (New South Wales, Queensland, Victoria, and Western Australia), Cook Islands, New Zealand, Papua New Guinea, and Solomon Islands

Predisposing Factors

The bacterium is able to survive in the soil (on plant debris) so that infestation remains between two crops.

High humidity and free water favor spread and penetration of the bacteria. Disease development is dependent on high temperatures, generally 25°C–30°C.

Host specialization has not definitely been proved in *E. chrysanthemi*, except in pv. *paradisiacal* (Dickey and Victoria, 1980; Dickey, 1981). The pathogen is ubiquitous and isolates from maize and potato seem to be rather polyphagous, while *Philodendron* and *Kalanchoe* do appear to be differential hosts for temperate isolates (Janse and Ruissen, 1988).

Economic Impact

The disease causes destruction of many flower and ornamental crops, particularly carnation and chrysanthemum in rooting beds. Losses are also recorded on different glasshouse ornamentals (*Saintpaulia ionantha*, *Kalanchoe*).

Control Measures
Control varies with the crop attacked. It entails strict attention to sanitation and plant hygiene in the nursery or glasshouse and usually a rigidly controlled propagation program to produce disease-free plants.

10.8.3 Fasciation of Chrysanthemum

Pathogen: *Rhodococcus fascians* (Tilford, 1936; Goodfellow, 1984)

Synonyms: *Rhodococcus rubropertinctus*
   Bacterium fascians (Tilford; Lacey 1939)
   Phytomonas fascians (Tilford, 1936)
   Pseudobacterium fascians (Tilford; Krasil'nikov 1949)
   Corynebacterium fascians (Tilford, 1936; Dowson 1942)

Common names: Fasciation (leafy gall), witches' broom syndrome, and cauliflower disease of ornamentals

Symptoms
Clusters of short, spindly or swollen, fleshy shoots develop at a node on the main stem. These shoots are dwarfed, with misshapen leaves, and may be at, below, or near the soil line. This mass of fascinated growth resembles small witches' broom or a cauliflower-like head and may reach a diameter of an inch to several inches. The main stem of an affected plant sometimes appears to grow normally but may be stunted. Blossoming is reduced. The flowers are also infected due to fascination pathogen (Photo 10.11). The roots on a diseased plant are sometimes short with swollen areas.

Hosts
The main hosts are sweet pea, carnation, chrysanthemum, dahlia, geranium (florists, ivy, zonal), gladiolus, nasturtium, petunia, and pyrethrum.

   Other ornamental plants reported as hosts include African violet, angel's trumpet, aster, baby's breath, Begonia, Buddleia, butterfly flower, cardamine, coral bells, crassula, delphinium, euphorbia, European cranberry bush, forsythia, hebe, hollyhock, impatiens, *Kalanchoe* larkspur (rocket), lily (Easter, regal), marigold (African, French), mullein (common, moth), nicotiana (flowering tobacco), phlox, physalis, *Piqueria*, primrose, rhododendron, schizanthus, Shasta daisy, asparagus (Sprenger), sweet William, and wallflower.

PHOTO 10.11 Symptoms of fasciation in chrysanthemum flower. (Photo courtesy of Perduejn, Wikimedia.)

Geographical Distribution
This kind of infection is widely distributed in the following regions:

Europe: Belgium, Czechia, Denmark, Estonia, France, Germany, Hungary, Italy, Latvia, the
   Netherlands, Norway, Russia, Slovakia, Sweden, Ukraine, and the United Kingdom
Asia: India and Iran
Africa: Egypt
America: Canada, Colombia, Mexico, and the United States
Oceania: Australia and New Zealand

Disease Cycle
The causal organism grows on the surface or in the outer few cell layers of a susceptible, young plant and stimulates the growth of shoots from normally dormant buds at the base of the plant. The bacterium can survive in the soil from one season to another, particularly in the greenhouse. How long the bacterium survives in the soil in the absence of a host plant is not known. It is not likely to be able to over season outdoors in Illinois.

The bacterium is spread by (1) taking cuttings from diseased stock plants, (2) planting infected seed, and (3) transferring the organism on hands or tools.

Economic Impact
Rhodococcus fascians can cause tremendous economic losses that consistently limit profitable production of certain ornamental crops. In 1950, for example, the pathogen was reported to be a limiting factor in commercial culture of Shasta daisy (*Leucanthemum* × *Superbum*) 'Esther Read' in California and to occur on certain other ornamentals in most seasons (Baker, 1950). Twenty-five years later, *R. fascians* was still prevalent in the same areas and, in some fields, was present on virtually all plants (Oduro, 1975). In contrast, the pathogen appears to occur sporadically on other ornamental crops grown in different production systems. Carnations, reportedly one of the less susceptible species, were affected by the disease in one nursery in England for a period of 10–12 years but at an incidence of only about 5% (Lacey, 1939). In chrysanthemums, incidence varied from 0.5% to 77.7% (Williams, 1933, 1934; Pape, 1938), depending on the cultivar. In *Dahlia*, incidence was as great as 100% (Faivre-Amiot, 1967). More recently, the disease caused by *R. fascians* can be responsible for serious recurring losses to the nursery industry. One wholesale nursery estimated losses over a 1-year period as great as US$1 million due to lost sales, recall of infected plants, and time lost in propagating and maintaining material that was eventually destroyed. A second nursery estimated US$394,000 annual loss due to *R. fascians*. Other nurseries have had to destroy all plants of a particular cultivar or had to stop growing particular species or cultivars altogether due to persistent, severe problems with *R. fascians*.

Control Measures
1. Carefully remove and destroy (burn) all infected plants.
2. Immediately after handling a diseased plant, do the following: (a) Wash hands thoroughly with soap and hot water and (b) sterilize all knives and other tools. Dip or swab them with a disinfectant such as a fresh solution of household bleach (one part of liquid bleach in five parts of water), 70%–95% grain or rubbing (wood) alcohol, Lysol, or Listerol household disinfectant.
3. Take cuttings only from healthy stock plants that have been inspected and found to be free of disease. Plant disease-free seed as the bacterium can be transmitted on the seed coat.
4. In greenhouses, plant in soil that has been sterilized by steam (80°C for 30 min or 75°C for an hour). Also sterilize all pots, boxes, and other plant containers. Be careful to avoid introducing the causal organism into the soil or onto plant containers that have been sterilized previously.

5. Rotate with nonsusceptible plants for one growing season or longer. The practices outlined earlier in addition to common sense hygiene should eliminate fasciation as a problem in growing flowers in greenhouses or outdoors.

10.9 MARIGOLD

10.9.1 SOUTHERN BACTERIAL WILT IN MARIGOLD

Pathogen: *Ralstonia solanacearum*

Synonym: *Pseudomonas solanacearum*

Southern bacterial wilt, caused by *Ralstonia solanacearum*, is a widespread and destructive disease of numerous crops in the warm climates of the world. The bacterium responsible for causing Southern wilt is believed to have originated in Kenya and is now found in regions ranging from tropical to temperate throughout the world. It is a major disease of tobacco, tomato, potato, dahlia, geranium, hollyhock, hydrangea, marigold, nasturtium, zinnia, and others. Bacterial wilt is the most frequent disease problem in marigolds in North Carolina landscapes.

Symptoms

Southern wilt attacks quickly and with devastating impact on infected marigolds; therefore, it is critical that gardeners recognize its signs and symptoms. Initially, marigold plants may cease to thrive and appear stunted. In early stages, the marigold's leaves may turn light green, yellow or gray green. First, top foliage droops slightly, but soon the entire marigold plant wilts completely. Marigolds infected with Southern wilt will die within 2 weeks after the symptoms are first noticed. Southern wilt can sometimes be mistaken for bacterial blight (caused by *X. campestris*); however, marigolds infected with bacterial blight would have distinctive leaf spots, which is not the case with marigolds suffering from Southern wilt.

Pathogen

Ralstonia solanacearum is a soilborne pathogen that can overwinter unnoticed in the ground. It thrives in warm climates, especially favoring those with high soil moisture; however, *R. solanacearum* can flourish in many soils with widely ranging pH values. Beyond soil, *R. solanacearum* can survive in infected plant material or on wild host plants.

Control Measures

Southern bacterial wilt is a difficult disease to control. There are no chemicals that provide effective control before or after planting.

The only control is to avoid susceptible plants or to grow resistant cultivars. The following marigold cultivars are resistant to bacterial wilt: Cupid, Irish Lace, Papaya Crush, Pineapple Crush, Pumpkin Crush, Rusty Red, Sparky, Sparky Mix, Bonanza Yellow, Choice Mix, Copper Canyon, Cupid Mix, Fort Knox, Golden Harmony, Goldie, Gypsy Dancer, Naughty Marietta, Orange Lady, Senator Dirksen, and Tangerine Gem. If it is necessary to grow a susceptible cultivar in an area where the disease has been a problem in the past, plant the susceptible cultivar in another part of the area or plant them in new soil or a soilless media in planter boxes.

10.10 ZINNIA

10.10.1 BACTERIAL LEAF SPOT IN ZINNIA

Pathogen: *Xanthomonas campestris* pv. *zinnia*

Symptoms

Spots begin as dull gray water-soaked areas on the leaf. Numerous yellow or tan spots can develop across the leaf forming irregular dead areas (Photo 10.12). Spots turn brown and angular, which could

PHOTO 10.12 Bacterial spot on a zinnia leaf. (Photo courtesy of OSU Plant Clinic Image, 2013, pnwhandbooks.org.)

be confused with leaf spots of *Alternaria*. Small brown spots may form on the ray flowers, and if severe, flower heads can disfigure and decay completely.

Disease Cycle

X. campestris pv. *zinniae* can survive in dried zinnia leaves for as long as a year. They can reside on foliage for several months before initiating disease. They may also be seed borne. Infected plant material is probably the most important source of contamination. Bacteria can be spread from plant to plant by overhead or sprinkler irrigation.

Disease Management

No sprays are effective. Whenever possible, water in a manner that keeps the leaves dry to inhibit this seed-borne disease.

Cultural Practices

Practice strict sanitation and rigorous scouting to identify and remove infected plants in the early stages before the pathogen spreads.

- Start with pathogen-free seed. Seed can be treated with a dilute bleach solution (10%) for 15–30 min or hot water. Be sure to test a small batch of seed for sensitivity to these treatments.
- Plants with leaf spots should be discarded. Diseased plant debris should be removed from the growing area.
- Workers should wash their hands after handling diseased plants or soil.
- Handling of wet foliage should be avoided.
- Minimize splashing when watering.
- Increase spacing between plants.
- Rotate plantings to different beds each year.

Chemical Control
Bactericides are only marginally effective, so focus on cultural control tactics:

- CuPRO 2005 T/N/O at 0.75–2 lb/A (or 1–2 Tbsp/1000 sq ft) when new growth is present
- Phyton-27 at 1.3–2.5 oz/10 gal water

Biological Control
- Bayer Advanced Natria Disease Control RTU (*Bacillus subtilis* strain QST 713) is registered for the home garden. Active ingredient is a small protein.
- Cease or Rhapsody (*B. subtilis* strain QST 713) at 2–8 quarts/100 gal water. Active ingredient is a small protein. Efficacy in the Pacific Northwest is unknown although it was effective in one trial in Maryland.

10.11 *DAHLIA*

10.11.1 BACTERIAL WILT IN *DAHLIA*

Pathogen: *Erwinia chrysanthemi*

Synonyms: *Pectobacterium chrysanthemi* and *Dickea chrysanthemi* (Burkholder, McFadden, and Dimock, 1953)

Symptoms
Erwinia chrysanthemi can affect any plant organ such as roots, stems, leaves, and storage organs, depending on plant species and environmental conditions. The resulting symptoms vary from soft rot to wilting.

Disease Cycle
Disease caused by *E. chrysanthemi* usually occurs in a severe form on succulent plant organs at high temperatures, a humid atmosphere, and high nitrogen levels. The symptoms are usually destructive at temperatures at or above 28°C. However, the disease may be very mild, or the plants remain symptomless, at temperatures below 20°C.

Systematically infected stock plants, or externally infested seed materials, become important sources of primary infection in vegetative propagation (*Dahlia, Dianthus, Dieffenbachia, Musa* spp., and *Solanum tuberosum*). The disease may be spread by knives or fingers during farming practices such as pruning, cutting, and harvesting (Burkholder et al., 1953; McGovern et al., 1985; Serfontein et al., 1991).

10.11.2 CROWN GALL IN *DAHLIA*

Pathogen: *Agrobacterium tumefaciens*

Symptoms
It is characterized by the abnormal growth of the tubers and at the base of the plant. The plants are stunted with spindly shoots.

This bacterium swims through the soil and accumulates around the roots of the dahlia plant.

Control Measures
Plants, tuber, and all infected material should be destroyed, and dahlia should be rotated to another location for at least 3 years while soil is fumigated. Grow nonwoody plants such as cereals or legumes in the space before returning dahlia to the area.

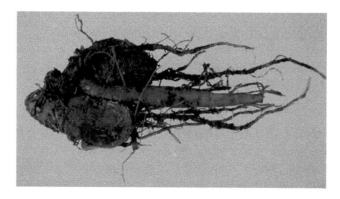

PHOTO 10.13   Dahlia tuber with bacterial soft rot. (Photo courtesy of Melodie Putnam, 2010. pnwhandbooks. org.)

10.11.3 Bacterial Soft Rot in *Dahlia*

Pathogen: *Erwinia carotovora*

Symptoms

Tuber rot is common in storage. The infected tubers become darken and soften (Photo 10.13) with the production of foul smell. Such tubers are unsuitable for planting.

Control Measures

Destroy the infected tubers. Use healthy tubers only for planting.

10.12 GERANIUM

10.12.1 Bacterial Blight in Geranium

Pathogen: *Xanthomonas campestris* pv. *pelargonii*

Geraniums (*Pelargonium* species) are the most popular flowering pot plants grown in the United States. Plant diseases are one of the key limiting factors to the high-quality production of geranium in the home and commercially. Three bacterial diseases are known to attack geraniums.

Of these, bacterial blight is the only one of these diseases currently known to occur in the Midwest. This disease is a major problem of geraniums in Illinois during warm, humid weather. It has caused losses of 10%–100%, depending on the cultivar.

Symptoms

The symptoms are often masked by low winter temperatures. Disease symptoms usually appear in warm, sunny weather when the plants are under moisture stress. Bacterial blight attacks the leaves and stems of the common garden or florist's geranium (*Pelargonium hortorum*) as well as the ivy geranium (*Pelargonium peltatum*).

Leaves infected with bacterial blight show two types of symptoms:

1. Small, round, water-soaked spots form on the underside of the leaf. Several days later, the spots develop into angular, slightly sunken, well-defined lesions about 1/16–1/8 in. (up to 3 mm) in diameter that may be surrounded by a diffuse yellow halo. The spots eventually turn dark brown to black (Photo 10.14) and become hard and dry. Leaves quickly wilt, die, and may fall off or hang on the plant for a week or more. The bacteria in these leaf lesions may migrate into the water-conducting tissues and move down through the petiole and into the stem, eventually causing the plant to die from the stem rot phase of the disease.

PHOTO 10.14 Symptoms of bacterial blight on geranium leaves. (Photo courtesy of R. Wick, UMass, University of Massachusetts Amherst, negreenhouseupdate.info.)

2. A wilting of the leaf margin occurs in some cultivars, resulting in large angular (V-shaped), yellow, or dead areas bounded by the veins. These leaves soon wither and hang on the petiole or drop off. Other factors, such as potash deficiency and Botrytis blight, may cause angular, dead patches, in geranium leaves, but the wilted condition is always associated with bacterial blight infections.

The stem rot phase of the disease is commonly called "black rot" by growers. Water-conducting tissues within the branches and stems of infected plants turn brown to black, 2–4 weeks after infection. At this stage, one or more leaves on the branch will usually wilt and develop angular dead areas. The bacteria soon spread from the water-conducting tissue to other stem tissues. The stem rapidly turns a dull gray to blackish brown. The roots are blackened but not decayed.

Plants gradually become defoliated, except for a few dwarfed leaves at the tips of the upright blackened branches. The stem and branches rapidly blacken and shrivel into a dry rot with all the tissues destroyed, except for the fibers and epidermis. Some plants may partially recover from initial infection and produce new, healthy-looking branches. Since these branches harbor the bacterium, such plants usually die within 3–6 months.

Cuttings infected with bacterial blight fail to root, the stems slowly rot, turn a dull blackish brown from the base upward, and eventually die. The leaves wilt, often showing the typical symptoms described earlier. Cuttings serve as a source of infection in the propagation bed.

Disease Cycle
Infection occurs through plant openings or through wounds that are created when cuttings are taken from stock plants. A cutting knife that is not disinfected is one of the surest means of transmitting the bacteria. The bacterium survives in water-conducting and other tissues of infected cuttings that are taken from symptomless stock plants. Infected plants are symptomless during cool weather when the temperatures are below 21°C or at very high temperatures in the range of 32°C–38°C. The bacteria are very infectious and are spread easily from plant to plant through plant contact and contaminated propagation media, by splashing during hose watering or misting, cultural operations, contaminated cutting knives, and insects. The organism has been isolated from whiteflies (*Trialeurodes vaporariorum*), which may transmit the bacteria to healthy geraniums. The organism can survive in moist soil from 3 to 6 months or until infected plant residues are completely decayed. Symptoms and disease spread occur most rapidly at temperatures of 21°C–29°C and in plants overfed with nitrogen fertilizer.

Control Measures

Resistance to bacterial blight is available in geranium varieties Lady Washington and Martha Washington geraniums (*Pelargonium domesticum*), as well as *Pelargonium acerifolium*, *Pelargonium* "Torento," *Pelargonium tomentosum*, and *Pelargonium scarboroviae*.

10.12.2 SOUTHERN BACTERIAL WILT IN GERANIUM

Pathogen: *Ralstonia solanacearum*

Synonym: *Pseudomonas solanacearum*

Southern bacterial wilt caused by *Pseudomonas solanacearum* was first reported in geranium in 1979. Although it is not a major disease in geraniums, it could pose a threat to geranium growers in the South and the Midwest due to the exchange of cuttings and plants from these geographical regions. The common geranium (*Pelargonium hortorum*) has been found to be susceptible to this disease. No geranium cultivars are resistant.

Symptoms

The initial symptoms include wilting of the lower leaves followed by chlorosis and finally necrosis of the leaves. The stem eventually turns brown and then black at the soil line, and finally, the plant collapses when 1/2–1 in. of the basal stem is infected (Photo 10.15). The roots can also become necrotic. Flowers fail to open normally. Stem sections reveal vascular discoloration. The symptoms of southern bacterial wilt always progress upward from the soil line, whereas those of bacterial blight move in either direction up or down the stem. It is unknown if infected plants can remain symptomless.

Disease Cycle

Ralstonia solanacearum is a soilborne bacterium that may persist in soil for 6 years or more in the absence of a suitable host plant. The pathogen survives and spreads within susceptible plants and cuttings. Infection normally occurs through the roots. The organism is transmitted on infected cutting knives, in soils, and other planting media, in crop debris, by splashing or flowing water and by insects. *Ralstonia solanacearum* has a very wide host range.

Symptoms develop rapidly at 23°C–39°C and plants die within 2–3 weeks under these conditions, whereas it may take 5 weeks for similar development of bacterial blight and wilt.

PHOTO 10.15 Symptoms of southern bacterial wilt in geranium plant. (Photo courtesy of R. Wick, UMass, University of Massachusetts Amherst, Amherst, MA, negreenhouseupdate.info.)

10.12.3 Bacterial Leaf Spot in Geranium

Pathogen: *Pseudomonas cichorii*

A bacterial leaf spot caused by *Pseudomonas cichorii* has been known to only attack geranium in the past few years. It has been reported only in Florida, but due to the exchange of plant material, the disease is a potential threat to the florist industry in the Midwest. This bacterium also has a wide host range.

Symptoms

The appearance of leaf spot symptoms varies, depending on the weather conditions. Plants growing outside exposed to rains exhibit dark brown to black, irregularly shaped areas 5–10 mm or larger (Photo 10.16). The margins of the spots appear water soaked a few hours after a rain. The lesions may extend or enlarge along the veins and may coalesce until the entire leaf may eventually die and curl up.

Under less favorable conditions for disease development, such as when plants are exposed to dews or only occasional overhead watering, the lesions that develop are sunken on both the upper and lower leaf surfaces. These spots may have tan centers and may be slightly raised. A lesion may have a dark margin and a yellow halo that may be as wide as the diameter of the lesion. The lesions commonly enlarge in wet weather but are checked by dry weather. Eventually, most spots turn dark brown to black, and large lesions, formed by the merging of smaller lesions, may appear to have "eyes" 1–2 mm in diameter.

Flower clusters are susceptible and individual flowers fail to open when infection occurs in the bud stage. Young seedlings can also become infected, causing a soft, dark, wet decay of the developing leaves.

Disease Cycle

The bacterium is thought to be carried on clothes, hands, sprayers, and other equipment. Environmental parameters specific for this disease on geranium are presently unknown. Free moisture on the leaves and rainy periods are known to promote disease development.

Control Measures

The control measures outlined will control all three diseases: bacterial blight or wilt, southern bacterial wilt, and bacterial leaf spot.

1. Disease-Free Planting Material

Purchase disease-free cuttings or plants from a reputable nursery or garden supply store. Florists should obtain CVI (culture-virus-indexed) cuttings from a commercial propagator who has

PHOTO 10.16 Bacterial leaf spot on geranium leaf. (Photo courtesy of R. Wick, UMass, University of Massachusetts Amherst, Amherst, MA, negreenhouseupdate.info.)

established disease-free planting stock and who practices strict sanitary measures or uses culture-indexed cuttings to set up a mother block system. Start with new stock plants each year; do not keep them from year to year.

2. Sanitary Measures
Follow strict sanitary procedures to prevent the contamination of healthy plants.

1. Immediately remove and destroy all infected plants, preferably by burning. Remove all old leaf and plant debris.
2. Florists should use an automatic watering system. If this is impossible, hang the host so that the nozzle does not touch the greenhouse floor. Avoid overwatering and sprinkling or misting water on the foliage.
3. When propagating, select a raised bench away from other areas of geranium production. The bench surface should be periodically wiped or sprayed with a fresh household bleach (Clorox, Purex, Sunny Sol) solution prepared by adding one part of bleach to nine parts of clean tap water. Also disinfect any equipment used to transport, hold, or work with geraniums.
4. Steam the propagating medium, soil mix, clay pots, and tools for 30 min at 82°C or 72°C for 1 h.
5. Wash hands thoroughly with soap and hot running water before handling plants. Take cuttings by (1) breaking off succulent tips of disease-free stock plants or (2) sterilizing the cutting knife or razor blade by dipping in a 70% solution of rubbing alcohol and flaming or soaking for 5 min in a liquid bleach solution (add one part of bleach to nine parts of water). Knives or razor blades should be changed between stock plants. Do not dip cuttings in liquid solutions. Place cuttings in clean, sterile flats lined with new newspapers.
6. After establishment, take cuttings from flats and plant them directly into steam-treated soil mix in clean pots.
7. Maintain good cultural conditions and ventilation. Space the plants to avoid plant contact. The greenhouse bench surface should be covered with woven wire mesh.
8. Control whiteflies and other insects to lessen spread of bacteria.
9. Avoid forcing the plants too rapidly, especially during warm (21°C–29°C) and humid weather. Maintain a balanced fertility based on a soil test.
10. In a field, flower bed, or nursery, wait at least a year before replanting with geraniums.

Chemical Control
Infrequent sprays containing basic copper sulfate or other fixed copper fungicide may aid in control of bacterial leaf spot caused by *P. cichorii*. Frequent use (every 7–10 days), however, causes a surface glazing of the leaves and a stunting of growth.

10.13   GERBERA

10.13.1   BACTERIAL BLIGHT AND LEAF SPOT IN GERBERA

Pathogen: *Pseudomonas cichorii*

Bacterial leaf spot in Gerbera, caused by *P. cichorii*, was recorded for the first time on Gerbera in Victoria in 2008.

Bacterial leaf spot tends to be a problem not only during warm weather with periods of heavy rain but also where overhead watering is practiced.

P. cichorii has been recorded (Miller and Knauss, 1974) to have a wide host range including hibiscus, cyclamen, vinca, chrysanthemum, impatiens, lettuce, and chicory.

PHOTO 10.17 Symptoms of bacterial blight and leaf spot in gerbera plant. (Photo courtesy of Florida Division of Plant Industry, Florida Department of Agriculture and Consumer Services, Tallahassee, FL, Bugwood.org.)

Symptoms

Symptoms of this disease are large black spots concentrated at the base of the plant. The spots often begin at the leaf margin but may also occur randomly. The spots are soft when tissue is wet and sunken and brittle when leaves are dry (Photo 10.17). From the leaf, the bacterium can move through the petiole and into the stem, resulting in a canker. The sepals of infected flower buds will become brown to black, and up to several inches of pedicel may be killed.

Disease Management

Key disease management practices include the following:

1. Plant pathogen-free seed and cultivars or resistant varieties.
2. Maintain good sanitation.
3. Avoid overhead irrigation or handle plants when they are wet.
4. Once plants become infected with bacteria, it is best to rogue infected plants.

10.14 BIRD OF PARADISE

10.14.1 BACTERIAL WILT IN BIRD OF PARADISE

Pathogen: *Pseudomonas* sp.

Bird of paradise is susceptible to southern bacterial wilt. Southern bacterial wilt is caused by a soil-borne bacterium that can survive in the soil for more than 6 years. This bacterium normally infects plants through their roots but can be transmitted via infected planting media, plant debris, water, insects, and infected gardening tools.

Symptoms

The first signs of southern bacterial wilt are wilting and yellowing of the leaves, while the base of the plant begins to turn brown or black near the soil line. The symptoms of this bacterium begin at the soil line and progress upward through the plant.

Control Measures

Once the disease has infected the white bird of paradise, there is little that can be done to save the plant, which will probably need to be removed and destroyed to prevent the bacterium from spreading.

10.15 HIBISCUS

10.15.1 Bacterial Leaf Spot in Hibiscus

Pathogen: *Pseudomonas cichorii*

There is a conspicuous bacterial leaf spot disease that occurs on the most commonly grown flowering hibiscus, *Hibiscus rosa-sinensis* L. The disease is found mainly in high rainfall areas of Hawaii such as near Kurtistown in the Puna district on the island of Hawaii. It is one of the several distinct bacterial leaf spots in hibiscus found globally.

In Florida, where these bacterial diseases have previously been studied, they are most common during hibiscus propagation in nurseries, as the high moisture and humidity necessary during nursery propagation favor them. In the landscape, however, the diseases develop less often (Chase, 1986). Florida, however, does not receive the high levels of precipitation as received in some areas of Hawaii. For example, Hilo and the Puna district can receive 180 in. (457 cm) of rainfall per year. This high level of rainfall favors infection and development of bacterial diseases.

Symptoms

Lesions range from 2 to 10 mm in diameter and are usually surrounded by a two-color border, which is black next to the tan-colored, necrotic center of the lesion and dark purplish red next to the black margin. A diffuse yellowish halo usually surrounds the whole lesion (Photo 10.18). The lesions have a similar appearance on the abaxial (lower) leaf surface; they are medium brown with a dark border and pale yellow halo. The purplish or reddish marginal tinge (caused by the production of an anthocyanin pigment) and the grayish-white color in the lesion centers tend to be absent from these surfaces. The distinctive purplish or reddish tinge may also be absent from adaxial (upper) surfaces, depending on the hibiscus cultivar. The central necrotic portions of the lesions are tan to whitish and sometimes fall out, creating a "shot-hole" appearance. Bacterial growth is limited by leaf veins, and lesions tend to be angular in shape. The spots can appear on young leaves that have not yet fully expanded, but they reach their maximum size on fully mature leaves. Severe spotting may cause premature defoliation. Other parts of the hibiscus plant (e.g., flowers, petioles, stems) are not susceptible to the disease.

Disease Cycle and Epidemiology

Dispersal of bacteria from lesions occurs primarily by splashing or windblown rain. The pathogen infects through wounds and natural openings such as leaf stomata, hydathodes, broken trichomes, and cracks in the cuticle. After penetration, the bacteria multiply between the plant cells and destroy cell walls with enzymes. The cellular contents serve as food for the bacterial population.

PHOTO 10.18   Bacterial leaf spot in hibiscus plant. (Photo courtesy of Scot Nelson, flickr.com.)

The optimum temperature for the growth of *P. cichorii* and for disease development is from 21°C to 27°C. The pathogen can survive for a period of time on fallen leaves. The disease cycle of bacterial leaf spot in hibiscus is illustrated in Figure 10.1.

Disease Management
Bacterial leaf spot in hibiscus can be managed by using a combination of the approaches listed here.

Pathogen-Free Cuttings
Start new planting using pathogen-free cuttings.

Resistant Cultivars
None of the 10 cultivars of *Hibiscus rosa-sinensis* tested for resistance to *P. cichorii* showed any significant resistance to the pathogen or disease (Chase et al., 1987).

Location
Avoid planting susceptible hibiscus in high rainfall areas such as the Puna district on the island of Hawaii.

Fertilizer
Fertilizer has been shown to affect the severity of many bacterial diseases of ornamental plants. Both the amount and type of fertilizer can affect plant disease development. However, the relationship needs to be determined for each disease. Therefore, observe the effects of fertilizer practices on the expression of disease symptoms in the garden and modify the practices accordingly to realize reduced disease severity.

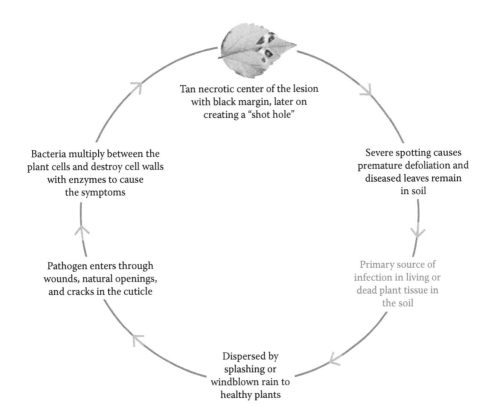

Tan necrotic center of the lesion with black margin, later on creating a "shot hole"

Bacteria multiply between the plant cells and destroy cell walls with enzymes to cause the symptoms

Severe spotting causes premature defoliation and diseased leaves remain in soil

Pathogen enters through wounds, natural openings, and cracks in the cuticle

Primary source of infection in living or dead plant tissue in the soil

Dispersed by splashing or windblown rain to healthy plants

FIGURE 10.1   Disease cycle of bacterial leaf spot in hibiscus caused by *Pseudomonas cichorii*.

Alternate Hosts

As *P. cichorii* has a potentially wide host range, it is important to understand that other plants or crops in the agroecosystem may be affected and require management. For a list of potential hosts of *P. cichorii*, please refer to a publication by CAB International (2009).

Sanitation

Sanitation is the periodic removal of diseased plant organs such as leaves from the vicinity of healthy leaves. Routinely prune and destroy branches with heavily diseased leaves. Pick up and destroy fallen leaves, as *P. cichorii* can survive there. Remove and destroy diseased cuttings from nurseries or young transplants.

Bactericides

Although pesticides, including Junction*, Camelot*, Kocide®, Phyton-27, and Actigard®, may be available, foliar sprays of bactericides will not usually control this disease in high rainfall areas. Always check a pesticide label for instructions on use of the product and whether it is right for your situation.

Moisture Management (Relative Humidity)

Long periods of leaf wetness and high relative humidity values favor infection and disease development. Use the following practices to reduce disease severity by minimizing leaf wetness and relative humidity in the plant canopy.

Irrigation

Avoid overhead sprinkler irrigation. Reduce leaf wetness by using drip irrigation.
   Limit water applications.

Soil Drainage

Ensure that soils in which hibiscus grow drain adequately. This can lower relative humidity around the plant and reduce infection and disease severity.

Weed Control

Trim weeds around hibiscus plants to reduce relative humidity.

Plant Spacing, Intercropping

Use wide plant spacing between hibiscus plants to decrease the probability of bacterial movement among plants. Place nonhost plants between hibiscus plants to block the spread of bacteria between plants.

Windbreaks

A breeze provides aeration of canopies, allowing leaves to dry rapidly after a rainfall and therefore avoid planting hibiscus near windbreaks.

Shade

Avoid planting hibiscus in heavily shaded areas.

Row Orientation

If planting hibiscus in a row, try to orient the row perpendicular to the direction of prevailing trade winds. This will help plant foliage to dry uniformly within the row and reduce spread of the bacteria among plants by wind-driven rain.

10.16 HORTENSIA

10.16.1 BACTERIAL BUD BLIGHT IN HORTENSIA

Pathogen: *Pseudomonas viridiflava*

Bud blight in Hortensia was observed on the growing bud of Hortensia plant when these are exposed to cold climatic condition. The disease was reported for the first time by Gardan and Borkar in 1984 in France.

Symptoms

The bud blight in Hortensia appears on the growing young buds of Hortensia plant. The infected bud turns water soaked and blackish in color. The bacterial infection on the growing bud progresses as much as 10 cm on the stem bearing the bud. The affected stem turns watery and pulpy black and eventually dies. In case of heavily infected plants, they do not sprout again and the whole plant is killed.

Disease Cycle

The pathogen is known to occur on tomato (Burki, 1973), Haricot, and Abricotier (Gardan et al., 1973). However, the *P. viridiflava* infecting hortensia was not pathogenic on tomato but infected petits pois. *P. viridiflava* is mostly an epiphytic bacteria, but under high humidity and at 16°C–18°C temperature may cause disease on some host plants.

Control Measures

Not worked out.

10.17 HYACINTH

10.17.1 BACTERIAL SOFT ROT IN HYACINTH

Pathogen: *Pectobacterium carotovorum*

Synonym: *Dickeya zeae*

The disease is widely distributed but infrequent.

Symptoms

Flowers do not form or open irregularly and rot. Flower stalks frequently rot at the base. The top of the bulb may be rotten.

Disease Management

Cultural Practices

Spread and development of the disease can be controlled only by the application of an integrated strategy involving the use of pathogen-free propagation material, disinfection of equipment, and other hygienic measures:

- Store bulbs under dry, well-ventilated conditions. Do not heat or freeze.
- Plant only sound bulbs.
- Avoid overwatering.

10.17.2 YELLOW DISEASE IN HYACINTH

Pathogen: *Xanthomonas hyacinthi* (Wakker, 1883; Vauterin, Hoste, Kersters, and Swings, 1995)

Synonym: *Xanthomonas campestris* pv. *hyacinthi*

Symptoms

When the infected bulb is used for planting, yellow discoloration of the vascular tissue and surrounding parenchyma can be observed in the scales and bottom plate. From heavily diseased bulbs, no plants develop or those plants suddenly wilt and die. Bulbs rot completely and have an unpleasant smell. In others, dark streaks following the veins in the leaves develop from the leaf base.

Primary infection in the field (usually not before May) gives rise to leaf spots and dark streaks along the veins at the margin (hydathode infection) or other parts of the leaf. Leaf tips may turn black and wither (also in secondary infections). No vectors have been described, but mechanical transmission by man is easily possible (Kamerman, 1975).

Hosts
The following are to be the hosts: Hyacinth (*Hyacinthus orientalis*), *Scilla tubergeniana*, *Eucomis autumnalis* (pineapple lily), and *Puschkinia scilloides* (Janse and Miller, 1983; van der Tuin et al., 1995).

Geographical Distribution
This disease is widely distributed in countries such as the Netherlands, Australia, Finland, France, Hungary, Ireland, Italy, Japan, Poland, Romania, Russian Federation, Sweden, the United Kingdom, the United States, and the former Yugoslavia.

Control Measures
In early springtime volunteers should be removed. Visitors should not be allowed to enter the fields. Further field inspections should trace diseased plants. The leaves of these plants and those present in the vicinity should be removed and destroyed by hydrochloric acid or covered with sand. Bulbs and surrounding soil can be treated with diquat or 5% formalin. Surrounding plants can be treated with kasugamycin (if permitted). Alkyl dimethyl benzyl ammonium chloride (if permitted) can be applied when leaf symptoms occur but are not completely effective. Bulbs from diseased spots should be harvested last and separately and destroyed or heat treated. After harvest, disinfection of machines and implements should take place and bulbs should be dried rapidly.

For bulbs, a heat treatment has been found to be quite effective. Bulbs are stored first for a period of 4 weeks at 30°C, thereafter for 2 weeks at 38°C, with a final treatment for 3 days at 44°C. Subsequently bulbs are stored at 28°C–30°C (Kruyer and Vreeburg, 1981). To avoid the negative effects of the heat treatment, sufficient ventilation is essential. Before planting, bulbs should be carefully inspected. The susceptibility of cultivars is different, and resistance breeding has been moderately successful. A classification system has been developed to produce yellow disease-free hyacinths (OEPP/EPPO, 1998).

10.17.3 INTERNAL BROWNING IN HYACINTH

Pathogen: *Erwinia rhapontici* (Millard, 1924; Burkholder, 1948)

Synonym: *Pectobacterium rhapontici*

Symptoms
Pinkish brown discoloration of inner scales of *Hyacinthus* bulb is observed.

Main Hosts
Bulbous plants such as hyacinth (*H. orientalis*) and onion (*Allium cepa*) are the main hosts, but *Erwinia rhapontici* has also been reported from *Allium sativum* (garlic) and *Hippeastrum* (amaryllis). *Amaranthus hybridus* (smooth pigweed) was reported as a host from Mexico and the bacterium found to be seed transmitted. *E. rhapontici* has also been reported to be the cause of foot and crown rot of *Rheum rhaponticum* and *Rheum hybridum* (rhubarb) and pink seed (pale pinkish brown-to-bright pink discoloration throughout the seed coat) of *Triticum aestivum* (wheat), *Pisum sativum* (pea), and *Phaseolus vulgaris* (bean) in the United States. Pink seed has also been described from *Lens culinaris* (lentil) and *Cicer arietinum* (chickpea) in Canada (Huang et al., 2002). Furthermore, *Cyclamen persicum* (cyclamen), *Dianthus* (carnation), *Armoracia rusticana* (horseradish), and *Morus* (mulberry) have been reported as hosts.

Geographical Distribution
This infection is evidently present in countries such as Belgium, Canada, the former Czechoslovakia, France, Iran, Israel, Italy, Japan, Korea, Lithuania, Malaysia, the Netherlands, Norway, Poland, Ukraine, the United Kingdom, and the United States.

Importance

E. rhapontici is an opportunistic pathogen, only causing problems under adverse conditions for the plant, and remains incidental in bulbaceous plants. The seed infections give problems of quality, not so much of yield loss.

Disease Cycle

The pathogen is spread by bulbs, rhubarb crowns used for propagation and infected seeds. Plant-to-plant infection may occur by insects spreading the bacterium, and infection with the nematode *Anguillulina dipsaci* enhances the occurrence of crown rot of rhubarb. This nematode probably also transmits the organism. Wounding is necessary during procedures connected with cultivation and vegetative propagation.

Control Measures

E. rhapontici can be readily transmitted via transplanting infected crown bulbs. In establishing new plantings, only disease-free crown bulbs should be selected. Documented movement has occurred when infected crowns were used to establish new plantings. New crown bulbs must not be replanted in areas where the disease has previously been observed. Evidence exists that root- and foliage-feeding insects can move the bacteria from infected to uninfected plants. Therefore, good insect management will also reduce localized spread. Healthy seeds and use of healthy crowns or bulbs and early spraying with insecticides to reduce insect populations on the hosts are the only way for prevention or control.

10.18   ORNAMENTAL PALM

10.18.1   BACTERIAL LEAF BLIGHT IN FISHTAIL PALM

Pathogen: *Acidovorax avenae* subsp. *avenae*

Species of single- and cluster-stemmed fishtail palms in the genus *Caryota* from India and Sri Lanka to Southeast Asia, northern Australia, and the Solomon Islands and in Hawaii (the Burmese fishtail palm) is widely grown in landscapes and is sometimes exported for use in interior spaces or as houseplants.

   Caryota mitis, the Burmese fishtail palm or clustered fishtail palm, is in the Arecaceae family and occurs naturally from Burma to the Malay Peninsula, Java, and the Philippines. These palms are grown in greenhouses, nurseries, and outdoors in landscapes within warm regions of the United States, such as Florida and Hawaii. Fishtail palms are also used as interior-scape plants and houseplants in some locations. Outdoors they can grow in large clusters of plants, from 12 to 40 ft in height. The Burmese fishtail palm (*Caryota mitis*) is prone to damage caused by bacterial pathogen, under high rainfall or overhead irrigation, causing a leaf striping blight of leaves.

   Bacterial leaf blight is a relatively new plant disease described and reported by plant pathologists in 1978. The disease has been in Hawaii since 1995, according to the CTAHR Agricultural Diagnostic Service Center in Hilo.

Symptoms

Initial symptoms are small, water-soaked, and translucent to light-yellow to light-brown banded areas running along and around leaf veins. Mature lesions develop a brown to black color and may have a chlorotic (yellow) halo; lesions range from a minimum of 1–2 mm wide and extend in length up to the entire length of the affected leaf (Photo 10.19). Initial infections often occur through hydathodes at leaf tips or margins.

Host Range

Reported hosts of *Acidovorax avenae* subsp. *avenae* include corn (*Zea mays*) and rice (*Oryza sativa*). In Iran, 12 isolates of *A. avenae* subsp. *avenae* isolated from rice were reportedly pathogenic on

PHOTO 10.19   Bacterial leaf blight infection on fishtail palm leaf. (Photo courtesy of Scot Nelson, flickr.com.)

corn, sorghum, barley, oat, sugarcane, barnyard millet, Italian millet, wheat, barnyard grass, and Johnson grass, indicating a wide host range (Rostami and Taghavi, 2001). It is unknown if the pathogen affects these hosts in Hawaii.

Disease Cycle and Epidemiology
The most important environmental factor for infection and disease development on *C. mitis* leaves in Hawaii is free water resulting from high rainfall, frequent dew, or overhead irrigation. Splashing water disperses the pathogen from leaf to leaf or between plants and enhances its survival. The bacteria can enter wounds or natural openings in palm leaves, probably most frequently through hydathodes at the leaf margins. Infections and disease symptoms appear and move along and near leaf veins, resulting in the black stripe symptoms. Symptoms of this disease can appear about 5–10 days after inoculation in greenhouses with air temperature ranging from 18°C to 32°C. The pathogen reproduces and survives primarily within infected *Caryota* leaf tissues.

The disease is relatively common in Hawaii. It is most likely to occur in palm nurseries where *C. mitis* plants are clustered together and irrigated by overhead sprinklers or in high rainfall areas.

Disease Management
- Eliminate overhead irrigation.
- Irrigate in the morning instead in the evening.
- Grow plants under cover and protected from frequent rainfall.
- Use preventive sprays of copper-containing or antibiotic pesticides.
- Remove symptomatic leaves and destroy them; remove and destroy entire plants if they are severely affected.
- Provide good air circulation around plants to allow leaf drying after they become wet.
- Do not purchase, sell, or distribute diseased plants.
- Increase spacing among plants in nurseries or greenhouses.
- Intercrop *C. mitis* only with nonsusceptible plants (i.e., not with reported hosts of the bacterium)
- Do not transplant symptomatic plants into landscapes.

REFERENCES

Alvarez, A., Lipp, R., and B. Bushe. 1989. Resistance of bacteria to antibiotics used for control of anthurium blight. In: Fernandez, J.A. and W.T. Nishijima (eds.), *Proceedings of the Second Anthurium Blight Conference*, Hawaii Institute of Tropical Agriculture and Human Resources, University of Hawaii, Honolulu, HI, pp. 11–12.

Alvarez, A. and C. Mizumoto. 2001. Bioprotection and stimulation of aroids with phylloplane bacteria. *Phytopathology*, 91, S3.

Alvarez, A., Norman, D., and R. Lipp. 1991. Epidemiology and control of anthurium blight. In: Alvarez, A.M., Deardorff, D.C., and K.B. Wadsworth (eds.), *Proceedings of the Fourth Anthurium Blight Conference*. Hawaii Institute of Tropical Agriculture and Human Resources, University of Hawaii, Honolulu, HI, pp. 12–18.

Baker, K.F. 1950. Bacterial fasciation disease of ornamental plants in California. *Plant Dis. Rep.*, 34, 121–126.

Burkholder, W.H., McFadden, L.H., and A.W. Dimock. 1953. A bacterial blight of chrysanthemums. *Phytopathology*, 43, 522–525.

CAB International. 2009. *Pseudomonas cichorii* ((Swingle 1925) Stapp 1928), bacterial blight of endive. Crop Protection Compendium 2009 No. AQB CPC record, Sheet 1965.

Chase, A.R. 1986. Comparisons of three bacterial leaf spots of *Hibiscus rosa-sinensis*. *Plant Dis.*, 70, 334–336.

Chase, A.R. 1988. Chemical and nutritional aspects of controlling *Xanthomonas* diseases on Florida ornamentals. In: Alvarez, A.M. (ed.), *Proceedings of the First Anthurium Blight Conference*. Hawaii Institute of Tropical Agriculture and Human Resources, University of Hawaii, Honolulu, HI, pp. 32–34.

Chase, A.R., Osborne, L.S., Yuen, J.M.F. et al. 1987. Effects of growth regulator chlormequat chloride on severity of three bacterial diseases on 10 cultivars of *Hibiscus rosa-sinensis*. *Plant Dis.*, 71, 186–187.

Chase, A.R., Stall, R.E., Hodge, N.C. et al. 1992. Characterization of *Xanthomonas campestris* strains from aroids using physiological, pathological, and fatty acid analyses. *Phytopathology*, 82, 754–759.

Cho, H.R., Lim, J.H., Yun, K.J. et al. 2005. Virulence variation in 20 isolates of *Erwinia carotovora* subsp. *carotovora* on Zantedeschia cultivars in Korea. *Acta Hortic.*, 673, 653–659.

Cook, J.R. 1988. Biological control: Some concepts, and the potential for application to bacterial blight of *Anthurium*. In: Alvarez, A.M. (ed.), *Proceedings of the First Anthurium Blight Conference*. Hawaii Institute of Tropical Agriculture and Human Resources, University of Hawaii, Honolulu, HI, pp. 35–36.

Dickey, R.S. 1981. *Erwinia chrysanthemi*: Reaction of eight plants to strains from several hosts and to strains of other *Erwinia* species. *Phytopathology*, 71, 23–29.

Dickey, R.S. and J.I. Victoria. 1980. Taxonomy and emended description of strains of *Erwinia isolated* from *Musa paradisiaca Linnaeus*. *Int. J. Syst. Bacteriol.*, 30, 129.

Dickey, R.S. and C.H. Zumoff. 1987. Bacterial leaf blight of *Syngonium* caused by a pathovar of *Xanthomonas campestris*. *Phytopathology*, 77, 1257–1262.

Dowson, W.J. 1942. On the generic name of the gram positive bacterial plant pathogens. *Trans. Br. Mycol. Soc.*, 25, 311–314.

EPPO/CABI. 1996a. *Phialophora cinerescens*. In: Smith, I.M., McNamara, D.G., Scott, P.R., and M. Holderness (eds.), *Quarantine Pests for Europe*, 2nd edn. CAB International, Wallingford, U.K.

EPPO/CABI. 1996b. *Erwinia chrysanthemi*. In: Smith, I.M., McNamara, D.G., Scott, P.R., and M. Holderness (eds.), *Quarantine Pests for Europe*, 2nd edn. CAB International, Wallingford, U.K.

Fernandez, J.A., Tanabe, M.J., Moriyasu, P. et al. 1989. Biological control. In: Fernandez, J.A. and W.T. Nishijima (eds.), *Proceedings of the Second Anthurium Blight Conference*. Hawaii Institute of Tropical Agriculture and Human Resources, University of Hawaii, Honolulu, HI, pp. 27–29.

Fernandez, J.A., Tanabe, M.J., Moriyasu, P. et al. 1990. Biological control. In: Alvarez, A.M. (ed.), *Proceedings of the Third Anthurium Blight Conference*. Hawaii Institute of Tropical Agriculture and Human Resources, University of Hawaii, Honolulu, HI, pp. 41–43.

Fernandez, J.A., Tanabe, M.J., Wolff, W.J. et al. 1991. Biological control. In: Alvarez, A.M., Deardorff, D.C., and K.B. Wadsworth (eds.), *Proceedings of the Fourth Anthurium Blight Conference*. Hawaii Institute of Tropical Agriculture and Human Resources, University of Hawaii, Honolulu, HI, pp. 28–30.

Fukui, H., Alvarez, A.M., and R. Fukui. 1998. Differential susceptibility of *Anthurium* cultivars to bacterial blight in foliar and systemic infection phases. *Plant Dis.*, 82, 800–806.

Fukui, R., Fukui, H., and A.M. Alvarez. 1999. Comparisons of single versus multiple bacterial species on biological control of *Anthurium* blight. *Phytopathology*, 89, 366–373.

Funnel, K.A. 1993. Zantedeschia. In: Hertogh, A. and M. Le Nard (eds.), *Physiology of Flowering Bulbs*. Elsevier, Amsterdam, the Netherlands, pp. 638–704.

Gardan, L. and S.G. Borkar. 1985. Observation de dessechement de bourgeons d''Hortensia du a *Pseudomonas viridiflava*. P.H.M. *Revue Horticole*, 253, 48–49.

Gardan, L., Prunier, J.P., Luisetti, J. et al. 1973. Responsibilite de diverse *Pseudomonas* dans le deperissement bacterien de l' abricotier en France. *Rev. Zool. Agric. Pathol. Veg.*, 4, 112–120.

Goodfellow, M. 1984. Reclassification of *Corynebacterium fascians* (Tilford) Dowson in the genus *Rhodococcus*, as *Rhodococcus fascians* comb. nov. *Syst. Appl. Microbiol.*, 5, 225–229.

Hayward, A.C. 1972. A bacterial disease of *Anthurium* in Hawaii. *Plant Dis. Rep.*, 56, 904–908.

Higaki, T., Imamura, J.S., and D. Moniz. 1992. Nutritional and cultural effects on bacterial blight of *Anthurium*. In: Delate, K.M. and C.H.M. Tome (eds.), *Proceedings of the Fifth Anthurium Blight Conference*. Hawaii Institute of Tropical Agriculture and Human Resources, University of Hawaii, Honolulu, HI, pp. 44–45.

Higaki, T., Imamura, J., Tanabe, M. et al. 1990. Nutritional and cultural effects on *Anthurium* bacterial blight. In: Alvarez, A.M. (ed.), *Proceedings of the Third Anthurium Blight Conference*. Hawaii Institute of Tropical Agriculture and Human Resources, University of Hawaii, Honolulu, HI, pp. 7–11.

Huang, H.C., Erickson, R.S., Yanke, L.J. et al. 2002. First report of pink seed of common bean caused by *Erwinia rhapontici*. *Plant Dis.*, 86, 921.

Huang, T.C. 1990. Characteristics and control of *Pseudomonas cattleyae* causing brown spot of Phalaenopsis orchid in Taiwan. *Plant Prot. Bull.*, 32: 327 (abstract).

Janse, J.D. 2006. *Phytobacteriology: Principles and Practice*. CAB International, Wallingford, U.K.

Janse, J.D. and H. Miller. 1983. Yellow disease in Scilla tubergeniana and related bulbs caused by *Xanthomonas campestris* pv. *hyacinthi*. *Neth. J. Plant Pathol.*, 89, 203–206.

Janse, J.D. and M.A. Ruissen. 1988. Characterization and classification of *Erwinia chrysanthemi* strains from several hosts in the Netherlands. *Phytopathology*, 78, 800–808.

Jones, J.B. and A.W. Engelhard. 1984. Crown and leaf rot of statice incited by a bacterium resembling *Pseudomonas caryophylli*. *Plant Dis.*, 68, 338–340.

Joubert, J.J. and S.J. Truter. 1972. A variety of *Xanthomonas campestris* pathogenic to *Zantedeschia aethiopica*. *Neth. J. Plant Pathol.*, 78: 212–217.

Kamemoto, H. and A. Kuehnle. 1989. Breeding for blight resistance: A progress report. In: Fernandez, J.A. and W.T. Nishijima (eds.), *Proceedings of the Second Anthurium Blight Conference*. Hawaii Institute of Tropical Agriculture and Human Resources, University of Hawaii, Honolulu, HI, p. 10.

Kamemoto, H. and A. Kuehnle. 1996. *Breeding Anthurium in Hawaii*. University of Hawaii Press, Honolulu, HI.

Kamemoto, H., Kuehnle, A., and J. Kunisaki etal. 1990. Breeding for bacterial blight resistance in *Anthurium*. In: Alvarez, A.M. (ed.), *Proceedings of the Third Anthurium Blight Conference*. Hawaii Institute of Tropical Agriculture and Human Resources, University of Hawaii, Honolulu, HI, pp. 45–48.

Kamerman, W. 1975. Biology and control of *Xanthomonas hyacinthi* in hyacinths. *Acta Hortic.*, 47, 99–105.

Krasil'nikov, N.A. 1949. *Opredelitel bakterii I aktinomitsetov (Guide to the Bacteria and Actinomycetes)*. Izdatelstvo AkademiI Nauk SSSR, Moskva-Leningrad, USSR.

Krejzar, V., Mertelík, J., Pánková, I. et al. 2008. *Pseudomonas marginalis* associated with soft rot of *Zantedeschia* spp. *Plant Prot. Sci.*, 44, 85–90.

Kruyer, C.J. and P.J.M. Vreeburg. 1981. How can hyacinths best be stored after heat treatment? *Bloembollencultuur*, 92, 224–225.

Kuehnle, A.R., Chen, F.C., Jaynes, J.M. et al. 1992. Engineering blight resistant *Anthurium*: A progress report. In: Delate, K.M. and C.H.M. Tome (eds.), *Proceedings of the Fifth Anthurium Blight Conference*. Hawaii Institute of Tropical Agriculture and Human Resources, University of Hawaii, Honolulu, HI, pp. 17–18.

Kuehnle, A.R., Chen, F.C., and J.M. Jaynes. 1993. Status of genetically engineered *Anthurium*. In: Delate, K.M. and E.R. Yoshimura (eds.), *Proceedings of the Sixth Hawaii Anthurium Industry Conference*. Honolulu, HI, pp. 7–8.

Kuehnle, A.R., Chen, F.C.N., Sugii, N. et al. 1991. Engineering of blight resistance in *Anthurium*. In: Alvarez, A.M., Deardorff, D.C., and K.B. Wadsworth (eds.), *Proceedings of the Fourth Anthurium Blight Conference*. Hawaii Institute of Tropical Agriculture and Human Resources, University of Hawaii, Honolulu, HI, pp. 42–43.

Kuehnle, A.R., Fujii, R., Chen, F.C. et al. 2004a. Peptide biocides for engineering bacterial blight tolerance and susceptibility in cut flower *Anthurium*. *Hortic. Sci.*, 39, 1327–1331.

Kuehnle, A.R., Fujii, T., Mudalige, R. et al. 2004b. Gene and genome mélange in breeding of *Anthurium* and Dendrobium Orchid. *Acta Hort. (ISHS)*, 651, 115–122.

Lacey, M.S. 1939. Studies in Bacteriosis XXIV. Studies on a bacterium associated with leafy galls, fasciations and "cauliflower" disease of various plants. Part III. Further isolations, inoculation experiments and cultural studies. *Ann. Appl. Biol.*, 26, 262–278.

Lee, Y.A. and K.P. Chen. 2002. First report of bacterial soft rot of white flowered calla lily caused by *Erwinia chrysanthemi* in Taiwan. *Plant Dis.*, 86, 1273.

Lee, Y.A., Chen, K.P., and Y.C. Chang. 2005. First report of bacterial leaf blight of white-flowered calla lily caused by *Xanthomonas campestris* pv. *zantedeschia* in Taiwan. *Plant Pathol.*, 54, 239.

Lipp, R.L., Alvarez, A.M., Benedict, A.A. et al. 1992. Use of monoclonal and pathogenicity tests to characterize strains of *Xanthomonas campestris* pv. *dieffenbachiae* from Aroids. *Phytopathology*, 82, 677–682.

McCulloch, L. 1924. A bacterial blight of gladioli. *J. Agric. Res.*, 27, 225–229.

McGovern, R.J., Horst, R.K., and R.W. Dickey. 1985. Effect of plant nutrition on susceptibility of *Chrysanthemum morifolium* to *Erwinia chrysanthemi*. *Plant Dis.*, 69, 1086–1088.

Mikicinski, A., Sobiczewski, P., Sulikowska, M. et al. 2010. Pectolytic bacteria associated with soft rot of calla lily (*Zantedeschia* spp.) tubers. *J. Phytopathol.*, 158, 201–209.

Miller, J.W. and J.F. Knauss. 1974. Bacterial blight of Gerbera Daisy. *Plant Pathol.*, Circular No. 139.

Mills, H.A. 1989. Cultural practices and *Anthurium* nutrition. In: Fernandez, J.A. and W.T. Nishijima (eds.), *Proceedings of the Second Anthurium Blight Conference*. Hawaii Institute of Tropical Agriculture and Human Resources, University of Hawaii, Honolulu, HI, pp. 40–42.

Nishijima, W.T. 1988. *Anthurium* blight: An overview. In: Alvarez, A.M. (ed.), *Proceedings of the First Anthurium Blight Conference*. Hawaii Institute of Tropical Agriculture and Human Resources, University of Hawaii, Honolulu, HI, pp. 6–8.

Nishijima, W.T. 1994. Diseases. In: Higaki, T., Lichty, J.S., and D. Moniz (eds.), *Anthurium Culture in Hawaii*. HITHAR Research Extension Series 152. College of Tropical Agriculture and Human Resources, University of Hawaii, Honolulu, HI, pp. 13–18.

Nishijima, W.T. and M. Chun, 1991. Chemical control of *Anthurium* blight. In: Alvarez, A.M., Deardorff, D.C., and K.B. Wadsworth (eds.), *Proceedings of the Fourth Anthurium Blight Conference*. Hawaii Institute of Tropical Agriculture and Human Resources, University of Hawaii, Honolulu, HI, pp. 21–23.

Nishijima, W.T. and D.K. Fujiyama. 1985. Bacterial blight of *Anthurium*. Commodity Fact Sheet AN-4(A). Institute of Tropical Agriculture and Human Resources, University of Hawaii, Honolulu, HI.

Nishiyama, J., Kobayashi, T., and K. Azegami. 1988. Bacterial wilt of statice caused by *Pseudomonas caryophylli*. *Ann. Phytopathol. Soc. Jpn.*, 54, 444–452.

Norman, D.J., Alvarez, A., and R. Lipp. 1993. Latent infections of *Xanthomonas campestris* pv. *dieffenbachiae* in tissue-cultured *Anthurium*. In: Delate, K.M. and E.R. Yoshimura (eds.), *Proceedings of the Sixth Hawaii Anthurium Industry Conference*. Hawaii Institute of Tropical Agriculture and Human Resources, University of Hawaii, Honolulu, HI, pp. 12–16.

Norman, D.J. and A.M. Alvarez. 1994. Latent infections of *in vitro Anthurium* caused by *Xanthomonas campestris* pv. *dieffenbachiae*. *Plant Cell Tissue Organ Cult.*, 39, 55–61.

Norman, K.A., Polyn, S.M., Detre, G.J. et al. 2006. Beyond mind-reading: Multi-voxel pattern analysis of fMRI data. *Trends Cogn. Sci.*, 10, 424–430.

Oduro, K.A. 1975. Factors affecting epidemiology of bacterial fasciation of *Chrysanthemum maximum*. *Phytopathology*, 65(6), 719–721.

OEPP/EPPO. 1978. Data sheets on quarantine organisms No. 55, *Pseudomonas caryophylli*. *Bull. OEPP/ EPPO Bull.*, 8(2).

OEPP/EPPO. 1990. Specific quarantine requirements. EPPO Technical Documents No. 1008.

OEPP/EPPO. 1991. Certification schemes. Pathogen-tested material of carnation. *Bull. OEPP/EPPO Bull.*, 21, 279–290.

OEPP/EPPO. 1998. EPPO Standards PM 4/23 (1) Classification scheme for hyacinth. *Bull. OEPP/EPPO Bull.*, 28, 235–242.

OEPP/EPPO. 2004. Diagnostic protocols for regulated pests, *Xanthomonas axonopodis* pv. *dieffenbachiae*. *Bull. OEPP/EPPO Bull.*, 34, 183–186.

OEPP/EPPO. 2008. Fiches informativessur les organismes de quarantaine. In: Data sheets on quarantine pests. *Bull. OEPP/EPPO Bull.*, 38, 441–449.

Onozaki, T., Ikeda, H., Yamaguchi, T. et al. 2002. 'Carnation Nou No.1', a carnation breeding line resistant to bacterial wilt (*Burkholderia caryophylli*). *Hortic. Res. (Jpn.)*, 1(1), 13–16.

Pape, H. 1938. Eine noch wenig beachtete Krankheit der Zierpflanzen. *Der Blumen-und Pflanzenbau die Gartenwelt*, 42, 384–386.

Rostami, N. and S.M. Taghavi. 2001. Etiology, distribution and phenotypical characterization of different isolates of the *Acidovorax avenae* subsp. *avenae* causal agent of bacterial stripe of rice in Fars and Kohgiluyeh and Boyer ahmad provinces. *Iran. J. Plant Pathol.*, 37, 19–23.

Sakai, D.S. 1990. The effect of nitrate and ammonium fertilizer on the contents of *Anthurium* guttation fluid. In: Alvarez, A.M. (ed.), *Proceedings of the Third Anthurium Blight Conference*. Hawaii Institute of Tropical Agriculture and Human Resources, University of Hawaii, Honolulu, HI, pp. 18–19.

Sakai, D.S. 1991. The effect of nitrogen fertilizer levels on amino compounds in guttation fluid of *Anthurium* and incidence of bacterial blight. In: Alvarez, A.M., Deardorff, D.C., and K.B. Wadsworth (eds.), *Proceedings of the Fourth Anthurium Blight Conference*. Hawaii Institute of Tropical Agriculture and Human Resources, University of Hawaii, Honolulu, HI, pp. 51–52.

Sakai, W.S., Okimura, S., Hanohano, T. et al. 1992. A detailed study of nitrogen fertilization, glutamine production, and systemic blight on *Anthurium* cultivars *Ellison Onizuka* and *Calypso*. In: Delate, K.M. and C.H.M. Tome (eds.), *Proceedings of the Fifth Anthurium Blight Conference*. Hawaii Institute of Tropical Agriculture and Human Resources, University of Hawaii, Honolulu, HI, pp. 47–50.

Sathyanarayana, N., Reddy, O.R., and R.L. Rajak. 1998. Interception of *Xanthomonas campestris* pv. *dieffenbachiae* on *Anthurium* plants from the Netherlands. *Plant Dis.*, 82, 262.

Serfontein, S., Logan, C., Swanepoel, A.E. et al. 1991. A potato wilt disease in South Africa caused by *Erwinia carotovora* subspecies *carotovora* and *Erwinia chrysanthemi*. *Plant Pathol.*, 40(3), 382–386.

Shehata, S.A. 1992. Supply-demand and market analysis of the cut-flower industry: A focus on the Hawaiian *Anthurium* industry. In: Delate, K.M. and C.H.M. Tome (eds.), *Proceedings of the Fifth Anthurium Blight Conference*. Hawaii Institute of Tropical Agriculture and Human Resources, University of Hawaii, Honolulu, HI, pp. 35–38.

Shehata, S.A. and W.T. Nishijima. 1989. The impact of *Anthurium* blight on the profitability of the industry. In: *Proceedings of the Second Anthurium Blight Conference*, Hilo, HI, pp. 17–19.

Snijder, R.C. and J.M. van Tuyl. 2002. Evaluation of test to determine resistance of *Zantedeschia* spp. (Araceae) to soft rot caused by *Erwinia carotovora* subsp. *carotovora*. *Eur. J. Plant Pathol.*, 108, 565–571.

Stovold, G.E., J. Bradley, and P.C. Fahy. 2001. *Acidovorax avenae* subsp. *cattleyae* (*Pseudomonas cattleyae*) causing leaf spot and death of Phalaenopsis orchids in New South Wales. *Aust. Plant Pathol.*, 30, 73–74.

Tanabe, M., Fernandez, J., Moriyasu, P. et al. 1992. *Anthurium* in vitro triple indexing. In: Delate, K.M. and C.H.M. Tome (eds.), *Proceedings of the Fifth Anthurium Blight Conference*. Hawaii Institute of Tropical Agriculture and Human Resources, University of Hawaii, Honolulu, HI, pp. 8–11.

Tilford, P.E. 1936. Fasciation of sweet peas caused by *Phytomonas fascians* n. sp. *J. Agric. Res.*, 53, 383–394.

Trujillo, G. and Y. Hernández. 1999. Bacterial spot in orchid. *Fitopatal. Venezoel.*, 12, 4–8.

Van der Tuin, W.R., Spit, B.E., Nahumury, T. et al. 1995. *Xanthomonas campestris* pv. *hyacinthi* in *Eucomis autumnalis* and *Puschkinia* species. Verslagen en Mededelingen nr. 177, Annual Report 1994, Plant Protection Service, Wageningen, the Netherlands, pp. 22–23.

Vanneste, J. 1996. Biological control of soft rot on calla lily and potatoes. http://www.hortnet.co.nz/publications/science/jvann2.htm (verified September 20, 2009).

Williams, P.H. 1933. Leafy gall of Chrysanthemum. *Rev. Appl. Mycol.*, 12, 698–699.

Williams, P.H. 1934. Leafy gall of Chrysanthemum. *Rev. Appl. Mycol.*, 13, 638.

Wright, P.J. 1998. A soft rot of calla (*Zantedeschia* spp.) caused by *Erwinia carotovora* subspecies *carotovora*. *NZ J. Crop Hortic. Sci.*, 24, 331–334.

Wrighte, P.J., Burge, G.K., and C.M. Triggs. 2002. Effects of cessation of irrigation and time of lifting of tubers on bacterial soft rot of calla (*Zantedeschia* spp.) tubers. *NZ J. Crop Hortic. Sci.*, 30, 265–272.

Wright, P.J., Clark, G.E., and J. Koolaard. 2005. Growing methods and chemical drenches control calla soft rot. *Acta Hortic.*, 673, 769–774.

SUGGESTED READING

Alvarez, A., Lipp, R., Norman, D. et al. 1990. Epidemiology and control of *Anthurium* blight. In: Alvarez, A.M. (ed.), *Proceedings of the Third Anthurium Blight Conference*. Hawaii Institute of Tropical Agriculture and Human Resources, University of Hawaii, Honolulu, HI, pp. 27–30.

Faivre-Amiot, A. 1967. Quelques observations sur la presence de Corynebacterium fascians (Tilford) Dowson dans les cultures maraicheres et florales en France. *Phytiatrie-Phytopharmacie*, 16, 165–176.

Fujii, T.M., Alvarez, A., Fukui, R. et al. 2002. Effect of transgenic anthuriums producing the Shiva-1 lytic peptide on beneficial bacteria. *Phytopathology*, 92, S27.

Ngamau, K., Mugai, E.N., and B.N. Ng'ayu. 2008. Effect of irrigation and mulch on the incidence of Erwinia soft rot, flower and tuber production of Zantedeschia 'Black Magic' and 'Florex Gold'. *Acta Hortic.*, 766, 193–198.

11 Bacterial Diseases of Forage Crops

11.1 ALFALFA

11.1.1 BACTERIAL LEAF SPOT IN ALFALFA

Pathogen: *Xanthomonas campestris* pv. *alfalfae*

Synonyms: *Phytomonas alfalfae* and *Xanthomonas alfalfae*

Symptoms

Diseased seedlings are often killed or stunted at high temperatures. Initially, small, water-soaked spots appear as chlorotic areas on leaflets. The spots enlarge to irregular-shaped lesions (2–3 mm in diameter) with chlorotic margin that are pronounced on the underside of leaflets (Photo 11.1). Eventually, the lesions become light yellow to tan, often with a lighter center, and have a translucent, papery texture. Lesions often glisten because of the dried bacterial exudates on their surface. Diseased leaves defoliate prematurely. Stem lesions are water soaked initially, and then they turn brown or black.

Disease Cycle

The bacterium survives in infected residue that has been incorporated into soil or is lying on the soil surface, in hay, and in debris associated with seed. During warm wet weather, bacteria are splashed or blown onto leaves and enter through small wounds that have been made by any means. During dry weather, bacteria may enter leaves through wounds made by windblown soil particles. The bacteria multiply inside the leaf and frequently ooze to the leaf surface, where they may be splashed by rain or rubbed by leaf-to-leaf contact onto adjacent healthy leaves.

The disease is favored by extended periods of hot, rainy, windy weather. Optimum growth of the bacterium occurs at 27°C–32°C. The causal bacterium overwinters in crop debris and seed. It is spread by wind and rain, insects, all types of equipment, and by infected forage.

Control Measures
- Grow resistant cultivars. Plants within a cultivar differ in susceptibility.
- Sow in the spring.

11.1.2 BACTERIAL WILT IN ALFALFA

Pathogen: *Clavibacter* (*Corynebacterium*) *michiganensis* subsp. *insidiosus*

Bacterial wilt is a very destructive disease of susceptible alfalfa varieties that are 3 years or older. The disease occurs wherever the crop is grown in the United States, except in arid areas without irrigation. Damage seldom occurs during the first 2 years following seeding.

The bacterium that causes the disease has contaminated soils in many parts of Illinois. Bacterial wilt is favored by cool temperatures and abundant moisture, usually becoming most severe in low, poorly drained areas. Plants weakened by bacterial wilt are more susceptible to winterkill than healthy plants. Crown injuries may also increase disease levels.

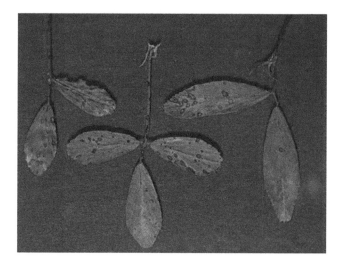

PHOTO 11.1 Bacterial leaf spot of alfalfa. (Photo courtesy of National Agriculture and Food Research Organization [NGRI], Tsukuba, Japan, naro.affrc.go.jp.)

Symptoms

Usually appear in the third or fourth year after planting. Foliar symptoms are most evident on regrowth after cutting. Infected plants are more susceptible to winterkill, particularly in southern areas such as Creston and Okanagan Falls.

Symptoms become apparent as the stand gets older (≥3 years). Infected plants are stunted and have a yellow-green color. In severe cases, the plant has spindly stems with small, distorted leaves. Infected plants that are stressed by water, heat, or both will wilt or die and are scattered throughout the stand. Infection stresses the plant and increases its susceptibility to winterkill. Cutting the taproot in half (in cross section) will show a light brown to yellow discoloration of the vascular tissue near the outer edge.

Bacterial wilt causes a stunting and yellowing of the entire plant. Growth is slow. Diseased plants are dwarfed, with a bunchy growth resulting from numerous, spindly, shortened stems and small, light green to yellow leaflets. Leaflets are commonly rounded at the tip and tend to curve upward (a condition called "mouse-leaved"). Affected plants may wilt during the heat of the day and recover temporarily during the cool of the night. Plants may wilt and die rapidly during warm, dry weather. At first, only the tips of the stems droop. This is followed by a more or less complete wilting and finally by the death of the infected plant. Stunting is most evident during regrowth following cutting. Progressively less growth is produced after each cutting. Infected plants usually die beginning in midsummer and into the next hay year. Severely diseased plants rarely survive the winter. Once infection has occurred, susceptible plants generally do not recover.

A sure sign of bacterial wilt is a yellow to dark golden brown discoloration in the outer vascular tissue of the taproot when the bark is peeled. This discoloration is in sharp contrast to the creamy white color of healthy roots. If an infected taproot is cut across just below the crown, scattered yellowish to brownish dots are visible or a ring of discolored tissue is usually evident. As the disease progresses, the entire stele becomes discolored.

The disease symptoms and death of alfalfa are probably due to the water-conducting vessels being plugged by the bacteria and due to the production of a bacterial toxin (a glycopeptide).

Geographical Distribution

Bacterial wilt was first recognized as a distinct disease in Illinois and Wisconsin in 1924. Bacterial wilt has been reported in Canada, Mexico, Chile, Europe, the USSR, the Near East, Japan, Australia, and New Zealand.

Bacterial wilt has recently occurred at damaging levels in the southern Okanagan and Creston valleys on susceptible varieties in British Columbia. It also occurs in central and northern British Columbia but does not appear to cause economic damage. In the 1950s, the disease caused serious problems in the southern half of the province. Bacterial wilt has historically been one of the most important forage diseases, not only in Ontario but also anywhere where forages are grown. The bacterial wilt pathogen is endemic in New York soils, and all adapted varieties need to possess a high level of bacterial wilt resistance.

Disease Cycle

The causal bacteria survive in living or dead alfalfa plant tissue in the soil. The bacteria have survived in dry plant tissue or seed for 10 years or more in the laboratory. The bacterial wilt organism is spread in the field by surface water, tillage equipment, mower sickles, infected hay, and animals. A long-distance spread most likely occurs by means of seed and hay. The infection of plants commonly occurs during cool, wet weather in spring and early summer. The bacteria enter plants through wounds in the roots and crowns produced by winter injury and animals in the soil, or through the cut ends of stems as a result of mowing or grazing. In advanced stages of the disease, bacteria multiply rapidly in crown and stem tissues and are released into the surrounding soil water. The disease cycle of bacterial wilt of alfalfa is illustrated in Figure 11.1.

Control Measures

1. The only practical control is to grow adapted, wilt-resistant varieties. This is especially true for an alfalfa stand that is to be maintained for 3 years or more. Varieties range from completely susceptible to highly resistant. No alfalfa variety is immune to bacterial wilt.

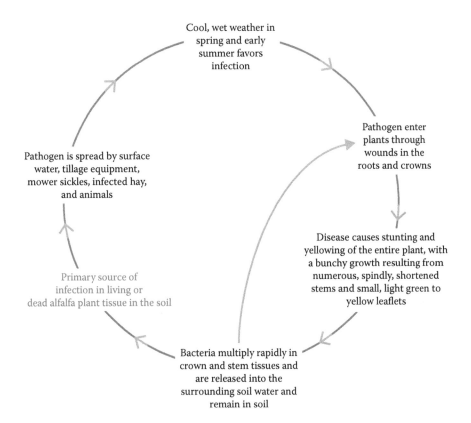

FIGURE 11.1   Disease cycle of bacterial wilt of alfalfa *Xanthomonas campestris* pv. *alfalfae*.

Moderately susceptible varieties may survive for 3 years after seeding. Moderately resistant ones are productive up to 5 years. Varieties vary somewhat in their resistance, but usually remain productive for 5–10 or more years. Stands are usually reduced by factors other than bacterial wilt. For a listing of currently recommended alfalfa varieties that are resistant to bacterial wilt, refer *Illinois Agricultural Pest Management Handbook* (revised annually).

2. Do not plant alfalfa in poorly drained soils.
3. For best yield, harvest at the late-bud to first-flower stage and every 30–40 days for succeeding harvests. This results in a rather rapid loss of plants in wilt-susceptible varieties. Less intensive harvesting schedules reduce the loss rate of susceptible plants.
4. Harvest young stands before old stands when using the same equipment. Harvest fields showing wilt symptoms last. Clean equipment with steam before moving from field to field, especially where wilt is present.
5. Mow only when the foliage is dry.
6. Use a program of high, balanced fertility based on a soil test to help maintain plant vigor.
7. Grow other crops for 2 or 3 years before reseeding a field with alfalfa.
8. Reduce injury to crowns (livestock movement, equipment, etc.), which provide entry wounds and may weaken plants.

11.1.3 Alfalfa Dwarf

Pathogen: *Xylella fastidiosa*

The first report of "bacteria-like bodies" associated with a plant caused by *Xylella fastidiosa*, referred to as alfalfa dwarf (AD), was reported in the 1920s (Weimer, 1937).

Symptoms

Plants infected with the bacterium *X. fastidiosa* are stunted. They often have small bluish-green leaflets and fine short stems. Taproots are normal in size but when cut open, roots appear abnormally yellowish in color with dark streaks of dead tissue scattered throughout. In newly infected plants, the yellowing is mostly in a ring beginning under the bark. Unlike bacterial wilt, there are no gummy pockets underneath the bark. Plants become stunted and eventually die.

The major symptoms include gradual decline in rate of regrowth after cutting. Affected plants have as little as a third the normal regrowth after cutting as healthy plants. Diseased plants have smaller, darker (often with a bluish color), and slightly more upright leaflets than healthy plants. If the taproot of chronically infected plants is cut diagonally, the wood of chronically infected plants is slightly yellow with streaks or flecks of dark brown, dead wood instead of the normal white color without dark streaks. There is a long incubation period—a year or more—between infection and the first appearance of symptoms. The symptoms are subtle, and other pathogens or physical or chemical factors may produce plants with the same appearance as AD. This makes diagnosis by symptoms alone difficult and unreliable. As a result, many cases of AD may not be recognized. The chief effect of AD is that a high incidence of the disease may shorten the life span of a typical alfalfa hay field. The slow regrowth of plants with AD may make hay production unprofitable. The incidence of AD was noticeably less in alfalfa seed fields than in hay fields (Weimer, 1933). This is perhaps because the less frequent irrigation and harvest schedules for seed fields decrease the amount of grass weeds that are so important to supporting populations of the vectors. A conceptual model of the epidemiology of AD within hay fields has been presented in Purcell and Frazier (1985).

History and Distribution

Alfalfa dwarf was first described in southern California in the 1920s (Weimer, 1931). Interestingly, the first photographs of the causal bacterium *X. fastidiosa* were published in 1936 (Weimer, 1936), but no claim was made as to their role in causing disease. It is unfortunate that this clue was not pursued a few years later when Pierce's disease of grape was found to be associated with AD

(Hewitt and Houston, 1941). Instead Pierce's disease and AD were then considered to be virus diseases, an erroneous conclusion that persisted for almost 40 years (Davis et al., 1978).

Alfalfa dwarf has been reported from Imperial Valley to as far north as Madera County in northern California. It is mysterious why AD does not seem to occur farther north in California, despite the abundance of alfalfa plantings and the common and often abundant populations of vector insects throughout northern California. Further studies may show strain differences between isolates of *X. fastidiosa* from grape and isolates from AD diseased plants. Vineyards with a high incidence of Pierce's disease were often located next to alfalfa fields with a high incidence of AD disease (Hewitt and Houston, 1941). Both diseases were later shown to have been caused by the same pathogen (Hewitt et al., 1946), which was then thought to be a virus. The same strains of *X. fastidiosa* that cause AD can cause Pierce's disease of grape and almond leaf scorch (Thomson et al., 1978).

Alfalfa dwarf was reported from Rhode Island, based on symptoms (Stoner, 1958). However, its occurrence outside California has not been definitively confirmed. It is possible that AD is an unrecognized limiting factor for commercial alfalfa production in regions of North America, such as the Gulf Coast, where *X. fastidiosa* precludes grape production by causing Pierce's disease.

Because of the difficulty of recognizing AD, the relatively low value of alfalfa hay, tolerance and resistance of modern alfalfa varieties to AD, and other unknown factors, AD has been considered an "obscure" disease. This status probably accounts for the very small amount of research on AD since the 1940s. Since then *X. fastidiosa* has been cultured from alfalfa with AD, and isolates of *X. fastidiosa* have been shown to react in serological tests in similar ways as isolates from grape and almond. In addition, AD isolates of *X. fastidiosa* were shown to cause Pierce's disease and almond leaf scorch disease.

Disease Cycle

The main vectors are assumed to be grass-feeding sharpshooter leafhoppers, the green sharpshooter, and the red-headed sharpshooter. The impact of the disease is that alfalfa stands with dwarf disease have to be replanted much sooner than normal. Symptoms of AD differ sharply from Pierce's disease and other leaf scorch diseases caused by *X. fastidiosa*. Alfalfa plants decline slowly after infection. They are slower to grow after cutting and have smaller, darker colored leaves and stems compared to healthy plants. A diagonal slice of the taproot reveals that the woody tissue of the root has a yellowish color, with streaks of brown dead wood.

Dwarf is not recognized as an economic disease of alfalfa. However, the bacterium that causes AD, *X. fastidiosa*, is the same pathogen that causes Pierce's disease of grapes, a very important grape disease in California. The role that alfalfa plays in the epidemiology of Pierce's disease is important. Leafhoppers, including the blue-green sharpshooter, spread the disease from alfalfa to grapes. Increased levels of Pierce's disease in grape vineyards located adjacent to alfalfa have been documented in the San Joaquin Valley.

Disease Management

Minimize the attractiveness of an alfalfa stand to sharpshooters by preventing the growth of grasses. This, in turn, will reduce the transfer of the bacterial pathogen between alfalfa and grape vineyards. When possible, avoid planting alfalfa adjacent to grape vineyards in areas where Pierce's disease is prevalent.

11.1.4   ALFALFA SPROUT ROT

Pathogen: *Erwinia chrysanthemi*

Symptoms

Radicals are translucent yellow as they emerge from the seed. In 24–48 h, seeds stop growing and turn into yellowish, odiferous mass that contains numerous bacteria. Disease initially occurs in a few trays, but it may spread throughout all trays in a few days.

Disease Cycle

The bacterium survives up to two weeks on inoculated dried seed. Air, the water supply, or greenhouse workers may introduce bacteria into a sprouting house. Spread within a tray is by seed-to-seed contact. Rot is most severe in high moisture and when the temperature is 28°C and above. Rot is less severe below 21°C.

Control Measures

*   Control temperature in the sprouting house.
*   Practice sanitation. Bacteria are likely to survive in water remaining in tanks used for soaking seeds.
*   Soak seeds for 2 h in 0.5% sodium hypochlorite or calcium hypochlorite. However, hypochlorite is not currently registered for alfalfa seed treatment in the United States.

11.2   FORAGE GRASSES

11.2.1   Bacterial Blight or Leaf Streak of Forage Grasses

Bacterial blights occur widely in Illinois on forage grasses during warm, wet weather. The damage fluctuates considerably from year to year. The forage yield and its quality suffer during a local or general epidemic (epiphytotic). Affected leaves may wither and drop prematurely. Plants may be stunted. When infection is severe, the heads (spikes and panicles) are blighted and are killed as though injured by frost.

The bacteria attacking forage grasses are divided into two taxonomic groups, *Pseudomonas syringae* and *Xanthomonas campestris*.

Symptoms on Sudan Grass and Johnson Grass

1. Bacterial Leaf Stripe

Pathogen: *Pseudomonas syringae* subsp. *andropogonis*

It is widely distributed and destructive on many types of sorghums (grain, forage, and sweet or sorgo), Sudan grass, and broomcorn as well as on several other related grasses.

Long, narrow stripes (1/4 in. × 9 in. or more) form on the leaves. These stripes are initially water soaked, irregular, and bounded by veins. These areas soon dry and turn brick red, dark purplish red, reddish brown, or tan to dark brown depending on the grass variety. The color is continuous throughout the lesion. Later, the stripes elongate and fuse to form irregular blotches that cover a large part of the leaf surface and extend into the leaf sheath. Stalks and floral structures may show similar but more restricted lesions.

Bacterial exudate, the same color as the stripe, forms in droplets and dries down to reddish crusts or scales over the lesions, especially on the lower leaf surface. Severely infected leaves dry and wither.

Infection occurs primarily through wounds produced by wind or insects and to a lesser extent through stomata. The bacterium is seed borne and survives in plant residue. Leaf stripe first appears about mid-summer and continues until plant maturity. Disease development is favored by periods of warm, moist weather with an optimum of 22°C–28°C. The maximum is about 38°C and the minimum about 6°C.

2. Bacterial Leaf Streak

Pathogen: *Xanthomonas campestris* subsp. *holcicola*

It is widely distributed on sorghums—including Sudan grass and Johnson grass. Other plants that may become infected include corn, pearl millet, and foxtail millet.

Narrow, water soaked, and translucent (1/8 in. × 1–6 in. long streaks), with red or brown borders and irregular blotches of color interrupting the streaks form on the leaves. Irregular, oval blotches develop later as lesions enlarge and merge covering much of the leaf blade. The blotches have tan centers and red or brown margins. When infection is severe, leaves may wither and drop early. Abundant yellow-to-cream droplets of bacterial exudate form on the lesions. This exudate dries to a thin, white-to-cream scale that is lighter than the bacterial leaf stripe exudate.

Damp or rainy weather with a temperature of 28°C is optimum for disease development (maximum, about 37°C; minimum 4°C).

3. Bacterial Eyespot or Leaf Spot

Pathogen: *Pseudomonas syringae* subsp. *syringae*

It occurs more commonly than bacterial leaf streak, but is not as widespread. Besides sorghums, Sudan grass, and Johnson grass, hosts include broomcorn, pearl millet, foxtail millet, field or dent corn, sweet corn, and popcorn.

Round to elliptical water-soaked spots form on lower leaves. The lesions soon dry, becoming papery, and light tan with red or brown borders. Diseased areas may exudate and merge forming large, irregular blotches that kill the whole leaf. Bacterial eyespot can be distinguished from both bacterial stripe and bacterial streak by the absence of streaking and bacterial exudate. Eyespot appears soon after seedlings emerge in the spring and progresses with plant development throughout the entire growing season, gradually spreading upward from the lower leaves on the plant.

Bacterial eyespot is favored by moist weather and cool temperatures of 13°C.

Symptoms of Bacterial Blight on Other Grasses

1. Bacterial Blight or Chocolate Spot

Pathogen: *Pseudomonas syringae* subsp. *coronafaciens*

It is widely distributed in Illinois and is a serious disease of smooth bromegrass. Other grass hosts include quackgrass, wild ryegrasses, wheat grasses, prairie June grass, and perennial ryegrass.

The lesions on the leaves and sheaths are circular to elliptical and are initially water soaked. Lesions enlarge and merge to form chocolate-to-purplish brown blotches, usually with a light-colored "halo." The blighted areas may involve the entire leaf blade and sheath. As infected leaves wither and die, the lesions fade to a rusty-brown color. Bacterial exudate is not normally present. Lesions on the pedicels and heads (panicles) are smaller and more restricted than those on the blade and sheath. When the disease is severe, the upper nodes may be killed, and in such plants, the panicles wither and die as though injured by frost.

2. Bacterial Stripe Blight or Leaf Streak

Pathogen: *Xanthomonas campestris* subsp. *translucens*

It is widely distributed on a number of grass and cereal hosts, including timothy, various brome-grasses (including smooth brome, mountain brome, rescue grass, rattlesnake chess, Japanese chess, soft chess, and ripgut grass), quackgrass, barley, wild barleys, wheat, rye, and oats.

Leaf lesions are usually long, narrow stripes, interrupted fine lines, or blotches that at first are water soaked. Later, the lesions may elongate, dry, and turn yellow to dark brown with isolated, translucent areas. Finally, they may become brownish black with small, golden areas. Water-soaked to black stripes develop on the stems. Frequently, the top leaves and inflorescence are blighted. Infections on the floral bracts (glumes) are dark brown, fine stripes to elongated spots. When attacked, the kernels are shriveled and brown.

Bacterial exudate is sometimes conspicuous on diseased leaf, leaf sheath, and inflorescence tissues. When dry, the droplets form yellowish, resinous "beads" or dry flakes. Numerous lesions usually result in a slow yellowing and death of the leaf, starting at the tip. Where severe, stripe blight may reduce the yields of forage and of grain.

Disease development is optimum in rainy or damp weather when the temperature is about 26°C with a maximum of about 38°C.

Disease Cycles

The disease cycles of the various blight-producing bacteria are very similar. They overwinter in or on seed, in crop residues, and in the soil. The bacteria are spread from plant to plant by the wind, splashing rain, and dew and by sucking as well as chewing insects. Mowing and grazing when the foliage is wet also spread the bacteria. Young grass tissues are infected through stomata (natural openings) and wounds.

New leaves become infected during wet periods when the temperature is favorable. Primary infections usually occur during the seedling stage. Secondary infections occur on younger tissues throughout the growing period.

The bacteria embedded in the exudate on infected tissues remain dormant during dry periods and resume active development when the weather again becomes favorable. The bacteria remain viable for 2 years or more within dried crop residues and seed.

Control Measures

1. Sow only certified, disease-free seed of improved, well-adapted cereal and grass cultivars: whenever available.
2. Grow cultivars that are resistant to bacterial blight when they are available and are recommended: Resistant cultivars of wheat, barley, sorghum, Sudan grass, and smooth brome-grass are available. Refer Illinois Agricultural Pest Management Guide Handbook for suggested cultivars to plant. Sorgos as a class are more susceptible to bacterial stripe than are the grain sorghums and Sudan grass. The Kafirs and broomcorn varieties are resistant to bacterial leaf streak.
3. Treat cereal and grass seed, where feasible, with a suggested fungicide: Seed treatment helps prevent the introduction of bacteria, which are carried on the seed, to new fields.
4. Avoid disease buildups by rotating grasses with nongrass crops, such as soybeans and forage legumes, for at least 2 years: Cut early and remove from the field any hay crop that becomes heavily infected. This practice reduces losses in hay quality and removes inoculum that may threaten future cuttings. Avoid leaving a heavy mat of hay on the grass during warm, moist weather. Leaf-blighting diseases are seldom destructive in frequently cut or closely grazed pastures.
5. When erosion is not a hazard, plow to neatly clean all cover crops, severely infected stands, volunteer cereals or grasses, and plant residues. Where practical, seed a mixture of forage species.
6. Keep down weed grasses by cultural or chemical means.
7. Use a careful, controlled burning of dead grass in early spring, which may be warranted if pastures are severely affected. This ancient practice destroys organic matter and kills leaf blighting bacteria and fungi in overwintering leaves and crop residues. Check local EPA regulations about open burning.
8. Practice balanced fertility, based on a soil test. Maintain a high level of potash (K20) and avoid excessive rates of fertilizers high in nitrogen.

11.2.2 Leaf Blotch of Sudan Grass

Pathogen: *Pantoea ananas* and *Pantoea stewartii*

This disease of Sudan grass (*Sorghum sudanense*) was observed in commercial fields in Imperial Valley of California.

Symptoms

Symptoms include light-colored necrotic streaks and white or tan irregular blotches, often associated with reddish-purple to dark-brown margins (Photo 11.2). *Pantoea ananas* was consistently isolated

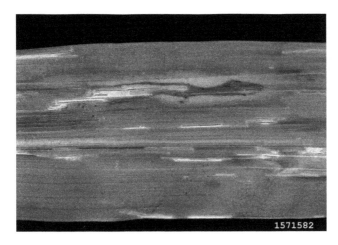

PHOTO 11.2 Symptoms of leaf blotch of Sudan grass. (Photo courtesy of Gerald Holmes, California Polytechnic State University, San Luis Obispo, CA, Bugwood.org.)

from the blotches with reddish margins, while *Pantoea stewartii* or mixtures of both species were isolated from necrotic streaks without reddish margins.

Symptoms in inoculated plants appeared as early as two and as late as twenty days after inoculation, depending on the inoculum level, methods of inoculation, temperature, and available moisture. The initial symptoms caused by inoculations with both bacteria were similar, but as symptoms progressed, *P. ananas* was associated with white streaks or irregular necrotic blotches often surrounded by a reddish or purplish hue. *Pantoea stewartii* was associated with light-colored necrotic streaks. A synergistic or antagonistic relationship was not observed between the two pathogens in co-inoculations.

Host Range

Both bacteria caused disease on sorghum and Sudan grass at similar levels of severity. *Pantoea ananas* was also pathogenic on corn and oat. *Pantoea stewartii* from Sudan grass was pathogenic on corn but did not cause wilting that was observed with Stewart's wilt strains of *P. stewartii* from corn.

The Sudan grass strains of *P. stewartii* also infect oat and triticale, whereas the Stewart's wilt strains did not. Both *P. ananas* and *P. stewartii* from Sudan grass grew at relatively high temperatures (43°C and 37°C, respectively) and caused disease at elevated temperatures and conditions of relative humidity similar to those in the Imperial Valley during late summer when epidemics of the disease were common.

11.2.3 Halo Blight of Bromegrass

Pathogen: *Pseudomonas syringae* pv. *atropurpurea* (Reddy & Godkin) Young, Dye & Wilkie
The disease occurs mainly in the warm regions.

Symptoms

The disease produces at first water-soaked spots in leaves and then the lesions become brown and oval to spindle-shape or irregular and surrounded with yellow halos. The lesion expands to long stripe when the disease advances and finally it might invade the head spikes and seeds (Photo 11.3).

PHOTO 11.3   Field symptoms of halo blight on bromegrass. (Photo courtesy of National Agriculture and Food Research Organization [NARO], Tsukuba, Japan.)

11.2.4   BACTERIAL BROWN STRIPE OF PASPALUM

Pathogen: *Acidovorax avenae* subsp. *avenae* (Manns, 1909) Willems et al.

Symptoms

Lesions are at first ash white, small oval in a young leaf, and gradually it extends along leaf veins and becomes reddish brown. The lesion part becomes thin like a paper and splits easily.

Many of the bacteria were detected from the infected leaf, and they disperse and transmit by wind and rain. The causal bacteria parasitize many kinds of gramineous plants, such as corn, rice, foxtail millet, African millet, and millet.

11.2.5   BACTERIAL WILT OF STYLOSANTHES HUMILIS

Pathogen: *Pseudomonas solanacearum*

Bacterial wilt due to *Pseudomonas solanacearum* has been found in some *Stylosanthes humilis* (*Townsville stylo*) pastures in the northern, higher rainfall region of the Northern Territory of Australia. The disease was found each year during the mid-wet season but usually caused only moderate damage to affected pastures.

The wilt disease of Townsville stylo pastures was first noticed during the 1965–1966 wet season (J. B. Heaton, unpublished reports), and it has been observed in the mid-wet season (February) each year since then. An outbreak of the disease was noticed in February 1969.

The disease has been found at sites located approximately 10, 30, 45, and 60 miles south of Darwin. The average wet season rainfalls of the surrounding areas vary from 56 in. at Darwin to 47 in. at Adelaide River (72 miles south of Darwin), and in each case January and February are the months of highest rainfall. The occurrence of the disease was not restricted to a particular soil type or situation, and it was found to recur at certain sites that apparently had little in common. Affected pastures had been established for several years, in one case on cleared, cultivated flat land, and in another among uncleared native vegetation on a stony slope.

Symptoms

Symptoms are most obvious in vigorous, dense stands of Townsville stylo at the flowering stage. Occasionally plants wilted suddenly, the leaves became dry and silvery green, and later shed as the plant died. In the early wilting stage the stem and roots of the plant appeared normal, with no external lesions or rotting. In newly dead plants however the root epidermis and the fibrous roots were easily peeled or rubbed off, leaving a slimy surface that blackened on drying. Near the end of the wet season, the wilting symptoms were often masked by signs of water stress or senescence, but diseased plants could still be detected on close examination.

It was difficult to assess the loss of production due to this disease in the 1969 wet season because wilted plants were usually scattered singly throughout the pasture, although some affected pastures also had numerous small patches of dead, leafless stalks, which may have been foci of infection. However, an outbreak of the disease in 1967 caused yellowing, wilting, and death of plants in 10%–15% of the affected area (J. B. Heaton, unpublished report). In most cases the disease symptoms had caused some concern to the landholders, who then drew attention to its occurrence.

Pathogen

The preliminary results showed that the bacterium could proliferate in Townsville stylo plants and that it was likely to be the causal agent of the wilting disease. The bacterial morphology and the colonial characteristics indicated that it was a species of *Pseudomonas*, and the disease symptoms were typical of *P. solanacearum* infections.

Cultures of the bacterium isolated from a field case of wilt and from an experimentally infected plant were subsequently identified by the Commonwealth Mycological Institute, England, as *P. solanacearum* biotype 4.

Control Measures

As given in Section 11.2.

REFERENCES

Davis, M.J., Purcell, A.H., and S.V. Thomson. 1978. Pierce's disease of grapevines: Isolation of the causal bacterium. *Science*, 199, 75–77.

Hewitt, W.B. and B.R. Houston. 1941. Association of Pierce's disease and alfalfa dwarf in California. *Plant Dis. Rep.*, 25, 475–476.

Hewitt, W.B., Houston, B.R., Frazier, N.W. et al. 1946. Leafhopper transmission of the virus causing Pierce's disease of grape and dwarf of alfalfa. *Phytopathology*, 36, 117–128.

Manns, T.F. 1909. The blade blight of oats—Bacterial disease. *Ohio Agric. Res. Stn. Res. Bull.*, 210, 91–167.

Purcell, A.H. and N.W. Frazier. 1985. Habitats and dispersal of the principal leafhopper vectors of Pierce's disease in the San Joaquin Valley. *Hilgardia*, 53, 1–32.

Stoner, W.N. 1958. Field symptoms indicate occurrence of "alfalfa dwarf" or "Pierce's disease" virus in Rhode Island. *Plant Dis. Rep.*, 42, 573–580.

Thomson, S.V., Davis, M.J., Kloepper, J.W. et al. 1978. Alfalfa dwarf: Relationship to the bacterium causing Pierce's disease of grapevines. In: *Proceedings of the Third International Congress of Plant Pathology*, August 16–23, 1978, Munich, Germany, Vol. 3, p. 65 (abstract).

Weimer, J.L. 1931. Alfalfa dwarf, a hitherto unreported disease. *Phytopathology*, 21, 71–75.

Weimer, J.L. 1933. Effect of environmental and cultural factors on the dwarf disease of alfalfa. *J. Agric. Res.*, 47, 351–368.

Weimer, J.L. 1936. Alfalfa dwarf, a virus disease transmissible by grafting. *J. Agric. Res.*, 53, 333–347.

Weimer, J.L. 1937. Effect of the dwarf disease on the alfalfa plant. *J. Agric. Res.*, 55, 87–104.

SUGGESTED READING

Willems, A., Goor, M., Thielemans, S. et al. 1992. Transfer of several phytopathogenic *Pseudomonas* species to *Acidovorax* as *Acidovorax avenae* subsp. *avenae* subsp. nov. comb. nov., *Acidovorax avenae* subsp. *citrulli*, *Acidovorax avenae* subsp. *cattleyae*, and *Acidovorax konjaci*. *Int. J. Syst. Bacteriol.*, 42, 107–119.

12 Bacterial Diseases of Plantation Crops

12.1 TEA

12.1.1 BACTERIAL CANKER IN TEA

Pathogen: *Xanthomonas campestris* pv. *theicola*
  Xanthomonas gorlencovianum
The disease is found in Japan.

Symptoms

It is a relatively new cankerous disease found on the leaves and stems of tea plants. Small, water-soaked spots appear at the beginning, then expand and turn into brown hard spots with cracks at the top. Such spots became typical canker lesions later (Photo 12.1).

Control Measures

Not worked out but copper-containing bactericide with antibiotic may reduce the disease incidence.

12.1.2 BACTERIAL SHOOT BLIGHT IN TEA

Pathogen: *Pseudomonas avellanae* pv. *theae*
Bacterial shoot blight (BSB) in tea is a major bacterial disease in Japan. It occurs mainly in low temperature season, and lesion formation by the bacterial pathogen is enhanced by both low temperature and presence of nucleation-active bacterium. Low temperature is the most important environmental factor influencing the incidence of this disease.

Symptoms

The bacterium produces deep brown necrotic lesions on leaf blade, petioles, and stem. However the bacterium may induce diverse symptoms. On tea leaves, brown shallow necrosis limited to lower epidermis and few layers of subepidermal spongy parenchyma cells is observed. The infected leaf blade, petiole, and stem of young shoot give shoot blight appearance.

Control Measures

Copper bactericides are commercially used for the control of BSB. Copper bactericide with kasugamycin had the longest residual efficacy than other bactericides. Copper bactericide with kasugamycin at initial incidence of BSB and regular application of copper bactericide once a month after initial incidence control the outbreak of BSB.

12.1.3 CROWN GALL OF TEA

Pathogen: *Agrobacterium tumefaciens*
The disease prevails in Japan.

Symptoms

The general symptoms are formation of gall at the crown region of tea plant. Severe infection girdles the stem and affects the growth and new shoot and leaf formation in tea plant.

(a)

(b)

PHOTO 12.1   Bacterial canker of tea: (a) healthy field and (b) canker spots on diseased tea leaf. (Photo courtesy of S.G. Borkar and vitabio.com.)

Disease Cycle
As described in other crown gall diseases.

Control Measures
As described in other crown gall diseases.

12.2   COFFEE

12.2.1   BACTERIAL BLIGHT OF COFFEE

Pathogen: *Pseudomonas syringae* pv. *garcae*
Bacterial blight has been a very serious problem in coffee in Ethiopia. The incidence of the disease is estimated at 70%–80%. The disease is also reported from Kenya and East and West Rift valley of Africa.

Symptoms
Circular to oval brown spots with concentric rings in the center of the leaves are found on all young coffee trees. The severe infection causes blighting of leaves and plant.

Predisposing Factors
The disease occurs where wet and cold conditions prevail.

Control Measures
Not worked out but copper bactericide with antibiotics may reduce the incidence of disease.

12.3   RUBBER

12.3.1   CROWN GALL OF RUBBER

Pathogen: *Agrobacterium tumefaciens*
Crown gall is a bacterial infection affecting the roots and stems of rubber trees, caused by *Agrobacterium tumefaciens*.

Symptoms
Characterized by swollen areas along the stems, crown gall can cause large, disfiguring masses to appear (Photo 12.2), which are deadly to the plant. The roots can also be affected and are then inhibited from absorbing nutrients and moisture.

(a)   (b)

PHOTO 12.2 Crown gall on rubber plant: (a) healthy plant and (b) disease plant with gall at crown level. (Photo courtesy of SiewLian Chan, freeimages.com and somasleep.wordpress.com.)

Control Measures
Unfortunately, there is no treatment for crown gall, and infected plants must be destroyed to prevent the spread of the bacterium.

12.3.2 *Xanthomonas* Leaf Spot of Rubber

Pathogen: *Xanthomonas campestris*
Xanthomonas leaf spot is one of the most common diseases found in rubber trees.

Symptoms
It is characterized by tiny oozing spots that appear on the leaves, eventually causing them to yellow and die. The spots rapidly grow in size and are often bordered by bright yellow discolorations that spread between the leaf veins.

Control Measures
Copper-based bactericides can keep the infection from spreading if applied early, although they can harm healthy tissue if overused. *Xanthomonas* leaf spot can be prevented if excessive watering is avoided and fertilizer is used sparingly.

12.4 ARECA NUT

12.4.1 Bacterial Leaf Stripe of Areca Nut

Pathogen: *Xanthomonas campestris* pv. *arecae*
Bacterial leaf stripe is a disease of areca nut palm in Karnataka, India.

Symptoms
The symptoms are purely parenchymatous in nature causing water-soaked linear lesions parallel to the midrib of the leaflet. The lesions are covered with abundant creamy white bacterial exudates on the undersurface, which is a striking feature of the disease. The entire leaflet in a frond

may be affected resulting in complete or partial blighting. In severe cases, entire crown may be affected. When growing buds are affected, death of palm takes place. The disease is aggressive during monsoon. Younger palms (3–5 years old) are highly susceptible.

Predisposing Factors and Epidemiology
Disease outbreaks are confined to the monsoon season, that is, July–October. The disease intensity and severity are high when there are more than 10 rainy days per month in the period from July to October, with an average of 130 mm of rain during the month. High temperatures reduce disease occurrence but the temperature range of 17.5°C–25.5°C combined with high rainfall, causes a heavy disease incidence.

Control Measures
Spraying or stem injection with tetracycline group of antibiotic at 500 ppm concentration is effective.

12.5 CASSAVA

12.5.1 BACTERIAL BLIGHT OF CASSAVA

Pathogen: *Xanthomonas axonopodis* pv. *manihotis* (Bondar) Vaulerin et al.

Synonyms: *Xanthomonas campestris* pv. *manihotis* (Bondar) Dye
Xanthomonas manihotis (Bondar) Starr

Common names: Bacterial blight of cassava
Xanthomonas axonopodis pv. *manihotis* causes bacterial blight of cassava. Originally discovered in Brazil in 1912, the disease has followed cultivation of cassava across the world. Among diseases that afflict cassava worldwide, bacterial blight causes the largest losses in terms of yield.

Large variations in the predominance and severity of the various symptoms are observed according to the location, season, aggressiveness of the occurring bacterial strains and the cassava cultivars (Maraite, 1993).

In cassava, symptoms vary in a manner that is unique to this pathogen. Symptoms include blight, wilting, dieback, and vascular necrosis. A more visible diagnostic symptom in cassava with *X. axonopodis* infection is angular necrotic spotting of the leaves often with a chlorotic ring encircling the spots. These spots begin as distinguishable moist, brown lesions normally restricted to the bottom of the plant until they enlarge and coalesce often killing the entire leaf. A further diagnostic symptom often embodies itself as pools of gum exudate along wounds and leaf cross veins. It begins as a sappy golden liquid and hardens to form an amber-colored deposit.

Symptoms

On Leaves
Leaves show dark-green to blue, water-soaked, angular spots (1–4 mm in diameter), limited by veinlets and irregularly distributed on the lamina. With time they frequently extend and coalesce along the veins or the edges of the leaf; the central portion turns brown, and the water-soaked part often becomes surrounded by a chlorotic halo. The lesions appear as translucent spots when viewed against the light. Under a magnifying glass, small droplets of exudate oozing from the central portion of the lesion are visible on the lower surface of the leaves. The droplets, which first glisten creamy-white and later yellow, are easily dissolved by rain or dew; on drying, they form a thin scale. Under favorable conditions (young leaves, high soil and air humidities), water-soaked pinpoint spots develop, scattered around young angular spots. The surrounding part of the lamina turns light brown, and within 2–3 days, extensive areas of the leaflets become withered not only toward the tip or border of the leaflet but also toward the base. The affected parts show light-brown and green zonations as if they had been superficially burnt. These necrotic areas are not translucent, no bacterial exudate is observed and bacteria are absent or present in

only very limited amounts at the borders of the extending blight lesions. A severe attack leads to premature drying and shedding of the leaves.

On Stem and Growing Point

Under conditions of high humidity, infection may spread through the vascular bundles from the leaflets to the petiole and twigs or stems, with the formation of black and dark-brown streaks as well as exudation drops along the pathway of progression. These parts may also become infected directly through wounds, which may be due to removal of leaves for consumption or insect punctures.

On the un-lignified twig or stem, a dark-green to black water-soaked area develops around the infection point. Large gummy exudation drops appear some distance away from the infection point, in the axis of vascular bundles, and one or a small number of leaves, located on the same side, show a sudden loss of turgidity, followed by rapid wilting and shriveling. Afterward the base of the petiole collapses, but the dried leaves generally remain attached for some time. All leaves located above those showing the first symptoms wilt progressively. Finally the un-lignified tip dies, appearing as a wick on the withered stem end, giving the "candle" symptom, while new shoots grow out lower down the stem. As the infection progresses toward the base of the stem, these shoots often also become wilted, leading to plant dieback. In the infected shoots, xylem vessels are brownish.

Under the microscope, the vessels appear obstructed by bacteria, tyloses, and mucilaginous substances. Lytic pockets develop around the protoxylem. The spread of these pockets causes rupture of the xylem ring, development of lytic pockets in the phloem, and later rupture of the fiber ring in the cortical collenchyma. Externally, the latter pockets become visible as dark-green water-soaked spots and small black streaks, corresponding to altered laticifers. These spots swell, rupture, and extrude a sticky white-yellow gum. In fully lignified stems or branches only internal vascular browning is visible. Infection may spread more than 50 cm below any external visual symptom.

On Roots

Only rarely does infection reach the roots in some very susceptible cultivars, whose swollen roots may show dry, rotted spots around the dead vascular strands (Lozano, 1986).

On Fruits

On the green capsules, water-soaked expanding spots can also be observed. Heavily infected seeds from such fruit may be deformed, with corrugation of the testa and necrotic areas on the cotyledons and endosperm.

Host

Manihot esculenta

Geographical Distribution

Asia: India, Indonesia, Japan, Malaysia, Philippines, Southeast Asia, Taiwan, Thailand

Africa: Africa, Benin, Cameroon, Central African Republic, Comoros, Congo, Cote d'Ivoire, Ghana, Madagascar, Malawi, Mali, Mauritius, Niger, Nigeria, Rwanda, South Africa, Sudan, Tanzania, Togo, Uganda, Zaire

America: Argentina, Barbados, Brazil, Caribbean, Colombia, Cuba, Dominican Republic, French Guiana, Mexico, Nicaragua, Panama, South America, Trinidad and Tobago, Venezuela

Oceania: Guam, Micronesia, Palau

Disease Cycle

Xanthomonas axonopodis pv. *manihotis* is a vascular and foliar pathogenic bacteria. It normally enters its host plants through stomatal openings or hydathodes. Wounds to stems have also been noted as a means of entry. Once inside its host, *X. axonopodis* enzymatically dissolves barriers to the plant's vascular system and so begins a systemic infection. Because of its enzyme's inability to break down highly lignified cell walls, this pathogen prefers to feed on younger tissues and often follows xylem vessels into developing buds and seeds. Seeds that have been invaded by a high number of bacteria are sometimes deformed and necrotic, but assays have shown a high percentage of infected seeds are asymptomatic carriers. In moist conditions, *X. axonopodis* pv. *manihotis* has been shown to survive asymptomatically for up to 30 months without new host tissue but is a poor survivor in soil. It persists from one growing season to the next in infected seeds and infected clippings planted as clones in fields. Once one cassava plant is infected, the whole crop is put at risk to infection by rain splash, contaminated cultivation tools, and foot traffic. These are effective methods of transmission because they cause wounds to healthy cassava plants, and *X. axonopodis* uses these wounds as an entry point.

The bacterium is spread to new areas in infected, symptomless stem cuttings and seed (Persley, 1979). Within the crop, spread is mostly by rain splash. Infection requires 12 h at 90%–100% relative humidity with temperatures of 22°C–26°C. The bacterium remains viable for many months in stems and gum, renewing activity in wet periods. Entry occurs through stomata or wounds and via the vascular tissues to other parts of the plant, including seeds (Lozano, 1986). In addition to rainfall, wide fluctuations between night and day temperatures, in the range 15°C–30°C, affect disease severity. The disease cycle of bacterial blight of cassava is illustrated in Figure 12.1.

Favorable Conditions

Xanthomonas axonopodis pv. *manihotis* excels in a humid subtropical to tropical climate. Cassava cultivation primarily takes place in these climates across the world, and *X. axonopodis* has followed it. It has been confirmed up and down Latin America into North America, Sub-Saharan Africa,

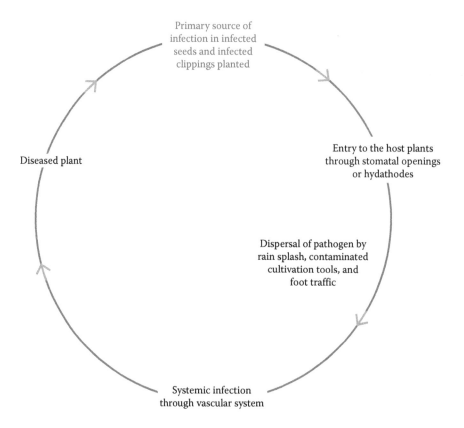

FIGURE 12.1 Disease cycle of bacterial blight of cassava caused by *Xanthomonas axonopodis* pv. *manihotis*.

Southeast Asia/India, and even Polynesia. As this geographical distribution would suggest, this pathogen requires a humid environment and warm habitat for pathogenicity. Nonpathogenic, epiphytic survival of *X. axonopodis* has been demonstrated under field conditions of high relative humidity, high rainfall, and high cloud cover/sun obstruction. These conditions match those of the early rainy seasons in the pathogen's (and cassava's) eco-region. The favorable conditions described allow colonial growth and eventual swarm behavior to enter hydathodes, stomata, or wounds.

Importance and Economic Impact
Cassava is a staple food of the human diets in developing countries in the tropics. In 2007, global production was 228 million tons, with 52% coming from Africa. It is estimated that cassava accounts for 37% of total calories consumed by humans in Africa. It has been further estimated that it provides the sixth most calories of any crop worldwide. These numbers would probably be even more impressive if bacterial blight were eradicated. The estimates for destruction caused by *X. axonopodis* pv. *manihotis* vary widely, but studies have shown that one infected transplant can result in 30% loss of yield in one growing cycle and up to 80% by the third cycle if no control measures are taken. There have been a number of historical outbreaks of bacterial blight. Zaire lost 75% of its tuber yield and almost all its protein-rich leaf yield every year in the early 1970s, while parts of Brazil lost 50% of tuber yield in 1974.

In 1973, one year after the first report of cassava bacterial blight in Nigeria the estimated yield losses were 75% (Ezelio, 1977). Crop losses as high as 90%–100% were observed in some parts of Uganda, 2 years after the disease was first recorded (Otim-Nape, 1980). In Zaire, the epidemics between 1971 and 1973 in the Kasaï and Bandundu provinces led to severe starvation because cassava roots and leaves are the staple foods in these areas (Maraite and Meyer, 1975). In field experiments with *X. axonopodis* pv. *manihotis*-free and *X. axonopodis* pv. *manihotis*-infected stem cuttings, Otim-Nape (1983) observed a reduction in fresh tuber yield from 40.1 to 26.6 t/ha. By comparing the yield of susceptible clones to that of resistant ones under natural infection in Colombia during 1974–1982, Umemura and Kawano (1983) observed an 18%–92% yield reduction, depending on locality, planting time, and degree of simultaneous infection by *Elsinoë brasiliensis*. Based on a CIAT survey of five major cassava-producing zones in Colombia, the estimated reduction in national production was, nevertheless, 6.64% in 1973 (Lourido, 1974).

Control Measures
Losses can be greatly reduced by a combination of measures taken within the perspective of Integrated pest management (IPM) (Lozano, 1986).

Regulatory Control
Great care must be taken while introducing germplasm in areas where cassava bacterial blight has not yet occurred. Vegetative propagated material must be introduced as meristem culture multiplied *in vitro* and certified disease-free. Botanical seed should originate from areas unfavorable for disease development, be heat-treated and planted in quarantine. Details can be found in the FAO/IBPGR technical guidelines for the safe movement of cassava germplasm (Frison and Feliu, 1991).

Cultural Control and Sanitary Methods
In areas where cassava bacterial blight is already widespread, disease incidence can be reduced by the use of clean planting material. Cuttings should be taken only from plantations that have been found to be free of the disease by inspections at the end of the rainy season. In cases of sporadic occurrence of the disease, great care must be taken in collecting cuttings only from healthy plants and from the most lignified portion of the stem, up to 1 m from the base, combined with visual inspection for the absence of vascular browning. Tools should be regularly disinfected using a bactericide.

Infected clones can be cleaned by rooting bacteria-free stem tips in conditions unfavorable for infection or by meristem cultures *in vitro*.

Crop rotation and fallowing proved very successful when the new crop was planted with uninfected cuttings. All infected plant debris and weeds on which epiphytic survival may occur should be removed and burned or incorporated into the soil. Rotation or fallowing should last at least one rainy season.

In some areas, planting toward the end of the rainy season, instead of the beginning, will delay epidemic development during the growing period and hence help reduce yield losses. Intercropping cassava with maize or melon has been reported to reduce cassava bacterial blight significantly (Ene, 1977).

In potassium-deficient soils, increasing the potassium content of the leaves by fertilization tends to reduce disease severity (Odurukwe and Arene, 1980).

Seed Treatments
Soaking infested botanical seeds in hot water at 60°C for 20 min, followed by drying in shallow layers at 30°C overnight or at 50°C for 4 h, reduced the number of bacteria to less than the minimum detectable level without appreciably reducing germination (Persley, 1979a).

The FAO/IBPGR technical guidelines for the safe movement of cassava germplasm recommend visual inspection of the seeds, density selection, followed by treatment of the seeds by immersing them in water and heating in a microwave oven at full power until the water temperature reaches 73°C and then immediately pouring the water off. If a microwave oven is not available, a dry heat treatment for 2 weeks at 60°C is recommended (Frison and Feliu, 1991). A subsequent thiram dust treatment reduces seed reinfestation. Lozano and Nolt (1989) mentioned 77°C instead of 73°C for the microwave treatment.

Resistance Cultivars
Clear differences in host-plant resistance occur, especially with regard to stem infection and wilt; use of resistant genotypes is a major control strategy. Resistance to cassava bacterial blight appears to be due to several genes, mainly with additive effects but also to some extent with nonadditive effects; resistance appears to be recessive to susceptibility (Hahn, 1979; Umemura and Kawano, 1983). A variation in aggressiveness, but no clear-cut pathogenic specialization, is observed among *X. axonopodis* pv. *manihotis* isolates from various countries and also among those from a single country (Maraite et al., 1982; Alves and Takatsu, 1984). A strong genotype–environment interaction is often observed. Host-plant resistance is sustained by adequate fertilization.

Ninety-three varieties of *M. esculenta* were assessed by amplified fragment length polymorphisms (AFLPs) for genetic diversity and for resistance to *X. axonopodis* pv. *manihotis*. AFLP analysis was performed using two primer combinations, and a 79.2% level of polymorphism was found. The phenogram obtained showed between 74% and 96% genetic similarity among all cassava accessions analyzed. The results demonstrate that resistance to *X axonopodis* pv. *manihotis* is broadly distributed in cassava germplasm and that AFLP analysis is an effective and efficient means of providing quantitative estimates of genetic similarities among cassava accessions (Sanchez et al., 1999).

Biological Control
Foliar application of *Pseudomonas fluorescens* and *Pseudomonas putida* has been shown significantly to reduce leaf infection by *X. axonopodis* pv. *manihotis* (Lozano, 1986). However, biological control has not yet gained practical acceptance.

REFERENCES

Alves, M.L.B. and A. Takatsu. 1984. Variability in *Xanthomonas campestris* pv. *manihotis*. *Fitopatol. Bras.*, 9(3), 485–494.

Ene, L.S.O. 1977. Control of cassava bacterial blight (CBB). *Trop. Root Tuber Crops Newslett.*, 10, 30–31.

Ezelio, W.N.O. 1977. Control of cassava bacterial blight in Nigeria. Nigeria, 1976. In: Persley, G., Terry, R.E., and R. MacIntyre (eds.), *Report Workshop on Cassava Bacterial Blight*. International Development Research Centre, Ottawa, Ontario, Canada, pp. 15–17.

Frison, E.A. and E. Feliu. 1991. *FAO/IBPGR Technical Guidelines for the safe Movement of Cassava Germplasm.* Food and Agriculture Organization of the United Nations/International Board for Plant Genetic Resources, Rome, Italy.

Hahn, S.K. 1979. Breeding of cassava for resistance to cassava mosaic disease (CMD) and bacterial blight (CBB) in Africa. In: Maraite, H. and J.A. Meyer (eds.), *Diseases of Tropical Food Crops, Proceedings of an International Symposium*, UCL, 1978, Université Catholique de Louvain, Louvain-la-Neuve, Belgium, pp. 211–219.

Lourido, L.C. 1974. Unametodologia para estimar los beneficios y los costosesperados en un programa de investigacionagricolaaplicada: el anublo bacterial en la yuca. Thesis Econ., Universidad de los Andes, Bogota, Colombia.

Lozano, J.C. 1986. Cassava bacterial blight: A manageable disease. *Plant Dis.*, 70(12), 1089–1093.

Lozano, J.C. and B.L. Nolt. 1989. Pest and pathogens of cassava. In: Kahn, R.P. (ed.), *Plant Protection and Quarantine.* Vol. II: *Selected Pests and Pathogens of Quarantine Significance.* CRC Press Inc., Boca Raton, FL, pp. 169–182.

Maraite, H. 1993. *Xanthomonas campestris* pathovars on cassava: Cause of bacterial blight and bacterial necrosis. In: Swings, J.G. and E.L. Civerolo (eds.), *Xanthomonas.* Chapman & Hall, London, U.K., pp. 18–24.

Maraite, H. and J.A. Meyer, 1975. *Xanthomonas manihotis* (Arthaud-Berthet) Starr, causal agent of bacterial wilt, blight and leaf spots of cassava in Zaire. *PANS*, 21(1), 27–37.

Maraite, H., Weyns, J., Yimkwan, O. et al. 1982. Physiological and pathogenic variations in *Xanthomonas campestris* pv. *manihotis*. In: Lozano, J.C. (ed.), *Proceedings of the Fifth International Conference on Plant Pathogenic Bacteria.* Centro Internacional de Agricultura Tropical Cali Colombia, Palmira, Colombia, pp. 358–368.

Odurukwe, S.O. and O.B. Arene. 1980. Effect of N, P, K fertilizers on cassava bacterial blight and root yield of cassava. *Trop. Pest Manag.*, 26(4), 391–395.

Otim-Nape, G.W. 1980. Cassava bacterial blight in Uganda. *Trop. Pest Manag.*, 26(3), 274–277.

Otim-Nape, G.W. 1983. The effects of bacterial blight on germination, tuber yield and quality of cassava. In: *Proceedings of the 10th International Congress of Plant Protection*, British Crop Protection Council, Brighton, U.K., p. 1205.

Persley, G.J. 1979a. Studies on the epidemiology and ecology of cassava bacterial blight. In: Terry, E.R., Persley, G.J., and S.C.A. Cook (eds.), *Cassava Bacterial Blight in Africa: Past, Present and Future.* Report of an inter-disciplinary workshop, IITA, Ibadan, Nigeria, 1978. Centre for Overseas Pest Research, London, U.K., pp. 5–7.

Sanchez, G., Restrepo, S., Duque, M.C. et al. 1999. AFLP assessment of genetic variability in cassava accessions (*Manihot esculenta*) resistant and susceptible to the cassava bacterial blight (CBB). *Genome*, 42(2), 163–172.

Umemura, Y. and K. Kawano. 1983. Field assessment and inheritance of resistance to cassava bacterial blight. *Crop Sci.*, 23(6), 1127–1132.

SUGGESTED READING

Ando, Y., Hamaya, E., Takikawa, Y. et al. 1986. Variation in symptoms of bacterial shoot blight of tea plant. *Jpn. J. Phytopathol.*, 52(3), 478–483.

CABI. 2015. *Xanthomonas axonopodis* pv. *manihotis* (cassava bacterial blight). (June 11, 2015). Retrieved July 20, 2015, from http://www.cabi.org/isc/datasheet/56952.

Korobko, A. and E. wondimagegne. 1997. Bacterial blight of coffee (*Pseudomonas syringae* pv. *Garcae*) in Ethiopia. *Pseudomonas syringae* pathovars and related pathogens developments in Plant Pathology. *Dev. Plant Pathol.*, 9, 538–541.

Kumar, S.N.S. 1983. Epidemiology of bacterial leaf stripe disease of Arecanut palm. *Trop. Pest Manag.*, 20(3), 249–253.

Persley, G.J. 1979b. Studies on the survival and transmission of *Xanthomonas manihotis* on cassava seed. *Ann. Appl. Biol.*, 93(2), 159–166.

Rao, Y.P. and S.K. Mohan. 1970. A new bacterial leaf stripe disease of Arecanut in Mysore state. *Indian Phytopath.*, 23(4), 702–704.

Uehara, K., Arai, K., Nonaka, T. et al. 1980. Canker of tea a new disease and its causal bacterium *Xanthomonas campestris* pv. *theaecola*. *J. Bull. Faculty Agric.*, 30, 17–21.

Vauterin, L., Hoste, B., Kersters, K. et al. 1995. Reclassification of *Xanthomonas*. *Int. J. Syst. Bacteriol.*, 45(3), 472–489.

13 Bacterial Diseases of Forest Trees

13.1 POPLAR

13.1.1 BACTERIAL LEAF SPOT AND CANKER IN POPLAR

Pathogen: *Pseudomonas syringae* pv. *syringae* van Hall

Symptoms

Leaf spotting: Cankers on twigs, branches, and stems of mainly 1- to 2-year-old plants with associated dieback and occasional whole-plant death where cankers form low on the stem are the general symptoms of this disease.

Yellow flecks (1–3 mm diameter) appear on upper and lower leaf surfaces followed by dark-brown spots of irregular outline on both leaf surfaces (Photo 13.1). Spots may remain discrete, or merge forming larger blotches. Terminal and side shoot dieback with blackened buds and collapsed to desiccate foliage. Rough fissures or sunken, blackened lesions on stems (cankers) appear (Photo 13.2). Bacteria ooze from cankers, which, when dry, appear as a white deposit on the stem.

Host

    Populus spp.: All species present in New Zealand are susceptible but leaf spotting is most prevalent on *Populus ciliata, P. deltoides, P. szechuanica, P. trichocarpa, P. yunnanensis, and their hy*brids.

    Salix spp.: All species present in New Zealand are susceptible. Stem and twig dieback is more common than leaf spotting on these species.

    Alnus cordata—Leaf spotting and twig dieback.

Geographical Distribution

Throughout New Zealand, particularly in localities with high summer rainfall.

Disease Development

Leaf symptoms first appear in spring when temperatures are rising and moisture levels are high. Throughout the growing season, during periods of continuous rainfall (>3 days), spots are produced on new leaves. For example, a severe attack of *Pseudomonas syringae* occurred on foliage and stems of poplars growing in a nursery at Aokautere during a 4-month period (November 1985 to February 1986), when rainfall was 64% (194 mm) more than average, and it rained on 52 of those 120 days. Depending on the cultivar, spots may either remain discrete or merge, forming extensive blotches. In poplars leaf spotting is generally more common than stem cankering and is often seen on plants that do not exhibit any other symptoms.

Formation of stem and twig cankers also depends on high moisture levels; however, symptoms vary, and sometimes stems may be roughly fissured or have sunken black lesions oozing bacteria. Sometimes the whole stem is affected, killing the plant; more often only the top one-third to one-half of the plant, or only lateral branches, is involved. When there is extensive shoot and twig dieback with blackened dead foliage the condition is referred to as "blast." New shoots grow from below the infected areas on stems and branches. Although stem fissures and cankers may heal and the plant continues to grow, this often leaves a weak point that makes the plant susceptible to breakage in high winds.

PHOTO 13.1   Bacterial leaf spot on poplar leaf. (Photo courtesy of (NZFFA) New Zealand Farm Forestry Association, NZFFA.org.nz.)

PHOTO 13.2   Bacterial canker on poplar shoot. (Photo courtesy of (NZFFA) New Zealand Farm Forestry Association, NZFFA.org.nz.)

Some severe outbreaks of the disease occur when sudden frosts follow a warmer wet period. Approximately 50% of *P. syringae* isolates, tested from poplars and willows in New Zealand were capable of causing ice-nucleation. Ice-nucleating bacteria initiate the formation of ice crystals within host cells during frosts, and the combination of bacteria and frost causes more extensive tissue disruption than if either factor were present independently.

Economic Importance

Losses due to this disease are generally not great. Outbreaks of disease due to *P. syringae* on 1- and 2-year-old nursery-grown poplars are sporadic and very dependent on prevailing weather conditions. In rare instances, damage can be severe when *P. syringae* is combined with late frosts.

Control

Control of disease is generally not warranted. Traditionally, copper-based inorganic compounds have been used to reduce bacterial populations but with only limited success. There is potential for biological control with competitive, nonpathogenic strains of other *Pseudomonas* species.

13.1.2 CROWN GALL IN POPLAR

Pathogen: *Agrobacterium tumefaciens*

Symptoms

Crown gall caused by *Agrobacterium tumefaciens* can be a major nursery problem for the production of 1-year-old rooted plants of white poplars. Large galls may form at ground level causing girdling of the main stem, poor growth, restricted shoot growth, and toppling. Large galls may occasionally form on the trunk of older tree at the crown level or at first 1-2 m height from the crown level.

Crown gall infections appear as round swellings with rough, wrinkled bark usually confined to the root collar and lower stem (Photo 13.3). The galls are irregularly shaped and have deep fissures. Galls are hard when young and later become spongy. During moist weather, the galls may ooze a slimy substance, consisting of the infectious bacteria.

Disease Cycle

Crown gall bacterium exists in the soil and infects trees through wounds. The bacterium can be spread by ground-inhabiting insects, rain splash, irrigation, cultivation, pruning, and movement of infested soil particles. The bacterium is also spread from infected to healthy trees when infected trees are pruned or otherwise cut and the contaminated tools are used to prune healthy trees. When the infected tree parts are sloughed into the soil, the natural cycle begins anew.

Economic Importance

The galls themselves may be considered unsightly. More important is that galls disrupt nutrient and water transport in the vascular tissues, which results in poor growth of young trees. Affected stems are weakened at the points of infection and can be invaded by organisms that cause discoloration and decay resulting in stem breakage.

PHOTO 13.3 Formation of gall at crown region of poplar tree. (Photo courtesy of William Jacobi, Colorado State University, Bugwood.org.)

Disease Management

- Do not plant in soil that is known to be infested by the bacterium. This is especially important when locating nurseries.
- Do not propagate infected plant material.
- Destroy all infected plant material.
- Do not move soil from areas known to be infected with the bacterium, either on plant material or equipment.
- If infected trees are pruned or cut. Disinfect contaminated equipment before pruning healthy trees.
- A possible biocontrol is to inoculate cuttings with a non-gall-forming competitor (*A. radiobacter*) prior to planting in contaminated soils.

13.2 TEAK

13.2.1 BACTERIAL WILT OF TEAK

Pathogen: *Pseudomonas* sp.

The disease has wide distribution in the tropics and warmer parts of the temperate zone. The wilt pathogen has a wide host range in plants belonging to 17 families of which *Solanaceous* families are most susceptible. Wilt of teak seedlings has been recorded from the Philippines, Malaysia, and Burma.

Bacterial wilt of teak was recorded in young plantations, varying from 6-month-old to 2-year-old in India at Thundathil (Malayattoor Div.), Kariampanny, Kannoth (Wynad Div.), Mullachal (Thiruvananthapuram Div.), and Gurunathanmannu, Angamoozhi (Ranni Div.). The incidence of the diseases varied from <1% to ca. 20% in a given area of the plantation, maximum being at Thundathil.

Symptoms

The disease manifests during warm and wet period, especially just after the onset of monsoon showers. The infection usually occurred through injury and occasionally through lenticels. The symptoms, characteristic of vascular wilt disease, are expressed in the following dry period. The bacterium causes systemic infection of vessels, which show necrosis and discoloration. Initially the bottom leaves of the affected plants turn yellow and show scorching and browning of leaf tissues in between the veins. The upper leaves, near the apex, become flaccid and droop. Such plants wilt and fail to survive. The roots of the affected plants get discolored and decay. The vascular tissue of the wilted plants shows characteristic browning, which could be easily seen in the stem or at the leaf scar (point of attachment of petiole to the stem). When the cut edge of the affected tissue was immersed in a clean glass of water, an off-white streak of ooze, emerging within a few seconds, confirmed systemic bacterial infection.

Control Measures

Once a plant is wilted, it cannot be cured. However, with proper sanitation methods and precautions, the development and spread of the disease can be checked to a considerable extent (Bakshi, 1975).

Management Practices

1. High soil moisture and low waterlogging in the nursery favor the bacterial wilt in teak. Suitable site should be selected for nurseries with good drainage. Low land areas should be avoided.
2. The affected plants should be dug out and burnt.
3. Root injuries during weeding or transplanting constitute the chief avenues for the entry of parasite. Weeding may be minimized by using chemical fertilizers instead of farm yard manure as the latter encourages weed growth.

4. In sick nursery beds, the soil should be sterilized by treatment with formalin, about 2 weeks prior to planting.
5. Nurseries where the disease occurs for successive years should be abandoned for some time and a new nursery site for teak on unbroken forest soil should be selected in which the possible introduction of contaminated soil should be avoided.

13.3 EUCALYPTUS

13.3.1 BACTERIAL BLIGHT OF EUCALYPTUS

Pathogen: *Pantoea ananatis*
Bacterial leaf spots caused by *Xanthomonas* sp. also occur, but under field conditions *Pantoea ananatis* is the major problem.

Bacterial blight and dieback has been reported in Uganda previously (Roux, 1999; Nakabonge et al., 2003; Roux and Slippers, 2007) and is known to occur on Eucalyptus in many countries (Roux, 1999; Old et al., 2003). In South Africa, the disease has been recorded on Eucalyptus clones, hybrids, and species. The disease generally affects trees <2 years old (Roux, 1999). Differences in susceptibility of *Eucalyptus grandis* clones have been reported.

In 1998, a severe bacterial disease appeared in a single nursery in KwaZulu/Natal on ramets of an *E. grandis* × *E. nitens* (GN) hybrid clone. The disease subsequently spread to other nurseries and commercial plantations and was reported from different *Eucalyptus* sp., hybrids, and clones. It also records a serious new disease problem affecting one of the most widely planted forest trees in South Africa and elsewhere in the world.

Host Range
Pineapple fruitlets, Sudan grass, cantaloupe fruit, sugarcane, onions, and honeydew melons. The *P. ananatis* isolates from eucalypts have the ability to infect a number of Eucalyptus clones, hybrids, and species including *E. grandis*, *E. saligna*, *E. dunnii*, *E. nitens*, *E. smithii*, *E. grandis* × *E. camaldulensis* (GC), and *E. grandis* × *E. urophylla* (GU). This is of considerable concern as these represent some of the most crucial planting stock on which forestry in South Africa is based.

Symptoms
Typical symptoms of bacterial blight include tip dieback and leaf spots of young leaves. The leaf spots are initially water soaked and often coalesce to form larger lesions (Photo 13.4). The pathogen appears to spread from the leaf petiole into the main leaf vein and from there to the adjacent tissue. Thus lesions on the leaf are often concentrated along the main veins. Leaf petioles become necrotic, which results in premature abscission of the leaves. Trees assume a scorched appearance in the advanced stages of the disease and, after repeated infections become stunted. In humid conditions bacterial exudates are often evident on diseased tissue. Due to the resultant formation of many new growing tips and endemic shoots, the trees have a bushy appearance. Highly susceptible species, hybrids, and clones exhibit a combination of dieback and blight symptoms while those more tolerant show only leaf spot symptoms.

Disease Cycle
Bacterial blight and dieback on Eucalyptus is more prevalent in areas in South Africa where the temperatures are relatively low (between 20°C and 25°C) and high relative humidity. The means of entry of this pathogen into its host has yet to be established. However, in nurseries where vegetative propagation is practiced, the bacterium enters the cut surfaces of cuttings and reduces their ability to root by nearly 100%.

Management Strategies
Bacterial blight and dieback has become a serious problem in nurseries and young plantations throughout South Africa. Not only is the bacterium infecting cuttings but also roots in the nursery.

PHOTO 13.4 Bacterial blight symptoms on Eucalyptus leaf. (Photo courtesy of Rivadalve C. Gonçalves, Douglas Lau, José R. Oliveira, Luiz A. Maffia, Júlio C.M. Cascardo, Acelino C. Alfenas. 2008, www.scielo.br.)

This seriously hinders the ability of forestry companies to produce vegetative material for rooting. There are, however, significant differences in susceptibility among *E. grandis* clones, and this provides an excellent opportunity for the selection of tolerant material. Development of management strategies to reduce the impact of this disease is now a priority. A rapid screening technique to detect this bacterium is needed, and commercially important clones should be tested to determine their level of tolerance to this disease.

13.3.2 BACTERIAL WILT OF EUCALYPTUS

Pathogen: *Ralstonia solanacearum*. Smith

Synonyms: *Bacillus solanacearum* (Smith)
            Burkholderia solanacearum (Smith) Yabuuchi et al.

The impact of bacterial wilt disease can be high, especially during the wet season and in sites located in highly humid and soggy sites. Up to 30% tree mortality has been reported in severely infected plantations of *Eucalyptus urophylla* in northern Vietnam at 1 year after planting (Old et al., 2003).

Symptoms

The pathogen is soilborne, and disease symptoms develop shortly after planting. Affected trees are often scattered through the stand and show wilting, leaf drop, stem death, and reduced growth rate (Photo 13.5). Vascular discoloration commonly occurs, roots die, and basal cankers may be found on affected trees. Symptomatic trees usually wilt and die. Stem sections, cut through discolored vascular tissue exude bacterial masses if incubated moist in plastic bags for 24 h or suspended in water.

Host

Eucalypt-infecting strains are all Race I (the race with a broadest host range) and either biovar I (in South America) or biovar 3 (Asia and Australia). These biovars also attack a wide range of hosts

PHOTO 13.5   Bacterial wilt of Eucalyptus. (Photo courtesy of Dr. Huang from APP, China, Forestry and Agricultural Biotechnology Institute 2015.)

including many woody species, for example, casuarina, olive, teak, neem, cassava, and cashew. Eucalypt hosts recorded so far include *Eucalyptus camaldulensis, C. citriodora, E. grandis, E. leizhou, E. pellita, E. propinqua, E. saligna, E. urophylla*, and *E. grandis*.

Geographical Distribution
Bacterial wilt is widespread throughout tropical, subtropical, and warm temperate regions of the world. On *Eucalyptus*, the disease is recorded in Brazil, China, Indonesia, Taiwan, Thailand, Vietnam, South Africa, Uganda, and Australia.

Disease Cycle
Local splash-dispersal within plantations from cankers on infected eucalypt stems seems likely. Sequences and lengths of crop rotations may also be a factor in the incidence of bacterial wilt in *Eucalyptus*. For example, experience in Vietnam and eastern Thailand suggested that wilt of clonal plantings of *E. camaldulensis* occurred more commonly when eucalypts were planted in fields formerly growing cassava.

Economic Impacts
In severely affected plantations of *E. urophylla* in northern Vietnam, tree mortality was as high as 30% one year after planting. In a plantation of 13-month-old *Eucalypt pellita* examined in northern Queensland in 1996, significant numbers of 1-year-old trees suffered bacterial wilt. Eighteen months later, no newly diseased trees were found and some recovery of trees was evident through healing of basal cankers. In older trees, such cankers may be invaded by secondary pathogens causing stem and butt rot with increased incidence of wind-throw.

Management Practices
Bacterial wilt can originate from infected nursery stock, although for *Eucalyptus, R. solanacearum* is rarely reported as a nursery problem. Different *Eucalyptus* sp. undoubtedly vary in their susceptibility to disease, with *E. pellita* and *E. urophylla* often being susceptible. Variation in resistance to bacterial wilt has been identified for a range of eucalypt species in Brazil. Clonal variation in susceptibility also occurs, with evidence from South Africa that the disease is restricted to some interspecific hybrid clones. The circumstantial evidence for enhanced disease incidence following cassava is of interest, and other crop sequences may influence pathogen populations. No control practices for bacterial wilt

of eucalypts have been evaluated, nor can any be recommended. For example, cutting of diseased trees would not be effective as the pathogen would remain in infected roots and infested soil. However, stringent weed control followed by a fallow period may greatly reduce the incidence of the disease (Roux, 1999). Similarly, good silvicultural practices may reduce the impact of this disease in nurseries. Proper drainage of soggy sites such as the Kyembogo mother garden is necessary to reduce soil moisture and humidity that enhance outbreaks of bacterial wilt and dieback. Reduction in watering frequency during wet seasons and timely cutting of ramets may reduce losses caused by the disease.

13.4 BAMBOO

13.4.1 BACTERIAL WILT OF BAMBOO

Pathogen: *Erwinia sinocalami* sp. nov.
The bacterial wilt of bamboo shoot was reported in Taiwan at Chushan in 1954 and subsequently appeared in main bamboo cultivation areas of Nantou, Yucanlin, and Chiayi of Taiwan Island.

Symptoms
The disease is characterized by appearance of pink or brown, elliptic or irregular lesions on the margin of outer sheath at a sprouting shoot. The spots are large and form a series of bands across the sheath as the disease progress.

Yellowing of the foliage and premature defoliation occurs. The basal part of the culm becomes discolored and shriveled. The rhizomes and roots of the affected culm show browning and necrosis; leading to severe infection to death of the culm.

Geographical Distribution
A bacterial wilt disease of bamboo has been reported on *Dendrocalamus latiflorus* from Taiwan, China.

Disease Cycle
Erwinia sinocalami is very close to *E. amylovora* and *E. carotovora*, which cause various diseases in different hosts. The soilborne bacterial propagule enters the host tissue through wounds and infects the roots and rhizome tissues; infection causes necrosis of the host tissue and affects the conductive system of the plant.

Control
Selection of good planting sites and resistant bamboo species have been suggested for managing the disease.

13.5 PINE

13.5.1 STEM CANKER OF PINE

Pathogen: *Pseudomonas syringae* pv. *syringae*

Symptoms
Cankers (sunken patches of dead tissue) form on the stem of young plants and may cause their death. Cankers are found on lower stem. These range from small, sunken patches to large, distorted, resinous areas, which may girdle the stem. This condition is generally confined to 1-year-old trees. Foliage fades in color and wilts prior to death of plant.

Bacterial stem canker occurs only in winter, and its incidence is thought to be dependent on a particular sequence of weather conditions. *Pseudomonas syringae* pv. *syringae* has the property of being able to act as an ice nucleus and to initiate the formation of ice crystals. This causes more damage to plant tissue than normal frosts. *Pseudomonas syringae* pv. *syringae* can also invade such damaged tissue. The disease is therefore favored by a period of frosts suddenly following a time of mild, wet weather, which has prevented plants from hardening off. This type of stem canker is most frequently found on plants growing in hollows and on flats, which are prone to cold-air ponding. Trees growing on grassy, fertile sites appear more susceptible than that on clear ground, probably because the presence of ground cover can lower the temperature by as much as 4°C.

Cankers caused by *Pseudomonas syringae* pv. *syringae* commonly occur on the lower stem of plants below the bottom whorl of branches, though they may sometimes be found higher up the stem. They range in size from small, sunken patches of dead tissue only a few millimeters in diameter to large, distorted, resinous areas, which in severe cases completely girdle the stem, resulting in wilting of foliage and finally the death of the plant. The small cankers, which have little or no effect on the growth and vitality of the tree, and the larger ones, which do not produce resin, become more apparent when the bark is removed to reveal the dead tissue beneath. Affected plants may also show signs of frost damage, such as needle scorching or dead buds. Bacterial stem canker is generally confined to 1-year-old plants though 2-year-old plants are sometimes susceptible. Plants that do not die in the few months following the initial infection seldom show any further deterioration. With the onset of spring growth, callus tissue begins to cover the dead tissue of the cankered part of the stem, and after three or four growing seasons have passed, even trees that have been three-quarters girdled show no signs of the previous damage. Badly cankered trees are, however, prone to breakage at the site of the canker in the first year after infection. In those cases where cankers develop in the mid-stem region, branches below the canker may take over to form new leaders, giving the tree a poor form.

Host
Pinus radiata

Geographical Distribution
Bacterial stem canker has been confirmed in both North and South Islands from Bay of Plenty south.

Economic Importance
The disease is of only sporadic occurrence and is not considered to be of any great importance. Severe losses have, however, occurred, for example, in the winter of 1977 two sites in the central North Island with a combined area of approximately 200 ha had 35% infection, and 50% of infected plants died as a result of stem girdling.

Control
Direct control of the pathogen is not practical. However, on fertile, frost-prone sites, removal of ground cover around individual plants may reduce the risk of bacterial canker by preventing additional temperature lowering during frosts.

13.6 NEEM

13.6.1 BACTERIAL WILT OF NEEM TREE

Pathogen: *Pseudomonas solanacearum* biovar 3
The disease is reported from Queensland as slow wilting of neem tree following a prolonged dry period.

Symptoms
Black to brown discoloration in the root and stem with a color rot developing in the advanced stages.

The isolate was pathogenic to neem tree and tomato but not to native white cedar.

13.6.2 BACTERIAL BLIGHT OF NEEM TREE

Pathogen: *Pseudomonas azadirachtae*

Symptoms

A bacterial disease of neem causing angular leaf spot, shot hole, and vein blight symptoms was found to occur in a severe form in parts of Rajasthan (India). The causal bacterium cause shot hole. The development of vein blight symptoms might play an important role in the perpetuation of disease from season to season. Leaf spot, blight, and shot hole are the main symptoms in neem foliage (Desai et al., 1966; Srivastva and Patel, 1969).

Predisposing Factors

Frequent rains favored the disease spread. Optimum temperature for disease progress is 22°C–27°C. Bacteria in infected leaves survive in a temperature of 0°C–65°C for 10 days. The Bacterium is host specific.

13.6.3 BACTERIAL LEAF SPOT OF NEEM TREE

Pathogen: *Xanthomonas azadirachtae*

Leaf spot of neem caused by *Xanthomonas azadirachtae* was reported from India by Charkravarti and Gupta (1975).

Symptoms

The symptoms appear on leaves as small translucent, water-soaked lesion, which turn brownish. These lesions gradually increase in size and coalesce together covering the larger leaf area. The diseased tissue frequently falls off giving shot holes. The coalesce lesions are often delimited by veins and veinlets giving angular appearance. The infected leaves eventually become chlorotic dry and defoliate prematurely. Leaflets showing blight symptoms have a narrow band of translucent water-soaked area on either or both sides of midveins or veins arising from them. The pathogen is detected in the vascular bundles of the petiole of blighted leaves. Petioles also have water-soaked streaks full of bacteria.

Geographical Distribution

India (Rajasthan, Maharashtra)

Predisposing Factors

Frequent rains favor the disease spread. Disease appears at a temperature between 22°C and 27°C, while it spreads during rains. The leaf becomes susceptible to infection when it is half the normal size and remains so till senescence.

Control Measures

Control measures for bacterial infections on neem are not yet worked out.

13.7 LEUCAENA

13.7.1 POD ROT OF LEUCAENA

Pathogen: *Pseudomonas fluorescens* biotype II

Symptoms

Under humid conditions, pods and seeds rot rapidly reduce seed production. *Leucaena esculenta*, *L. leucocephala*, and *L. pulverulenta* were more susceptible than *L. diversifolia* and *L. shannoni* in inoculation studies. Seed infection with *P. fluorescens* can be as high as 95%.

Geographical Distribution

Brazil, Colombia, Mexico and Panama, and recently in Guatemala

Control Measures

Not worked out

13.8 CASUARINA

13.8.1 BACTERIAL WILT OF CASUARINA

Pathogen: *Pseudomonas solanacearum*
The pathogen also causes wilt on tomato, eggplant, and capsicum indicating that the pathogen belongs to race 1 and biovar III of *P. solanacearum*.

Symptoms
Bacterial wilt of casuarinas is characterized by yellowing of bottom needles, which gradually proceeds upward and finally it results in wilting and death. The disease was recorded on 2-year-old plantation in Kasargod forest range of Calicut forest division, Kerala, India.

Yellowing of foliage and wilting and death have been reported in China and India (Liang and Chen, 1982). Bacterial wilt of *Casuarina equisetifolia* sapling also been recently reported to cause mortality in young plantation in Kerala (Mohammed et al., 1991). The pathogen mostly attacks the seedlings. Common symptoms include sudden wilt and shriveling of leaves. The root tissue of wilted plants shows brown discoloration of vascular bundles (Brown, 1968).

Geographical Distribution
India, Mauritius, and China

Control Measures
Not worked out

13.9 MORUS

13.9.1 LEAF SPOT OF MORUS

Pathogen: *Pseudomonas mori*

Symptoms
Leaf spot and elongated lesions are observed on old twigs of *Morus alba*, *M. indica*, *and M. nigra*. Bacterial infection results in stunting and dieback of diseased plants. In wet weather, bacteria ooze from infected twigs (Peace, 1962).

Geographical Distribution
Australia, Brazil, East and South Africa, Europe, Japan, Korea, North America, and Turkey

Control Measures
Not worked out

13.10 FLAME OF FOREST

13.10.1 LEAF SPOT OF FLAME OF FOREST

Pathogen: *Xanthomonas butae*

Symptoms
Few to numerous minute, water-soaked areas with brown center and pale yellow halo on leaves; injured stem and rachis of tree are formed. The spot become round to angular and get black with the progress of the disease. In severe infection, the entire leaf surface covered by disease results in premature defoliation (Bhatt et al., 1955).

Geographical Distribution
India (Maharashtra)

Control Measures
Not worked out

13.11   WILLOW AND ALDER

13.11.1   BACTERIAL TWIG BLIGHT OF WILLOW

Pathogen: *Pseudomonas syringae* pv. *syringae*
             Pseudomonas saliciperda

Willow trees are susceptible to bacterial blight, where the disease affects the shoots and leaves of trees. Overwintering on bark, the disease invades the willow tree through openings, such as the tree's natural stomata or wounds. The disease travels throughout the tree through the system that carries water, becoming active in areas of new growth. Encouraged by cool, wet weather and deterred by warm weather, this disease is most prevalent in spring.

Symptoms
Blighted twigs and branches dieback, and leaves turn brown and wilt. Brown streaks occur in sections of affected wood. Severe defoliation may occur. Cankers with longitudinal cracks have also been reported.

Predisposing Factors
Pseudomonas syringae pv. *syringae*, overwinters in twig and branch cankers. Two common genetic traits increase the bacteria's ability to cause disease. Most produce a powerful plant toxin, syringomycin, which destroys plant tissues as bacteria multiply in a wound. Bacteria also produce a protein that acts as an ice nucleus, increasing frost wounds that bacteria easily colonize and expand. The bacteria are ubiquitous and easily found on healthy tissues.

Bacterium overwinters in the cankers, so young leaves are infected as soon as they unfold. The damage can be confused with frost injury.

Disease Management
- Cut out affected twigs and branches if practical.
- Make chemical applications in fall after leaves drop to protect leaf scars.
- Spray with bordeaux 8-8-100 in fall after leaves drop.
- Copper-Count-N at 1 quart/100 gal water.
- CuPRO 2005 T/N/O at 0.75–3 lb/A (or 1–3 Tbsp/1000 sq ft) dormant or at 0.75–2 lb/A when new growth is present.
- Junction at 1.5 lb/100 gal water.

13.11.2   CROWN GALL OF WILLOW

Pathogen: *Agrobacterium tumefaciens*

Symptoms
Willow trees are at risk for crown gall, a bacterium that causes galls to form on the roots and stems. Numerous galls can cause stunting, discoloration, and dieback. Crown gall can also make the tree susceptible to secondary tree diseases that enter decaying galls.

Crown gall is a nursery disease. Large, rough, woody swellings or galls on the lower part of the stem and crown of the plant are formed. The infected plants may be deformed, stunted, or even killed. Weeping willow is susceptible.

Control Measures
No practical control is known for this disease.

13.11.3 Bacterial Leaf Spot and Canker of Willow and Alder

Pathogen: *Pseudomonas syringae* pv. *syringae* van Hall

Symptoms
Leaf spotting, cankers on twigs, branches, and stems of mainly 1- to 2-year-old plants with associated dieback and occasional whole-plant death where cankers form low on the stem are the general symptoms of this disease.

Yellow flecks (1–3 mm diameter) appear on upper and lower leaf surfaces followed by dark-brown spots of irregular outline on both leaf surfaces. Spots may remain discrete or merge forming larger blotches. Terminal and side shoot dieback with blackened buds and collapsed, desiccate foliage. Rough fissures or sunken, blackened lesions appear on stems (cankers). Bacteria ooze from cankers, when dry, appears as a white deposit on the stem.

Leaf symptoms first appear in spring when temperatures is rise and moisture levels are high. Throughout the growing season, during periods of continuous rainfall (>3 days), spots are produced on new leaves. Formation of stem and twig cankers also depends on high moisture levels. Symptoms vary, stems may be roughly fissured or have sunken black lesions oozing bacteria. Sometimes the whole stem is affected, killing the plant; more often only the top one-third to one-half of the plant, or only lateral branches, is involved. Although stem fissures and cankers may heal and the plant continues to grow, this often leaves a weak point that makes the plant susceptible to breakage in high winds. Some severe outbreaks of the disease occur when sudden frosts follow a warmer wet period.

Approximately 50% of *P. syringae* isolates, tested from willows in New Zealand, were capable of causing ice nucleation. Ice-nucleating bacteria initiate the formation of ice crystals within host cells during frosts, and the combination of bacteria and frost causes more extensive tissue disruption than if either factor were present independently.

Control Measures
Control of disease is generally not warranted. Traditionally, copper-based inorganic compounds have been used to reduce bacterial populations but with only limited success.

13.12 OAK

13.12.1 Bacterial Leaf Scorch of Oak

Pathogen: *Xylella fastidiosa*

Bacterial leaf scorch has devastated many landscape and shade trees in Kentucky's urban forests in recent years. Especially hard hit has been the mature pin oaks lining many urban streets. First diagnosed in the United States in the early 1980s, this epidemic shows no signs of abating.

Bacterial leaf scorch (BLS) is a systemic disease caused by the bacterium *Xylella fastidiosa*, which invades the xylem (water- and nutrient-conducting tissues) of susceptible trees. It is most commonly seen in pin, red, shingle, bur, and white oaks but can also affect elm, oak, sycamore, mulberry, sweet gum, sugar maple, and red maple. Xylem-feeding leafhoppers and spittlebugs spread the bacterium from tree to tree. Transmission between trees through root grafts has also been reported. There is no cure for this disease; it is chronic and potentially fatal.

Symptoms
Symptoms of BLS are described as marginal leaf burn (Photo 13.6) and are very similar to drought stress symptoms. In addition to marginal leaf burn, there is a defined reddish or yellow border separating the necrosis from green tissue. Symptoms are more noticeable in late summer after hot, dry conditions, but symptoms can be expressed all year around.

PHOTO 13.6 Bacterial leaf scorch of oak. (Photo courtesy of Edward L. Barnard, Florida Department of Agriculture and Consumer Services, Bugwood.org.)

Symptoms first appear on one branch or section of branches and on the oldest leaves. Each year symptoms will reoccur and progress to other parts of the tree.

Browning of the oldest leaves along their margins begins in mid- to late summer on one branch or a few branches on inner and lower portions of the tree. A wavy, reddish-brown band sometimes develops between the brown and green tissue of the leaf. The browning of leaves progresses to include more leaves toward the ends of branches. Branches and eventually the entire tree dies.

The first noticeable symptom is premature browning of leaves in midsummer. Symptoms worsen throughout late summer and fall. Leaf margins turn brown, beginning with the older leaves and moving outward, spreading to leaves toward the branch tip. In most, but not all infected trees, browned, dead areas of the leaf are separated from green tissue by a narrow yellow border. The browned leaves may drop from the tree. Infected trees leaf-out normally the following year, with leaves on a few more branches turning prematurely brown in late summer. Symptoms become progressively worse over a period of 3–8 years, until the entire tree turns brown prematurely. The lack of green, chlorophyll-producing leaves year after year leads to twig, branch, and limb death due to continual defoliation.

Bacterial leaf scorch can easily be mistaken for oak wilt except for the following:

1. The cycle of bacterial leaf scorch repeats and becomes worse over a long period of time. Oak wilt and Dutch elm disease are both capable of killing susceptible trees within a matter of months.
2. There is no streaking of the sapwood with bacterial leaf scorch.
3. In bacterial leaf scorch, the leaf browning develops from the leaf edges and works toward the mid-vein, whereas browning tends to happen in a more overall, uniform manner with oak wilt.

Bacterial leaf scorch can also be mistaken for drought and heat stress. However, damage by bacterial leaf scorch begins in old leaves and spreads to the branch tips, with browning around the leaf edges. Damage due to environmental stresses tends to cause overall browning to the canopy and to individual leaves. Trees tend to react to environmental stresses soon after damaging conditions occur, whereas bacterial leaf scorch is unique in its timing. Leaf browning is generally not noticed until midsummer and intensifies through late summer and fall.

The only way to confirm the diagnosis of BLS is through laboratory analysis. The best time to test for the presence of this disease is in late summer or early fall, when the bacterial count is at its highest.

Host Range and Location

Bacterial leaf scorch is found throughout much of the eastern and southern United States. In Kentucky, it is present in landscape trees in many urban areas, including Paducah, Madisonville,

Owensboro, Bowling Green, Somerset, Louisville, and Lexington. This disease has not been detected in Kentucky's forest trees in oaks, especially pin oak and red oak, and in sycamore in Kentucky. It is also occasionally found here in red maple, sugar maple, silver maple, London plane, hackberry, mulberry, elm, and sweet gum.

Disease Cycle

Infected leafhoppers and spittlebugs feed on the succulent, terminal shoots of susceptible host trees, transmitting the bacteria. Xylem vessels become clogged with bacterium as it travels within, multiplying, and infecting other parts of the tree. There are no viable control options for the insect vectors. The cold-sensitive bacteria overwinter in protected areas within the xylem of the tree, and their populations begin to climb again as the next growing season progresses.

This bacterium is spread by leafhoppers and treehopper insects, although it does not appear to spread from tree to tree very rapidly. Nevertheless, in some neighborhoods where the disease has been present for many years, a high proportion of mature oaks may show symptoms of BLS. Little is known about which of these leafhopper vectors are active in Kentucky. There is some evidence that *X. fastidiosa* is present in symptomless shrubs, grasses, and weeds in the landscape. Thus, leafhoppers may not necessarily pass the disease from tree to tree but may acquire the bacterium from other hosts. The pathogen infects the xylem where it partially blocks the flow of water to the leaves, resulting in leaf scorch symptoms. Some researchers working with this disease suggest that leaf scorch symptoms are more severe during times when other stresses are placed on the tree. Timing of symptom development in mid- to late summer in urban trees is often associated with various moisture and heat stresses occurring that season.

Favorable Conditions

The disease is spread by leafhoppers, spittlebugs, and through root contact with neighboring trees. Since the pathogen is harbored within insects, warm temperatures and high populations of leafhoppers and spittlebugs are conducive for BLS.

Control Measures

Once a host becomes infected with BLS there is no cure. Efforts to reduce disease spread by vector control have not been effective. Maintaining the health of the tree through proper mulching and irrigation practices may delay and suppress symptoms by assuring adequate moisture availability for the plant. Secondary pests, including canker diseases, borers, and bark beetles should be monitored and controlled as needed. There is no data to suggest that pruning diseased limbs or immediate removal of diseased trees reduces the incidence of new infections.

Remove severely infected trees and replant using resistant species. Control weeds (to minimize insect populations) and ensure proper tree fertility and irrigation to maintain health and vigor.

Integrated Pest Management Strategies

1. *Maintain plant vigor.* There is no cure for the disease. Keeping susceptible trees healthy and thriving can help them resist infection and survive longer once they are infected.
2. *Practice good sanitation.* Branches that have died due to BLS should be routinely removed. Infected trees that are in a severe state of decline should also be removed. Disinfect pruning tools with a 10% bleach solution between pruning cuts.
3. *Plant resistant species.* In areas where BLS has occurred, avoid planting highly susceptible trees.
4. *Antibiotic injections*: Oxytetracycline (OTC) root flare injections applied in spring can reduce bacterium levels and delay symptoms by a couple of weeks. They are expensive, need to be reapplied each year, and possible damage resulting from long-term use is unknown. A certified arborist should be contacted while considering injections. Trees in the early stages of BLS respond best to yearly OTC injections. The suppression of BLS by the use of OTC treatments is only temporary.

13.12.2 Bacterial Wetwood or Slime Flux of Oak

Pathogen: *Various bacteria*

Symptoms

A sour odor is often associated with wetwood as water-soaked wood with large numbers of dead bacteria begin to break down. The buildup of bacterial populations within the tree causes fermentation resulting in internal gas pressure of up to 60 lb/in.^{2}. Foliage sometimes wilts and branches may dieback. However, most of the time, wetwood is a minor problem that leaves a vertical streak on the bark where pressurized liquid escapes out of wounds (Photo 13.7). Many times, secondary fungi and bacteria infect the surface liquid and create a slimy texture on the bark.

Dark streaks of sap, usually foul smelling, ooze from holes or cracks in the bark (Photo 13.8). The heartwood is discolored dark brown. Pin oaks are especially prone to wetwood.

PHOTO 13.7 Bacterial wetwood and slimeflux on oak tree. (Photo courtesy of Randy Cyr, Greentree, Bugwood.org.)

PHOTO 13.8 Aspen/poplar (Populus spp.) Cottonwood tree showing symptoms of infection with bacterial wetwood. (Photo courtesy of William Jacobi, Colorado State University, Bugwood.org.)

Favorable Conditions

Bacteria that cause wetwood tolerate low oxygen and are often found in soils and on plant surfaces. Bacteria enter through assorted wounds above and below the soil line. The bacteria may lay dormant during the greatest periods of growth and become active in mature or older tissues.

Control Measures

There are no known controls for bacterial wetwood. A 10% bleach solution may be used cosmetically to clean stains off the bark.

13.13   ELM

13.13.1   BACTERIAL LEAF SCORCH OF ELM

Pathogen: *Xylella fastidiosa*

This disease is caused by the bacterium *X. fastidiosa*, which infects and clogs the water-conducting tissues of the tree. Infection by this bacterium causes a slow decline over many years. Once a tree is infected, symptoms recur annually.

Symptoms

Symptoms of scorch are irregular browning along the leaf margin with a yellow border between green and scorched leaf tissue (Photo 13.9). Older leaves on a branch are affected first (Photo 13.10).

PHOTO 13.9   Symptoms of bacterial leaf scorch on elm leaves. (Photo courtesy of Sandra Jensen, Cornell University, Bugwood.org.)

PHOTO 13.10   Symptoms of bacterial leaf scorch on elm leaves. (Photo courtesy of Brian Olson, Oklahoma State University, Bugwood.org.)

Damage is initially observed on single branches and later spreads to the entire crown, the oldest leaves being affected first. Leaves appear brown along margin, with a yellow halo. Symptoms appear in summer and early fall. No discoloration in sapwood. No discoloration of inner bark. No wintergreen odor.

Control Measures
The control measures are the same as applied for leaf scorch of oak.

13.13.2 Bacterial Wetwood or Slime Flux of Elm

Pathogen: *Enterobacter cloacae*

Bacterial wetwood, also called slime flux, is a major bole rot of trunk and branches of trees. Slime flux has been attributed to bacterial infection in the outer sapwood and inner heartwood area of the tree. The bacterial infection is normally associated with wounding or environmental stress. The bacterium *Enterobacter cloacae* is determined to cause wetwood in elm, but numerous other bacteria have been associated with this condition in other trees such as cottonwood, willow, ash, maple, birch, hickory, beech, oak, sycamore, cherry, and yellow poplar.

Symptoms
A tree with slime flux is water soaked and "weeps" from visible wounds and even from healthy looking bark (Photo 13.11). The "weeping" may be a good thing as it has a slow, natural draining effect on a bacterium that needs a dark, damp environment. A tree with this bole rot is trying its best to compartmentalize the damage.

This bacterium alters wood cell walls, causing moisture content of the wood to increase. One interesting thing is that the weeping liquid is fermented sap, which is alcohol based and toxic to new wood.

Control Measures
Several United States Forest Service books say not to bore holes to drain the rotting wood as it will further spread the bacterium. There is some debate about this practice. Actually, nothing can stop further rot except the tree's ability to isolate the spot by growing good wood around the diseased portion.

Using an insecticide will not help prevent the rot going inside. One can see secondary insects feeding on sap and the rotting remains but they do not affect the disease process. It is not as though they spread the infection. Don't waste money on spraying for insects.

PHOTO 13.11 Bacterial wetwood and slime flux of elm. (Photo courtesy of William Jacobi, Colorado State University, Bugwood.org.)

13.14 WILD CHERRY

13.14.1 Bacterial Canker of Wild Cherry

Pathogen: *Pseudomonas syringae* pv. *morsprunorum*

Bacterial canker is one of the most important diseases of wild cherry (*Prunus avium* L.). It is a major limitation for timber production from wild cherry. Susceptible trees include *P. avium* (wild cherry) and its ornamental and fruiting varieties and other *Prunus* sp. including plums, peaches, and apricots.

Symptoms

Symptoms include cankers on twigs, branches, and/or trunk, gum exudation, dieback, and leaf spots.

Symptoms usually consist of scattered shoots, which fail to flush in the spring, but sometimes this can affect entire branches or even whole young trees. Close inspection will reveal large areas of dead and dying bark, usually girdling shoots or branches, and an amber-colored gum also exudes from affected areas.

This disease can be caused by two pathovars of *Pseudomonas syringae*: pv. *morsprunorum* (*Psm*) and pv. *syringae* (*Pss*). Recently *Pss* and/or intermediate forms between *P. syringae*: pv. *morsprunorum* and pv. *syringae* were also found in sweet and wild cherry in the United Kingdom during a survey of woodland plantations and nurseries in 2000/01 (Vicente et al., 2004).

Disease Cycle

During the summer the bacteria that cause the cankering can be found on the cherry leaves causing brown spots. Cracks develop around the brown spots, so the discolored tissue falls out to leave the so-called shot-hole symptom. The bacterium also infects the bark via scars left after leaf fall and through any injuries. It is apparently inactive during winter but in early spring can grow rapidly in the bark causing cankers.

Disease Management

Cultural control tactics supplemented with chemical control can be followed. Traditional recommendations encourage the first spray in October before rains and again in early January. Copper-based products have not worked well under conditions favorable for disease development.

13.15 ASH

13.15.1 Bacterial Canker of Ash

Pathogen: *Pseudomonas syringae* ssp. *savastanoi* pv. *fraxini*

Symptoms

First symptoms are red-brown, more or less lens-shaped blisters on the bark of younger branches or the stems of young trees. These occur at wounds, at feeding sites of the twig miner *Prays fraxinella*, and at lenticels. Early stages of infection can only be detected by close inspection, whereas the later stages are more conspicuous. The blister surface is ultimately split by vertical and lateral cracks to reveal brown-black, proliferated, necrotic bark tissue. The swollen bark tissue forms into large excrescences that contain bacterial cavities surrounded by cork layers (formed as part of the plant defense mechanism) and necrotic, black bark parenchyma. The cambium is destroyed, leading to the formation of open wounds. Proliferated, thickened bark is still present at the margins (Photo 13.12).

Secondary spread of the disease, perhaps promoted by frost damage, produces long, vertical cracks on the trunk, sometimes more than 1 m in length. Infection makes the trees vulnerable to attack by other pests and diseases, such as bark beetles and fungi, for example, *Nectria galligena*, exacerbating dieback. The growth of young trees may be severely restricted and eventually they

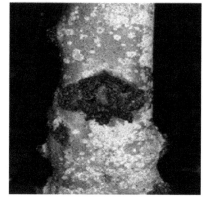

PHOTO 13.12   Bacterial canker of ash tree. (Photo courtesy of Thomas L.Cech, BFW Austria.)

may die. Larger trees only show restricted growth of branches and deformation of the wood, especially when the open canker form is prevalent (van Vliet, 1931; Riggenbach, 1956; Janse, 1981, 1982a). The limited necrotic excrescences on *Fraxinus excelsior* and *Olea europaea* caused by pv. *fraxini* strains are considered to be the result of the low amounts of IAA produced and the absence of cytokinin production (Alvarez et al., 1998; Iacobellis et al., 1998).

Host
Ash

Geographical Distribution
Europe and Russia

Disease Cycle
The bacterium can apparently persist for many years in the thickened, infected bark, slowly spreading into surrounding healthy bark each growing season. In spring or early summer, bacteria ooze out of infected bark as a yellowish slime, but little is known about how infection occurs.

Prevention and Control
Unless trees are grown for wood production or are significant as ornamental plants, control is not generally necessary. Complete control would be by eradication of infected trees. However, because the disease is usually not severe, pruning to remove seriously affected branches is practically effective. Chemical control of *P. savastanoi* pv. *fraxini* is not practicable. An evaluation of susceptibility of species and cultivars showed that *F. americana*, *F. pennsylvanica*, and *F. excelsior* cultivars "Geesink" and "Atlas" and clone 04.72 were found to have low susceptibility, whereas other *F. excelsior* cultivars or clones had moderate to high susceptibility to the pathogen (van Dam and van der Voet, 1991).

13.16   OLEANDER

13.16.1   Leaf Scorch of Oleander

Pathogen: *Xylella fastidiosa*

Oleander leaf scorch is found mainly in southern California. It is caused by the bacterium *X. fastidiosa*, which is the same species (although a different strain) that causes Pierce's disease of grapevines and almond leaf scorch. The strain of *X. fastidiosa* that causes oleander leaf scorch will not cause Pierce's disease, so removing oleanders will not reduce the source of *X. fastidiosa* that can affect grapes. As with other diseases caused by *X. fastidiosa*, the bacterium is vectored by insects, primarily sharpshooters, which feed on the water-conducting tissue (xylem) of the plant.

This disease was first noticed on oleanders in the Palm Springs-Indio area of Riverside County and in Tustin (Orange County) in the early 1990s and has spread to other parts of southern California including Santa Barbara, Ventura, San Diego, San Bernardino, and Los Angeles counties. Evidence to date suggests that the disease resulted from the introduction of a strain of *X. fastidiosa*, which was new to California. The disease has now been reported across the southern United States. While the disease has not yet been recorded north of Santa Barbara County, it is believed that it could spread north through California's Central Valley and along the coast where the glassy-winged sharpshooter is established. Oleanders affected by this disease decline and then die, usually within 3–5 years of the first symptoms.

Symptoms

Symptoms can be expressed year-round, although they may be more noticeable in late spring and summer; they develop more quickly in warm weather. Leaves on one or more branches may yellow and begin to droop; soon the margins of the leaves turn a deeper yellow or brown, and the leaves eventually die (Photo 13.13). As the disease progresses, more branches of the plant are affected and the plant dies. Symptoms are much more severe and develop more rapidly in hot interior valleys than in cooler coastal areas.

Symptoms of this disease are often confused with those caused by drought. However, under limited water conditions leaves on all branches of a healthy plant yellow and droop at the same time. Drought-stressed leaves yellow uniformly or along the central leaf vein, whereas in leaf scorch disease yellowing of leaves progresses from the tip or margins of leaves inward. Also, unless the drought is severe, the plant recovers when watered. An oleander infected with *X. fastidiosa* will not recover when watered because bacteria plug the xylem tubes and limit the flow of water to the affected branches.

Marginal browning of leaves can also be associated with salt or boron toxicity. In these cases leaves usually do not droop, and the symptoms are more noticeable in older leaves.

Plant diagnostic laboratories that test for Pierce's disease in grapevines can also detect *X. fastidiosa* in infected oleander. A soil or tissue test can help determine if the symptoms are caused by an excess of minerals. If salt toxicity is the problem, plants will improve if salts are leached through the soil and below the root zone, whereas no improvement will be seen in plants infected with leaf scorch bacteria.

PHOTO 13.13 Leaf scorch symptoms on oleander leaves. (Photo courtesy of Pompilid, wikipedia.org.)

Bacterial Transmission

Xylella fastidiosa growth in plants is limited to the xylem and is spread from plant to plant by xylem-feeding insect vectors. The dominant vector in southern coastal California is the glassy-winged sharp-shooter (*Homalodisca vitripennis* [*formerly H. coagulata*]), which was first identified in California in 1990. This insect acquires the bacteria from infected plants while feeding on the xylem sap. The bacteria replicate in the sharpshooter's mouth, so that once a sharpshooter adult acquires the bacteria, it is infectious for life. Nymphs lose the bacteria when they molt. When the sharpshooter moves to another plant to feed, it takes bacteria with it in its mouthparts and deposits them into the next host plant. When the bacteria enter the xylem, they can then multiply and spread throughout the plant to such an extent that they greatly reduce the movement of water within the plant.

In the Coachella Valley, the smoke-tree sharpshooter (*H. liturata*) is more abundant than the glassy-winged sharpshooter and is the most likely vector of oleander leaf scorch bacterium. Other sharpshooters such as the blue-green sharpshooter (*Graphocephala atropunctata*) may also spread the bacteria from plant to plant. The blue-green sharpshooter is often common in irrigated ornamental landscapes in coastal California.

Vector Identification and Biology

The glassy-winged sharpshooter is nearly 1/2 in. long with transparent wings; it is larger than most other sharpshooters found in California. The insect is dark brown on top and has a slightly lighter underside. The top of the arrowhead-shaped head is stippled with ivory or yellow dots. A large white spot is frequently present on each forewing of reproductively mature females, although the insect may rub these spots off when it lays eggs. The glassy-winged sharpshooter excretes large amounts of liquid when feeding, and on heavily infested plants, excrement gives the leaves or fruit a white-washed appearance. The smoke-tree sharpshooter is similar in appearance to the glassy-winged sharpshooter, but its head has distinct yellow wavy lines instead of dots.

There are two generations per year of glassy-winged sharpshooter in California. Glassy-winged sharpshooters overwinter as adults and lay eggs in spring. The first generation develops from late spring to early summer, with egg-laying beginning in midsummer. The second generation matures into the overwintering adults starting in late summer.

Disease Management

There is no known cure for oleander leaf scorch. Pruning out the part of the plant showing symptoms may help to prolong the appearance of the oleander tree or shrub but will not save the plant. The bacteria by then have already moved throughout the plant via the xylem, and limbs that show symptoms are only the first to become affected. Research indicates that some cultivars of oleander may express symptoms to lesser degrees than others and may live longer than other varieties when infected.

Because of the year-round abundance of the glassy-winged sharpshooter, currently available insecticides are not effective in stopping the spread of the disease. The best management may be early removal of plants infected with the oleander leaf scorch bacteria to reduce the source of inoculum, but there are no experimental data to validate this method. Although only a few plant species have been tested as hosts of the oleander leaf scorch strain of *X. fastidiosa*, it is possible that other plant species may harbor the bacteria without showing disease symptoms.

13.17 FRUIT-BEARING FOREST TREES

13.17.1 Fire Blight of Fruit-Bearing Forest Trees

Pathogen: *Erwinia amylovora*

Fire blight is a serious disease of apple and pear. This disease occasionally damages fruit-bearing forest trees like cotoneaster, crabapple, hawthorn, mountain ash, ornamental pear, firethorn, plum quince, and spiraea and can affect many parts of a susceptible plant but generally noticed first on damaged leaves.

Symptoms

The name "fire blight" describes the most characteristic symptoms of this disease: a brown-to-black scorched appearance of twigs, flowers, and foliage. It is usually seen first in spring when blossoms and fruit spurs appear water soaked, wilted, and shriveled and finally turn brown to black. Shoot blight occurs when infections begin at shoot tips, moving rapidly down the shoots and then to limbs and trunk. Crabapple leaves turn brown. Frequently, the tip of the blighted shoot bends over and resembles a shepherd's crook. The bacterium *E. amylovora* overwinters in trunk and branch cankers. In the spring, the bacteria resume multiplication when temperature is above 17°C.

Disease Control

A combination of pruning, reduced fertilization, and chemicals can help control fire blight. Prune and remove all stems showing symptoms as they first appear, and cut back into the healthy portion of both stems and limbs. Too much fertilizer application will cause rapid new growth, which is most susceptible to the blight. Moderate fertilizer to reduce rapid tree or shrub growth. Over-pruning can have the same effect; hence one should refrain from heavy pruning. The antibiotic streptomycin is the most effective spray material for controlling fire blight. It will prevent but not control infection on use of streptomycin in spring during bloom.

13.17.2 CROWN GALL OF FRUIT-BEARING FOREST TREES

Pathogen: *Agrobacterium tumefaciens*

Symptoms

Galls form on roots and stems especially at the root collar or root crown region. Aerial galls are common on highly susceptible plants like poplar, willow, and euonymus. Young plants with numerous galls tend to be stunted and predisposed to drought damage or winter injury. Floral display or fruit production may be suppressed. Damage is most severe when galls encircle the root crown, but few plants are killed by crown gall alone. Severely diseased plants are subject to attack by secondary pathogens that enter through decaying galls.

Geographical Distribution

Crown gall occurs around the world, causing economically significant damage to fruit and nut trees and ornamental plants. *Agrobacterium tumefaciens* has the broadest host range of any bacterial plant pathogen. More than 600 plant species in over 90 families are susceptible, although relatively few species sustain significant damage. Some of the common hosts include apple, blackberry, cherry, euonymus, forsythia, peach, pear, plum, poplar raspberry, rhododendron, and willow.

Disease Cycle

Crown gall bacteria are dispersed in soil or irrigation water, on horticultural implements, and on or within plants. Plants with latent infection are a big problem, because dormant nursery stock that became infected at the time of harvest may not develop galls until after planting in a new location. Tumor development is most favored by temperatures around 21°C. Temperature above 30°C prevents transformation of normal cells to tumor cells, but does not prevent gall growth after transformation.

Disease Management

If only a few galls are present, cut off and destroy the stems on which they occur. Dip pruning shears in a 10% bleach solution between prunes to prevent the spread of the bacteria. Heavily infected plants should be removed and destroyed.

13.17.3 BACTERIAL LEAF SPOTS OF FRUITS-BEARING FOREST TREES

Pathogen: The most common bacterial pathogens that cause leaf spots include *Pseudomonas* and *Xanthomonas*.

Symptoms
Bacterial leaf spots may appear similar to fungal leaf spots, so it may be difficult to distinguish between the two. Leaf spots caused by bacteria are often initially light green and look water soaked. Later, these leaf spots turn brown or black and may have definite margins.

Disease Cycle
Bacteria are often splashed from the soil onto wet foliage, where they enter a leaf through stomata or wounds. Thereafter, the bacteria spread from leaf to leaf when plants are watered in nurseries or during rainy periods.

Disease Management
Fungicides containing copper may be effective against BLS, but copper may burn the foliage of some plant species. To prevent spread of the bacteria, avoid overhead irrigation in nurseries. Pick off and destroy infected leaves.

REFERENCES

Alvarez, F., García de los Ríos, J.E., Jimenez, P. et al. 1998. Phenotypic variability in different strains of *Pseudomonas syringae* subsp. *savastanoi* isolated from different hosts. *Eur. J. Plant Pathol.*, 104(6), 603–609.

Bakshi, B.K. 1975. *Forest Pathology. Principles and Practice in Forestry*. Controller of Publications, New Delhi, India, pp. 134–147.

Bhatt, V.V., Patel, M.K., and M.J. Thirumalachar. 1955. Two new *Xanthomonas* species on legumes. *Indian Phytopathol.*, 8, 136–142.

Brown, F.G. 1968. *Pests and Diseases of Forest Plantation Trees*. Clarendon Press, Oxford, U.K., p. 1330.

Desai, S.G., Gandhi, A.B., and M.K. Patel. 1966. A new bacterial leaf spot and blight of *Azadirachta indica*. *Indian Phytopathol.*, 19, 322–323.

Iacobellis, N.S., Caponero, A., and A. Evidente. 1998. Characterization of *Pseudomonas syringae* ssp. *savastanoi* strains isolated from ash. *Plant Pathol.*, 47(1), 73–83.

Janse, J.D. 1981. The bacterial disease of ash *Fraxinus excelsior* caused by *Pseudomonas syringae* subsp. *Savastanoi* pv. *fraxini*. 11. Etiology and taxonomic considerations. *Eur. J. Forest Pathol.*, 11, 425–438.

Janse, J.D. 1982a. *Pseudomonas syringae* subsp. *savastanoi* (ex Smith) subsp. *nov.*, nom. rev., the bacterium causing excrescences on Oleaceae and *Nerium oleander* L. *Int. J. Syst. Bacteriol.*, 32, 166–169.

Liang, Z.E. and B.Q. Chen. 1982. Preliminary study on the susceptibility of *Casuarina equisetifolia* to *Pseudomonas solanacearum* and its relation to the permeability of the cell membrane and peroxidase isoenzymes. *J. South China Agric. Coll.*, 3(2), 28–35.

Nakabonge, G., Coutinho, T.A., Roux, J. et al. 2003. Bacterial blight of eucalyptus tree in Ugania. *Tree Protect. News*, 5, P1–P16.

Old, K.M., Wingfield, M.J., and Z.Q. Yuan. 2003. *A Manual of Diseases of Eucalypts in South-East Asia*. Australian Centre for International Agricultural Research (ACIAR), Canberra, Australia. Center for International Forestry Research, Bogor, Indonesia.

Peace, T.R. 1962. *Pathology of Trees and Shrubs with Special Reference to Britain*. Clarendon Press, Oxford, U.K., p. 753004.

Riggenbach, A., 1956. Untersuchungüber den Eschenkrebs. *Phytopathol. Z.*, 27, 1–40.

Roux, J. 1999. Pathology survey of plantation forest trees in southern Uganda. A report prepared for the Ugandan Forestry Department and SFD. Tree Pathology Cooperative Programme, University of Pretoria, South Africa.

Roux, J. and B. Slippers, 2007. Entomology and pathology survey with particular reference to *Leptocybe invasa*, July 23–26. Report submitted to the Sawlog Production grant Scheme (SPGS) Uganda.

Roux, J., Wingfield, M.J., Wingfield, B.D. et al. 1999. A serious new disease of Eucalyptus caused by *Ceratocystis fimbriata* in Central Africa. *Forest Pathol.*, 30, 175–184.

van Dam, B.C. and H. Van Der Voet. 1991. Testing *Fraxinus excelsior, Fraxinus americana*, and *Fraxinus pennsylvanica* for resistance to *Pseudomonas syringae* subsp. *savastanoi* pv. *fraxini. Eur. J. Forest Pathol.*, 21(6–7), 365–376.

van Vliet, J.I. 1931. Ash cankers and their structure. Thesis, University of Utrecht, Hollandia Drukkerij, Baarn, Utrecht, the Netherlands, p. 73.

Vicente, J.G., Alves, J.P., Russell, K. et al. 2004. Identification and discrimination of *Pseudomonas syringae* isolates from wild cherry in England. *Eur. J. Forest Pathol.*, 110, 337–351.

SUGGESTED READING

Anonymous (2015) Bacterial canker of ash (*Pseudomonas savastanoi* pv. *fraxini*). (n.d.). Retrieved July 20, 2015, from http://www.plantwise.org/KnowledgeBank/Datasheet.aspx?dsid=45003.

Anonymous, 2015. *Xanthomonas axonopodis* pv. *manihotis* (cassava bacterial blight). June 11, 2015. Retrieved July 20, 2015, from http://www.cabi.org/isc/datasheet/56952.

Blua, M.J., Phillips, P., and R.A. Redak. 1999. A new sharpshooter threatens both crops and ornamentals. *Calif. Agric.*, 53(2), 22–25.

Chakravarti, B.P. and D.K. Gupta. 1975. A nonpigmented strain of *Xanthomonas azadirachtii* Moniz and Raj causing leaf spot of neem (*Azadirachta indica* A. Juss.). *Curr. Sci.*, 44, 240–241.

Diatloff, A., Wood, B.A., and D.G. Wright. 1993. Bacterial wilt of neem tree caused by *Pseudomonas solanacearum. Austr. Plant Pathol.*, 22(1), 1.

Haworth, R.H. and A.G. Spiers. 1988. Characterisation of bacteria from poplars and willows exhibiting leaf spotting and stem cankering in New Zealand. *Eur. J. Forest Pathol.*, 18, 426–436.

Janse, J.D. 1982b. The bacterial disease of ash (*Fraxinus excelsior*), caused by *Pseudomonas syringae* subsp. *savastanoi* pv. *fraxini*. III. Pathogenesis. *Eur. J. Forest Pathol.*, 12, 218–213.

Kam, M. de. 1982. Damage to poplar caused by *Pseudomonas syringae* in combination with frost and fluctuating temperatures. *Eur. J. Forest Pathol.*, 12, 203–209.

Mohamed Ali, M.I., Anunadha, C.S., and J.K. Sharma. 1991. Bacterial wilt of Casuarina equisetifolia in India. *Eur. J. Forest Pathol.*, 21(4), 234–238.

Purcell, A.H. 2006. *Xylella fastidiosa* Web site. University of California, Berkeley, CA. Accessed April 2006. http://nature.berkeley.edu/xylella/index.html.

Ramstedt, M., Astrom, B., and H.A. von Fricks. 1994. Dieback of poplar and willow caused by Pseudomonas syringae in combination with freezing stress. *Eur. J. Forest Pathol.*, 24, 305–315.

Roux, J., Coutinho, T.A., Mujuni Byabashaija, D., Wingfield, M.J. (2001). Diseases of plantation Eucalyptus in Uganda. *South African Journal of Science*, 97: 16–18.

Srivatsava, S.K. and P.N. Patel. 1969. Epidemiology of bacterial leaf spot blight and shot hole disease of neem in Rajasthan. *Indian Phytopathol.*, 22, 237–244.

Srivastva, S.K. 1970. Symptoms of bacterial disease of neem. *PANS Pest Articles and Summaries*, 16(3), 518–521.

Tsing-che Lo, Dati-Wuchen and Jeng-Sheng Huang. 1966. A New Disease (Bacterial Wilt) of Taiwan Giant Bamboo 1. Studies on the Causal Organism (*Erwinia sinocalami* sp. Nov.). *Botanical Bulletin of Academia Sinica*. 7(2):14–22.

14 Bacterial Diseases of Lawn Grasses

14.1 TURFGRASS

14.1.1 BACTERIAL WILT IN TURFGRASS

Pathogen: *Xanthomonas translucens* pv. *poae*
Bacterial wilt was a major problem in cultivar *Toronto* of creeping bentgrass, but continues to be an issue on annual bluegrass as well in many regions of North America.

Outbreaks tend to occur during extended periods of rainfall. The disease can be devastating if prolonged rainfall is followed by bright sunny days and warm weather. The disease is most devastating where sand top dressing programs are implemented.

Symptoms
Bacterial wilt is characterized by tiny red-copper-colored spots first appearing about the size of a dime. As more plants die, spots become larger (Photo 14.1). Small, yellow leaf spots, or streaked tan to dark brown spots, or dark green water-soaked lesions, or shriveled blue to dark green leaves, and yellow elongated leaves are all symptoms that have been associated with bacterial wilt. Numerous small, pit-like or speckled spots, about 0.25–0.75 in. (0.5–2 cm) in diameter, may develop on greens. The patches of dead wilted plants can be observed in lawns of turfgrass.

Control Measures
Not worked out. However, copper bactericidal spray may stop further spread of infection in lawns.

14.1.2 BACTERIAL WILT AND DECLINE OF CREEPING BENTGRASS

Pathogen: *Xanthomonas campestris* pv. *graminis*
Bacterial wilt and decline occurs from late spring through summer and early fall during warm, sunny weather. Infection and disease development is favored by closely mowed turf and warm, sunny days with cool-to-warm nights and heavy rainfall or watering.

Symptoms
The leaves on individual plants start to wilt from the tip down. Within a few days the entire leaf wilts, turns blue-green, and becomes shriveled and twisted. At this early stage, root and crown tissues appear white and healthy. The leaves soon become reddish brown followed by a discoloration of the roots and crown. Death and decomposition of the entire plant soon follow. Symptoms of bacterial wilt and decline can be confused with anthracnose, red leaf spot, and leaf smuts.

The disease spreads through an infected golf green, killing susceptible plants in just a few days, while apparently immune or highly resistant grass cultivars and species remain unaffected. A diseased golf green characteristically has an uneven mottled appearance with areas of green resistant grass among the withered and dead diseased turf. When a recently infected and freshly cut leaf or stem is examined in a drop of water under a light microscope (at about 100×) a white "cloud" of bacteria can be seen oozing from the cut surface within a few seconds.

PHOTO 14.1   Bacterial wilt of turfgrass. (Photo courtesy of Frank S. Rossi, Cornell University, Ithaca, NY, turfnet.com.)

Host

Include perennial and annual (Italian) ryegrasses, bluegrasses, tall and meadow fescues, orchard-grass, timothy, quackgrass, and possibly others. Preliminary evidence in the United States indicates that the bacterium may be adapting itself to other grasses, such as bluegrasses.

Disease Cycle

The bacterium overwinters in diseased plants and thatch and is disseminated by rain splash or flowing water, by physical transmission on mowers, hoses, other turfgrass equipment, or shoes, and by planting infected sprigs, sod, or plugs. The bacteria enter healthy grass plants through wounds such as leaf tips cut by mowers. Originally, the bacteria are located in leaf tissue but soon occupy the water-conducting vessels (xylem) in the roots and crown where they quickly multiply. Masses of bacteria in the vessels prevent normal movement of water within infected plants causing them to wilt and die. The disease cycle of bacterial wilt and decline of creeping bentgrass is illustrated in Figure 14.1.

Control Measures

Chemical Control

Disease reduction has been achieved on Toronto creeping bentgrass greens using the antibiotic oxytetracycline (sold as Mycoshield, C. Pfizer Corp). Preventative control appears possible when the turf is drenched with antibiotic and at least 50 gal of water per 1000 ft^2. Applications are needed at 3–4-week intervals throughout the spring and fall when the disease is active. Directions on the container label should be carefully followed. Chemical control, however, is presently too expensive and time-consuming to be practical over a long period.

Cultural Practices

The use of one or more resistant cultivars should eliminate the need to apply repeated antibiotic drenches to susceptible turf. The disease is usually most severe where the bentgrass is cut at 1/8–3/16 of an inch, less severe at 1/4 in., and uncommon on the collars of greens cut higher than 1/4 in.

In the future, bacterial wilt and decline is likely to infect other bentgrass cultivars as well as other species of turfgrass. The best long-term solution to the problem is to use resistant turfgrasses; however, none are currently being developed for use in the United States.

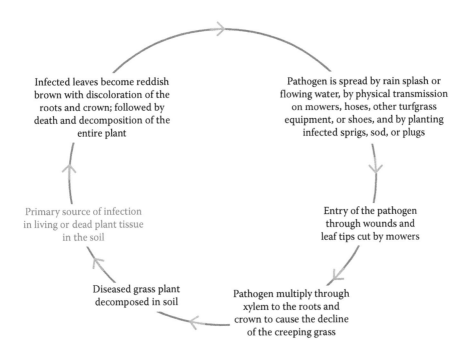

Infected leaves become reddish brown with discoloration of the roots and crown; followed by death and decomposition of the entire plant

Pathogen is spread by rain splash or flowing water, by physical transmission on mowers, hoses, other turfgrass equipment, or shoes, and by planting infected sprigs, sod, or plugs

Primary source of infection in living or dead plant tissue in the soil

Entry of the pathogen through wounds and leaf tips cut by mowers

Diseased grass plant decomposed in soil

Pathogen multiply through xylem to the roots and crown to cause the decline of the creeping grass

FIGURE 14.1 Disease cycle of bacterial wilt and decline of creeping bentgrass caused by *Xanthomonas translucens* pv. *poae.*

SUGGESTED READING

Anonymous. 2015. Bacterial Wilt *Xanthomonas translucens* pv. *poae*. (n.d.). Retrieved June 22, 2015, from http://www.msuturfdiseases. net/details/_/bacterial_wilt_12/.

Arielle Chaves and Nathaniel Mitkowski. 2013. Virulence of *Xanthomonas translucens* pv. *poae* Isolated from Poa annua. *Plant Pathol J.* 29(1): 93–98. doi: 10.5423/PPJ.NT.08.2012.0127.

15 Bacterial Pathogens of Phytosanitary Risk in International Trade of Seed and Planting Stock

15.1 SIGNIFICANCE OF INTERNATIONAL TRADE OF SEED AND PLANTING STOCKS

The introduction of plant pathogens in new areas is largely influenced by international exchange of seed and planting stock. In fact, global trade and exchange has contributed to the dispersal of many pathogens into regions of the world where they previously did not exist (Zadoks, 2008). Also, the movement of people from and between low- and middle-income countries, carrying their own food and dodging border controls, may contribute to the spread of pathogens.

The study of international trade in agricultural products has developed rapidly over the past 50 years. In the 1960s, the disarray in world agriculture caused by domestic price support policies became the focus of analytical studies. These followed attempts to measure the distortions caused by policies in developing countries and to model their impact on world agricultural markets.

> A staggering $1.1 trillion worth of agricultural products are traded internationally each year, with food accounting for more than 80 per cent of the total. In this more and more globalized world, we need to increase our efforts to protect food security and the environment, and ensure safe trade from pests of plants. A failure to monitor the spread of plant pests and diseases can have disastrous consequences on agricultural production and food security for millions of poor farmers. Learning from past experiences, preventions are the first line of defense against plant pests and diseases and they have also proven the most cost effective ways.
>
> **FAO (2015)**

FAO estimates that between 20% and 40% of global crop yields are reduced each year due to the damage wrought by plant pests and diseases.

Intercontinental spread of emerging plant diseases is one of the most serious threats to world agriculture (Mazzaglia et al., 2012). Kiwifruits (*Actinidia chinensis and Actinidia deliciosa*) are economically important crops that are grown in several EPPO countries (by order of importance in production: Italy, Greece, France, Portugal, and Spain). In Japan and Korea, bacterial canker has become one of the most serious limiting factors for cultivating kiwifruit. In Italy, it is estimated that the economic losses (including impact on trade) due to *Pseudomonas syringae* pv. *actinidiae* have reached €2 million (EPPO, 2009). Although its precise route of entry into New Zealand remains unknown, *Pseudomonas syringae* pv. *actinidiae* (Psa) impact on the kiwifruit sector is clear: up to $885 million in losses more than 15 years (Mazzaglia et al., 2012; McCann, 2014). Control strategies are being developed against the disease and include preventive measures (e.g., good fertilization, avoidance of overhead irrigation, disinfection of pruning equipment, pruning, and destruction of diseased parts), regular inspections of the orchards for disease symptoms, and the use of healthy planting material. Chemical control has been implemented in Japan (e.g., with copper compounds and antibiotics), but this has lead to the appearance of resistant strains. It seems desirable to better understand the biology of *P. syringae* pv. *actinidiae* in order to develop adequate control strategies in areas where it occurs and to avoid its further spread in Europe (EPPO, 2009).

In the Apulia region at the southernmost tip of Italy several thousand hectares of olive plantations were affected by *Xylella fastidiosa*, also known as olive leaf scorch. The disease's impact comes on top of a particularly bad year for Spanish and Italian olive growers in 2014 due to pests and the weather, with harvests in Italy down by 40%–50%. Spain and Italy account for 70% of Europe's olive output, leading to warnings that olive oil prices rise. European commission (EU) earmarked €7.5 m (£5.9 m) for fighting several pests, including *Xylella*. Some €751,000 of this went to Italy, with the Italian government providing the same amount (Neslen, 2015). It is highly probable that the Apulian isolate originates from Central America because the European countries annually import from Costa Rica millions of ornamental plants belonging to several species, which rarely undergo appropriate phytosanitary inspections at the point of entry into the EU. The recent interceptions of *X. fastidiosa*, by the Dutch, French, German, and Italian Customs and in a number of nurseries, on ornamental coffee plants imported from Costa Rica and Honduras provide further evidence in support of this hypothesis (Digiaro and Valentini, 2015).

EPPO considers *X. fastidiosa* as an A1 quarantine pest (OEPP/EPPO, 1989) and it is also a quarantine pest for Comite Regional de Sanidad Vegetal del Cono Sur (COSAVE). In the EPPO region, it is clear that the grapevine strain of *X. fastidiosa* has the potential to kill large numbers of grapevines and to make areas unfit for growing *V. vinifera*. Its vectors in North America do not occur in the EPPO region, but vector capacity is so nonspecific that one could certainly expect European Cicadellinae (e.g., *Cicadella viridis*) or Cercopidae to transmit the bacterium if introduced. The main danger in the long term is that *X. fastidiosa* could become established in natural vegetation, which would then act as a reservoir for infection of vineyards. It is less likely that Pierce's disease would become a problem in the production of planting material, for it is easily detected and rapidly self-eliminating. Nevertheless, infected planting material could introduce the disease to new areas. *Xylella fastidiosa* is only likely to establish in the warmer parts of the EPPO region such as the southern Iberian and Italian peninsulas and the lowlands of Greece, which have winter temperatures approaching those of the southern United States. However, its potential natural range in Europe may depend on the biology of potential vectors and is accordingly rather difficult to assess. The South American strain on citrus also presents a major risk, the climatic conditions of Mediterranean countries seem favorable for its development. The reported damage in Brazil suggests potential damage even greater than for the grapevine disease. No critical evaluation of possible European vectors has yet been made, but the arguments developed earlier for grapevine should certainly also apply to citrus. The peach strain is relatively less important, but *X. fastidiosa* still presents a definite danger for peaches, almonds, oaks, and, *sensu lato*, for other fruit and amenity trees (Dunez, 1981).

Black rot of crucifers (*Xanthomonas campestris* pv. *campestris*) believed to have been introduced to India with seeds imported from Holland, and other European countries after World War II, prevailed for some years on the hills and then spread to the plains and became established in Indian seed stocks, especially in West Bengal.

15.2 INTERNATIONAL AGREEMENT ON PLANT PROTECTION

International plant protection convention, the first effort toward international agreement on Plant Protection was made in 1914 under the auspices of the International Institute of Agriculture in Rome. This was followed by an International Convention of Plant Protection by over 50 member countries of the institute in 1919, and certain agreements regarding the issue and acceptance of phytosanitary certificates were finalized. The project received a setback due to World War II and was later on revived by the FAO. In post-war period International action in Plant Protection and particularly in plant quarantine was encouraged by FAO with the establishment of the International Plant Protection Convention in 1951. This agreement was constituted with the purpose of securing common and effective action to prevent the introduction and spread of pests and diseases of plants and plant products so as to encourage governments to take all steps necessary to implement its prevention (Ling, 1953).

A national health standard is illegal under the SPS agreement if the WTO decides that it is not "based on scientific principles and is maintained without sufficient scientific evidence." In making this judgment, the WTO examines the extent to which the country has done a scientific assessment of the risk to "human, animal, or plant life or health."Article 5 provides that "Members shall ensure that such measures are not more trade-restrictive than required to achieve their appropriate level of sanitary or phytosanitary protection, taking into account technical and economic feasibility" (WTO, 2000).

The WTO under the SPS agreement may force a nation to choose between lowering its health standards for humans, animals, or plants and paying an international penalty. The penalty can take the form of either compensating the foreign government whose exports to the nation are limited by the stricter standard or permitting that country to impose additional trade restrictions on exports from the nation with the more protective health standards.

Quarantine regulations are mainly aimed at excluding pests and diseases, including phytopathogenic bacteria from a territory. These regulations may be laws, orders, or decrees that limit the import of plants or plant products and specify the pathogens of interest. More specifically, quarantines facilitate the isolation and inspection of plants and plant products for prohibited organisms. This is only used for small-scale importations, or germplasm sent to post-entry quarantine stations for breeding or research purposes. Inspection at the point of import is not adequate due to latent symptom development and undetectable levels of bacteria on planting material. If the host plants are not prohibited, then phytosanitary regulations are based on ensuring that the imported plants are symptomless at the point of origin, as certified by the National Plant Protection Organization (NPPO) of the exporting country. An important principle for ensuring that commodities are pathogen-free is the concept of the pest-free area, which is assigned when a disease has not been observed in a certain country or part of that country. A pest-free area can be created and officially recognized according to international standards, and commodities can be freely exported from it. A similar concept is the "protected zone" that also must be free of a particular bacterium, but must be especially protected from introduction by stricter phytosanitary measures than adjoining areas. Examples of protected zones can be found in Canada and Europe for *Clavibacter michiganensis* subsp. *sepedonicus* and *Erwinia amylovora*, respectively. If a pest-free area cannot be established or maintained, phytosanitary regulations may allow plants or plant products to originate from a production site that has been free from the particular bacterium for a defined period of time. Quarantine regulations should also include the eradication or containment of introduced bacterial pathogens. Additionally, they should be used to ensure the production of healthy planting material by certification schemes and standard production protocols.

Other phytosanitary measures applied to host plants include requirements that the previous generation of the plants, or seeds of the commodity be free of the pathogen, and that planting materials be tested, or subjected to physical or chemical eradicative treatment. Furthermore, the exchange of bacterial strains from culture collections is subject to greater regulation, mostly at the national level, but also through the World Federation for Culture Collections.

International action in plant protection, and particularly in plant quarantine, was encouraged by the FAO with the establishment of the International Plant Protection Convention in 1951. The text of the convention was revised in 1979 and again in 1997; 111 countries are now signatories of the convention. The aim of the convention is to promote and harmonize as far as possible plant quarantine legislation, regulations, and associated practices. The 1997 revised text emphasizes cooperation and exchange of information toward the objective of global harmonization. It includes an agreement with the World Trade Organization on the application of sanitary and phytosanitary measures (WTO-SPS agreement), and this has led to the establishment of International Standards for Phytosanitary Measures (ISPMs). Signatories to the convention are obliged to provide information on the pests present in their countries and to provide official contact points to facilitate information exchange.

Regional action is needed to prevent a pathogen or pest absent from a whole area from being introduced into any part of the area, as its entry into one territory will endanger neighboring countries. Sovereign states, under the International Plant Protection Convention and the Sanitary and Phytosanitary (SPS) Agreement, have the right and responsibility for preparing phytosanitary regulations to protect themselves against exotic plant pests, including bacterial diseases and pathogens. The Interim Commission on Phytosanitary Measures, established by the International Plant Protection Convention (IPPC), develops International Standards on Phytosanitary Measures. These standards establish the principles and procedures of phytosanitary measures, ensuring that countries develop their measures consistently and fairly. In addition, most countries belong to Regional Plant Protection Organizations (RPPOs) that coordinate and harmonize the phytosanitary actions of their member countries on a regional basis.

With regard to bacterial pathogens, the European Plant Protection Organization (EPPO) advises its members which bacteria should be quarantine pests, dividing them into A1 quarantine pests (those absent from the region and against which all countries are recommended to take phytosanitary action) and A2 quarantine pests (those present in some parts of the region and of concern to only some countries). The European Community (EU) issues Control Directives for some quarantine pathogens that include obligatory official testing schemes, standardized and harmonized for the whole EU region (Anonymous, 1998).

Regional agreements and organizations have been created to safeguard the interests of groups of neighboring countries with similar plant protection problems. This may be associated with closer trade and political linkages (Ebbels, 1993). Regional action is particularly desirable to prevent a pathogen or pest absent from a whole area from being introduced into any part of the area, as its entry into one territory would immediately endanger neighboring countries.

15.3 WORLD ORGANIZATIONS DEALING WITH PLANT QUARANTINE ISSUES

The major organizations that deal with plant quarantine activities in their respective regions are as follows:

1. Asia and Pacific Plant Protection Commission (APPPC)
2. The Caribbean Plant Protection Commission (CPPC)
3. Comite de Sanidad Vegetal del Cono Sur (COSAVE).
4. Communidad Andina (CAN)
5. The European and Mediterranean Plant Protection Organization (EPPO)
6. The Inter-African Phytosanitary Council (IAPSC)
7. The North American Plant Protection Organization (NAPPO)
8. Organismo Internacional Regional de Sanidad Agropecuaria (OIRSA)
9. Near East Plant Protection Organization (NEPPO)
10. The European Union (EU)
11. Pacific Plant Protection organization (PPPO)

15.3.1 MAJOR ROLES AND ACTIVITIES OF THESE ORGANIZATIONS

1. *Preparation of international standards for phytosanitary measures*: International Standards for Phytosanitary Measures (ISPMs) are prepared by the Secretariat of the International Plant Protection Convention (IPPC) as part of the United Nations Food and Agriculture Organization's global program of policy and technical assistance in plant quarantine. This program makes available to FAO members and other interested parties these standards, guidelines, and recommendations to achieve international harmonization of phytosanitary measures, with the aim to facilitate trade and avoid the use of unjustifiable measures as barriers to trade.

2. *Application of phytosanitary standards*: International Standards for Phytosanitary Measures are adopted by contracting parties to the IPPC through the Commission on Phytosanitary Measures. ISPMs are the standards, guidelines, and recommendations recognized as the basis for phytosanitary measures applied by Members of the World Trade Organization under the Agreement on the Application of Sanitary and Phytosanitary Measures. Noncontracting parties to the IPPC are encouraged to observe these standards.

3. *Review and amendment of phytosanitary standards*: International Standards for Phytosanitary Measures are subject to periodic review and amendment. The next review date for each standard is 5 years from their endorsement, or such other date as may be agreed upon by the Commission on Phytosanitary Measures. Standards will be updated and republished as necessary. Standard holders should ensure that the current edition of standards is being used.

4. *Distribution of phytosanitary standards*: International Standards for Phytosanitary Measures are distributed by the Secretariat of the International Plant Protection Convention to IPPC contracting parties, plus the Executive/Technical Secretariats of the RPPOs:

The structure of individual plant protection organization, its member countries, agreements, and the quarantine bacterial plant pathogens are described next.

15.4 ORGANIZATIONAL SET UP, ACTIVITIES, AND QUARANTINE PLANT BACTERIAL PATHOGENS IN JURISDICTION OF ORGANIZATIONS

15.4.1 ASIA AND PACIFIC PLANT PROTECTION COMMISSION: ASIA-PACIFIC PLANT PROTECTION ORGANIZATION

The Asia-Pacific Plant Protection Commission (APPPC) was established in 1956 with 24 member countries. The Commission administers the Regional Plant Protection Agreement for Asia and the Pacific. It reviews the plant protection situation at the national level in member countries and also at the regional level. Coordinating and promoting development of regional plant protection systems, assisting member countries to develop effective plant protection regimes, setting standards for phytosanitary measures, and facilitating information sharing are among its key objectives. The APPPC meets at least once in every 2 years. Twenty-five countries in Asia and Pacific region are members of APPPC.

Member Countries of APPPC in Asia
Afghanistan, Bangladesh, Bhutan, Brunei Darussalam, Cambodia, China, DPR Korea, India, Indonesia, Iran, Japan, Kazakhstan, Lao PDR, Malaysia, Maldives, Mongolia, Myanmar, Nepal, Pakistan, Philippines, Republic of Korea, Russian Federation, Singapore, Sri Lanka, Thailand, Timor-Leste, Uzbekistan, Vietnam.

Member Countries of APPPC in Pacific
Australia, Cook Islands, Fiji, France, Kiribati, Marshall Islands, Micronesia, Nauru, New Zealand, Niue, Palau, Papua New Guinea, Samoa, Solomon Islands, Tonga, Tuvalu, United States, Vanuatu.

Structure and Organization
The Plant Protection Agreement for Asia and Pacific Region is an intergovernmental treaty and administered by the APPPC. The Commission consists of representatives of all member countries and elects among them a Chairperson who serves for a period of 2 years. The director general of Food and Agriculture Organization appoints and provides the secretariat that coordinates, organizes, and follows up the work of the Commission. The Commission, according to its provisions convenes a meeting at least once in every 2 years with participation to all member countries. For implementation of the agreement, the Commission has established four standing committees, namely,

APPPC Standard Committee
APPPC Standing Committee on Plant Quarantine
APPPC Standing Committee on Pesticide Management
APPPC Standing Committee on IPM

Agreement
The Plant Protection Agreement for the Asia and Pacific Region (formerly the Plant Protection Agreement for South-East Asia and Pacific Region) was approved by the 23rd Session of the FAO Council in November 1955 and entered into force on July 2, 1956. The FAO Council approved amendments to the Agreement in 1967, 1979, 1983, and 1999. Some of these amendments have entered into force for all Contracting Governments, while others only with respect to the Contracting.

In Asia and Pacific region the bacterial plant pathogens for phytosanitary risk are as follows:

A1 Bacterial Pathogen	*Curtobacterium flaccumfaciens* pv. *flaccumfaciens*
A2 Bacterial Pathogen	*Clavibacter michiganensis* subsp. *insidiosus*
	Clavibacter michiganensis subsp. *michiganensis*
	Clavibacter michiganensis subsp. *nebraskensis*
	Clavibacter michiganensis subsp. *sepedonicus*
	Erwinia amylovora
	Pantoea stewartii
	Pseudomonas syringae pv. *tabaci*
	Ralstonia solanacearum race 2
	Xanthomonas albilineans
	Xanthomonas axonopodis pv. *manihotis*
	Xanthomonas axonopodis pv. *vasculorum*

15.4.2 Caribbean Plant Protection Commission

The Caribbean Plant Protection Commission (CPPC) was established in 1967 with 23 member countries including South and Central American countries bordering the Caribbean with a purpose: To strengthen intergovernmental cooperation in plant quarantine in the Caribbean area in order to prevent the introduction of destructive plant pests and diseases and to preserve the existing plant resources of that area.

Member Countries of CPPC

Barbados, Colombia, Costa Rica, Cuba, Dominica, Dominican Republic, France, Grenada, Guyana, Haiti, Jamaica, Mexico, the Netherlands, Nicaragua, Panama, Saint Kitts and Nevis, Saint Lucia, Suriname, Trinidad and Tobago, the United Kingdom, the United States, Venezuela (Bolivarian Republic)

The bacterial plant pathogen for phytosanitary risk in the CPPC region are as follows:

A1 Bacterial Pathogen	*Curtobacterium flaccumfaciens* pv. *flaccumfaciens*
	Liberibacter africanus
	Liberibacter asiaticus
	Xanthomonas axonopodis pv. *cajani*
	Xanthomonas hortorum pv. *carotae*
A2 Bacterial Pathogen	*Clavibacter michiganensis* subsp. *michiganensis*
	Dickeya chrysanthemi
	Pseudomonas marginalis pv. *marginalis*
	Ralstonia solanacearum
	Ralstonia solanacearum race 2
	Xanthomonas axonopodis pv. *citri*
	Xanthomonas axonopodis pv. *dieffenbachiae*
	Xanthomonas axonopodis pv. *manihotis*
	Xanthomonas oryzae pv. *oryzae*

15.4.3   COMITE REGIONAL DE SANIDAD VEGETAL DEL CONO SUR (COSAVE)

The Plant Protection Committee (or COSAVE) is a Regional Plant Protection Organization (RPPO) set up under the International Plant Protection Convention (IPPC) with a view to resolve phytosanitary problems of common interest to its member countries and strengthen regional integration.

COSAVE was established in 1980 as an ad hoc committee initiated by the Directors of Plant Protection of the five member states. It was created by agreement signed on March 9, 1989 between the Governments of Argentina, Brazil, Chile, Paraguay, and Uruguay. The first accession to the agreement was formalized in March 2013, date on which the Government of Peru made the deposit of an instrument of accession with the Ministry of Foreign Affairs of the Eastern Republic of Uruguay, becoming therefore an integral part of the organization. Bolivia is participating in the activities of the organization and with the later stages of the formal accession process.

Member Countries of COSAVE

Argentina, Brazil, Chile, Paraguay, Uruguay.

The present name and constitution were agreed in Montevideo in 1989 and by 1991 had been ratified by all member states. The administrative structure and regulations were approved during 1991 and regular activities started in 1992. The COSAVE Presidency and the Technical Secretary are intended to rotate every 2 years among the member states, but sometimes it is convenient for the Technical Secretary to remain in a different country to the Presidency. Finance is through annual subscriptions from member states and is administered by the Inter-American Institute for Cooperation in Agriculture (IICA). COSAVE is now a recognized RPPO within the framework of the IPPC. Its main objectives are to prevent the introduction and spread of agricultural pests, to minimize their impact on agricultural production, and to harmonize phytosanitary measures in order to facilitate regional and international trade in plants and plant products. Since the World Trade Organization Agreement on the Application of Sanitary and Phytosanitary Measures (WTO-SPS) in 1994, COSAVE has started to develop regional standards to harmonize phytosanitary regulations and procedures.

COSAVE formulates standards that are not obligatory but provide general guidelines for the community legislation developed by MERCOSUR for the commercial sector.

The bacterial plant pathogens for phytosanitary risk in the COSAVE region are as follows:

A1 Bacterial Pathogen	*Clavibacter michiganensis* subsp. *sepedonicus*
	Curtobacterium flaccumfaciens pv. *flaccumfaciens*
	Erwinia salicis
	Liberibacter africanus
	Liberibacter asiaticus
	Xanthomonas axonopodis pv. *citrumelo*
	Xanthomonas oryzae pv. *oryzae*
	Xanthomonas oryzae pv. *oryzicola*
	Xanthomonas populi
	Xylophilus ampelinus
A2 Bacterial Pathogen	*Pantoea stewartii*
	Ralstonia solanacearum
	Xanthomonas axonopodis pv. *citri*
	Xylella fastidiosa

15.4.4 Comunidad Andina (CAN)

The Andean Community (Spanish: Comunidad Andina, CAN) is a customs union comprising the South American countries of Bolivia, Colombia, Ecuador, and Peru. The trade bloc was called the Andean Pact until 1996 and came into existence when the Cartagena Agreement was signed in 1969. Its headquarters are in Lima, Peru.

Current Members
Bolivia (1969), Colombia (1969), Ecuador (1969), Peru (1969)

Associate Members
Argentina (2005), Brazil (2005), Paraguay (2005), Uruguay (2005), Chile (2006)

Observer Country
Spain

Former Full Members
Venezuela (1973–2006), joined MERCOSUR
Chile (full member 1969–1976, observer 1976–2006, associate member since 2006)

Goals of CAN
1. Promote the balanced and harmonious development of member countries under equitable conditions through integration and economic and social cooperation.
2. Accelerate growth and employment generation work for the people of the member countries.

3. Facilitate the participation of member countries in the process of regional integration, with a view to the gradual formation of a Latin American common market.
4. Reduce external vulnerability and improve the position of the member countries in the international economic context.
5. To strengthen subregional solidarity and reduce the differences in development among the member countries.
6. Seek an enduring improvement in the standard of living of the people of the subregion.

Strategic Agenda: Guiding Principles of CAN
1. Taking a realistic and historic opportunity virtues and limits of Andean integration.
2. Preserving the common Andean heritage, consolidating the achievements of 40 years of integration.
3. Respecting the diversity of approaches and visions that form the foundation of community coexistence.
4. Promote the development of the Andean market and trade by developing new opportunities for economic inclusion and social solidarity.
5. Progress in reducing the asymmetries within member countries through initiatives that promote economic and social development.
6. Develop the comprehensive nature of the integration process.
7. Deepening physical and border integration among member countries.
8. Promoting Amazon issues in the Andean integration process.
9. Promote citizen participation in the integration process.
10. Rate and assume unity in cultural diversity.
11. Sustainably enhance the biodiversity resources of the member countries.
12. Strengthen the institutions of the Andean Integration System to improve coordination and efficiency.
13. Strengthen regional cooperation on security issues.
14. Strengthen common foreign policy.
15. Generate practical mechanisms for coordination and convergence between integration processes.

The bacterial plant pathogens for phytosanitary risk in the CAN region are as follows:

A1 Bacterial Pathogen	*Clavibacter michiganensis* subsp. *sepedonicus*
	Xanthomonas axonopodis pv. *citri*
A2 Bacterial Pathogen	Nil

15.4.5 European and Mediterranean Plant Protection Organization

The European and Mediterranean Plant Protection Organization (EPPO) (Organisation Européenne et Méditerranéenne pour la Protection des Plantes [OEPP]) is an intergovernmental organization responsible for international cooperation in plant protection in the European and Mediterranean region. In the sense of the article IX of the FAO International Plant Protection Convention (IPPC), it is the Regional Plant Protection Organization for Europe. Founded in 1951 with 15 member governments, now it has 50 member governments including nearly every country of Western and Eastern Europe and the Mediterranean region.

EPPO Member Countries (upto 2014)

1. Albania	17. Greece	34. the Netherlands
2. Algeria	18. Guernsey	35. Norway
3. Austria	19. Hungary	36. Poland
4. Azerbaijan	20. Ireland	37. Portugal
5. Belarus	21. Israel	38. Romania
6. Belgium	22. Italy	39. Russia
7. Bosnia and	23. Jersey	40. Serbia
Herzegovina	24. Jordan	41. Slovakia
8. Bulgaria	25. Kazakhstan	42. Slovenia
9. Croatia	26. Kyrgyzstan	43. Spain
10. Cyprus	27. Latvia	44. Sweden
11. Czechia	28. Lithuania	45. Switzerland
12. Denmark	29. Luxemburg	46. Tunisia
13. Estonia	30. Macedonia (FYROM)	47. Turkey
14. Finland	31. Malta	48. United Kingdom
15. France	32. Moldova	49. Ukraine
16. Germany	33. Morocco	50. Uzbekistan

Aims of EPPO
- To protect plant health in agriculture, forestry, and the uncultivated environment
- To develop an international strategy against the introduction and spread of pests (including invasive alien plants) that damage cultivated and wild plants, in natural and agricultural ecosystems
- To encourage harmonization of phytosanitary regulations and all other areas of official plant protection action
- To promote the use of modern, safe, and effective pest control methods
- To provide a documentation service on plant protection

The bacterial plant pathogens for phytosanitary risk in the EPPO region are as follows:

A1 Bacterial Pathogen	*Acidovorax citrulli*
	Liberibacter africanus
	Liberibacter americanus
	Liberibacter asiaticus
	Liberibacter solanacearum
	Pseudomonas syringae pv. *pisi*
	Xanthomonas axonopodis pv. *allii*
	Xanthomonas axonopodis pv. *citri*
	Xanthomonas axonopodis pv. *citrumelo*
	Xanthomonas hyacinthi
	Xanthomonas oryzae pv. *oryzae*
	Xanthomonas oryzae pv. *oryzicola*
	Xanthomonas populi
	Xylella fastidiosa

(Continued)

A2 Bacterial Pathogen	*Burkholderia caryophylli*
	Clavibacter michiganensis subsp. *insidiosus*
	Clavibacter michiganensis subsp. *michiganensis*
	Clavibacter michiganensis subsp. *sepedonicus*
	Curtobacterium flaccumfaciens pv. *flaccumfaciens*
	Dickeya dianthicola
	Erwinia amylovora
	Pantoea stewartii
	Pseudomonas syringae pv. *actinidiae*
	Pseudomonas syringae pv. *persicae*
	Pseudomonas syringae pv. *pisi*
	Ralstonia solanacearum
	Ralstonia solanacearum race 1
	Ralstonia solanacearum race 2
	Ralstonia solanacearum race 3
	Xanthomonas arboricola pv. *corylina*
	Xanthomonas arboricola pv. *fragariae*
	Xanthomonas arboricola pv. *pruni*
	Xanthomonas axonopodis pv. *dieffenbachiae*
	Xanthomonas axonopodis pv. *phaseoli*
	Xanthomonas axonopodis pv. *poinsettiicola*
	Xanthomonas axonopodis pv. *vesicatoria*
	Xanthomonas fragariae
	Xanthomonas hyacinthi
	Xanthomonas populi
	Xanthomonas translucens pv. *translucens*
	Xanthomonas vasicola pv. *holcicola*
	Xylophilus ampelinus

The bacterial pathogens on alert list are as follows:

Xanthomonas arboricola pv. *fragariae*
Xanthomonas hyacinthi
Xanthomonas populi
Erwinia pyrifoliae
Pseudomonas syringae pv. *aesculi*
Pseudomonas syringae pv. *pisi*

15.4.6 INTER-AFRICAN PHYTOSANITARY COUNCIL

The member countries of Inter-African phytosanitary council are as follows:

Algeria, Angola, Benin, Botswana, Burkina Faso, Burundi, Cameroon, Cape Verde, Central African Republic, Chad, Comoros, Congo, Democratic Republic of the Cote d'Ivoire, Djibouti, Egypt, Equatorial Guinea, Eritrea, Ethiopia, Gabon, Gambia, Ghana, Guinea, Guinea-Bissau, Kenya, Lesotho, Liberia, Libya, Madagascar, Malawi, Mali, Mauritania, Mauritius, Morocco, Mozambique, Niger, Nigeria, Rwanda, Sao Tome and Principe, Senegal, Seychelles, Sierra Leone, Somalia, Sudan, Swaziland, Tanzania, Togo, Tunisia, Uganda, Zaire, Zambia, Zimbabwe

The Inter-African Phytosanitary Council of the African Union (AU-IAPSC) holds its meeting by its Steering Committee and General Assembly every 2 years. While the statutory members of the Steering Committee are all U's REC, those of the General Assembly are all the National Plant Protection Organizations of African Union member countries.

The bacterial plant pathogens of phytosanitary risk in IAPSC region are as follows:

A1 Bacterial Pathogen	*Pantoea stewartii*
	Xanthomonas albilineans
	Xanthomonas fragariae
	Xylella fastidiosa
	Xylophilus ampelinus
A2 Bacterial Pathogen	*Clavibacter michiganensis* subsp. *michiganensis*
	Erwinia amylovora
	Liberibacter africanus
	Liberibacter asiaticus
	Pseudomonas syringae pv. *pisi*
	Ralstonia solanacearum
	Xanthomonas arboricola pv. *pruni*
	Xanthomonas axonopodis pv. *citri*
	Xanthomonas oryzae pv. *oryzae*
	Xanthomonas oryzae pv. *oryzicola*
	Xanthomonas translucens pv. *translucens*
	Xanthomonas vasicola pv. *holcicola*

15.4.7 NORTH AMERICAN PLANT PROTECTION ORGANIZATION

The North American Plant Protection Organization (NAPPO) is the phytosanitary standard-setting organization recognized by the North American Free Trade Agreement (NAFTA). It was created in 1976 as a regional organization of the International Plant Protection Convention (IPPC) of the Food and Agriculture Organization (FAO) of the United Nations.

Member Countries
Canada, Mexico, United States

NAPPO Mandate
1. Global Mandate

The global mandate comes from Article IX of the New Revised Text (1997) of the Plant Protection Convention (IPPC) of the Food and Agriculture Organization (FAO) of the United Nations. The main activity under this mandate is to cooperate with the IPPC Secretary in achieving the objectives of the Convention and, where appropriate, cooperating with the Secretary and the Commission in developing international standards.

The goal of the IPPC is to protect the world's cultivated and natural plant resources from the introduction and spread of plant pests while minimizing interference with the international movement of goods and people.

2. Regional Mandate

The regional mandate for NAPPO was formalized by Canada, the United States, and Mexico in a Cooperative Agreement signed in 1976 at the Minister/Secretary of Agriculture level. The NAPPO Constitution and Bylaws confirm that NAPPO is accountable to the Minister/Secretary of Agriculture in NAPPO member countries.

NAPPO Mission
NAPPO's mission is to provide a forum for public and private sectors in Canada, the United States, and Mexico to collaborate in the development of science-based standards intended to protect agricultural, forest, and other plant resources against regulated plant pests, while facilitating trade and participate in related international cooperative efforts.

Strategic Goals of NAPPO

A satellite composite image of North America.

In order to accomplish this mission, the following strategic goals have been established:

- Protecting plant resources and the environment
- Capacity building
- Communicating results
- Building partnerships
- Offering an effective dispute settlement mechanism
- Working with sound management practices
- Working with a stable funding base

The bacterial plant pathogens of phytosanitary risk in NAPPO region are as follows:

A1 Bacterial Pathogen: *Clavibacter michiganensis* subsp. *sepedonicus*
A2 Bacterial Pathogen: Not mentioned

The bacterial pathogens kept on alert list are as follows:

Dickeya solani
Erwinia salicis
Liberibacter asiaticus
Pantoea ananatis
Pseudomonas lignicola
Xanthomonas acernea
Xanthomonas axonopodis pv. *manihotis*
Xanthomonas axonopodis pv. *vasculorum*
Xanthomonas populi
Xylella fastidiosa
Xylophilus ampelinus

15.4.8 ORGANIZATION INTERNATIONAL REGIONAL DE SANIDAD AGROPECUARIA

The International Regional Organization for Plant and Animal Health (Organismo Internacional Regional de Sanidad Agropecuaria, OIRSA), an intergovernmental organization founded in 1953, provides technical assistance to the ministries and departments of agriculture and livestock of nine member states in Central America: Belize, Costa Rica, the Dominican Republic, El Salvador, Guatemala, Honduras, Nicaragua, Mexico and Panama.

The organization plays an important role in disease and pest control throughout Central America, protecting and strengthening agricultural, forestry, and aquaculture-related development by enhancing production capacity and the safety of crops and agricultural products.

The bacterial plant pathogens of phytosanitary risk in OIRSA region are as follows:

A1 Bacterial Pathogen: Not mentioned
A2 Bacterial Pathogen: Not Mentioned

15.4.9 NEAR EAST PLANT PROTECTION ORGANIZATION

The Organization for Plant Protection in the Middle East (NEPPO) is a regional intergovernmental organization that encourages cooperation, within the framework of the IPPC, in particular its Article IX establishing regional organizations for the protection of plants. It is a platform for member countries to develop and implement a regional strategy and standards for the protection of plants. Since its establishment on January 9, 2009, 11 countries participate actively. Three countries signed the Agreement but have not yet ratified.

Background

The lack of an RPPO in the Near East Region has a negative impact on the regional collaboration in the area of plant protection, in particular the development of a regional strategy to monitor and control the transboundary plant pests (including diseases and weeds). Pest outbreaks are an increasing trend and threat to the region in the last 20 years (particularly in the light of increasing international trade in agricultural plants and plant products), which require close attention from the region at minimum in the form of information exchange and the initiation of an effective regional monitoring and control strategy.

In response to a request made by the Near East Regional Commission on Agriculture at its third session held in Nicosia (Cyprus) from September 11 to 15, 1989, and following a recommendation made by a technical consultation held in Rome from April 14 to 16, 1992, a conference of plenipotentiaries on the establishment of the Near East Plant Protection Organization (NEPPO), organized by FAO, was held in Rabat, Morocco, from February 16 to 18, 1993. The conference was attended and signed by 17 countries from the FAO Near East region.

The NEPPO was established on January 9, 2009. The first Governing Council was held in Rabat (Kingdom of Morocco) in October 2010, during which NEPPO established its structure, adopted rules and procedures, elected its President and Vice-president and nominated its Executive Director.

NEPPO was formally recognized as an RPPO under the IPPC on March 23, 2012. There are now 10 RPPOs under the IPPC and Article IX of the New Revised Text of the IPPC that states RPPOs

- Shall function as the coordinating bodies in the areas covered, shall participate in various activities to achieve the objectives of this Convention and, where appropriate, shall gather and disseminate information
- Shall cooperate with the Secretary in achieving the objectives of the Convention and, where appropriate, cooperate with the Secretary and the Commission in developing international standards

Member states of NEPPO are

		Signed but not ratified
1. Algeria	7. Morocco	
2. Egypt	8. Pakistan	1. Iran
3. Jordan	9. Sudan	2. Mauritania
4. Iraq	10. Syria	3. Yemen
5. Libya	11. Tunisia	
6. Malta		

NEPPO Objectives

The objectives of NEPPO shall be primarily to promote international and regional cooperation in the region in strengthening plant protection activities and capabilities with the aim of

1. Controlling pests of plants in an appropriate manner
2. Preventing the introduction and spread of pests
3. Facilitating safe trade

The bacterial plant pathogens of phytosanitary risk in NEPPO region are as follows:

A1 Bacterial Pathogen	*Curtobacterium flaccumfaciens* pv. *flaccumfaciens*
	Pseudomonas syringae pv. *pisi*
A2 Bacterial Pathogen	*Xanthomonas axonopodis* pv. *manihotis*
	Xanthomonas axonopodis pv. *vasculorum*

15.4.10 The European Union

The European Union (EU) is a union of 28 independent states based on the European Communities and founded to enhance political, economic, and social cooperation. Formerly known as European Community (EC) or European Economic Community (EEC). The EU was formed on November 1, 1993.

European Union rules on plant health aim to protect crops, fruits, vegetables, flowers, ornamentals, and forests from harmful pests and diseases (harmful organisms) by preventing their introduction into the EU or their spread within the EU. This aim helps to

1. Contribute to sustainable agricultural and horticultural production through plant health protection and
2. Contribute to the protection of public and private green spaces, forests, and the natural landscape

Council Directive 2000/29/EC provides the basis for this aim. The general principles are based upon provisions laid down in the International Plant Protection Convention (IPPC).

Directive 2000/29/EC is supported by a number of Control Directives and Emergency Measures in order to meet this aim, the EU

1. Regulates the introduction of plants and plant products into the EU from countries outside the EU
2. Regulates the movement of plants and plant products within the EU
3. Imposes eradication and containment measures in case of outbreaks, and co-finances them
4. Places obligations on countries outside the EU that want to export plants or plant products to the EU

Member States of the EU (Year of Entry)

1. Austria (1995)	15. Italy (1958)
2. Belgium (1958)	16. Latvia (2004)
3. Bulgaria (2007)	17. Lithuania (2004)
4. Croatia (2013)	18. Luxembourg (1958)
5. Cyprus (2004)	19. Malta (2004)
6. Czech Republic (2004)	20. the Netherlands (1958)
7. Denmark (1973)	21. Poland (2004)
8. Estonia (2004)	22. Portugal (1986)
9. Finland (1995)	23. Romania (2007)
10. France (1958)	24. Slovakia (2004)
11. Germany (1958)	25. Slovenia (2004)
12. Greece (1981)	26. Spain (1986)
13. Hungary (2004)	27. Sweden (1995)
14. Ireland (1973)	28. United Kingdom (1973)

On the road to EU membership | *Potential candidates*

Candidate countries

1. Albania
2. Montenegro
3. Serbia
4. The former Yugoslav Republic of Macedonia
5. Turkey

Potential candidates

1. Bosnia and Herzegovina
2. Kosovo[a]

[a] This designation is without prejudice to positions on status and is in line with UNSCR 1244/99 and the ICJ opinion on the Kosovo declaration of independence.

Bacterial plant pathogens of phytosanitary risk in EU region are as follows:

A1 Bacterial Pathogen	*Liberibacter africanus*
	Liberibacter americanus
	Liberibacter asiaticus
	Liberibacter solanacearum
	Pantoea stewartii
	Xanthomonas axonopodis pv. *citri*
	Xanthomonas oryzae pv. *oryzae*
	Xanthomonas oryzae pv. *oryzicola*
	Xylella fastidiosa
A2 Bacterial Pathogen	*Clavibacter michiganensis* subsp. *insidiosus*
	Clavibacter michiganensis subsp. *michiganensis*
	Clavibacter michiganensis subsp. *sepedonicus*
	Dickeya dianthicola
	Erwinia amylovora
	Pseudomonas syringae pv. *persicae*
	Ralstonia solanacearum
	Xanthomonas arboricola pv. *pruni*
	Xanthomonas axonopodis pv. *phaseoli*
	Xanthomonas axonopodis pv. *vesicatoria*
	Xanthomonas fragariae
	Xanthomonas vesicatoria
	Xylophilus ampelinus

15.4.11 PACIFIC PLANT PROTECTION ORGANIZATION

The Pacific Plant Protection Organization (PPPO) council meets every 3 years but the PPPO executive committee comprising of representatives from the three different geographical subregions, Melanesia, Micronesia, and Polynesia plus Australia and New Zealand meet annually. Guam is the current chair of PPPO with vice chairs, New Zealand and Australia. The PPPO executive committee members are Cook Islands, Tonga, New Caledonia, Solomon Islands, Federated States of Micronesia, and Nauru.

The Secretariat of the Pacific Community hosts the PPPO Secretariat as it is the regional organization that hosts all member countries and is involved in providing assistance in plant protection and quarantine to member countries since its inception in 1947. Non–IPPC contracting parties in the region such as the French, the United States, and New Zealand territories are also invited to participate in PPPO meetings for their information and to implement standards.

In the standard-setting process such as the draft ISPM workshops, member countries meet to discuss and endorse standards and discuss ways to implement the standards. The PPPO secretariat also facilitates information exchange among its member countries.

At present, the member states of PPPO are as follows:

1. American Samoa	13. Niue
2. Australia	14. Northern Mariana Islands
3. Cook Islands	15. Palau
4. Fiji	16. Papua New Guinea
5. French Polynesia	17. Pitcairn
6. Guam	18. Samoa
7. Kiribati	19. Solomon Islands

(Continued)

8. Marshall Islands	20. Tokelau
9. Micronesia	21. Tonga
10. Nauru	22. Tuvalu
11. New Caledonia	23. Vanuatu
12. New Zealand	24. Wallis and Futuna Islands

Bacterial plant pathogens of phytosanitary risk in PPPO are as follows:

A1 Bacterial Pathogen	Nil
A2 Bacterial Pathogen	*Leifsonia xyli*
	Pantoea stewartii
	Ralstonia solanacearum
	Xanthomonas albilineans
	Xanthomonas axonopodis pv. *citri*
	Xanthomonas axonopodis pv. *manihotis*
	Xanthomonas axonopodis pv. *vasculorum*

REFERENCES

Anonymous. 1998. Interim testing scheme for the diagnosis, detection and identification of *Ralstonia solanacearum* (Smith) Yabuuchi et al. in potatoes. Annex II to the Council Directive 98/57/EC of 20 July 1998 on the control of *Ralstonia solanacearum* (Smith) Yabuuchi et al. Publication 97/647/EC. *Off. J. Eur. Commun.*, L 235, 8–39.

Digiaro, M. and F. Valentini. 2015. The presence of *Xylella fastidiosa* in Apulia region (Southern Italy) poses a serious threat to the whole Euro-Mediterranean region. *Watch Lett.*, 33, CIHEAM-Bar: http://ciheam.org/images/CIHEAM/PDFs/ Publications/LV/WL33/08%20-%20Valentini.pdf.

Dunez, J. 1981. Exotic virus and virus-like diseases of fruit trees. *Bull. OEPP/EPPO Bull.*, 11, 251–258.

Ebbels, D.L. 1993. Preface, x–xi. In: Ebbels, D.L. (ed.), *Plant Health and the European Single Market*. BCPC Monograph No. 54, BCPC, Farnham, U.K., 416pp.

EPPO. 2009. *Pseudomonas syringae* pv. *actinidiae*. (2009). Retrieved July 4, 2015, from http://www.eppo.int/QUARANTINE/Alert_List/bacteria/P_syringae_pv_actinidiae.htm.

FAO. 2015. Keeping plant pests and diseases at bay: Experts focus on global measures: Phytosanitary standards for trade in plants and plant products come under review. FAO News Article, March 16, 2015, Rome, Italy.

Ling, L. 1953. International plant protection convention—Its history, objectives and present status. *FAO Plant Prot. Bull.*, 1(5), 65–68.

Mazzaglia, A., Studholme, D.J., Taratufolo, M.C. et al. 2012. *Pseudomonas syringae* pv. *actinidiae* (PSA) isolates from recent bacterial canker of kiwifruit outbreaks belong to the same genetic lineage. *PLoS ONE*, 7(5), e36518.

McCann, H. (November 13, 2014). Biosecurity: the lessons from Psa. Massey University. Retrieved from http://www.massey.ac.nz/massey/about-massey/news/article.cfm?mnarticle_uuid=87AFEB49-FBC1-C414-CC79-F55E6B968F7D).

Neslen, A. Europe's olive trees threatened by spread of deadly bacteria. *The Guardian*. January 8, 2015. Accessed May 9, 2015. http://www.theguardian.com/environment/2015/jan/08/europes-olive-trees-threatened-spread-deadly-bacteria.

OEPP/EPPO. 1989. Data sheets on quarantine organisms No. 166, *Xylella fastidiosa*. *Bull. OEPP/EPPO Bull.*, 19, 677–682.

WTO. 2000. The WTO Agreement on Sanitary and Phytosanitary Measures—Weakening Food Safety Regulation to Facilitate Trade: WTO: Linking with Development: Sponsored by the South-North Federation on Developmental Perspectives of the World Trade Organization; By: Bruce Silver glade, Director of Legal Affairs Center for Science in the Public Interest, May 25, 2000, Amsterdam, the Netherlands.

Zadoks, J. 2008. The potato murrain on the European continent and the revolutions of 1848. *Potato Res.*, 51(1), 5–45.

SUGGESTED READING

Addoh, P.G. 1977. The International Plant Protection Convention: Africa. *FAO Plant Prot. Bull.*, 25, 164–166.

Anonymous. NAPPO Phytosanitary Alert System. Accessed August 9, 2015. http://www.pestalert.org/opr_search.cfm.

Berg, G.H. 1977. The International Plant Protection Convention: Central America and the Caribbean. *FAO Plant Prot. Bull.*, 25, 160–163.

Campbell, F.T. 2001. The science of risk assessment for phytosanitary regulation and the impact of changing trade regulations. *BioScience*, 51(2), 148–153.

Campos, H. 1998. The Ten Commandments of the SPS Agreement of the WTO. Retrieved July 22, 2015, from http://www.fao.org/ag/againfo/resources/documents/Vets-l-2/7eng.htm.

EPPO Global Database. (n.d.). Retrieved July 24, 2015, from http://gd.eppo.int/.

Josling, T., Anderson, K., Schmitz, A. et al. 2010. Understanding international trade in agricultural products: One hundred years of contributions by agricultural economists. *Am. J. Agric. Econ.*, 92(2): 424–446.

Mathys, G. 1977. European and Mediterranean Plant Protection Organisation. *FAO Plant Prot. Bull.*, 25, 152–156.

Reddy, D.B. 1977. The International Plant Protection Convention: Plant Protection Committee for the South East Asia and Pacific Region. *FAO Plant Prot. Bull.*, 25, 157–159.

Sanitary and phytosanitary measures. (n.d.). Retrieved July 22, 2015, from https://www/wto/org/English/tratop_e/sps_issues_e.htm.

Schaad, N.W., Frederick, R.D., Shaw, J. et al. 2003. Advances in molecular-based diagnostics in meeting crop biosecurity and phytosanitary issues. *Annu. Rev. Phytopathol.*, 41, 305–324.

The WTO and the International Plant Protection Convention (IPPC). (n.d.). Retrieved July 22, 2015, https://www.wto.org/english/thewto_e/coher_e/wto_ippc_e.htm.

16 Sanitary and Phytosanitary Measures Agreement

16.1 INTERNATIONAL PLANT PROTECTION CONVENTION

FAO International Plant Protection Convention (IPPC) is a multilateral treaty for international cooperation in plant protection. The Convention makes provision for the application of measures by governments to protect their plant resources from harmful pests (phytosanitary measures), which may be introduced through international trade. The IPPC is deposited with the director general of the FAO and is administered through the IPPC Secretariat located in FAO's Plant Protection Service. The IPPC was first adopted in 1951 and has been amended twice, most recently in 1997.

The WTO's SPS Agreement states that "to harmonize sanitary and phytosanitary measures on as wide basis as possible, Members shall base their sanitary or phytosanitary measures on international standards, guidelines or recommendations." The Agreement names the IPPC for plant health standards.

The revision of the IPPC, agreed in 1997, and which entered into legal force on October 2, 2005, represents an updating of the Convention to reflect contemporary phytosanitary concepts and the role of the IPPC in relation to the Uruguay Round Agreements of the WTO, particularly the SPS Agreement. The SPS Agreement identifies the IPPC as the reference organization developing international standards for plant health (phytosanitary) measures. The standards of IPPC have also proved an important reference point for the dispute settlement mechanism of the WTO, for example, Japan—Measures Affecting the Importation of Apples DS245.

IPPC work includes standards on pest risk analysis, requirements for the establishment of pest-free areas, and others that give specific guidance on topics related to the SPS Agreement.

The Secretariat of the IPPC is located at the FAO headquarters in Rome.

The IPPC requires its contracting parties to make arrangements to issue phytosanitary certificates certifying compliance with the phytosanitary regulations of other contracting parties. This standard describes an export certification system to produce valid and credible phytosanitary certificates. Exported consignments certified under these systems should meet the current phytosanitary requirements of the importing country.

The basic elements of the phytosanitary certification process include

- Ascertaining the relevant phytosanitary requirements of the importing country (including import permits if required)
- Verifying that the consignment confirms to those requirements at the time of certification—Issuing a phytosanitary certificate

The requirements for a certification system to fulfill these functions comprise the following:

- Legal authority
- Management responsibility, including resources, documentation, communication, and review mechanism

Since 1948, national food safety and animal and plant health measures, which affect trade, were subject to General Agreement on Tariffs and Trade (GATT) rules. Article I of the GATT, the most-favored nation clause, required nondiscriminatory treatment of imported products from different foreign suppliers, and Article III required that such products be treated no less favorably than domestically produced goods with respect to any laws or requirements affecting their sale. These rules applied, for instance, to pesticide residue and food additive limits, as well as to restrictions for animal or plant health purposes.

The GATT rules also contained an exception (Article XX:b) which permitted countries to take measures "necessary to protect human, animal, or plant life or health," as long as these did not unjustifiably discriminate between countries where the same conditions prevailed, nor were a disguised restriction to trade. In other words, where necessary, for purposes of protecting human, animal, or plant health, governments could impose more stringent requirements on imported products than they required of domestic goods.

In the Tokyo Round of multilateral trade negotiations (1974–1979), an *Agreement on Technical Barriers to Trade* was negotiated (the 1979 TBT Agreement or "Standards Code"). Although this agreement was not developed primarily for the purpose of regulating sanitary and phytosanitary measures, it covered technical requirements resulting from food safety and animal and plant health measures, including pesticide residue limits, inspection requirements, and labeling. Governments that were members of the 1979 TBT Agreement agreed to use relevant international standards (such as those for food safety developed by the Codex) except when they considered that these standards would not adequately protect health. They also agreed to notify other governments, through the GATT Secretariat, of any technical regulations that were not based on international standards. The 1979 TBT Agreement included provisions for settling trade disputes arising from the use of food safety and other technical restrictions.

16.2   SANITARY AND PHYTOSANITARY MEASURES

16.2.1   THE WTO AGREEMENT ON SANITARY AND PHYTOSANITARY MEASURES

The *Agreement on the Application of Sanitary and Phytosanitary Measures* (known as the "SPS Agreement") entered into force with the establishment of the World Trade Organization (WTO) on January 1, 1995. It concerns the application of food safety and animal and plant health regulations.

The SPS agreement discusses the Final Act of the Uruguay Round of Multilateral Trade Negotiations, signed in Marrakesh on April 15, 1994. This agreement and others contained in the Final Act, along with the GATT as amended (the General Agreement on Tariffs and Trade [GATT], as revised in 1994, which is part of the WTO Agreements; GATT 1994 includes the original General Agreement, which is known as GATT 1947), are part of the treaty that established the WTO. The WTO superseded the GATT as the umbrella organization for international trade.

The WTO Secretariat has prepared the text to assist public understanding of the SPS Agreement. It is not intended to provide legal interpretation of the agreement.

16.2.2   THE SANITARY AND PHYTOSANITARY MEASURES AGREEMENT

The Agreement on the Application of Sanitary and Phytosanitary Measures sets out the basic rules for food safety and animal and plant health standards.

It allows countries to set their own standards. But it also says regulations must be based on science. They should be applied only to the extent necessary to protect human, animal, or plant life or health. And they should not arbitrarily or unjustifiably discriminate between countries where identical or similar conditions prevail.

Member countries are encouraged to use international standards, guidelines, and recommendations where they exist. However, members may use measures that result in higher standards if there is scientific justification. They can also set higher standards based on appropriate assessment of risks so long as the approach is consistent, not arbitrary.

The agreement still allows countries to use different standards and different methods of inspecting products.

In the SPS Agreement, sanitary and phytosanitary measures are defined as any measures applied

- To protect human or animal life from risks arising from additives, contaminants, toxins, or disease-causing organisms in their food
- To protect human life from plant- or animal-carried diseases
- To protect animal or plant life from pests, diseases, or disease-causing organisms
- To prevent or limit other damage to a country from the entry, establishment, or spread of pests

These include sanitary and phytosanitary measures taken to protect the health of fish and wild fauna, as well as of forests and wild flora.

Measures for environmental protection (other than as defined earlier), to protect consumer interests or for the welfare of animals are not covered by the SPS Agreement. These concerns, however, are addressed by other WTO agreements (i.e., the TBT Agreement or Article XX of GATT, 1994).

16.2.3 KEY FEATURES OF SPS AGREEMENT

All countries take measures to ensure that food is safe for consumers and to prevent the spread of pests or diseases among animals and plants. These sanitary and phytosanitary measures can take many forms, such as requiring products to come from a disease-free area, inspection of products, specific treatment or processing of products, setting of allowable maximum levels of pesticide residues, or permitted use of only certain additives in food. Sanitary (human and animal health) and phytosanitary (plant health) measures apply to domestically produced food or local animal and plant diseases, as well as to products coming from other countries.

16.2.3.1 Protection or Protectionism

Sanitary and phytosanitary measures, by their very nature, may result in restrictions on trade. All governments accept the fact that some trade restrictions may be necessary to ensure food safety and animal and plant health protection. However, governments are sometimes pressured to go beyond what is needed for health protection and to use sanitary and phytosanitary restrictions to shield domestic producers from economic competition. Such pressure is likely to increase as other trade barriers are reduced as a result of the Uruguay Round agreements. A sanitary or phytosanitary restriction that is not actually required for health reasons can be a very effective protectionist device, and because of its technical complexity, a particularly deceptive and difficult barrier to challenge.

The Agreement on Sanitary and Phytosanitary Measures (SPS) builds on previous GATT rules to restrict the use of unjustified SPS measures for the purpose of trade protection. The basic aim of the SPS Agreement is to maintain the sovereign right of any government to provide the level of health protection it deems appropriate, but to ensure that these sovereign rights are not misused for protectionist purposes and do not result in unnecessary barriers to international trade.

16.2.3.2 Justification of Measures

The SPS Agreement, while permitting governments to maintain appropriate sanitary and phytosanitary protection, reduces possible arbitrariness of decisions and encourages consistent decision-making. It requires that sanitary and phytosanitary measures be applied for no other purpose than that of ensuring food safety and animal and plant health. In particular, the agreement clarifies which factors should be taken into account in the assessment of the risk involved. Measures to ensure food safety and to protect the health of animals and plants should be based as far as possible on the analysis and assessment of objective and accurate scientific data.

16.2.3.3   International Standards

The SPS Agreement encourages governments to establish national sanitary and phytosanitary measures consistent with international standards, guidelines, and recommendations. This process is often referred to as "harmonization." The WTO itself does not and will not develop such standards. However, most of the WTO's member governments (132 at the date of drafting) participate in the development of these standards in other international bodies. The standards are developed by leading scientists in the field and governmental experts on health protection and are subject to international scrutiny and review.

International standards are often higher than the national requirements of many countries, including developed countries, but the SPS Agreement explicitly permits governments to choose not to use the international standards. However, if the national requirement results in a greater restriction of trade, a country may be asked to provide scientific justification, demonstrating that the relevant international standard would not result in the level of health protection the country considered appropriate.

16.2.3.4   Adapting to Conditions

Due to differences in climate, existing pests or diseases, or food safety conditions, it is not always appropriate to impose the same sanitary and phytosanitary requirements on food, animal, or plant products coming from different countries. Therefore, sanitary and phytosanitary measures sometimes vary, depending on the country of origin of the food, animal or plant product concerned. This is taken into account in the SPS Agreement. Governments should also recognize disease-free areas that may not correspond to political boundaries, and appropriately adapt their requirements to products from these areas. The agreement, however, checks unjustified discrimination in the use of sanitary and phytosanitary measures, whether in favor of domestic producers or among foreign suppliers.

16.2.3.5   Alternative Measures

An acceptable level of risk can often be achieved in alternative ways. Among the alternatives and on the assumption that they are technically and economically feasible and provide the same level of food safety or animal and plant health, governments should select those that are not more trade restrictive than required to meet their health objective. Furthermore, if another country can show that the measures it applies provide the same level of health protection, these should be accepted as equivalent. This helps ensure that protection is maintained while providing the greatest quantity and variety of safe foodstuffs for consumers, the best availability of safe inputs for producers, and healthy economic competition.

16.2.3.6   Risk Assessment

The SPS Agreement increases the transparency of sanitary and phytosanitary measures. Countries must establish sanitary and phytosanitary measures on the basis of an appropriate assessment of the actual risks involved, and, if requested, make known what factors they took into consideration, the assessment procedures they used and the level of risk they determined to be acceptable. Although many governments already use risk assessment in their management of food safety and animal and plant health, the SPS Agreement encourages the wider use of systematic risk assessment among all WTO member governments and for all relevant products.

16.2.3.7   Transparency

Governments are required to notify other countries of any new or changed sanitary and phytosanitary requirements that affect trade, and to set up offices (called "Enquiry Points") to respond to requests for more information on new or existing measures. They also must open to scrutiny how they apply their food safety and animal and plant health regulations. The systematic communication of information and exchange of experiences among the WTO's member governments provide a better basis for national standards. Such increased transparency also protects the interests of consumers, as well as of trading partners, from hidden protectionism through unnecessary technical requirements.

A special Committee has been established within the WTO as a forum for the exchange of information among member governments on all aspects related to the implementation of the SPS Agreement. The SPS Committee reviews compliance with the agreement, discusses matters with potential trade impacts, and maintains close cooperation with the appropriate technical organizations. In a trade dispute regarding a sanitary or phytosanitary measure, the normal WTO dispute settlement procedures are used, and advice from appropriate scientific experts can be sought.

16.3   SANITARY AND PHYTOSANITARY AGREEMENT

16.3.1   What Is New in the SPS Agreement?

Because sanitary and phytosanitary measures can so effectively restrict trade, GATT member governments were concerned about the need for clear rules regarding their use. The Uruguay Round objective to reduce other possible barriers to trade increased fears that sanitary and phytosanitary measures might be used for protectionist purposes.

The SPS Agreement was intended to close this potential loophole. It sets clearer, more detailed rights and obligations for food safety and animal and plant health measures that affect trade. Countries are permitted to impose only those requirements needed to protect health that are based on scientific principles. A government can challenge another country's food safety or animal and plant health requirements on the grounds that they are not justified by scientific evidence. The procedures and decisions used by a country in assessing the risk to food safety or animal or plant health must be made available to other countries upon request. Governments have to be consistent in their decisions on what is safe food and in responses to animal and plant health concerns.

16.3.2   The Scope and Differences in SPS and TBT Agreement

The scope of the two agreements is different. The SPS Agreement covers all measures whose purpose is to protect

- Human or animal health from food-borne risks
- Human health from animal- or plant-carried diseases
- Animals and plants from pests or diseases

(whether or not these are technical requirements)

The TBT (Technical Barriers to Trade) Agreement covers all technical regulations, voluntary standards, and the procedures to ensure that these are met, except when these are sanitary or phytosanitary measures as defined by the SPS Agreement. It is thus the type of measure that determines whether it is covered by the TBT Agreement, but the purpose of the measure that is relevant in determining whether a measure is subject to the SPS Agreement.

Technical Barriers to Trade measures could cover any subject, from car safety to energy-saving devices, to the shape of food cartons. To give some examples pertaining to human health, TBT measures could include pharmaceutical restrictions, or the labeling of cigarettes. Most measures related to human disease control are under the TBT Agreement, unless they concern diseases that are carried by plants or animals (such as rabies). In terms of food, labeling requirements, nutrition claims and concerns, quality and packaging regulations are generally not considered to be sanitary or phytosanitary measures and hence are normally subject to the TBT Agreement.

On the other hand, by definition, regulations that address microbiological contamination of food, or set allowable levels of pesticide or veterinary drug residues, or identify permitted food additives, fall under the SPS Agreement. Some packaging and labeling requirements, if directly related to the safety of the food, are also subject to the SPS Agreement.

The two agreements have some common elements, including basic obligations for nondiscrimination and similar requirements for the advance notification of proposed measures and the creation of information offices ("Enquiry Points"). However, many of the substantive rules are different. For example, both agreements encourage the use of international standards. However, under the SPS Agreement the only justification for not using such standards for food safety and animal/plant health protection is scientific arguments resulting from an assessment of the potential health risks. In contrast, under the TBT Agreement governments may decide that international standards are not appropriate for other reasons, including fundamental technological problems or geographical factors.

Also, sanitary and phytosanitary measures may be imposed only to the extent necessary to protect human, animal, or plant health, on the basis of scientific information. Governments may, however, introduce TBT regulations when necessary to meet a number of objectives, such as national security or the prevention of deceptive practices. Because the obligations that governments have accepted are different under the two agreements, it is important to know whether a measure is a sanitary or phytosanitary measure, or a measure subject to the TBT Agreement.

16.3.3   Transparency in SPS Agreement

The transparency provisions of the SPS Agreement are designed to ensure that measures taken to protect human, animal, and plant health are made known to the interested public and to trading partners. The agreement requires governments to promptly publish all sanitary and phytosanitary regulations, and, upon request from another government, to provide an explanation of the reasons for any particular food safety or animal or plant health requirement.

All WTO member governments must maintain an Enquiry Point, an office designated to receive and respond to any requests for information regarding that country's sanitary and phytosanitary measures. Such requests may be for copies of new or existing regulations, information on relevant agreements between two countries, or information about risk assessment decisions.

Whenever a government is proposing a new regulation (or modifying an existing one) which differs from an international standard and may affect trade, they must notify the WTO Secretariat, who then circulates the notification to other WTO member governments (over 700 such notifications were circulated during the first 3 years of implementation of the SPS Agreement). The notifications are also available to the interested public. Alternatively, notifications can be requested from the Enquiry Point of the country that is proposing the measure.

Governments are required to submit the notification in advance of the implementation of a proposed new regulation, so as to provide trading partners an opportunity to comment. The SPS Committee has developed recommendations on how the comments must be dealt with.

In cases of emergency, governments may act without delay, but must immediately notify other Members, through the WTO Secretariat, and also still consider any comments submitted by other WTO member governments.

The SPS Agreement explicitly recognizes the right of governments to take measures to protect human, animal, and plant health, as long as these are based on science, are necessary for the protection of health, and do not unjustifiably discriminate among foreign sources of supply. Likewise, governments will continue to determine the food safety levels and animal and plant health protection in their countries. Neither the WTO nor any other international body will do this.

The SPS Agreement does, however, encourage governments to "harmonize" or base their national measures on the international standards, guidelines, and recommendations developed by

WTO member governments in other international organizations. These organizations include, for food safety, the joint FAO/WHO Codex Alimentarius Commission; for animal health, the Office International des Epizooties; and for plant health, the FAO IPPC. WTO member governments have long participated in the work of these organizations including work on risk assessment and the scientific determination of the effects on human health of pesticides, contaminants, or additives in food; or the effects of pests and diseases on animal and plant health. The work of these technical organizations is subject to international scrutiny and review.

One problem is that international standards are often so stringent that many countries have difficulties in implementing them nationally. But the encouragement to use international standards does not mean that these constitute a floor on national standards, nor a ceiling. National standards do not violate the SPS Agreement simply because they differ from international norms. In fact, the SPS Agreement explicitly permits governments to impose more stringent requirements than the international standards. However, governments that do not base their national requirements on international standards may be required to justify their higher standard if this difference gives rise to a trade dispute. Such justification must be based on an analysis of scientific evidence and the risks involved.

Three different types of precautions are provided for in the SPS Agreement. First, the process of risk assessment and determination of acceptable levels of risk implies the routine use of safety margins to ensure adequate precautions are taken to protect health. Second, as each country determines its own level of acceptable risk, it can respond to national concerns regarding what are necessary health precautions. Third, the SPS Agreement clearly permits the precautionary taking of measures when a government considers that sufficient scientific evidence does not exist to permit a final decision on the safety of a product or process. This also permits immediate measures to be taken in emergency situations. There are many examples of bans on the production, sale, and import of products based on scientific evidence that they pose an unacceptable risk to human, animal, or plant health. The SPS Agreement does not affect a government's ability to ban products under these conditions.

It is accepted in the SPS Agreement that food safety and animal and plant health regulations do not necessarily have to be set by the highest governmental authority and that they may not be the same throughout a country. Where such regulations affect international trade, however, they should meet the same requirements as if they were established by the national government. The national government remains responsible for implementation of the SPS Agreement and should support its observance by other levels of government. Governments should use the service of nongovernmental institutions only if these comply with the SPS Agreement.

The SPS Agreement allows countries to give food safety, animal and plant health priority over trade, provided there is a demonstrable scientific basis for their food safety and health requirement. Each country has the right to determine what level of food safety and animal and plant health it considers appropriate, based on an assessment of the risks involved.

Once a country has decided on its acceptable level of risk, there are often a number of alternative measures that may be used to achieve this protection (such as treatment, quarantine, or increased inspection). In choosing among such alternatives, the SPS Agreement requires that a government use those measures that are no more trade restrictive than required to achieve its health protection objectives, if these measures are technically and economically feasible. For example, although a ban on imports could be one way to reduce the risk of entry of an exotic pest, if requiring treatment of the products could also reduce the risk to the level considered acceptable by the government, this would normally be a less trade-restrictive requirement.

Since the GATT began in 1948, it has been possible for a government to challenge another country's food safety and plant and animal health laws as artificial barriers to trade. The 1979 TBT Agreement also had procedures for challenging another signatory's technical regulations, including food safety standards and animal and plant health requirements. The SPS Agreement makes more explicit not only the basis for food safety and animal and plant health requirements that affect trade, but also the basis for challenges to those requirements. While a nation's ability to establish legislation is not restricted, a specific food safety or animal or plant health requirement can be challenged by another country on the grounds that there is no sufficient scientific evidence supporting the need for the trade restriction. The SPS Agreement provides greater certainty for regulators and traders alike, enabling them to avoid potential conflicts.

The WTO is an intergovernmental organization and only governments, not private entities or nongovernmental organizations, can submit trade disputes to the WTO's dispute settlement procedures. Nongovernmental entities can, of course, make trade problems known to their government and encourage the government to seek redress, if appropriate, through the WTO.

By accepting the WTO Agreement, governments have agreed to be bound by the rules in all of the multilateral trade agreements attached to it, including the SPS Agreement. In the case of a trade dispute, the WTO's dispute settlement procedures encourage the governments involved to find a mutually acceptable bilateral solution through formal consultations. If the governments cannot resolve their dispute, they can choose to follow any of several means of dispute settlement, including good offices, conciliation, mediation, and arbitration. Alternatively, a government can request that an impartial panel of trade experts be established to hear all sides of the dispute and to make recommendations.

In a dispute on SPS measures, the panel can seek scientific advice, including by convening a technical experts group. If the panel concludes that a country is violating its obligations under any WTO agreement, it will normally recommend that the country bring its measure into conformity with its obligations. This could, for example, involve procedural changes in the way a measure is applied, modification or elimination of the measure altogether, or simply elimination of discriminatory elements.

The panel submits its recommendations for consideration by the WTO Dispute Settlement Body (DSB), where all WTO member countries are represented. Unless the DSB decides by consensus not to adopt the panel's report, or unless one of the parties appeals the decision, the defending party is obliged to implement the panel's recommendations and to report on how it has complied. Appeals are limited to issues of law and legal interpretations by the panel.

Although only one panel was asked to consider sanitary or phytosanitary trade disputes during the 47 years of the former GATT dispute settlement procedures, during the first 3 years of the SPS Agreement 10 complaints were formally lodged with reference to the new obligations. This is not surprising as the agreement clarifies, for the first time, the basis for challenging sanitary or phytosanitary measures that restrict trade and may not be scientifically justified. The challenges have concerned issues as varied as inspection and quarantine procedures, animal diseases, "use by" dates, the use of veterinary drugs in animal rearing, and disinfection treatments for beverages. Dispute settlement panels have been requested to examine four of the complaints; the other complaints have been or are likely to be settled following the obligatory process of bilateral consultations.

The decision to start the Uruguay Round trade negotiations was made after years of public debate, including debate in national governments. The decision to negotiate an agreement on the application of sanitary and phytosanitary measures was made in 1986 when the Round was launched. The SPS negotiations were open to all of the 124 governments that participated in the Uruguay Round. Many governments were represented by their food safety or animal and plant health protection officials. The negotiators also drew on the expertise of technical international organizations such as the FAO, the Codex, and the OIE.

Developing countries participated in all aspects of the Uruguay Round negotiations to an unprecedented extent. In the negotiations on sanitary and phytosanitary measures, developing countries were active participants, often represented by their national food safety or animal and plant health experts. Both before and during the Uruguay Round negotiations, the GATT Secretariat assisted developing countries to establish effective negotiating positions. The SPS Agreement calls for assistance to developing countries to enable them to strengthen their food safety and animal and plant health protection systems. FAO and other international organizations already operate programs for developing countries in these areas.

GATT was an intergovernmental organization and it was governments that participated in GATT trade negotiations; neither private business nor nongovernmental organizations participated directly. But as the scope of the Uruguay Round was unprecedented, so was the public debate. Many governments consulted with both their public and private sectors on various aspects of the negotiations, including the SPS Agreement. Some governments established formal channels for public consultation and debate while others did on a more ad hoc basis. The GATT Secretariat also had considerable contact with international nongovernmental organizations as well as with the public and private sectors of many countries involved in the negotiations. The final Uruguay Round results were subject to national ratification and implementation processes in most GATT member countries.

The SPS Agreement established a Committee on Sanitary and Phytosanitary Measures (the "SPS Committee") to provide a forum for consultations about food safety or animal and plant health measures that affect trade, and to ensure the implementation of the SPS Agreement. The SPS Committee, like other WTO committees, is open to all WTO Member countries. Governments that have an observer status in the higher level WTO bodies (such as the Council for Trade in Goods) are also eligible to be observers in the SPS Committee. The Committee has agreed to invite representatives of several international intergovernmental organizations as observers, including Codex, OIE, IPPC, WHO, UNCTAD, and the International Standards Organization (ISO). Governments may send whichever officials they believe appropriate to participate in the meetings of the SPS Committee, and many send their food safety authorities or veterinary or plant health officials.

The SPS Committee usually holds three regular meetings each year. It also holds occasional joint meetings with the TBT Committee on notification and transparency procedures. Informal or special meetings may be scheduled as needed.

During its first year, the SPS Committee developed recommended procedures and a standardized format for governments to use for the required advance notification of new regulations. More than 700 notifications of sanitary and phytosanitary measures were submitted and circulated by the end of 1997. The Committee considered information provided by governments regarding their national regulatory procedures, their use of risk assessment in the development of sanitary and phytosanitary measures, and their disease-status, notably with respect to foot-and-mouth disease and fruit-fly. In addition, a considerable number of trade issues were discussed by the SPS Committee, in particular with regard to bovine spongiform encephalopathy (BSE). As required by the SPS Agreement, the SPS Committee developed a provisional procedure to monitor the use of international standards. The SPS Committee is continuing to work on guidelines to ensure consistency in risk management decisions, in order to reduce possible arbitrariness in the actions taken by governments. In 1998, the SPS Committee again reviewed the operation of the SPS Agreement.

The SPS Agreement helps ensure, and in many cases enhances, the safety of their food as it encourages the systematic use of scientific information in this regard, thus reducing the scope for arbitrary and unjustified decisions. More information will increasingly become available to consumers as a result of greater transparency in governmental procedures and on the basis for their food safety, animal and plant health decisions. The elimination of unnecessary trade barriers allows consumers to benefit from a greater choice of safe foods and from healthy international competition among producers.

Specific sanitary and phytosanitary requirements are most frequently applied on a bilateral basis between trading countries. *Developing countries* benefit from the SPS Agreement as it provides an international framework for sanitary and phytosanitary arrangements among countries, irrespective of their political and economic strength or technological capacity. Without such an agreement, developing countries could be at a disadvantage when challenging unjustified trade restrictions. Furthermore, under the SPS Agreement, governments must accept imported products that meet their safety requirements, whether these products are the result of simpler, less sophisticated methods or the most modern technology. Increased technical assistance to help developing countries in the area of food safety and animal and plant health, whether bilateral or through international organizations, is also an element of the SPS Agreement.

Exporters of agricultural products in all countries benefit from the elimination of unjustified barriers to their products. The SPS Agreement reduces uncertainty about the conditions for selling to a specific market. Efforts to produce safe food for another market should not be thwarted by regulations imposed for protectionist purposes under the guise of health measures.

Importers of food and other agricultural products also benefit from the greater certainty regarding border measures. The basis for sanitary and phytosanitary measures that restrict trade are made clearer by the SPS Agreement, as well as the basis for challenging requirements that may be unjustified. This also benefits the many processors and commercial users of imported food, animal, or plant products.

16.4   DEVELOPING COUNTRIES AND SPS AGREEMENTS

16.4.1   Difficulties Faced by Developing Countries in Implementing the SPS Agreement, Assistance Received, and Special Provisions for Developing Countries

Although a number of developing countries have excellent food safety and veterinary and plant health services, others do not. For these, the requirements of the SPS Agreement present a challenge to improve the health situation of their people, livestock, and crops, which may be difficult for some to meet. Because of this difficulty, the SPS Agreement delayed all requirements, other than those dealing with transparency (notification and the establishment of Enquiry Points), until 1997 for developing countries, and until 2000 for the least developed countries. This means that these countries are not required to provide a scientific justification for their sanitary or phytosanitary requirements before that time. Countries that need longer time periods, for example, for the improvement of their veterinary services or for the implementation of specific obligations of the agreement, can request the SPS Committee to grant them further delays.

Many developing countries have already adopted international standards (including those of Codex, OIE, and the IPPC) as the basis for their national requirements, thus avoiding the need to devote their scarce resources to duplicate work already done by international experts. The SPS Agreement encourages them to participate as actively as possible in these organizations, in order to contribute to and ensure the development of further international standards that address their needs.

One provision of the SPS Agreement is the commitment by members to facilitate the provision of technical assistance to developing countries, either through the relevant international organizations or bilaterally. FAO, OIE, and WHO have considerable programs to assist developing countries with regard to food safety, animal and plant health concerns. A number of countries also have extensive bilateral programs with other WTO Members in these areas. The WTO Secretariat has undertaken a program of regional seminars to provide developing countries (and those of Central and Eastern Europe) with detailed information regarding their rights and obligations stemming from this agreement. These seminars are provided in cooperation with the Codex, OIE, and IPPC, to ensure that governments are fully aware of the role these organizations can play in assisting countries to meet their requirements and fully enjoy the benefits resulting from the SPS Agreement. The seminars are open to participation by interested private business associations and consumer organizations. The WTO secretariat also provides technical assistance through national workshops and to governments through their representatives in Geneva.

16.5 PRINCIPLES OF SPS RELATED TO INTERNATIONAL TRADE

16.5.1 GENERAL PRINCIPLES

16.5.1.1 Sovereignty

With the aim of preventing the introduction of quarantine pests into their territories, it is recognized that countries may exercise the sovereign right to utilize phytosanitary measures to regulate the entry of plants and plant products and other materials capable of harboring plant pests.

16.5.1.2 Necessity

Countries shall institute restrictive measures only where such measures are made necessary by phytosanitary considerations, to prevent the introduction of quarantine pests.

16.5.1.3 Minimal Impact

Phytosanitary measures shall be consistent with the pest risk involved, and shall represent the least restrictive measures available, which result in the minimum impediment to the international movement of people, commodities, and conveyances.

These are based on the following principles:

1. *Modification*: As conditions change, and as new facts become available, phytosanitary measures shall be modified promptly, either by inclusion of prohibitions, restrictions, or requirements necessary for their success, or by removal of those found to be unnecessary.
2. *Transparency*: Countries shall publish and disseminate phytosanitary prohibitions, restrictions, and requirements and, on request, make available the rationale for such measures.
3. *Harmonization*: Phytosanitary measures shall be based, whenever possible, on international standards, guidelines, and recommendations developed within the framework of the IPPC.
4. *Equivalence*: Countries shall recognize as being equivalent to those phytosanitary measures that are not identical but that have the same effect.
5. *Dispute settlement*: It is preferable that any dispute between two countries regarding phytosanitary measures be resolved at a technical bilateral level. If such a solution cannot be achieved within a reasonable period of time, further action may be undertaken by means of a multilateral settlement system.

16.5.2 SPECIFIC PRINCIPLES

1. *Cooperation*: Countries shall cooperate to prevent the spread and introduction of quarantine pests and to promote measures for their official control.
2. *Technical authority*: Countries shall provide an official Plant Protection Organization.
3. *Risk analysis*: To determine which pests are quarantine pests and the strength of the measures to be taken against them, countries shall use pest risk analysis methods based on biological and economic evidence and, wherever possible, follow procedures developed within the framework of the IPPC.
4. *Managed risk*: Because some risk of the introduction of a quarantine pest always exists, countries shall agree to a policy of risk management when formulating phytosanitary measures.
5. *Pest-free areas*: Countries shall recognize the status of areas in which a specific pest does not occur. On request, the countries in whose territories the pest-free areas lay shall demonstrate this status based, where available, on procedures developed within the framework of the IPPC.
6. *Emergency action*: Countries may, in the face of a new and/or unexpected phytosanitary situation, take immediate emergency measures on the basis of a preliminary pest

risk analysis. Such emergency measures shall be temporary in their application, and their validity will be subjected to a detailed pest risk analysis as soon as possible.

7. *Notification of noncompliance*: Importing countries shall promptly inform exporting countries of any noncompliance with phytosanitary prohibitions, restrictions, or requirements.

8. *Nondiscrimination*: Phytosanitary measures shall be applied without discrimination between countries of the same phytosanitary status, if such countries can demonstrate that they apply identical or equivalent phytosanitary measures in pest management. In the case of a quarantine pest within a country, measures shall be applied without discrimination between domestic and imported consignments.

SUGGESTED READING

Addoh, P.G. 1977. The International Plant Protection Convention: Africa. *FAO Plant Protect. Bull.*, 25, 164–166.

Berg, G.H. 1977. The International Plant Protection Convention: Central America and the Caribbean. *FAO Plant Protect. Bull.*, 25, 160–163.

Campbell, F.T. 2001. The science of risk assessment for phytosanitary regulation and the impact of changing trade regulations. *BioScience*, 51(2), 148–153.

Campos, H. 1998. The Ten Commandments of the SPS Agreement of the WTO. Retrieved July 22, 2015, from http://www.fao.org/ag/againfo/resources/documents/Vets-l-2/7eng.htm.

Josling, T., Anderson, K., Schmitz, A. et al. 2010. Understanding international trade in agricultural products: One hundred years of contributions by agricultural economists. *Am. J. Agric. Econ.*, 92(2), 424–446.

Mathys, G. 1977. European and Mediterranean Plant Protection Organisation. *FAO Plant Protect. Bull.*, 25, 152–156.

NAPPO Phytosanitary Alert system. (n.d). Retrieved from http://www.pestalert.org/opr_search.cfm. Accessed September 22, 2015.

Reddy, D.B. 1977. The International Plant Protection Convention: Plant Protection Committee for the South East Asia and Pacific Region. *FAO Plant Protect. Bull.*, 25, 157–159.

Sanitary and phytosanitary measures. Retrieved July 22, 2015, from https://www/wto/org/ English/tratop_e/sps_issues_e.htm.

The WTO and the International Plant Protection Convention (IPPC). Retrieved July 22, 2015, from https://www.wto.org/english/thewto_e/coher_e/wto_ippc_e.htm. Accessed September 22, 2015.

World Trade Organization (WTO). https://www.wto.org/english/res_e/booksp_e/analytic_index_e/cusval_e.htm. Accessed August 22, 2015.

17 Efforts Made and Success Stories

Management of Bacterial Plant Pathogens

To manage the economically important bacterial plant disease pathogens across the world, serious efforts have been made with partial or full success. The success in the bacterial disease management depends on the seriousness and efforts of the farmers, cultivators, nurseryman, gardeners, government agencies, seed companies, local citizens, and others who are in the chain of crop production, crop husbandry, or plant users. Per se the seed companies should see that the seeds they produce and sell are free from bacterial plant pathogens. The nurseryman should see that the nursery stock they raise and sell are free from any bacterial plant pathogens and their latent infection. In case local citizens detect the bacterial disease in their backyard or surrounding places, they must either make effort to control this or inform the local government agencies, so as to control it. Many a times the bacterial plant pathogen is found to infect the alternate host of minor importance, which must be destroyed so as to minimize the damage it may cause to their main crop host plant. Control of disease by individual cultivator or farmer sometimes does not help much as the inoculum is already present in the neighbor's field to cause second attack. In such case, the joint control program may be carried out. The involvement of government agencies in such case is of great significance in the successful management of the disease.

The efforts made by government agencies, cultivators, and others may help either to eradicate the bacterial plant pathogen or not allow to establish it to cause substantial losses. In the regime of open market access and intercontinental trade of seed and planting stock, several bacterial plant pathogens are likely to cross the state boundaries to establish themselves in new areas.

17.1 INTERNATIONAL INTRODUCTION OF BACTERIAL PLANT PATHOGENS

The bacterial plant pathogens that were introduced in foreign countries but were not established are given in Table 17.1.

Other bacterial pathogens that were introduced in foreign countries and were eradicated are given in Table 17.2.

The eradication of the disease and its bacterial pathogen may be either complete or substantial as the disease reappear after some years with new race or biotype of the pathogen.

Some of the efforts made in the eradication of bacterial plant pathogen/disease are enumerated in the following text.

17.2 STORIES OF ERADICATION OF BACTERIAL PLANT PATHOGENS

17.2.1 *XANTHOMONAS AXONOPODIS* PV. *CITRI*

This bacterium causes citrus canker disease, which is considered to be a primary disease in citrus industry. The disease and pathogen have probably originated in Southeast Asia (as the canker lesions were noted on herbarium specimens collected in India as early as 1827) and continues to

TABLE 17.1

Bacterial Pathogens Introduced but Not Established

Sr. No.	Bacterial Pathogens	Diseases	Hosts	Countries
1.	*Xanthomonas translucens* pv. *translucens*	Black chaff	Wheat	Belgium, France Sweden, Bulgaria
2.	*Xanthomonas oryzae* pv. *oryzae*	Bacterial leaf blight	Rice	Southern Russia
3.	*C. flaccumfaciens* pv. *flaccumfaciens*	Bacterial tan spot	Soybean	Greece Hungary
4.	*X. arboricola* pv. *pruni*	Bacterial leaf spot	Apricot	Cyprus
5.	*Xanthomonas fragariae*	Angular leaf spot	Strawberry	Israel Ecuador
6.	*Clavibacter michiganensis* subsp. *michiganensis*	Bacterial wilt and canker	Tomato	The United Kingdom China
7.	*Xanthomonas vesicatoria*	Bacterial spot	Tomato	Azerbaijan Kazakhstan
8.	*Xanthomonas euvesicatoria*	Bacterial spot	Capsicum	Azerbaijan, Germany Kazakhstan
9.	*X. axonopodis* pv. *phaseoli*	Common blight	Beans	Czech Republic Israel
10.	*E. chrysanthemi*	Soft rot and wilt	Chrysanthemum	Finland

TABLE 17.2

Bacterial Pathogens Introduced but Eradicated

Sr. No.	Bacterial Pathogen	Disease	Host	Countries
1.	*X. axonopodis* pv. *citri*	Citrus canker	Citrus	South Africa: Mozambique; Florida (1933), the United States (1947), Strain A reappeared in Florida in 1986; Uruguay (A strain under eradication, while B strain eradicated since 1985); Queensland Australia
2.	*Xanthomonas fragariae*	Angular leaf spot	Strawberry	Africa: Reunion, Chile; Australia: New South Wales, Victoria; India: Mahabaleshwar
3.	*Clavibacter michiganensis* subsp. *michiganensis*	Bacterial wilt and canker	Tomato	Norway Portugal
4.	*E. pyrifoliae*	Fruit blight	Strawberry	EPPO Region
5.	*P. syringae* pv. *aesculi*	Bacterial canker	*Aesculus hippocastanum*	EPPO Region
6.	*P. syringae* pv. *pisi*	Bacterial blight	Pea	EPPO Region
7.	*Xanthomonas hyacinthi*	Bacterial blight	Hyacinth	EPPO Region
8.	*Xanthomonas populi*	Bacterial canker	Poplar	EPPO Region

increase its geographical range in spite of strict regulations imposed by many countries to prevent the introduction of this bacterial pathogen. Citrus canker presently occurs in more than 30 countries in Asia, the pacific and Indian Ocean Islands, South America, and the United States. In the United States, the citrus canker was first described in 1915 in the Gulf state, which included seven southern states, as a result of a shipment of infected trifoliate orange nursery stock imported from Japan.

Citrus canker was found in Florida in 13 locations from 1985 to 1992. Through extensive inspection and tree removal, eradication was believed to have been achieved. However, the disease reemerged in commercial plantation in Manatee country of Florida state in June 1997, where eradication efforts had previously been taken place. This outbreak has largely been suppressed by destruction of several hundred acres of infected commercial citrus plantations.

A new and extensive outbreak was reported in urban Miami of Florida State in 1995. The original Miami outbreak consisted of approximately 14 square miles of infected residential properties when first discovered but had expanded to more than 202 square miles in 1998. By 2005 nearly all Florida counties with commercial citrus range have had one or more outbreaks, and about 10% of the commercial acreage has been removed in an attempt to eradicate the diseases.

The disease also appeared earlier this century in South Africa and Australia; however, it was reportedly eliminated in these countries through nursery and orchard inspections, quarantine, and the onsite burning of infected trees. Subsequent epidemics have occurred in Australia, Argentina, Uruguay, Brazil, Oman, Saudi Arabia, and Reunion Island. In some locations, eradication efforts have been attempted and failed. In others, particularly Australia, Brazil, and Florida, active eradication campaigns continue.

The Florida Department of Agriculture and Consumer Services and the U.S. Department of Agriculture, Animal and Plant Health and Inspection Service are currently engaged in what may be the largest single regulatory agriculture program to eradicate a plant disease like citrus canker ever undertaken in the history of the United States, known as citrus canker eradication program (CCEP); this is an attempt to mitigate the serious consequences the disease would have on the $8.5 million Florida commercial citrus industry (Schubert et al., 2001). The state witnessed several outbreak of this disease, the first in 1912, second in 1986, third in 1995, fourth in 1997, and fifth in 1999, and the cost of CCEP exceeds $200 million (Mavrodieva et al., 2004).

Goto (1992) reported that the disease once reported to be eradicated from Australia, New Zealand, South Africa, and the United States during 1980s was again reported in Australia as well as in Mexico in 1981 and in Florida in 1984.

17.2.2 *Xanthomonas campestris* pv. *visicatoria*

The bacterium causes leaf spot and blight on tomato and capsicum and is observed to be a major disease on tomato in Nashik region of Maharashtra, India. Tomato cultivation in Nashik District occupies more than 14,000 ha of land around the year with a net profit of Rs. 0.1 million/ha as 60% of tomatoes are transported to other parts of Indian states and abroad. Infection of the bacterium *X. campestris* pv. *visicatoria* was consistently appearing on tomato cultivation in the area since 1990, but during kharif 1992, tomato crop succumbed heavily to the disease throughout the cultivated area due to favorable epiphytotic condition (26°C–30°C temperature, 93%–97% Relative Humidity (RH) coupled with rain/downpour and cloudiness), and within 10–15 days, whole tomato crop was devastated forcing the farmers to have had only one picking as against four to five normal picking. Most of the tomato hybrids under cultivation were susceptible to the disease. Market surveys revealed that there was 40% decline in the tomato supply to the market yard due to the disease, and the losses estimated were around INR 190 million, thus assuming the disease an epidemic.

Estimating the gravity of this bacterial disease, the pathogen *X. campestris* pv. *vesicatoria* was isolated and various agrochemicals were tested against the bacterial pathogen for their effectiveness at Agriculture Research Station, Niphad, Nashik District, by Professor S.G. Borkar. These effective agrochemicals were recommended for the management of this disease. A documentary film was made on this disease and its management, which was the first documentary film on bacterial plant disease in India and probably in the world. The Indian phytopathological society awarded a commendation certificate to this film under teaching aid during its annual conference held at Coimbatore in 1993. The film was shown at different locations in the districts to create awareness about this disease among the tomato cultivators, which subsequently helped them control this disease. Since then, the region did not have another epidemic of this disease.

17.2.3 ERWINIA AMYLOVORA

The bacterium causes fire blight on apple and pear, the most important fruit crops of temperate region. It is generally believed that fire blight was originated on wild hosts presumably Crataegus in the north-eastern United States and has been described after the import and cultivation of European apple and pear varieties in the United States (van der Zwet and Keil, 1979). The first description of the disease outside the United States was in New Zealand (1919). In Europe, fire blight was first described in the United Kingdom (Kent) in 1957 and in subsequent years the spread of the disease was recorded in Northern, Western, and Central Europe. In the year 1998, all countries belonging to the European Union (except Portugal) had fire blight on pears, apples, or ornamentals, either wide-spread (England, Belgium, Germany) or localized (France, Switzerland) or in restricted spots, under control and local eradication (Spain, Italy, Austria). It is said that Western Europe has been invaded by fire blight in the second half of the twentieth century. However, even today, wide areas of Europe (Italy, Spain, and the South-East of France) remain free of fire blight. Fire blight also invaded a large area around the Mediterranean Sea and most probably spread from an initial outbreak detected in the Nile delta region of Egypt in 1964. The disease was later found in Greece (Crete), Israel, Turkey, Lebanon, Iran, and countries of Central Europe. The introduction and infection of *E. amylovora* in Egypt and England have resulted in one continuous zone of fire blight area, which encompasses most of the Western Europe and most of the Mediterranean region. Only countries in North Africa seem to be free of fire blight; although the disease has recently been described in Morocco.

A number of unconfirmed reports of fire blight (China, India, Korea, Saudi Arabia, Vietnam, Colombia) may rely on misdiagnosis or insufficient description of the causal agent (confusion of fire blight with pear-blast symptoms caused by *Pseudomonas syringae* pv. *syringae* or with other *Erwinia* species reported on Asian pear). It must also be remembered that *E. amylovora* is a quarantine organism (list A2 OEPP), and the economic consequences of a declaration of the presence of fire blight in a country may have costly consequences for the international trade of this country: it cannot be ruled out that the list of actually "infected" countries is slightly longer than the list of officially declared areas.

In most cases, attempts to eradicate the pathogen in newly infected countries only slow down the spread of the disease. Until fire blight is again detected in Australia, this country might be the only case where eradication has been successful. Fire blight–like symptoms were detected on cotoneaster in the Royal Botanic Gardens, Melbourne, Victoria, in April 1997, and diagnostic tests confirmed that the causal organism was *E. amylovora* (Rodoni et al., 1999). An intensive eradication program was undertaken, and national surveys conducted for 3 years following the detection of *E. amylovora* have confirmed the absence of the disease in all states of Australia (Rodoni et al., 2002). There has been no positive detection of *E. amylovora* in New South Wales.

Large areas of the world are still free of fire blight (South America, most of Africa and Asia), in spite of the fact that susceptible cultivars of European and American origin are grown in these areas and that potentially susceptible host plants may be common in the environment (EPPO/CABI, 1998).

17.2.4 ERWINIA PYRIFOLIAE

The bacterium causes strawberry fruit blight. The first finding of *E. pyrifoliae* in strawberry (*Fragaria X ananassa*) was informed by the NPPO of the Netherlands recently to the EPPO Secretariat. *E. pyrifoliae* is closely related to *E. amylovora* (EPPO A2 List) and was initially described in 1999 in Korea on *Pyrus pyrifolia* (Asian or nashi pear) causing symptoms resembling those of fire blight (EPPO RS 99/134). In Japan, *Erwinia* isolates from *Pyrus ussuriensis* or *Pyrus communis* were found to be closely related to *E. pyrifoliae* but it is not entirely clear whether they were distinct species or not from *E. pyrifoliae*. In the Netherlands, symptomatic strawberry plants collected from two different locations (glasshouse commercial crops) were received in June and October 2013 for diagnosis. The identity of the bacterium was confirmed on December 23, 2013, by the National Reference Centre (biochemical and molecular tests of pure cultures followed by pathogenicity tests on *F. ananassa* cv. 'Elsanta'). Affected strawberry plants showed intense blackening of immature fruits, fruit calyx, and attached stems. In many cases, fruits were also heavily distorted. No symptoms were observed on the leaves. The discoloration was obvious inside young fruits, which presented an intense darkening/blackening of the fruit tissue at the edges, and the fruit tissue was extremely shiny in the middle. Symptomatic fruits were unmarketable. The origin of the introduction of *E. pyrifoliae* in strawberry crops in the Netherlands is unknown. Since affected crops have already been removed, no further phytosanitary measures have been taken at the affected fruit companies. The pest status of *E. pyrifoliae* in the Netherlands is officially declared as: "Present, only in some areas, only in protected cultivation."

17.2.5 *Pseudomonas syringae* pv. *aesculi*

The bacterium causes canker on *Aesculi* spp. The first report of *P. syringae* pv. *aesculi* (formerly EPPO Alert List) in the regions of Vienna and Niederösterreich was communicated by the NPPO of Austria recently to EPPO Secretariat. In Vienna, the bacterium was found in a private garden on 14 trees of *Aesculus x carnea* cv. 'Briotii' which had been purchased from another EU Member State. In Niederösterreich, the bacterium was detected on *Aesculus hippocastanum* trees in a public place where new trees of different origins have been replanted several times from 2005 to 2009. The origin of the infection is unknown. In both regions, the Regional Plant Protection Services have ordered the infected trees to be uprooted and destroyed. The pest status of *P. syringae* pv. *aesculi* in Austria is officially declared as: "Transient, actionable, under eradication."

17.2.6 *Ralstonia solanacearum*

The bacterial wilt pathogen has a very wide host range. In Austria, annual systematic official surveys for the occurrence of *R. solanacearum* (EPPO A2 List) have been conducted since 1995. In 2008, the bacterium was first detected during routine testing in one lot of ware potatoes (*Solanum tuberosum* cv. 'Ditta') from a supermarket in Kärnten. Investigations showed that the infested lot originated from a producer located in Niederösterreich and had been grown from certified seed potatoes (which had tested negative before planting). The infested lot of ware potatoes was only marketed in Austria. The possible source of contamination remained unknown. In 2009, the Austrian NPPO had informed the EPPO Secretariat of the first finding of *R. solanacearum* in Niederösterreich (EPPO RS 2009/083). Phytosanitary measures were taken in accordance with the EU Directive 98/57/EC. As none of the subsequent surveys detected the bacterium, the NPPO of Austria now considers that *R. solanacearum* has been eradicated. The pest status of *R. solanacearum* in Austria is officially declared as: "Absent: pest eradicated."

17.2.7 *Candidatus Liberibacter asiaticus*

This bacterium infects citrus plants causing symptoms known as huanglongbing. In Argentina during recent routine surveys, *Candidatus Liberibacter asiaticus* (EPPO A1 List) have been detected in the province of Corrientes. During 2014, *Candidatus Liberibacter asiaticus* was detected in one citrus plant growing in an orchard in the area of Mocoretá (Departmento Monte Caseros). During 2015, the bacterium was reported in one citrus plant at a private residence in the locality of Wanda (Departamento Puerto Iguazú). More recently, the pathogen was detected in three citrus plants in commercial orchards in the Departamento General Manuel Belgrano and in five plants growing in a private residence near Puerto Iguazú. In all cases, infected plants were destroyed and additional intensive surveys were carried out in their surroundings. These surveys did not detect other positive cases. The situation of *Candidatus Liberibacter asiaticus* in Argentina is described as "absent, eradicated."

17.2.8 *Xanthomonas arboricola* pv. *corylina*

This is an important bacterial pathogen of hazelnut crop. *X. arboricola* pv. *corylina* is an EPPO A2 quarantine organism (OEPP/EPPO, 1986), but is not considered of quarantine concern by any other regional plant protection organization due to the reason that hazelnut is not grown in the regions of other regional plant-protection organizations. *X. arboricola* pv. *corylina* is still of very limited distribution in the EPPO countries where it has been introduced. Other countries, however, while they may not be major hazelnut producers, are still at risk.

The NPPO of Germany informed the EPPO Secretariat of recent outbreaks of *X. arboricola* pv. *corylina* (EPPO A2 list) in Bayern and Baden-Württemberg. The EPPO Secretariat had no data on the previous occurrence of this disease in Germany. On August 24 and November 15, 2005, the plant-protection service of Bayern notified the occurrence of *X. arboricola* pv. *corylina* in five companies producing hazelnut fruits and one case in natural environment. The affected companies had supplied the infected material by another one in Baden-Württemberg, where the disease was also detected.

The infected parts of the diseased plants have been removed. At Baden-Württemberg, the infected lot was put under quarantine. In spring 2006, all lots of *Corylus avellana* from the company concerned was tested for *X. arboricola* pv. *corylina*. The pest status of *X. arboricola* pv. *corylina* in Germany was officially declared as: "Transient: actionable."

17.2.9 *Curtobacterium flaccumfaciens* pv. *flaccumfaciens*

This bacterium causes bacterial wilt in *Phaseolus* beans. Following the first report of its occurrence in 1920, *C. flaccumfaciens* pv. *flaccumfaciens* has become one of the most important bacterial diseases of beans in the United States, causing up to almost total losses in some years. More recently, however, it has become very much less important and has indeed not been reported on beans since the early 1970s (Hall, 1991). In soybeans, the disease was not reported in the United States until 1975 and is of rather minor importance (Sinclair and Backman, 1989). In the EPPO region, it is important on beans in Turkey but causes only minor losses in other countries.

Control was effected by using disease-free seeds and crop rotations. Seeds grown in dry climates are usually free from infection and are, therefore, recommended for distribution.

EPPO has listed *C. flaccumfaciens* pv. *flaccumfaciens* as an A2 quarantine pest (OEPP/EPPO, 1982,a), and CPPC and IAPSC also consider it of quarantine significance. Because of its very less importance currently in its area of origin, the quarantine status of the pathogen will be reviewed within EPPO. From its existing distribution, the disease seems most likely to be important in the southern part of the EPPO region, where *Phaseolus* spp. are widely grown. It is not present in the western Mediterranean countries and not established in most eastern Mediterranean countries. The disease does not seem important enough on soybeans to merit any special attention on this crop. EPPO recommends that consignments of seeds of *Phaseolus vulgaris* imported from infested countries should come from an area where the disease does not occur or from a crop that was found free from the disease during the growing season (OEPP/EPPO, 1990).

17.2.10 *Erwinia chrysanthemi* (*Dickeya chrysanthemi*)

The pathogen causes destruction of many flower and ornamental plants and particularly carnation and chrysanthemum in rooting beds. Losses are also recorded on different glasshouse ornamentals (*Saintpaulia ionantha, Kalanchoe*), as well as in potato and Dahlia tuber production.

E. chrysanthemi is listed as an A2 quarantine pest by EPPO (OEPP/EPPO, 1982,b). In 1980, EPPO specified that the quarantine pests *E. chrysanthemi* comprise the pathovars *dianthicola* and *chrysanthemi* (OEPP/EPPO, 1988). The major means of infection of host crops, at least in some countries, was the use of infected planting material, which, according to current trade practice, was mostly imported, and protection against this pathogen was perceived as a plant quarantine problem.

It is now accepted within EPPO that the risk from *E. chrysanthemi* can be adequately covered by national nuclear-stock certification schemes for the crops concerned and that *E. chrysanthemi* will be deleted from the EPPO A2 list as soon as such schemes have been agreed for carnations (OEPP/EPPO, 1991) and chrysanthemums. No other regional plant-protection organization considers *E. chrysanthemi* to be a quarantine pest.

EPPO recommends (OEPP/EPPO, 1990) that plantlets of carnations or chrysanthemums should be from mother plants found free of *E. chrysanthemi* (by testing in the case of carnations). However, now that nuclear-stock certification schemes are recommended for carnations (OEPP/EPPO, 1991) and chrysanthemums (in preparation), it is simpler to recommend that planting material of these crops should originate from such schemes.

17.2.11 *XANTHOMONAS AXONOPODIS* PV. *PUNICAE*

This bacterial pathogen causes a serious disease on pomegranate. Bacterial blight is a devastating disease of this crop reported from major pomegranate-growing areas of Maharashtra, Karnataka, and Andhra Pradesh in India. The infection of the bacterium *X. axonopodis* pv. *punicae* was so severe that most of the pomegranate orchards in Jat and Sangola areas of Solapur district have been destroyed and uprooted by the farmers. The economic losses due to this disease were estimated to the tune of INR 23,183 million, and the contribution of the Indian government to combat the disease was INR 10,000 million during 2003–2008.

Based on the laboratory and field study conducted at the Department of Plant Pathology, Mahatma Phule Agriculture University, Rahuri, under the leadership of Professor S.G. Borkar, a protocol was formulated and validated on farmers field with 100% success in the management of this disease. Those contract farmers who followed the protocol step by step without missing any component of the protocol could control the disease in their field and are still managing their crop without bacterial incidence. The recommendations/protocol for the control of the disease should be applied in totality to keep the pathogen under check.

17.2.12 BANANA *XANTHOMONAS* WILT

Banana *Xanthomonas* wilt (BXW) or banana bacterial wilt or enset wilt is a bacterial disease caused by *X. campestris* pv. *musacearum* (Tushemereirwe et al., 2004). After being originally identified on a close relative of banana, *Ensete ventricosum*, in Ethiopia in the 1960s (Yirgou and Bradbury, 1968), BXW emanated in Uganda in 2001 affecting all types of banana cultivars. Since then, BXW has been diagnosed in Central and East Africa including banana-growing regions of Rwanda, Democratic Republic of the Congo, Tanzania, Kenya, Burundi, and Uganda, despite control efforts.

It reached epidemic levels in Uganda in 2001. Analysis from Uganda indicated that, if unchecked, the disease could result in national cumulative losses of $4 billion more than a 5-year period.

BXW symptoms include the wilting of leaves, premature ripening of bunches and rotting of fruit, and death of the plant—damage is up to 100%. Contaminated tools, insects, and even birds spread the bacteria, but long distance transmission is often man-induced through the movement of planting material carrying latent infections. All banana cultivars are susceptible to the disease, and BXW has to be controlled through eradication or field management practices that reduce pathogen spread.

In Uganda, participatory management campaigns were initially successful in reducing infections, and in pilot villages where Farmers' Field Schools were introduced by FAO in 2006, the disease was practically eliminated. Recently, however, the disease is showing resurgences in areas where it was previously controlled, probably resulting from a form of fatigue and the lack of sustained support system and incentives to the farming communities—including availability of clean planting material. Control practices, including destruction of infected plants, removal of male buds, and use of clean tools are physically onerous. Besides, more information on the disease epidemiology and spread is needed.

Global concern arose over the livelihoods of African banana farmers and the millions relying on bananas as a staple food when the disease was at its worst between the years 2001 and 2005. It was estimated that in Central Uganda from 2001 and 2004, there was a 30%–52% decrease in banana yield due to BXW infection (Castellani, 1939). Although extensive management of the disease outbreaks has helped reduce the impact of BXW, even today BXW continues to a pose a real problem to the banana farmer of Central and East Africa. They say a journey of 1000 miles begins with one step. This is the story of the long walk toward the elimination of BXW disease that is threatening to wipe out the valued banana crop in East Africa. Bananas and plantains are important sources of food for more than 100 million people in Sub-Saharan Africa as well as income for more than 50 million smallholder farmers. Kalyebara et al. (2006) reported that if BXW is not controlled, Uganda stands to lose an estimated $295 million worth of banana output valued at farm gate prices.

Since the disease was discovered, banana farmers have used several cultural methods such as sanitary measures to try and control its spread without much success. As they say necessity is the mother of invention, scientists have had no option but to explore a lasting solution to this malady. This led to the birth of the Banana Bacterial Wilt project, which develops bacterial wilt–resistant bananas for use by smallholder farmers in the Great Lakes region. The BXW project is a public–private partnership that brings together the African Agricultural Technology Foundation (AATF), the International Institute of Tropical Agriculture (IITA), and the National Agricultural Research Organisation (NARO) of Uganda. In 2004, IITA and NARO scientists started developing transgenic bananas resistant to BXW in a joint project funded by Gatsby Charitable Foundation. AATF joined the project in 2005 and successfully negotiated for two genes (*pflp* and *hrap*) from Academia Sinica, Taiwan, for use in the project free of royalty. The three organizations agreed to work together and, therefore, signed a tripartite agreement in 2009 to guide the partnership especially on roles and responsibilities. IITA and NARO were responsible for developing, transforming, and evaluating transgenic bananas that are resistant to BXW, while AATF was mandated to coordinate the partnership and provide expertise in management of intellectual property and regulatory issues. The project then commenced with funding from AATF in addition to support from Gatsby Charitable Foundation and USAID. The developmental research work is now being carried out in both Uganda and Kenya by Dr. Leena Tripathi of IITA and Dr. Wilberforce Tushemereirwe of NARO. The project uses modern tools of biotechnology to develop BXW-resistant banana varieties through genetic modification of East Africa's farmer preferred banana varieties.

The progress made so far in the project is positive following the successful transformational work in the laboratories and subsequent conduct of the first confined field trial in NARO's Kawanda field station in October 2010. The scientists are looking forward to a victorious war against the BXW once the product complies with the necessary regulatory requirements in the project countries.

There is hope that most countries in Sub-Saharan Africa will put in place enabling legislations to allow for the development and growing of genetically modified agricultural products to curb food production constraints in Africa. It is estimated that the first BXW-resistant varieties from these efforts could get to farmers in 6–7 years depending on research results and regulatory approvals in each of the project countries.

REFERENCES

Castellani, E. 1939. Su un marciume dell' Ensete. *Agr. Colon.*, 33, 297–300.

EPPO/CABI. 1998. Map 257. In: *Distribution Maps of Quarantine Pests for Europe*. (edited by Smith, I. M. and Charles, L. M. F.). Wallingford, UK: CAB International, xviii + 768 pp.

Goto, M. 1992. Citrus canker. In: Kumar, J., Chaube, H.S., Singh, U.S., and A.N. Mukhopadhyay (eds.), *Plant Diseases of International Importance*, Vol. III: *Diseases of Fruit Crops*. Prentice Hall, Englewood Cliffs, NJ, pp. 170–208.

Hall, R.J.B. (ed.). 1991. *A Compendium of Bean Diseases*. American Phytopathological Society, St. Paul, MN.

Kalyebara, M.R., Ragama, P.E., Kagezi, G.H. et al. 2006. Economic importance of the banana bacterial wilt in Uganda. Special issue. Banana Bacterial Wilt in Uganda: A disease that threatens livelihoods. *Afr. Crop Sci. J.*, 14(2), 93–103.

Mavrodieva, V., Levy, L., and D.W. Gabriel. 2004. Improved sampling methods for real time polymerase chain reaction diagnosis of citrus canker from field samples. *Phytopathology*, 94(1), 61–68.

OEPP/EPPO. 1982a. Data sheets on quarantine organisms No. 48, *Corynebacterium flaccumfaciens*. *Bull. OEPP/EPPO Bull.*, 12(1).

OEPP/EPPO. 1982b. Data sheets on quarantine organisms No. 53, *Erwinia chrysanthemi*. *Bull. OEPP/EPPO Bull.*, 12(1).

OEPP/EPPO. 1986. Data sheets on quarantine organisms No. 134, *Xanthomonas campestris* pv. *corylina*. *Bull. OEPP/EPPO Bull.*, 16, 13–16.

OEPP/EPPO. 1988. A1 and A2 lists of quarantine pests. Specific quarantine requirements. EPPO Publications Series B No. 92.

OEPP/EPPO. 1990. Specific quarantine requirements. EPPO Technical Documents No. 1008.

OEPP/EPPO. 1991. Certification schemes. Pathogen-tested material of carnation. *Bull. OEPP/EPPO Bull.*, 21, 279–290.

Rodoni, B., Gardner, R., Gile, R. et al. 2002. National surveys did not detect *Erwinia amylovora* on host p ants in Australia. *Acta Hortic.*, 590, 39–45.

Rodoni, B., Kinsella, M., Gardner, R. et al. 1999. Detection of *Erwinia amylovora*, the causal agent of fire blight, in the Royal Botanic Gardens, Melbourne, Australia. *Acta Hortic.*, 169–170.

Schubert, T.S., Rizvi, S.A., Sun, X. et al. 2001. Meeting the challenge of eradicating citrus canker in Florida-Again. *Plant Dis.*, 85, 340–356.

Sinclair, J.B. and P.A. Backman. 1989. *A Compendium of Soybean Diseases*, 3rd ed. American Phytopathological Society, St. Paul, MN.

Tushemereirwe, W., Kangire, A., Ssekiwoko, F. et al. 2004. First report of *Xanthomonas campestris* pv. *musacearum* on banana in Uganda. *Plant Pathol.*, 53(6), 802.

van der Zwet, T. and H.L. Keil. 1979. Fire blight, a bacterial disease of Rosaceous plants. In: *Agriculture Handbook*. Science and Education Administration USDA, Beltsville, MD, 200pp.

Yirgou, D. and J.F, Bradbury. 1968. Bacterial wilt of Enset (*Ensete ventricosum*) incited by *Xanthomonas musacearum* sp. n. *Phytopathology*, 58, 111–112.

SUGGESTED READING

Gottwald, T.R. 2000. Citrus canker. *Plant Health Instruct*. DOI: 10.1094/PHI-I-2000 1002-01. Internet.

Kim, W.S., Gardan, L., Rhim, S.L. et al. 1999. *Erwinia pyrifoliae* sp. *nov.*, a novel pathogen that affects Asian pear trees (*Pyrus pyrifolia Nakai*). *Int. J. Systemat. Bacteriol.*, 49, 899–906.

OEPP/EPPO. 1986. Data sheets on quarantine organisms No. 134, *Xanthomonas arboricola* pv. *corylina*. *Bull. OEPP/EPPO Bull.*, 16, 13–16.

Outbreaks of *Xanthomonas arboricola* pv. *corylina* in Germany. EPPO Global Database. 2006. Retrieved August 6, 2015, from https://gd.eppo.int/reporting/article-887.

Samson, R., Poutier, F., Sailly, M. et al. 1987. Caractérisation des *Erwinia chrysanthemi* isolées de *Solanum tuberosum* etd'autresplantes-hôtesselon les biovars et sérogroupes. *Bull. OEPP/EPPO Bull.*, 17, 11–16.

Schubrt, T.S., Gottawald, T.R., Graham, J.H. et al. 2001. Meeting the challenges of eradicating citrus canker in Florida again. *Plant Dis.*, 85(4), 340–356.

SENASA. 2015-06-02. Misiones: erradicación de plantaspositivas a la presenciadel HLB en Puerto Iguazú y General Belgrano. http://www.senasa.gov.ar/contenido.php?to=n&in=1897&ino=0&io=30513.

Index

A

Acibenzolar-*S*-methyl (A-S-M), 80–81, 100, 170, 315, 418
Acidovorax avenae subsp. *avenae*
    bacterial brown stripe of paspalum, 488
    bacterial leaf blight and stalk rot of maize, 35
        common names, 35
        control measures, 36
        disease cycle, 36
        geographical distribution, 36
        host, 36
        symptoms, 35–36
        synonyms, 35
    bacterial leaf blight in fishtail palm, 473
        disease cycle, 474
        disease management, 474
        epidemiology, 474
        host range, 473–474
        symptoms, 473
    red stripe and top rot of sugarcane, 87–88
Acidovorax avenae subsp. *cattleyae, see* Leaf spot in
        orchids
Acidovorax avenae subsp. *citrulli, see* Bacterial fruit
        blotch of watermelon
Actinidia chinensis, 531
Actinidia deliciosa, 531
Aerobic bacterial library (TABA50), 257
Aesculus hippocastanum, 166, 565
African Agricultural Technology Foundation (AATF),
        208, 568
African countries
    banana wilt pathogen, 1
    blackarm of cotton, 113
    cassava bacterial pathogen, 1
Aglaonema, 443–444
    A. commutatum, 444
    A. crispum, 444
    A. pictum, 456
Agri-Mycin 17, 100, 126
Agri-Mycin 77, 100, 202, 455
Agri-Mycin 455, 500
Agrobacterium radiobacter, 151, 504
Agrobacterium radiobacter K84 strain, 138, 163, 181, 327
Agrobacterium rhizogenes, 135
Agrobacterium rubi, 133
Agrobacterium spp., 151
Agrobacterium tumefaciens, 5
    black knot of grapevine, 200
    crown gall and cane gall of raspberry, 247–248
    crown gall in *Dahlia,* 461–462
    crown gall in poplar, 503–504
    crown gall in rose, 433–435
    crown gall of almond, 163
    crown gall of apple and pear, 133, 135
        disease cycle, 136
        disease management, 136–139
        epidemiology, 136
        host, 135

        predisposing factors, 134
        symptoms, 133–134
        Ti plasmid and virulence genes, 136
    crown gall of apricot, 142–143
    crown gall of avocado, 260–261
    crown gall of fruit-bearing forest trees, 523
    crown gall of rubber, 492–493
    crown gall of tea, 491–492
    crown gall of walnut, 151–153
    crown gall of willow, 512
Agrobacterium vitis, see Crown gall of grapevine
Agropyron repens, 132
Agrostis avenacea, 22
Ag streptomycin, 100
Alfalfa
    bacterial leaf spot, 479–480
    bacterial wilt, 479–482
    dwarf, 482–483
    sprout rot, 483–484
Allium cepa, 415
Allium cepa var. *ascalonicum,* 415
Allium fistulosum, 415
Allium porrum, 415
Allium sativum, 415
Allium schoenoprasum, 415
Almond
    bacterial canker, 161
    bacterial spot
        disease cycle, 156–157
        disease management, 157
        epidemiology, 156
        pathogen, 155
        symptoms, 155–156
        synonyms, 155
    crown gall, 163
    hyperplastic canker, 162–163
    leaf scorch
        disease management, 160–161
        disease spread, 159
        geographical distribution, 159
        GWSS effect, 160
        pathogen, 158
        symptoms, 158–159
        varietal susceptibility, 161
        vectors, 159–160
Alternaria brassicae, 355
Ambarella *(Spondias cytherea* or *Spondias dulcis),* 201
Amelanchier (June berry), 118
Amplified fragment length polymorphisms (AFLPs), 498
Anacardiaceae, 201–202
Anguina tritici, 22–23
Angular leaf spot of cucumber, 377–378
Angular leaf spot of cucurbits, 388–390
Angular leaf spot of strawberry
    common names, 249
    control measures, 253–254
    detection and inspection methods, 252–253
    disease cycle, 251–252